Analog Circuit Techniques

With Digital Interfacing

To Jeanette. Now I know why...

Quotes

Theory

When you are a Bear of Very Little Brain and
you Think of Things, you sometimes find that a
Thing which seemed very Thingish inside you is
quite different when it gets out into the open
and has other people looking at it.

Practice

"That's funny," said Pooh, "I dropped it
on the other side," said Pooh, "and it came out
on this side! I wonder if it would do it again?"
And he went back for some more fir cones.
It did. It kept on doing it.

Simplicity

Rabbit is clever. Rabbit has brain.
I suppose that is why he never
understands anything.

Winnie the Pooh's little book of Wisdom.
Methuen, 1999.

Couldn't use these because Disney has the US rights! - see Preface.

***

hence came forth Maul, a giant. This Maul did use to spoil young pilgrims with sophistry. *Bunyan.*

Analog Circuit Techniques

With Digital Interfacing

T. H. Wilmshurst

Previously Reader in Electronics
Department of Electronics and Computer Science
University of Southampton

Neil
regards.
Trevor. W.

OXFORD AUCKLAND BOSTON JOHANNESBURG MELBOURNE NEW DELHI

Newnes
An imprint of Bufterworth-Heinemann
Linacre House, Jordan Hill, Oxford OX2 8DP
225 Wildwood Avenue, Woburn, MA 01 801-2041
A division of Reed Educational and Professional Publishing Ltd

First published 2001

British Library Cataloguing in Publication Data

A catalogue record for this book is available from the British Library

ISBN 0 7506 55094 X

Printed and bound in Great Britain

Contents

Preface viii

1. Basic circuit techniques 1

1.1 Resistor 1
1.2 Capacitor 3
1.3 Inductor 4
1.4 Transformer 5
1.5 Network theorems 7
1.6 Nodal analysis 10
1.7 AC theory 11
1.8 Fourier analysis 13
1.9 Exercises 15

2. Simple operational amplifier circuits 18

2.1 Operational amplifier 18
2.2 Non-inverting negative-feedback amplifier 19
2.3 Inverting feedback amplifier 20
2.4 Voltage-follower 21
2.5 Current-follower 21
2.6 Adder 22
2.7 Subtractor 22
2.8 Integrator 22
2.9 Differentiator 23
2.10 Exercises and experiments 23

3. Operational amplifier limitations 26

3.1 Finite op-amp gain 26
3.2 Op-amp non-linearity 27
3.3 Voltage offset 27
3.4 Bias current 28
3.5 Op-amp input impedance 29
3.6 Op-amp output impedance 30
3.7 Feedback resistor values 31
3.8 Common-mode response 31
3.9 Op-amp frequency response 32
3.10 Feedback amp frequency and step response 33
3.11 Slewing 34
3.12 Differentiator resonance 35
3.13 Photo detector amplifier 36
3.14 Exercises and experiments 37

4. Op-amp and other switching circuits 40

4.1 Frequency counter 40
4.2 Schmidt trigger 40
4.3 Comparator 42
4.4 Triangular and square wave 43
4.5 Precision 555 type astable 44
4.6 Monostable 44
4.7 Precision rectifier 46
4.8 Precision envelope detector 47
4.9 Peak-and-valley detectors 47
4.10 Sampling gates 48
4.11 Auto zero 49
4.12 DC restorer 50
4.13 Exercises and experiments 51

5. Semiconductor diode and power supplies 55

5.1 Semiconductors 55
5.2 Semiconductor diode 57
5.3 Incremental diode model 60
5.4 Diode reverse breakdown 61
5.5 DC power supplies 63
5.6 Switched-mode power supply 64
5.7 Diode capacitance and switching time 66
5.8 Exercises and experiments 69

6. Single-stage common-emitter amplifier 71

6.1 Bipolar transistor 72
6.2 Voltage gain 73
6.3 Collector output resistance 74
6.4 Base current 75
6.5 Hybrid-π transistor model 76
6.6 Frequency response 79
6.7 Biasing circuits 81
6.8 Exercises and experiments 83

7. Emitter-follower and other circuits 86

7.1. Low frequency operation of emitter-follower 86
7.2. Emitter-follower capacitances 88
7.3. Resonant capacitive loading 90
7.4. Common-base amplifier 91
7.5. Cascode amplifier 91
7.6. Long-tailed pair 92
7.7. Bipolar inverter switch 96

7.8. Bipolar analogue shunt switch 98
7.9 Exercises 100

8. Basic MOS circuits 102

8.1 MOS operation (descriptive) 102
8.2 MOS operation (quantitative) 104
8.3 Common-source voltage amplifier 105
8.4 Source-follower 108
8.5 Common-gate stage 111
8.6 Cascode 111
8.7 CMOS switch 113
8.8 Exercises 116

9. MOS-based operational amplifiers 117

9.1 LTP difference-mode transconductance 118
9.2 LTP common-mode transconductance 119
9.3 LTP frequency response 121
9.4 Cascode tail 122
9.5 Cascode current mirrors 123
9.6 Cascode OTA 124
9.7 Folded cascode 125
9.8 Two-stage Miller op-amp 126
9.9 OTA-based feedback amplifier 128
9.10 Supply-independent bias source 129
9.11 Exercises 130

10. Junction field-effect transistor 132

10.1 JFET operation 132
10.2 Common-source voltage amplifier 132
10.3 JFET capacitances 134
10.4 Frequency response 135
10.5 Temperature coefficient 135
10.6 JFET applications 135

11. Single-section passive RC filters 137

11.1 Simple RC low-pass 137
11.2 Simple RC high-pass 141
11.3 Pole-zero diagrams 142
11.4 Laplace transform 145
11.5 Exercises and experiments 145

12. Simple LC resonator-based filters 149

12.1 Series resonator narrow-band band-pass filter 149
12.2 Parallel resonator 152
12.3 Narrow-band band-stop filter 153
12.4 Notch filters 154
12.5 Second-order low- and high-pass filters 155
12.6 LC impedance matching 157
12.7 Broad-band band-pass and band-stop filters 157
12.8 Inductor specifications 158
12.9 LC oscillator 159
12.10 Exercises and experiments 159

13. Active RC filters 164

13.1 Low- and high-pass Butterworth filters 164
13.2 Linear-phase filters 166
13.3 Narrow-band band-pass filter 167
13.4 Notch filter 169
13.5 Notch-enhanced low-pass 170
13.6 Switched-capacitor filters 170
13.7 OTA-C filters 172
13.8 MOSFET-C filter 175
13.9 Exercises and experiments 177

14. Modulator and demodulator applications 180

14.1 AM radio link 180
14.2 Reversing-switch modulator 181
14.3 Reversing-switch demodulator 182
14.4 Single-sideband suppressed-carrier radio link 183
14.5 Frequency-changer 184
14.6 Exclusive-OR 185
14.7 Exercises 185

15. Long-tailed pair modulator circuits 187

15.1 Long-tailed pair modulator 187
15.2 LTP large signal analysis 188
15.3 Transposed input LTP 189
15.4 Four-quadrant modulator 189
15.5 Gilbert cell 189
15.6 Full range analogue multiplier 190
15.7 Detector circuits 191
15.8 Exercises and experiments 191

16. RF amplifiers and frequency-changers 193

16.1 RF and VHF amplifiers 194
16.2 Envelope distortion 195
16.3 Dual-gate MOS 198
16.4 Balanced frequency-changers 199
16.5 JFET frequency-changers 199
16.6 CMOS switching modulator 200
16.7 Diode ring modulator 200
16.8 Microwave modulator 201
16.9 Exercises 201

17. Noise 204

17.1 Thermal noise 204
17.2 Shot noise 205
17.3 1/f noise and drift 206
17.4 Dual source noise model 207
17.5 Bipolar transistor amplifier 209
17.6 Long-tailed pair noise 210
17.7 FET noise 211
17.8 Effect of feedback on noise 212
17.9 Photodiode amplifier 213
17.10 Exercises 215

18. Interference and its prevention 216

18.1 Electrostatic screening 216
18.2 Magnetic screening 218
18.3 Eddy-current screening 219
18.4 Ground loops 219
18.5 Decoupling 221
18.6 Exercises and experiments 222

19. Phase-locked loop and oscillators 224

19.1 PLL function 224
19.2 System diagram 225
19.3 Loop stability, lock range and capture range 226
19.4 Motor phase-locking 227
19.5 Noise-optimised measurements 230
19.6 Voltage-controlled oscillator 232
19.7 PLL frequency synthesisers 234
19.8 Direct digital synthesis 236
19.9 Exercises 237

20. Analogue-digital data conversion 240

20.1 DAC function 240
20.2 Switched output ladder DAC 241
20.3 Switched input ladder DAC 241
20.4 High-speed DAC (0800 type) 242
20.5 Multiplying DAC 244
20.6 DAC glitches 244
20.7 Successive-approximation ADC 245
20.8 Flash ADC 245
20.9 Dual-ramp ADC 246
20.10 Oversampling (sigma-delta) ADC 247
20.11 Oversampling DAC 250
20.12 Sampling gate and anti-aliasing 251
20.13 Exercises and experiments 251

21. SPICE and the audio power amplifier 255

21.1 Power amplifier 255
21.2 SPICE bipolar transistor model 258
21.3 Thermal stability 259
21.4 Final stage linearity 261
21.5 Final stage frequency response 263
21.6 Overall frequency response 264
21.7 Overall linearity 266
21.8 Exercises 267

22. Transmission lines and transformers 269

22.1 Line response to step input 269
22.2 Digital signal transmission 270
22.3 Transmission line equations 272
22.4 Sine wave transmission 273
22.5 Reflectometer 275
22.6 Balun 276
22.7 Transmission line transformer 277
22.8 Exercises 279

23. S-parameters and the Smith chart 283

23.1 y- and h-parameters 283
23.2 s-parameters 284
23.3 Impedance matching 286
23.4 Smith chart 286
23.5 Amplifier matching using the Smith chart 288
23.6 s-parameter measurement 289
23.7 Exercises 290

24. Optoelectronics and other subjects 292

24.1 Fiber-optic data link 292
24.2 Light emitting diode (LED) 293
24.3 Semiconductor laser 294
24.4 Photo-diode 295
24.5 Opto-isolator 297
24.6 Piezo-electric filters 297
24.7 Electrostatic microphones 300
24.8 Bandgap voltage reference 301
24.9 Current feedback op-amp 302
24.10 Thyristor (SCR) 304

Index 307

Preface

This text derives initially from the notes for the course on Analog Electronics given to students taking the MSc courses in Electronics and Microelectronics in the Department of Electronics and Computer Science of the University of Southampton. These are students normally with a first degree in an allied subject such as Mathematics, Physics or Electrical Engineering. As such, a feature is the coverage of a large amount of material in a relatively short space, hopefully good value for money!

With the material much the same as for the Analog component of the three-year undergraduate course in Electronic Engineering, the book has been expanded to suit this readership also, in particular Chapter 1 on Basic Circuit Techniques.

With some of the material given to research chemists and to undergraduates taking the Physics-with-Electronics, and Chemistry-with-Electronics courses, the interests of the experimental scientist have also been covered.

With all of the above courses including laboratory work, that given in the final exercises section of most chapters is thoroughly 'road tested'.

All of the numeric calculations have been checked, by hand with the calculator, and by simple BASIC programs.

Use as reference text

With the book intended for use in this mode as much as for the above teaching support, careful attention has been given to indexing, and to the content of chapter, section, and sub-section headings.

SPICE

The book is not a SPICE manual, but the audio power amplifier in Chapter 21 provides an excellent example of the use of this method of circuit simulation.

Analog-to-digital conversion (ADC)

As the title implies, coverage includes examples of both this and the complementary digital-to-analog converter (DAC). The now popular oversampling ADC and DAC are included, with experimental exercises for each.

Optoelectronics

With this subject of great current interest, the last chapter includes a brief coverage of the topic, with particular emphasis on the fibre-optic digital data link.

Source texts

My source of knowledge has mainly been lectures and hands-on experience in the company of many experts in the various areas of the subject, rather than from extensive reading. Exceptions have been the material on MOS devices and s-parameters where I am heavily indebted to the specialist texts quoted. All chapters, however, have required some degree of reference, with the texts quoted in the Reference section.

Acknowledgements

Working now at home, I am most grateful for the help and patience of past colleagues whom I have repeatedly phoned at inconvenient moments for confirmation of various points, in no particular order, Bill Redman-White (oversampling ADC) Arthur Brunnschweiler (Microelectronics), David Stewart and Steve Braithwaite (Communications), Bob Greef (Physical Chemistry), Ed Zaluska (Digital Electronics), and my son Richard (Nortel at Paignton, (Opto-Electronics).

I also gratefully acknowledge the free three-month loan of the full version of PSpice from MicroSim and the generous gift of a complete copy of SpiceAge from Those Engineers. With both used in Chapter 21, the earlier subsection work was mainly with the freely available evaluation version of PSpice, with the later 'whole system' simulations with SpiceAge. For the two column format of the book, it was convenient to Coreldraw hand trace both sets of graphs.

Notation

Reference to all equations is by the use of square brackets. For example, for an equation in view

'with I_d as in [5.2] the first term is usually dominant',

while for the equation not in view

'with $\quad I_d = I_{ds}[\exp(qV_d/kT) - 1] \quad$ as in [5.2]

the first term is usually dominant'

$\approx$ means 'is approximately equal to' as usual
$\sim$ means 'is roughly equal to' from say $\times 1/3$ to $\times 3$.

...and finally

Milne's 'bear of little brain' was fascinated to see that every time he dropped in a fir cone 'on the other side' it came out 'on this side'. This is the first approach to our subject but some can't leave it at that, they want to know how it works. This book is for them.

Winnie the Pooh's little book of wisdom, Methuen, 1999.

1

Basic circuit techniques

Summary

This chapter introduces the passive components and basic analytical tools that are used in the following chapters. The components are the resistor, capacitor, inductor and transformer. Active components such as the bipolar transistor are covered later.

For a text on analog circuit techniques, it is appropriate to start with the familiar hydraulic analog of the flow of current through a resistor in Fig. 1.1, and the possibly less familiar analogs of the capacitor, inductor and resonator that follow. After this the basic network theorems such as Thévenin and superposition are introduced, extending to the method of nodal analysis, here for resistor-only networks.

Reactive components

The trouble with the inductor and capacitor is that they change the shape of an applied signal. For example, with I=CdV/dt for the capacitor, and for the applied current I constant then, instead of the resulting voltage V being constant as for the resistor, for the capacitor V is a steadily rising ramp.

An important exception is the sine wave, for which differentiation or integration changes the phase and the amplitude but not the sinusoidal shape, making calculation of the circuit waveforms less difficult. The corresponding techniques are known as AC theory, the sine wave being an 'alternating current'.

Fourier analysis

All periodic waveforms, and some others, can be resolved into sinusoidal components, by the method of Fourier analysis. The effect of the network on each of the input components can then be calculated, and the resulting set of output components recombined to give the actual output waveform.

The Laplace transform, a method for dealing with waveforms that cannot be resolved into Fourier components, is covered in Chapter 11.

1.1 Resistor

For the resistor R in Fig. 1.1(a), the hydraulic (water) analog is as in (b), a section of pipe offering resistance to flow. With $I \propto V$ then by Ohm's law

$$I=V/R \tag{1.1}$$

With the 'conductance' G=1/R, these two relations need be taken as memorised, to avoid tedious repetition.

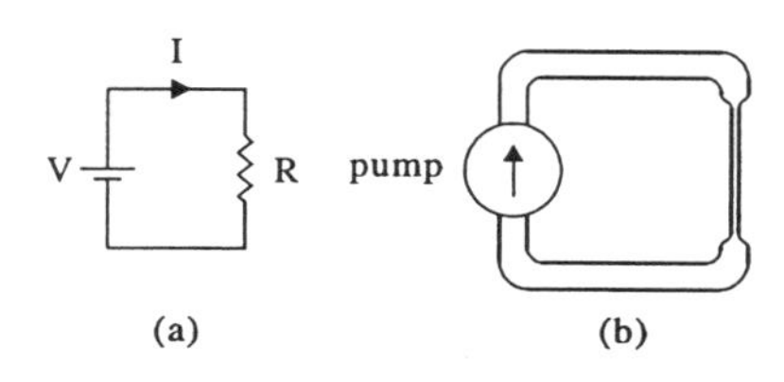

Fig. 1.1 Resistor connected to constant voltage supply.
(a) Circuit diagram.
(b) Water analog.
Voltage V ⇔ pump pressure.
Current I ⇔ flow rate.
R ⇔ pipe resistance.

The unit of resistance is the ohm (Ω). For the widely used metal-film resistor, R typically ranges from 1Ω to 10^7Ω.

As for normal practice, we use the notation in Table 1.1 to represent the various orders of ten. Here the range of 1Ω to 10^7Ω for the metal-film resistor would be stated as from 1Ω to 10MΩ.

The convention applies equally to all other variables, for example 3×10^{-2}V is 30mV.

The unit of conductance is the siemens (S), with 1S=1/1Ω, 1/1kΩ=1mS, etc.

Preferred values

The 'preferred values' for manufactured components are logarithmically graded. For resistance R=10, 15, 22, 33, 47, 68, and 100Ω, then continuing, 150, 220, 330Ω, etc.

Here the increment is $10^{1/6} \approx 1.5$. For higher precision the increment is $10^{1/12}$, giving 10, 12, 15, 18, etc.

Values beyond the 1Ω and 10MΩ limits for the metal-film resistor exist using other methods of manufacture.

Pico	Nano	Micro	Milli
1pΩ	1nΩ	1µΩ	1mΩ
10^{-12}Ω	10^{-9}Ω	10^{-6}Ω	10^{-3}Ω

	Kilo	Mega	Giga
1Ω	1kΩ	1MΩ	1GΩ
10^{0}Ω	10^{3}Ω	10^{6}Ω	10^{9}Ω

Table 1.1 Symbolic representation of orders of ten, as for resistance.

Variable resistor

These are normally in the form of a resistive strip, with a movable contact in addition to the two end contacts. The strip is in the form of a nearly complete circle for front-panel knob operation. For the preset 'trimmer', the strip may be either circular or linear. The linear trimmer is screw driven, giving more turns to cover the range than the circular device. Both are screwdriver operated.

Power

The resistor dissipates power P as heat. With P=IV then

$$P=V^2/R \quad (1.2) \quad \text{or} \quad P=I^2R \quad (1.3)$$

The unit of power is the watt (W: James Watt). A manufactured resistor has a power rating, the maximum continuous power it is able to dissipate without overheating. 0.6W is typical for the commonly used metal-film resistor.

Sine wave

This is defined as in Fig. 1.17. The vector of unit length in (a) rotates at the angular frequency ω to trace sin(θ)=sin(ωt) on the vertical axis as in (b). The associated cos(ωt) is traced on the horizontal axis. A vector which rotates in this way is known as a 'phasor'. The sine wave is the natural output of the rotating power alternator giving

$$V = \hat{V}\sin(\omega t) \qquad (1.4)$$

where $\hat{V}$ is the peak value of V. With ω the frequency in radians/sec then

$$\omega=2\pi f \qquad (1.5)$$

where f is the frequency in hertz (Hz = cycles/sec). This relation is in constant use and so also needs be taken as memorised.

RMS value

This is a concept first encountered in the supply industry, where some areas were once supplied with 'direct current' (DC) and others with AC. Here the requirement was to state the supply voltage in a way that the power delivered to a resistive load could be calculated from $P=V^2/R$ as in [1.2] regardless of the type of supply.

With $\hat{V}_{ac}$ the peak value of the AC supply voltage V_{ac} then the instantaneous power P_{ac} is only equal to $\hat{V}_{ac}^2/R$ at the peak of V_{ac}. The mean power $\overline{P_{ac}}$ is less. Thus the value of V_{ac} quoted is $\tilde{V}_{ac}$, that for which [1.2] does apply, giving

$$\overline{P_{ac}} = (\tilde{V}_{ac})^2/R$$

With the instantaneous

$$P_{ac}=V_{ac}{}^2/R$$

then

$$\overline{P_{ac}} = \overline{V_{ac}^2}/R,$$

where $\overline{V_{ac}^2}$ is the 'mean square' value of V_{ac}.

Thus

$$\tilde{V}_{ac} = \sqrt{\overline{V_{ac}^2}} \qquad (1.6)$$

the 'root-mean square' value of V_{ac}.

With

$$V = \hat{V}\sin(\omega t)$$

as in [1.4] then

$$\overline{V_{ac}^2} = \hat{V}_{ac}^2\overline{\sin^2(\omega t)} = \hat{V}_{ac}^2/2$$

giving

$$\tilde{V}_{ac} = \hat{V}_{ac}/\sqrt{2} \qquad (1.7)$$

With the stated 240V mains voltage the rms value $\tilde{V}_{ac}$, then the peak value **$\hat{V}_{ac}$ =340V.**

> **Problem** For the three-phase distribution system the phasors for the three 'lines' are spaced at 120° intervals. Draw the resulting phasor diagram and thus show that, for the stated 240V rms between line and 'neutral', the rms line-to-line voltage is 415V.

Noise fluctuation

The rms value of [1.6] is a valid indicator of the power transferred to the load for any time-varying voltage or current that has a definable mean, in particular for the random noise fluctuation of Chapter 17.

1.2 Capacitor

Fig. 1.2(a) shows the capacitor C connected to a constant voltage supply. At its simplest, the capacitor is two metal plates separated by an insulator. Current cannot flow through it, but charge Q_c, in the form of conducting electrons, can be added to one plate and removed from the other, giving

$$Q_c = CV \tag{1.8}$$

where C is the 'capacitance' of the component. C is proportional to the plate area and inversely proportional to the separation. While no current can flow through the capacitor, a current will flow in the outer circuit while the charge is being established.

Water analog

The water analog of the capacitor is the elastic membrane stretched across the pipe in Fig. 1.2(b). No water flows through the membrane, but there is a flow through the adjoining pipework as pressure is applied to stretch the membrane.

If two shutters are inserted across the pipe to isolate the stretched membrane, this is equivalent to disconnecting the capacitor once it has been charged. The charge is then held until the wire ends are shorted together (the two shuttered ends of the pipe bent round to meet and the shutters removed). Then an instantaneous and theoretically infinite current (a current 'impulse') flows as the capacitor discharges.

Alternatively the capacitor may be connected to a resistor R (attach narrow pipe – remove shutters). The discharge then takes the non-zero time T where, as shown in Section 11.1, T=CR.

The water analogy extends to the breakdown of the insulator (rupture of the membrane) for excessive V.

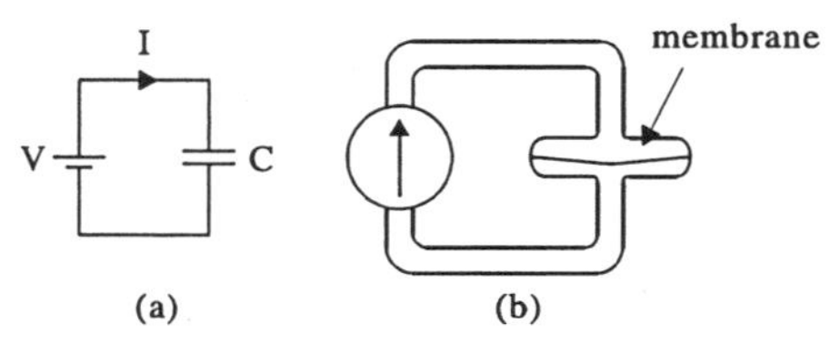

Fig. 1.2 Capacitor C connected to constant voltage supply.
(a) Circuit diagram.
(b) Water analog.
Voltage V ⇔ pump pressure.
Charge Q=CV ⇔ volume displaced by membrane.
C ⇔ membrane elasticity.

Capacitor current

With the current I at a point in a circuit defined as the rate dQ/dt at which charge Q moves past that point, and with the capacitor charge Q_c as in [1.8], then for the present capacitor

$$I = C\,dV/dt \tag{1.9}$$

For the alternating

$$V = \hat{V}.\sin(\omega t) \tag{1.10}$$

then

$$I = \hat{V}\omega C.\cos(\omega t) \tag{1.11}$$

Thus for an applied AC voltage there is a proportional AC current (a pump of alternating pressure will drive an alternating flow, as the membrane is alternately stretched in one direction and the other). In short, a capacitor is a component that will pass an AC but not a DC current.

Reactance

From [1.10] and [1.11]

$$\hat{V} / \hat{I} = 1 / \omega C \tag{1.12}$$

Here $1/\omega C$ is somewhat equivalent to R for the resistor, except for the resistor I and V are in phase, while for the capacitor I_{in} leads V_{in} by 90°. Thus $1/\omega C$ is termed the 'reactance' of C, with the symbol for reactance X. With X_c the reactance of the present C, we write

$$|X_c| = 1/\omega C \tag{1.13}$$

rather than $X_c = 1/\omega C$, since $1/\omega C$ is only the magnitude of X_c. A representation of X_c which gives both the magnitude and the 90° phase difference is given in Section 1.7.

Much as G=1/R for the resistor, for the 'susceptance' B of a reactive component B=1/X. This too needs to be memorised.

Example. For C=100nF and f=10kHz then $\mathbf{|X_c| = 159\Omega}$.

Values

The unit of capacitance is the farad (F: Michael Faraday). For manufactured devices C ranges from approximately 1pF to one farad, with the usual preferred values. For a capacitor formed of two circular plates (Compton, 1990)

$$C \approx \varepsilon_o \varepsilon_r A/d \tag{1.14}$$

where ε_o is the permittivity of free space, ε_r is the relative permittivity of the dielectric, A is the area of one face of one of the plates, d is the plate separation, and d<<D where D is the plate diameter.

Example. For $A = 1cm^2$, d=1mm, air as the dielectric, and with $\boldsymbol{\varepsilon_o = 8.85 \times 10^{-12}}$ then $\mathbf{C \approx 0.885pF}$.

Electrolytic capacitor

With $C \propto 1/d$ as above then for $C > 1\mu F$ the capacitor will normally be of the 'electrolytic' type, where the very thin insulating film is developed electrochemically within the device. Normally the electrolytic deposition will allow a mean voltage of only one polarity across the capacitor, that used to form the film. The other polarity will reverse the process and destroy the film. Any capacitor has a voltage rating, the value above which the dielectric may break down.

High-k capacitor

These use a type of dielectric for which the dielectric constant is extremely high, allowing values of C up to about 100nF with the capacitor still in the form of two metal plates separated by a single layer of dielectric. The conducting 'plates' are two metal films deposited on a disc of the high-k material. The high-k capacitor is particularly suitable for high-frequency applications where the parasitic series inductance L_p needs to be as low as possible.

Polyester type construction

For a capacitor of the above kind of value using a normal dielectric, the construction will be in the form of two long strips of conducting foil separated by a strip of flexible insulating material (e.g. polyester). The assembly is then rolled up for compactness, making L_p a good deal larger.

Imperfections

Apart from the larger L_p, the conventional capacitor approximates very well to the ideal. When an alternating voltage is applied to a capacitor to give a current of proportional amplitude then for the ideal capacitor no power is dissipated. Energy is simply accepted over one half of the cycle and returned over the next. Unfortunately the same is not true for the high-k component for which, although the series L_p is minimal, there is also a small effective series R_p. This becomes progressively more important as the frequency f of the applied voltage increases.

It is also difficult to manufacture the high-k component to an exact value. The quoted tolerance thus tends to be relatively large.

Finally, for the high-k the variation of C with temperature (the 'temperature coefficient') is higher than for other types.

The high-k capacitor is thus best suited to power decoupling applications, where the applied voltage is essentially DC, the exact value (and variations in it) do not matter, and a low series inductance is a prime requirement.

Variable capacitors

These exist in several forms. The knob-controlled type normally consists of two sets of interleaved plates, each in the form of a filled half circle, and on a common axis. One set of plates is then rotated into the other, as far as is needed for the required C.

For a pre-set 'trimmer' the sets of plates are cylindrical, the one rotating on a screw into the other. Alternatively two sets of rectangular plates are separated by slips of plastic insulator (often mica). The plates are springy and are closed on to the dielectric using a screw, again to the degree needed for the required C.

Problem By integrating the instantaneous power $V \times I$ flowing into a capacitor C over one cycle of the applied voltage, show that, with the phase of I and V differing by 90°, the mean power transferred over one cycle of the sinusoidal applied voltage is zero.

Similarly show that, regardless of the charging waveform, when C is charged to V_c the stored energy $E = CV^2/2$.

1.3 Inductor

The inductor is simply a coil of wire, sometimes with a core of iron or ferrite. If a current I is passed through the coil this sets up a magnetic flux Φ. If I changes then so does Φ, to develop the voltage

$$V = N d\Phi/dt \tag{1.15}$$

where N is the number of turns on the coil. The unit of flux is the weber (Wb), which is such that if the flux linking a coil of one turn is changing at the rate of one Wb/sec then the induced voltage will be one volt. With $V \propto d\Phi/dt$ as in [1.15] and $\Phi \propto I$ then

$$V = L dI/dt . \tag{1.16}$$

where the constant L is termed the 'inductance' of the component. The inductor is thus the complement of the capacitor. It develops no voltage for constant I, while the capacitor allows no current for constant V.

Water analog

The analog of the inductor is as in Fig. 1.3(b). The flywheel with non-leaking paddles requires pressure (V) to make the speed (I) change, but for constant speed (I) no pressure (V) is needed.

With the inductor connected as in (a), the constant V gives constant dI/dt. Thus I increases at a steady rate. Correspondingly for the flywheel, constant pressure (V) gives constant acceleration and thus a steadily increasing flow rate (I).

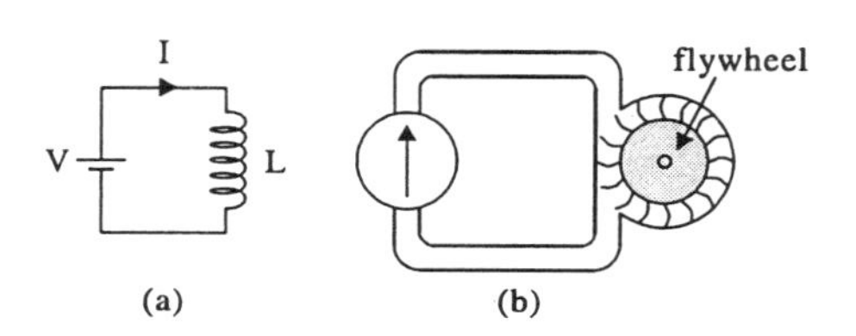

Fig. 1.3 Inductor L connected to constant voltage supply.
(a) Circuit diagram.
(b) Water analog.
Voltage $V = LdV/dt$ ⇔ pump pressure.
Current I ⇔ flow rate.
L ⇔ flywheel moment of inertia.

Current cut-off

If the pipes connecting to the moving wheel are shuttered, and if the shuttering is sufficiently abrupt, then either shutter or paddles or pipe will be shattered. This happens for the inductor too. As the wire is broken then briefly $dI/dt=-\infty$. With $V=LdI/dt$ as in [1.16] then $V=-\infty$, to be checked by the generation of a spark across the opened contacts, the well-known basis of the ignition system in a car engine.

If a resistor R is connected across the point where the wire is to be broken then V is limited to the value IR where I is the current that was flowing before the break. Thereafter I decays at the rate $dV/dt=-V/L$, to fall to zero after a time $T=L/R$.

Reluctance

With $\Phi \propto NI$ then
$$\Phi=NI/\mathsf{R} \quad (1.17)$$

where the 'reluctance' R of the magnetic circuit is analogous to the resistance R in an electric circuit. With $V=Nd\Phi/dt$ as in [1.15] and $V=LdI/dt$ as in [1.16] then

$$L=N^2/\mathsf{R} \quad (1.18)$$

R is not always easy to calculate but for an inductor wound on a toroidal core (Compton, 1990)

$$\Phi=NI(\mu_o\mu_r a/d) \quad (1.19)$$

where μ_o is the permeability of free space, μ_r is the relative permeability of the core material, a is the cross-sectional area (csa) of the toroid, and d is the distance round the toroid. From [1.17] and [1.19] then $\mathsf{R}=d/(\mu_o\mu_r a)$. With L as in [1.18] then

$$L=\mu_o\mu_r N^2 a/d \quad (1.20)$$

The unit of inductance is the henry (H). With $V=LdI/dt$ as in [1.16], 1H is the inductance for which $dI/dt=1A/sec$ gives $V=1Volt$.

Example For ten turns on a ferrite toroid of $\mu_r=1000$, 4cm diameter, and 40mm² csa, and with $\mu_o=4\pi\times10^{-7}$
L=126μH.

Reactance X_L

With $V=LdI/dt$ as in [1.16] for the inductor, and for $I=\hat{I}\sin(\omega t)$, then $V=\omega L.\hat{I}\cos(\omega t)$ giving

$$\hat{V}/\hat{I}=\omega L \quad (1.21)$$

With X_L the reactance of the inductor then

$$|X_L|=\omega L \quad (1.22)$$

now with V leading I by 90°, rather I leading V as for the capacitor.

Example For a 1mH inductor with f=50kHz
$|X_L|=314\Omega$.

Resonance

Suppose that the wheel in Fig. 1.4 is rotated manually to stretch the membrane and then released. Assuming no power loss, then what follows will be an oscillation which continues indefinitely (simple harmonic motion – SHM). The same occurs for the LC circuit, if the capacitor is initially charged before being connected to the inductor.

If L and C in (a) are connected in series, with I an applied sinusoidal current, then the voltages across the two components will be in anti-phase, V_L leading I by 90° and V_c lagging. With $|X_L|=\omega L$ and $|X_c|=1/\omega C$ then at the frequency ω_o for which the two are equal the total voltage, and thus the total impedance, is zero.

With here a finite current for zero applied voltage, this is precisely the condition for the above SHM, so the 'resonant frequency'

$$\omega_o=1/\sqrt{LC} \quad (1.23)$$

is both the frequency for zero impedance, and also that of the undamped simple harmonic oscillation that can occur for no resistance in the circuit.

Example For L=2mH and C=1nF then **f_o=113kHz.**

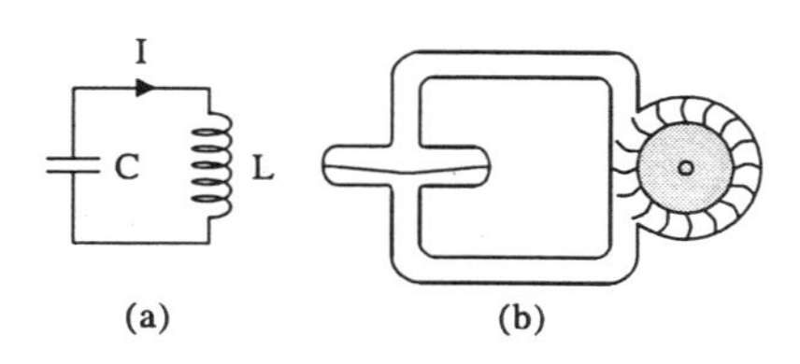

Fig. 1.4 LC resonator.
(a) Circuit diagram.
(b) Water analog.

1.4 Transformer

The transformer is essentially an inductor with two windings, usually on an iron or ferrite core. It is widely used in the power distribution industry for converting one voltage to another, ideally without loss of power. At a lower level, this is as the transformer in the DC power supply of Section 5.5, where the 240V of the AC mains is converted to the lower voltage needed to drive the rectifier. The transformer is also used for altering signal voltage levels, and the associated impedances.

Voltage transformation

Fig. 1.5(a) shows the normal symbol for the transformer. With Φ the core flux, and for the winding senses shown

$$V_1=N_1 d\Phi/dt \quad (1.24)$$

and
$$V_2=N_2 d\Phi/dt \quad (1.25)$$

giving
$$V_2/V_1=N_2/N_1 \quad (1.26)$$

Magnetising current

I_1 has two components, I_{12} incurred by the load current I_2, and the magnetising current I_m needed to establish Φ.

With $V=LdI/dt$ as in [1.16] for the inductor and for $I_2=0$ then

$$V_1=L_1dI_m/dt \qquad (1.27)$$

With V_m sinusoidal then $I_m(rms)=V_1(rms)/\omega L_1$.

Example For the mains $V_1=240V$ rms at 50Hz, and the typical $L_1=5H$, **$I_m=153mA$ rms.**

Load current

With $V_1=N_1d\Phi/dt$ as in [1.24] then Φ is determined entirely by the applied V_1. Thus when I_2 flows the resulting I_{12} must be such as to produce the flux Φ_{12} which exactly cancels $\Phi_2=N_2I_2/R$. With $\Phi_{12}=N_1I_{12}/R$ then

$$I_{12}/I_2=N_2/N_1 \qquad (1.28)$$

With $I_1=I_m+I_{12}$, the transformer can then be modelled as in Fig. 1.5(b).

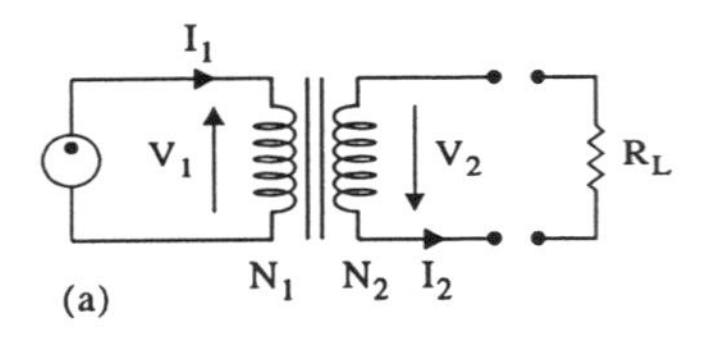

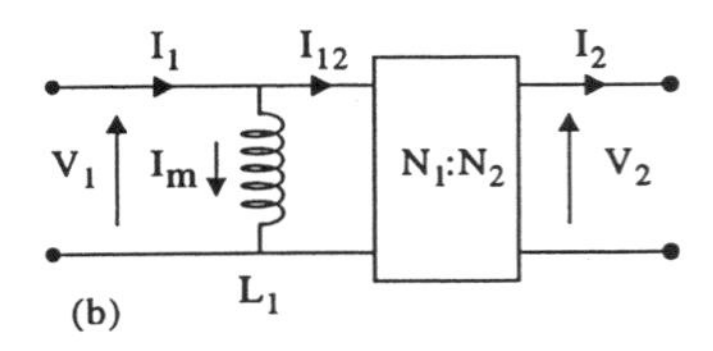

Fig. 1.5 Use of transformer to couple source to load.
(a) Circuit diagram.
(b) Circuit equivalent with turns ratio N_1/N_2.

Power

For the transformer as so far presented there is no power loss. With I_m purely inductive then, as for C, there is no power dissipation, giving the input power $P_{in}=V_1I_{12}$. With $P_{out}=V_2I_2$, $V_2/V_1=N_2/N_1$ as in [1.26], and with $I_{12}/I_2=N_2/N_1$ as in [1.28], then $P_{in}=P_{out}$.

Impedance transformation

With Z_1 the impedance presented by the transformer primary then $Z_1=\omega L_1//R_1$ where $R_1=V_1/I_{12}$ is the component due to I_2. With again [1.26] and [1.28] then

$$R_1=R_L(N_1/N_2)^2 \qquad (1.29)$$

With ωL_1 in parallel it is clear why the transformer can only be used with V_1 alternating. Otherwise $\omega L_1=0$ giving blown fuses!

Power matching

For a DC voltage source of output impedance R_s connected to a load R_L, the load power P_L is highest when $R_L=R_S$. With $P_L=V_L^2/R_L$ where V_L is the load voltage, and with E the open-circuit source voltage then

$$P_L = E^2\left(\frac{R_L}{R_L+R_s}\right)^2 \Big/ R_L \qquad (1.30)$$

By differentiation, or from the symmetry of the expression in the form

$$P_L = \frac{E^2}{R_s}\left(\frac{1}{\sqrt{R_L/R_s}+\sqrt{R_s/R_L}}\right)^2 \qquad (1.31)$$

it is confirmed that P_L is maximum for $R_s=R_L$

Available source power

With the resulting $\qquad P_{av} = E^2/(4R_s) \qquad (1.32)$

essentially the same for an AC signal source, the transformer can be used to power match a differing R_L and R_S.

Example With R_1 as in [1.29], to match an 8Ω loudspeaker to the $R_s=10k\Omega$ of a valve audio power amplifier, **$N_1/N_2=35.4$.**
Here $N_1/N_1=35$ or 36 would be used.

> **Problem** Use some gear wheels to extend the inductor water analog of Fig. 1.3(b) to one for a transformer.

Imperfections

For a real transformer the windings have a small ohmic resistance. For the circuit model these simply appear as R_{w1} and R_{w2} in series with the windings. These do incur a power loss, as do the eddy currents and hysteresis losses in the core.

It has also been assumed that all of the flux induced by one winding is linked to the other. For the types of transformer considered so far this is nearly so. The 'mutual inductance' for less than 100% linkage is discussed below.

Construction

For the type of transformer used in a mains supplied DC power supply the core is traditionally of soft iron and in the form of a squared-off figure of eight set on its side. To overcome the problems of winding, this is done first on a suitable bobbin, with one winding on top of the other. The core is then composed of a large number of sheet soft iron stampings, alternately in the shape of letters E and T. These are threaded alternately into the bobbin, to fill it. The stampings are insulated from each other to reduce core eddy-current losses. More recently a ferrite toroid is favoured, methods having been found for winding this with the necessary large number of turns for the 240V primary.

Current sampling transformer

Fig. 1.6 shows a method of sampling the line current I_1, ideally without changing it, and with no electrical contact. One use here is as a current-sensing oscilloscope probe, with the toroid split and sprung so that it can be clipped over the current carrying wire. V_2 is then applied to the scope input.

For R_L disconnected the ferrite ring adds X_1 to the self-inductance of the line, with the resulting change in I_1 possibly significant. With R_L connected, and with $N_1=1$, then the impedance $R_1=R_L/N_2^2$ is placed in parallel with X_1. This can be made as low as is required by appropriate choice of R_L and N_2.

Example For $R_L=200\Omega$ **N=10** gives $\mathbf{R_1=2\Omega}$.

An important feature of the arrangement is that it gives $V_2 \propto I_1$, in contrast to $V_2 \propto dI_1/dt$ for the secondary open. However, with $X_L=\omega L_2$ in parallel with R_L, this imposes a low-frequency limit, at the frequency ω_L for which $\omega_L L_2=R_L$. Thus

$$\omega_L=R_L/L_2 \qquad (1.33)$$

Example For $L_2=126\mu H$ as previously calculated, and with $f_L=R_L/2\pi L_2$, then $\mathbf{f_L=253kHz}$.

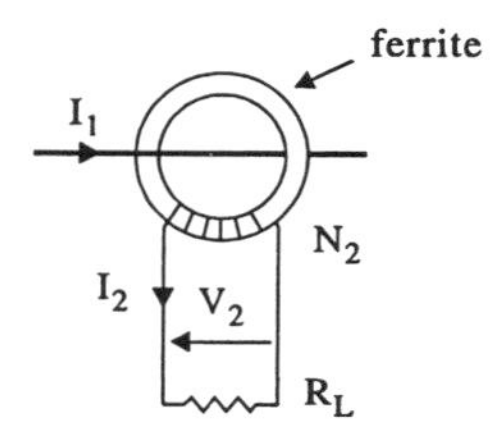

Fig. 1.6 Current sensing transformer.

Mutual inductance

It is assumed above that the core flux Φ is the same for both primary and secondary. This is nearly so for the soft-iron figure-of-eight core assembly used for the mains-driven DC power supply, and also nearly so for the ferrite toroid. It is much less so when there is no ferromagnetic core, as for higher frequency applications. Here it is most suitable to view the transformer as two inductances L_1 and L_2 coupled by the 'mutual inductance' M.

For full linkage, with $\Phi=NI/R$ as in [1.17], and Φ_1 and Φ_2 the components of Φ due to I_1 and I_2, then

$$\Phi_1=N_1I_1/R \ldots(a) \qquad \Phi_2=N_2I_2/R \ldots(b) \qquad (1.34)$$

and for the winding senses shown

$$\Phi=\Phi_1-\Phi_2 \qquad (1.35)$$

With $V_1=N_1 d\Phi/dt$ as in [1.24] and $V_2=N_2 d\Phi/dt$ as in [1.25] then

$$V_1=L_1 dI_1/dt - M dI_2/dt \quad \ldots(a)$$
$$V_2=L_2 dI_2/dt - M dI_1/dt \quad \ldots(b) \qquad (1.36)$$

where

$$L_1=N_1^2/R \quad \ldots(a)$$
$$L_2=N_2^2/R \quad \ldots(b)$$
$$M=\sqrt{L_1L_2} \quad \ldots(c) \qquad (1.37)$$

For less than full linkage

$$M=k\sqrt{L_1L_2} \qquad (1.38)$$

where k is the 'coupling factor', with k=1 for full linkage.

1.5 Network theorems

For this and the next section the networks contain only resistors, voltage sources, and current sources. Extension to the reactive C and R follows in Section 1.7.

Simple potentiometer

One of the simplest resistive networks is the voltage-dividing potentiometer of Fig. 1.7. With

$$I=E/(R_1+R_2) \quad \text{and} \quad V_o=I/R_2$$

then

$$V_o=E\frac{R_2}{R_1+R_2} \qquad (1.39)$$

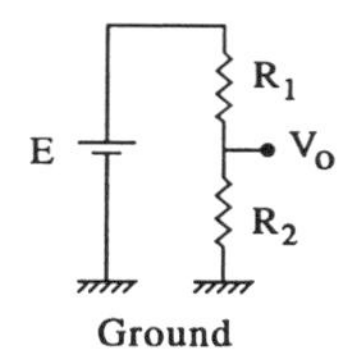

Fig. 1.7 Potentiometer

Ground connection

Notice the two symbols labeled 'ground'. These represent the common terminal to which all of the other voltages are referred. Thus V_o represents the voltage between the terminal so labeled and ground. Usually this point is connected to the metal box in which the circuit is enclosed.

As here, the symbol is often used several times over, to reduce the number of wires that have to be drawn.

Parallel resistors

This equally simple resistive network is shown in Fig. 1.8. With G_p the total parallel conductance,

$$I_p = I_1 + I_2, \quad I_1 = EG_1, \quad I_2 = EG_2, \quad I_p = EG_p$$

then, rather obviously,

$$G_p = G_1 + G_2 \qquad (1.40)$$

Also $$R_p = 1/G_p \qquad (1.41)$$

Alternatively, from these relations

$$R_p = \frac{1}{1/R_1 + 1/R_2}$$

giving

$$R_p = \frac{R_1 R_2}{R_1 + R_2} \qquad (1.42)$$

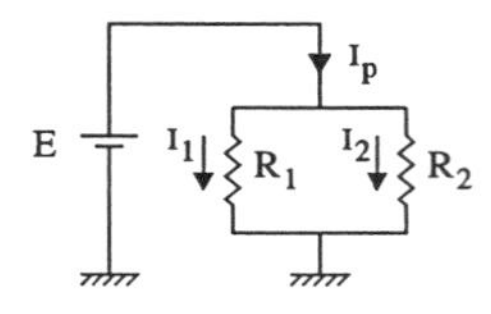

Fig. 1.8 Parallel resistors.

Example Calculate R_p for $R_1 = 3k\Omega$ and $R_2 = 5k\Omega$.

With $G_p = 1/R_1 + 1/R_2$

then $G_p = 1/3k\Omega + 1/5k\Omega = 533mS$

giving $\mathbf{R_p = 1.88k\Omega.}$

Alternatively with R_p as in [1.42]

$$R_p = (3k\Omega \times 5k\Omega)/(5k\Omega + 3k\Omega)$$

giving the same result.

The parallel R_p in Fig. 1.8 is often written as $R_1//R_2$ or $R_{1//2}$, with the representation also used for other variables combined in this reciprocal fashion.

Potentiometer of Fig. 1.9

Here

$$V_O = E\frac{R_{2//3}}{R_1 + R_{2//3}} \quad \text{which with} \quad R_{2//3} = \frac{R_2 R_3}{R_2 + R_3}$$

gives

$$V_O = E\frac{R_2 R_3}{R_1 R_2 + R_2 R_3 + R_3 R_1} \qquad (1.43)$$

Example Calculate V_o for

$E = 10V$, $R_1 = 1k\Omega$, $R_2 = 3k\Omega$ and $R_3 = 2k\Omega$.

With $G_{2//3} = 1/3k\Omega + 1/2k\Omega = 833\mu S$

then $R_{2//3} = 1.2k\Omega$.

With $V_o = ER_{2//3}/(R_{2//3} + R_1)$

then $V_o = 10V \times 1.2k\Omega/[1.2k\Omega + 1k\Omega]$

giving $\mathbf{V_o = 5.45V}$

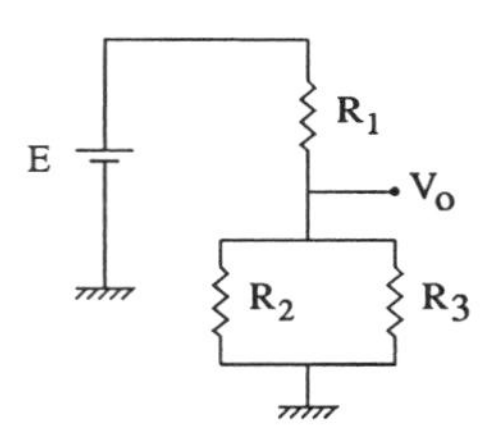

Fig. 1.9 Resistor network giving V_O by viewing R_1 and $R_2//R_3$ as potentiometer.

Three parallel resistors

With $R_p = R_1//R_2//R_3$

then $$R_p = 1/G_p \qquad (1.44)$$

where $$G_p = G_1 + G_2 + G_3 \qquad (1.45)$$

Alternatively

$$R_p = \frac{1}{1/R_1 + 1/R_2 + 1/R_3}$$

giving

$$R_p = \frac{R_1 R_2 R_3}{R_1 R_2 + R_2 R_3 + R_3 R_1} \qquad (1.46)$$

Example Calculate R_p for

$R_1 = 1k\Omega$, $R_2 = 500\Omega$ and $R_3 = 1.5k\Omega$.

We use [1.44] and [1.45], rather than [1.46].

With G_p as in [1.45] then

$$G_p = 1/1k\Omega + 1/500\Omega + 1/1.5k\Omega = 3.66mS.$$

With R_p as in [1.45] then $R_p = 1/3.66mS$
giving $\mathbf{R_p = 273\Omega.}$

While the two approaches were of comparable complexity for calculation for two parallel resistors, for the present three, calculation using [1.46] is the more complex. The disparity increases further for larger numbers of components.

Thevenin's theorem

It is normally possible to solve a circuit problem by several of the present methods. It is thus wise to do so by at least two, to check for errors. Thus we shall introduce

Thévenin's theorem by showing how it can be used to confirm [1.43] for Fig. 1.9.

The circuit is first divided as in Fig. 1.10(a,b). The Thevenin equivalent of (a) is then as in (c), with E_T the value of V_o with R_3 disconnected. Also R_T is the resistance presented by (a) at the point labelled V_o but with E=0 (short circuit). Thus for (a)

$$E_T = ER_2/(R_1+R_2) \quad (1.47)$$

and

$$R_T = R_1//R_2 \quad (1.48)$$

Then for (c,d) $V_o = E_T R_3/(R_T+R_3)$ (1.49)

giving V_o is as in [1.43].

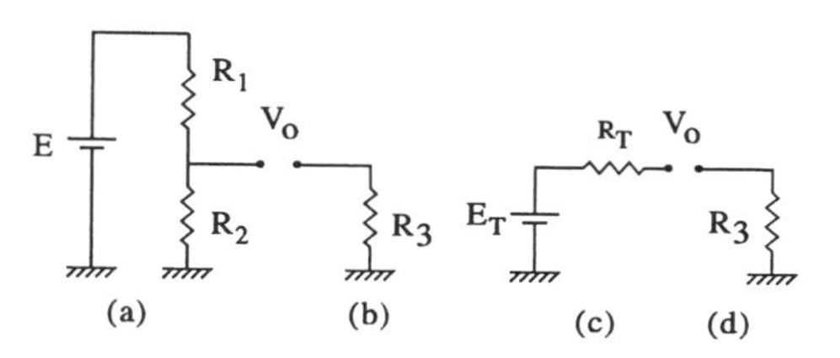

Fig. 1.10 Circuit of Fig. 1.9 divided as for Thévenin equivalent. (a,b) Division of circuit. (c) Thévenin equivalent of (a). (d) R_3.

> **Example** Calculate V_o for
>
> $E=10V$, $R_1=1k\Omega$, $R_2=3k\Omega$ and $R_3=2k\Omega$
>
> in Fig. 1.9 using the Thévenin method.

With E_t as in [1.47]

$$E_t = 10V \times 3k\Omega/(1k\Omega+3k\Omega)$$

giving $E_t=7.5V$.

With R_T as in [1.48]

$$G_T = 1/1k\Omega + 1/3k\Omega = 1.33mS$$

giving $R_T=750\Omega$.

With V_o as in [1.49] then

$V_o=7.5V\times2k\Omega/(2k\Omega+750\Omega)$ confirming $V_o=5.45V$.

Superposition

Fig. 1.11 is the common example for the remaining methods. For superposition the expression for V_o is first written for V_o with E_1 as normal and $E_2=0$, i.e. E_2 replaced by a short circuit. Next the expression is written with E_2 as normal and $E_1=0$. Finally the two expressions are added, to give

$$V_o = E_1 \frac{R_{2//3}}{R_1 + R_{2//3}} + E_2 \frac{R_{1//3}}{R_2 + R_{1//3}} \quad (1.50)$$

and thus

$$V_o = \frac{E_1R_2R_3 + E_2R_1R_3}{R_1R_2 + R_2R_3 + R_3R_1} \quad (1.51)$$

> **Example. Superposition** Calculate V_o in Fig. 1.11 using the method of superposition, given
>
> $E_1=8V$, $E_2=5V$, $R_1=1k\Omega$, $R_2=2k\Omega$, $R_3=3k\Omega$.

With $G_{2//3}=1/2k\Omega+1/3k\Omega=0.833mS$

then $R_{2//3}=1.2k\Omega$.

With $G_{1//3}=1/1k\Omega+1/3k\Omega=1.33mS$

then $R_{1//3}=750\Omega$.

With

$$V_o=E_1R_{2//3}/(R_1+R_{2//3})+E_2R_{1//3}/(R_2+R_{1//3})$$

then

$$V_o=8V\times1.2k\Omega/(1k\Omega+1.2k\Omega)$$
$$+5V\times50\Omega/(2k\Omega+750\Omega)$$

giving **V_o=5.73V.**

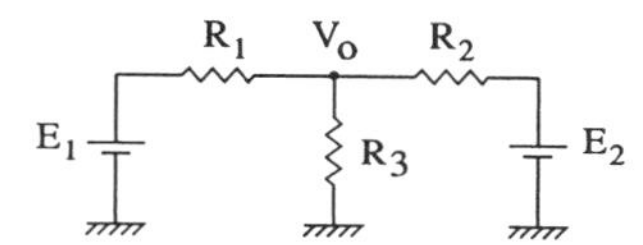

Fig. 1.11 Circuit to be solved by various methods below.

> **Example. Thévenin** Confirm the above result using the Thevenin method.

With the circuit divided as in Fig. 1.12, the Thévenin voltage

$$E_T=E_1R_2/(R_1+R_2)+E_2R_1/(R_1+R_2)$$

giving

$$E_T=8V\times2k\Omega/(1k\Omega+2k\Omega)+5V\times1k\Omega/(1k\Omega+2k\Omega)$$

and thus $E_T=7V$.

With $G_T=1/R_1+1/R_2$

then $G_T=1/1k\Omega+1/2k\Omega=1.5mS$

giving $R_T=667\Omega$.

With $V_o=E_T\times R_3/(R_3+R_T)$

then $V_o=7V\times3k\Omega/(3k\Omega+667\Omega)$

confirming **V_o=5.73V.**

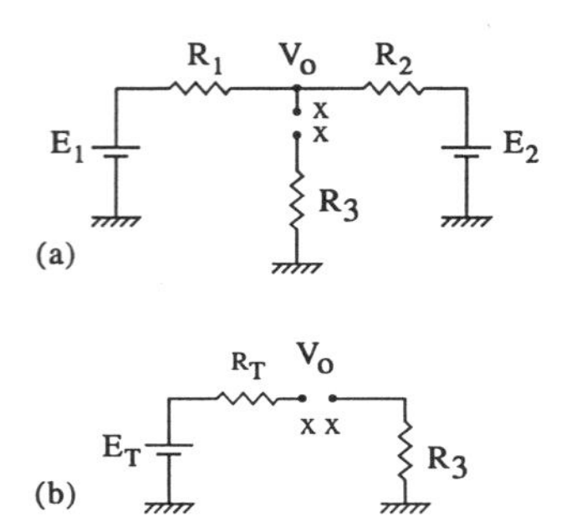

Fig. 1.12 Division of circuit in Fig. 1.11 for solution by Thévenin's theorem.

Ideal voltage and current sources

Fig. 1.13 shows the symbols used to represent these sources. An ideal voltage source is one for which the voltage V is independent of the current passing through the source.

An ideal current source is one where the current I is independent of the voltage across the source.

The symbol in (c) is that for a battery. This is an ideal voltage source where the voltage does not vary with time, as well as not varying with I as in (a). It is traditional to represent the voltage by E (electromotive force) rather than V.

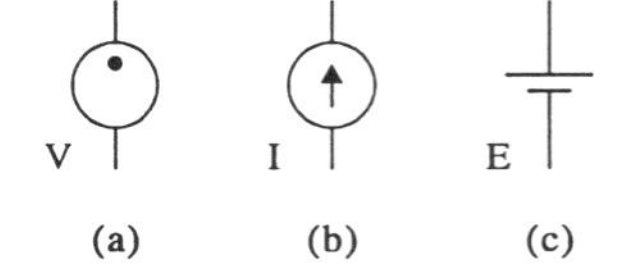

Fig. 1.13
(a) Ideal voltage source.
(b) Ideal current source.
(c) Ideal DC voltage source.

For (a) and (b), V and I may or may not vary with time. The dot in (a) marks the terminal which is positive when V is positive. The arrow in (b) shows the direction of the current when I is positive. This is not to imply that say in (a) the upper terminal is always positive. V might be negative and then the upper terminal would also be negative, similarly for the current in (b).

When a circuit contains both ideal current and ideal voltage sources then for superposition these are all set to zero apart from the source being considered. The voltage sources are replaced by short circuits as before, while the current sources are replaced by open circuits.

Norton's theorem

Fig. 1.14 shows how the circuit of Fig. 1.11 is solved using Norton's theorem. As for Fig. 1.12 for Thévenin's theorem, the circuit is split into R_3 and the rest of the circuit. Then, instead of the centre point being left open to give E_T, the point is grounded to give the Norton current I_N. The circuit can then be represented as in Fig. 1.14(b), with the Norton resistance R_N the same as R_T, here $R_1//R_2$.

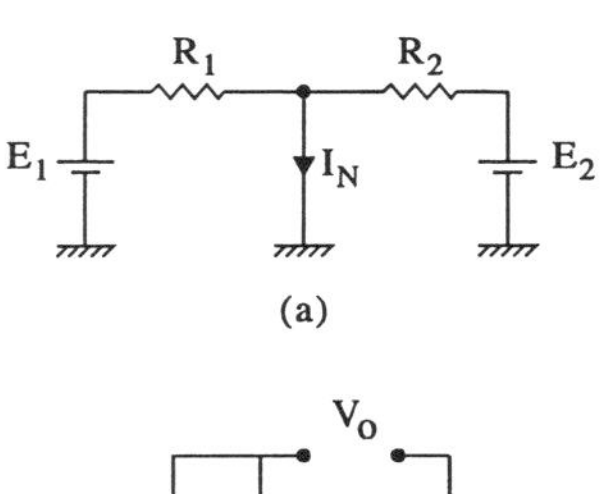

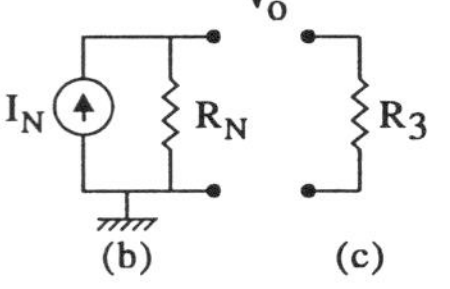

Fig. 1.14 Use of Norton's theorem to solve circuit of Fig. 1.11.
(a) Derivation of Norton current I_N.
(b) Norton equivalent of (a).
(c) R_3.

> **Example. Norton** Confirm the result of the last two examples by the Norton method.

With $\quad I_N=E_1/R_1+E_2/R_2=8V/1k\Omega+5V/2k\Omega$

then $I_N=10.5mA$.

With $\quad V_o=I_N/(G_1+G_2+G_3)$

then $\quad V_o=10.5mA/(1/1k\Omega+1/2k\Omega+1/3k\Omega)$

confirming **$V_o=5.73V$.**

For this circuit the Norton method is the simplest.

> **Problem** The above confirmations are numeric. Using Thévenin and then Norton, derive the expression for V_o to confirm [1.51].

1.6 Nodal analysis

Single-node circuit

Fig. 1.15 shows Fig. 1.11 drawn as for nodal analysis. To obtain the expression for V_o the method is as follows. First the expression is written for I_{out}, the current that will flow away from the V_o node with V_o as labeled and the voltages at the adjoining nodes E_1 and E_2 both zero, giving

$$I_{out}=V_o(G_1+G_2+G_3) \qquad (1.52)$$

Next the expression is written for I_{in}, the current that will flow into the V_o node for $V_o=0$ and E_1 and E_2 as labeled, giving

$$I_{in}=E_1G_1+E_2G_2 \qquad (1.53)$$

With $I_{in}=I_{out}$ then

$$V_o = \frac{E_1G_1 + E_2G_2}{G_1 + G_2 + G_3} \quad (1.54)$$

With $R_1=1/G_1$, etc., then [1.51] is again confirmed.

Even for a circuit as simple as that above, the nodal analysis gives the required solution as quickly as any of the other methods. For circuits of greater complexity nodal analysis is nearly always the best choice.

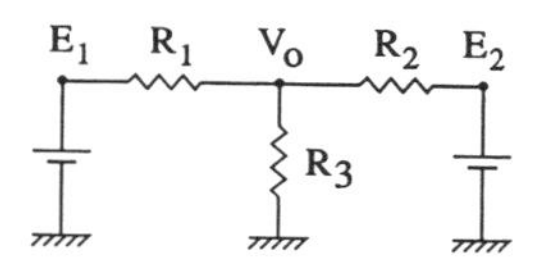

Fig. 1.15 Circuit of Fig. 1.11 shown as for nodal analysis.

Two-node circuit

For Fig. 1.15 there is only one unknown nodal voltage, and hence one equation to solve. Fig. 1.16 shows an example where there are two unknown voltages V_a and V_b. Here the two equations become as in the table below.

Node	I_{out}	=	I_{in}
A	$V_a(G_1+G_2+G_3)$	=	$E_1G_1+V_bG_3$
B	$V_b(G_3+G_4+G_5)$	=	$V_aG_3+E_5G_5$

giving

$$V_a = \frac{E_1G_1(G_3 + G_4 + G_5) + E_2G_3G_5}{(G_1 + G_2 + G_3)(G_4 + G_5) + G_3(G_1 + G_2)} \quad (1.55)$$

$$V_b = \frac{E_1G_1G_3 + E_2G_5(G_1 + G_2 + G_3)}{(G_3 + G_4 + G_5)(G_1 + G_2) + G_3(G_4 + G_5)} \quad (1.56)$$

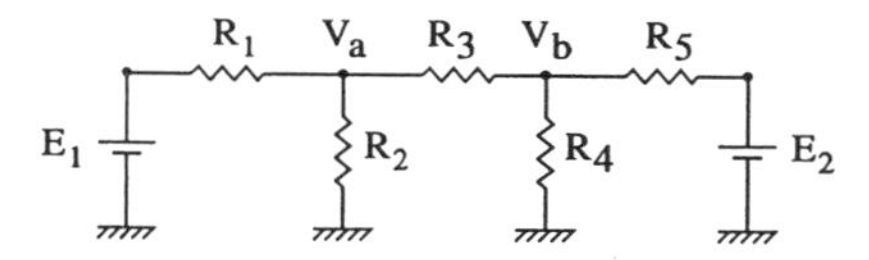

Fig. 1.16 Circuit with two unknown nodal voltages.

Linearity

While the above methods can be extended to include the C and L components as in the next section, they cannot be applied when a 'non-linear' component such a diode is included. For say a resistor R in series with an ideal voltage source E then V=E–IR, with the plot of V against I a straight line. Thus the circuit and the components are 'linear'. For the diode, in contrast, the I-V relation is curved, in fact being exponential. Both transistor and diode are essentially non-linear devices becoming effectively linear for sufficiently low-level signals. The appropriate 'incremental' or 'small-signal' methods are introduced in Chapter 5.

1.7 AC theory

We have now considered a number of networks containing ideal voltage and current sources and resistors, and also the combination of L and C to form a resonator. What we have not yet covered is a network containing both resistive and reactive components. Fig. 1.18(a) shows R and L in series, with the need to calculate the resulting impedance Z. Were these both resistors then V would simply be the sum of the two component voltages, but now V_L and V_r are no longer in phase and this must be allowed for.

Phasors

With the method based on the phasor diagrams of Fig. 1.18(b) and (c), Fig. 1.17 shows what is meant by a 'phasor'.

With θ as shown then sin(θ) and cos(θ) are as also shown. Thus as the emphasised arrow (the phasor) rotates then the sine wave is traced on the vertical axis and the cosine wave on the horizontal.

Normally the rotation is at the constant rate ω=dθ/dt, to give sin(ωt) on the vertical axis and cos(ωt) on the horizontal. Also, as already noted, ω=2πf where ω is the angular velocity in radians/sec and f is the frequency in Hz (cycles/sec).

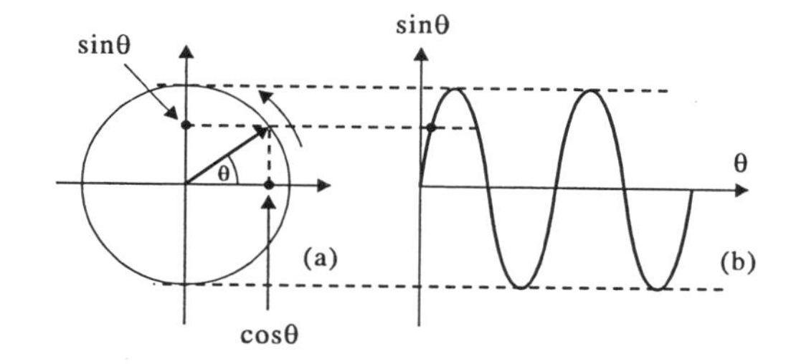

Fig. 1.17 Derivation of sin(θ) from rotating phasor of unit length. (a) Phasor. (b) sin(θ).

Phasor diagram

Next Fig. 1.18(c) shows the phasor tracing $\hat{I}\cos(\omega t)$ on the horizontal axis. This is the reference phasor for which θ=0 when t=0, the time for which the diagram is drawn.

With V_r in phase with I then the phasor tracing V_r is as in (b), with that tracing V_L 90° ahead. Thus the phasor tracing V becomes as shown. With

$$\hat{V}_r = \hat{I}R \quad \text{and} \quad \hat{V}_L = \hat{I}\omega L \qquad \text{as in [1.21],}$$

then

$$\hat{V} = \hat{I}\sqrt{R^2 + (\omega L)^2} \quad (1.57)$$

With the impedance Z of the network having both magnitude and phase then the magnitude

$$|Z| = \sqrt{R^2 + (\omega L)^2} \quad (1.58)$$

and with the phase $\phi_z = \phi$ shown then

$$\phi_z = \tan^{-1}(\hat{V}_L / \hat{V}_r) \quad (1.59)$$

giving $$\phi_z = \tan^{-1}(\omega L/R) \quad (1.60)$$

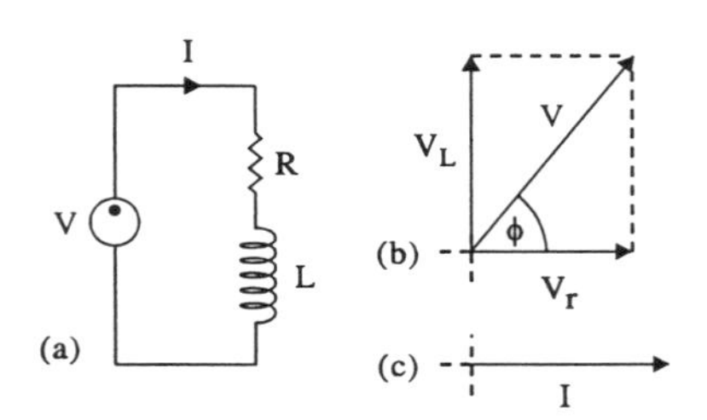

Fig. 1.18 Phasor diagram for parallel R and C.
(a) Circuit diagram.
(b) Phasors tracing voltages on horizontal axis.
(c) Phasor tracing current I on horizontal axis.

The j-operator

This is a useful way of avoiding the need to keep drawing phasor diagrams. Consider the unlikely number $\sqrt{-1}$. This is nowhere on the scale from $-\infty$ through zero to $+\infty$ and so is termed 'imaginary'. Indeed an entire imaginary scale is needed for the variable $\sqrt{-1} \times y$ where y ranges from $-\infty$ to $+\infty$.

Normal numbers, in contrast, are termed 'real'. If the real scale is laid on the X-axis and the imaginary along the Y then we have the 'complex plane'. Any point on the plane then represents a 'complex number' having a real and an imaginary part. For example, for X=3 and Y=5 then the complex number is $3+\sqrt{-1}\times 5$. Here $\sqrt{-1}$ is awkward to keep writing so we use 'j' instead. Thus the present number is 3+j5.

Comparing say 7 and j7 then j7 is 7 rotated through 90°. Also $j \times j \times 7 = -7$ which is 7 rotated through 180°. Further $j^3 \times 7$ is $-j7$, a rotation through 270°, and $j^4 \times 7 = 7$, a rotation through 360°. Thus j is a rotation operator. With X_L the reactance of L then

$$X_L = j\omega L \quad (1.61)$$

signifying that the magnitude $|X_L|$ of X_L is ωL, as previously established, and now also that V_L leads I by 90°. Conversely the susceptance B_L of L becomes

$$B_c = 1/j\omega L \quad (1.62)$$

with I lagging behind V_L by 90°. Thus the complex impedance

$$Z = R + 1/j\omega L \quad (1.63)$$

Plotting Z on the complex plane then gives

$$|Z| = \sqrt{R^2 + (\omega L)^2} \quad \text{and} \quad \phi_z = \tan^{-1}(\omega L/R),$$

thus confirming [1.58] and [1.60].

> **Example. Series impedance** Calculate $|Z|$ and ϕ_z for R=10kΩ in series with L=2mH with the signal frequency f=500kHz.

With $\omega L = 2\pi \times 500\text{kHz} \times 2\text{mH}$

then $\omega L = 6.28\text{k}\Omega$

With $|Z|$ as in [1.58] then

$$Z = 1/[10\text{k}\Omega^2 + 6.28\text{k}\Omega^2]^{1/2}$$

giving **$|Z| = 11.8\text{k}\Omega$.**

With ϕ_z as in [1.60] then

$$\phi_z = \tan^{-1}(\omega L/R) = \tan^{-1}(6.28\text{k}\Omega/10\text{k}\Omega)$$

giving **$\phi_z = 32.1°$.**

Parallel R and C

We now derive the expression for Z, for R and C in parallel.
With $|X_c| = 1/\omega C$ as in [1.13] and I leading V by 90° as in [1.11], then

$$X_c = 1/j\omega C \quad (1.64)$$

Also with the susceptance $B_c = 1/X_c$ then $B_c = j\omega C$ (1.65)

Thus the complex 'admittance' $Y = G + j\omega C$ (1.66)

Plotting Y on the complex plane then

$$|Y| = \sqrt{G^2 + (\omega C)^2} \ldots(a) \quad \phi_y = \tan^{-1}(\omega CR)\ldots(b) \quad (1.67)$$

However, it is the complex impedance

$$Z = \frac{1}{G + j\omega C} \quad (1.68)$$

which we seek. Here, as shown in Exercise 1.1, if for the general complex numbers a, b, and y,

$$y = a/b$$

then $$|y| = |a|/|b| \ldots(a) \quad \phi_y = \phi_a - \phi_b \ldots(b) \quad (1.69)$$

Thus $$|Z| = 1\Big/\sqrt{G^2 + (\omega C)^2} \quad \ldots(a)$$

$$\phi_z = -\tan^{-1}(\omega CR) \quad \ldots(b) \quad (1.70)$$

> **Example. Parallel impedance** Calculate $|Z|$ and ϕ_z for R=10kΩ in parallel with C=10nF for a signal of frequency f=500Hz.

With $\omega C = 2\pi \times 500 \times 10\text{nF}$ then $\omega C = 31.4\mu\text{S}$.

With $|Z|$ as in [1.70](a)

$$|Z| = 1/[(1/10k\Omega)^2 + (31.4\mu S)^2]^{1/2}$$

giving **$|Z| = 9.54k\Omega$.**

With ϕ_z as in [1.70](b) then

$$\phi_z = -\tan^{-1}(\omega CR) = -\tan^{-1}(31.4\mu S \times 10k\Omega)$$

giving **$\phi_z = -17.4^o$.**

Equivalent series impedance

The above parallel Z can be resolved into the real and imaginary components

$$Z_r = |Z|\cos(\phi) \;...(a) \quad Z_i = j|Z|\sin(\phi_z) \;...(b) \qquad (1.71)$$

corresponding to the equivalent series resistance and reactance. Here

$$Z_r = 95.4k\Omega \times \cos(-17.4^o), \quad Z_i = 9.54k\Omega \times \sin(-17.4^o)$$

giving **$Z_r = 9.10k\Omega$ and $Z_i = -2.86k\Omega$**

Normalisation

With the above results only numeric, for the full analytic expressions for Z_r and Z_i the process of 'normalisation' is needed. Here for $y = a/b$ as in [1.69](a) then

$$y = \frac{(a_r + ja_i)}{(b_r + jb_i)} \qquad (1.72)$$

Next both numerator and denominator are multiplied by $b_r - jb_i$, the 'complex conjugate' of b to give

$$y = \frac{(a_r + ja_i)}{(b_r + jb_i)} \times \frac{(b_r - jb_i)}{(b_r - jb_i)} = \frac{(a_r + ja_i) \times (b_r - jb_i)}{(b_r^2 + b_i^2)} \qquad (1.73)$$

and thus

$$y_r = \frac{(a_r b_r + a_i b_i)}{(b_r^2 + b_i^2)} \;...(a) \quad y_i = j\frac{a_i b_r - a_r b_i}{(b_r^2 + b_i^2)} \;...(b) \qquad (1.74)$$

Example. Normalisation Confirm the above values of Z_r and Z_i using the process of normalisation.

With $Z = 1/(G + j\omega C)$

as in [1.68], and for y in [1.72] equal to Z, then

$$a_r = 1, \quad a_i = 0, \quad b_r = G, \quad b_i = j\omega C$$

giving

$$Z_r = \frac{G}{G^2 + (\omega C)^2} \;...(a)$$

$$Z_i = \frac{-j\omega C}{G^2 + (\omega C)^2} \;...(b) \qquad (1.75)$$

With $G = 1/10k\Omega = 100\mu S$ and $\omega C = 31.4\mu S$

then $Z_r = 100\mu S/(100\mu S^2 + 31.4\mu S^2)$

confirming **$Z_r = 9.10k\Omega$.**

Also $Z_i = -31.4\mu S/(100\mu S^2 + 31.4\mu S^2)$

confirming **$Z_i = -j2.85k\Omega$.**

For more complex expressions where

$$y = \frac{a \times b \times c...}{d \times e \times f...}$$

then

$$|y| = \frac{|a||b||c|...}{|d||e||f|...} \;...(a)$$

$$\phi_x = (\phi_a + \phi_b + \phi_c...) - (\phi_d + \phi_e + \phi_f...) \;...(b) \qquad (1.76)$$

as also shown in Exercise 1.1.

1.8 Fourier analysis

The mathematician Fourier showed that essentially any periodic waveform of frequency f could be resolved into the mean plus a series of sine and cosine waves of frequency $f_n = nf$, where n is integer and ranges from 0 to ∞. The two for $n=1$ combine to constitute the 'fundamental', while the remainder are termed 'harmonics'. Thus for the periodic voltage V

$$V = A_o + \sum_{n=1}^{\infty} A_n \sin(\omega_n t) + B_n \cos(\omega_n t) \qquad (1.77)$$

where $\omega_n = n \times 2\pi f$ with n integer.

Distortion

This discovery is useful for at least two reasons. For the example of a real audio amplifier, this always gives some degree of distortion. Thus for a sine wave input, the output is a distorted sine wave which can be resolved into the Fourier series.

Next, the intelligence component of speech ranges from 300 to 3.4kHz while the response of the healthy ear extends to about 20kHz. Thus, by including a low-pass filter cutting off at 3.4kHz, those Fourier components resulting from the distortion which are above 3.4kHz will be removed, thus improving the intelligibility.

Circuit analysis

The other main value of the method is that, as stated above, The analysis of a circuit containing capacitors and inductors is much easier for a sinusoidal signal than for other types of waveform. Thus a non-sinusoidal input waveform can be resolved into its Fourier components, the effect of the circuit on each component calculated, and the resulting output components summed to give the full output waveform. The procedure is rarely carried out just like that, but it does form the basis of the methods of Chapters 11 to 13 that are used for more complex circuits.

Procedure

With A_o the average and $T=1/f$ then

$$A_o = \frac{1}{T}\int_0^T V.dt \qquad (1.78)$$

while

$$A_n = \frac{2}{T}\int_0^T V.\sin(\omega_n t)\ ...(a)$$

$$B_n = \frac{2}{T}\int_0^T V.\cos(\omega_n t)\ ...(b) \qquad (1.79)$$

Here [1.78] is the normal procedure for deriving the mean, while [1.79](a) for the sine components is confirmed in Exercise 1.2.

Exercise 1.3 shows the procedure applied to a square-wave, giving

$$V = \frac{4E}{\pi}\sum_{n=1}^{\infty}\frac{1}{n}\sin(\omega_n t) \qquad (1.80)$$

> **Problem** Either by sketching or by a short computer program, first display the first two non-zero components in [1.80] and their sum. Then add the third component. Thus show how the result progressively approaches the original square wave.

Symmetry

The series for the above square wave has several notable features.

(i) The mean is zero, as is obvious.
(ii) There are no cosine terms.
(iii) There are only odd-order sine terms.

(ii) is little more than a matter of timing. Were the square wave to start at $t=T/4$ rather than $t=0$ then cosine terms would appear. (i) and (iii), however, are more significant, being a feature of the type of non-linearity needed to convert a sine wave to the square wave.

Fig. 1.19 shows the transfer function needed to do this, and Fig. 1.20 the more general 'anti-symmetric' transfer function. As will shortly be shown, for any such anti-symmetric function (i) and (iii) obtain.

Symmetric transfer function

However the corresponding features of the output spectrum for the symmetric transfer function in Fig. 1.21(a) are simpler to derive, so we consider these first.

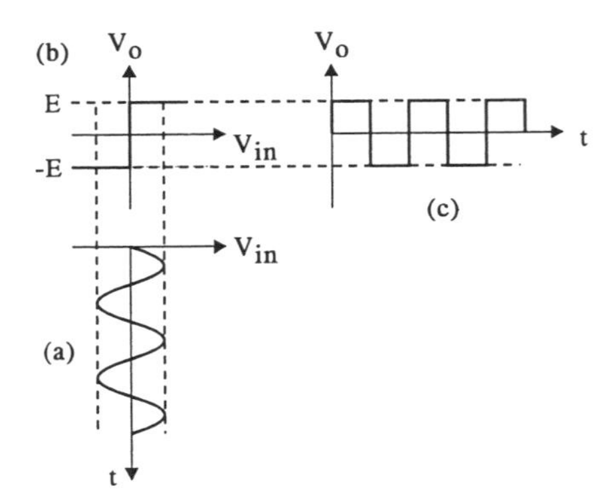

Fig. 1.19 Conversion of sine wave to square wave.
(a) Sine wave input V_{in}.
(b) Transfer function.
(c) Square wave output V_o.

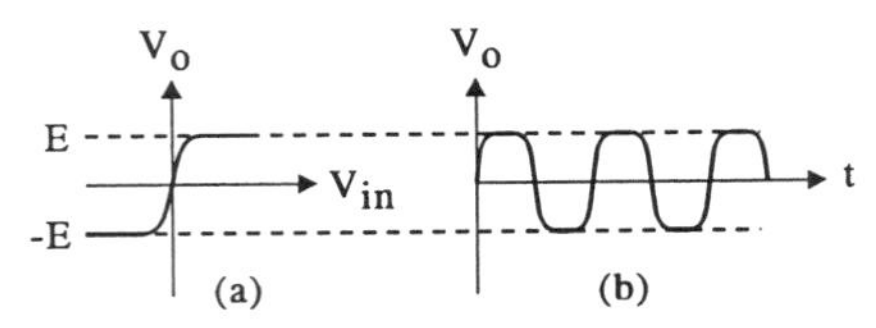

Fig. 1.20 Effect on V_o of more general type of anti-symmetric transfer function in place of that in Fig. 1.19(b).
(a) Transfer function.
(b) V_o waveform for V_{in} still the sine wave of Fig. 19(a).

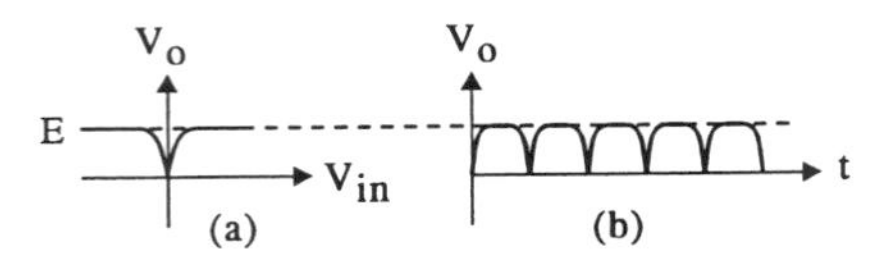

Fig. 1.21 Effect of symmetric transfer function in place of that in Fig. 1.19(b).
(a) Transfer function.
(b) V_o waveform for V_{in} still the sine wave of Fig. 19(a).

Here
(i) the average is non-zero,
(ii) with V_o periodic at twice the input frequency, there is no fundamental and no odd-order harmonics.

Anti-symmetric transfer function

For the general anti-symmetric transfer of Fig. 1.20, Fig. 1.22(c) shows one cycle of the V_o waveform, and (b) the $\sin(\omega_4 t)$ used in

$$A_4 = \frac{2}{T}\int_0^T V_o.\sin(\omega_4 t).dt$$

as in [1.79] to obtain A_4.

Here (b) for the first half cycle repeats over the second, while V_o also repeats but with sign reversal. Thus $A_4=0$.

With $\cos(\omega_4 t)$ also repeating over each half-cycle without sign reversal then $B_4=0$ also. Thus there is neither a sine nor a cosine component for n=4. The same is so for any even value of n.

In contrast, Fig. 1.22(a) shows $\sin(\omega_5 t)$ in [1.79]. This too repeats over the second half-cycle but now with sign reversal, cancelling the effect of the V_o reversal. Thus A_5 is non-zero.

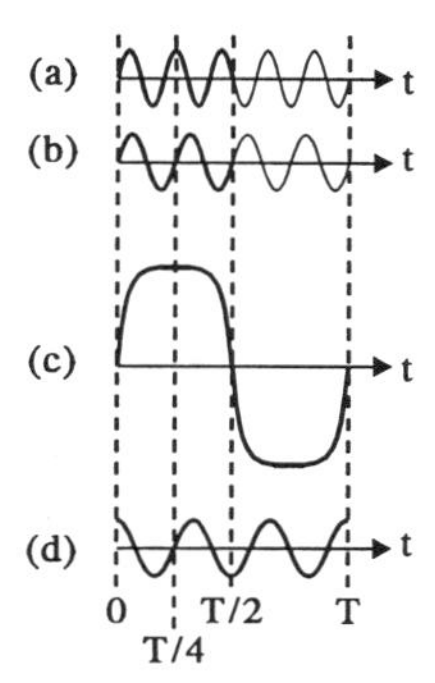

Fig. 1.22 One cycle of waveform in Fig. 1.20(b) as convolved with various sine and cosine components in Fourier analysis. (a) $\text{Sin}(\omega_5 t)$. (b) $\text{Sin}(\omega_4 t)$. (c) V_o. (d) $\text{Cos}(\omega_3 t)$.

Absence of cosine components

To complete the picture, Fig. 1.22(d) shows an odd-order cosine term, $\cos(\omega_3 t)$. Although this has the sign reversal over the second half-cycle of (c), it is also anti-symmetric about T/4, while the first half-cycle of (c) is symmetric about T/4. Thus the integral over the first half-cycle is zero. This is why there are no B_n terms at all in the Fourier series for the square wave.

Convolution

Fig. 1.23 shows more specifically how the formation of

$$\int_{-\infty}^{\infty} y_1 \times y_2 . dx = 0$$

when one y is symmetric and the other anti-symmetric. Here the $y_1 \times y_2$ for the left-hand pair of dots is equal and opposite to that for the right-hand pair. The integral is termed the 'convolution' of y_1 with y_2, sometimes written as $y_1 \otimes y_2$. Here $y_1 \otimes y_2=0$.

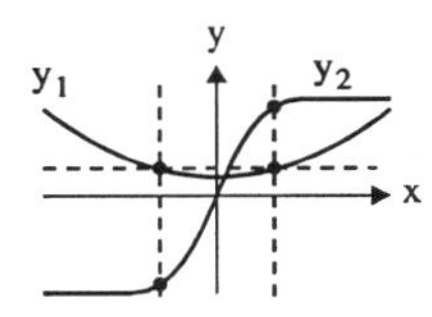

Fig. 1.23 Convolution of symmetric and anti-symmetric functions $y_1(x)$ and $y_2(x)$.

Push-pull amplifier

As stated, the absence of cosine terms from the series is merely a matter of timing. Of greater experimental importance is the lack of even-order harmonics for the anti-symmetric transfer function. This is exploited in amplifier design by giving the circuit anti-symmetry, as for the push-pull audio power amplifier of Chapter 21. A third harmonic is more likely to be out of the audio hearing range than a second.

Frequency doubler

Sometimes a sine wave output is required that is twice the frequency of the sine wave input. With the symmetric transfer function of Fig. 1.21 producing no fundamental then it becomes easier to extract the required component by band-pass filtering.

Sawtooth

Some periodic waveforms could not be the result of the distortion of a sine wave by any of the above types of transfer function, whatever their symmetry. The sawtooth of Fig. 1.24 is an example. Such waveforms may have both odd- and even-order harmonics. Exercise 1.4 shows the Fourier components for the sawtooth waveform of to be $\mathbf{A_o=E/2, \quad A_n=E/n\pi, \quad B_n=0.}$

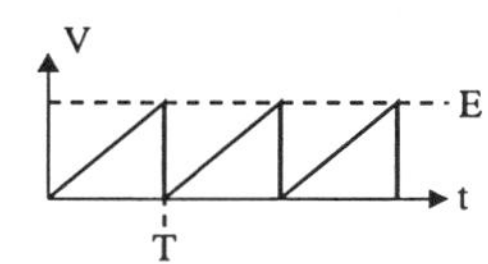

Fig. 1.24 Sawtooth wave.

Reference

Compton, A. J. 1990: *Basic Electromagnetism and its Applications*, Chapman and Hall.

1.9 Exercises

Exercise 1.1 Magnitude and phase for composite fraction

Confirm as stated in [1.76] that for

$$y = \frac{a \times b \times c\ldots}{d \times e \times f\ldots} \quad (1.81)$$

$$|y| = \frac{|a||b||c|\ldots}{|d||e||f|\ldots} \quad (1.82)$$

and

$$\phi_x = (\phi_a+\phi_b+\phi_c\ldots) - (\phi_d+\phi_e+\phi_f\ldots) \quad (1.83)$$

Product With

$$a=|a|(\cos\phi_a+j\sin\phi_a),$$

etc., for the simpler $y=a\times b$

then $y=|a|(\cos\phi_a+j\sin\phi_a)\times|b|(\cos\phi_b+j\sin\phi_b)$
giving

$$y=|a|\times|b|[\cos\phi_a\cos\phi_b-\sin\phi_a\sin\phi_b+j(\cos\phi_a\sin\phi_b+\sin\phi_a\cos\phi_b)]$$

and thus

$$y=|a|\times|b|[(\cos(\phi_a+\phi_b)+j(\sin(\phi_a+\phi_b)].$$

With also $y=|y|[\cos(\phi_y)+j\sin(\phi_y)]$

then $|y|=|a|\times|b|\ldots(a)\quad \phi_y=\phi_a+\phi_b\ldots(b)\qquad (1.84)$

Similarly for $y=(a\times b)\times c$

$$|y|=|a|\times|b|\times|c|\ldots(a)\quad \phi_y=\phi_a+\phi_b+\phi_c\ldots(b)\qquad (1.85)$$

and so on.

Quotient With

$$y=\frac{|a|(\cos\phi_a+j\sin\phi_a)}{|b|(\cos\phi_b+j\sin\phi_b)}$$

for $y=a/b$ then, normalising,

$$y=\frac{|a|(\cos\phi_a+j\sin\phi_a)|b|(\cos\phi_b-j\sin\phi_b)}{|b|^2}$$

giving

$$y=\frac{|a|}{|b|}\big(\cos(\phi_a-\phi_b)+j(\sin\phi_a-\phi_b)\big).$$

Thus $|y|=|a|/|b|$ and $\phi_y=\phi_a-\phi_b$,

which with the above relations extends to confirm [1.76].

Exercise 1.2 Fourier analysis procedure

Confirm that

$$A_n=\frac{2}{T}\int_0^T V.\sin(\omega_n t)$$

as in [1.79] gives the value of the Fourier sine component of frequency ω_n.

With

$$V=A_o+\sum_{n=1}^{\infty}A_n\sin(\omega_n t)+B_n\cos(\omega_n t)$$

the Fourier series for the periodic function V as in [1.77], then for

$$I_a=\frac{2}{T}\int_0^T V.\sin(\omega_n t)$$

$$I_a=\frac{2}{T}\int_0^T\left(A_o+\sum_{m=1}^{\infty}A_n\sin(\omega_m t)+B_n\cos(\omega_m t)\right)\times\sin(\omega_n t)$$

With $\int_0^T A_o.\sin(\omega_n t).dt=0$

for all n then

$$I_a=\frac{1}{T}\int_0^T\left(\sum_{m=1}^{\infty}A_n[\cos\{(\omega_m-\omega_n)t\}-\cos\{(\omega_m+\omega_n)t\}]\right)$$

$$+\frac{1}{T}\int_0^T\left(\sum_{m=1}^{\infty}A_n[\sin\{(\omega_n-\omega_m)t\}-\sin\{(\omega_n+\omega_m)t\}]\right)$$

Here every component integral is zero for $m\neq n$. The second cos term is also zero for $m=n$, leaving only the first which is equal to A_n. Thus $I_a=A_n$, confirming the above relation.

Exercise 1.3 Fourier analysis of square wave

Derivation Derive the Fourier components for a square wave V of peak value E and frequency f.

with $A_o=\frac{1}{T}\int_0^T V.dt$ as in [1.78]

then

$$A_o=\frac{1}{T}\int_0^{T/2}E.dt+\frac{1}{T}\int_{T/2}^T -E.dt$$

Thus $A_o=0$, as is obvious for the square wave.

With $A_n=\frac{2}{T}\int_0^T V.\sin(\omega_n t)$

as in [1.79(a)] then

$$A_n=\frac{2}{T}\int_0^{T/2}E.\sin(\omega_n t).dt+\frac{2}{T}\int_{T/2}^T(-E).\sin(\omega_n t).dt$$

giving

$$A_n=\frac{2}{T}\left[-E\frac{\cos(\omega_n t)}{\omega_n}\right]_0^{T/2}+\frac{2}{T}\left[E\frac{\cos(\omega_n t)}{\omega_n}\right]_{T/2}^{T}$$

which is zero for n even and $4E/(n\pi)$ for n odd.

Finally as in [1.79(b)]

$$B_n = \frac{2}{T}\int_0^{T/2} E.\cos(\omega_n t).dt + \frac{2}{T}\int_{T/2}^{T} (-E).\cos(\omega_n t).dt$$

giving

$$B_n = \frac{2}{T}\left[E\frac{\sin(\omega_n t)}{\omega_n}\right]_0^{T/2} + \frac{2}{T}\left[-E\frac{\sin(\omega_n t)}{\omega_n}\right]_{T/2}^{T}$$

which is zero for all n. Thus it is confirmed that

$$V = \frac{4E}{\pi}\sum_{n=1}^{\infty}\frac{1}{n}\sin(\omega_n t)$$

as in [1.80] where $\omega_n = 2\pi f_n$, $f_n = nf$ and n is odd.

Exercise 1.4 Fourier analysis of sawtooth wave

> **Derivation** Derive the Fourier components for the sawtooth waveform of Fig. 1.24

With $$A_o = \frac{1}{T}\int_0^T V.dt$$ as in [1.78]

and here $$V = Et/T$$

then $$A_o = \frac{E}{T^2}\int_0^T t.dt = \frac{E}{T^2}\left[\frac{T^2}{2}\right]_0^T$$

and thus, as is obvious, **$A_o = E/2$.**

With $$A_n = \frac{2}{T}\int_0^T V.\sin(\omega_n t)$$

as in [1.79(a)] and with

$$V = Et/T$$

then

$$A_n = \frac{2E}{T}\int_0^T t.\sin(\omega_n t).dt$$

giving

$$A_n = \frac{2E}{T}\left[\frac{\sin(\omega_n t)}{(\omega_n)^2} - \frac{t.\cos(\omega_n t)}{\omega_n}\right]_0^T$$

and thus **$A_n = E/n\pi$.**

With $$B_n = \frac{2}{T}\int_0^T V.\cos(\omega_n t)$$

as in [1.79(b)] then

$$B_n = \frac{2E}{T}\int_0^T t.\cos(\omega_n t).dt$$

giving

$$B_n = \frac{2E}{T}\left[\frac{\cos(\omega_n t)}{(\omega_n)^2} + \frac{t.\text{cin}(\omega_n t)}{\omega_n}\right]_0^T$$

and thus **$B_n = 0$.**

2

Simple operational amplifier circuits

Summary

The operational amplifier (op-amp) of Fig. 2.1(b) is the active element in almost all analog electronic circuitry operating with signal frequencies of up to ~1MHz.

The device is not intended to be used alone but with some degree of 'negative feedback'. With this applied many of the minor imperfections of the op-amp are suppressed, to give an excellent implementation of the required function.

The first and most obvious function, or 'operation', is that of amplification, with the feedback amplifier gain set by two resistors forming the feedback network.

Other operations described in the chapter are addition, subtraction, integration, differentiation, the voltage-follower and the current-follower.

In this chapter the op-amp is largely taken as ideal, with the experimental work and associated calculations in the last section designed to avoid all but the most obvious limitations. The full range of limitations is the topic of the next chapter.

2.1 Operational amplifier

The op-amp in Fig.2.1(b) differs from the normal 'single-ended' amplifier of (a) in that there are two signal input terminals. For (a)

$$V_o = AV_{in} \tag{2.1}$$

while for the op-amp

$$V_o = A_o(V_{i+} - V_{i-}) \tag{2.2}$$

Differential amplifier

The op-amp is said to be a 'differential amplifier' because, while for the normal amplifier the response of V_o is to the difference between the single input and ground, for the differential amplifier the response is to the difference between the two inputs, otherwise regardless of the values relative to ground.

Ideal op-amp

In addition to the differential response, for the ideal op-amp

- $A_o = \infty$,
- no current is drawn into either input terminal,
- V_o is independent of I_o drawn from the output node,
- the response of V_o to the differential input is immediate.

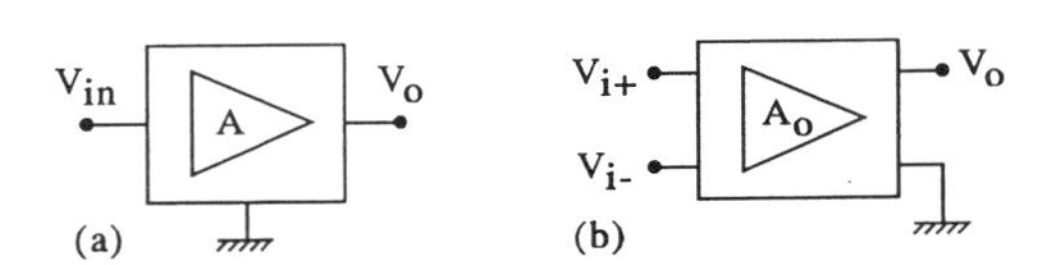

Fig. 2.1 Voltage amplifier types.
(a) Conventional non-differential voltage amplifier.
(b) Operational amplifier.

Power supplies

As shown in Fig. 2.2(a), the op-amp has two components, the microcircuit chip and the dual voltage supply V_+ and V_-, with the values typically ±15V.

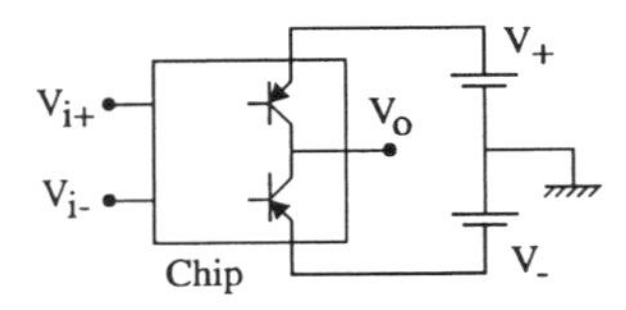

Fig. 2.2(a) Operational amplifier composed of integrated-circuit chip and dual power supply V_+ and V_-. Block diagram.

Voltage transfer function

With the voltage transfer function as in Fig. 2.2(b), this differs from the ideal of [2.2] in two respects.

(i) There is slight curvature (somewhat exaggerated).
(ii) The range of V_o is limited to the 'saturation' values V_{s+} and V_{s-}.

These, together with the finite A_o (typically 10^4 to 10^5), are the only limitations that will be evident in this chapter.

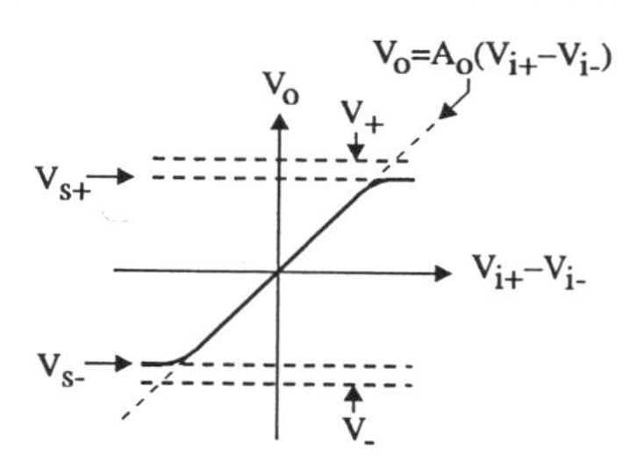

Fig. 2.2(b) Voltage transfer function for op-amp of (a).

2.2 Non-inverting negative-feedback amplifier

There are two types of op-amp-based negative-feedback amplifier, the inverting type of Fig. 2.7, and the present non-inverting type of Fig. 2.3. Here the feedback fraction

$$\beta=R_1/(R_1+R_2) \quad (2.3)$$

of V_o is 'fed-back' to the inverting input. With this in such a sense as to reduce the effect of the applied V_{in}, the feedback is negative. Positive feedback is sometimes used, but not in this chapter.

Formal analysis

We now derive the expression for the feedback amplifier gain

$$A_f=V_o/V_{in} \quad (2.4)$$

With the signal input currents for the ideal op-amp zero,

$$V_{i-}=\beta V_o \quad (2.5)$$

With $V_o=A_o(V_{i+}-V_{i-})$ as in [2.2], and with $V_{i+}=V_{in}$,

$$A_f=A_o/(1+A_o\beta) \quad (2.6)$$

With $A_o=\infty$ for the ideal op-amp then

$$A_f=1/\beta=(R_1+R_2)/R_1 \quad (2.7)$$

Feedback resistors

Here A_f depends only on the feedback resistors, which is one of the major advantages of the circuit. In principle this allows A_f to be set to any value =>1 by choosing the appropriate resistor values.

Thus essentially only one type of microcircuit op-amp chip needs to be manufactured, one for which A_o is as high as possible. Otherwise a different type of op-amp would be needed for each required value of A_f.

Also it is easier to make an op-amp chip for which A_o is high, than one for which A_o is lower, but with the value accurately determined.

Reduction of differential input

From the above relations

$$V_{i+}-V_{i-}=V_{in}/(1+A_o\beta) \quad (2.8)$$

which for $A_o=\infty$ gives

$$V_{i+}=V_{i-} \quad (2.9)$$

This is an extremely important result, applying to all of the negative-feedback circuits to be described. As a starting point, it invariably simplifies the analysis. With [2.9] giving $\beta V_o=V_{in}$ for the present circuit, then $A_f=1/\beta$ as in [2.7] follows immediately.

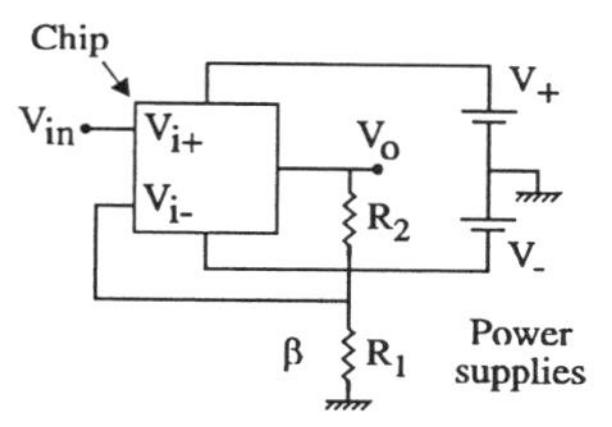

Fig. 2.3 Op-amp of Fig. 2.2(a) connected as a non-inverting negative feedback amplifier.

Descriptive account

The function of the op-amp is to compare V_{i-} with V_{i+} and to adjust V_o to the value needed to make the two equal.

Simplified circuit diagram

Fig. 2.4 shows the circuit of Fig. 2.3 as normally presented. With the dual power supply always required, normally powering more than one op-amp, the supply is 'taken as read' for simplicity.

Also the 'non-inverting' and 'inverting' signal input terminals are abbreviated, from V_{i+} and V_{i-} to just + and −. This does not necessarily mean that the voltages are positive or negative. The symbols merely indicate which terminal is which.

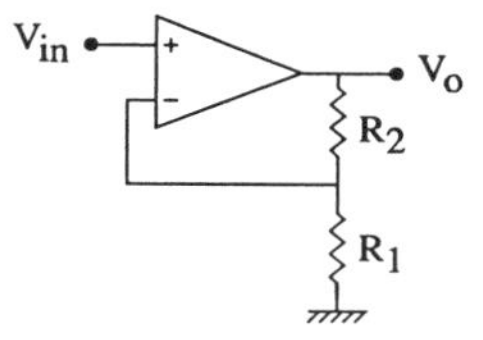

Fig. 2.4 Simplified circuit diagram for non-inverting feedback amplifier of Fig. 2.3.

'Missing' ground connection

Notice that the op-amp chip in Fig. 2.3 omits the essential output ground terminal in Fig. 2.1(b). Indeed the TL081 bi-fet op-amp used in all the experimental work has no pin so labelled.

The reason is that the entity referred to as the 'op-amp' must comprise both the microcircuit chip and its power supply, as in Fig. 2.2. Thus the op-amp ground terminal is the centre point of the dual power supply, not on the chip. The simplified representation of Fig. 2.4 evades this point.

Loop gain

With $A_f = A_o/(1 + A_o\beta)$ as in [2.6] then the requirement for A_f to approach the required $1/\beta$ is that $A_o\beta >> 1$. Here $A_o\beta$ is the feedback 'loop gain' A_L. With the loop broken as in Fig. 2.5 then $A_L = V_b/V_a$ giving

$$A_L = -A_o\beta \qquad (2.10)$$

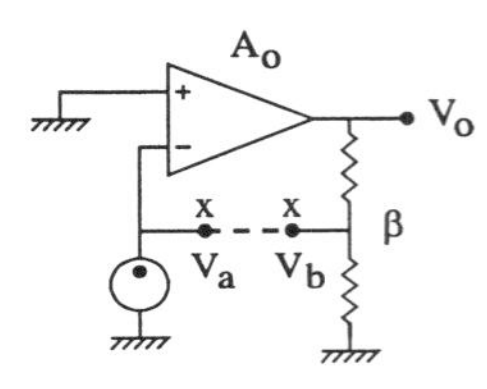

Fig. 2.5 Circuit showing feedback-amplifier loop-gain $A_L = A_o\beta$.

Actual and required feedback-amplifier gain

With the feedback-amplifier gain

$$A_f = A_o/(1 + A_o\beta)$$

as in [2.6] then for

$$A_o\beta >> 1, \quad A_f \approx 1/\beta$$

while for

$$A_o\beta << 1, \quad A_f \approx A_o,$$

as plotted in Fig. 2.6. This is as expected. For $A_o\beta >> 1$ then the feedback amplifier gain A_f is close to that required, but A_f can never be greater than the gain A_o of the op-amp without feedback.

Logarithmic plotting

With the log plotting of Fig. 2.6:

- distance is proportional to ratio,
- the ratio 1:1 is represented by zero distance,
- a ratio <1 is represented by a negative distance.

Thus the ratio $(A_o)/(1/\beta) = A_o\beta$ becomes as shown, and we see again that the point at which A_f switches from $\approx 1/\beta$ to $\approx A_o$ is that at which the loop gain $A_o\beta = 1$.

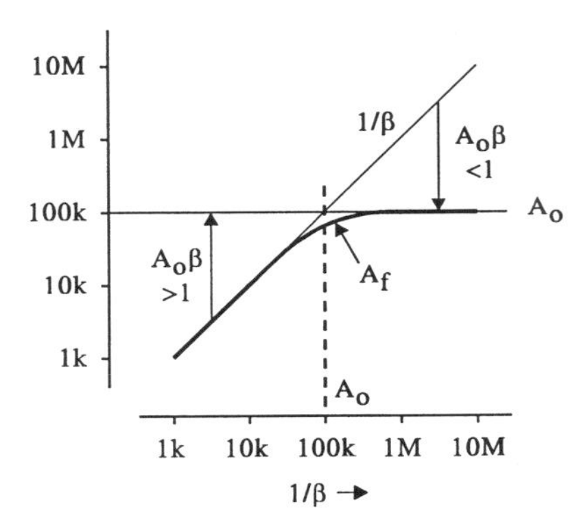

Fig. 2.6 Variation of actual feedback amplifier gain A_f with required gain $1/\beta$ (log axes).

Cascading

Where values of A_f are required which are not small compared with A_o it becomes necessary to cascade two or more feedback stages, each with a suitably reduced $1/\beta$.

Exercise 2.1 gives an example with associated experimental work for the non-inverting negative-feedback amplifier. Exercise 2.2 gives an example of cascading.

2.3 Inverting feedback amplifier

The inverting feedback amplifier of Fig. 2.7 has much the same properties as the non-inverting type, except that A_f is negative rather than positive.

Virtual ground

As before, the negative feedback gives

$$V_{i+} = V_{i-} \quad \text{for} \quad A_o = \infty,$$

much simplifying the analysis. Now, however, $V_{i+} = 0$, so V_{i-} becomes zero also, causing the point to be referred to as a 'virtual ground'.

With $V_{i-} = 0$ as the starting point then $I = V_{in}/R_1$. For zero op-amp input current then $V_o = -IR_2$ giving

$$A_f = -R_2/R_1 \qquad (2.11)$$

With

$$A_f = (R_1 + R_2)/R_1$$

as in [2.7] for the non-inverting amplifier then, as well as the inversion, $|A_f|$ for the present circuit is less than before by one.

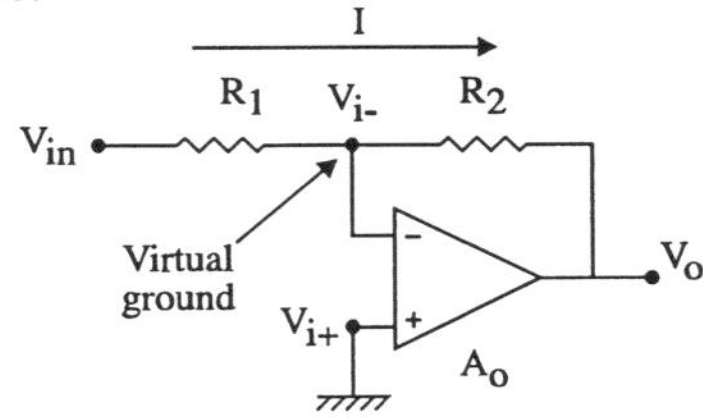

Fig. 2.7 Inverting negative-feedback amplifier.

Exact analysis

For zero op-amp input current, and with the resistor network 'feed-forward' and 'feed-back' fractions

$$\alpha=R_2/(R_1+R_2) \ldots\text{(a)} \quad \beta=R_1/(R_1+R_2) \ldots\text{(b)} \qquad (2.12)$$

then by superposition

$$V_{i-}=\alpha V_{in}+\beta V_o \qquad (2.13)$$

With $V_o=A_o(V_{i+}-V_{i-})$

as in [2.2] for the op-amp, with $V_{i+}=0$, and with $A_f=V_o/V_{in}$, then

$$A_f=-A_o\alpha/(1+A_o\beta) \qquad (2.14)$$

and

$$V_{i-}=V_{in}\alpha/(1+A_o\beta) \qquad (2.15)$$

With $V_{i-}=0$ for $A_o\beta=\infty$ in [2.15] then the virtual ground condition is confirmed.

With $A_f=-\alpha/\beta$ for $A_o\beta=\infty$

in [2.14], and with $\alpha/\beta=R_2/R_1$

from [2.12], then $A_f=-R_2/R_1$,

confirming [2.11]. In both cases the required condition is that the loop-gain $A_o\beta >> 1$.

Input impedance

For $V_{i-}=0$ then the input impedance of the inverting feedback amplifier is R_1. This is an important difference from the non-inverting amplifier, which ideally draws no current from the signal source and so has an infinite input impedance. Exercise 2.3 gives experimental work on the inverting negative-feedback amplifier.

2.4 Voltage-follower

This, in Fig. 2.8(b), is the non-inverting feedback amplifier of Fig. 2.4 with $R_1=\infty$ and $R_2=0$. With

$$A_f=(R_1+R_2)/R_1$$

as in [2.7] then $A_f=1$, so V_o 'follows' V_{in}.

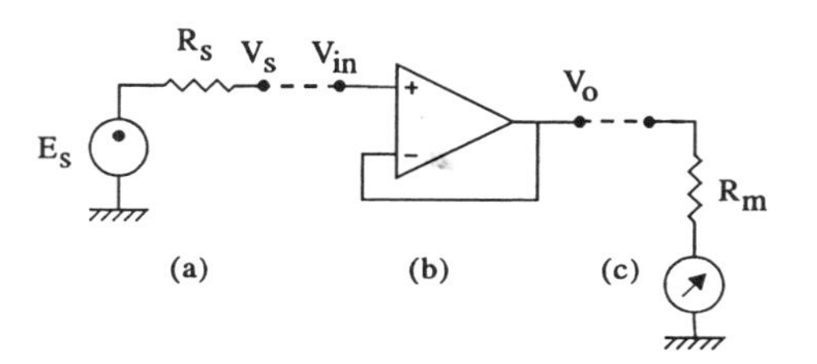

Fig. 2.8 Use of voltage-follower to isolate load from signal source.
(a) Signal source of impedance R_s.
(b) Voltage-follower.
(c) Meter of impedance R_m.

Fig. 2.8 shows the follower being used to prevent the loading of the source in (a) by the load in (c), allowing the meter in (c) to correctly register the open-circuit value E_s of V_s. Otherwise the loss factor $R_m/(R_m+R_s)$ results.

Exercise 2.4 gives an example with an associated experiment.

Clearly the non-inverting feedback amplifier gives the same degree of isolation where gain is also required.

2.5 Current-follower

This, shown in Fig. 2.9(b), is the 'dual' of the voltage-follower. There the requirement was to measure the voltage V at a point without drawing the load current that might otherwise alter V. Here the requirement is to measure the loop current I_s without developing the voltage that would result were the meter resistance R_m to be inserted directly into the loop.

With the connection as shown then only the virtual ground is connected into the loop, giving ideally no reduction in I_s. With $V_{i-}=0$ for the virtual ground,

$$V_o=-I_sR_f \qquad (2.16)$$

so V_o 'follows' I_s, to be recorded on the meter.

Current-follower view

For many of the present negative-feedback circuits the last section can be viewed as a current-follower. One such is the inverting feedback amplifier of Fig. 2.7, as follows.

- The virtual ground gives $I=V_{in}/R_1$.
- The current-follower converts I to $V_o=-IR_2$, giving $V_o/V_{in}=-R_2/R_1$.

Exercise 2.5 gives an example with an associated experiment.

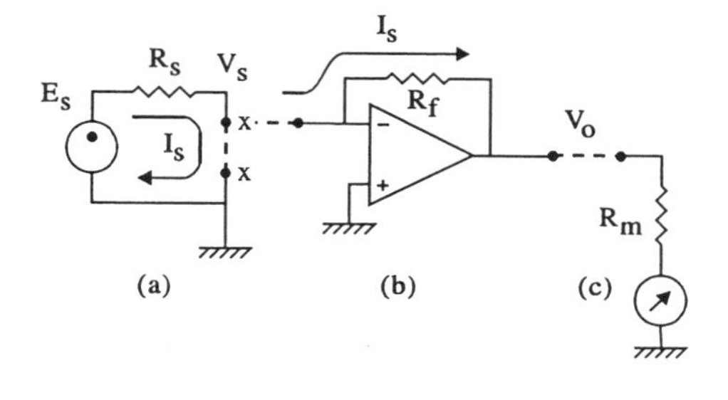

Fig. 2.9 Use of current-follower to convert circuit current I_s to meter voltage V_o without altering I_s.
(a) Loop carrying I_s.
(b) Current-follower.
(c) Meter.

2.6 Adder

Fig. 2.10 shows the feedback adder. With $V_{i-}=0$ for the virtual ground,

$$I=V_{in1}/R_1+V_{in2}/R_2+V_{in3}/R_3$$

With $V_o=-IR_f$ for the current-follower then

$$V_o=-\left(\frac{V_{in1}}{R_1}+\frac{V_{in2}}{R_2}+\frac{V_{in3}}{R_3}\right)R_f \qquad (2.17)$$

For all the R equal then the addition is simple, apart from the sign inversion, while weighting is possible by varying R_1 to R_3. For variable adjustment down to zero, as for example in a multi-channel sound mixer, the value of the input R for zero response is ∞, requiring the arrangement of Fig. 2.11.

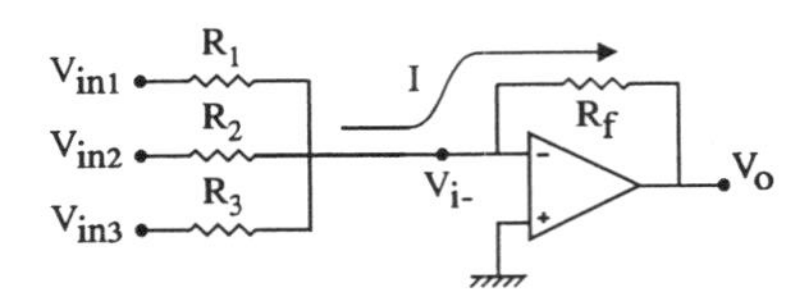

Fig. 2.10 Feedback adder

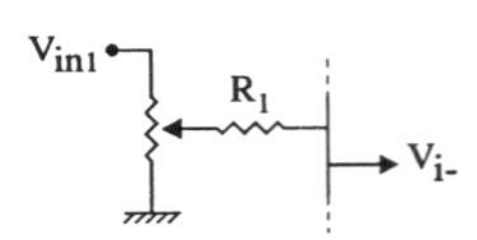

Fig. 2.11 Modification to adder allowing channel sensitivity to be reduced to zero.

Exercise 2.6 gives an example, an associated experiment, and a problem to solve.

2.7 Subtractor

The basic circuit structure of both the inverting and the non-inverting feedback amplifier is that of Fig. 2.12. The only difference is as to which input terminal the signal is applied, and which grounded. For signals applied to both inputs at once then

$$V_o=V_{in+}\frac{R_1+R_2}{R_1}-V_{in-}\frac{R_2}{R_1} \qquad (2.18)$$

the beginnings of a subtractor.

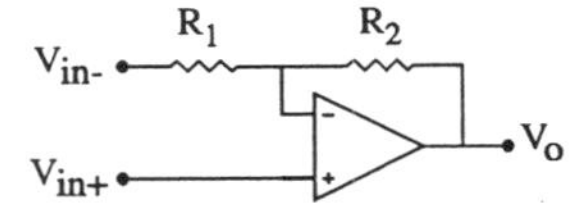

Fig. 2.12 Differential feedback amplifier with unequal responses to V_{in+} and V_{in-}.

The higher weighting for the non-inverting input is easily corrected as in Fig. 2.13, giving for $R_3=R_1$ and $R_4=R_2$

$$V_o=(V_{in+}-V_{in-})\frac{R_2}{R_1} \qquad (2.19)$$

For unity scaling all the R can be made equal. Exercise 2.7 gives appropriate experimental work.

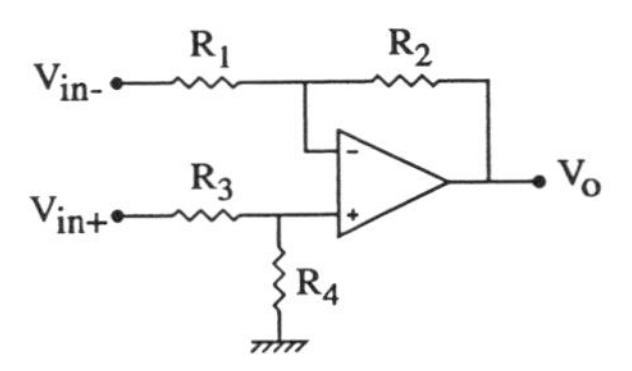

Fig. 2.13 Feedback subtractor obtained by modifying the circuit of Fig. 2.12 to make the responses to V_{in+} and V_{in-} of equal magnitude.

2.8 Integrator

The essential element in the analog integrator is the capacitor. For C in Fig. 2.14, the charge $Q=CV_c$. With

$$Q|_t=Q|_{t=0}+\int_0^t Idt$$

then, as required,

$$V_c|_t=V_c|_{t=0}+\frac{1}{C}\int_0^t Idt \qquad (2.20)$$

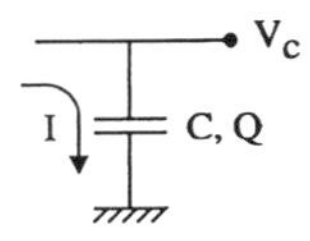

Fig. 2.14 Capacitor C as integrator.

Fig. 2.15 shows how the capacitor might be used to form the integral of the current pulse in (a) over the period from $t=0$ to t_1, with the following sequence.

- The Reset switch is briefly closed at some time prior to $t=0$, to make $V_c=0$.
- With the Reset switch open, the Run/Hold switch is closed at time $t=0$, allowing the integration to proceed.
- At $t=t_1$ the Run/Hold switch is opened, with V_c thereafter remaining constant as the record of the integral.

A limitation here is that the final V_c cannot be observed without discharging C. As soon as the monitoring oscilloscope or meter in (c) is connected then the discharge occurs.

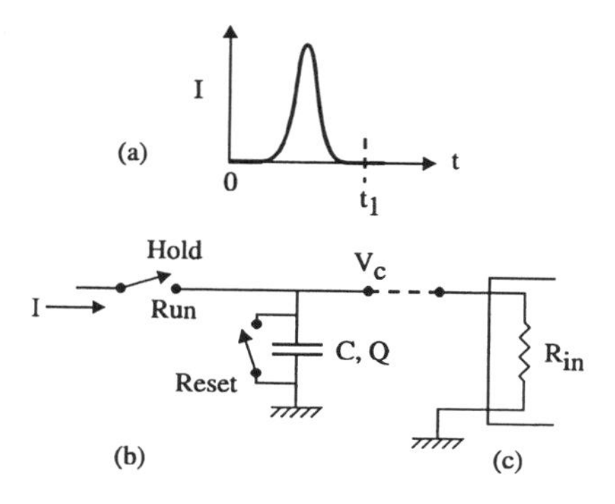

Fig. 2.15 Integration of current pulse by capacitor.
(a) Current pulse I.
(b) Capacitor C.
(c) Load R_{in}.

Feedback integrator

While the above loading of C could be resolved by interposing a voltage-follower, it remains that the input to such operational circuits is normally a voltage rather than a current. The op-amp-based feedback integrator of Fig. 2.16 overcomes both of these limitations.

With the circuit as in Fig. 2.16 then the feedback component is C. For as long as V_o is within the operating range, this is still an effective way of maintaining the virtual ground. Operation is as follows.

- For the Run switch closed, and with the virtual ground, then $I=V_{in}/R$.
- For the capacitor $$V_c|_t = V_c|_{t=0} + \frac{1}{C}\int_0^t I dt \qquad (2.21)$$
- With the virtual ground then $V_o=-V_c$, giving

$$V_o|_t = V_o|_{t=0} - \frac{1}{CR}\int_0^t V_{in} dt \qquad (2.22)$$

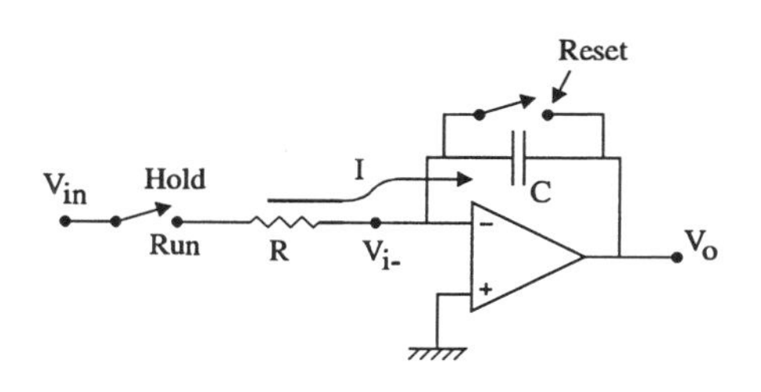

Fig. 2.16 Feedback integrator

Hold mode

With the Run switch opened at $t=t_1$ then V_o is held as before, but with now no current path through C, V_o does not change. Any current drawn by an output load is provided by the op-amp.

Reset mode

The Reset switch operates by shorting the capacitor, making $V_c=0$. With V_{i-} for the virtual ground then $V_o=0$ also.

From another viewpoint, the circuit becomes an inverting feedback amplifier with $R_2=0$ and thus $A_f=0$.

Exercise 2.8 gives five examples related to the feedback integrator, all verified by experimental work.

2.9 Differentiator

With [2.20] for the capacitor written in the differential form,

$$I=CdV_c/dt \qquad (2.23)$$

Thus the capacitor can also be used as a differentiator, with the voltage V_c as the input and the current I as the output.

For the circuit of Fig. 2.17, with the virtual ground, then V_{in} is applied as V_c, while the current-follower converts the resulting I to the proportional

$$V_o=-CRdV_{in}/dt \qquad (2.24)$$

Oscillatory transient response

Unlike the integrator which works well, the present differentiator circuit will not operate adequately without further components. The response of an ideal differentiator is an alternating series of impulses (sketch), while that of the real circuit might be expected to be the corresponding series of alternating narrow pulses.

Exercise 2.9 demonstrates experimentally that the actual response for the circuit is a series of weakly damped sinusoidal transients. Appropriate corrective measures are described in the next chapter.

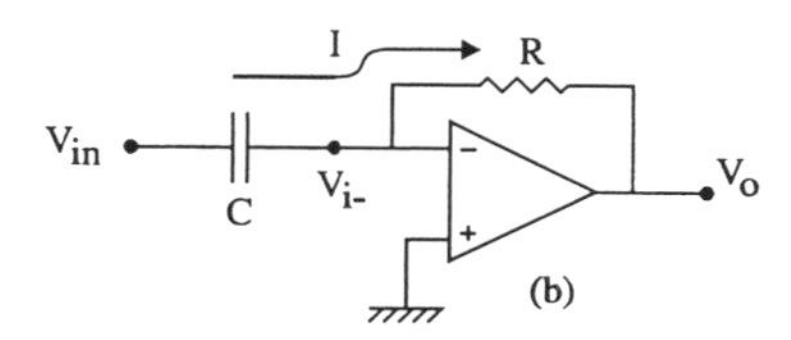

Fig. 2.17 Feedback differentiator, needing modification to function correctly.

2.10 Exercises and experiments

Exercise 2.1 Non-inverting negative feedback amplifier (Fig. 2.3)

> **Calculation** For the reasons given in Section 3.7, 10kΩ is suitable for R_1. Calculate the value of R_2 then required for the feedback amplifier gain $A_f=11$.

With $A_f=(R_1+R_2)/R_1$
as in [2.7] then

$$R_2=R_1(A_f-1)$$

giving $\mathbf{R_2=100k\Omega.}$

Experiment Connect the circuit shown using a TL081 op-amp and the above feedback resistors. A Vero or Experimentor 'breadboard' is recommended.

For V_+ and V_- use a dual regulated power supply set to ±15V. As explained in Section 18.5, a ceramic 100nF capacitor is needed between each supply and ground, to prevent oscillation.

Using a suitable signal generator, apply a 1V peak-to-peak 100Hz sine wave as V_{in}. Compare the V_{in} and V_o waveforms on a dual-channel oscilloscope. Thus confirm that the waveforms are in phase and that $A_f=11$. Next increase the input amplitude until the output exceeds the saturation limits.

Repeat with the scope in the X-Y mode, with X connected to V_{in} and Y to V_o. This shows the transfer function of the feedback amplifier, which is somewhat like that of Fig. 2.2(b) for the op-amp but with much better linearity between the saturation limits.

Exercise 2.2 Cascading of feedback amplifiers

Design Devise a suitable negative-feedback circuit giving $A_f=10^6$, where the op-amp gain $A_o=10^5$.

Since the required $A_f>A_o$ this can not be done with one op-amp. Using two, the required A_f for each is 10^3. For each then the loop-gain

$$A_o\beta=10^5\times(1/10^3)=100$$

which is normally considered adequate.

Exercise 2.3 Inverting negative-feedback amplifier

Experiment Repeat the experiments of Exercise 2.1 using the same component values but with the inverting configuration of Fig. 2.7. Check that input and output are now in anti-phase and that $|A_f|$ is now reduced from 11 to 10.

Exercise 2.4 Voltage-follower (Fig. 2.8)

Example Calculate the error factor α_{err} for $R_s=10k\Omega$ and $R_m=1k\Omega$ with the follower omitted.

With $V_o=E_sR_m/(R_m+R_s)$ and $\alpha_{err}=V_o/E_s$
then $\alpha_{err}=R_m/(R_m+R_s)$
giving $\mathbf{\alpha_{err}=\times 0.0909.}$

Experiment Set E_s to 10V p-p at 100Hz using the scope for measurement.

Increase R_s to approximately 10kΩ by adding a 10kΩ series resistor (the signal generator output impedance is typically 50Ω).

Directly connect a 1kΩ resistor as R_m and measure V_o using the scope. Thus confirm the above value of α_{err}.

Interpose the voltage-follower to confirm that the error is avoided.

Exercise 2.5. Current-follower (Fig. 2.9)

Example Calculate the error factor α_{err} for $R_s=1k\Omega$ and $R_m=10k\Omega$ with the follower omitted.

With $I_s=E_s/R_s$ for the link shorted, and

$$I_s=E_s/(R_s+R_m)$$

for R_m included then

$$\alpha_{err}=R_s/(R_s+R_m)$$

giving $\mathbf{\alpha_{err}=0.0909.}$

Experiment Connect the circuit of (a) and (b) with $R_s=R_f=10k\Omega$, and apply the previous 10V p-p, 100Hz, as E_s.

With $V_o=-I_sR_f$ as in [2.16] and $I_s=E_s/R_s$ for the virtual ground, confirm that the registered value of I_s is correct.

For R_m as above, and without the follower, confirm the above α_{err}.

Problem As an alternative, the current meter may be connected in series with R_f. Explain why no allowance needs to be made here when calculating the value of R_f. Is R_f needed at all?

Exercise 2.6 Adder (Fig. 2.10)

Example For $R_1=10k\Omega$, $R_2=22k\Omega$, $R_3=2.2k\Omega$ and $R_f=10k\Omega$, give the numerical expression for V_o in terms of the V_{in}.

With

$$V_o = -\left(\frac{V_{in1}}{R_1} + \frac{V_{in2}}{R_2} + \frac{V_{in3}}{R_3}\right)R_f$$

as in [2.17] then

$$V_o=-(V_{in1}+0.45V_{in2}+4.5V_{in3})$$

Experiment Connect the above circuit and apply V_{in}=1V p-p at 100Hz to each input in turn to confirm that the V_o is as expected. Do the unused inputs need to be grounded?

Exercise 2.7 Subtractor (Fig. 2.13)

Experiment Set each of the R values to 10kΩ and apply a 1V p-p 100Hz sine wave to each of the two inputs in turn, with the other grounded. Check that the voltage gain is +1 and –1 as expected.

Parallel the two inputs and see how close V_o comes to the expected zero.

How do the resistor tolerances effect the result?

Exercise 2.8 Integrator (Fig. 2.16)

Calculation With the required integration constant T_i=1s, electrolytic capacitors leak and the largest value of C for convenient size is 1μF. Calculate the value of the associated R.

With T_i=CR in [2.22], **R=1MΩ.**

DC input Predict the V_o waveform for V_{in}=+15V.

With V_o as in [2.22],

$$dV_o/dt=-V_{in}/(CR)$$

Thus V_o is a ramp falling at the rate **dV_o/dt=–15V/s.**

Square wave input Predict the V_o waveform for V_{in} a 10V peak-to-peak square wave of f_{in}=1Hz.

For V_{in}=+5V, dV_o/dt=–5V/s.

This lasts for 500ms to give a change of 2.5V. Thus V_o will be a triangular wave of 2.5V p-p at 10Hz.

Sine wave input Predict the V_o waveform for V_{in} a 10V peak-to-peak sine wave of f_{in}=10Hz.

With V_o as in [2.22],

$$V_o\big|_t = V_o\big|_{t=0} - \frac{1}{CR}\int_0^t \hat{V}_{in}\sin(2\pi ft).dt$$

giving $$V_o\big|_t = V_o\big|_{t=0} - \frac{\hat{V}_{in}}{2\pi fCR}[\cos(2\pi ft) - 1]$$

Thus, relative to the starting value, we have a cosine wave of amplitude

$$\hat{V}_{in}/(2\pi fCR)=80mV.$$

Thus **V_o=160mV p-p.**

X-Y display What will be the scope X-Y display be for the above sine wave input? For what value of the input frequency f_{in} will this be a circle?

With V_o lagging V_{in} by 90° and inverted, the spot will trace an unskewed ellipse, rotating in the clockwise sense. This becomes a circle for R=1/2πfC, giving **f_{in}=0.159Hz.**

Experiment Connect the integrator of Fig. 2.16.

Use the +15V supply for V_{in} and 'reset' the integrator by shorting the capacitor. Remove the short and confirm that V_o varies as in the above calculation.

Apply the square wave as above and confirm that V_o is the predicted triangular wave, after first resetting.

Note that the base line has a slow drift of constant rate. Contributing factors are:
- the + and – levels of the square wave will not be quite the same,
- the periods in the + and – states will not be exactly equal. For either case saturation eventually occurs.

Increase the frequency to 10Hz and confirm the expected ×1/10 decrease in V_o amplitude.

Apply the above sine wave and confirm the predicted V_o, paying particular attention to the phase relation between input and output.

Switch to the x-y display as above and confirm the predicted effect of varying f_{in}.

Exercise 2.9 Differentiator (Fig. 2.17)

Experiment Connect using R=10kΩ and C=10nF. Apply a 100mV p-p square wave of f_{in}=2kHz and note the series of alternating oscillatory transients in place of the expected alternating train of impulses.

Connect a 220pF capacitor across R and note the improvement.

3

Operational amplifier limitations

Summary

The limitations to be discussed in this chapter are as follows.

- The op-amp voltage-gain A_o is less than the ideal of infinity.
- The variation of output voltage V_o with differential input voltage $V_{id}=V_{i+}-V_{i-}$ is not quite linear.
- For the ideal op-amp the output voltage $V_o=0$ for $V_{id}=0$. For the real device there is a slight offset.
- The ideal op-amp draws no current into either signal input terminal. For the real device there is the small 'bias current' I_b, approximately the same for both signal inputs, added to which is a yet smaller component proportional to V_{id}, constituting a finite input impedance.
- For the ideal op-amp V_o is independent of the current I_o drawn. For the real device there is a small variation. Also the range of V_o is reduced with loading.
- For the ideal op-amp V_o only responds to V_{id}, and not at all to a 'common-mode' component V_{ic} applied to both signal input terminals at once. For the real device there is a small response.
- For the ideal op-amp the response of V_o to V_{id} is immediate. For the real device the response-time T_r obtains, with the corresponding upper limit f_r to the frequency response.
- In addition to T_r, there is an upper limit to the rate at which V_o can change.
- Some feedback circuits can either oscillate or show a strong resonance without suitable damping. The feedback differentiator, the photodetector amplifier, and the voltage-follower with high signal source impedance are cases discussed.

Value calculations are cumulative with the results set bold to the right-hand margin. The example is normally the non-inverting negative-feedback amplifier of Fig. 3.1, or the equivalent inverting configuration.

For both the starting values for the cumulative calculations are

$$\mathbf{A_o=10^5, \quad R_1=10k\Omega, \quad R_2=100k\Omega}$$

The final section gives experimental work and further calculations

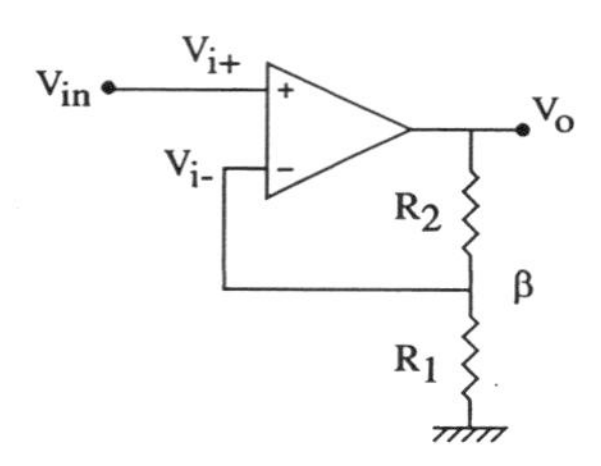

Fig. 3.1 Non-inverting feedback amplifier.

3.1 Finite op-amp gain

With A_o the voltage-gain of the op-amp in Fig. 3.1

$$V_o = A_o(V_{i+} - V_{i-}) \tag{3.1}$$

with ideally $A_o=\infty$. For the non-inverting feedback amplifier shown, [2.6] gives the feedback amplifier gain A_f as

$$A_f = A_o/(1+A_o\beta) \tag{3.2}$$

where the feedback fraction

$$\beta = R_1/(R_1+R_2) \tag{3.3}$$

With $\Delta A_f=A_f-1/\beta$ the error in A_f due to the loop-gain $A_o\beta$ being finite then

$$\Delta A_f = -1/[(1+A_o\beta)\beta] \tag{3.4}$$

or more significantly

$$\Delta A_f/A_f = -1/(A_o\beta) \tag{3.5}$$

Sensitivity

By differentiation of [3.2], for a small change δA_o in A_o,

$$\frac{\delta A_f / A_f}{\delta A_o / A_o} = \frac{-1}{1 + A_o\beta} \qquad (3.6)$$

Here δA_o may be due to production spread or varying temperature.

3.2 Op-amp non-linearity

Fig. 3.2 shows how negative-feedback improves upon the linearity of the op-amp. First (a) shows the transfer function of the op-amp without feedback, with the non-linearity exaggerated to make the point.

Next (c) shows the variation of the incremental gain

$$a_o = dV_o/d(V_{i+} - V_{i-})$$

with $(V_{i+} - V_{i-})$. With the required gain $1/\beta$ also shown then only between the intersection points is the loop-gain $A_o\beta > 1$. Thus only here does the incremental feedback amplifier gain a_f approximate to the required $1/\beta$.

Thus, shown to a somewhat expanded vertical scale in (d), $a_f \approx 1/\beta$ between the intersection points, falling to follow the much reduced a_o elsewhere.

With a_f the slope of the transfer function of the feedback amplifier, the transfer function becomes as in (b), with the excellent linearity retained up to very close to the saturation values V_{s+} and V_{s-}, where $A_o\beta$ falls below unity.

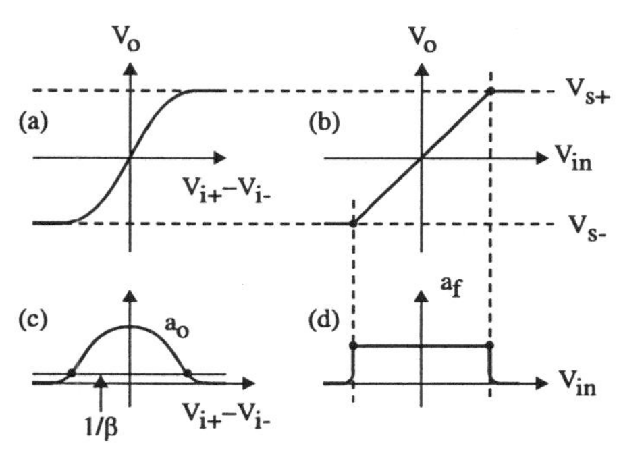

Fig. 3.2 Op-amp non-linearity.
(a) Op-amp transfer function with non-linearity exaggerated.
(b) Improved linearity of feedback amplifier.
(c) Incremental op-amp gain $a_o = dV_o/d(V_{i+} - V_{i-})$.
(d) Incremental feedback-amplifier gain $a_f = dV_o/dV_{in}$.

Notation

The lower-case a_o and a_f were used above to distinguish the incremental values from the upper-case 'large-signal' A_o and A_f. This is a convention which will be followed where appropriate for most of the remaining text. This, however, is the only point in the present chapter where the distinction needs to be made, so the upper-case symbols will continue to be used for the rest of the chapter.

3.3 Voltage Offset

Fig. 3.3(a) shows the effect of the input offset voltage v_{ofo} of a real op-amp on the transfer function, as modelled in (b). Here V_{ofo} is the value of $(V_{i+} - V_{i-})$ for which $V_o = 0$. With $V_o = A_o(V_{i+} - V_{i-})$ as in [3.1] for the ideal op-amp, now

$$V_o = A_o(V_{i+} - V_{i-} - V_{ofo}) \qquad (3.7)$$

For the two types of op-amp to be considered the specified V_{ofo} is the same

Bipolar 741 V_{ofo}=5mV, Bi-fet TL081 V_{ofo}=5mV.

V_{ofo} may be set to zero by the offset-correcting potentiometer R_{of}, external to the op-amp chip as in (c).

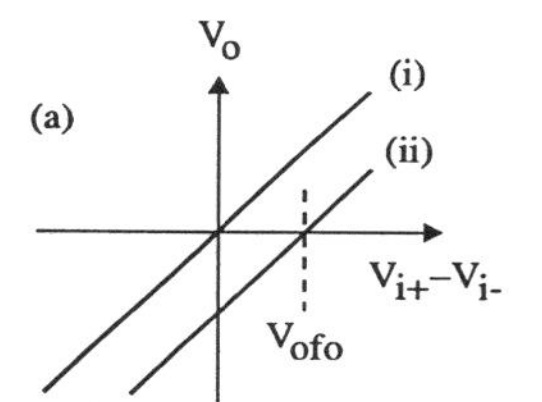

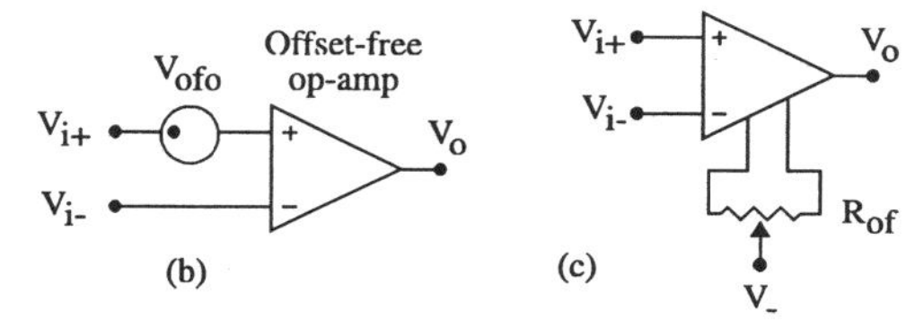

Fig. 3.3 Op-amp voltage offset V_{ofo} (of: offset, o: op-amp).
(a) Op-amp transfer function (i) without offset, (ii) with offset.
(b) Circuit model for offset.
(c) Offset-correcting potentiometer.

Non-inverting feedback amplifier voltage offset

With V_{off} the feedback amplifier offset (of: offset, f: feedback), it is clear from Fig. 3.4(a) that $V_{off} = V_{ofo}$.

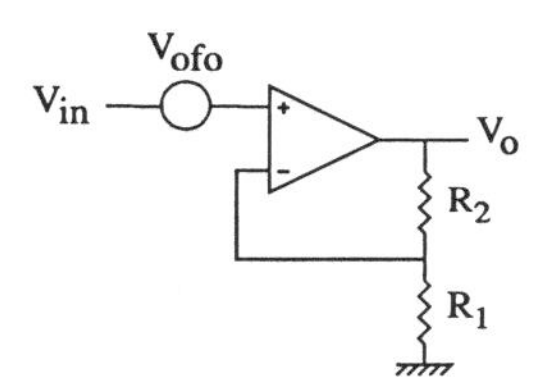

Fig. 3.4(a) Feedback amplifier offset due to op-amp offset V_{ofo} for non-inverting feedback amplifier.

Inverting feedback amplifier voltage offset

For the inverting feedback amplifier of Fig. 3.4(b), with V_{off} the value of V_{in} for $V_o=0$,

with $V_{i-}=\alpha V_{in}-\beta V_o$

where the 'feed-forward' fraction

$$\alpha=R_2/(R_1+R_2) \quad (3.8)$$

with $V_{i+}-V_{i-}=V_{ofo}$,

and with $V_{i+}=0$,

then $$V_{off}=-V_{ofo}/\alpha \quad (3.9)$$

The reason for this being larger than $V_{off}=V_{ofo}$ for the non-inverting configuration is that V_{in} is no longer applied directly as V_{i+}, but is attenuated by $\times\alpha$.

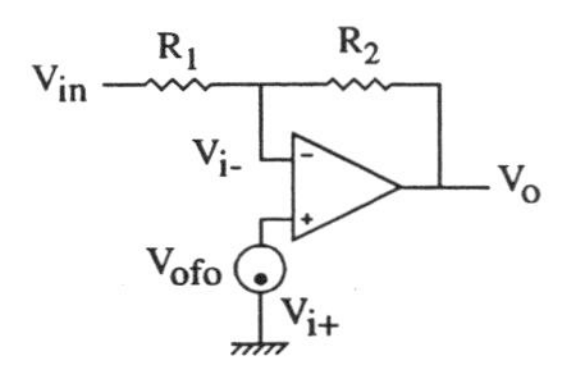

Fig. 3.4(b) Feedback amplifier offset due to op-amp offset V_{ofo} for inverting feedback amplifier.

Example For $R_1=R_2$ then for the non-inverting and inverting circuits V_{off}=5mV and 10mV respectively. Exercise 3.1 gives experimental verification.

3.4 Bias current

Fig. 3.5(b) shows the essentials of the 'long-tailed pair' which is the basis of the input stage of the bipolar transistor-based op-amp. Here the currents I_{b+} and I_{b-} have to be provided by the signal input circuit and so constitute the op-amp input 'bias currents'. To the first approximation these are independent of the signal input voltages V_{i+} and V_{i-} and so can be represented by the ideal current sources in (a).

One of the major advantages of the bi-fet op-amp, with its JFET input stage, is the much reduced I_b. Here

Bipolar 741 I_b=50nA, Bi-fet TL081 I_b=50pA.

Feedback amplifier offset voltage due to bias current

For the non-inverting feedback amplifier of Fig. 3.6, I_{b+} has no effect on V_{off}.

However, with V_{off} the value of V_{in} for $V_o=0$ then for $V_o=0$ I_{b-} develops $V_{i-}=-I_{b-}R_p$ where $R_p=R_1//R_2$. Thus

$$V_{off}=-I_{b-}R_p \quad (3.10)$$

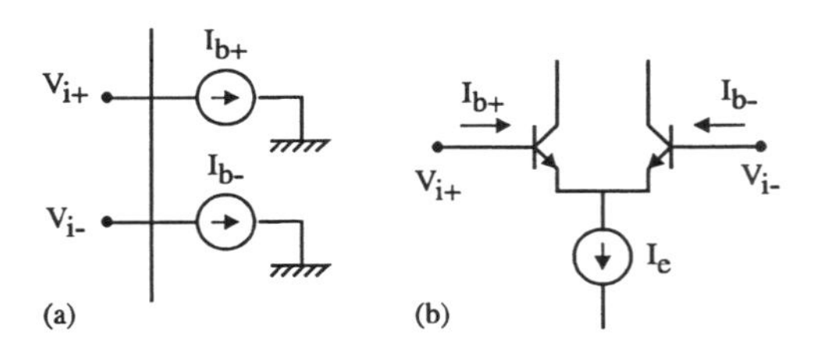

Fig. 3.5 Op-amp bias currents.
(a) Circuit model.
(b) Input circuit for long-tailed pair differential amplifier.

Critical resistance

Without zero adjustment of V_{ofo}, it is of value to reduce R_p to the point where this component of V_{off} is small compared with that due to V_{ofo}. For the two equal

$$R_p=V_{ofo}/I_{b-} \quad (3.11)$$

giving

Bipolar 741 R_p=100kΩ, Bi-fet TL081 R_p=100MΩ.

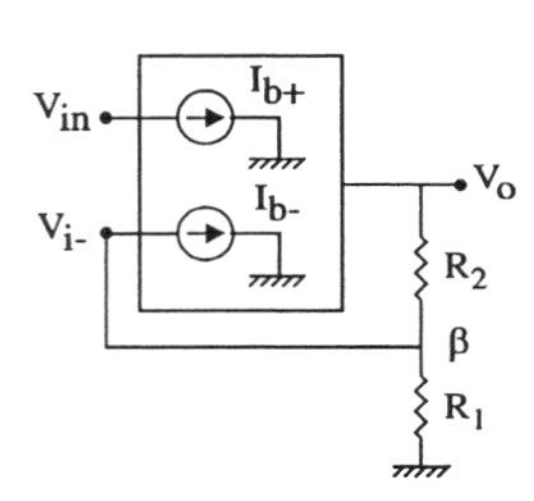

Fig. 3.6 Input bias currents for non-inverting feedback amplifier.

Inverting configuration

Here, as for the effect of V_{ofo}, V_{off} is increased by the factor $1/\alpha=(R_1+R_2)/R_2$ to give

$$V_{off}=I_{b-}R_1 \quad (3.12)$$

Bias current compensation

Fig. 3.7 shows the usual method of correcting for the I_b. If the added $R'_p=R_p$ and the I_b are equal then the component of V_{off} due to the I_b is zero.

Unfortunately the matching of the two I_b is seldom better than 20% and so, instead of eliminating this component of V_{off}, it is merely reduced, by approximately ×1/5.

With R'_p included and the offset potentiometer adjusted, what matters is the temperature-dependence of V_{off}. With θ the temperature, then [3.11] for the upper limit of R_p becomes

$$R_p=|(dV_{ofo}/d\theta)|/|d(I_{b+}-I_{b-})/d\theta| \quad (3.13)$$

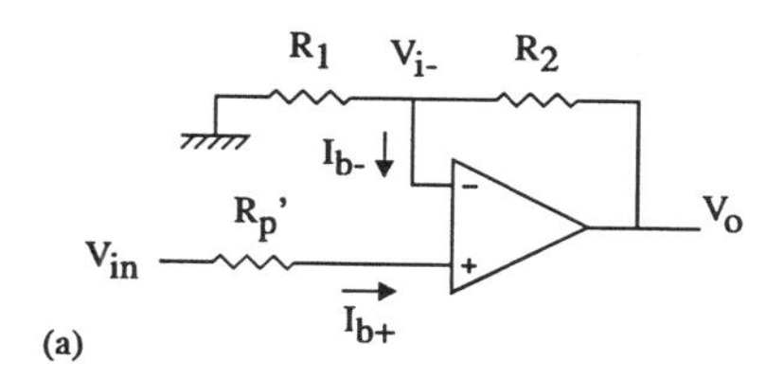

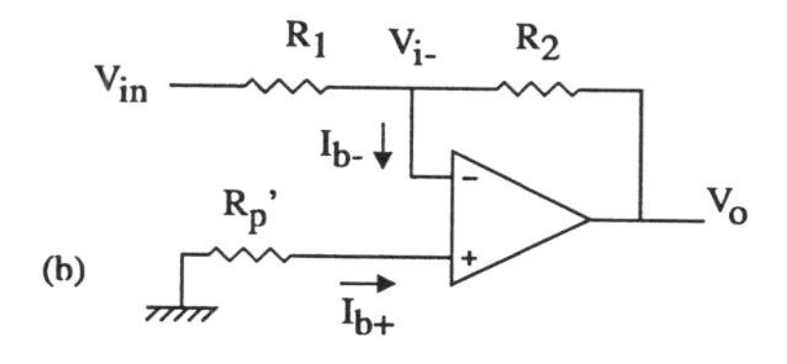

Fig. 3.7 Bias current compensating resistor R'_p.
(a) Non-inverting feedback amplifier.
(b) Inverting feedback amplifier.

Bipolar 741 With the specified

$$dV_{ofo}/d\theta = 5\mu V/°$$

and $\quad d(I_{b+} - I_{b-})/d\theta = 0.5nA/°$

then $\quad$ $\mathbf{R_p = 10k\Omega}$.

Bi-fet TL081 While the specified

$$dV_{ofo}/d\theta = 20\mu V/°$$

the variation of I_b with temperature is exponential, with $I_{b+} - I_b = 5pA$ doubling for each rise of 20°.

With the critical R_p thus varying with the temperature change, we calculate for a 20° rise. With

$$\Delta V_{ofo} = R_p \Delta(I_{b+} - I_{b-}), \qquad \Delta(I_{b+} - I_{b-}) = 5pA,$$

and $\Delta V_{ofo} = 400\mu V$ then $\quad$ $\mathbf{R_p = 80M\Omega}$.

3.5 Op-amp input impedance

The small differential change in the I_b of Fig. 3.5 that does occur with V_{id} can be represented as in Fig. 3.8, with R_{io} the incremental op-amp input impedance (i: input, o: op-amp).

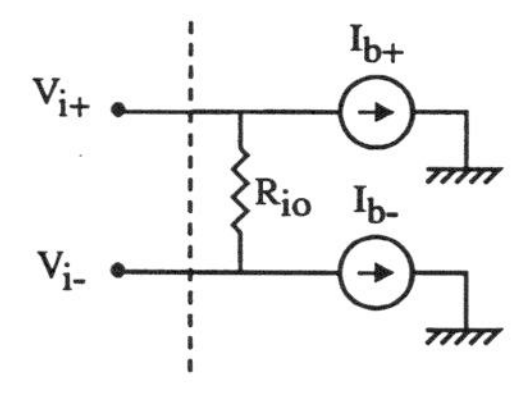

Fig. 3.8 Addition of op-amp input impedance R_{io} to model of op-amp input circuit.

The representation implies that a common change in the V_i produces no change in the currents. This is not a general feature of a three-terminal connection to a network, but is a good approximation for the long-tailed pair and its derivatives. Typical values are

Bipolar 741 $R_{io} = 1M\Omega$ Bi-fet TL081 $R_{io} = 1G\Omega$.

These differ by approximately the same factor as for the I_b above.

Feedback amplifier input impedance

With R_{if} the input impedance of the feedback amplifier (i: input, f: feedback) then for the non-inverting feedback-amplifier of Fig. 3.9 the effect of negative feedback is to increase R_{if} from the value R_{io} without feedback.

With $\quad V_{i-} \approx \beta V_o \quad$ for $\quad R_{io} >> R_1 // R_2$,

and with $\quad V_o = A_o(V_{i+} - V_{i-}) \quad$ as in [3.1]

then
$$R_{if} \approx R_{io}(1 + A_o\beta) \tag{3.14}$$

With the increase by approximately $\times A_o\beta$ with $A_o\beta$ the loop gain, this is due to the reduction in $V_{i+} - V_{i-}$ imposed by the feedback.

Bias current

With these added, the Norton equivalent of the circuit presented to the signal source is R_{if}, in parallel with the Norton current I_N which flows to ground with the input terminal grounded. With

$$I_N = I_{b+} - I_{b-} \times G_{io}/(G_{io} + G_p) \tag{3.15}$$

and normally $G_p >> G_{io}$ then $I_N \approx I_{b+}$.

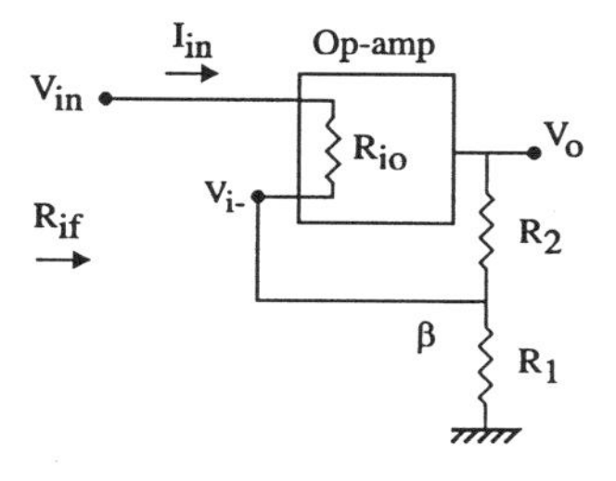

Fig. 3.9 Circuit for determining feedback amplifier input impedance R_{if} due to R_{io}.

Inverting configuration

With the input impedance ideally equal to R_1, this circuit has no pretence to a high input impedance, so R_{io} is of little consequence.

3.6 Op-amp output impedance

For the ideal op-amp the load current has no effect on V_o, while the real op-amp has the output impedance R_{oo} (o: output, o: op-amp) as in Fig. 3.10. There

$$V_o = A_o(V_{i+} - V_{i-}) - I_oR_{oo} \quad (3.16)$$

with typically **$R_{oo}=50\Omega$.**

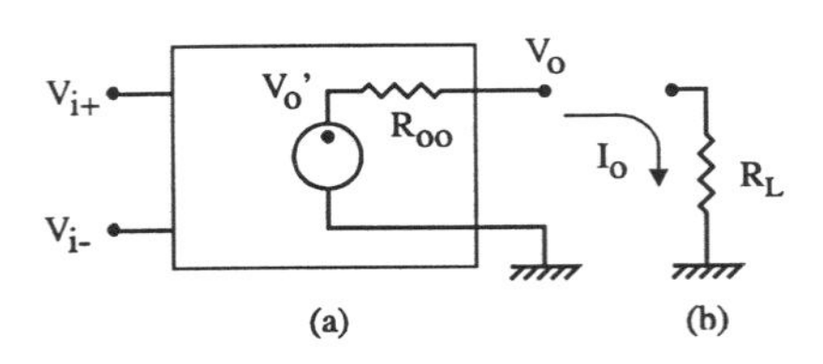

Fig. 3.10 Circuit model for op-amp output impedance R_{oo}.
(a) Op-amp with $V_o' = A_o(V_{i+} - V_{i-})$.
(b) Load resistor.

Feedback amplifier

Negative feedback reduces the effect of R_{oo}, to give $R_{of} << R_{oo}$ where R_{of} (o: output, f: feedback) is the feedback amplifier output impedance. Fig. 3.11 shows the non-inverting feedback amplifier with R_{oo} included. Here the test current I_t is inserted with $V_{in}=0$ and the resulting V_o determined, to give

$$R_{of} = V_o/I_t.$$

For I_β relatively small,

$$V_o \approx V_o' + I_tR_{oo}$$

With $V_o' = A_o(V_{i+} - V_{i-})$

$$V_{i-} = \beta V_o \quad \text{and} \quad V_{i+} = 0$$

then $R_{of} \approx R_{oo}/(1 + A_o\beta)$ (3.17)

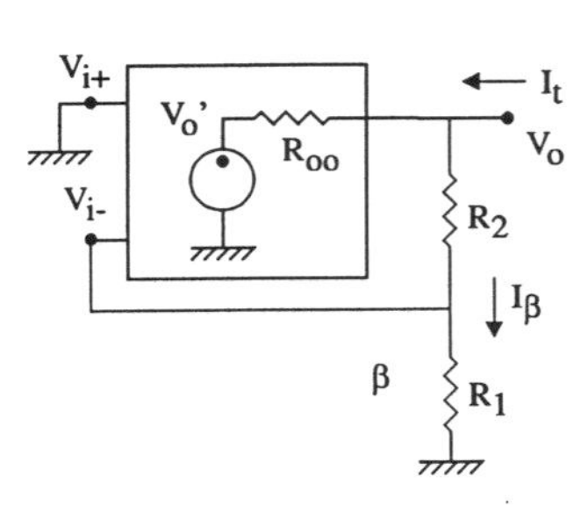

Fig. 3.11 Diagram for determining output impedance R_{of} of feedback amplifier due to R_{oo} by applying test current I_t.

Output range

As well as determining R_{of}, R_{oo} can also significantly reduce the voltage output range of the amplifier. Fig. 3.12 is an approximation to the output circuit of a typical op-amp and gives an idea of what determines R_{oo}. For $V_o > 0$, I_o is supplied via Q_1, while for $V_o < 0$ it is Q_2 that provides the supply.

The R_{LIM} shown are current limiting resistors. These avoid destruction of the op-amp when the V_o terminal is shorted to ground. Thus R_{oo} will be the sum of one of the R_{LIM} and the associated transistor impedance.

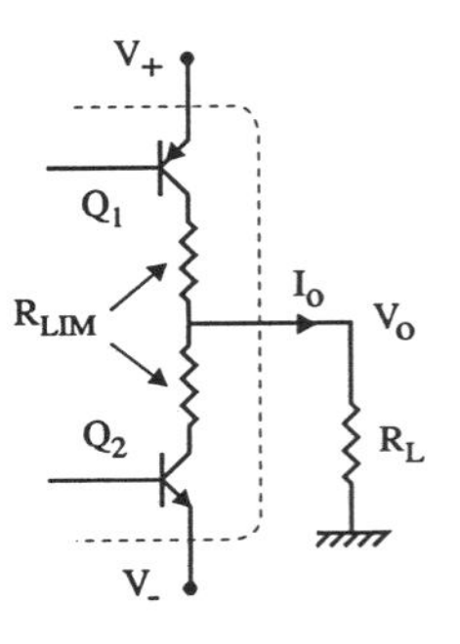

Fig. 3.12. Approximate op-amp output circuit for determining V_o range.

Fig. 3.13(a,b) show the effect on the op-amp transfer function for the load R_L a few times greater than R_{oo}. Here both the slope and saturation limits are reduced.

Then, as in (c,d) with a suitably compressed horizontal scale, the feedback suppresses the changes in slope but makes no difference to the reduced saturation limits.

In the experimental Exercise 3.2 it is found that the range reduction is somewhat larger than for $R_{oo}=50\Omega$. The reason here is that 50Ω is the incremental value, which would more correctly be represented by r_{oo}. The large-signal R_{oo} in this instance is greater.

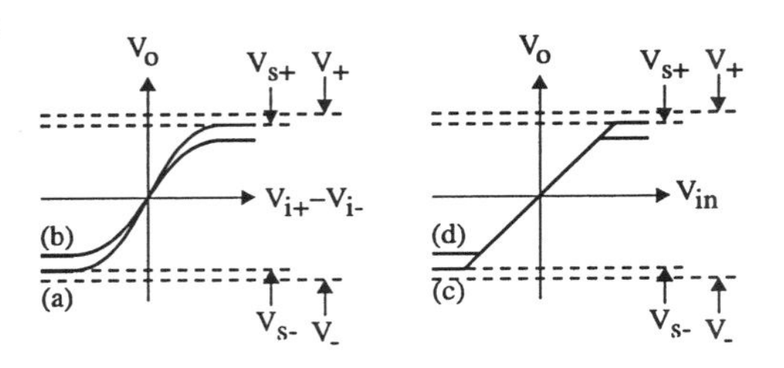

Fig. 3.13 Effect of load resistor R_L on amplifier transfer functions.
(a) Op-amp without feedback showing non-linearity and output saturation voltages V_{s+} and V_{s-}.
(b) As (a) but with R_L.
(c) Feedback amplifier without R_L.
(d) As (c) but with R_L.

3.7 Feedback resistor values

For the feedback amplifier the choice of absolute value for the feedback resistors is wide. The upper limit is that at which the thermal sensitivity due to the bias current begins to significantly increase the total, as in [3.13]. As calculated, the transition is at $R_p=R_1//R_2=10k\Omega$ for the bipolar op-amp and $\approx 80M\Omega$ for the bi-fet.

The lower limit is where R_1+R_2 begins to significantly reduce the output voltage range, with $R_1+R_2>1k\Omega$ a rough estimate. The present $R_1=10k\Omega$ and $R_2=100k\Omega$ conform well.

AC coupled feedback amplifier

The constraints are sometimes more demanding when the amplifier is AC coupled as in Fig. 3.14. Here the objective is to introduce a lower cut-off frequency, usually to accept a low-amplitude AC component while rejecting a much larger DC component (see Chapter 11 on Filters). With the AC coupling the concept of an input offset is inappropriate, because no applied DC V_{in} can bring V_o to zero. However the op-amp V_{ofo} and I_b will produce an offset in V_o. With

$$V_{i-}=V_o-I_{b+}R_2$$

and with $\quad V_{i-}=-V_{ofo} \quad \text{for} \quad A_o=\infty$

then

$$V_o=I_{b-}R_2-V_{ofo} \qquad (3.18)$$

With the cut-off frequency

$$f_f=1/(2\pi CR_1) \qquad (3.19)$$

there is a limit to the largest value of leak-free capacitor that can be made of reasonable size, with $\sim 1\mu F$ the practical limit. With this setting a lower limit to R_1, and with the required gain equal to $-R_2/R_1$, the limit to R_2 is even higher.

In Exercise 3.3 it is shown that for $f_f=1Hz$ and the required gain -100 then $R_1\approx 30M\Omega$, giving $V_o\approx 1.5V$ for the bipolar 741. The corresponding 1.5mV for the bi-fet TL081 is much to be preferred.

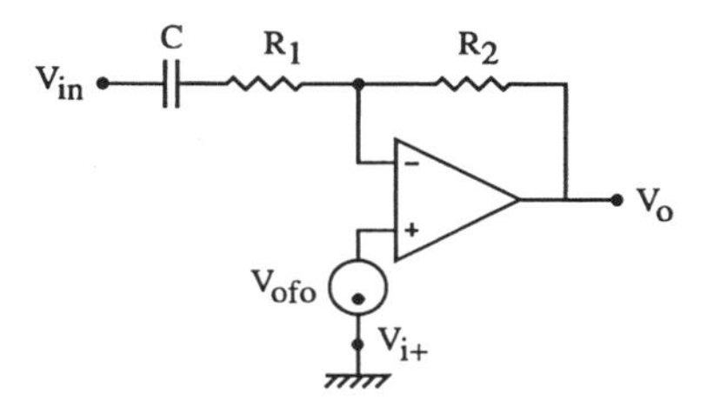

Fig. 3.14 AC-coupled inverting feedback amplifier for consideration of output offset.

3.8 Common-mode response

While for the ideal op-amp

$$V_o=A_o(V_{i+}-V_{i-})$$

for the real op-amp

$$V_o=A_{o+}V_{i+}+A_{o-}V_{i-} \qquad (3.20)$$

with $|A_{o-}|$ closely equal to $|A_{o+}|$ but only ideally exactly so.

A simple test of the degree to which the magnitudes of the A_o differ is to apply a common signal to both of the inputs. Here for the ideal op-amp V_o remains zero. The small response that occurs for the real op-amp is termed the 'common-mode' response.

It is helpful here to resolve the values of V_{i+} and V_{i-} into the 'common-mode' and 'difference-mode' input components V_{ic} and V_{id} as in Fig. 3.15. The ideal amplifier then responds to V_{id} but not to V_{ic}. More generally

$$V_o=A_cV_{ic}+A_dV_{id} \qquad (3.21)$$

where A_c and A_d are the common- and difference-mode gains of the amplifier. With $A_c=0$ for the ideal device, the recognised figure of merit is the 'common-mode rejection ratio'

$$CMR=A_d/A_c \qquad (3.22)$$

CMR is usually quoted in decibels with the typical value

CMR=80dB.

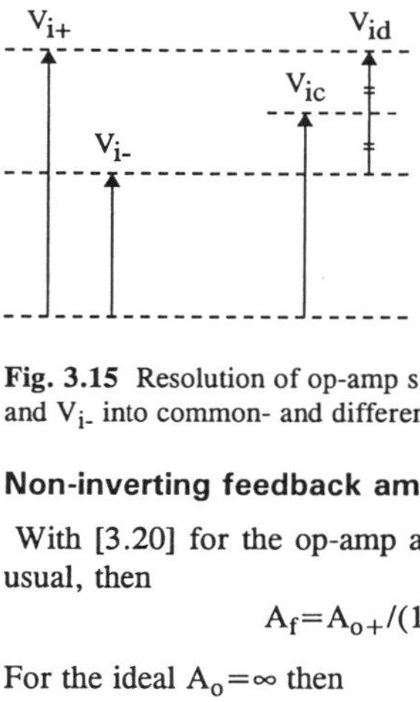

Fig. 3.15 Resolution of op-amp signal input components V_{i+} and V_{i-} into common- and difference-mode components.

Non-inverting feedback amplifier

With [3.20] for the op-amp and the rest of the circuit as usual, then

$$A_f=A_{o+}/(1-A_{o-}\beta).$$

For the ideal $A_o=\infty$ then

$$A_f=-A_{o+}/A_{o-}/\beta$$

rather than the required $1/\beta$, making the mismatch of the A_o significant.

Inverting feedback amplifier

Here the mismatch is of no consequence. With $V_{i+}=0$ then A_{o+} is eliminated from [3.20], making it of no concern whether $|A_{o+}|$ is the same as $|A_{o-}|$ or not.

Subtractor

A circuit for which CMR is of potential importance is the op-amp-based subtractor, particularly in the ground-loop cancelling applications of Chapter 18. For the inverting input of the subtractor the CMR is of no effect, for the same reasons as for the inverting feedback amplifier, while for the non-inverting input the error factor $-A_+/(A_-)$ again obtains. Thus the op-amp CMR is transferred to the subtractor. With CMR typically 80dB the typically 1% tolerance in the resistor values will vastly outweigh this. However, a small adjustment of one of the resistor values can correct for both effects.

CMR measurement

From [3.22], measurement of CMR is essentially a matter of measuring A_c and A_d. Strictly A_d should be measured by applying a difference-only input signal, i.e. one where $V_{i-}=-V_{i+}$. This, however, is a little pedantic, because for a typical real op-amp A_c is vastly below A_d making A_+, $-A_-$ and A_d all very nearly equal in magnitude. Thus one tends to measure say A_+, by applying V_{i+} with V_{i-} zero, and then to take this as the value of A_d. Direct measurement of A_c is simpler, requiring only the two inputs to be connected before application of the signal.

Op-amp

The above applies well enough when measuring the CMR of a feedback circuit, such as the subtractor, but cannot be applied directly to commonly used op-amps such as the TL081 without feedback. This is due to the small amount of positive feedback internal to the chip which is encountered in Section 3.3.

The solution is to make the measurement using the normal subtractor circuit of Fig. 2.13 but with only a very little negative feedback applied, just enough to comfortably override the chip positive feedback.

This implies making $R_2 >> R_1$ and $R_4 >> R_3$. A ratio of 10^4 is suitable, causing the subtractor gain to have this value. With these values the effect of the typically 1% tolerance for the resistor values is suppressed $\times 1/10^4$. Exercise 3.4 gives calculations and experimental work.

3.9 Op-amp frequency response

The real op-amp has a non-zero response-time due to the circuit capacitances. While the negative feedback can considerably speed the response, it also becomes possible for the circuit to oscillate or show a strong resonance.

To explain the possible instability, Fig. 3.16 first shows the op-amp as composed of three stages of identical gain ×30 and each having the same cut-off $f_{tr}=3$MHz (tr: transistor). Here

$$A_o=A_{oz}/(1+jf/f_{tr})^3 \quad (3.23)$$

where A_{oz} is the zero frequency value of A_o.

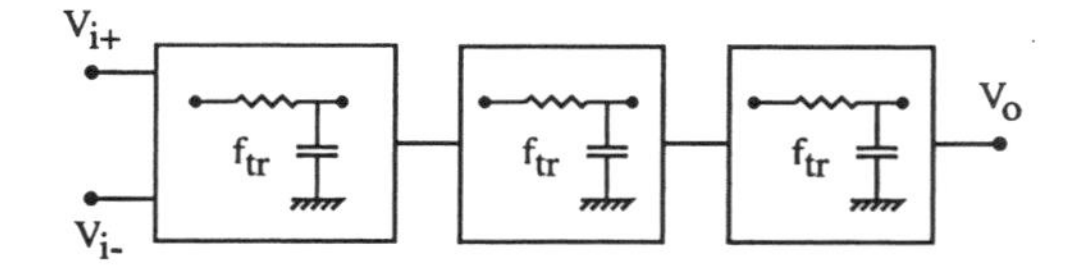

Fig. 3.16 Breakdown of op-amp into three component stages.

With the amplitude-frequency and phase-frequency responses as in Fig. 3.17, then the phase-shift ϕ_o reaches $-180°$ at f_{180} making the feedback positive rather than negative. If the feedback loop gain $A_o\beta > 1$ at this frequency then the loop will oscillate, approximately at f_{180}.

Alternatively if $A_o\beta$ is only a little less than one at f_{180}, there will be a marked resonance in the frequency response, and the damped oscillatory transient or 'ring' in the step response.

For the circuit shown, at f_{180} $|A_o|$ is only less than A_{oz} by a factor of about six. This is the limiting value of $A_o\beta$ for stability, which is far too small to give the above benefits of negative feedback.

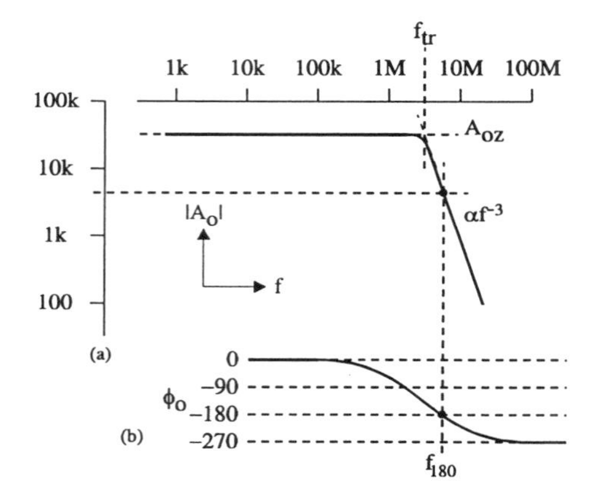

Fig. 3.17 Frequency response of three-stage op-amp of Fig. 3.16.

The solution here is to arrange for the cut-off of one of the filters to be much lower than any of the others. The frequency responses then become as in Fig. 3.18. Here at the point where ϕ_o has reached $-180°$, A_o has fallen to below unity. Thus even if 100% feedback is applied, as for the voltage follower, the loop still remains stable.

Usually it is the second stage that provides most of the voltage gain A_o, with the associated Miller feedback imposing the dominant cut-off. The last stage is essentially a voltage-follower, to prevent the following load from reducing the second stage gain. For the first stage the emphasis is on correct differential operation, i.e. high CMR, rather than voltage gain.

In the diagram f_u is the frequency for unity gain and f_o the op-amp cut-off frequency. Typical values for f_u are

Bipolar 741, f_u=1MHz, Bi-fet TL081, f_u=3MHz.

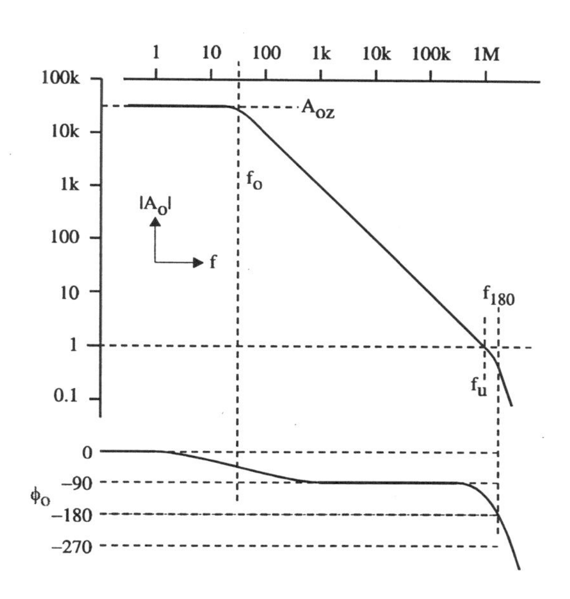

Fig. 3.18 Modification to frequency response of Fig. 3.17 for one of the three filter time constants is much increased.

3.10 Feedback amplifier frequency and step response

Fig. 3.19 shows the familiar non-inverting negative feedback amplifier, with the main filter in the op-amp included.

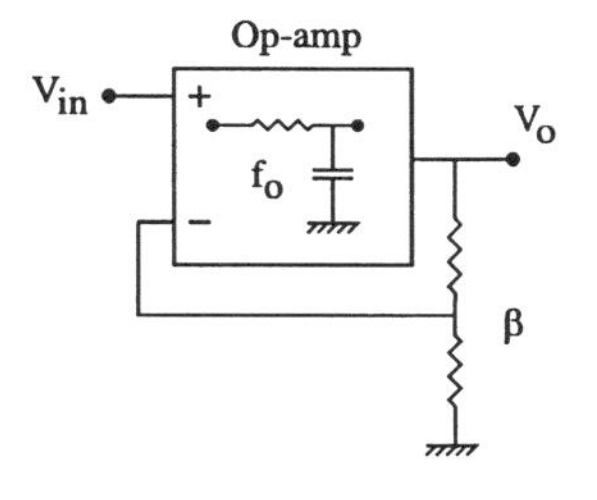

Fig. 3.19 Non-inverting feedback amplifier showing effect of f_o in Fig. 3.18.

Fig. 3.20 shows the resulting variation of $|A_o|$ with frequency as in Fig. 3.18, together with the required gain $1/\beta=10$. With the loop gain $A_o\beta$ as also shown, $|A_f|$ follows $1/\beta$ up to f_1 and then falls to follow $|A_o|$.

From the diagram

$$f_1=\beta f_u \qquad (3.24)$$

so there is a trade-off between $1/\beta$ and f_1. More precisely, with

$$A_f=A_o/(1+A_o\beta) \quad \text{and} \quad A_o=A_{oz}/(1+jf/f_o)$$

then

$$A_f=A_{oz}/(1+A_{oz}\beta+jf/f_o).$$

From Fig. 3.18 $f_o=f_u/A_{oz}$ giving, with $A_{oz}\beta >> 1$

$$A_f \approx (1/\beta)/(1+jf/f_1) \qquad (3.25)$$

Thus when both high A_f and f_1 are required this will require cascading of two or more feedback stages.

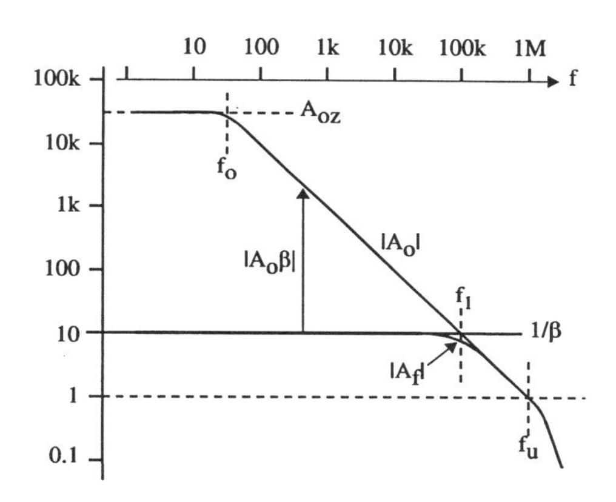

Fig. 3.20 Frequency response curves for non-inverting feedback amplifier of Fig. 3.19.

Step response

Apart from the region above f_u, the frequency response of $|A_f|$ in Fig. 3.20 is essentially that of the simple low-pass filter. Thus the response time

$$T_1=1/(2\pi f_1) \qquad (3.26)$$

For 100% feedback this becomes

$$T_u=1/(2\pi f_u) \qquad (3.27)$$

Thus

$$T_1=T_u/\beta \qquad (3.28)$$

giving the expected trade-off between required gain $1/\beta$ and T_1. The following quoted values of T_u accord approximately with the above, remembering that, in reality, there is more than just the one cut-off beyond f_u.

Bipolar 741 T_o= 150ns, Bi-fet TL081 T_o=50ns.

Exercise 3.5 gives calculations confirmed by experimental work.

Production spread

With the typical production spread of 10:1 for A_{oz}, and with $f_u=A_{oz}f_o$, it is fortunate that f_u remains constant, with the spread in f_o.

3.11 Slewing

The 'slew rate' is the maximum rate at which the output V_o can change. For the step response of Fig. 3.21(a) the slew-rate is not exceeded so the response time is the above T_1. But as the input amplitude is increased the slew limit to dV_o/dt is exceeded, giving the response of (c), rather than that of (b), (b) being merely (a) with increased vertical scaling. Thus slew effectively increases the small-signal response time T_1. Typical values for the slew rate are

Bipolar 741 0.5V/μs, Bi-fet TL081 13V/μs.

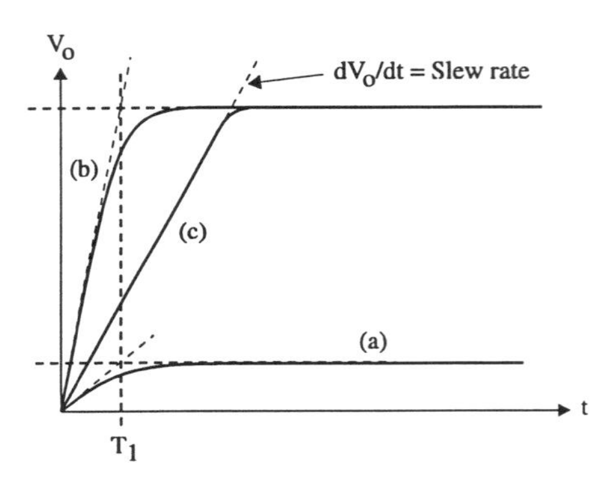

Fig. 3.21 Slewing in the step response of an op-amp-based feedback amplifier.
(a) Small-signal response giving no slewing increase to the response time T_1.
(b) Large-signal response without slewing.
(c) Large-signal response with slewing.

Fig. 3.22 shows a very approximate model of the op-amp which exhibits the slewing mechanism. This time the amplifier is divided into two sections, and we assume the gain to be equally divided with $A_1=A_2=200$, giving $A_o=4\times10^4$.

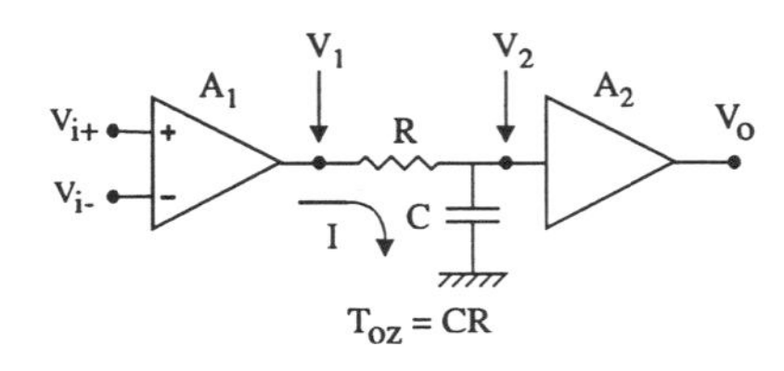

Fig. 3.22 Approximate circuit model for op-amp slewing.

The dominant low-pass filter is also shown and this has the response time $T_o=5$ms, corresponding to the usual op-amp cut-off frequency $f_o=30$Hz.

Suppose now that a square wave is applied to the differential input $V_{i+}-V_{i-}$, of sufficient amplitude to saturate A_1. This will cause the first stage output voltage V_1 to describe the square wave shown in Fig. 3.23. The limits of the square wave are the saturation limits of the stage, which are taken as being the same as those of the second stage, the op-amp saturation limits V_{s+} and V_{s-}. The filter then converts the square wave to that shown for V_2.

Also shown are the values of V_2 at which the second stage saturates. Thus V_o becomes as shown below, with the rate of the near linear rise between the V_o saturation points the slew rate.

Thus broadly slewing is the result of saturation before the dominant filter, while the normal saturation limits are set by the final output stage. As will be seen later, the dominant filter section is actually set by Miller feedback, which modifies the present account somewhat.

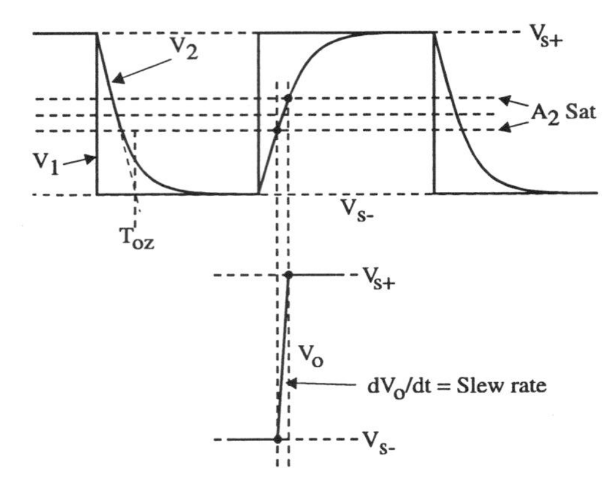

Fig. 3.23 Waveforms for slewing model of Fig. 3.22.

Calculation

The slew rate can be calculated from the above model. For the region in Fig. 3.23 corresponding to the linear rise in V_o we have

$$V_1=V_{s+} \quad \text{and} \quad V_2 \ll V_1$$

giving $I\approx V_{s+}/R$. With

$$I=CdV_2/dt \quad \text{and} \quad T_o=CR$$

then

$$dV_o/dt \approx A_2V_{s+}/T_o \tag{3.29}$$

With the typical values of V_s, T_o, etc., for the 741 as shown in Exercise 3.6, the model confirms the above $dV_o/dt=0.5$V/μs quoted for the device. The exercise includes experimental work.

3.12 Differentiator resonance

Fig. 3.25 shows the differentiator of Section 2.9 which, without the added damping resistor R_d, gives the previously observed resonance and ringing step response. In the absence of R_d and with the signal source impedance zero, the feedback fraction

$$\beta = X_c/(R+X_c) \quad \text{where} \quad X_c = 1/j\omega C,$$

giving
$$\beta = 1/(1+j\omega CR) \tag{3.30}$$

With this the response of a simple low-pass filter of cut-off

$$f_\beta = 1/2\pi CR \tag{3.31}$$

the feedback loop contains two significant low-pass sections, that of cut-off f_o in the op-amp, and now the external filter of cut-off f_β.

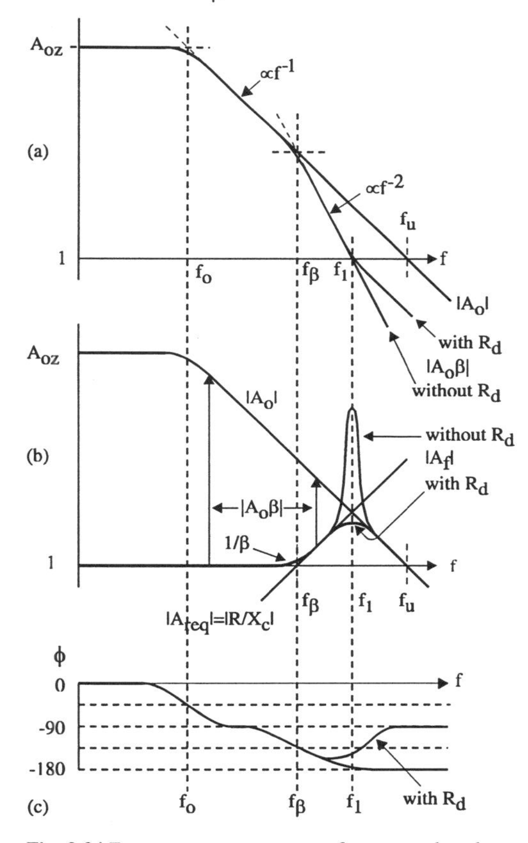

Fig. 3.24 Frequency response curves for op-amp-based differentiator of Fig. 3.25.
(a) Open-loop amplitude-frequency responses,
$|A_o|$ for op-amp alone,
$|A_o\beta|$ for op-amp with feedback network with and without R_d.
(b) Comparison of amplitude-frequency response curves for op-amp $|A_o|$ and required differentiator gain A_{req} showing resonance in differentiator gain A_f at cross over.
(c) Phase-frequency curves for loop gain $A_o\beta$.

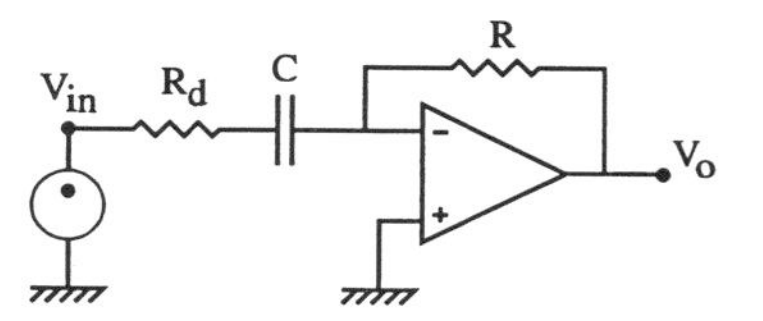

Fig. 3.25 Op-amp-based differentiator with damping resistor R_d.

The result is to make the amplitude-frequency response for the loop gain $A_o\beta$ to be as in Fig. 3.24(a) without R_d, with the associated phase-frequency response as in (c). Here at the frequency f_1 for which $|A_o\beta| = 1$ the phase lag is very nearly 180°. It is this which gives the unwanted resonance.

Complex plane plots

Fig. 3.26 gives further insight. This is the trace of $A_o\beta$ on the complex plane as f varies from 0 to ∞. First (a) gives the view for f in the region of the op-amp cut-off f_o. Here $|A_o\beta|$ is largest when f=0 and falls by $\times 2^{-1/2}$ at f_o, with the phase ϕ then lagging by 45°.

As f advances beyond f_o then ϕ swings back further, to lag by 135° at f_β (not shown). Ultimately as ϕ approaches ∞ the lag approaches 180°, becoming very close to 180° at f_1.

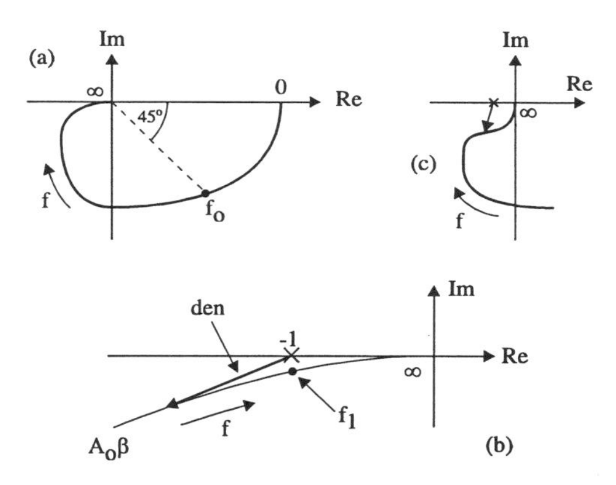

Fig. 3.26 Complex plane plots showing the variation of $A_o\beta$ with signal frequency f for differentiator of Fig. 3.26.
(a) Full frequency range from 0 to ∞.
(b) Expanded amplitude scale in the region of f_1 in Fig. 3.25 showing how magnitude of the denominator of A_f exhibits a sharp dip at f_1.
(c) Modification to (b) when damping resistor R_d is added.

Next (b) shows the last stages of (a) to a much expanded scale. The closed-loop differentiator gain is given just as for the inverting feedback amplifier of Section 2.3 but with R_1 replaced by X_c and R_2 by R. Thus as always

$$A_f = \frac{-A_o\alpha}{1+A_o\beta} \tag{3.32}$$

with now

$$\alpha = R/(R+X_c) = R/(R+1/j\omega C)$$

and β as in [3.30].

Consider next the denominator of [3.32] with f in the region of f_1 as in (b), which may be written as $A_o\beta-(-1)$. With the -1 as plotted the denominator becomes the arrow representing the difference between $A_o\beta$ and the -1 point. With ϕ here very close to $-180°$, the magnitude of the denominator undergoes a very sharp dip at f_1. With the numerator of A_f moderately 'well behaved' at this point, the overall result is the sharp resonant peak in $|A_f|$.

Frequency response

Fig. 3.24(b) shows a number of plots.

- $|A_o|$ copied from (a).
- $1/\beta$ with β as in [3.30].
- The resulting $|A_o\beta|$ as shown.
- $|A_{req}|$ for the required differentiator gain $A_{req}= R/X_c$.

With f_1 the frequency at which the loop gain $A_o\beta$ falls below unity, the feedback-differentiator gain A_f follows A_{req} up to f_1 and falls to follow A_o thereafter. Thus before f_1 we have the required differentiator response $\propto f$, while beyond f_1 the response becomes $\approx 1/f$, that of an integrator. With ϕ very close to $-180°$ at the transition, this is where the resonance occurs.

Resonant frequency

With distances in (b) representing ratios on the log axes used, then

$$f_1/f_\beta = f_u/f_1,$$

giving

$$f_1 = (f_u f_\beta)^{1/2} \qquad (3.33)$$

Q-factor

The Q-factor of the resonance is that of the series combination of C and R at f_1, giving

$$Q = f_1/f_\beta \qquad (3.34)$$

Damping resistor

R_d dampens the resonance by being set to equal X_c at f_1, giving

$$R_d = 1/2\pi f_1 C \qquad (3.35)$$

This has the effects shown dashed in Fig. 3.24(a) and (c), and also in the complex plane plot Fig. 3.26(c), thus avoiding the sharp dip in DEN in (b).

In Exercise 3.7 it is shown that for C=10nF, R=10kΩ and f_u=3MHz then f_β=1.6kHz, f_1=69.3kHz, Q=44.6 and the required C_d=230pF. These values are confirmed by experimental work.

Step response

With the resonance thus removed, the oscillatory transient in the step response becomes so heavily damped that only the first half cycle is evident. Thus we have a series of alternating pulses, of width approximately $1/2f_1$ with one for each transition of the input square wave. This is much more like the series of alternating impulses for the ideal differentiator than is the previous series of oscillatory transients.

3.13 Photodetector amplifier

Fig. 3.27 shows the usual arrangement for monitoring the output from a photo diode. Here light falling on the diode causes the diode current I_d to flow, with I_d proportional to the intensity of the light. I_d then passes into the current-follower, to produce the output voltage $V_o=-I_dR$.

The function of the reverse bias E_b is merely to reduce the diode capacitance C_{di}. Apart from this, the circuit would function in the same way.

With or without E_b, the equivalent circuit for the diode is merely the capacitor C_{di}. For the damping C_{da} absent, with the op-amp input capacitance C_{in} in parallel with C_{di}, and with R, then the feedback coupling of V_o to the op-amp input is via the simple low-pass filter so formed. Thus again with the internal op-amp low-pass we have the same possibility for resonance as for the differentiator.

With no possibility here of damping by the previous R_d then damping has to be by C_{da} shown.

In Exercise 3.8 it is shown that for the typical C_d=5pF, C_{in}=15pF, R=100kΩ and f_u=3MHz then $f_\beta \approx$ 80kHz, the resonant frequency f_1=490kHz, the Q of the resonance ≈6 and the required $C_{da} \approx$ 3pF.

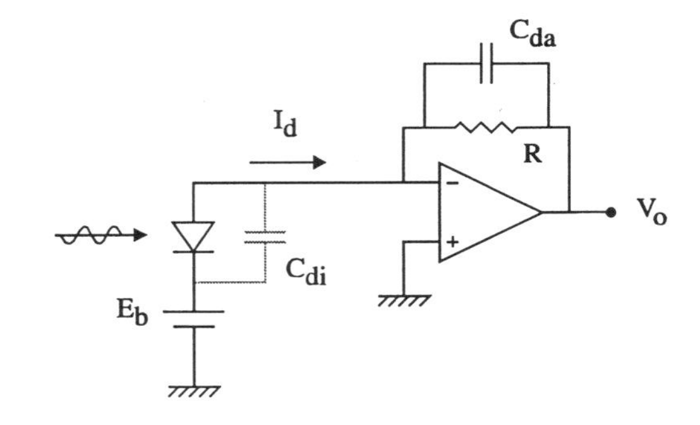

Fig. 3.27 Photodetector amplifier with damping capacitor C_d.

Voltage-follower with high signal source impedance

This circuit too has the potential for resonance. Here the high source impedance R_s and the op-amp C_{in} apparently form a high-pass filter in the feedback network. However, the op-amp differential input is across C_{in}, so the effect is as for the equivalent low-pass, of cut-off

$$f_c = 1/(2\pi C_{in} R_s)$$

For say C_{in}=15pF and R_s=1MΩ then f_c=10.6kHz, well below f_u. However, the effect is moderated by the stray capacitance across R_s.

Capacitive loading of voltage-follower

Often the voltage-follower is used to buffer the output when driving a relatively long line. This is another case where resonance can occur.

The zero-frequency output impedance r_{of} of any such negative-feedback amplifier $\propto 1/A_o$. Then beyond the op-amp cut-off f_o, $|A_o| \approx 1/f$ giving $|z_{of}| \approx f$. With this essentially as for an inductor, the effective inductance can resonate with the line capacitance. The theory is as presented for the emitter-follower in Section 7.3.

Reference

Coughlin, R.F. and Driscoll, F. F. 1982: *Operational Amplifiers and Linear Integrated Circuits*, Prentice Hall.

3. 14. Exercises and experiments

Exercise 3.1 Voltage offset

Calculation For the non-inverting and inverting feedback amplifiers of Fig. 3.4(a) and (b), $R_1 = R_2$=10kΩ and the op-amp input voltage offset V_{ofo} is the typical 5mV. Calculate the resulting feedback amplifier input offset voltage V_{off}.

Non-inverting From Fig. 3.4(a) $V_{off} = V_{ofo}$ giving **V_{off}=5mV.**

Inverting With $V_{off} = -V_{ofo}/\alpha$

as in [3.9] where $\alpha = R_2/(R_1+R_2)$

as in [3.8] then α=1/2 giving $V_{off} = -2V_{ofo}$ and thus **V_{off}=−10mV.**

Experiment. Op-amp voltage-offset Using a suitable potential divider, apply a small adjustable DC potential to the TL081 op-amp input with no feedback. Thus determine V_{ofo} from the response of V_o.

You will almost certainly find it impossible to set V_o to zero. Instead V_o swings slowly but inexorably to one or the other of the two saturation values V_{s+} or V_{s-}. The reason for this is that the op-amp chip has a degree of internal positive feedback, presumably due to some kind of leakage across the chip.

The internal β is very small but nevertheless enough to cause the effect. This turns the device into a Schmidt trigger, which is described in the next chapter.

Check that the effect ceases as soon as any normal degree of negative feedback is applied (e.g. R_1=100Ω, R_2=100kΩ). Then V_o can be set to zero with no difficulty.

With $R_1 = R_2$=10kΩ, confirm the values calculated above for the two configurations.

Exercise 3.2 Op-amp output impedance

Experiment. Loading of transfer function For the circuit of Fig. 3.9 with the usual R_1=10kΩ, R_2=100kΩ and TL081 op-amp, use the scope X-Y display to plot the variation of V_o with V_{in} for a sine wave input of f_{in}=100Hz. Thus confirm the unloaded transfer function of Fig. 3.13(c).

In turn shunt R_L=10kΩ, 1kΩ and 100Ω across the output to confirm the loaded transfer function of (d). Check that it is for $R_{oo} \approx 220Ω$ that the range is halved, giving the large-signal $R_{oo} \approx 200Ω$, rather than the above stated 50Ω, the incremental value.

Exercise 3.3 Output offset voltage for AC-coupled negative-feedback amplifier due to bias current.

Calculation. Resistor values For the AC-coupled feedback amplifier of Fig. 3.14 we require the low frequency cut-off f_f=0.5Hz. Calculate R_1 and R_2 for C=1μF with A_f=100 well above the cut-off.

With $R_1 = 1/(2\pi C f_f)$ as in [3.19]

then **R_1=321kΩ.**

With $A_f = (R_1+R_2)/R_1$

then **R_2=31.7MΩ.**

Calculation. Output offset For the above circuit, calculate the offset component of the output voltage V_o for the bipolar 741 (I_b=50nA) and the bi-fet TL081 (I_b=50pA). Assume the op-amp offset V_{ofo} for each device is zero.

With $V_o = I_b R_2 - V_{ofo}$ as in [3.18]

741 V_o=1.58V, TL081 V_o=1.58mV.

Exercise 3.4 Common-mode response

Calculation. Common- and difference-mode parameters For the two-input feedback amplifier of Fig. 2.12, without R_3 and R_4 in Fig. 2.13, $R_1=1k\Omega$ and $R_2=100k\Omega$. Calculate

- $A_+=V_o/V_{in+}$, and $A_-=V_o/V_{in-}$,
- the difference-mode gain A_d,
- the common-mode gain A_c,
- the common-mode rejection ratio CMR.

With $A_+=(R_1+R_2)/R_1$
then **$A_+=101$.**

With $A_-=-R_2/R_1$
then **$A_-=-100$.**

With $A_d=V_o/V_{id}$

for $V_{i+}=V_{id}/2$ and $V_{i-}=-V_{id}/2$

and with $V_o=A_+V_{i+} + A_-V_{i-}$ as in [3.20]

then $A_d=(A_+-A_-)/2$
giving **$A_d=100.5$.**

With $A_c=V_o/V_{ic}$

for $V_{i+}=V_{i-}=V_{ic}$

and with $V_o=A_+V_{i+}+A_-V_{i-}$ as in [3.20]

then $A_c=A_++A_-$
giving **$A_c=1$.**

With $CMR=A_d/A_c$ then **$CMR\approx40.0dB$.**

Calculation. Resistor value tolerance What effect will a 1% tolerance have on A_c and A_d?

With $A_c=A_++A_-=1$

as above then there is no effect.

With $A_d=(A_+ - A_-)/2$

as above, and with A_+ and A_- also as above then

$$A_d=R_2/R_1+1/2$$

With $R_2>>R_1$ then the worst case error in A_d will be 2%.

Experiment. Common- and difference-mode gain Connect the above circuit and for $f_{in}=100Hz$ confirm that $|A_+|\approx|A_-|=100$ with the expected signs. Then apply V_{in} to both V_{i+} and V_{i-} together, as V_{ic} and confirm that $A_c=1$.

Exercise 3.5 Feedback amplifier frequency and step response

Calculation. Closed-loop cut-off For a non-inverting feedback amplifier the required gain $A_{req}=1/\beta=30$ and for the TL081 op-amp the unity gain frequency $f_u=3MHz$. Calculate the closed-loop cut-off frequency f_1 in Fig. 3.20 and the associated response time T_1.

With $f_1=\beta f_u$ as in [3.24] then **$f_1=100kHz$.**

With $T_1=1/(2\pi f_1)$ as in [3.26]

then **$T_1=1.59\mu s$.**

Experiment. Slewing With the 741-based inverting feedback amplifier with $1/\beta=-11$, apply a square-wave input to observe the step responses of Fig. 3.21.

Confirm that slewing is more likely to be a significant factor when the required feedback amplifier gain is low than when it is high, by setting $1/\beta$ to say 1 and then to 100. Why is this so?

Exercise 3.7 Differentiator resonance

Experiment. Oscillatory step response and resonance Connect the differentiator circuit of Fig. 3.24 with $C=10nF$, $R=10k\Omega$ and without R_d.

Drive with a 1V p-p 500Hz square wave and observe how, instead of the required series of alternating impulses at each transition of the square wave, V_o consists of a corresponding series of oscillatory transients, with the oscillation frequency approximately 67kHz.

Switch to a sine wave input and vary the signal frequency through 67Hz to observe the corresponding resonant peak.

Calculation Calculate the value of the first transition frequency f_β.

With $f_\beta=1/2\pi CR$ as in [3.31]

then **$f_\beta=1.6kHz$.**

Calculation. Resonant frequency Calculate f_1 for the op-amp $f_u=3MHz$.

With $f_1=(f_uf_\beta)^{1/2}$ as in[3.33]

then **$f_1=69.3kHz$.**

Calculation Calculate the Q-factor of the resonance.

With $Q = f_1/f_\beta$ as in [3.34]

then **$Q = 43.3$.**

Experiment. Detailed frequency response With V_{in} a sine wave, repeat the above frequency response measurements, first relocating the resonance at $f_1 \approx 70$kHz.

Confirm that for $f < f_1$ the response is essentially as required for the differentiator. Here $|A_f| = f/f_\beta$ with V_o leading V_{in} by 90° (apart from the sign reversal incurred by the inverting configuration.

Confirm that for $f > f_1$ the response approaches that of an integrator, with $|A_f| \approx f_u/f$ and V_o now lagging V_{in} by 90°. This is now just the response of A_o, with the feedback of insignificant effect.

Calculation. Damping resistor Calculate the value of R_d needed to equal X_c at f_1 and so damp the resonance.

With $R_d = X_c$ at f_1 then

$$R_d = 1/(2\pi f_1 C)$$

giving **$R_d = 231$.**

Experiment. Damped response Add $R_d = 220\Omega$ and confirm that the frequency response becomes as in Fig. 3.24(b) but without the resonance. Increase R_d further and check the lowering of the central region to give a flat top.

Return R_d to 220Ω and measure the step response. Confirm that the resulting alternating single pulses are of width equal to one half cycle for f_1, i.e. 313μs.

Calculation. Damping capacitance Show that connection of a capacitor C_d across R, as in Fig. 3.27, is an equally valid way of damping the resonance. Calculate the required C_d.

With $X_d = R$ at f_1 then

$$C_d = 1/(2\pi f_1 R)$$

giving **$C_d = 231$pF.**

This accords with the 220pF used previously.

Exercise 3.8. Photodetector amplifier

For the photodetector amplifier of Fig. 3.27 the diode capacitance $C_d = 5$pF, the typical op-amp input capacitance $C_{in} = 15$pF, R=100kΩ and the op-amp $f_u = 3$MHz.

Calculation. Resonance and damping Calculate f_β and the resonant frequency f_1, the Q of the resonance, and the required value of the damping capacitor C_d.

With $f_\beta = 1/2\pi CR$

as in [3.31] for the differentiator then **$f_\beta = 80.2$kHz.**

With $f_1 = (f_o f_\beta)^{1/2}$

as in [3.33] then **$f_1 = 490$kHz.**

With $Q = f_1/f_\beta$

as in [3.34] then **$Q = 6.12$.**

For critical damping $X_d = R$ at f_1 giving

$$1/(2\pi f_1 C_d) = R.$$

Thus **$C_d = 3.27$pF.**

Experiment Connect as in Fig. 3.27 but with the diode and bias supply simulated by a 5pF capacitor. Loosely couple the V_{i-} terminal to a 2kHz square wave signal generator, by connecting a wire to the generator output and placing the end of the wire close to the V_{i-} point.

Check that the frequency of oscillatory transient is the above f_1, and that the frequency response shows the corresponding resonance.

The above $C_d = 3.27$pF is small enough for the 0.5pF stray capacitance for R, and the typically 2pF track-to-track capacitance of the Experimentor board to make up most of it, considerably damping the resonance even before the external capacitor is connected.

4

Op-amp and other switching circuits

Summary

The previous two chapters covered the use of the op-amp as a linear device, where the saturation values are normally avoided. This chapter describes some circuits where either the op-amp or its close cousin the comparator switch from one saturation state to the other.

With the same symbol normally used for the comparator as for the op-amp, as in Fig. 4.1, the comparator has largely the same structure as the op-amp.

However, the switching speed for the comparator is considerably higher than for the op-amp. Also the output voltage levels are those required by the digital logic that the comparator frequently has to drive, again as in Fig. 4.1. Thus most frequently, but not invariably, it is the comparator that is used.

The op-amp or comparator switching circuits to be described are usually a refinement of a simpler circuit using only passive components. For example, the precision rectifier of Section 4.7 is a refinement of the simple detector consisting of using just one diode and one resistor. In such cases the simpler circuit is presented first.

The final section gives experimental work with associated example calculations for virtually all of the circuits described.

4.1 Frequency counter

The frequency counter of Fig. 4.1 is a typical example of the use of the comparator as the interface between the analogue and digital parts of a circuit. Here the function of the circuit is to indicate the frequency f_{in} of a sine wave applied as V_{in}.

The function of the comparator here is to give logic output 1 (normally +5V) for $V_{i+} > V_{i-}$ and logic 0 (normally 0V) otherwise.

The counter is a digital register giving the parallel output D_o which increments each time the clock input (marked by the triangle) changes from logical 0 to logical 1 (usually 0V to 5V). With these components the sequence is as follows.

- D_o is set to zero by briefly applying logic 1 to the counter Reset input.
- Enable is set to logic 1 to allow the count to proceed.
- After an accurately timed interval T, Enable is returned to logic 0, causing the value of D_o to remain constant.

Here $f_{in} = D_o/T$, conveniently giving $D_o = f_{in}$ for T = 1s.

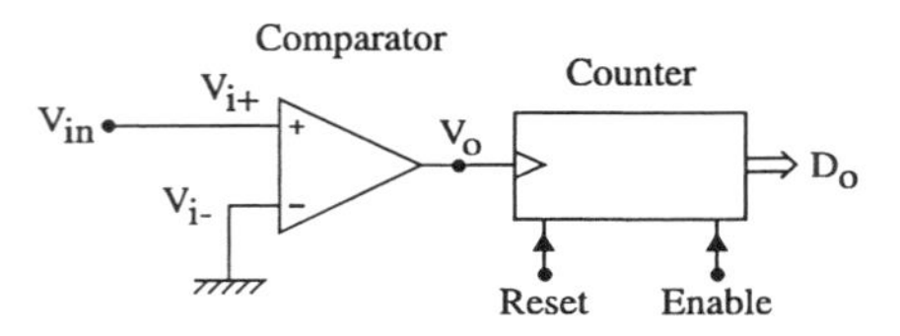

Fig. 4.1 Use of comparator in a frequency counter.

4.2 Schmidt trigger

Without further modification, the above frequency counter would register many more zero crossings of V_{in} than for the applied sine wave. This is because there is always some degree of noise present, to give the effect shown in Fig. 4.2.

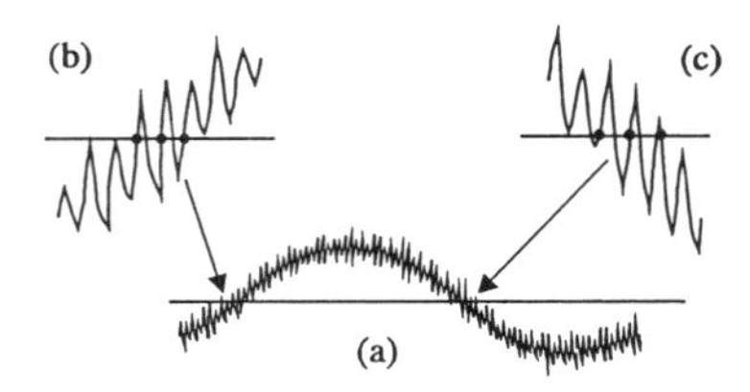

Fig. 4.2 Diagram showing how noise causes additional zero crossings to the one required per signal cycle.
(a) One cycle of noisy signal.
(b,c) Magnified sections of (a).

The solution here is to add positive feedback to the comparator as in Fig. 4.3(a), turning it into a 'Schmidt Trigger' circuit. We shall consider first the response of the op-amp based Schmidt, extending to the comparator shortly.

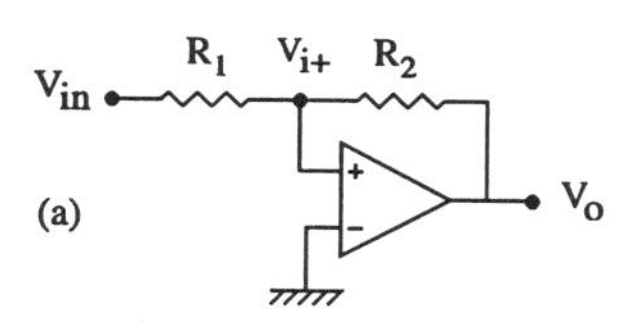

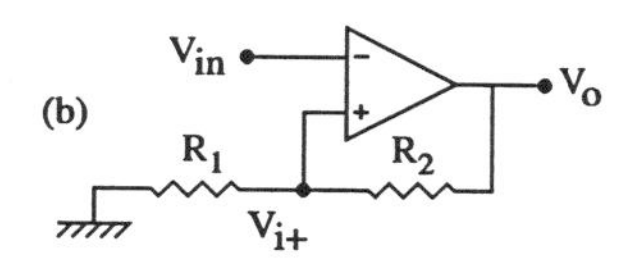

Fig. 4.3 Schmidt trigger circuits
(a) Non-inverting.
(b) Inverting.

Bistable

With the positive feedback, the op-amp has two stable states, making it a 'bistable' circuit. With V_{s+} and V_{s-} the op-amp output saturation voltages, and for $V_{in}=0$, then V_o is either V_{s+} or V_{s-}. In each case the sign of βV_o fed back as V_{i+} is such as to maintain the existing state.

Threshold splitting

For V_{in} initially low and rising, as in Fig. 4.2, then $V_o=V_{s-}$. With the positive feedback then the value at which the circuit changes state is the threshold value

$$V_{th+}=-(R_1/R_2)V_{s-} \qquad (4.1)$$

which is above zero.

Once V_{th+} has been crossed then V_o becomes V_{s+}, shifting the threhold to

$$V_{th-}=-(R_1/R_2)V_{s+} \qquad (4.2)$$

which is below zero.

Thus the circuit to switch back to the first state V_{in} must decrease significantly from the value V_{th+} at which the first switch occurred.

Thus by making the threshold splitting

$$\Delta V_{th}=V_{th+}-V_{th-}$$

greater than the noise amplitude as in Fig. 4.4, the multiple zero-crossings of Fig. 4.2 are avoided.

Inverting Schmidt

With the inverting Schmidt as in Fig. 4.3(b), notice that, apart from the reversal of the op-amp inputs, the configuration for the non-inverting Schmidt is as for the inverting feedback amplifier, and that for the inverting Schmidt is as for the non-inverting feedback amplifier.

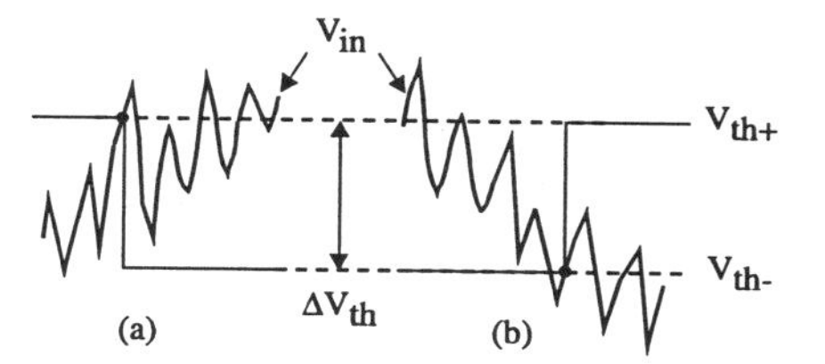

Fig. 4.4 Waveforms showing how non-inverting Schmidt trigger of Fig. 4.3(a) avoids unwanted noise-induced zero crossings.
(a) Rising zero crossing of sine wave.
(b) Falling zero crossing of sine wave.

Op-amp positive feedback

We now see why, with the usual small degree of on-chip positive feedback, it not possible to bring V_o to zero for the op-amp with no external feedback. As V_{i+}, $-V_{i-}$, or the offset-correcting potentiometer is adjusted then V_o simply snaps from one saturated state to the other. Here 'snap' is not too accurate a term, since with the large internal time constant, the progress is actually rather slow, albeit unstoppable.

Threshold adjustment

Fig. 4.5 shows how threshold adjustment can be incorporated into the Schmidt circuits of Fig. 4.3(a) and (b).

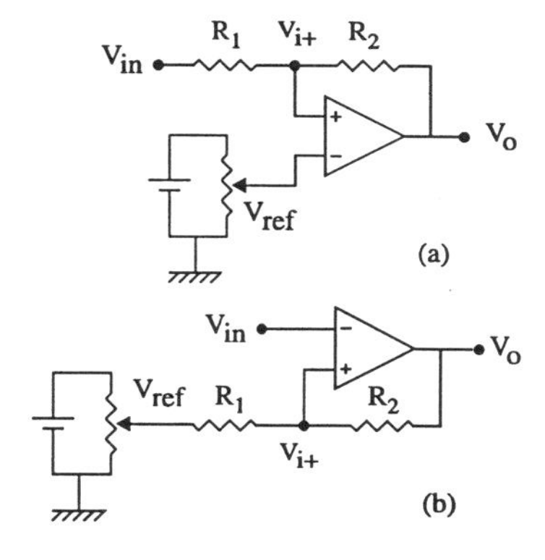

Fig. 4.5 Incorporation of adjustable threshold into Schmidt trigger circuit.
(a) Non-inverting.
(b) Inverting.

Threshold levels

We now derive the expressions for the V_{th} in terms of R_1, R_2 and V_{ref}. Then the equations are manipulated to give the values of R_1, etc., for the required V_{th}. The first is for analysis and the second for design, a recurring transition throughout the text. For simplicity we take the impedance of the adjustable V_{ref} source to be zero.

Threshold levels for non-inverting Schmidt

For Fig. 4.5(a) the V_{th} are the values of V_{in} at which $V_{i+}=V_{ref}$. Thus

$$V_{ref} = V_{th-}\frac{R_2}{R_1+R_2} + V_{s+}\frac{R_1}{R_1+R_2} \quad (4.3)$$

and

$$V_{ref} = V_{th+}\frac{R_2}{R_1+R_2} + V_{s-}\frac{R_1}{R_1+R_2} \quad (4.4)$$

giving

$$V_{th-} = V_{ref}\frac{R_1+R_2}{R_2} - V_{s+}\frac{R_1}{R_2} \quad (4.5)$$

and

$$V_{th+} = V_{ref}\frac{R_1+R_2}{R_2} - V_{s-}\frac{R_1}{R_2} \quad (4.6)$$

The separation ΔV_{th} between the V_{th} and the mean $\overline{V}_{th}$ thus become

$$\Delta V_{th} = \frac{R_1}{R_2}(V_{s+} - V_{s-}) \quad (4.7)$$

and

$$\overline{V}_{th} = V_{ref}\frac{R_1+R_2}{R_2} \quad (4.8)$$

Threshold levels for inverting Schmidt

From Fig. 4.3(b), the V_{th} are the values of V_{in} for which $V_{in}=V_{i+}$. Thus

$$V_{th+} = V_{ref}\frac{R_2}{R_1+R_2} + V_{s+}\frac{R_1}{R_1+R_2} \quad (4.9)$$

and

$$V_{th-} = V_{ref}\frac{R_2}{R_1+R_2} + V_{s-}\frac{R_1}{R_1+R_2} \quad (4.10)$$

giving

$$\Delta V_{th} = \frac{R_1}{R_1+R_2}(V_{s+} - V_{s-}) \quad (4.11)$$

and

$$\overline{V}_{th} = V_{ref}\frac{R_2}{R_1+R_2} \quad (4.12)$$

For both $\overline{V_{th}}$ it is assumed that $V_{s-}=-V_{s+}$, which is usually approximately so.

Design values

By manipulation of the above relations, for the non-inverting Schmidt

$$R_1/R_2=\Delta V_{th}/(V_{s+} - V_{s-}) \quad (4.13)$$

and

$$V_{ref} = \overline{V}_{th}\bigg/\left(1+\frac{\Delta V_{th}}{V_{s+} - V_{s-}}\right) \quad (4.14)$$

while for the inverting Schmidt

$$R_1/R_2=\Delta V_{th}/(V_{s+} - V_{s-} - \Delta V_{th}) \quad (4.15)$$

and

$$V_{ref} = \overline{V}_{th}\bigg/\left(1-\frac{\Delta V_{th}}{V_{s+} - V_{s-}}\right) \quad (4.16)$$

Oscilloscope triggering

The above technique for avoiding unwanted noise-induced zero crossings is used in oscilloscope triggering. Here the signal ultimately fed to the scope Y-plates is also used as the V_{in} for a Schmidt circuit. The Schmidt output is then used to trigger the scope time base. The Schmidt has an adjustable threshold, which is usually labelled the 'Trigger Level' control. The effect is to set an adjustable voltage 'level' on the screen. The time base is then triggered when the displayed signal crosses the level. It is also possible to select whether it is on the rising or falling crossing that the triggering occurs, usually by interposing an inverter, and selecting either the inverter input or output.

Without the positive feedback, any high frequency noise on the signal will cause the scope to trigger on both rising and falling crossings, regardless of which switch setting is used. This produces an annoying double image.

What is the effect when the signal amplitude falls below the level splitting caused by the positive feedback?

With digital circuitry present the noise spikes can still exceed the Schmidt threshold. Here an 'LF' setting adds a low-pass filter which smears the spikes out.

Bistable, monostable, astable

As above, a 'bistable' circuit is one where the output has two possible stable states, rather like a penny that can be laid one way up or the other.

In contrast the 'monostable' is stable in one of its two states until triggered. It then enters the other state for a period of time T_m determined by the monostable circuit, to return to the stable state thereafter.

Finally the 'astable' repeatedly alternates between its two allowed states without any external input.

Apart from the precision monostable described below, the monostable is usually implemented as a member of a family of digital logic gates, with the astable formed by a closed loop comprising two monostables.

4.3 Comparator

Fig. 4.6(a) and (b) compare the outputs of the op-amp and comparator, showing how the comparator, together with the 'pull-up' resistor R_{pu}, gives the required $V_o=V_{cc}$ or 0V. This is regardless of the value of V_{cc}, though this is almost invariably the +5V shown. The corresponding bipolar values for the op-amp are clearly incorrect and possibly damaging.

For the comparator V_{s+} in [4.3] and [4.9] must be replaced by V_{cc}, and R_{pu} added to R_2. Also V_{s-} in [4.4] and [4.10] becomes zero.

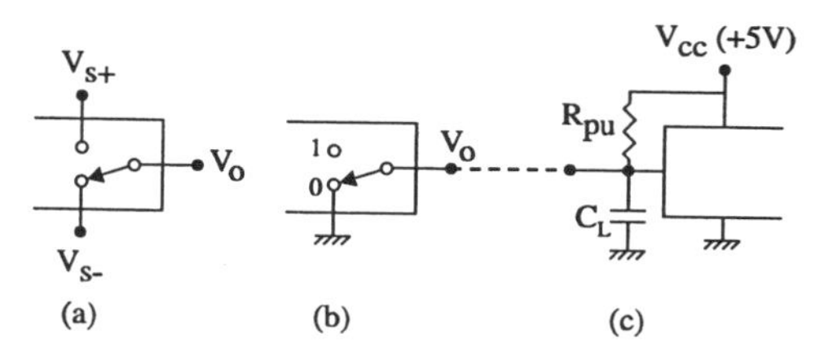

Fig. 4.6 Comparison of op-amp and comparator
(a) Op-amp output switching.
(b) Comparator output switching.
(c) Digital logic driven by comparator and pull-up resistor.

Switching speed

With the comparator not requiring the internal filtering needed by the op-amp to maintain stability with negative feedback, the switching speed is much higher. However, with the line capacitance C_L, together with the logic circuit input capacitance C_{in}, the upward transition is subject to the time constant $R_{pu}C_L$, and thus is usually slower than the downward transition. Exercise 4.1 gives calculations and experimental work on the comparator and its use with the frequency counter of Fig. 4.1.

Hysteresis

With the transfer function for the non-inverting op-amp-based Schmidt with $V_{ref}=0$ as in Fig. 4.7, this is said to have 'hysteresis'. How would the function differ for the comparator-based circuit? Exercise 4.2 demonstrates the hysteresis experimentally.

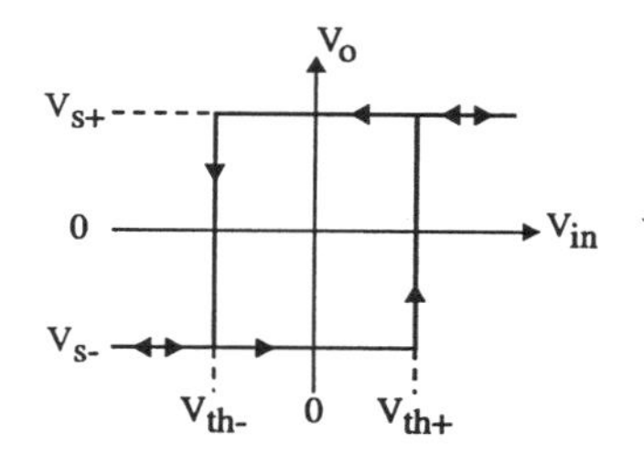

Fig. 4.7 Hysteresis in transfer function of op-amp-based non-inverting Schmidt of Fig. 4.3(a).

4.4 Triangular and square wave generator

This is shown in Fig. 4.8, with the waveforms as for the op-amp based Schmidt. For the saturation voltages V_{s+} and V_{s-} symmetric as shown, then the threshold splitting

$$\Delta V_{th}=2V_{s+}(R_1/R_2).$$

For the integrator with $V_a=V_{s+}$

$$dV_b/dt=-V_{s+}/CR$$

With also $\quad dV_b/dt=-\Delta V_{th}/(T_o/2)$

then $\quad T_o=4CR(R_1/R_2) \quad (4.17)$

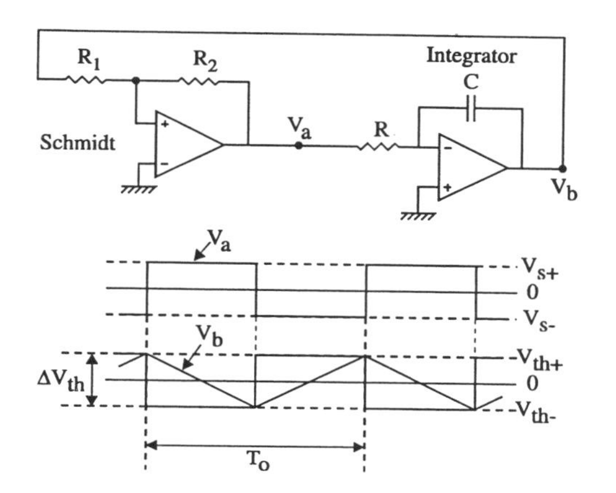

Fig. 4.8 Op-amp based triangular and square wave generator.

Sine wave adapter

With the adapter of Fig. 4.9(d) giving the corresponding sine wave, this is particularly useful for frequencies of ~ 1Hz and below, where the Wein oscillator of Section 13.3 becomes difficult to stabilise.

However, with the op-amp saturation levels neither well defined nor particularly stable, the above implementation merely demonstrates the principle. The single-chip 8083 microcircuit operates in the same way but with this limitation resolved.

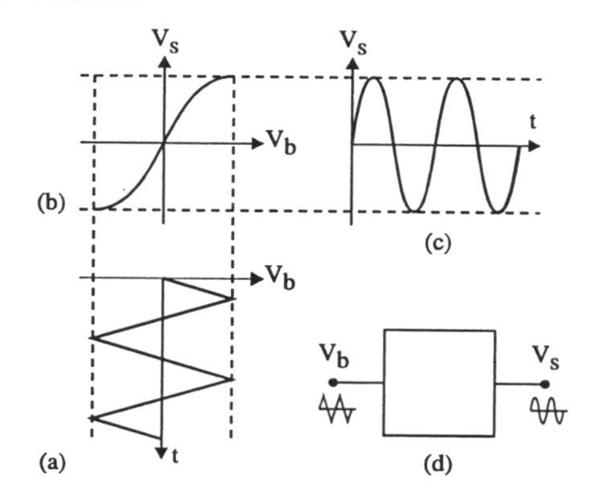

Fig. 4.9 Sine wave adapter for triangular and square wave generator of Fig. 4.8.
(a) Triangular input wave V_b.
(b) Adapter transfer function.
(c) Output sine wave.
(d) Adapter block diagram.

The main general limitation is the need to match the amplitude of V_b to the input of the transfer function of (d). Exercise 4.3 gives experimental work, etc.

Simplified square wave generator

For a simpler version of the above circuit, using only one op-amp, the integrator is replaced by the comparable low-pass RC filter, and the inverting version of the Schmidt is used. The limitation here is that the previously triangular wave has an exponential form.

4.5 Precision 555 type astable

With neither the op-amp nor the comparator output levels particularly well defined, and where only the square wave output is needed, a better implementation is that for the T555 timer chip shown in Fig. 4.10.

With the split threshold $V_{cc}/3$ and $2V_{cc}/3$, then $V_{cc}=5V$ is not well defined either. However, with V_{cc} supplying both the reference potential divider and the timing network, the output period T_o becomes independent of changes in V_{cc}. From the waveforms shown

$$T_1=C(R_1+R_2)\ln(2) \tag{4.18}$$

and

$$T_2=CR_2\ln(2) \tag{4.19}$$

giving

$$T_o=C(R_1+2R_2)\ln(2) \approx 0.7\times C(R_1+2R_2) \tag{4.20}$$

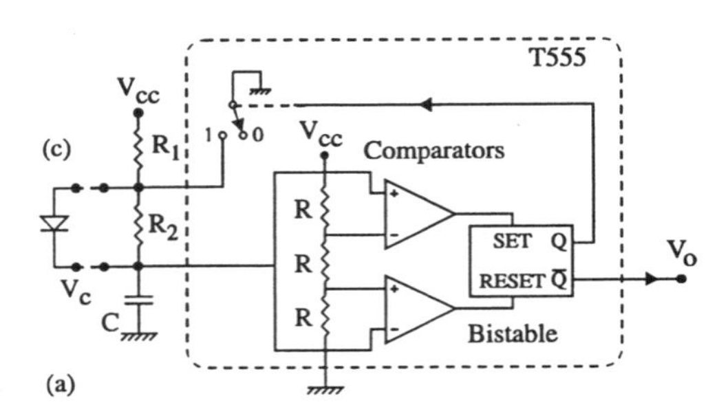

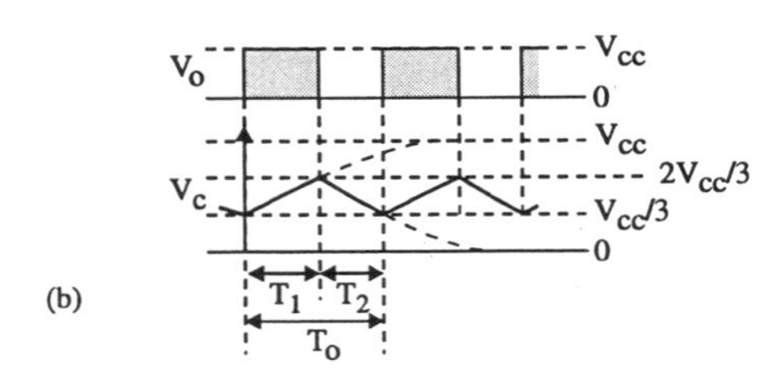

Fig. 4.10 T555-based precision astable.
(a) Block diagram.
(b) Waveforms.
(c) Diode needed for $T_1=T_2$.

1:1 duty cycle

So far the circuit is unable to give the often required 1:1 duty cycle, since always $T_1>T_2$. If however a diode is connected across R_2 as in (c) then this removes R_2 from [4.18], making $T_1=T_2$ if $R_1=R_2$. Here, however, slight adjustment is needed to allow for the 600mV drop across the diode when conducting.

4.6 Monostable

Fig. 4.11 shows a precision monostable based on the 555, the function of which is to output the pulse V_o of width T_m each time the circuit is triggered. The sequence is as follows.

Prior to triggering

- Here the bistable is in the 'set' state, giving $V_o=0$ and the switch as shown.
- For the lower comparator $V_{in}>V_{cc}/3$ so RESET=0.
- For the upper comparator $V_c=0$ giving $V_c<2V_{cc}/3$ and therefore SET=0.
- With both SET and RESET=0 then the bistable maintains its state.

On triggering

- With $V_{in}<V_{cc}/3$ during the brief trigger pulse, the bistable is reset, opening the switch.
- V_c then rises as shown, to reset the bistable when $V_c=2V_{cc}/3$. Thus

$$T_m=CR_1\ln(3) \approx 1.1CR_1 \tag{4.21}$$

End of output pulse

- With the bistable reset at this point, the switch closes, reducing V_c to zero and thus giving SET=0.
- With V_{in} previously having been raised then both SET and RESET are zero. Thus the state of the bistable, and indeed of the rest of the circuit, persists, waiting for the next trigger.

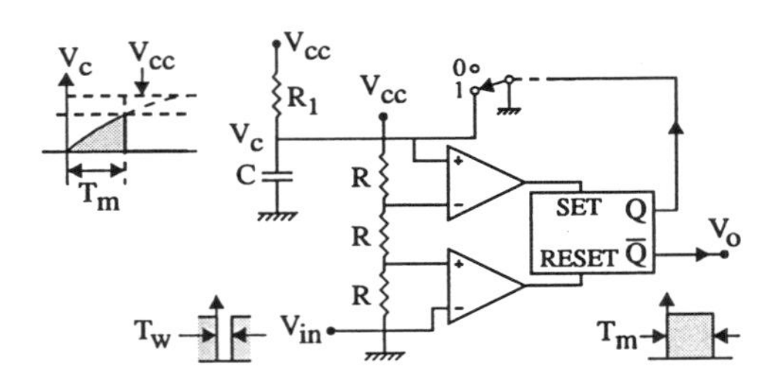

Fig. 4.11 T555-based precision monostable.

Trigger pulse narrowing

With early raising of V_{in} essential for the above sequence, where necessary, T_w can be reduced to T_t using the circuit of Fig. 4.12. This is essentially the simple high-pass filter, with the output biased to V_{cc}. The diode D is only needed because, for correct operation of the T555, the input must not exceed V_{cc} by more than 600mV.

For the type of monostable provided as a member of a digital logic family, the input pulse narrowing is based instead on the propagation delay of the on-chip logic gate.

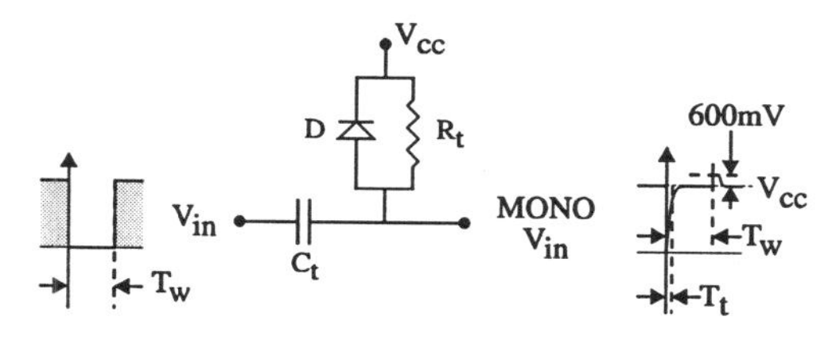

Fig. 4.12 Trigger pulse narrowing for monostable of Fig. 4.11.

Recovery time

At the end of the monostable output pulse in Fig. 4.11, the capacitor C is discharged through the switch. This, in reality, will be a transistor, so the discharge will take a finite time. Should the monostable be retriggered before this time has elapsed then the width of the following output pulse will be reduced. Thus the monostable has a finite recovery time, beyond which the reduction in subsequent pulse width is negligible.

Retriggerable monostable

If, for the above monostable, a further narrow triggering pulse is applied within the duration T_m of the output pulse, then the second trigger is ignored. This is because the effect of the second pulse would be to reset the bistable, but this is its state within the period T_m anyway. Such a monostable is said to be 'non-retriggerable', as distinct from the alternative 'retriggerable' type.

Fig. 4.13 makes the distinction. Here both types are triggered on the rising edge of the input pulse, and both ignore the falling edge.

Then for the non-retriggerable type in (b) the second triggering pulse is ignored, while for the retriggerable type in (c) the T_m pulse is restarted.

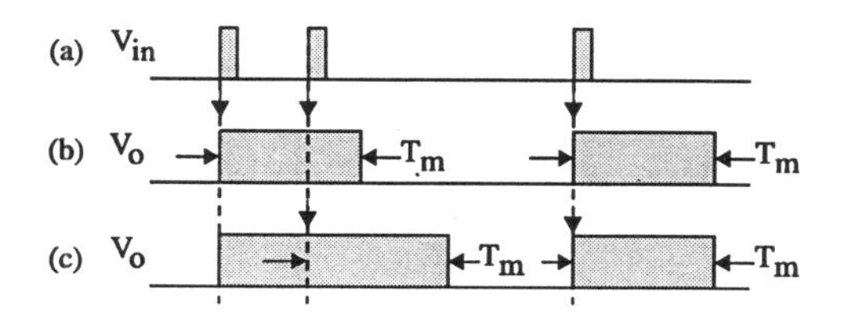

Fig. 4.13 Waveforms for non-retriggerable and retriggerable type monostables.
(a) Trigger pulses V_{in}.
(b) Output for non-retriggerable monostable.
(c) Output for retriggerable monostable.

Finally, with no retriggering involved, the response to the third trigger pulse is the same for both.

Elementary implementation

Fig. 4.14(a) shows the essentials of the retriggerable monostable. The main limitation here is that, unlike for Fig. 4.11, etc., T_w is added to the output pulse width.

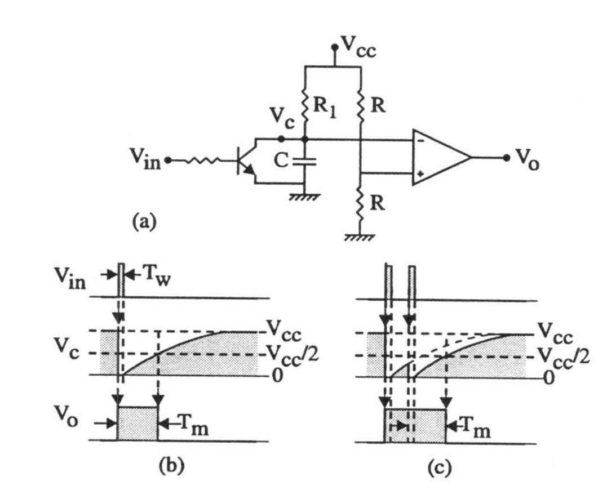

Fig. 4.14 Elementary retriggerable monostable.
(a) Circuit diagram.
(b) Response to single input pulse.
(c) Retriggering response.

A popular choice for a digital logic family type monostable is the HC123. This has two monostables on the chip. Each may be connected to be retriggerable or not, and the two can be connected in a ring to give the astable.

Missing pulse detector

A common application of the retriggerable monostable is illustrated in Fig. 4.15. Here the monostable input V_{in} is a periodic train of pulses, with the possibility of a missing pulse from time to time.

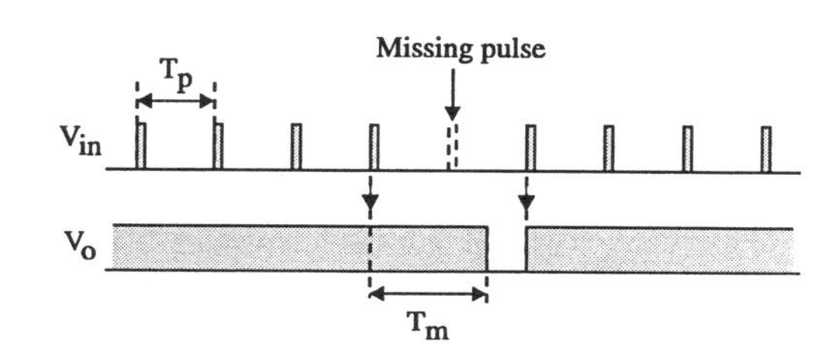

Fig. 4.15 Waveforms for retriggerable monostable used as missing pulse detector.

4.7 Precision rectifier

Fig. 4.16(a) shows the circuit of the simple half-wave diode rectifier, with the transfer function and waveforms as in (c) and (d), both for the ideal diode and the real diode.

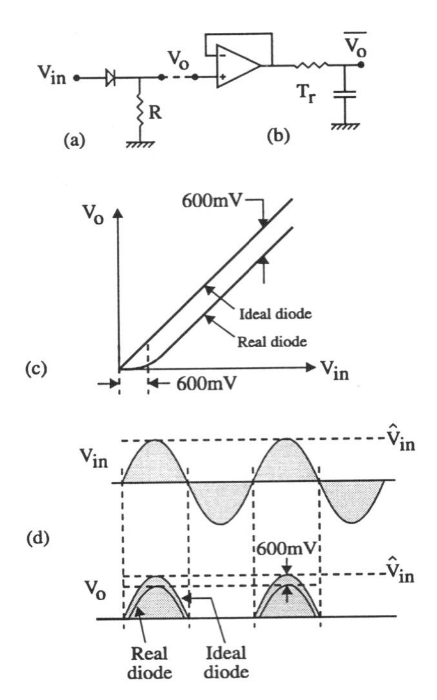

Fig. 4.16 Operation of simple half-wave rectifier showing imperfections due to diode.
(a) Rectifier.
(b) Low-pass filter with isolating voltage-follower.
(c) Rectifier transfer function.
(d) Waveforms.

For the real diode there is the ≈600mV needed to make the diode conduct, and also the curvature due to the exponential nature of the diode I-V relation.

Dynamic range

The 'dynamic range' D of a circuit is the ratio of the maximum to minimum input amplitude for which the circuit will operate satisfactorily. For the simple rectifier the minimum is thus ≈600mV. The corresponding maximum is set by the reverse breakdown voltage of the diode, which can be around 100V, giving D ≈ 166.

This figure will probably be reduced by saturation in the amplifier providing the input. If this is an op-amp-based circuit then the saturation voltages are about ±13V, giving D=13V/600mV and thus **D=21.7=26.7dB.**

This is the typical figure for the meter detector in the lab signal generator.

Precision circuit

Fig. 4.17 shows the op-amp-based precision rectifier, which considerably improves upon the dynamic range D. Here the two right-hand resistors are alternately switched in by the diodes according to the sign of V_{in}, thus maintaining the feedback. For the diodes considered ideal then the waveforms becomes as in (b).

For the precision circuit the value of V_{in} required for the appropriate diode to start conducting is reduced by × $1/A_o$, to $600mV/A_o$.

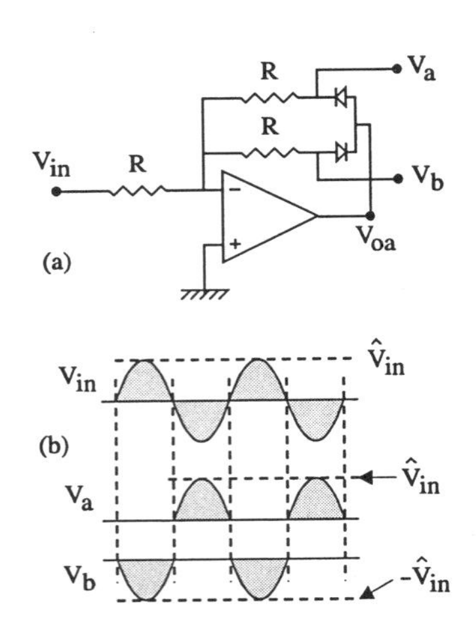

Fig. 4.17 Precision half-wave rectifier.
(a) Circuit diagram.
(b) Waveforms for diodes assumed ideal.

Example

With A_o typically 3×10^4, the minimum input amplitude $600mV/(3 \times 10^4) = 20\mu V$. With the V_s ±13V as usual then $D = 13V/20\mu V$, giving **$D \approx 650 \times 10^3 = 116dB$**

However, A_o varies with frequency, so that this value is only obtained below the op-amp cut-off frequency $f_o \approx 30Hz$. As f approaches the op-amp unity gain frequency f_u, so D approaches the unimproved figure of ≈20.

Full-wave precision rectifier

Full-wave precision rectification can be obtained using the adder of Fig. 4.18, for which

$$V_o = -(V_{in} + 2V_b) \tag{4.22}$$

For any of the above types of rectifier, the output can be averaged by the filter of Fig. 4.16(b), to give $\overline{V}_o$ proportional to the amplitude of V_{in}. For the precision types the voltage-follower is not needed. Exercise 4.4 gives calculations and experimental work on the above three types of rectifier.

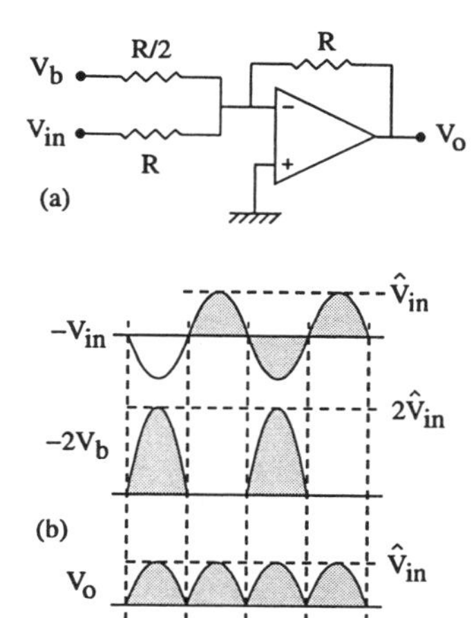

Fig. 4.18 Use of adder to convert half-wave precision rectifier of Fig. 4.17 to full-wave operation.

4.8 Precision envelope detector

Fig. 4.19(a) shows the simple 'envelope detector', with the waveforms for the ideal diode and for the amplitude of V_{in} constant, as in (b). Compare

- $\overline{V_o}$ here close to $\hat{V}_{in}$,
- $\overline{V_o} = 2\,\hat{V}_{in} / \pi$ for the precision full-wave rectifier,
- $\overline{V_o} = \hat{V}_{in} / \pi$ for the precision half-wave rectifier,
- $\overline{V_o}$ nearly $2\hat{V}_{in}$ for the high-efficiency envelope detector of Section 4.12.

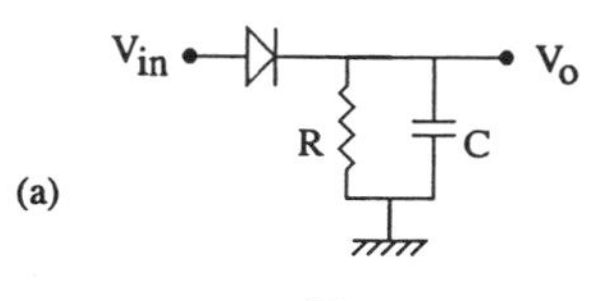

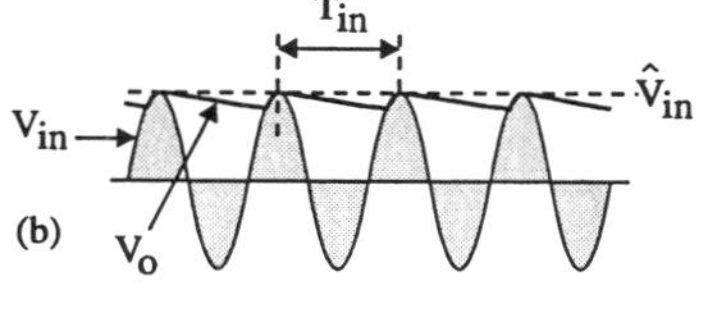

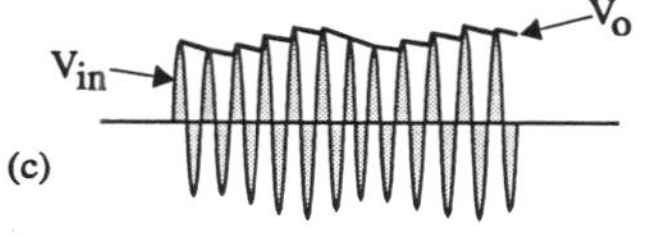

Fig. 4.19 Simple envelope detector, with waveforms for ideal diode.
(a) Circuit diagram.
(b) Waveforms for unmodulated sine wave input.
(c) Waveforms for amplitude modulated sine wave input.

In Fig. 4.19(b) the decay between peaks is with the time-constant CR, requiring $CR \gg T_{in}$ for low ripple. CR should not be made too long, however, or V_o will not follow the falling regions of the envelope properly for waveforms such as the amplitude modulated sine wave carrier in (c).

Precision circuit

With the imperfections of the real diode, the precision circuit of Fig. 4.20 is needed. Here, as for the precision rectifier, the dynamic range is increased by $\times A_o$. Also as there, V_a gives the positive output and V_b the negative. The inter-peak decay rate remains CR. Exercise 4.5 gives calculations and experimental work for both the simple and precision circuits.

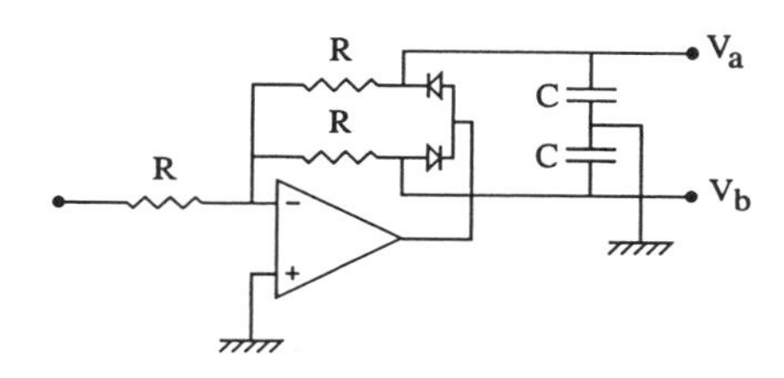

Fig. 4.20 Precision envelope detector.

4.9 Peak-and-valley detectors

The elementary peak detector of Fig. 4.21(a) is essentially the elementary envelope detector of Fig. 4.19(a) with R omitted, with Fig. 4.21(b) showing the corresponding valley detector.

The waveforms in (c) are for the ideal diode. Here V_p is the highest value of V_{in} after the switch is opened, and V_y the lowest.

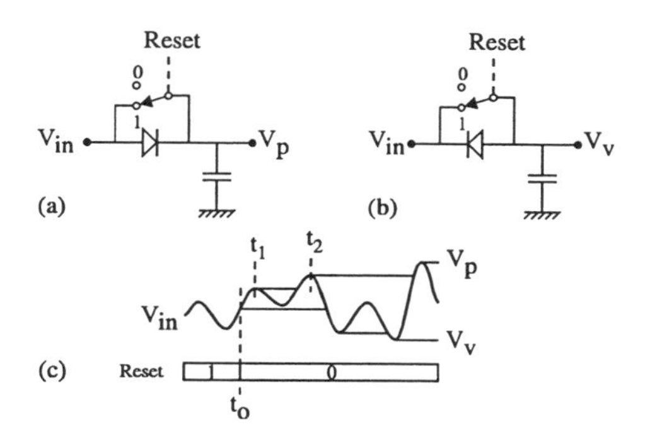

Fig. 4.21 Elementary peak and valley detectors.
(a) Peak detector.
(b) Valley detector.
(c) Waveforms for diodes ideal.

Precision circuit

The precision peak detector of Fig. 4.22 overcomes the limitations of the diodes, as usual to increase the dynamic range by $\times A_o$. The voltage-follower is included to prevent the loading of C by the following stage, etc., which would otherwise discharge C.

The components R_s and D_s are not essential but are added to improve the response time of the circuit. Without them, as soon as V_{in} falls below V_p, the feedback loop is broken and the output of the first op-amp falls to the negative saturation voltage V_{s-}. When next V_{in} rises above V_p, the op-amp output then has to recover from this extreme value, and takes a significant time to do so, being limited by the slew rate. With the extra two components, the op-amp output only falls to 600mV below V_p and so the recovery is much more rapid. Exercise 4.6 gives experimental work and calculations.

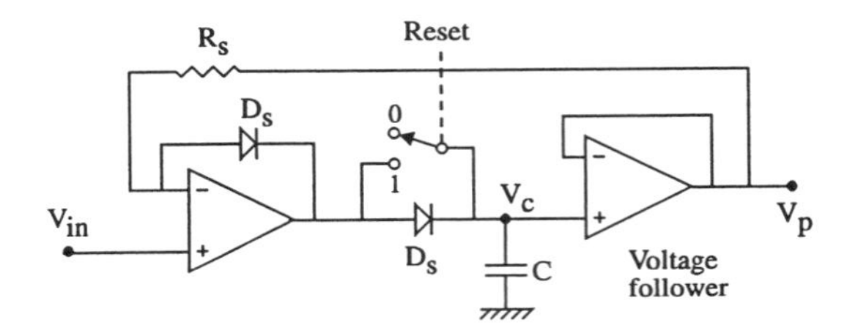

Fig. 4.22 Precision peak detector.

4.10 Sampling gates

The purpose of a sampling gate is to record the value of an input signal at some chosen point in time.

Track-and-hold

The 'track-an-hold' gate of Fig. 4.23 is the simplest type of sampling gate. With the switch initially closed then V_c, and hence V_o, follows (tracks) V_{in} exactly. With the switch opened at t_o then V_c becomes constant (holds) the value of V_{in} at t_o.

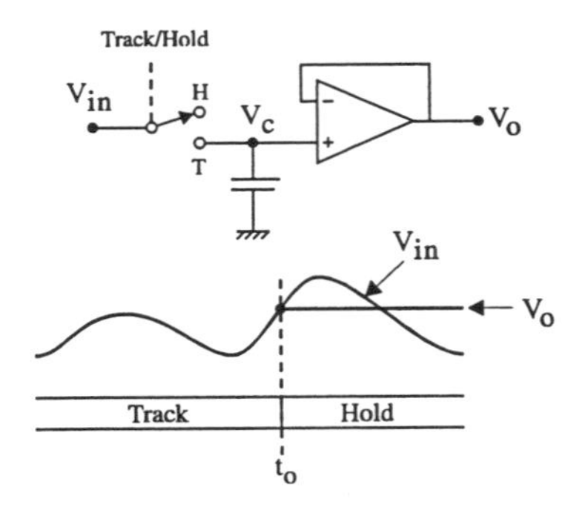

Fig. 4.23 Track-and-hold (sampling) gate.

Sample-and-hold

As shown in Fig. 4.24, this is the track-and-hold function with the tracking periods made vanishingly small. The function is sometimes referred to a 'point sampling', since what is held is the value of V_{in} at the point of sampling. However, the output of the track-and-hold is an equally valid point sample, the point being at the time at which the switch was opened. The only difference is that the held value is not maintained right up to the time of the next sample.

Frequently the gate is used to hold a value steady while a following ADC performs its conversion. Here the hold period only needs to be long enough for the conversion.

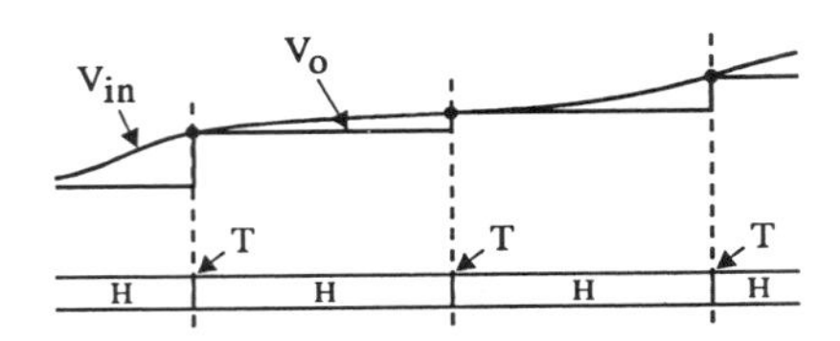

Fig. 4.24 Sample-and-hold function.

Switch resistance

In reality the switch will have the resistance R_s, as in Fig. 4.25(a). With the waveforms as in (b), then for V_c to complete its change to the new value of V_{in} requires the track period $T_t >> CR_s$.

Where T_t is fixed, and with R_s fixed also, the only way then to reduce CR is to reduce C.

The limitation here is that the bias current I_b of the associated voltage-follower causes V_c to drift over the hold period, with the drift rate $dV_c/dt = -I_b/C$.

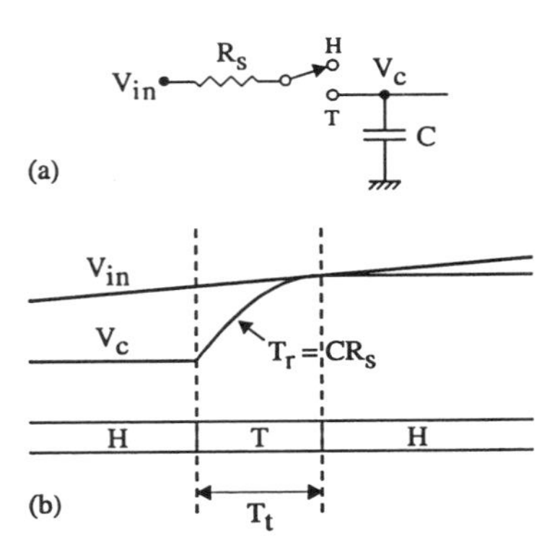

Fig. 4.25 Effect of series resistance R_s on track-and-hold gate of Fig. 4.23. H: hold, T: track.

Examples

With the typical I_b=50pA for the TL081, C=10nF and the holding period T_h=1ms, the change is by 5μV, which is normally negligible.

For the 741 with I_b=50nA the figure becomes 5mV, which is less acceptable.

Boxcar sampling gate

With R_s and C in Fig. 4.25 essentially a low-pass filter, then V_c in Fig. 4.25(b) at t_o is the usual exponentially tapered average of V_{in} over the period CR prior to t_o, but with the average now distributed over the periods when the switch is closed

For the boxcar gate of Fig. 4.26(a), CR is deliberately increased, to make the number N of samples included large for the purpose of noise averaging.

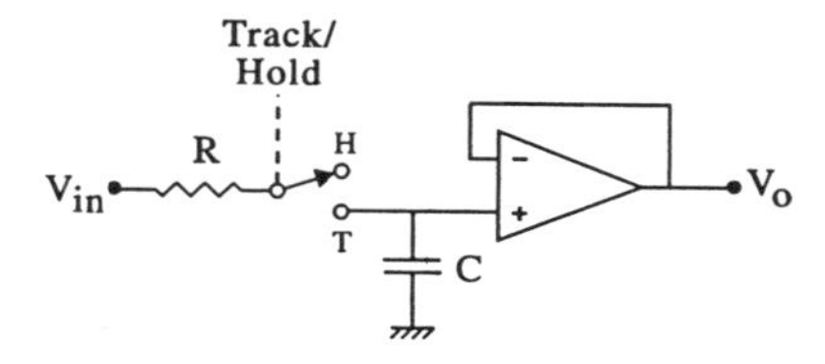

Fig. 4.26(a) Boxcar sampling gate.

With the waveforms as in (b), were the switch to be permanently closed then the filter memory function at time t would be as for M_1 shown. To be precise, with g(t) the filter impulse response, then $M_1=g(t-t')$ in the expression

$$V_o(t) = \int_{-\infty}^{t} g(t-t')V_{in}(t')dt' \qquad (4.23)$$

M_2 then shows the spreading of M_1 as the switch is normally operated, to give $N=CR_s/T_t$.

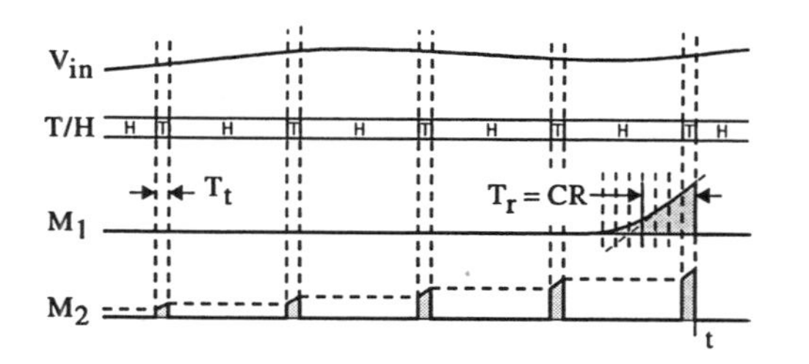

Fig. 4.26(b) Boxcar sampling waveforms.

Exercise 4.7 gives calculations and experimental work on each of the above types of sampling.

4.11 Auto Zero

Fig. 4.27(a) shows a circuit that is useful for correcting a severely drifting base line, while (c) shows the waveforms for a typical application, the recording of an electro cardiogram (ECG).

This is the potential developed between electrodes attached to two points on the body (e.g. the two arms), and generated by the signal that triggers the heartbeat. The resulting repeated complex is as for V_{in} shown. The relatively large slowly fluctuating base line derives from the muscle-trigger signals (Electro-Myogram, EMG).

Suppose, as is often the case, that it is required to digitise V_{in}. Apart from the base-line variations, an ADC resolution of eight bits (one in 265) would probably be adequate, or at the most 12 bits (one in 4096). However, with the base line variation several times greater in magnitude than the required signal, a proportionally higher resolution is required.

The problem is resolved by, in effect, resetting the zero once per beat. For the auto-zero circuit of (a), when the switch is closed then the voltage at the voltage-follower input will be made zero. After the switch has been opened, no capacitor current can flow and so the capacitor becomes effectively a DC supply of the correct value to remove the base line at the time the switch was opened. Thus the only drift that can accumulate is that arising over one beat, which is much less than when a single adjustment is made manually at the start of the series of beats.

High-pass filter

An alternative is to use the high-pass filter shown in (b). This produces an output at any time t which is the input at t, less the tapered average of V_{in} for the period CR immediately prior to t. If CR is made approximately equal to the beat period T_b, this too will reduce the drift to that occurring over one beat period only. However, the beat complex then becomes distorted, because CR is comparable with the duration of the complex.

The auto-zero method requires that the time of occurrence of the beats be known. This is best arranged by controlling the auto-zero from the computer receiving the digitised data.

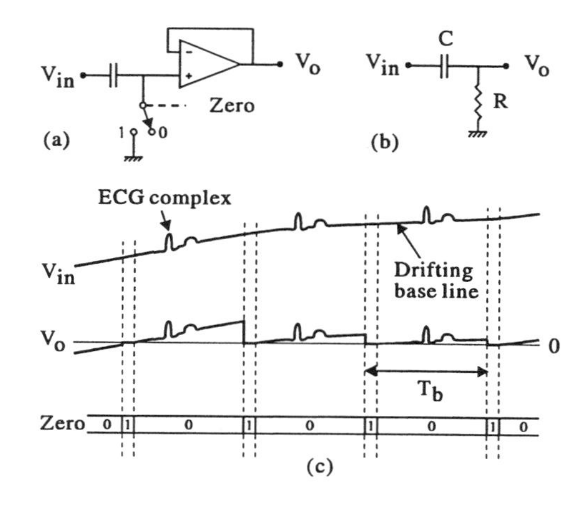

Fig. 4.27 Auto-zero circuit.
(a) Circuit diagram.
(b) High-pass filter for comparison.
(c) Waveforms for ECG signal with wavering base line.

4.12 DC Restorer

The DC restorer of Fig. 4.28(a) is the same as the auto-zero circuit, except that the switch is replaced by a diode. While the auto-zero sets V_o to zero at the time the switch is opened, for the restorer it is set to zero when V_{in} is minimum.

For the restorer, from A to B in (b) the diode conducts to maintain $V_o=0$, taking up the difference in C. Then at B conduction stops, to give the following waveform as shown. Thus the lower peaks of V_o are pinned to the zero line.

This result could also be obtained using a suitable DC level shifter. However, the restorer adjusts automatically to any increase in the signal amplitude.

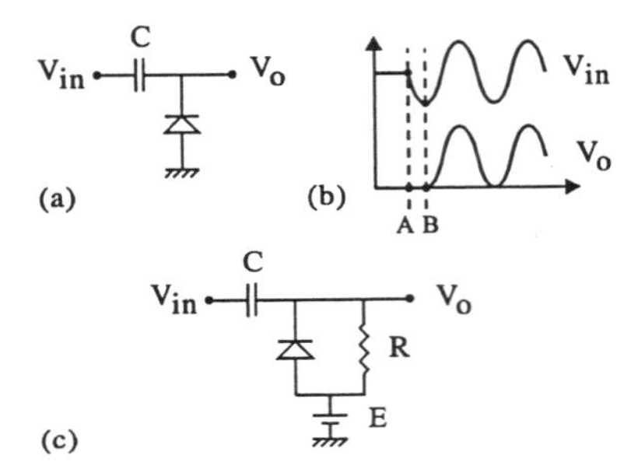

Fig. 4.28 DC restorer.
(a) Basic circuit.
(b) Waveforms.
(c) Addition of discharging resistor R and bias E.

Discharging resistor

So far the restorer does not automatically adjust to a decrease in the input amplitude. With the resulting waveforms as in Fig. 4.29, connection of a resistor R across the diode has the required effect.

Level shift

E in Fig. 4.28(c) allows the level to which V_o is pinned to be other than zero. Also V_o can be made to hang down from the pinning level instead of standing up on it, simply by reversing the diode.

The above waveforms are for the restorer diode ideal, with the real diode giving the usual limitations. Where the precision restorer is needed this can be provided by the precision envelope detector of Fig. 4.20, with the filter output $\overline{V_o}$ subtracted from V_{in}.

TV restorer

The classic application of the restorer is in the coupling of the video signal to the grid of the cathode ray tube in a television set. Here the video signal has to be applied between the grid and cathode, and both of these are at a potential of 20kV or so above ground.

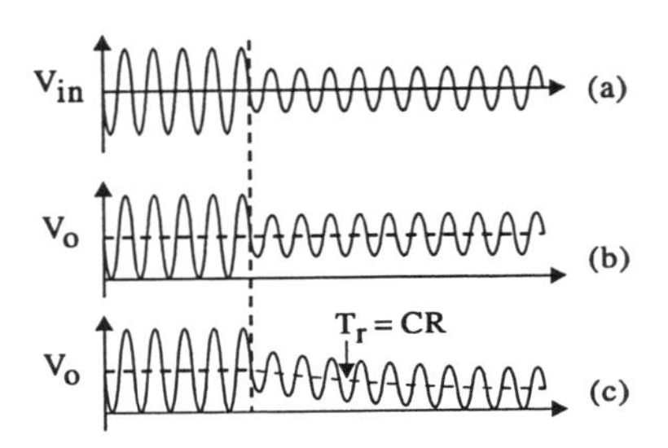

Fig. 4.29 Waveforms for DC-restorer with and without discharging resistor (for E=0).
(a) V_{in} with step reduction in input amplitude.
(b) V_o with no discharging resistor.
(c) V_o with discharging resistor.

Here the most convenient form of coupling is via a suitably high-voltage capacitor C.

Fig. 4.30 shows the video waveform as received. This comprises the time-base synchronising pulses and the picture content, proportional to screen intensity and varying as the spot scans across the screen.

A comparator with the threshold V_{th} derives the sync pulses, and a suitable clipper the video section.

Were the C to be formed into a high-pass filter, with R between the grid and cathode, then the black level applied to the grid-cathode would vary according to the picture content. But with the diode in Fig. 4.28(c) added to form the restorer, this is avoided.

The brilliance control sets E relative to the cathode voltage, such that the lowest value of the video waveform is at the threshold for blackness.

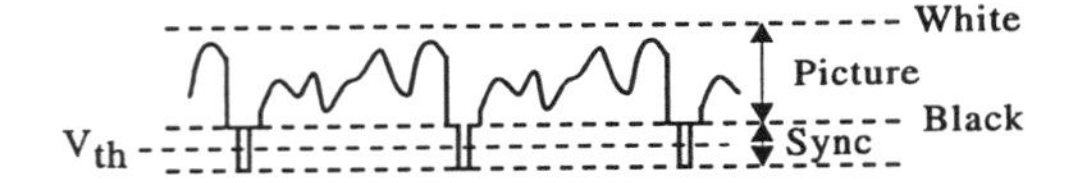

Fig. 4.30 TV video waveform.

High-efficiency envelope detector

Fig. 4.31 shows an extension to the DC-restorer of Fig. 4.28 which gives an envelope detector that is twice as efficient as the normal one, and therefore much more so than the simple half-wave rectifier. This is the DC restorer of Fig. 4.28 followed by the normal envelope detector. Exercise 4.8 gives experimental work on the DC restorer and present envelope detector.

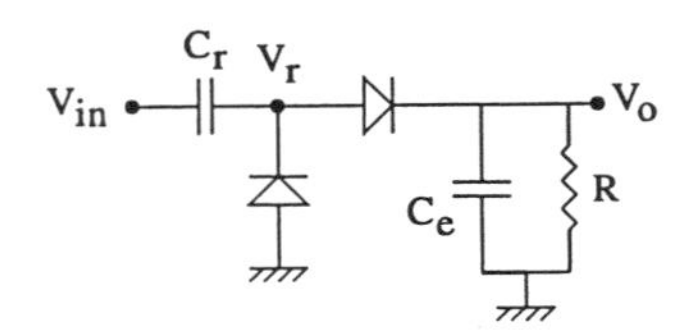

Fig. 4.31 High-efficiency envelope detector.

Reference

Coughlin, R. F. and Driscoll, F. F. 1982: *Operational Amplifiers and Linear Integrated Circuits*, Prentice Hall.

4.13 Exercises and experiments

Exercise 4.1. Comparator and counter.

Calculation For the comparator of Fig. 4.6 calculate the value of R_{pu} needed for the rise time T_r to be 1μs for $C_L=100pF$.

With $T_r=C_LR_{pu}$ then **$R_{pu}=10k\Omega$.**

Experiment. Comparator Connect the LM319 comparator as in Fig. 4.1 but without the counter or R_{pu}.

Apply $V_{in}=1V$ p-p at $f_{in}=1kHz$. Confirm that the output $V_o=0$ until $R_{pu}=10k\Omega$ in (c) is included.

Add $C_L=100pF$ and increase f_{in} to allow T_r to be measured. Use the ×10 scope probe and allow for the ≈ 10pF probe capacity. Note the relatively rapid decay, as C_L is discharged through the comparator output transistor.

Experiment. Counter Add a 12-bit HC4040 as counter. Apply $V_{in}=1V$ p-p at $f_{in}=10Hz$ and manually control Reset. Thus confirm for say T=1s that D_o is much in excess of the required ten.

Add 30% positive feedback as in Fig. 4.3(a) with $R_1=47k\Omega$ and $R_2=100k\Omega$ and confirm that the additional noise-induced crossings are avoided.

Exercise 4.2 Schmidt trigger

Calculation For the non-inverting Schmidt of Fig. 4.3(a) we require a threshold splitting ΔV_{th} of 1V, centred about zero. Derive suitable values for R_1 and R_2 where the op-amp is used, with $V_s=\pm 13V$.

For the reasons in Section 3.7, we choose **$R_1=10k\Omega$.**

With $R_1/R_2=\Delta V_{th}/(V_{s+}-V_{s-})$ as in [4.13]

then **$R_2=260k\Omega$.**

Experiment Connect the circuit using the TL081 op-amp. Use ±15V supplies to give the $V_s \approx$ the assumed ± 13V.

Using the preferred 270kΩ for R_2, apply a 100Hz sine wave and note how V_o does not respond until the input amplitude exceeds the threshold splitting.

For larger amplitudes compare the V_{in} and V_o waveforms checking that V_o is a square wave with the transitions where V_{in} crosses the thresholds.

Use the X-Y display to observe the hysteresis of Fig. 4.7.

Exercise 4.3 Triangular and square wave generator

Calculation For the circuit of Fig. 4.8, calculate suitable component values for V_b to be 10V p-p at 1kHz. Assume the V_s to be ±13V.

For the reasons in Section 3.7 choose **$R=R_1=10k\Omega$.**

With the p-p value of V_b equal to ΔV_{th} and

$$\Delta V_{th}=2V_{s+}(R_1/R_2)$$

then **$R_2=26k\Omega$.**

With

$$T_o=4CR(R_1/R_2)$$

as in [4.17] then **C=65nF.**

Experiment Connect the circuit as above with C the preferred 68nF and check the waveforms for frequency and amplitude.

Problem For the sine wave adapter of Fig. 4.9, sketch the V_s waveform for the amplitude of V_b
(i) below that shown (sharp peaks) and
(ii) above (flat tops).

Exercise 4.4. Precision rectifier

Calculation. Simple half-wave rectifier The 1N1418 signal diode is to be used in the following experiments. With this working best for $I_d \approx 1mA$, calculate a suitable value for R in the rectifier of Fig. 4.16(a), for V_{in} =20V p-p.

With the peak $V_{in}=10V$ and for $I_d=1mA$ then **R=10kΩ.**

Problem Show that for the ideal diode the average of the output V_o of the half-wave rectifier is $\hat{V}_{in}/\pi$, giving $2\hat{V}_{in}/\pi$ for the full-wave.

Experiment. Simple half-wave rectifier Connect the half-wave rectifier of Fig. 4.16(a), first using the 1N4148 diode and R=10kΩ only. Apply a 100Hz sine wave to confirm the V_{in} and V_o waveforms of (d).

Check that for $V_{in} \sim 1V$ then V_o does not start to increase until $\hat{V}_{in}$ exceeds ≈600mV. Note the rounding at the corners resulting from the exponential nature of the diode I-V curve.

Use the scope X-Y mode to see the transfer function of (c). Compare the ideal $V_o = V_{in}$ line for the ideal diode when conducting, by moving the V_o probe to the V_{in} point.

Add the voltage-follower and low-pass filter of (b) with R=10kΩ and C=10μF giving a 100ms response time. Confirm that $\overline{V_o} \approx \hat{V}_{in}/\pi$ less the 600mV diode voltage drop.

Calculation. Precision rectifier components For the circuit of Fig. 4.17 decide upon suitable components for operation at 100Hz.

For the TL081 the op-amp gain $A_o \approx 2 \times 10^5$ at 100Hz so this is suitable. For the reasons in Section 3.7 we choose R=10kΩ. The signal-type 1N1418 diode is designed for operating currents of a few mA so is also suitable.

Calculation. Precision rectifier dynamic range Calculate D for the circuit driven by an op-amp of saturation voltages ±13V.

The maximum peak output 13V and the corresponding minimum $600mV/A_o = 3\mu V$.

Thus D=13V/3μV giving **$D = 4.33 \times 10^6 = 133dB$.**

Experiment Connect the resulting circuit and check the waveforms shown for f_s=100Hz. Confirm that these obtain down to the typically 1mV limit of the scope.

Consider what the V_{oa} waveform will be and check by observation.

Increase f_s to 100kHz where $|A_o| \approx 30$ and confirm that the minimum level for correct operation is increased to $600mV/A_o = 20mV$.

Note also how capacitive effects such as that of the diodes begin to become significant.

Add the weighted adder of Fig. 4.18 to give full-wave rectification.

Why is an interposing voltage-follower not needed for the V_b input to the adder?

Exercise 4.5 Precision envelope detector

Calculation. Simple envelope detector For the circuit of Fig. 4.19, calculate the value of C required for a decay time T_d=10ms with the usual R=10kΩ.

With T_d=CR then **C=1μF.**

Calculate the resulting amplitude of the ripple component V_r of V_o for V_o=10V and the input signal frequency f_{in}=1kHz.

With $V_r = V_o T_{in}/T_d$ where $T_{in} = 1/f_{in}$ then **V_r=1V p-p.**

Experiment. Simple envelope detector Following the last experiment, revert to the scope y-t mode and increase f_{in} until an exponential tail begins to follow each V_o half sine wave. This is the discharge of the scope probe through R. Measure the decay time and thus calculate the scope probe capacity. This is very different for the ×1 and ×10 settings of the normal probe.

For f_{in}=1kHz, lengthen the tail with increasing values of added shunt C_p until the operation becomes that of the envelope detector.

Using the calculated C=1μF, check that V_o is ≈600mV below the peaks of V_{in}, also that the ripple is the calculated 1V p-p. Increase to 10μF for ×1/10 ripple reduction.

Calculation. Precision envelope detector For the circuit of Fig. 4.20 with the usual R=10kΩ, calculate the value of C needed for a T_d=1s decay.

With T_d=CR then **C=100μF.**

Calculate the amplitude of the resulting ripple component V_r of V_b for f_{in}=100Hz and V_b=10V.

With $V_r = V_b T_{in}/CR$ where $T_{in} = 1/f_{in}$, **V_r=100mV p-p.**

Experiment. Precision envelope detector Connect and drive the circuit as above and compare the waveforms of V_{in} and V_b.

Confirm that V_b is closely equal to $\hat{V}_{in}$ and that V_r is as calculated. Check that when V_{in} is abruptly removed then V_b decays with the 1s period.

Increase f_{in} to 1kHz and note the ×1/10 reduction in V_r.

Exercise 4.6 Peak and valley detectors

Experiment. Simple peak detector Connect the simple peak detector of Fig. 4.21(a) with the 1N4148 diode and C still 100μF. Apply V_{in} at 10V p-p and f_{in}=0.5Hz. Connect a wire link across the diode and check that V_p follows V_{in}. Remove the link and note how V_p continues to follow to the peak and then holds.

Repeat for V_{in}=2V p-p and note how V_p first becomes the expected 1V–600mV=400mV and then steps up very slowly towards 1V with much reduced diode current due to the exponential diode I-V relation. Reverse the diode for the valley detector.

Calculation. Output drift For the precision peak detector of Fig. 4.22 the input bias current I_b for the TL081 used is 50pA. Calculate the resulting output drift that will result for C=100μF.

With $dV_o/dt=-I_b/C$ then **$dV_o/dt=-0.5\mu V/s$**

Experiment. Precision peak detector Connect the circuit of Fig. 4.22 omitting R_s and D_s. Confirm that operation for f_s=0.5Hz is correct at least down to V_{in}=20mV p-p.

Confirm that the drift after the link has been removed is imperceptible, as calculated.

Reduce C to 0.1μF, disconnect the diode, and confirm the dV_o/dt, then of –0.5mV/s.

Exercise 4.7 Sampling gates

Fig. 4.32 shows the experimental system that will be used to demonstrate the sample-and-hold function. V_{in} is a 100Hz sine wave input which is not only the signal to be sampled but also provides the trigger for the sampling circuit. At the rising zero-crossing of V_{in} the leading edge of the comparator output V_c triggers the first monostable M_1. M_2 is then triggered on the trailing edge of Q_1 to output the short sampling pulse Q_2. In this way the sampling point can be adjusted by R_{m1} which controls the width of the M_1 output.

Calculation. Comparator threshold values The positive feedback around the comparator is needed to stop M_1 from being triggered also on the falling edge of V_{in}. We arrange ≈10% feedback by setting R_1=10kΩ and R_2=100kΩ. Calculate the threshold values V_{th+} and V_{th-} for R_{pu}=10kΩ.

For $V_c=V_{cc}$=5V

$$V_{th-}=V_{cc}\times(-R_1/(R_2+R_{pu})$$

giving **V_{th-}=–454mV.**

More simply, for V_o=0 **V_{th+}=0V**

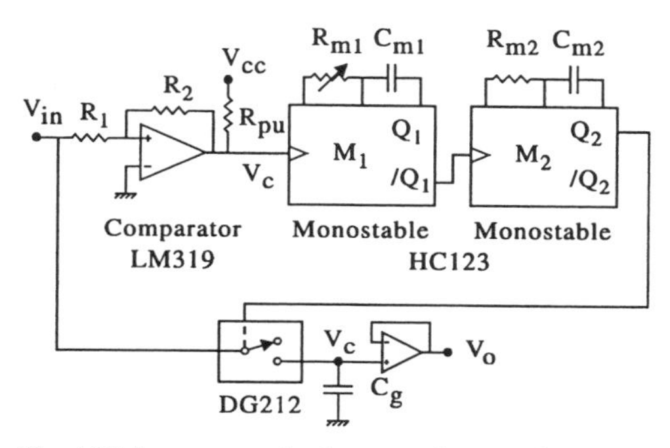

Fig. 4.32 Arrangement for demonstrating sampling gates.

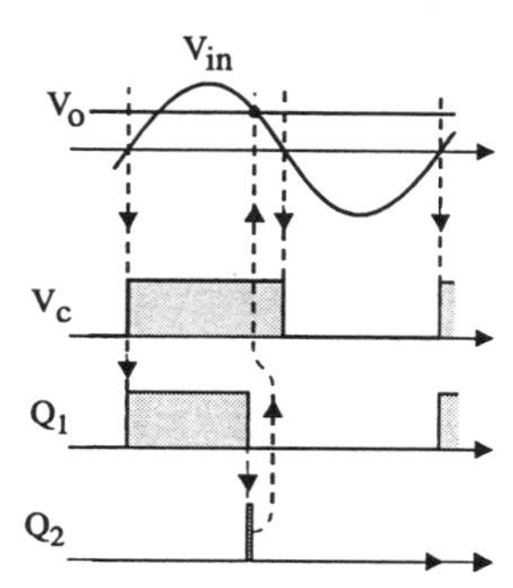

Fig. 4.33 Waveforms for sampling gate experiment.

Monostable pulse widths

From the elementary monostable of Fig. 4.14, the present monostable pulse width T_m is comparable with the associated C_mR_m, with the stated value $0.45C_mR_m$. It is also required that R_m < 2kΩ. We make the minimum R_{m1}=10kΩ and add a 100kΩ variable resistor to be able the vary T_{m1} up to the maximum of $T_{in}=1/f_{in}$.

Calculation For the above R_{m1} calculate the required value of C_{m1} for $f_{in}=100Hz$.

With $C_{m1}=T_{m1}/(0.45R_{m1})$

and with $R_{m1}=110k\Omega$ for $T_{m1}=T_{in}=10ms$
then **$C_{m1}=202nF$.**

Calculate the required value of C_{m1} for $R_{m2}=10k\Omega$.

As for T_{m1} **$C_{m2}=11.1nF$.**

Calculation. Gate capacitor The switch resistance R_{on} of the DG212 is specified as 125Ω. Calculate the maximum value of the gate capacitor C_g for adequate charging over one sampling pulse period T_{m2}.

For charging to $V_{in}\times\exp(-1)$

$T_{m2}=C_gR_{on}$

giving **$C_g=400nF$.**

For adequate charging we choose $C_g=C_{m2}$ less by ×1/4 to give better than 98% of full charge. Thus **$C_g=100nF$.**

Calculation. Drift rate Calculate the resulting drift rate for the op-amp bias current $I_b\approx 50pA$.

With $dV_o/dt=I_b/C_g$,
then **$dV_o/dt=500\mu V/s$.**

Experiment. Sample-and-hold Connect the comparator and M_1 without R_1 and R_2 and compare the waveforms of V_{in} and Q_1 for $V_{in}=10V$ p-p and $f_{in}=100Hz$.

Confirm that the triggering is mainly on the rising edge of V_{in} as shown, but also sporadically on the falling edge.

Add the feedback resistors to avoid the unwanted triggering. Confirm that T_{m1} is variable from 1ms to 10ms as required. Lower V_{in} to check that the triggering is at $V_{th+}=0V$.

Connect M_2 and check the Q_2 waveform as R_{m1} is varied. Confirm $T_{m2}=50\mu s$.

Connect the sampling gate and compare V_o with V_{in} as T_{m1} is varied. Increase C_{m2} to 100nF to see that V_o tracks V_{in} over T_{m2}. Over $T_{in}=10ms$ the present drift rate of 500μV/s is negligible.

Calculation. Boxcar sampling To convert the above arrangement for boxcar sampling, as in Fig. 4.26, calculate the value of R in Fig. 4.26 needed to spread the overall response time T_r to 1s, as for being able to view the effect. C=100nF as before. Here still $f_{in}=100Hz$.

With $T_r=CR\times(T_{in}/T_{m2})$ then **$R=50k\Omega$.**

Experiment. Boxcar sampling Add R_g and confirm the much slower response of V_o to changes in T_{m1} and to the amplitude of V_{in}.

Exercise 4.8 DC restorer

Calculation For the restorer of Fig. 4.28(c) we require a response time $T_r=1s$. For the usual $R=10k\Omega$, calculate the required C.

With $T_r=CR$ then **$C=100\mu F$.**

Experiment. Restorer With these values connect the circuit and observe the V_o waveform as the amplitude of V_{in} is abruptly changed. Check that the response is immediate for an increase but has the 1s lag to come down to the base line for the decrease.

Experiment. High efficiency envelope detector Connect the circuit of Fig. 4.31, with $R=10k\Omega$ and $C_r=C_e=1\mu F$.

Compare the V_{in} and V_o waveforms and also that at the restorer output V_r. Thus confirm that V_o is close to the peak-to-peak value of V_{in}, apart from the usual 600mV diode drop.

Note how much larger this is than the $\hat{V}_{in}/\pi$ for the filtered half-wave rectifier output (≈×6).

5

Semiconductor diode and power supplies

Summary

In Exercise 4.4 of the last chapter the semiconductor diode was found to be a device which passed current in one direction but not the other. Also

- for significant current to flow in the forward direction a voltage of ≈600mV is required,
- for applied voltages ≈600mV the transition from non-conduction to conduction is not quite abrupt,
- as the frequency of the input signal is increased there is evidence of the diode capacitance.

The first objective of this chapter is to give an understanding of these properties. The other is to describe the way the diode is used as the rectifier in various kinds of DC power supply. The final section is mainly example calculations and detailed derivations, but with one or two experiments.

5.1 Semiconductors

A semiconductor is a material with conducting properties intermediate between those of an insulator such as glass, and a conductor such as copper.

The most commonly used semiconductor today is silicon. Fig. 5.1 shows the atomic structure of a semiconductor. Here each atom has an outer shell with four electrons in it. These combine with the neighbouring electrons to constitute the complete 'covalent bonded' structure shown.

The structure has zero overall charge, because the negative charge of the electrons is exactly balanced by the positive charge of the lattice of nuclei. Also the 'complete' nature of the structure means that none of the electrons is available for conduction. Thus so far the material is an insulator.

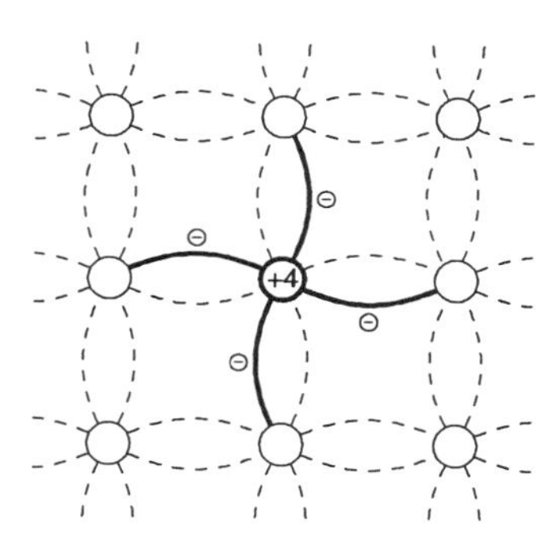

Fig. 5.1 Covalent bonding of Silicon.

Balls on a tray model

The mathematical model describing this situation most completely is quantum theory. Another more homely model, that conforms reasonably well with the findings of the quantum theory, is the 'balls on a tray' model of Fig. 5.2.

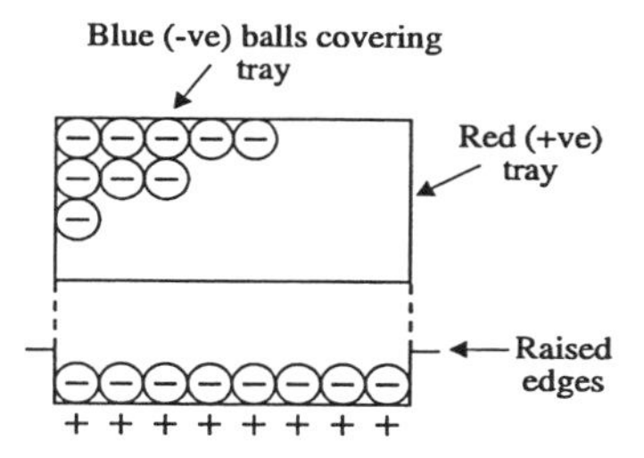

Fig. 5.2 Balls on a tray model of silicon semiconductor, locked in at absolute zero so no conduction.

Here the potentially mobile electrons are thought of as a layer of negatively charged balls placed on a positively charged tray which represents the fixed 'lattice' of silicon atoms. The balls are trapped by the edges of the tray and completely fill it. They are thus locked in and unable to move. Hence, even when a voltage is applied setting up an electric field in the Silicon, no current flows.

Intrinsic semiconductor

The above describes the situation at the absolute zero temperature of 0°K. For T non-zero the tray (lattice) vibrates with thermal energy and this causes a few of the balls (electrons) to be dislodged as in Fig. 5.3. Two mechanisms of conduction now become possible.

(i) The raised electrons are now no longer locked in and can move around at the upper level.

(ii) With a number of 'holes' left in the lower layer then a neighbouring electron can move into the hole leaving a space behind it, another can move into that space, and so on.

A positive charge imbalance exists at the hole position, wherever it is, and so the hole can most simply be thought of as a single moving positive charge carrier, although it is actually a series of electrons that are executing the movement.

Similarly, as shown, there is a negative charge imbalance at the position of each of the upper-layer electrons. At all other positions there is no charge imbalance, so the region is electrically neutral.

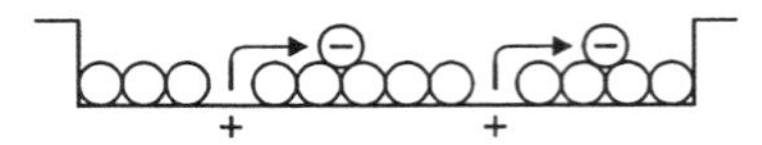

Fig. 5.3 Thermal excitation of balls (electrons).

Applied field

Fig. 5.4 shows how, when an electric field E is applied, the holes move in one direction and the electrons the other. However, both current components are in the direction of the field.

The overall conduction is 'ohmic', with I proportional to the applied voltage V.

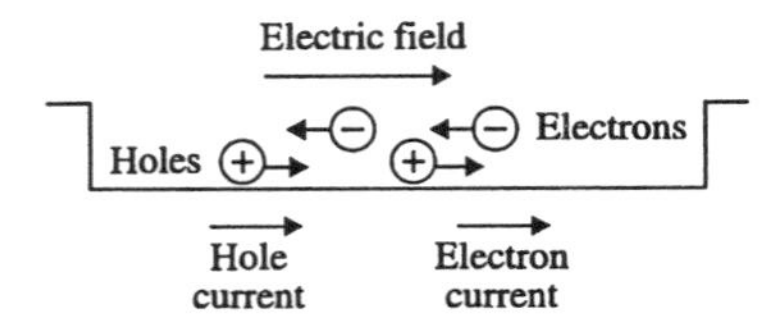

Fig. 5.4 Hole and electron movement and current for applied electric field.

Extrinsic semiconductor

Intrinsic semiconduction in silicon is very small, and conductivity levels are raised by doping with impurity atoms. The impurities are of two types, electron 'donor' atoms which give the 'N-type' of semiconductor, and electron 'acceptor' atoms which give the 'P-type'.

N-type

The N-type impurity atom is as shown in Fig. 5.5(a). Here the outer shell has five outer electrons, rather than the four of the silicon. Also the nucleus of the impurity atom has the required balancing charge of five units.

Only four of the outer electrons are required for the covalent bonding. Thus the fifth is much more easily detached from the parent atom than for one of the valence electrons in the host silicon.

Balls on a tray

Fig. 5.5(b) shows the situation at T=0K where the fifth electron remains bonded to its parent atom, giving as yet no conduction and a totally neutral charge distribution. (c), in contrast, shows the situation at room temperature. Here the thermal energy is sufficient to detach the electron, thus allowing conduction. The fixed parent atoms are then left with a net positive charge, while the mobile charge carriers, being electrons, have a negative charge.

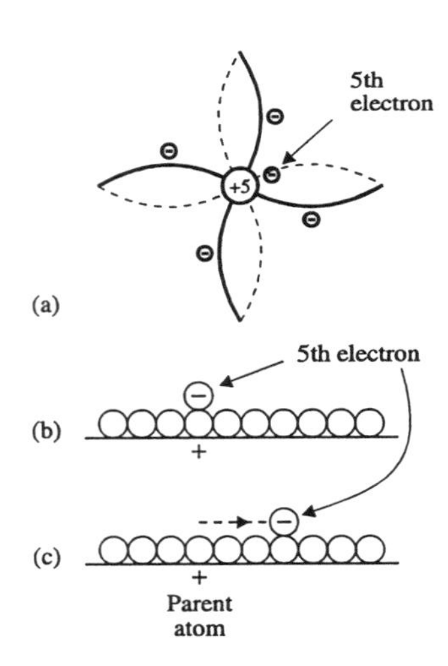

Fig. 5.5 Donor type impurity semiconductor.
(a) Covalent bond view.
(b) Balls on a tray view, T=0K.
(c) As (b) but at room temperature.

Relative conductivity

For silicon

- the energy required to free an electron (lift a ball to the top of the tray) is 1.2eV,
- the mean thermal energy at room temperature is 25meV
- the proportion of freed electrons varies exponentially with the energy ratio.

Thus for the undoped material the proportion of freed electrons is very small indeed.

For the impurity atoms, in contrast, the energy needed to free the fifth electron is smaller than the thermal energy and so virtually all of these are freed.

P-type

Fig. 5.5 shows the corresponding situation for P-type material. Here the doping atom in (a) has only three atoms in its outer shell. Hence it is relatively easy for an electron from a neighbouring atom to move in and, in effect, release a hole from the parent atom.

At T=0K in (b) the hole remains at the site of the parent atom. Thus again there is no conduction and a neutral charge distribution.

But in (c) thermal energy has released the hole (allowed a neighbouring ball to move in) and the freed hole is then available for conduction.

Again conductivity is much higher than for the intrinsic material, here because the energy needed to free the hole from its parent is small compared with the thermal energy, thus freeing virtually all of the previously trapped holes.

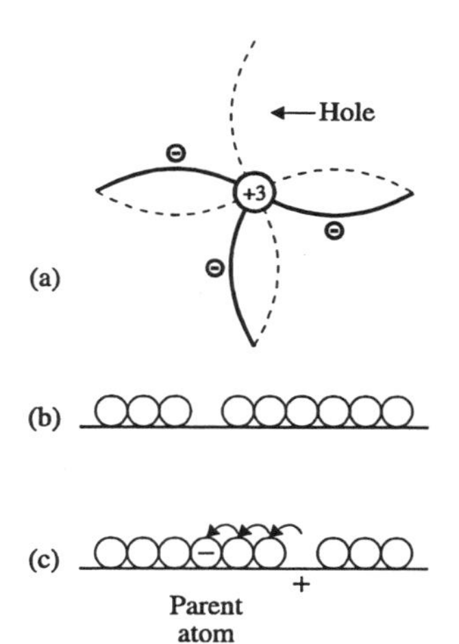

Fig. 5.6 Acceptor type impurity semiconductor.
(a) Covalent bond view.
(b) Balls on a tray view, T=0K
(c) As (b) but at room temperature.

Temperature sensitivity

Because for both of the extrinsic types of semiconductor virtually all of the available charge carriers are freed from their parents, increasing the temperature makes little difference to the conductivity. There is only a slight reduction due to the decrease in mobility resulting from the higher degree of lattice vibration.

In contrast, for intrinsic conductivity the number of carriers, and thus the conductivity, increases rapidly with temperature. Thus the process becomes highly temperature-dependent. However, for both extrinsic and intrinsic mechanisms at T=0K the conductivity is reduced to zero.

Carrier concentration

For the P-type material the typical doping concentration

$$\mathbf{P \sim 2 \times 10^{21}/m^3}.$$

With p_i and n_i the intrinsic hole and electron concentrations that existed prior to the doping,

$$\mathbf{n_i = p_i \approx 1.5 \times 10^{16}/m^3}.$$

Thus P is vastly in excess of the intrinsic concentrations, with the same excess for the N-type semiconductor.

For the doped material the opportunity for one of the intrinsically generated electrons to recombine is much increased, due to the increased hole concentration. Thus $n_p << n_i$, where n_p is the electron concentration in the P-type material. With the hole concentration $p_p \approx P$ then the factor by which n_p is reduced from n_i is $\times p_i/P$. With $p_i = n_i$ then

$$n_p \approx n_i^2/P \tag{5.1}$$

giving $\mathbf{n_p \approx 1 \times 10^{11}/m^3}$

With p_p and n_p the 'majority' and 'minority' carrier concentrations, then the minority concentration is less than that of the majority by approximately $\times 10^{-10}$, a very large difference indeed. It is thus surprising that both types play an important part in determining the above 600mV for the diode.

5.2 Semiconductor diode

Fig. 5.7(b) shows the circuit symbol for the semiconductor diode, with the arrow head indicating the direction of allowed current flow. The terms 'anode' and 'cathode' derive from its predecessor, the thermionic valve diode.

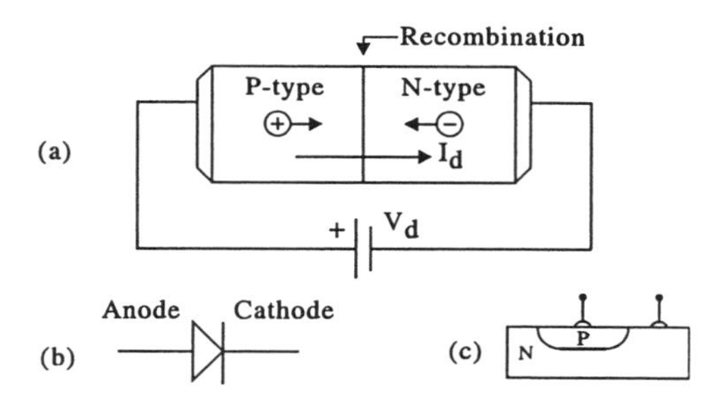

Fig. 5.7 Semiconductor diode.
(a) Essential structure with forward bias giving current flow.
(b) Circuit symbol.
(c) Normal planar construction.

The device is a piece of semiconductor material, usually silicon, half of which is doped to be N-type and the other P-type. (a) shows the essential structure and (c) a common method of construction. Starting with the N-type slab, P-type impurities are diffused into this at a higher concentration to make the region overall P-type. Metal pads are then added, for attachment of the connecting wires.

(a) also shows how the forward current is supported. With the polarity of the applied V_d as shown, the carriers both move towards the junction. There they recombine to effectively vanish, to be replaced by others at the two end junctions. Thus the current continues.

For the reverse polarity of V_d the carriers move away from the junction. Thus, apart from the initial movement, the current is prevented.

Depletion layer

Imagine the P- and N-type regions in Fig. 5.6(a) to be initially separate and then brought into contact, with the diode otherwise disconnected.

With thermal energy causing the carriers to diffuse across the junction, and with each type of entering carrier meeting its opposite, recombination occurs. With the recombined carriers ceasing to be mobile, and with the entering carriers retaining their charge, the charge dipole shown is established. Over the charged regions virtually no mobile carriers remain, which is why the region is known as the 'depletion layer'.

Associated with the charge dipole is the electric field E and the resulting potential step ψ_o shown in (b). Here (b) is termed the 'energy diagram' for the device, because it represents, in electron-volts, the variation of the potential energy ψ of an electron with position.

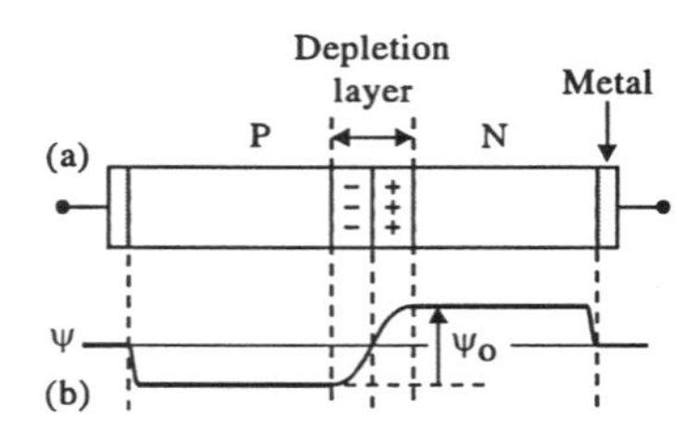

Fig. 5.8 Semiconductor diode on open circuit.
(a) Essentials of physical arrangement.
(b) Energy diagram resulting showing potential function ψ.

Junction potential

With typically $\psi_o \approx$ **900mV**
this is determined by the balance of two opposing mechanisms.

(i) Majority carrier flow With the polarity of ψ_o such as to oppose the majority flow, as the potential builds up the proportion of carriers having enough energy to traverse the layer decreases. With the

$$\text{energy distribution} \propto \exp[-E/(kT/q)]$$

where E is the energy and kT/q the thermal energy, then ψ_o only needs to be a few times kT/q to give a large reduction relative to the rate when the regions were first brought together.

(ii) Minority carrier flow The minority carriers also diffuse towards the junction, with the important difference that when they get there the polarity of the potential is such as to assist, rather than oppose, the passage. As a result the minority carrier flow is limited only by the diffusion process that takes them to the edge of the layer, and thus is virtually independent of the magnitude of the rising potential step.

Thus the final value ψ_o is that for which the initially vastly greater majority carrier flow is reduced to become equal to the minority flow.

With Boltzmann's constant $k=1.38\times10^{-23}$J/°, the electron charge $q=1.60\times10^{-19}$ Coulomb, and the standard absolute room temperature $T_o=290°$, for normal values of temperature T **kT/q≈25mV.**

With $\psi_o/(kT/q)=36$ for $\psi_o=900$mV,

and with the exponential energy distribution, this is sufficient to compensate for the $\times10^{10}$ difference in carrier densities.

Diode voltage current law

As the applied voltage V_d is increased from zero then the junction potential ψ_d is reduced from ψ_o to ψ_o-V_d. This makes no difference to the minority flow but much increases the majority flow, to constitute the dominant component of the diode current I_d.

Conversely, for V_d made negative then ψ_d is increased from ψ_o, soon to make the majority flow negligible. With the resulting diode current

$$I_d=I_{ds}[\exp(qV_d/kT)-1] \tag{5.2}$$

as plotted in Fig. 5.9 (Gibbons, 1966; Sze, 1981) the 'saturation' component I_{ds} corresponds to the minority flow, and the exponential term to the majority.

The majority variation is exponential because of the exponential energy distribution of the mobile carriers, making the proportion having enough thermal energy to traverse the opposing barrier vary exponentially with the height of the barrier.

Working values

While the plot of Fig. 5.9(a) is for $V_d \sim kT/q$, that of (b) is for $V_d \sim 600$mV, with the resulting 'working values' of I_d there vastly greater than those for (a).

For $I_d=1$mA at $V_d=600$mV, [5.2] gives $\mathbf{I_{ds} \sim 0.0378}$ **pA.** Thus for the working values, to a very good approximation indeed,

$$I_d \approx I_{ds}.\exp(qV_d/kT) \tag{5.3}$$

As shown in (b), I_d is reduced by ×1/10 each time V_d is reduced by (kT/q).ln(10)=57mV. With a further ×1/10 reduction for the next 57mV decrease then I_d soon becomes invisible on the working scale.

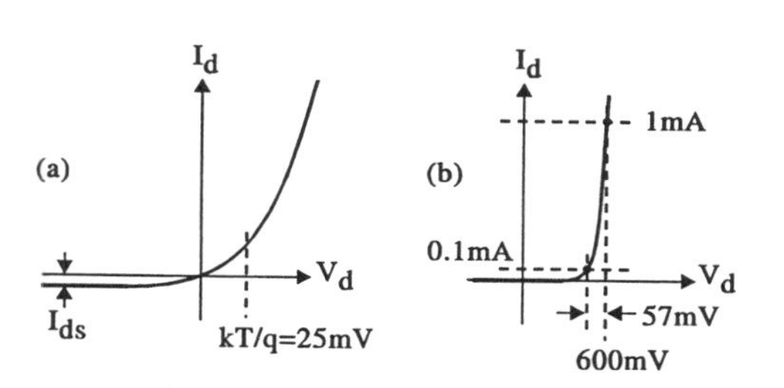

Fig. 5.9 Diode voltage current relations.
(a) V_d comparable with kT/q=25mV.
(b) V_d comparable with 600mV.

Carrier flow

Fig. 5.10(a) shows the majority carrier flow in the region of the junction for the diode forward biased, with the recombination in the regions shown, while (b) shows the much smaller minority flow. Here there is no recombination, since each entering carrier is of the majority type for the region entered.

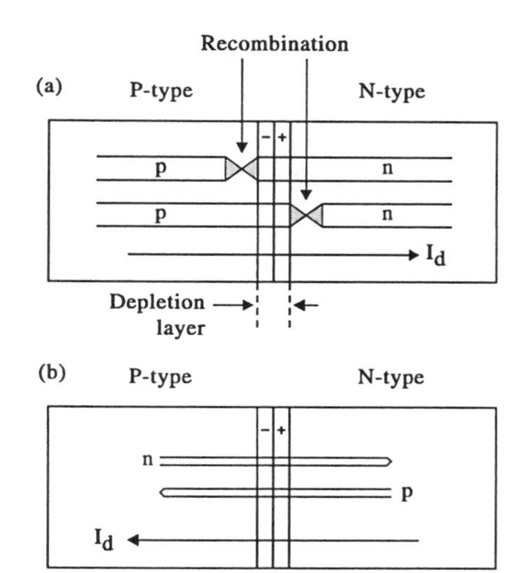

Fig. 5.10 Carrier flow for semiconductor diode.
(a) Forward bias, majority carriers.
(b) Reverse bias, minority carriers.

End connections

Fig. 5.8 also shows the voltage steps at the terminal connections that are necessary for V_d to equal zero for the diode disconnected. These junctions do not rectify in the same way as that at the centre because, while the change in doping concentration at the central junction is close to being abrupt, for the other two the doping is 'graded', with the grading over a distance which is large compared with that of the associated recombination region.

For the end connection to the N-type region the electrons constituting the majority flow simply pass from the connecting wire into the region. For the connection to the P-type region the process is more complex. Here the holes needed for the majority flow are the result of the generation of hole-electron pairs in the graded region, with the holes moving towards the depletion layer as required, and the electrons flowing back into the wire to complete the current path.

600mV diode model

Fig. 5.11(a) and (b) show the ideal diode with its voltage-current relation, while (c) and (d) show the changes allowing for the nominally 600mV voltage drop.

With the actual relation the exponential one of [5.2], it is this that accounts for the transition from non-conduction to conduction not being as abrupt as the 600mV model suggests.

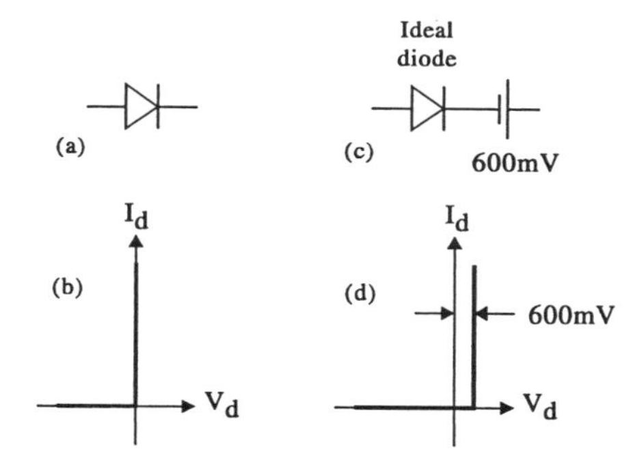

Fig. 5.11 Voltage current relations for simplified diode models.
(a) Ideal diode.
(b) Voltage/current relation for (a).
(c) 600mV diode model.
(d) Voltage/current relation for (c).

Bulk resistance

For ψ_o=900mV, and with the working V_d=600mV, then the junction potential ψ_d is not fully eliminated. Indeed with I_d increasing ×10 for each increment of 57mV in V_d, then for $\psi_d=0$, I_d=163A!

Long before this could happen the resulting reduced V_d/I_d, falls below the 'bulk resistance' R_b of the diode, the resistance of the semiconductor material apart from the depletion layer. Thereafter the $I_d - V_d$ relation becomes linearised as in Fig. 5.12(a).

Addition of R_b to the 600mV model of Fig. 5.11(c) as in Fig. 5.12(b) then gives the $I_d - V_d$ relation of Fig. 5.12(c), which is an adequate approximation to the relation of (a) for the real diode.

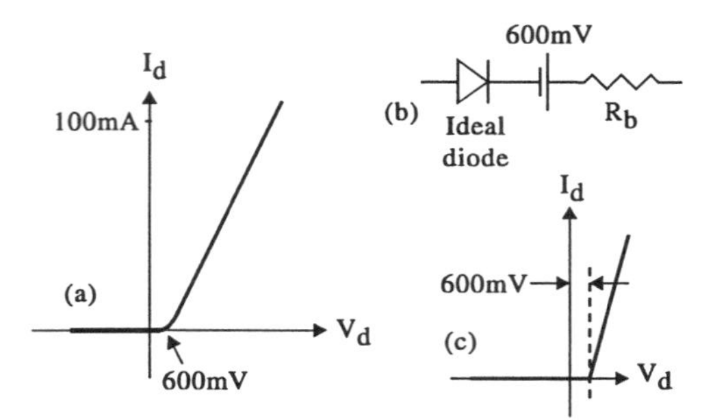

Fig. 5.12 Effect of diode bulk resistance R_b at higher values of I_d.
(a) $I_d - V_d$ relation for real diode.
(b) 600mV diode model of Fig. 5.11(c,d) with R_b added.
(c) $I_d - V_d$ relation for (b).

5.3 Incremental diode model

Fig. 5.13(b,c) shows the semiconductor diode used as a voltage-controlled attenuator. There are better ways of realising this function, but the example is a good one for introducing the important principle of 'incremental modelling' of a non-linear device. With the relatively low-amplitude V_s the signal input to the attenuator, the bias voltage E_b controls the impedance of the diode, and thus the transfer of V_s to the output V_d.

Diode

From (a) the incremental

$$i_d = v_d \times dI_d/dV_d$$

Thus dI_d/dV_d can be regarded as an incremental conductance g_d, or as the equivalent incremental resistance $r_d = 1/g_d$. With

$$I_d \approx I_{ds}\exp(qV_d/kT)$$

as in [5.3] then

$$g_d \approx I_d/(kT/q) \quad \ldots(a)$$

and

$$r_d \approx (kT/q)/I_d \quad \ldots(b) \qquad (5.4)$$

For say $I_d = 1\text{mA}$ **$r_d = 25\Omega$ and $g_d = 40\text{mS}$.**

With $v_d = i_d r_d$

this is represented diagramatically (modelled) as in (e).

Source and resistor R

With the section of the circuit in (b) linear, the components i_d of I_d and v_d of V_d due to V_s are the same as for $E_b = 0$, giving

$$v_d = V_s - i_d R \qquad (5.5)$$

This is modelled as in (d) to give (d,e) as the incremental, or 'small-signal' circuit model for the attenuator, in contrast to the 'large-signal' circuit of (b,c). From the incremental model, the attenuation $\alpha = v_d/V_s$ is given by

$$\alpha = r_d/(R + r_d) \qquad (5.6)$$

With $r_d \approx 1/I_d$ and I_d increasing with E_b then α is controlled by E_b.

Incremental values

Lower-case symbols are used as above for incremental values, and upper-case for normal large-signal values. This convention will be observed in all following chapters for which it is relevant. Otherwise the upper-case will be used. For example, at the 'operating point' P in (a), $g_d = dI_d/dV_d$ differs considerably from the large-signal $G_d = I_d/V_d$.

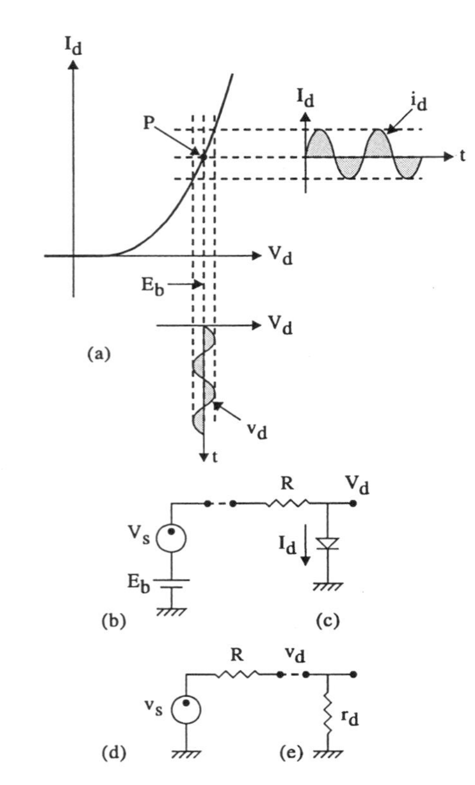

Fig. 5.13 Voltage-controlled attenuator.
(a) Waveforms and voltage current relation for diode.
(b) Linear section of circuit diagram.
(c) Non-linear section of circuit diagram (diode).
(d) Incremental circuit model for (b)
(e) Incremental circuit model for (c).

Limitations

The low permissible amplitude of V_s is a considerable limitation for the present circuit. Also for such circuits linear control is usually preferred, while the present control is both reciprocal and exponential. Better implementations are given in Chapter 15.

5.4 Diode reverse breakdown

For the reverse-biased diode, 'breakdown' ultimately occurs. The breakdown voltage V_{bk} is usually large compared with 600mV, so the $I_d - V_d$ curve becomes as in Fig. 5.14. There are two breakdown mechanisms. For light doping V_{bk} is relatively high, giving 'avalanche' breakdown. For higher doping then V_{bk} is lower. Here it is the Zener mechanism which dominates.

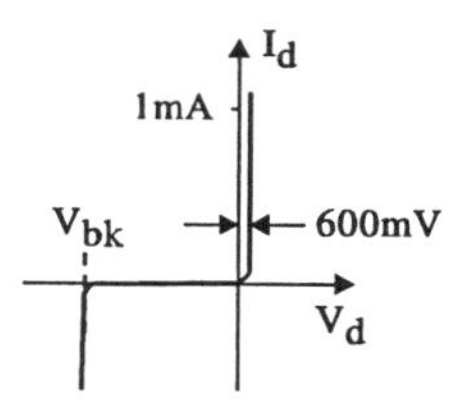

Fig. 5.14 Diode reverse breakdown.

Avalanche breakdown

With reverse bias, when a minority electron in the P-type region diffuses into the depletion layer it is swept across by the E field within the layer. During the journey the electron is repeatedly accelerated by the field, each time to collide with the thermally vibrating lattice to give up the energy gained. Eventually a value of V_d is reached for which the mean kinetic energy gained between collisions exceeds that needed to release another electron from the lattice on collision (1.2eV for silicon), and thus to create another hole-electron pair. The released electron is then accelerated to create yet another hole-electron pair, and so the 'avalanche' occurs.

Zener breakdown

Zener breakdown is a simpler mechanism, and occurs when E in the depletion layer is large enough to significantly perturb the wave functions that make up the crystal lattice, thus making it easier to detach a valence electron.

Breakdown voltage V_{bk} vs doping

Fig. 5.15 shows the profiles for the depletion layer variables, with ρ the charge density for the trapped carriers in the depletion region, E the resulting electric field and ψ the resulting electric potential.

With each profile the integral of the previous one, and with P and N the doping concentrations then, as shown in Exercise 5.1,

$$\psi_d \propto E_{max}^2(P^{-1}+N^{-1}) \qquad (5.7)$$

With E_{max} the value limited by breakdown, then the value of ψ_d at which breakdown occurs decreases with increased doping.

With $\psi_d=\psi_o - V_d$, then so also does $|V_{bk}|$, the value of $|V_d|$ at which breakdown occurs, with approximately

$$|V_{bk}| \propto (\text{concentration})^{-1/2} \quad \text{for} \quad |V_{bk}| >> \psi_o.$$

This however is moderated by a modest increase of ψ_o with doping, for which

$$\psi_o=(kT/q)\ \ln(PN/n_i^2)$$

(Gibbons, 1996), but the overall increase of V_{bk} remains.

Differential doping

With $$(\psi_o - V_{bk}) \propto (P^{-1}+N^{-1})$$

then to increase V_{bk} it is only necessary to reduce one of the dopings. This is confirmed by the diagram, in which most of ψ_d is accounted for by the more lightly doped half. This feature will be used to advantage later.

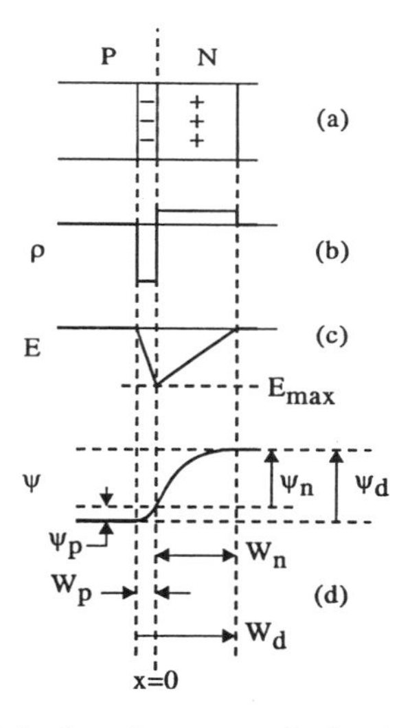

Fig. 5.15 Depletion layer for an unequally doped PN junction.
(a) Depletion layer.
(b) Charge density ρ=charge/vol.
(c) Electric field strength E.
(d) Electrostatic potential ψ.

Zener diode

This is simply a diode with V_{bk} set to the specified value, and is widely used in limiting and simple voltage-regulator circuits. A full range of such diodes is manufactured, with the preferred values V_z=1V, 1.2V 1.5V, 1.8V, 2.2V, etc. Such diodes are invariably called 'Zener' diodes, although only for the higher dopings (lower V_{bk}) is the breakdown predominantly Zener. For the others the breakdown is by the avalanche mechanism.

Zener voltage reference

Fig. 5.16 shows how the Zener diode can be used to produce a stable reference voltage V_o in the face of a varying supply voltage V_{in}. The symbol in (a) is the usual one for the Zener diode, with the breakdown current in the reverse direction from that of normal current flow.

First (c) shows the diode voltage-current curve (here I vs V_o) while (b) shows the relation $V_o = V_{in} - IR$ for the rest of the circuit. Then the actual I and V_o are those at the intersection point P.

For small changes in V_{in} then V_o changes only slightly relative to that at P. Over this range the diode curve is approximated well by the line shown in (d) for which

$$V_o = E_d + r_z I \tag{5.8}$$

where the incremental $r_z = dV_o/dI$ for the diode at P. Thus (a) with the approximation of (d) becomes as in the partially incremental model of (e). The factor dV_o/dV_{in} by which the circuit suppresses changes in V_{in} is then given by

$$dV_o/dV_{in} = r_z/(R + r_z) \tag{5.9}$$

The arrangement can also be used as a rudimentary stabilised power supply, provided that the current demand is comparable with the allowed Zener current. Exercise 5.2 gives experimental work.

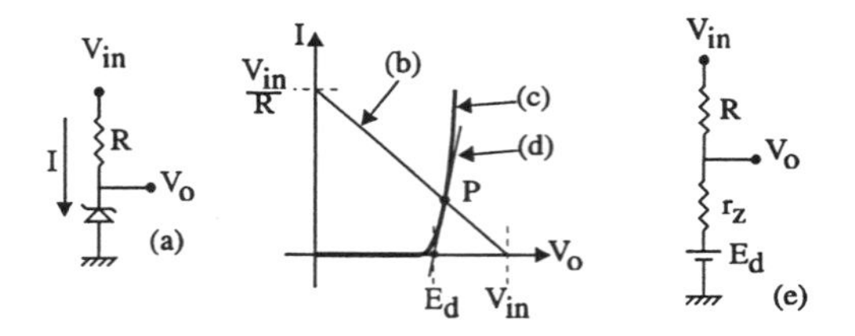

Fig. 5.16 Zener voltage reference.
(a) Circuit diagram.
(b) Relation $V_o = V_{in} - IR$ for linear section of (a).
(c) Voltage-current ($I - V_o$) relation for Zener diode in (a).
(d) Linear approximation to (c).
(e) Incremental circuit model.

Back diode

If the doping is increased to the point where the internal depletion-layer potential step ψ_o reaches 1.2V, this is the voltage corresponding to the energy in electron-volts needed to detach an electron from the lattice and create a hole-electron pair. With V_{bk} here zero then the Zener diode $I_d - V_d$ curve of Fig. 5.14 becomes as for the 'back diode', in Fig. 5.17.

The diode is so called because, provided that the normal turn-on voltage of 600mV is not exceeded, the device works as one which conducts with V_d negative but not with V_d positive, the reverse of the usual case. The device has the advantage of a much lower turn-on voltage and so is useful for rectifying low-level signals. It can also be made to operate at very high frequencies, including the microwave region, and is of low noise. Its obvious disadvantage is its low dynamic range.

Tunnel diode

With a yet further increase in doping, the result becomes as for the tunnel diode, also as plotted in Fig. 5.17. Here the brief region of negative slope constitutes a negative incremental resistance, allowing the device to be used as the basis of a low-dynamic range amplifier, or as an oscillator. The device is also of low noise but is rarely used today, because of the considerable advances in the design of MOSFET devices, particularly those based on the gallium arsenide type of semiconductor.

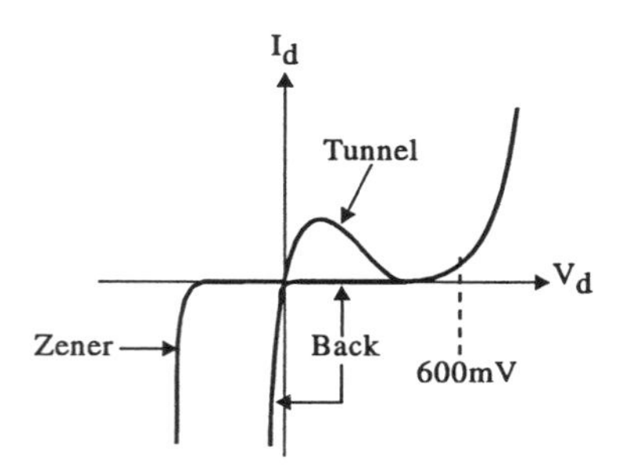

Fig. 5.17 Tunnel and back diode I_d–V_d curves compared with Zener.

Diode limiters

Fig. 5.18 shows a very simple diode limiter. Should prior high-pass filtering be needed, say to eliminate a DC component, this can be simply achieved by adding a suitable capacitor in series with the resistor.

With the waveforms assuming the 600mV diode model, for a real diode the transition points will be slightly curved and $|V_o|$ will rise slightly in the clipping regions.

Higher clipping levels can be provided by using two Zener diodes in series in place of the two parallel normal diodes.

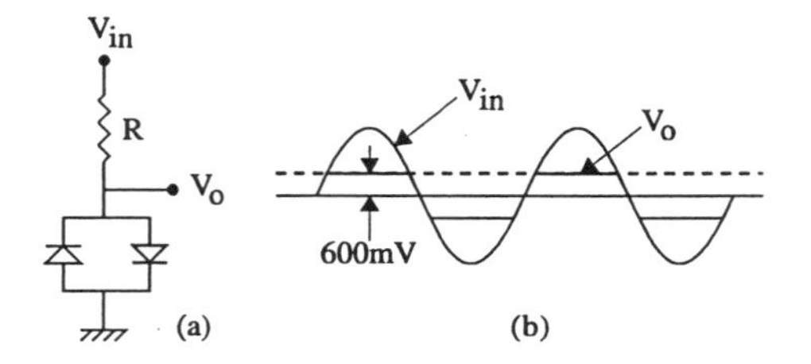

Fig. 5.18 Simple diode limiter.
(a) Circuit diagram.
(b) Waveforms for 600mV diode model.

5.5 DC power supplies

Fig. 5.19 shows one of the usual ways is which the diode is used as the rectifier in a DC power supply. With the waveforms as in (c), the operation is much as for the simple envelope detector of Fig. 4.19, except that with the full-wave rectification the 'reservoir capacitor' C is 'topped up' twice per input cycle, rather than once as for the half-wave circuit. Thus the ripple amplitude is reduced and the ripple frequency doubled.

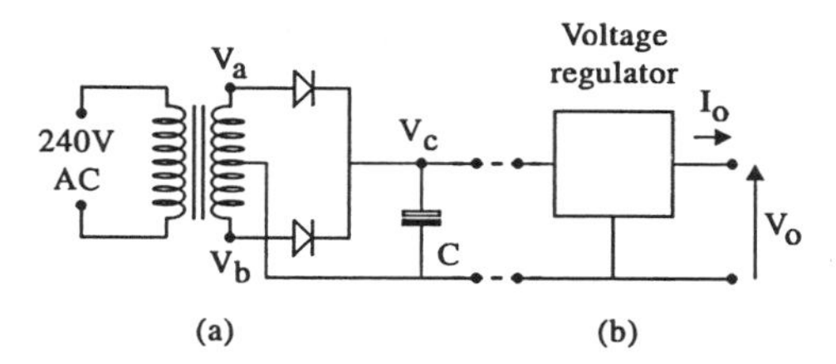

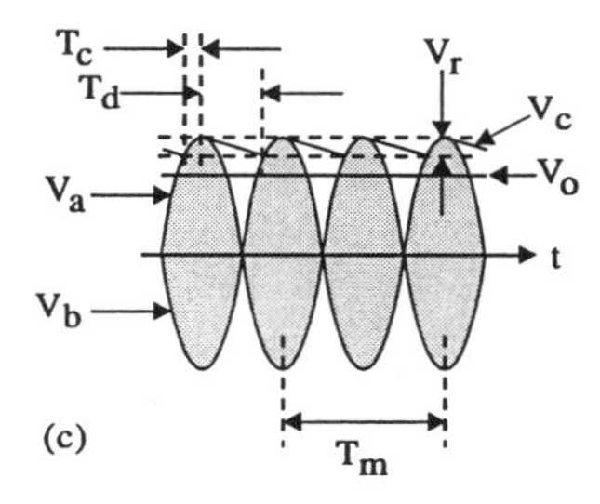

Fig. 5.19 Full-wave power rectifier and regulator.
(a) Transformer and rectifier.
(b) Voltage regulator.
(c) Waveforms for (a) with components assumed ideal.

Experimental result

The waveforms in Fig. 5.19 assume ideal diodes with no forward voltage drop, and also an ideal transformer with zero winding resistance. Fig. 5.20, in contrast, shows the experimentally derived V_c waveform for the supply detailed in Table 5.1.

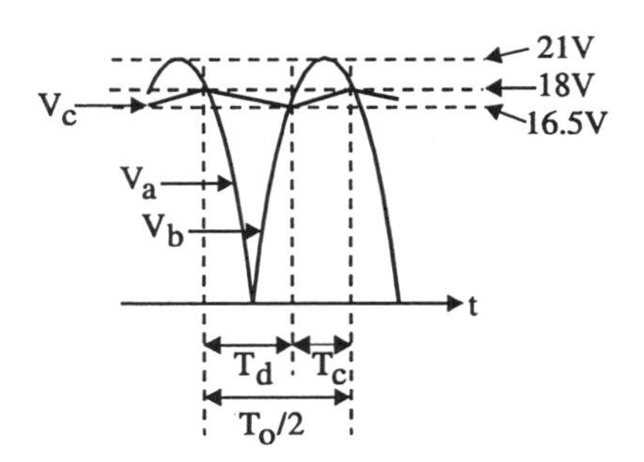

Fig. 5.20 Waveforms for an experimental power supply of the type of Fig. 5.19.

Transformer secondary voltage 15V-0-15V.
Secondary winding resistance R_w= 4.5Ω+4.5Ω.
Reservoir capacitor C=440μF.
Rectifier diodes 1N4001.
Voltage regulator type LM78L12
Regulator current drain I_{reg}=3mA.
Output voltage V_o=12V.
Load current I_o=100mA.
Voltage regulator requires V_c > 14.5V.
Mains period T_m=20ms.

Table 5.1 Details of DC power supply of Fig. 5.19.

The waveform is for the full 100mA load current, and is shown relative to the transformer secondary voltages on open circuit, for reference.

It is this kind of circuit where the SPICE simulation is particularly to be preferred. However, some useful analytical calculations can be made.

Peak secondary voltage

Transformer ratings are always for the RMS value. Thus $\hat{V}_a = \tilde{V} \times \sqrt{2}$ giving **$\hat{V}_a$ =21.2V.**

Ripple amplitude

With $dV_c/dt=(I_o+I_{reg})/C$

and with Fig. 5.20 showing the decay time $T_d \approx 6.5ms$, then

$$V_r=(I_o+I_{reg})T_d/C \quad (5.10)$$

Giving **$V_r \approx 1.52V$ p-p.**

Both the calculated values are close to those observed.

Diode peak current

With T_d and T_c the decay and conduction periods then, from the waveform, the diode conducts for about 40% of the time. With the mean I_d over this period

$$I_o \times (T_c+T_d)/T_c$$

and with the peak I_d approximately twice this then
peak $I_d \approx 571mA$.

Transformer and diode voltage drop

This is the reason for the 3V difference between the peak open-circuit V_a=21V and the peak V_c=18V. With the transformer drop

$$I_{d.peak} \times R_w=2.25V$$

this leaves 750mV for the diode, roughly as expected.

Peak inverse voltage

With V_{piv} the peak inverse voltage across the diode then

$$V_{piv} = V_{a.peak} + V_c$$

giving $\mathbf{V_{piv} = 39V}$.

This is within the 50V rating of the 1N4001, as is the $I_{d.peak}$ within the 1A limit for the device. However, most designers would use the equally priced 1N4002 rated at 100V (same price for each 5p).

Voltage regulator

Fig. 5.21 shows the essentials of the voltage regulator in Fig. 5.19(b). Here V_o is sampled by the right-hand potential divider and the result is compared with the Zener-derived reference voltage. Any difference is amplified and applied to the 'pass transistor' in such a way as to correct the difference. Thus V_o is maintained constant in the face of varying V_c, including the ripple.

Usually the circuit contains certain refinements such as arrangements for improving the immunity of the reference to changes in V_o, and current limiting circuitry.

The pass transistor needs $V_c - V_o$ to be only about 1V. However, the value will usually be a good deal higher, to allow for the change in V_c with I_o.

For a given I_o the power dissipated in the pass transistor is proportional to $V_c - V_o$, so adequate heat sinking is required if the voltage drop is large.

The present minimum $V_c = 16.5V$ is suitably above the required 14.5V for the regulator used.

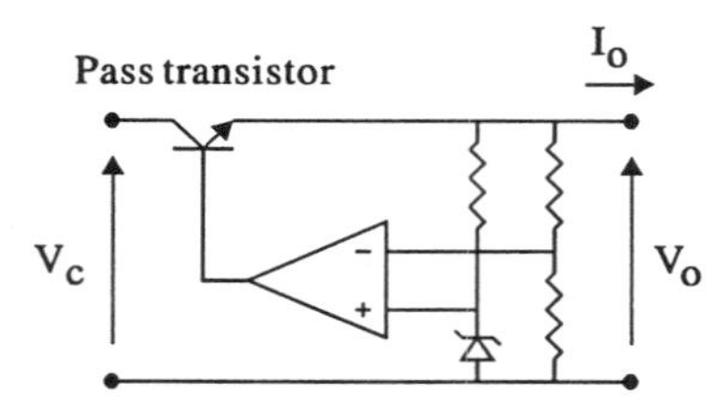

Fig. 5.21 Essentials of voltage regulator in Fig. 5.19.

Bridge rectifier

Fig. 5.22 shows a popular variant on the full-wave rectifier of Fig. 5.19. The main advantage of this arrangement is that full-wave rectification is obtained without the second half of the split transformer secondary. Since the transformer is likely to be the most expensive item in the supply, this is a useful saving.

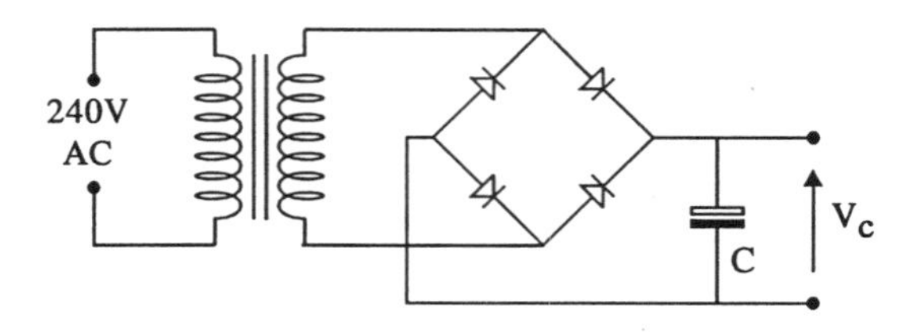

Fig. 5.22 Bridge rectifier-based power supply.

The only disadvantage is that the reservoir charging current must now pass through two diodes rather than one. This requires a slightly larger input voltage but, more importantly, gives twice the overall diode power loss.

5.6 Switched-mode power supply

A limitation of the conventional power supply of Fig. 5.19 is the size, weight and cost of the transformer. Even when no secondary current is drawn there is the 'magnetising' primary current I_m due to the finite impedance $2\pi f L_1$ of the primary winding. With f the mains frequency of 50Hz then L_1 needs to be several henries. This calls for a large ferrite toroid or heavy laminated iron core, with windings of many turns.

Example For $L_1 = 5H$, the mains $f = 50Hz$, and with $\tilde{I}_m = \tilde{V}_1 / 2\pi f L_1$, then $\mathbf{\tilde{I}_m = 153mA}$.

The switched-mode supply resolves this limitation as in Fig. 5.23(a). First the simple transformerless half-wave rectifier D_1 and C_1 convert the 240V RMS mains voltage to the DC voltage $V_{in} \approx 240V \times \sqrt{2} = 340V$.

V_{in} is then interrupted at the switching frequency f_{sw} normally ~30kHz, in effect increasing the mains frequency ×30kHz/50Hz. Thus L_1 can then be proportionally reduced, allowing a much smaller core, and fewer turns requiring less copper.

Interference

The switched-mode supply has one serious disadvantage. The abrupt switching action tends to generate interference, which can propagate back into the mains supply and elsewhere. This calls for exhaustive filtering and screening.

Flyback switched-mode supply

There are many varieties of switched-mode supply (Mohan et al, 1989), with the present flyback type typical. Prior to the switch closure V_1, V_2, I_1, I_2, and the transformer flux Φ, are all zero.

On closure $V_1 = V_{in}$. With $V_1 = L_1 dI_1/dt$ and V_{in} constant then I_1 rises at the constant rate V_{in}/L_1 as in (c), with Φ rising in proportion. The polarity is such that the proportional V_2 reverse biases D_2.

After the interval T_1 set by the controller, the switch is opened. Φ cannot change immediately without trauma, but now D_2 is forward biased to give the I_2 needed to initially maintain Φ at the previous value.

Throughout the entire cycle V_o is held approximately constant by C_2. With the diode conducting $V_2 \approx V_o + 600mV$ is then also nearly constant.

With $V_2 = -L_2 dI_2/dt$ then I_2 decreases at the essentially constant rate $-V_2/L_2$ in (d).

With I_2 unable to be negative, due to the diode, then at the point where I_2 falls to zero it remains so. With I_2 thus constant, and $V_2=-L_2dI_2/dt$ then $V_2=0$ also. With $\Phi \propto I_2$ then $\Phi=0$ also, so the transformer reverts to the inert state that obtained before the switch closed.

Output voltage V

With N_1 and N_2 the number of primary and secondary turns, and R the core reluctance, then during T_1

$$\Phi=N_1I_1/R$$

while during the conduction period T_c for the diode D_2

$$\Phi=N_2I_2/R$$

For continuity of Φ at the transition then

$$I_{2.max}=I_{1.max}(N_1/N_2)$$

For the switch closed

$$dI_1/dt=V_1/L_1$$

giving

$$I_{1.max}=V_{in}T_1/L_1$$

and thus

$$I_{2.max}=\frac{V_1T_1}{L_1}\times\frac{N_1}{N_2} \qquad (5.11)$$

With

$$V_2=-L_2dI_2/dt$$

and V_2 constant during T_c then

$$T_c=I_{2.max}L_2/V_2 \qquad (5.12)$$

Since the average of I_2 flowing into C must equal the virtually constant I_o flowing out of it then, with I_2 triangular,

$$I_{2.max}T_c/2=I_oT_{sw} \qquad (5.13)$$

From these relations

$$V_2=(V_1T_1)^2\times\frac{L_2}{L_1}\left(\frac{N_1}{N_2}\right)^2\times\frac{1}{2I_oT_{sw}} \qquad (5.14)$$

Control of V

With V_2, and thus $V_o \propto I_o^{-1}$, this is very poor regulation indeed, worse than the conventional supply of Fig. 5.19 without its regulator. However with $V_2 \propto T_1^2$ the controller can maintain V_2 constant by varying T_1.

With $\quad dV_o/dT_1 \propto T_1/I_o \quad$ and $\quad I_o \propto T_1^2$

for V_2 constant, then $\quad dV_o/dT_1 \propto I_o^{-1/2}$

With the control loop gain thus dependent on I_o, then some compensation is needed in the controller if stability is to be maintained.

Isolation

For reasons of safety there must be no electrical connection between the supply output and either of the mains terminals.

Thus the connection between controller and switch (shown dashed) must be through some form of isolator, either as an AC signal through a second transformer, or through the combination of light emitting diode and photodetector known as an opto-isolator.

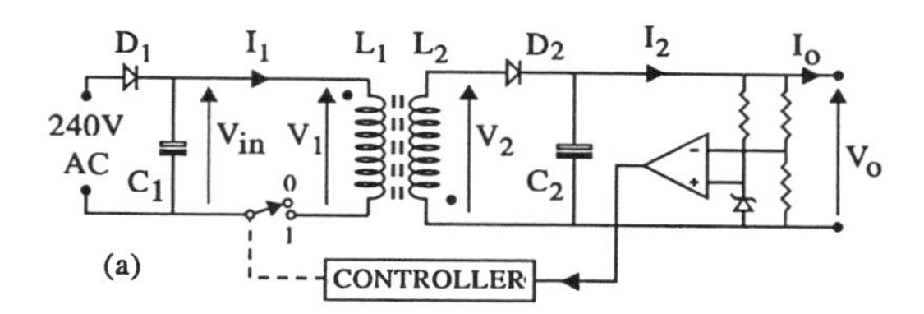

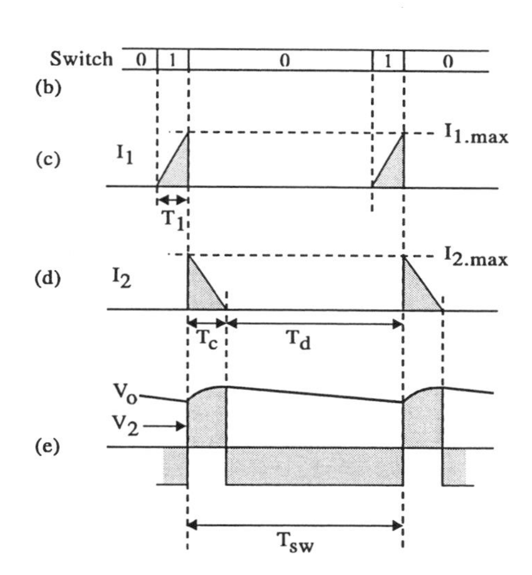

Fig. 5.23 Flyback type of switched-mode power supply with output voltage regulator.
(a) Circuit diagram.
(b-e) Waveforms for diodes considered ideal.

Primary inductance

With $L=N^2/R$ as in [1.18] then $L_1/L_2=(N_1/N_2)^2$. With the L as in [5.14] then

$$L_1=\frac{V_1^2T_1^2}{2I_oV_oT_{sw}} \qquad (5.15)$$

Example With the switching frequency $f_{sw}=30kHz$, $V_o=5V$ for $I_o=1A$ and with $T_1=10\mu s$ then $\mathbf{L_1=34.6mH}$,

a value easily realisable using a ferrite or powdered iron cored inductor. Finally, the arrangement eliminates the power loss in the pass transistor for the conventional system, which further improves efficiency.

5.7 Diode capacitance and switching time

Fig. 5.24 shows the effect of applying a 50kHz square-wave to the input of the simple diode rectifier. This is somewhat as for a capacitor in parallel with the diode, but with the charge transfer clearly a good deal more complex.

Depletion-layer capacitance

With no applied voltage the diode is effectively a capacitor, with the insulating depletion layer separating the two conducting undepleted regions.

For the rectifier, this remains so for the amplitude of V_{in} sufficiently low to avoid significant conduction, giving V_o as in (b). With C_j the depletion-layer capacitance (j: junction), then the time constant is C_jR as shown.

These waveforms are as for the experimental Exercise 5.3, but with the transient periods T_{s1}, etc., extended for clarity of annotation. The period marked C_jR was actually 500ns giving $\mathbf{C_j=50pF}$ for $\mathbf{R=10k\Omega}$.

'Stored charge' capacitance

This is the effect accounting for the much longer periods of length T_{s1} and T_{s2} in (c). The term is not entirely helpful because C_j also involves stored charge, as does any normal capacitance. However, the term is the usual one so we adopt it. The 'stored charge' effect is associated with the regions of recombination on either side of the depletion layer, as shown in Fig. 5.10, and in more detail in Fig. 5.25(b). The two components of the diode capacitance are best understood by considering the carrier flow at the two transitions of V_{in}.

Switch-on

As V_{in} is switched to +5V then the first process is the charging of C_j to the ≈600mV needed for the diode to conduct. This and what follows are best understood by considering the diode V_d waveform in (d).

The charging of C_j then gives V_d the usual low-pass response, with V_o the corresponding high-pass.

Once V_d has reached 600mV, then what follows is not evident from the waveforms. Here the majority carriers begin to flow across the depletion layer into the opposite conducting region, to recombine with the majority carriers there.

For say the flow of electrons from the N-type region to the P-type, there is a transfer of negative charge from the N-type region to the P. To maintain charge neutrality in both conducting regions, electrons have to flow into the N-type region via the connecting wire. Also electrons must flow from the other end connection into the wire, leaving the remaining holes from the hole-electron pair generation there to compensate for the invading electrons from across the depletion layer. This, and the corresponding effect for the holes flowing across the depletion layer in the other direction, constitute the 'stored-charge' component of the current flowing in the external circuit.

The effective capacitor becomes fully charged when the number of invading carriers is enough for the resulting recombination to account for the whole of the forward current.

With Q_s the total transfer of charge around the external circuit during the charging process then $Q_s=\int I_d dt$. Once the charging has been completed, and with T_r the recombination time, then

$$I_d=Q_s/T_r \tag{5.16}$$

Thus T_r is essentially the charging time.

During the whole of this charging process V_d remains ≈600mV, so the effect is barely discernible from the waveform.

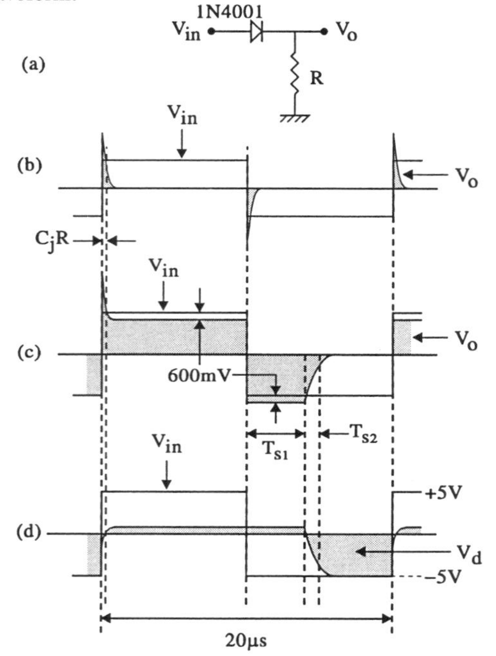

Fig. 5.24 Experimentally recorded switching response of simple rectifier for a square wave input.
(a) Circuit diagram.
(b) Waveforms for input amplitude below 600mV.
(c) V_o waveforms for input amplitude well above 600mV.
(d) V_d waveforms for (c).
(Displayed switching times somewhat increased for clarity of annotation, e.g. T_{s1}=2.8μs.)

Switch-off

It is here, as V_{in} is switched from +5V to –5V, that the above effects become evident. First Fig. 5.25(b) shows the minority concentrations for the diode conducting normally, prior to switch-off. With the carrier flow in the invaded regions at all times diffusion controlled then the minority concentration gradients at the edges of the depletion layer become as required for this purpose. The gradient then decreases with distance from the layer edge as the flow decreases, due to the decrease in concentration resulting from the recombination.

(c,d) then shows the two-stage process for switch-off. As V_{in} goes negative then the invading carriers begin to flow back across the depletion layer, now being swept across by the electric field in the layer.

With I_d reversed then the concentration gradient at the layer edge is also reversed, initially as in (c). As the invading carriers thus diffuse back to the edge and across the layer, the previous compensating charges are extracted via the end connections to constitute the external discharging current.

This is able to proceed at the same rate, keeping I_d constant and until the stage of (d) is reached.

Up to this point, with $V_d = V_{in} - I_dR$, and with I_d constant then so also is V_d, still at ≈600mV. But after (d) then the initial concentration gradient can no longer be maintained, so I_d is reduced, causing V_d at last to start falling towards the negative V_{in}.

By the time (d) has been reached most of the 'stored charge' has been extracted, leaving T_{s2} only a modest fraction of T_{s1}. With I_d as in [5.16], and with the displayed waveforms, then Exercise 5.4 estimates the recombination time **$T_r \approx 4.4\mu s$.**

Incremental charge storage capacitance

With the finally established

$$I_d = Q_s/T_r$$

as in [5.16] and with $V_d \approx 600mV$ largely independent of I_d then the usual

$$C_s = Q_s/V_d$$

has no single value, being $\propto I_d$. However, for an application such as the voltage-controlled attenuator of Fig. 5.13 the incremental c_s does have meaning, as an incremental capacitance across the diode, giving the response time $T_s \approx c_j r_r$ to a small-signal square wave input.

With $$I_d \approx I_{ds} \exp(qV_d/kT)$$

as in [5.3] and $$Q_s = I_dT_r$$

as in [5.16] then

$$Q_s = T_rI_{ds} \exp(qV_d/kT) \qquad (5.17)$$

With $c_s = dQ_s/dV_d$ then

$$c_s = T_rI_d/(kT/q) \qquad (5.18)$$

This too is proportional to I_d. However, with

$$r_d = (kT/q)/I_d$$

as in [5.4] then $c_sr_d = T_r$, which is independent of I_d.

Exercise 5.5 gives $c_s >> c_j = 50pF$ as **$c_s \approx 78nF$.**

Improvement of diode switching time

The 2.8μs 'turn-off' time in Fig. 5.24 for the 1N4001 diode makes it quite adequate for use in a 50Hz mains driven power supply, where the input period is 20ms, but not very suitable for use as say the final rectifier of a switched-mode power supply, where the switching frequency is in the region of 30kHz.

Further, when used as the simple rectifier of Fig. 5.24 (a) the situation cannot be improved by reducing R. True, the time constant C_jR can be so reduced, but at 500ns this is already so much less than the T_s as to make no significant difference. Also the T_s are essentially determined by T_r, which is a physical constant and therefore independent of operating conditions.

It is thus necessary to find other ways of improving the diode switching time. Here, however, we have a conflict. The switching speed can be reduced by increasing the doping, since this reduces the recombination time T_r. However, increasing the doping also reduces the reverse breakdown voltage V_{bk}, and for a power rectifier this may need to be large.

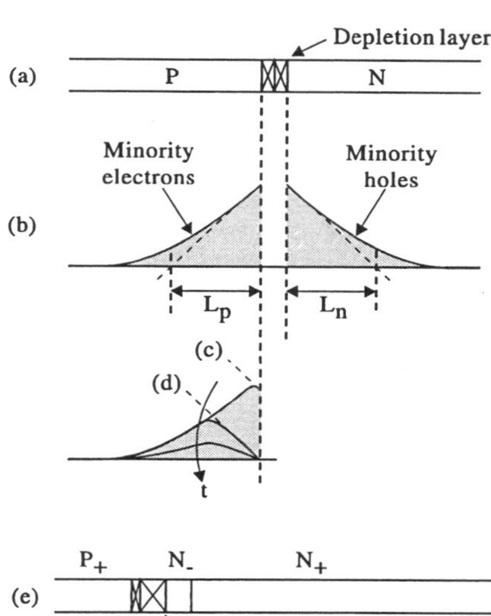

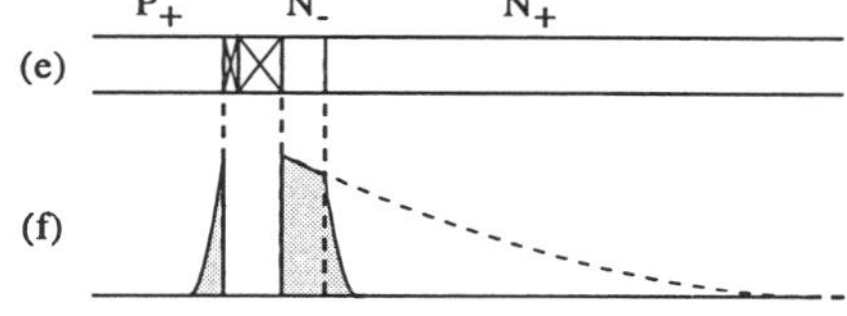

Fig. 5.25 Diode minority carrier concentrations.
(a) Doping type for normal diode.
(b) Minority carrier concentrations for (a) with diode in conduction.
(c) As LHS of (b) but shortly after switch-off.
(d) As (c) but at end of T_{s1} in Fig. 5.24.
(e) As (a) but with modifications to reduce stored charge.
(f) Minority carrier concentrations for (e) with diode in conduction.

Short base diode

Fig. 5.25(e,f) shows how this conflict is resolved. Consider first (a,b) for the normal diode. In principle it would be possible to use a low doping for the depletion layer and a high one elsewhere. This would give a wide depletion layer with the associated high V_{bk}, while a low T_r in the recombination regions.

However, it has been noted in connection with Fig. 5.15 that the doping only needs to be low on one side of the depletion layer to give a high V_{bk}. Thus in (e,f) we make the N-type doping initially light (N_-) to give the high V_{bk}, and the P-type heavy (P_+) to make T_{rp} low.

With T_{rn} still high, this is resolved by adding N_+ doping as close as is practical to the edge of the depletion layer as in (e,f). This reduces the area of the carrier concentration profile to almost equal that for the P region, giving the required reduction in stored charge.

The device can be fabricated as in Fig. 5.26(a). Starting with the heavily doped N_+ material as a slab, with the metal pad attached to the bottom, a thin layer of lightly doped N_--type is deposited epitaxially (from the vapour phase). Finally the heavily doped P_+-type is diffused into this such as to leave the required thin N_- region between the P_+ and the N_+.

Strapped transistor

Fig. 5.26(b) shows another arrangement for reducing the diode recovery time, with the device really a transistor with the collector and base strapped together. The stored charge Q_s is now only that in the narrow base region, which is normally made much smaller than the recombination length L_r, so the stored charge is reduced in proportion. The resulting recovery time T_{rec} is thus also proportionally reduced, to typically 1ns, a vast improvement on the 3.5μs for the 1N4001 power diode. Such a diode can be used up to about $1/2\pi T_{rec}=150$MHz.

Unfortunately the strapped transistor does not naturally give a high V_{bk}, due to the heavy doping required in the emitter (the upper N region). However, the arrangement is well suited to microcircuit fabrication and tends to be used for this purpose, with the emitter doping somewhat adjusted.

With the reduction in recovery time afforded by these methods, the effect of the junction capacitance c_j may well no longer be insignificant. For signal diodes where little power has to be dissipated, c_j may be reduced by simply reducing the area of the junction.

Schottky diode

This is a more recently developed device, which uses a metal semiconductor junction. Here the 'stored-charge' component is eliminated, leaving only c_j. With the area made suitably small then for signal processing applications the device can operate up to GHz frequencies.

With the forward voltage drop ≈400mV in place of the ≈600mV for the normal diode, this gives less power loss in power rectifier applications. A limitation of the device here is that its V_{bk} is only ~50V.

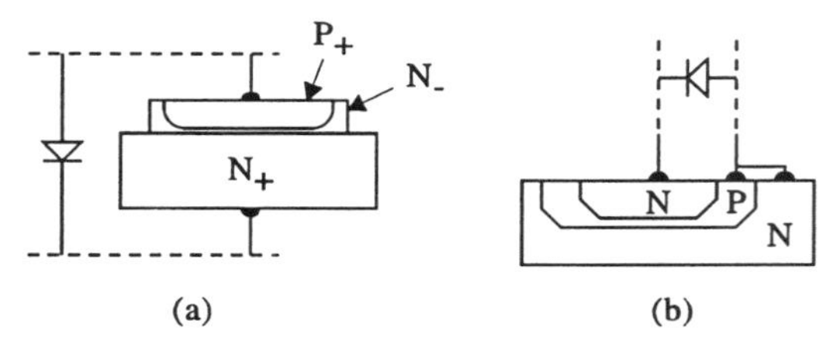

Fig. 5.26 Diode constructions reducing recovery time.
(a) Short-base diode, including symbol.
(c) Strapped transistor diode.

Varactor diode

While marginally reducing the upper frequency limit for use of the diode, the depletion-layer junction capacitance c_j can also be useful. As reverse bias is applied then the depletion layer widens, to increase the junction potential ψ_d by the applied bias. The increase in width then gives a reduction in capacitance, giving a voltage-variable capacitance.

The varactor is widely used in tuning and other applications, for example for channel selection in a TV receiver, tuning the LC circuit in the local oscillator.

From the details of the depletion layer shown in Fig. 5.15, Exercise 5.6, gives the incremental

$$c_j=c_{jo}(1-V_d/\psi_o)^{-1/2} \qquad (5.19)$$

where c_{jo} is the value of c_j when the applied voltage $V_d=0$. This relation is plotted in Fig. 5.27. For a more exact relation see [21.7].

c_{jo} is dependent on the area of the diode and typically ranges from 1 to 100pF. The above exercise gives for $c_{jo}=30$pF that as V_d varies from −10V to +600mV then c_j varies from 8.6pF to 52pF.

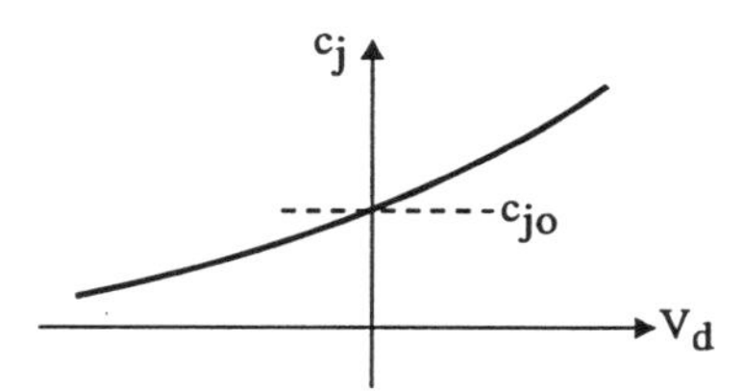

Fig. 5.27 Variation of semiconductor diode junction capacitance c_j with applied voltage V_d.

Hyper-abrupt junction diodes

The above range for a normal diode usually does not offer the degree of variation required. In such cases a diode based on the 'hyper-abrupt junction', a different doping profile, is available, at rather greater cost. This point is developed further in Section 19.6, in connection with the phase-locked loop.

Reference

Mohan, N., Undeland, T. M. and Robbins, W. P., 1995, *Power Electronics, applications and design*, John Wiley.

5.8 Exercises and experiments

Exercise 5.1. Diode depletion layer (Fig. 5.15)

> **Electric field-strength** Obtain the expressions for the electric field strength E(x) in (c) in terms of P, N, W_d and W_n.

Gauss' law gives $\varepsilon dE/dx = \rho(x)$ (5.20)

where ρ is the charge density, E is the electric field strength, and ε the permittivity.

P-type region From [5.20], with the trapped ρ_p negative and the electron charge q also negative, then $\rho_p = qP$

giving $E(x) = \varepsilon^{-1} qP(x + W_p)$ (5.21)

N-type region Similarly, with the trapped ρ_n positive then $\rho_n = -qN$ giving

$$E(x) = E_{max} - \varepsilon^{-1} qNx \quad (5.22)$$

> **Charge dipole** Show that the charges Q_p and Q_n are equal and opposite.

$E_{max} = E(0)$ in [5.21] gives

$$E_{max} = \varepsilon^{-1} qPW_p \quad (5.23)$$

$E(x) = 0$ for $x = W_n$ in [5.22] giving

$$E_{max} = \varepsilon^{-1} qNW_n \quad (5.24)$$

With the two E_{max} equal then

$$W_p P = W_n N \quad (5.25)$$

With $Q_p = qW_p P$ and $Q_n = -qW_n N$

then $Q_p = -Q_n$.

> **Potential step** Obtain the expressions for ψ_d in terms of P and N and the maximum field strength E_{max}.

With $\psi_p = -\int_{-W_p}^{0} E(x)dx$ and the triangular E(x) then

$$\psi_p = -E_{max} W_p/2 \quad (5.26)$$

Similarly

$$\psi_n = -E_{max} W_n/2 \quad (5.27)$$

giving

$$\psi_d = -(E_{max}/2)(W_p + W_n) \quad (5.28)$$

With the W as in [5.23] and [5.24] then

$$\psi_d = \frac{\varepsilon E_{max}^2}{2q}\left(\frac{1}{P} + \frac{1}{N}\right) \quad (5.29)$$

Exercise 5.2 Zener diode voltage reference

> **Experiment** For the circuit of Fig. 5.16, use a 6.2V Zener diode with R=1kΩ. With $V_{in} \approx 10V$, measure the suppression factor $\alpha_z = dV_o/dV_{in}$ and thus calculate the Zener r_z (typically 4Ω).
>
> Construct an op-amp-based adder circuit to superimpose a low-amplitude sine wave on the otherwise constant V_{in}. Thus confirm the value of α_z.

Exercise 5.3 Diode switching time

> **Experiment** Execute the measurements detailed in Section 5.7 and giving the waveforms of Fig. 5.24, being careful to use the scope ×10 probe to avoid smearing the transients. The V_d waveform is obtained by transposing R and the diode.
>
> For (c) and (d) it may not be possible to discern the initial transients of width C_jR. This is because, as shown in Exercise 5.6 below, c_j is considerably lower for the initial reverse bias of −5V than for V_d approaching the +600mV at which conduction starts. The initial period is too low for the normal scope response time. Also c_j is comparable with the scope probe 10pF input capacitance.
>
> Reduce R to 1kΩ and confirm that this makes little difference to the T_s as predicted, these being ~T_r.
>
> With R restored to 10kΩ, replace the 1N4001 diode by the popular 1N4148 signal dopde, probably one of the strapped transistor type. Here it becomes difficult to see the T_s transients at all.

Exercise 5.4 Diode recombination time

> **Calculation** The time T_{s1} in Fig. 5.24 taken to turn the diode off is 2.8μs. Calculate the approximate value of the recombination time T_r.

With the stored charge $Q_s = I_d T_r$

as in [5.16], then here $Q_s = I_{d+} T_r$

where I_{d+} is the value of I_d for the conducting half-cycle.

With α the fraction of Q_s extracted over T_{s1} then

$$\alpha Q_s = I_{d-}.T_{s1}$$

where I_{d-} is the discharging current, giving

$$T_r = (T_{s1}/\alpha \times (I_{d-}/I_{d+}).$$

With α estimated as 0.8 from the waveform of $V_o \propto I_d$ in (c) then **$T_r = 4.45\mu s$.**

Exercise 5.5 Incremental capacitance due to diode stored charge

> **Calculation** For the rectifier of Fig. 5.24 and with the diode fully conducting, calculate the component c_{ds} of the incremental diode capacity due to stored charge. $R = 10k\Omega$.

With $I_d = (V_{in+} - 600mV)/R$

then $I_d = 440\mu A$.

With $c_{ds} = T_r I_d/(kT/q)$

as in [5.18] and $T_r = 4.4\mu s$ as above then **$c_{ds} = 78.4nF$.**

Exercise 5.6 Varactor diode

> **Derivation** Derive the expression for the incremental junction capacitance c_j of a non-conducting semiconductor diode, in terms of the applied voltage V_d and the value c_{jo} of c_j for $V_d = 0$.

With A the junction area then

$$c_j = \varepsilon A/W_d$$

so we require W_d in terms of V_d, derived as follows.

With $\psi_p = -E_{max} W_p/2$ as in [5.26]

and $E_{max} = \varepsilon^{-1} qPW_p$ as in [5.23]

then $\psi_p = -\varepsilon^{-1} qPW_p^2/2$

Similarly $\psi_n = -\varepsilon^{-1} qNW_n^2/2$

With also $W_p P = W_n N$, as in [5.25],

and $\psi_d = \psi_p + \psi_n$

then $W_d = K\psi_d^{1/2}$

where
$$K = \sqrt{-\frac{2\varepsilon}{q}\left(\frac{N+P}{NP}\right)} \qquad (5.30)$$

giving, with c_j as above,

$$c_j = (\varepsilon A/K)\psi_d^{-1/2} \quad \text{and} \quad c_{jo} = (\varepsilon A/K)\psi_o^{-1/2}.$$

With $\psi_d = \psi_o - V_d$ then, as required,

$$c_j = c_{jo}(1 - V_d/\psi_o)^{-1/2} \qquad (5.31)$$

> **Calculation** the junction capacitance c_j of a signal diode for $V_d = 0$ is 30pF. The internal junction potential $\psi_o = 900mV$. Calculate the values of c_j for $V_d = +600mV$ and $-10V$.

$V_d = 600mV$. With c_j as in [5.31] then **$C_j = 52.0pF$.**

$V_d = -10V$. Similarly **$C_j = 8.62pF$.**

6

Single-stage common-emitter amplifier

Summary

Fig. 6.1(d) shows an elementary version of the single-stage common-emitter bipolar-transistor amplifier, based on the NPN transistor, and as suited for a simple laboratory demonstration of the operation.

Here (b) shows the transfer function relating the output V_o to the input V_b. Then (a), (b) and (c) show the way the transfer function converts the input waveform in (a) to the output waveform in (c). Here the magnitude of the signal component of V_o is considerably greater than that of V_b, making the circuit a voltage amplifier. The bias E_b is needed to find the appropriate position on the transfer function.

The transfer function and general operation differ from that of the op-amp in a number of ways.

- The input is non-differential, as in Fig. 2.1(a), rather than differential as for the op-amp in Fig. 2.1(b).
- The voltage gain is much lower, with the sign reversal ≈ -200.
- Significant current is drawn into the signal input terminal (the base), giving a much lower input impedance.
- The output impedance is much higher.
- The transfer function is exponential rather than linear.
- The output voltage range is unipolar (only positive).
- The cut-off frequency is much higher, with the corresponding response time therefore much lower.

For the transistor shown, E is the 'emitter', B is the 'base' and C the 'collector'. With B here the input terminal and C the output then it is E which is the grounded terminal 'common' to input and output circuit, as for the non-differential amplifier in Fig. 2.1(a).

Other configurations exist, the 'common-collector' and the 'common-base' but these are left for the next chapter. This chapter constitutes a fairly detailed look at the operation of the bipolar transistor generally, and its use in the common-emitter configuration.

Numeric calculations are cumulative, with the values set bold to the right-hand margin.

The last section gives experimental work covering the material presented, and confirming the calculated values. Some more detailed calculations are included.

According to facilities available to the reader, the SPICE simulation of the experimental work is an effective alternative.

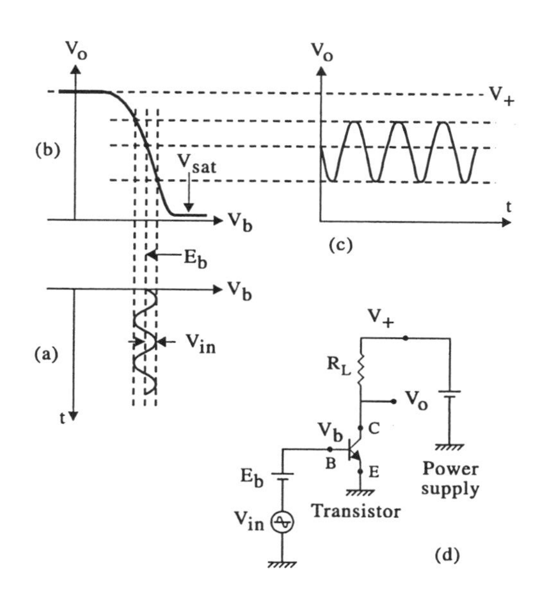

Fig. 6.1 Single-stage NPN bipolar-transistor common-emitter amplifier.
(a) Input waveform.
(b) Transfer function.
(c) Output waveform.
(d) Circuit diagram.

6.1. Bipolar transistor

There are two types of bipolar transistor, the NPN and the PNP. Fig. 6.2(a) shows the NPN device as produced by the normal planar fabrication process, while (b) gives a simplified view. For the PNP the N- and P-type regions are simply transposed.

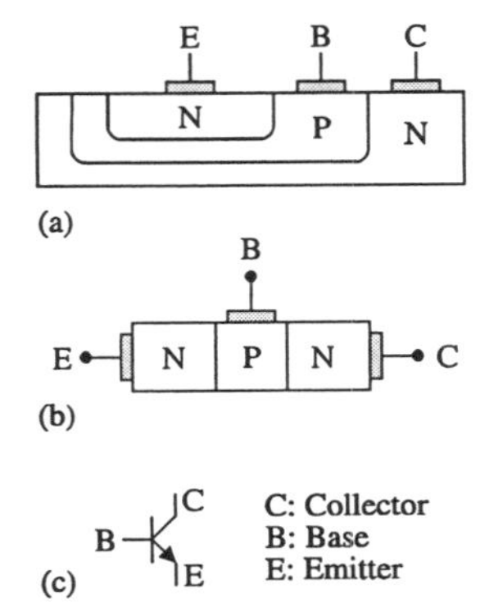

Fig. 6.2 NPN bipolar transistor.
(a) Planar construction.
(b) Essential structure.
(c) Circuit symbol.

Thick-base transistor

For the transistor to give voltage gain, the width of the base region must be small compared with that of the recombination region. However, we start by considering the carrier flow in Fig. 6.3(a), where the reverse is the case. The device is then much as for the two back-to-back diodes in (b). With the right-hand diode reverse-biased then no current flows, while for the forward-biased left-hand diode the flow is as shown.

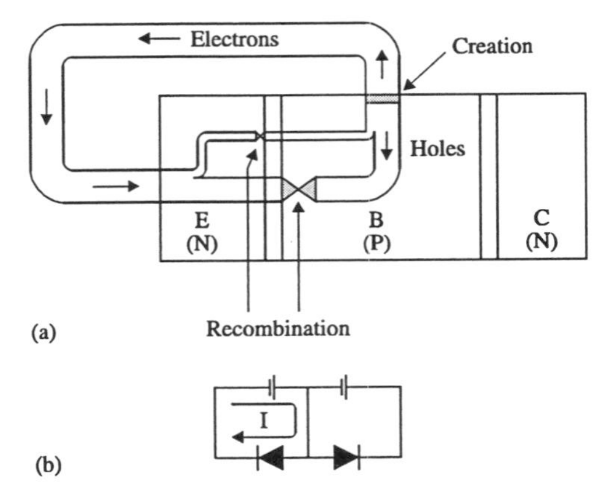

Fig. 6.3 Hole and electron flow for thick-base NPN transistor with forward BE bias and reverse BC bias.
(a) Carrier flow.
(b) Equivalent two-diode circuit.

The reason for the lower flow in (a) being the larger is that, for reasons given below, the emitter doping is made many times higher than that of the base.

The process requires the generation of holes and electrons at the base connection as shown. This is ensured by making the junction between the metal connecting pads and the semiconductor graded, rather than relatively abrupt as for the BE and BC junctions.

Thin-base transistor

With no collector current for the thick-base transistor, and thus clearly no voltage gain, Fig. 6.4 shows the thin-base equivalent, with the base width exaggerated the show the carrier flow.

The important difference is that, with the base width small compared with the recombination length, then most of the electrons injected into the base reach the collector before recombining. Here the polarity of the reverse-biased depletion layer is such as to rapidly sweep the arriving electrons across the layer and into the collector region.

With these the majority of the electrons injected, then the main electron flow becomes from emitter to collector, and round back to the emitter via the external CE circuit, rather than from emitter to base and back round the external BE circuit as previously.

There does remain the small base current I_b shown, partly due to the holes injected into the emitter from the base, and partly due to the small fraction of electrons injected by the emitter that recombine in the base. Until Section 6.4 I_b will be ignored.

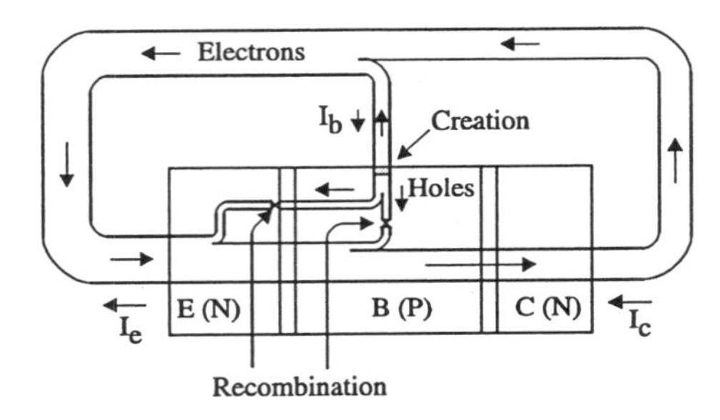

Fig. 6.4 Hole and electron flow for NPN transistor of Fig. 6.3 but with thin base.

Independence of collector voltage

As long as the CB junction remains reverse-biased then V_c has very little effect on I_c. Electrons reaching the edge of the depletion layer meet a downward rather than an upward step. The probability of falling over the edge of a cliff due to some process at the top is independent of the extent of the drop.

The process at the top is the diffusion of the electrons across the base, proportional to the concentration gradient and assisted by a weak electric field across the base. We shall later see that this independence of V_{cb} is the feature of the transistor needed for it to provide voltage gain.

Ideal dependent current source

The above mechanism makes the transistor approximate to an 'ideal dependent current source', as shown in Fig. 6.5(a). The source is 'ideal' in that I_c is independent of the collector-emitter voltage V_{ce}, and is 'dependent' in that I_c is controlled by V_{be}.

Variation of collector current with base voltage

The dotted diode (not usually shown) is a reminder that I_c is the current that would flow in the BE diode were it normal, as for the thick-base transistor. Now, instead of flowing in the external BE loop as I_b, the current flows in the external CE loop as I_c. Thus the $I_c - V_{be}$ relation in (b) corresponds to the $I_d - V_d$ relation for the diode in [5.2], to give

$$I_c = I_{co}[\exp(qV_{be}/kT) - 1] \qquad (6.1)$$

As for the diode, I_{co} is independent of V_{be}. Also for normal forward bias $I_c >> I_{co}$ giving

$$I_c \approx I_{co} \exp(qV_{be}/kT) \qquad (6.2)$$

Transfer function

The shape of the transfer function of Fig. 6.1(b) is thus explained. For $V_b=0$ and below then $I_c=0$ giving $V_o=V_+$. This continues until $V_b \approx 600$mV, after which the current I_c in Fig. 6.5(b) becomes significant, causing the proportional voltage drop I_cR_L across R_L. This continues until V_o reaches the 'saturation' value V_{sat}. This is a little past the point at which the CB junction ceases to be reverse-biased, at which much of the above picture breaks down.

6.2 Voltage gain

We now derive the expression for the incremental voltage gain

$$a_{ce} = dV_o/dV_b \qquad (6.3)$$

of the CE stage in Fig. 6.1.

Starting values

For the following series of calculations

$$\mathbf{V_+=10V, \quad I_c=1mA, \quad KT/q=25mV.}$$

Mutual conductance

Fig. 6.5(b) shows the variation of I_c with V_{be} as a further form of transfer function, giving the incremental

$$i_c = g_m v_{be} \qquad .4)$$

where the mutual conductance

$$g_m = dI_c/dV_{be} \qquad (6.5)$$

With I_c as in [6.2] then

$$g_m = I_c/(kT/q) \qquad (6.6)$$

giving $\mathbf{g_m = 40mA/V = 1/(25\Omega)}$.

With [6.4] expressed diagramatically as in (c) then this becomes the first incremental circuit model of the transistor.

Mid-point bias

With the supply V_+ and the load R_L as in (a), the large-signal

$$V_o = V_+ - I_cR_L \qquad (6.7)$$

For the 'mid-point' $V_o=V_+/2$ then $\mathbf{R_L=5k\Omega}$.

Incremental circuit model

With $dV_o/dI_c = -R_L$ the incremental

$$v_o = -i_cR_L \qquad (6.8)$$

With this in (d) the incremental circuit model for the load and supply then (c) and (d) constitute the full incremental circuit for the voltage amplifier, giving

$$a_{ce} = -g_mR_L \qquad (6.9)$$

and thus $\mathbf{a_{ce} = -200}$.

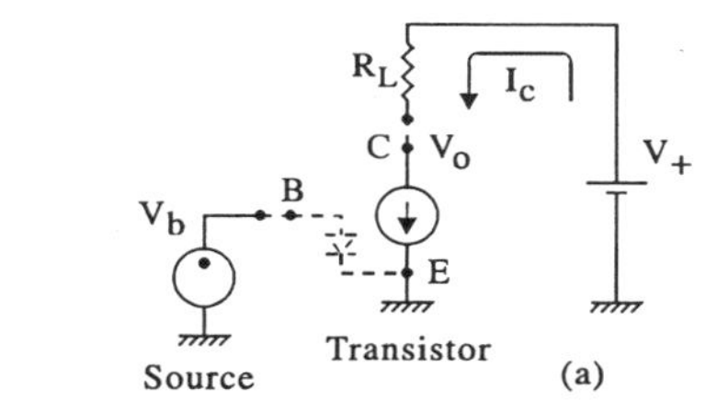

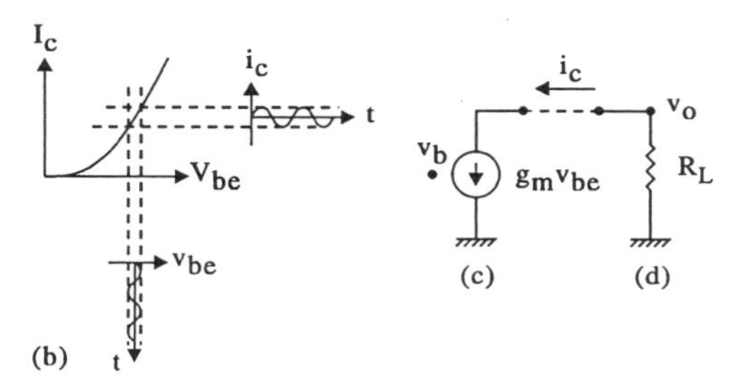

Fig. 6.5 Large-signal and incremental circuits for voltage amplifier of Fig. 6.1.
(a) Large-signal circuit diagram, showing transistor as ideal dependent current source of [6.2].
(b) Transistor $I_c - V_{be}$ relation.
(c) Incremental circuit model for transistor in (a).
(d) Incremental circuit model for supply V_+ and load R_L in (a).

Independence of I_c

With g_m as in [6.6] and R_L as in [6.9],

$$a_{ce}=-(V_+/2)/(kT/q) \qquad (6.10)$$

which is independent of I_c, since $g_m \propto I_c$ and $R_L \propto 1/I_c$. With kT/q the thermal energy of the electron, and V_+ the potential electron energy for the supply, then a_{ce} is the ratio of the two.

6.3 Collector output resistance

Early effect

So far it has been assumed that a change in the collector voltage V_c has no effect on the collector current I_c. In fact there is a small dependence, due the the Early effect.

- As V_c is increased this increase is imposed on the BC depletion layer, increasing its width.
- With the base width accordingly reduced, the proportion of electrons reaching the collector from the emitter is increased, thus increasing I_c.

Fig. 6.6 shows the variation of I_c with V_c for a number of values of V_{be}, with the typical Early voltage **V_{ea}=70V.**

Saturation voltage

The transition point in each characteristic corresponds to the saturation voltage V_{sat} in Fig. 6.1.

With $V_{be} \approx 600$mV, it is when V_c is equal to this value that the BC junction ceases to be reverse-biased. However, for the resulting current to be comparable with that for the initial I_c then the voltage across the junction must approach that across the BE junction to within a few times kT/q. Here typically **V_{sat}~150mV.**

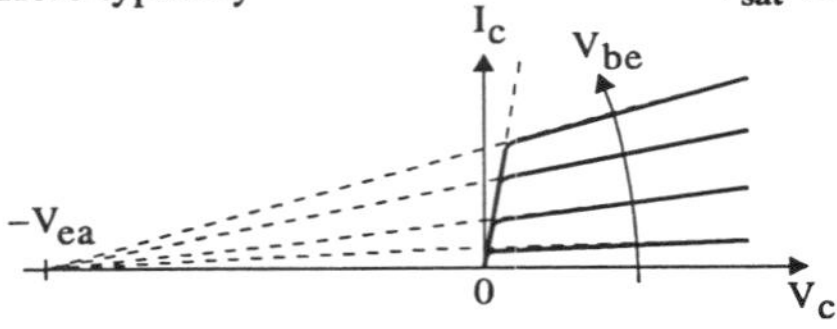

Fig. 6.6 Transistor output ($V_c - I_c$) curves for various values of V_{be}, showing Early voltage V_{ea}.

Incremental model

Previously the incremental $i_c=g_m v_{be}$ where $g_m=dI_c/dV_{be}$, but now

$$i_c= v_{be}g_m+v_{ce}/r_o \qquad (6.11)$$

where

$$g_m=(\partial I_c/\partial V_{be}) \qquad (6.12)$$

and

$$r_o=(\partial I_c/\partial V_{ce})^{-1} \qquad (6.13)$$

This extends the incremental transistor circuit model of Fig. 6.5(c) as in Fig. 6.7(a). Also the amplifier model of Fig. 6.5(c,d) is extended to that in Fig. 6.7(b).

Voltage gain

From the extended model

$$a_{ce}=-gm(R_L r_o)/(R_L+r_o) \qquad (6.14)$$

With $V_{ea} >> V_c$, and I_c and V_c as in Fig. 6.6 then

$$r_o \approx V_{ea}/I_c \qquad (6.15)$$

giving **r_o=70kΩ.**

With this large compared with R_L=5kΩ then the above calculated a_{ce}=−200 is not much reduced in magnitude.

Potential voltage gain

However, for $R_L >> r_o$

$$a_{ce} \approx -g_m r_o \qquad (6.16)$$

With $g_m=I_c/(kT/q)$

as in [6.6] and r_o as in [6.15] then

$$a_{ce} \approx -V_{ea}/(kT/q) \qquad (6.17)$$

giving **a_{ce}=−2800**

a large increase from the previous a_{ce}=−200.

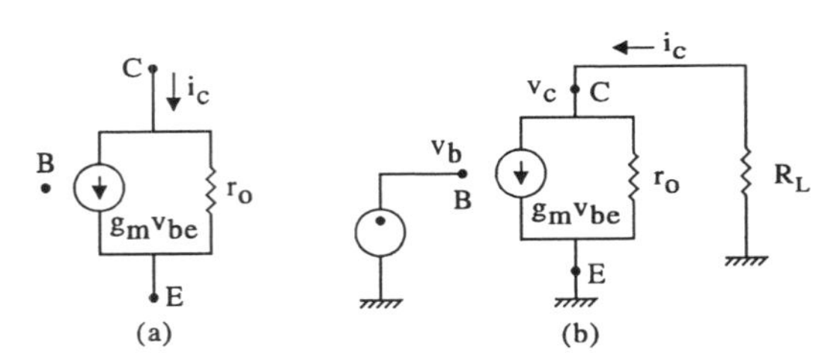

Fig. 6.7 Incremental model for transistor collector output impedance r_o.
(a) Transistor model.
(b) Incorporation into amplifier circuit.

Active load

The above ideal can be approached by using the active PNP load as in Fig. 6.8. Here r_L in [6.14] becomes the r_o of the active load. For the two equal then

$$a_{ce}=-g_m r_o/2$$

giving **a_{ce}=−1400.**

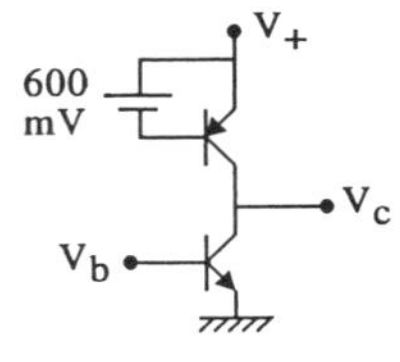

Fig. 6.8 Replacement of R_L in transistor amplifier by transistor to obtain increased voltage gain.

Criterion for voltage gain

With $r_o=(\partial I_c/\partial V_c)^{-1}$ as in [6.13], $g_m=\partial I_c/\partial V_b$ as in [6.12], and with $a_{ce}=-g_m r_o$ as in [6.16] for the potential voltage gain then

$$a_{ce} = -\frac{\partial I_c / \partial V_{be}}{\partial I_c / \partial V_{ce}} \tag{6.18}$$

Thus the feature of the transistor needed for voltage gain is $\partial I_c/\partial V_{be} >> \partial I_c/\partial V_{ce}$, as for the account of the thin-base transistor in Section 6.1.

With a_{ce} as in [6.17] then the ratio of I_c dependencies in [6.18] is equal to $V_{ea} \sim 70V$ divided by the electron thermal energy in electron-volts, $kT/q=25mV$.

6.4 Base current

In Fig. 6.4 the transistor base current I_b has two components, I_{b1} and I_{b2}. I_{b1} is due to the small fraction of electrons injected into the base from the emitter which recombine in the base region rather than reaching the collector. This fraction is reduced as far as possible by making the base thin. With

$$\beta_1=I_{b1}/I_c \tag{6.19}$$

then typically **$\beta_1=400$.**

I_{b2} is due to the holes injected into the emitter from the base. This component, being independent of the width of the base, is made relatively small by making the N-type doping of the emitter much higher than the P-type doping of the base. With

$$\beta_1=I_{b1}/I_c \tag{6.20}$$

then typically also **$\beta_2=400$.**

With the total transistor 'current gain'

$$\beta=I_c/I_b \tag{6.21}$$

and with $I_b=I_{b1}+I_{b2}$ then typically **$\beta=200$.**

However, there is usually a wide production spread in the value. For the BC182 currently being investigated $\beta=100$ to 480.

I_b–V_{be} relation

With $$I_c \approx I_{co}\exp(qV_{be}/kT)$$

as in [6.2] and $I_b=\beta I_c$ in [6.21] then

$$I_b \approx I_{bo}\exp(qV_{be}/kT) \tag{6.22}$$

with the constant

$$I_{bo}= I_{co}/\beta \tag{6.23}$$

[6.22] is plotted in Fig. 6.9(a), with V_b written for V_{be} since $V_e=0$.

Incremental input resistance

As for the variation of I_c with V_{be},

$$i_b=v_b(dI_b/dV_b) \tag{6.24}$$

Here dV_{be}/dI_b is termed the 'incremental input resistance' of the transistor which, to accord with the hybrid-pi model below, we term r_π. Thus

$$i_b=v_{be}/r_\pi \tag{6.25}$$

where

$$r_\pi=dV_{be}/dI_b \tag{6.26}$$

giving

$$r_\pi=(kT/q)/I_b \tag{6.27}$$

or with $\beta=I_c/I_b$ as in [6.21]

$$r_\pi=\beta(kT/q)/I_c \tag{6.28}$$

and thus **$I_b=50\mu A$, $r_\pi=5k\Omega$.**

With $$g_m=I_c/(kT/q)$$

as in [6.6] then

$$\beta=g_m r_\pi \tag{6.29}$$

From [6.28] then $r_\pi \propto I_c^{-1}$, as for $1/g_m$ in [6.6].

Incremental circuit model

For the incremental signal source voltage e_s,

$$v_b=e_s - i_b R_s \tag{6.30}$$

which is modelled in Fig. 6.9(b). With [6.25] as in (c) then

$$v_b=e_s r_\pi/(r_\pi+R_s) \tag{6.31}$$

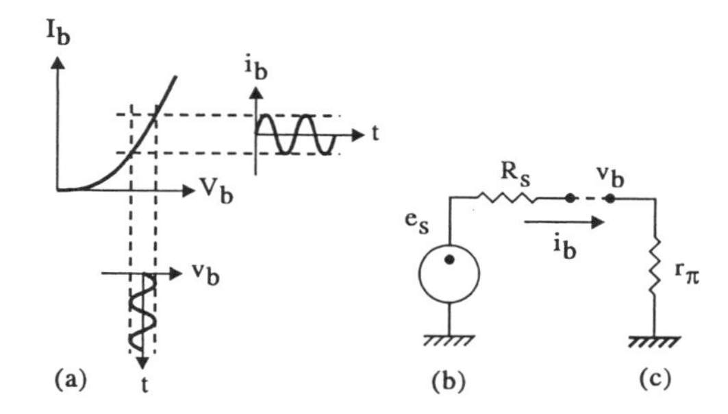

Fig. 6.9 Incremental model for transistor input.
(a) Transistor I_b–V_b relation and small-signal waveforms.
(b) Incremental model for signal source.
(c) Incremental model for transistor input.

Voltage gain

With r_π added, the amplifier incremental model of Fig. 6.7(b) becomes as in Fig. 6.10. With $a_{ce}=v_c/e_s$ then

$$a_{ce} = -\frac{r_\pi}{r_\pi + R_s} g_m \frac{R_L r_o}{R_L + r_o} \tag{6.32}$$

This is [6.14] for Fig. 6.7 with the input loss factor $r_\pi/(r_\pi+R_s)$ included.

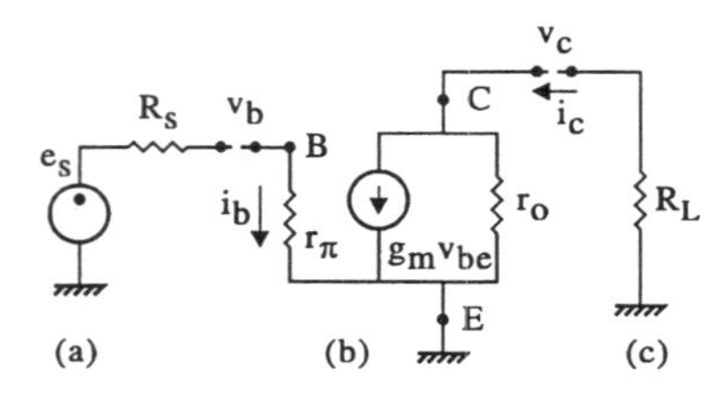

Fig. 6.10 Amplifier incremental circuit model including r_π.
(a) Signal source.
(b) Transistor.
(c) Load resistor.

Current drive

While the amplifier V_o/V_b transfer function of Fig. 6.1(b) is exponential and therefore basically non-linear, $I_c=\beta I_b$ as in [6.21] is fully linear. Thus if the signal source can be made to approach an ideal independent current source, rather than the low-impedance voltage source of the lab signal generator, then the resulting current drive should give linear operation over the full range of V_o, from V_+ to V_{sat}.

Experimental work

Exercise 6.1 gives extensive experimental work on the above low-frequency characteristics of the single-stage common-emitter transistor amplifier.

6.5 Hybrid-π transistor model

Fig. 6.11(a) shows the transistor capacitances c_π and c_μ, and also the 'spreading resistance' r_x resulting from the need to make the base thin. When these components are added, the incremental transistor model of Fig. 6.10(b) becomes the 'hybrid-π' model of Fig. 6.11(b).

Spreading resistance

With typically $\quad$ $\mathbf{r_x=150\Omega}$

then $\quad r_x << (r_\pi=5k\Omega)$

which is why r_x has so far been ignored. However, with

$$r_\pi=\beta(kT/q)/I_c$$

as in [6.28] then, for the higher current densities typical of power transistors, the significance of r_x is increased. r_x also increases in importance with signal frequency, as the impedances of c_π and c_μ decrease.

Diode capacitances

c_μ is for the BC depletion layer, with typically $\quad$ $\mathbf{c_\mu=5pF.}$

c_π has the two components $c_{\pi d}$ and $c_{\pi s}$, with $c_{\pi s}$ the base 'charge-storage' component and $c_{\pi d}$ the base-emitter depletion layer capacitance, where typically $\quad$ $\mathbf{c_{\pi d}=10pF.}$

As for the diode, $c_{\pi s} \propto I_c$, while c_μ and $c_{\pi d}$ are independent of I_c. However, both c_μ and $c_{\pi d}$ are somewhat dependent on the junction voltage, again as for the diode. Thus all of the capacitances are incremental, making the lower case 'c' appropriate.

With the area of c_μ in (a) larger than that of $c_{\pi d}$, the reason why $c_\mu < c_{\pi d}$ is that the reverse-bias of the base-collector junction makes the width of the associated depletion-layer the larger.

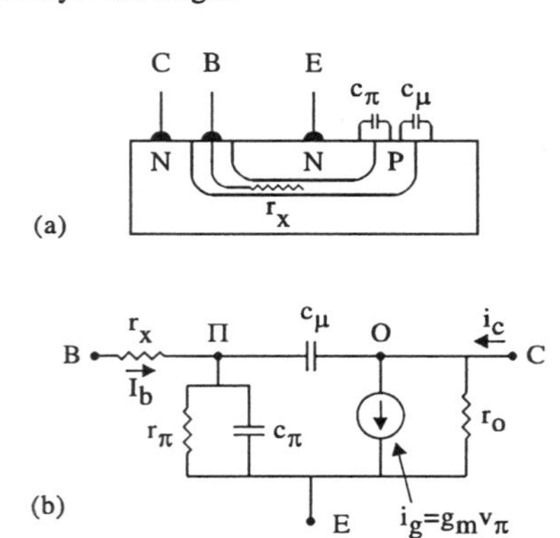

Fig. 6.11 Hybrid-π incremental model of transistor.
(a) Planar transistor construction showing spreading resistance r_x and capacitances c_p and c_m.
(b) Circuit model derived by adding components in (a) to model of Fig. 6.13(b).

Base charge-storage capacitance

Fig. 6.12(a) shows the distribution of electron concentration ρ in the base for normal operation. This is for the carrier movement entirely diffusion controlled, for which the carrier velocity is proportional to concentration gradient. The diagram assumes all of the electrons injected into the base to reach the emitter, making the current over the base, and therefore the slope of the concentration gradient, constant.

Charging

Despite the Q_{s-} shown for the passing electrons, the base region is electrically neutral, with the compensating Q_{s+} provided by an equal increase in the base hole concentration, and with the holes provided by an additional transient component of I_b. Only after the Q_{s-} and Q_{s+} have been established does most of I_e flow across the base to constitute I_c. Prior to this the current flow which establishes the charges is in the base-emitter loop.

Discharging

Similarly, when V_b is reduced to zero the carriers comprising Q_- flow back over the base-emitter junction and ultimately out via the emitter terminal. Simultaneously the charge-balancing Q_{s+} carriers are removed at the base terminal, to complete the discharging path, once more around the base-emitter loop. Only when this is complete does I_c cease. Thus the effect of the stored charges Q_{s+} and Q_{s-} is much as for those on the opposite plates of a capacitor. However, there is an important difference. For the normal capacitor $Q=CV$, a linear relation. For the transistor stored-charge the voltage–charge relation is highly non-linear, as will now be shown.

Base transit time

The time T_τ taken for an electron to traverse the base is independent of the electron concentration, with typically

$$T_\tau = 1\text{ns}.$$

With $$I_c = -Q_{s-}/T_\tau \qquad (6.33)$$

with $$I_c \approx I_{co}\exp(qV_\pi/kT) \qquad (6.34)$$

as in [6.2], and with $Q_{s+} = -Q_{s-}$

then $$Q_{s+} = T_\tau I_{co}\exp(qV_\pi/kT) \qquad (6.35)$$

as plotted in Fig. 6.12(b).

Incremental model

Fig. 6.12 also shows the waveform for the incremental q_{s+} when a low amplitude sine wave v_π is added to the V_π bias. Here

$$q_{s+} = v_\pi c_{\pi s} \qquad (6.36)$$

where

$$c_{\pi s} = dQ_{s+}/dV_\pi \qquad (6.37)$$

Differentiating [6.35] then

$$c_{\pi s} = T_\tau I_c/(kT/q) \qquad (6.38)$$

giving, for $I_c = 1\text{mA}$, $\quad c_{\pi s} = 40\text{pF},$

and also confirming that $c_{\pi s} \propto I_c$.

Here $\quad (c_{\pi s} = 40\text{pF}) > (c_{\pi d} = 10\text{pF})$

but with $c_{\pi d}$ independent of I_c this can be reversed by reducing I_c.

With the calculation of the values for the hybrid-π model thus completed, the results are summarised in Table 6.1.

r_x	r_π	g_m	$c_{\pi d}$	$c_{\pi s}$	c_μ	r_o
150Ω	5kΩ	1/25Ω	10pF	40pF	5pF	70kΩ

Table 6.1 Component values for hybrid-π transistor model of Fig. 6.11(b) for $I_c = 1\text{mA}$, $V_o = 5\text{V}$ and $\beta = 200$.

Unity gain frequency

Where two or more stages are cascaded then a value of concern is the signal frequency f_u at which the gain per stage falls to unity.

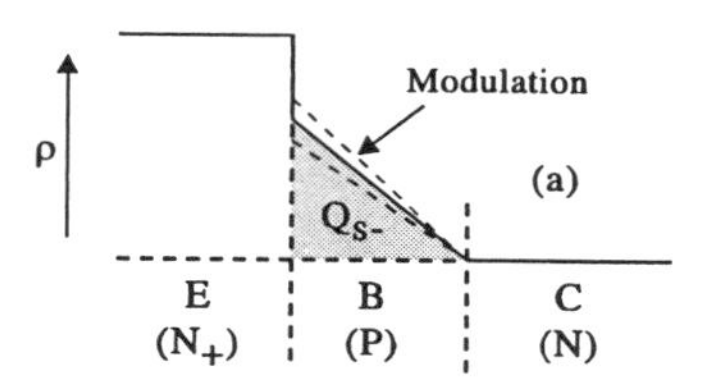

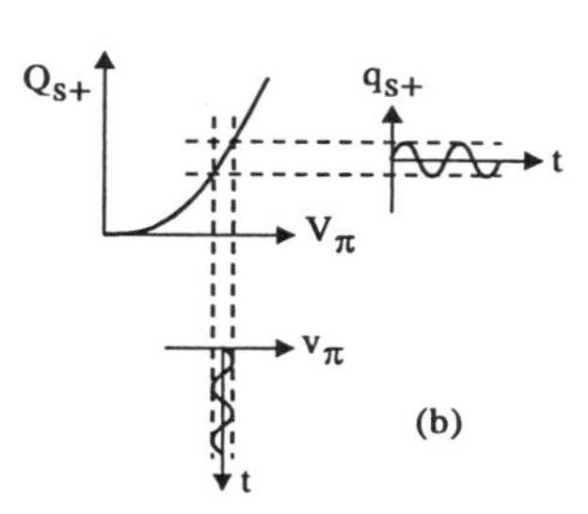

Fig. 6.12 Stored base charge Q_{s-} showing modulation due to v_p in (b).
(a) Electron concentration ρ for transistor carrying current I_c.
(b) Modulation of Q_{s+} by v_p.

Grouping the last half of one stage with the first half of the second, and with $x_\pi \ll (r_o$ and $r_\pi)$ at this frequency then r_o and r_π may be neglected. Also with c_π somewhat greater than c_μ, and now the feedback of little effect, for an initial estimate we ignore c_μ. With these approximations and with β' the current gain per stage

$$\beta' \approx g_m x_\pi$$

giving $$\beta' = g_m/(2j\pi f c_\pi) \qquad (6.39)$$

Thus $$f_u = g_m/(2\pi c_\pi) \qquad (6.40)$$

for which $\quad f_u = 127\text{MHz}.$

Base transit-time cut-off

With T_τ the base transit time, the associated cut-off frequency

$$f_\tau = 1/(2\pi T_\tau) \qquad (6.41)$$

With $$c_{\pi s} = T_\tau I_c/(kT/q) \qquad \text{as in [6.38]}$$

and $$g_m = I_c/(kT/q) \qquad \text{as in [6.6]}$$

then $$f_\tau = g_m/(2\pi c_{\pi s}) \qquad (6.42)$$

giving $\quad f_\tau = 159\text{MHz}.$

Thus f_τ is the value of f_u for $c_{\pi d} = 0$. With f_u the parameter of interest to the designer, beware that manufacturers frequently list the specified value of f_u as 'f_τ'.

π-network

While the network $r_\pi // c_\pi$ is conspicuous in the hybrid-π model, it is shown in the next section that the effect of $c_\pi = 50$pF tends to be dominated by that of the relatively small $c_\mu = 5$pF, much increased by the associated negative feedback.

However, the time constant $c_\pi r_\pi$ has some physical significance, and there are some circuits, in particular the emitter-follower of the next section, where the negative feedback is absent. Here $c_\pi r_\pi$ becomes the time constant of greatest importance.

Dependence of $c_\pi r_\pi$ on collector current

With f_π the transition frequency for the π-network

then $$f_\pi = 1/2\pi r_\pi (c_{\pi s} + c_{\pi d}) \quad (6.43)$$

With $$f_{\pi s} = 1/2\pi r_\pi c_{\pi s} \quad (6.44)$$

$$r_\pi \propto 1/I_c \quad \text{as in [6.28]}$$

$$c_{\pi s} \propto I_c \quad \text{as in [6.38]}$$

$c_{\pi d}$ independent of I_c,

and with $I_{cx} = (I_c \text{ for } c_{\pi s} = c_{\pi d})$

then $f_{\pi s}$ varies with I_c as in Fig. 6.13.

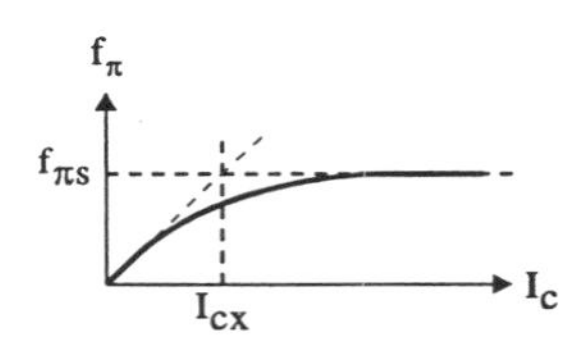

Fig. 6.13 Variation of $f_\pi = 1/2\pi r_\pi c_\pi$ with I_c.

From Exercise 6.2 $\quad$ **$I_{cx} = 250\mu A$, $f_{\pi s} = 795$kHz,**

and for $I_c = 1$mA $\quad$ **$f_\pi = 637$kHz.**

Base recombination time

With $r_\pi \propto 1/I_c$, and $c_{\pi s} \propto I_c$ as above, then $r_\pi c_{\pi s}$ is independent of I_c. As will now be shown for the NPN transistor, this is because $r_\pi c_{\pi s}$ is closely related to the base recombination time T_r.

As in Section 6.4, I_b has two components, I_{b1} and I_{b2}. Of these, I_{b1} results from recombination in the base of a small proportion of the electrons in transit from emitter to collector, while I_{b2} represents the flow of holes from base to emitter. Here

$$\beta_1 = I_{b1}/I_c \quad \text{and} \quad \beta_2 = I_{b2}/I_c.$$

With T_r and T_τ the base recombination and transit times

then $$\beta_1 = T_\tau / T_r.$$

With $$r_\pi = \beta(kT/q)/I_c \quad \text{as in [6.28]}$$

and $$c_{\pi s} = T_\tau I_c/(kT/q) \quad \text{as in [6.38]}$$

then $$r_\pi c_{\pi s} = \beta T_\tau$$

Apart from I_{b2} and the associated β_2 then

$$r_\pi c_{\pi s} = T_r.$$

Interruption of base connection

The two views of $c_\pi r_{\pi s}$, electrical and physical, can be compared for the effect of breaking the base connection while I_c is flowing.

Electrically, when the lead is broken then $c_{\pi s}$ discharges through r_π, taking the time $r_\pi c_{\pi s}$ do so, reducing I_c to zero in the process. Does $c_\pi r_{\pi s}$ change during this decay?

Physically, the broken lead provides no flow of holes to match the electrons in the base lost by recombination. Thus the carriers representing both Q_{s+} and Q_{s-} cease to be mobile, taking the recombination time T_r to do so.

Hole diffusion

With Q_{s+} equal to the Q_{s-} in Fig. 6.12, there is an equal concentration-gradient for the holes in the base. Why does this not cause a hole current of equal magnitude, cancelling the effect of the electron flow and making $I_c = 0$?

With the hole concentration much the greater, the effect of the weak electric field in the base is sufficient to cancel the hole diffusion current, while having relatively little effect on the electron current.

Charge control model

As noted previously, the higher current density for the power transistor considerably reduces r_π relative to r_x, tending to linearise the voltage input-output relation. The charge control model shows that this continues to be the case for VHF power transistors where usually $x_\pi < r_\pi$.

With $$Q_{s-} = -I_c T_\tau \quad (6.45)$$

as in [6.33] for Q_{s-} in Fig. 6.12(a), and with $Q_{s+} = -Q_{s-}$

then $$I_c = Q_{s+}/T_\tau.$$

This is the highly linear charge control relation, where Q_{s+} rather than the input voltage V_{in} is seen as the factor controlling I_c.

With Q_{s+} the charge transferred in and out of the base via the base terminal then

$$Q_{s+} = \int I_b dt$$

giving $$I_c = T_\tau^{-1} \int I_b dt \quad (6.46)$$

Despite being integral, this too is a fully linear relation. With V_{in} the input voltage, and to the extent that

$$(R_s + r_x) >> (r_\pi // x_\pi)$$

then $I_b \propto V_{in}$ giving the equally linear $I_c \propto \int V_{in}$.

6.6. Frequency response of single-stage C-E amplifier

The expression for the amplifier frequency response will now be derived from the incremental circuit of Fig. 6.14, with the result plotted in Fig. 6.15(a).

Fig. 6.14 is the incremental voltage-amplifier circuit of Fig. 6.10 with the transistor represented by the hybrid-π model of Fig. 6.11(b), and with the calculated values of Table 6.1 added for convenience.

With normally $r_o >> R_L$ then r_o has been omitted. Otherwise R_L can be replaced by $R_L//r_o$ in the final expression.

Similarly, with the signal source impedance R_s omitted, this can be added to r_x in the final expression.

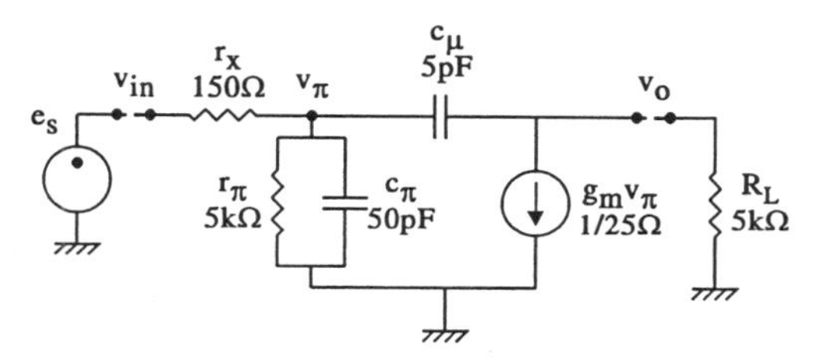

Fig. 6.14 Small-signal model for single-stage bipolar transistor amplifier using hybrid-π model for transistor.

For the nodal analysis there are two nodes, making the analysis a little tedious. With the general method covered in Chapter 1, this is the first case where frequency-dependent elements are present. Thus the analysis will be described in detail, including the necessary final approximations. From the diagram, the nodal equations become

Π-node $$v_\pi y_{\pi\pi} - v_o y_{\pi o} = e_s g_x \quad (6.47)$$

O-node $$v_\pi y_{\pi o} - v_o y_{oo} = 0 \quad (6.48)$$

where $$y_{\pi\pi} = g_x + g_\pi + j\omega(c_\pi + c_\mu) \quad (6.49)$$

$$y_{\pi o} = j\omega c_\mu \quad (6.50)$$

$$y_{o\pi} = -g_m + j\omega c_\mu \quad (6.51)$$

$$y_{oo} = G_L + j\omega c_\mu \quad (6.52)$$

Also, as usual,

$$g_x = 1/r_x, \quad g_\pi = 1/r_\pi \quad \text{and} \quad G_L = 1/R_L.$$

[6.47] and [6.48] can then be solved to give either v_π or v_o in terms of e_s, with our interest only in v_o. With the terms in [6.49] to [6.52] ultimately restored, and with

$$a_{ce} = v_o/e_s \quad \text{then}$$

$$a_{ce} = \frac{g_x\left(-g_m + j\omega c_\mu\right)}{(g_x + g_\pi)G_L + j\omega\left\{G_L\left(c_\pi + c_\mu\right) + c_\mu\left(g_x + g_\pi + g_m\right)\right\} - \omega^2 c_\pi c_\mu} \quad (6.53)$$

Low frequency gain

With a_{cez} the zero-frequency value of a_{ce} then

$$a_{cez} = \frac{-g_x g_m}{(g_x + g_\pi)G_L} \quad (6.54)$$

With $g_x = 1/150\Omega$ and $g_\pi = 1/5k\Omega$ then $g_x >> g_\pi$ giving

$$a_{cez} \approx -g_m R_L \quad (6.55)$$

giving $$\mathbf{a_{cez} \approx -200}$$

This is as for a_{ce} in [6.54], but now with the small effect of r_x included. From [6.53] and [6.54]

$$a_{ce} \approx a_{cez} \times \frac{1 + j\omega c_\mu / g_m}{1 + j\omega\left\{\dfrac{G_L(c_\pi + c_\mu) + c_\mu g_m}{(g_x + g_\pi)G_L}\right\} - \omega^2 \dfrac{c_\pi c_\mu}{(g_x + g_\pi)G_L}} \quad (6.56)$$

Transition at f_m

This occurs where the ω term in the denominator exceeds unity. Thus

$$f_m = \frac{1}{2\pi} \times \frac{G_L(g_x + g_\pi)}{G_L(c_\pi + c_\mu) + c_\mu g_m} \quad (6.57)$$

With $r_x << r_\pi$ in Fig. 6.14 and with a_{cez} as in [6.55]

then $$f_m \approx 1/[2\pi r_x\{c_\pi + (1 + |a_{cez}|)\, c_\mu\}] \quad (6.58)$$

From [6.56], for the region between f_m and the next transition

$$|a_{ce}| \propto f^{-1} \quad \text{as plotted.}$$

Transition at f_1

With this where the magnitude of the ω^2 term in the denominator of [6.53] exceeds that of the ω term, then

$$f_1 = \frac{1}{2\pi} \times \frac{G_L(c_\pi + c_\mu) + c_\mu g_m}{c_\pi c_\mu} \quad (6.59)$$

With $$c_\mu g_m >> G_L(c_\pi + c_\mu)$$

then $$f_1 \approx g_m/(2\pi c_\pi).$$

Here $f_1 = f_u$ in [6.40]. Again from [6.56], beyond f_1

$$|a_{ce}| \propto f^{-2} \quad \text{as plotted.}$$

Transition at f_μ

With this the frequency at which the ω term in the numerator exceeds unity then

$$f_\mu = g_m/2\pi c_\mu \quad (6.60)$$

Beyond f_μ then $|a_{ce}|$ reverts to being $\propto f^{-1}$, as plotted.

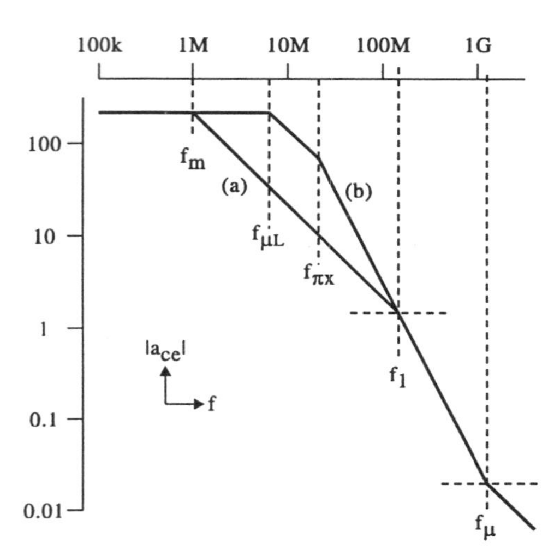

Fig. 6.15 Frequency response for circuit of Fig. 6.14.
(a) With feedback through c_m as normal.
(b) Without feedback through c_m.

With the above relations and the component values in Fig. 6.14

$$\mathbf{f_m = 1.01MHz, \quad f_1 = 127MHz, \quad f_\mu = 1.27GHz,}$$

as for the plot.

Miller effect

With $\quad f_m \approx 1/[2\pi r_x\{c_\pi + (1 + |a_{cez}|)c_\mu\}]$

as in [6.58], and with $c_\pi = 50pF$, $c_\mu = 5pF$ and $|a_{cez}| = 200$ then c_π is negligible. Thus the response is as if the source with r_x had been terminated in the Miller capacitance

$$c_m = (1 + |a_{cez}|)c_\mu \tag{6.61}$$

as in Fig. 6.16(a), for which $\quad \mathbf{c_m \approx 1nF.}$

(b) shows how this occurs. With

$$i_{in} = (v_\pi - v_o)/x_\mu, \quad v_o = a_{cez}v_\pi, \quad \text{and} \quad z_m = v_\pi/i_{in}$$

then $\quad z_m = x_\mu/(1 + |a_{cez}|)$

confirming c_m as in [6.61]

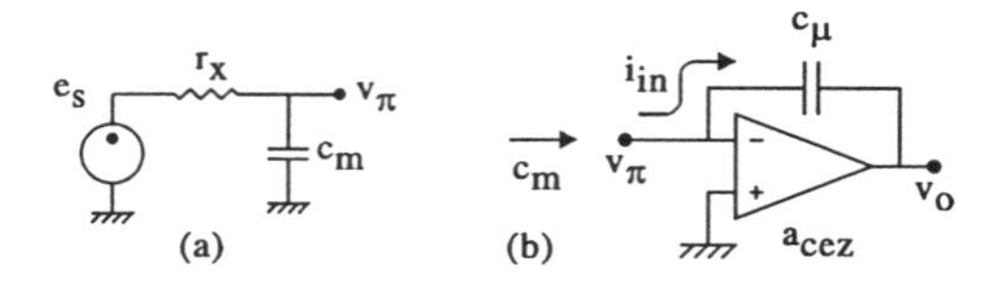

Fig. 6.16 Miller effect.
(a) e_s and r_x in Fig. 6.14 with remainder presenting as Miller capacitance c_m.
(b) Formation of c_m.

Response without feedback

Fig. 6.15(b) shows the frequency response without the feedback through c_μ, as if c_μ came separately across c_π and r_L. The first transition is where $R_L = x_\mu$, giving

$$f_{\mu L} = 1/(2\pi c_\mu R_L) \tag{6.62}$$

and thus $\quad \mathbf{f_{\mu L} = 6.37MHz.}$

With $c_\pi >> c_\mu$ and $r_x << r_\pi$ then the second transition $f_{\pi x}$ is where $r_x = x_\pi$ giving

$$f_{\pi x} = 1/2\pi r_x c_\pi \tag{6.63}$$

and thus $\quad \mathbf{f_{\pi x} = 21.2MHz.}$

Evidently f_m and f_1 are the frequencies between which the loop-gain > 1.

Increased source impedance

With r_x replaced by $R_s + r_x$ in [6.57] for f_m, the approximation

$$c_\mu g_m >> G_L(c_\pi + c_\mu)$$

remains valid, but the condition $g_x >> g_\pi$, now

$$(r_x + R_s) << r_\pi$$

may not. With [6.57] thus modified

$$f_m = \left\{2\pi c_m \frac{r_\pi(R_s + r_x)}{r_\pi + R_s + r_x}\right\}^{-1} \tag{6.64}$$

The simple input circuit of Fig. 6.16(a) then becomes as in Fig. 6.17. From this or [6.64], for the extreme where

$$(r_x + R_s) >> r_\pi$$

then $\quad f_m \approx 1/(2\pi r_\pi c_m) \quad (6.65)$

giving $\quad \mathbf{f_m \approx 31.7kHz}$

considerably lower than the 1MHz in Fig. 6.15(a), for $R_s = 0$.

Exercise 6.3 gives experimental work with further examples on the response-time of the above circuits with low R_s, and Exercise 6.4 for $R_s >> r_\pi$.

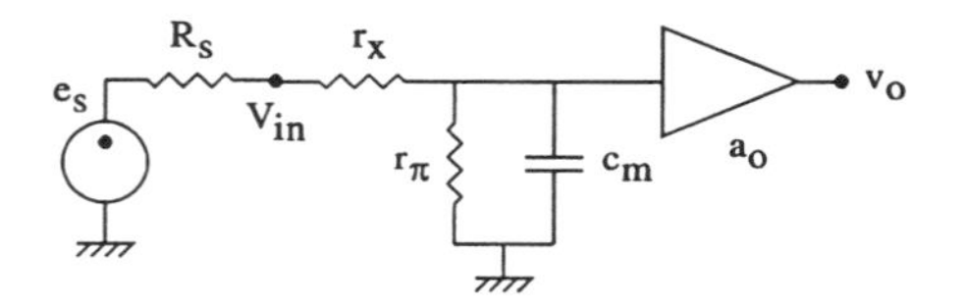

Fig. 6.17 Effect of non-zero signal source resistance R_s upon amplifier input circuit.

6.7 Biasing circuits

Current bias

As shown in Fig. 6.18, current bias is the first of the methods to be described for eliminating the inconvenient floating bias supply E_b in Fig. 6.1. For $\beta=200$, $V_+=10V$ and with $V_{be}\approx 600mV$ then **$R_b \approx 2M\Omega$.**

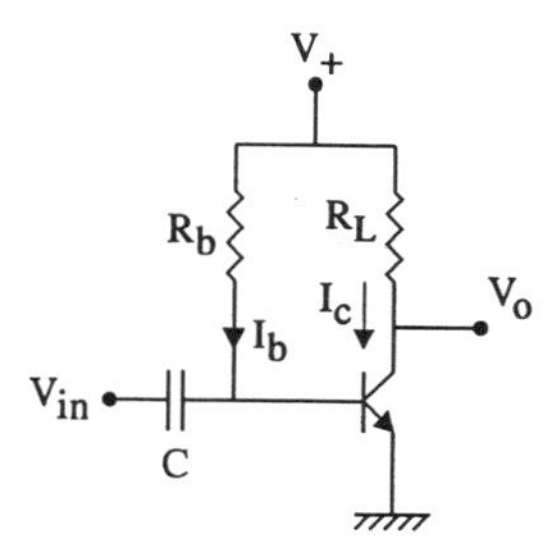

Fig. 6.18 Current bias for AC-coupled amplifier.

The main disadvantage of the method is the extremely wide production spread for β. $100>\beta>480$ for the BC182. Here for a production run a different R_b has to be provided for each circuit. Also the circuit is 'AC-coupled', i.e. not able to amplify a DC signal. However, the circuit remains the most convenient one for the following experimental work.

Voltage bias

Fig. 6.19 shows the original voltage bias circuit of Fig. 6.1.

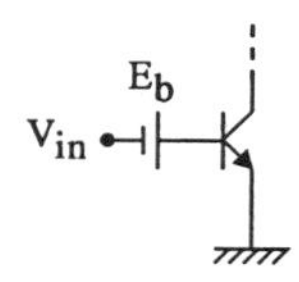

Fig. 6.19 Voltage bias.

Here I_c is independent of β, but unfortunately far too temperature-dependent to be acceptable. With

$$I_c \approx I_{co}\exp(qV_{be}/kT)$$

as in [6.2] then apparently I_c decreases with temperature. However, the reverse is the case, because I_{co} is twice as temperature dependent as the exponential term, and in the opposite direction. Here

$$I_{co} \propto T^3 \exp(-qV_g/kT)$$

(Massobrio and Antognetti, 1993) where V_g is the band-gap voltage (1.2V for S_i), corresponding to the energy needed to detach an electron from its initial location in the lattice (lift a ball from the tray to roll on top of the others) and so create an intrinsic hole-electron pair. With the temperature dependence of the exponential term dominating that of T^3 then

$$I_{co} \approx A\exp(-qV_g/kT) \tag{6.66}$$

with A assumed independent of T. Thus

$$I_c = A\exp\left(\frac{V_{be}-V_g}{kT/q}\right) \tag{6.67}$$

With $V_g=1.2V$ and $V_{be}\approx 600mV$, the reversal is confirmed.

Equivalent change in base-emitter voltage

Because of the exponential variation, it is usual to represent the temperature change ΔT as an equivalent change ΔV_{be} in V_{be}. With I_c as in [6.67] and for constant I_c

$$dV_{be}/dT=-(V_{be}-V_g)/T \tag{6.69}$$

giving **$\Delta V_{be}/\Delta T \approx 2mV/^o$.**

With I_c as in [6.67] then for an increase ΔV_{be} in V_{be}, I_c increases $\times\gamma$ where

$$\gamma=\exp[\Delta V_{be}/(kT/q)] \tag{6.70}$$

For $\Delta T=20^o$, and this equivalent to $\Delta V_{be}=40mV$, then **$\gamma=4.95$**

a totally unacceptable figure which is comparable with the effect of the $\times 480/100$ production spread of β for the current bias circuit.

Feedback bias

Fig. 6.20(a) shows a refinement to the above circuit that much reduces its temperature sensitivity. Here I_c passes through R_e to develop the voltage $V_e=I_cR_e$. With V_e placed in series with V_{in} as applied to the transistor V_{be}, this constitutes negative feedback, thus reducing the sensitivity of I_c to V_{in}, and hence also to temperature.

With the incremental circuit as in Fig. 6.20(b), the single-node (E) analysis gives

$$v_e = \frac{v_{in}(g_m+g_\pi)}{g_m+G_e+g_\pi} \tag{6.71}$$

With $i_c=g_m(v_{in}-v_e)$ and $g_m >> g_\pi$

then $$i_c \approx v_{in}g_m/(1+g_mR_e) \tag{6.72}$$

With the feedback loop gain

$$A_L=-g_mR_e \tag{6.74}$$

and for $g_mR_L >> 1$, then as expected, the sensitivity of I_c to V_{in} is reduced by approximately $\times$(loop gain)$^{-1}$ to give

$$i_c \approx v_{in}/R_e \tag{6.73}$$

instead of $i_c=v_{in}g_m$ without the feedback.

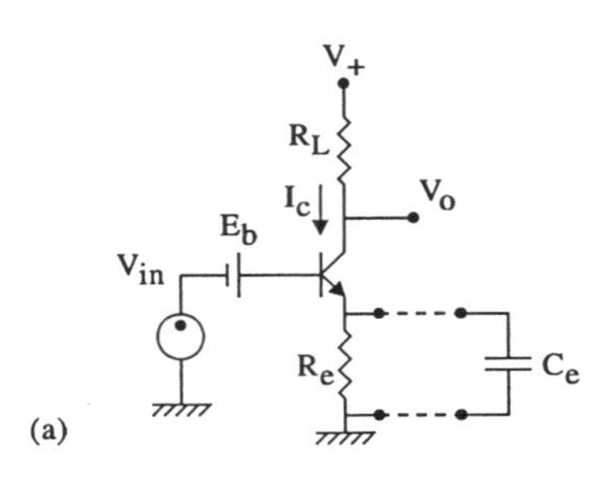

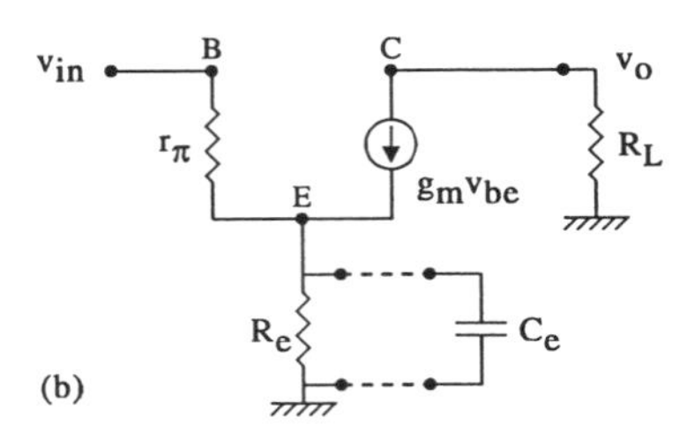

Fig. 6.20 Feedback bias.
(a) Circuit diagram.
(b) Elements of inherent negative-feedback loop.
(c) Incremental equivalent circuit of amplifier in (a).

With $V_e=I_cR_L$

and V_e limiting the V_o range, we limit V_e to 1V.

For $I_c=1$mA then **$R_e=1k\Omega$** giving **$A_L=40$.**

Also for V_o mid-way between V_e and V_+, **$R_L=4.5k\Omega$.**

Thermal sensitivity

Recalling the ≈×5 increase in I_c for a 20° change in temperature for voltage bias without R_e, now the previous 2mV/° temperature sensitivity is reduced ×$1/A_L=1/40$ to give the effective $\Delta V_{be}=1$mV. With $\Delta I_c=g_m\Delta V_{be}$, and with $g_m=40$mS then **$\Delta I_c=40\mu A$,**

an acceptable fraction of $I_c=1$mA. Exercise 6.5 includes experimental work on the thermal sensitivity of each of the above circuits.

Emitter capacitor

As a DC amplifier, the addition of R_e gives no real advantage, since the sensitivity to the DC input signal is reduced by the same factor as for the thermal drift. For DC amplification the op-amp with its long-tailed pair input stage is the best arrangement.

However, as an AC amplifier, the addition of C_e in the diagram will restore the signal gain to the value without feedback, while retaining the thermal stability afforded by R_e. With v_e as in [6.71], and with

$$v_o=-g_mR_Lv_{be}$$

then
$$\frac{v_o}{v_{in}}=\frac{-g_mR_LG_e}{g_m+g_\pi+G_e} \quad (6.75)$$

For present values $g_m >> (g_\pi+g_L)$

giving $v_o/v_{in}\approx -R_L/R_e$ (6.76)

and thus **$v_o/v_{in}\approx -4.5$.**

As usual for a negative-feedback circuit, this is determined by the resistors rather than the active device. Also for $R_e=0$, [6.75] reduces to the no-feedback value

$$v_o/v_{in}=-g_mR_L$$

Frequency response

The effect of adding C_e is obtained by adding the admittance $j\omega C_e$ of C_e to G_e in [6.75]. With

$$g_m >> (g_\pi+G_e)$$

then
$$\frac{v_o}{v_{in}}\approx -g_mR_L\frac{G_e+j\omega C_e}{g_m+j\omega C_e} \quad (6.77)$$

Thus the frequency response of $|V_o/V_{in}|$ becomes as in Fig. 6.21, with the two transition frequencies

$$f_1=1/(2\pi R_eC_e) \quad (6.78)$$

and

$$f_2=g_m/(2\pi C_e) \quad (6.79)$$

For $f_2=10$Hz then **$C_e=637\mu F$** giving **$f_1=0.25$H**

Potential divider derived feedback bias

Fig. 6.22 shows a way of avoiding the floating E_b still present in the feedback bias circuit of Fig. 6.20(a).

With $R_1/(R_1+R_2)=(V_e+V_{be})/V_+$

then $R_2/R_1=V_+/(V_e+V_{be})-1$ (6.80)

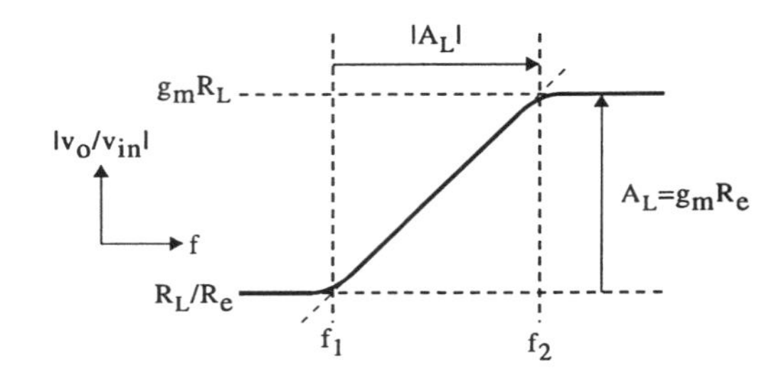

Fig. 6.21 Frequency response of voltage gain of amplifier in Fig. 6.20 with feedback bias (log axes).

With ΔV_b the acceptable voltage drop due to I_b then

$$\Delta V_b = I_b \times R_b \tag{6.81}$$

where

$$R_b = R_1 // R_2 // r_\pi \tag{6.82}$$

With f_b the cut-off frequency incurred by C_b then

$$f_b = 1/(2\pi R_b C_b) \tag{6.83}$$

With these relations Exercise 6.6 gives for f_b equal to the above f_2=10Hz that **C_b=22.1μF**

and with 5% the acceptable drop in I_c due to ΔV_b then **R_1=1kΩ, R_2=5.25kΩ.**

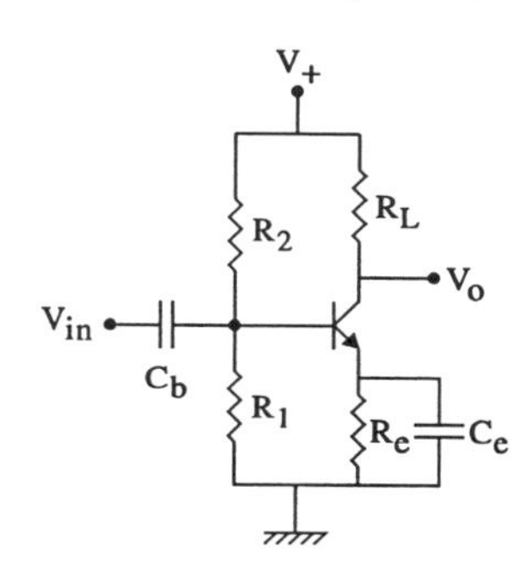

Fig. 6.22 Feedback bias as derived by potential divider.

> **Experiment. Operation with bias** The circuit of Fig. 6.23(a) is used to add adjustable bias to the input signal, thus avoiding the inconvenient floating E_b.
>
> As a preliminary, show that the circuit gives a bias range of 0 to ≈1V.

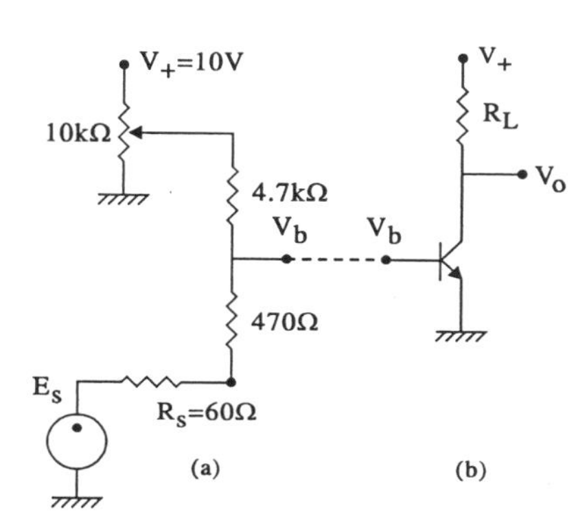

Fig. 6.23 Method of deriving adjustable base bias E_b from supply V_+ for experimental measurements.
(a) Arrangement for adding bias to signal.
(b) Amplifier.

References

Gibbons, J. F. 1966: *Semiconductor Electronics,* McGraw Hill.

Massobrio, G. and Antognetti, P. 1993: *Semiconductor Device Modelling with SPICE,* McGraw Hill.

6.8 Exercises and experiments

Exercise 6.1 Amplifier low-frequency characteristics

> **Experiment. Operation without base bias** Connect the circuit of Fig. 6.1(d) but with the signal generator connected directly to the transistor base, and with V_+=10V, R_L=4.7kΩ, and the transistor a BC182.
>
> Consider how, with the transfer function of (b), for signal amplitudes << (V_{be} ≈ 600mV) there will be negligible response, while for larger amplitudes gross distortion will occur. Apply a 100Hz sine wave as V_{in} and use the scope dual-channel facility to compare the input and output waveforms, and thus confirm the predictions.
>
> Switch the scope X-Y mode to view the V_b-V_o transfer function.

> With the transistor base disconnected, use the scope to confirm that V_b in (a) is the required signal E_s, with little reduction, with the above calculated bias added.
>
> With E_s=0, reconnect (b) and set the bias to the mid-point condition V_o=V_+/2.
>
> Increase the amplitude of E_s and thus repeat the above measurements but now with the base bias. Thus show that, for a sufficiently low amplitude of V_{in}, the amplifier gives distortion-free amplification. Confirm that a_{ce}=−200 as calculated.
>
> Change to the scope X-Y mode and increase E_s to again display the V_o-V_b transfer function. Confirm that the mid-point slope is equal to a_{ce}=−200.
>
> Reduce the amplitude of E_s to restore the linear operation. Observe how, as the bias is varied, $|a_{ce}|$ increases with decreasing V_o until the saturation point is met.
>
> Increase R_L to 47kΩ and reset for mid-point bias. Thus verify that a_{ce} is independent of I_c as in [6.10].

Experiment. Operation with increased signal source impedance For a source impedance $R_s >> r_\pi$ then the exponential transfer function gives way to the linear $I_c = \beta I_b$.

To give $R_s >> r_\pi = 5k\Omega$, connect a $100k\Omega$ resistor in series with the transistor base, with V_i the voltage at the resistor terminal remote from the base.

Repeat the previous experiments to confirm that the transfer function V_i vs V_o is now linear.

Confirm that the waveforms are now distortion-free, even for an output amplitude approaching the full output range. Why at this limit does the V_b waveform appear distorted? Does this matter?

Exercise 6.2 Transistor π-network calculations (Fig. 6.13).

Calculate I_{cx}, the value of I_c for $c_{\pi s} = c_{\pi d}$.

With $c_{\pi s} \approx I_c$ as in [6.38]

$c_{\pi s} = 40pF$ for $I_c = 1mA$

and $c_{\pi d} = 10pF$ then **$I_{cx} = 250\mu A$.**

Calculate the value of the component $f_{\pi s}$ of f_π due to the base stored charge.

With $f_{\pi s} = 1/2\pi r_\pi c_{\pi s}$ as in [6.44]

then **$f_{\pi s} = 795kHz$.**

Calculate the value of the total f_π.

With $f_\pi = 1/2\pi r_\pi (c_{\pi s} + c_{\pi d})$ as in [6.43]

then **$f_\pi = 637kHz$.**

Exercise 6.3 Amplifier response time for low signal source impedance

Calculation For the circuit of Fig. 6.16(a) with the signal source impedance R_s included, calculate the response time T_m for R_s the typical 50Ω of the lab signal generator.

With $T_m = (r_x + R_s)c_m$

where $c_m = (1 + |a_{cez}|)c_\mu$ as in [6.61]

then **$T_m = 200ns$.**

Experiment. Amplifier voltage gain and response time The simplicity of the current-bias circuit of Fig. 6.18 makes it the most suitable. However, the value of R_b has to be selected for the obtaining value of β. With $R_L = 4.7k\Omega$, $C = 10\mu F$, and with $V_+ = 10V$ as usual, choose R_b for mid-point bias.

From the resulting value of R_b, calculate β, I_b and hence $r_\pi = (kT/q)/I_b$.

Apply a sine wave of $f_{in} = 100Hz$ and suitably low amplitude and check that the voltage gain a_{cez} is the expected $-g_m R_L = -200$.

Switch to a square wave and increase f_{in} to allow the response-time T_m to be measured.

Thus calculate c_m and c_μ from the measured T_m and a_{cez}. The measured c_μ should be comparable with the typical $c_\mu = 5pF$.

Exercise 6.4 Amplifier response time for higher signal source impedance

Calculation For the circuit of Fig. 6.17, calculate the increased T_m for $R_s >> r_\pi$.

With now $T_m \approx r_\pi c_m$ **$T_m = 5\mu s$.**

Experiment For $R_s >> r_\pi = 5k\Omega$ add $47k\Omega$ to the signal generator output impedance, measure the resulting T_m, and compare with the above value.

Exercise 6.5 Thermal stability

Precision measurements here are not realistic. A hair-dryer and thermometer will give some idea.

Current bias With the circuit still as in Fig. 6.18, check that a 20° rise has very little effect on V_o, since β is not very temperature dependent.

Voltage bias For the simple circuit of Fig. 6.19, with no signal needed, the bias can be derived from V_+ using a potential divider, with the lower resistor $<< r_\pi = 5k\Omega$.

Set the divider for the mid-point $V_o = 5V$ and confirm that the effect of the hair-dryer is now gross, driving the transistor well into saturation.

Feedback bias calculation For the circuit of Fig. 6.20, calculate the change ΔI_c in I_c for the above 20° change in T.

$dV_{be}/dT = 2mV/^o$ without R_e is reduced $\times 1/|A_L|$ to

$$(1/40) \times 2mV/^o = 50\mu V/^o$$

For $\Delta T = 20^o$ then

$$\Delta V_{be} = 2mV/^o \times 20^o = 1mV$$

With $\Delta I_c = g_m \Delta V_{be}$ then **$\Delta I_c = 40\mu A$.**

Experiment Assemble the circuit (the capacitors are not needed) and confirm the above figure.

Exercise 6.6 Potential-divider-derived feedback bias circuit.

Resistor values For the circuit of Fig. 6.22, calculate suitable values of R_1 and R_2 for the change in I_c due to I_b to be <5%.

With $R_2/R_1 = V_+/(V_e + V_{be}) - 1$

as in [6.80] then $R_2/R_1 = 5.25$.

With $I_c = V_e/R_e$ then 5% change in I_c is 5% change in V_e giving $\Delta V_e = 50mV$.

With $A_L >> 1$ then also $\Delta V_b = 50mV$.

With $R_b = \Delta V_b/R_b$ as in [6.81] then $R_b = 1k\Omega$.

With $r_\pi = 5k\Omega$ and $R_2 >> R_1$

then **$R_1 = 1k\Omega$.**

would be acceptable, giving $R_2 = 5.25k\Omega$ and thus
$R_1 = 1k\Omega$, $R_2 = 5.25k\Omega$.

Capacitor C_b. Calculate C_b for $f_b = 10Hz$.

With $R_b = R_1 // R_2 // r_\pi$

as in [6.82] then $R_b = 719\Omega$.

With $C_b = 1/(2\pi f_b R_b)$

as in [6.83] then **$C_b = 22.1\mu F$.**

7

Emitter-follower and other circuits

Summary

The last chapter covered the normal common-emitter connection of the single-stage bipolar transistor amplifier. In this chapter we first cover another single-stage circuit, the 'emitter-follower'.

Here the input terminal is again the base, but now it is the emitter terminal that is the output, while the collector is the common terminal. Thus the emitter-follower is alternatively termed the 'common-collector' stage.

The term 'emitter-follower' refers to the way in which the signal component of the emitter terminal voltage follows that applied to the base, essentially as for the op-amp based voltage-follower of Section 2.4.

The remaining possibility is the 'common-base' connection, for which the emitter terminal is the input and the collector the output. This responds much as for the common-emitter, except that the input (emitter) impedance is much lower. Also the stage does not invert.

With the above the three single-stage configurations, next comes the 'cascode', which is a common-emitter stage followed by a common-base. This gives essentially the same voltage gain as for the common-source alone, but with a much higher upper frequency cut-off, due to the virtual elimination of the Miller feedback.

Next come the 'long-tailed pair' that forms the basis of first stage of almost all operational amplifiers, first in the elementary two-transistor form, and then with the addition of the current mirror and active tail needed to obtain an adequate common-mode rejection ratio.

V_+	kT/q	V_{be}	I_c	I_b	β
10V	25mV	600mV	1mA	5μA	200
g_m	r_π	R_x	c_π	c_μ	r_o
40mS	5kΩ	150Ω	50pF	5pF	70kΩ

Table 7.1 Stating values for calculations.

The bipolar transistor may also be used as a saturating switch. Here the factors determining the reponse time are somewhat more complex than for the incremental analysis of the last chapter.

With most of the numeric calculations in the course of the chapter, the Exercises section is relatively brief. As usual the calculated values are set bold with the sequence largely cumulative.

The starting values for the calculations are those derived in the last chapter for the BC182 operating at $I_c = 1mA$, as listed in Table 7.1.

7.1 Low frequency operation of emitter-follower

As for the voltage-follower, the usual function of the emitter-follower is to reduce the loading of a signal source. Fig. 7.1 shows the circuit used to reduce the loading of the common-emitter stage in (a) by the load R_L in (c).

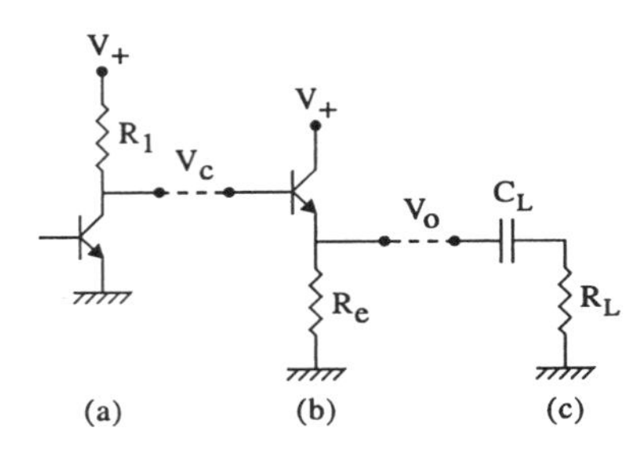

Fig. 7.1 Use of emitter-follower to reduce loading of common-emitter amplifier stage.
(a) Common-emitter stage.
(b) Emitter-follower.
(c) AC-coupled load.

Comparison with voltage-follower

For the ideal voltage-follower the voltage gain is unity, the input impedance is infinite, and the output impedance is zero. With the general signal source impedance R_s and load impedance R_L, the emitter-follower approaches these ideals as follows.

- The voltage gain a_{ef} is only a little short of unity.
- The input impedance r_{ie} is increased from the R_L that would obtain without the follower to $\approx\beta R_L$, with the lower limit of r_π for $R_L=0$.
- The output impedance r_{oe} is reduced from the R_s that would obtain without the follower to $\approx R_s/\beta$, with the lower limit of $\approx 1/g_m$ for $R_s=0$.

Incremental circuit

Fig. 7.2 gives the full incremental circuit for Fig. 7.1, with the signal source generalised in (a) and with the transistor in (b) represented by the hybrid-pi model of Fig. 6.11(b).

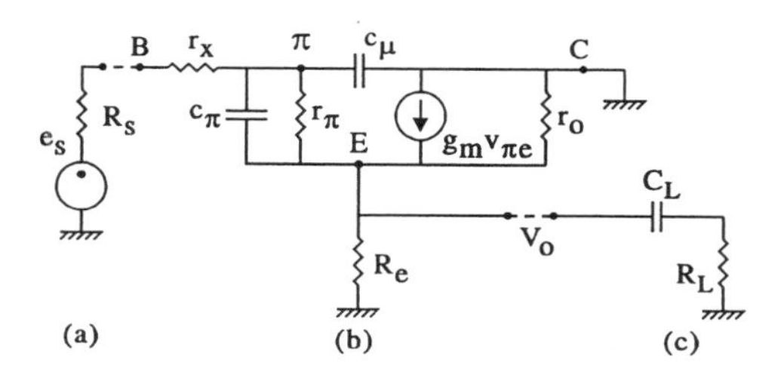

Fig. 7.2 Full small-signal equivalent circuit of emitter-follower.
(a) Signal source.
(b) Emitter-follower.
(c) Load.

For the present low-frequency analysis the circuit is simplified to that in Fig. 7.3, in which

- c_π and c_μ are omitted,
- the impedance of C_L is taken as zero,
- r_x is omitted, later to be added to R_s,
- r_o and R_e are omitted, later to be added in parallel with R_L.

Resistor values (Fig. 7.1)

Common-emitter stage With $R_1=(V_+/2)/I_c$

for mid-point bias of V_c then **$R_1=5k\Omega$.**

Emitter-follower With

$R_e=(V_c - V_{be})/I_c$ and $V_{be}=600mV$

then **$R_e=4.4k\Omega$.**

Voltage gain

From the two-node analysis (π and E) of Fig. 7.3, and with the follower voltage gain $a_{ef}=v_o/e_s$

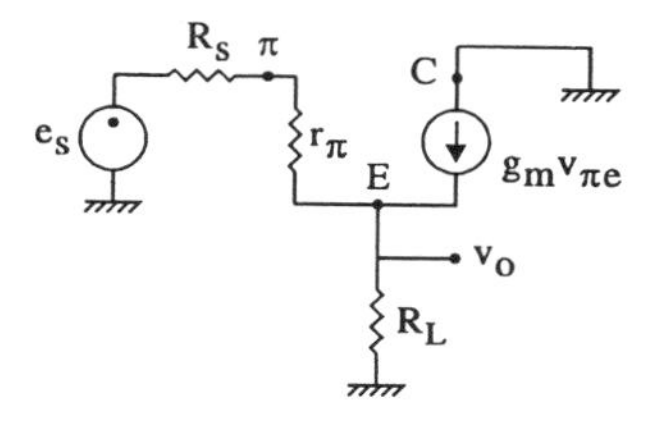

Fig. 7.3 Low-frequency version of Fig. 7.2 ignoring C_L.

then
$$a_{ef}=\frac{G_s(g_\pi + g_m)}{G_L(G_s + g_\pi) + G_s(g_m + g_\pi)} \quad (7.1)$$

With
$$\beta = g_m r_\pi \quad (7.2)$$
as in [6.29] then
$$a_{ef}=\frac{R_L}{R_L + \dfrac{R_s + r_\pi}{\beta+1}} \quad (7.3)$$

For $R_L=\infty$ then $a_{ef}=1$, while for the more realistic $R_L=R_e$ **$a_{ef}=0.994$.**

Output impedance

From the form of [7.3], the follower output impedance

$$r_{oe} = \frac{R_s + r_\pi}{\beta + 1} \quad (7.4)$$

With $\beta >> 1$, with $\beta=r_\pi g_m$, with r_x and R_e included, and with $r_o >> R_e$,

$$r_{oe} \approx [(R_s+r_x)/\beta + 1/g_m]//R_e \quad (7.5)$$

which is plotted against R_s in Fig. 7.4.

Lower limit For $R_s=0$ and with $r_x/\beta << 1/g_m$ and $R_e >> 1/g_m$ then $r_{oe} \approx 1/g_m$ giving, as plotted, **$r_{oe} \approx 25\Omega$.**

Upper limit For $R_s=\infty$ then $r_{oe}=R_e$ giving **$r_{oe}=4.4k\Omega$,** again as plotted.

With $\beta/g_m=r_\pi$ and $r_x << r_\pi$ in [7.5], the first transition is approximately where $R_s=r_\pi$, and the second where $R_s=\beta R_e$. It is between these limits that $r_{oe} \approx R_s/\beta$.

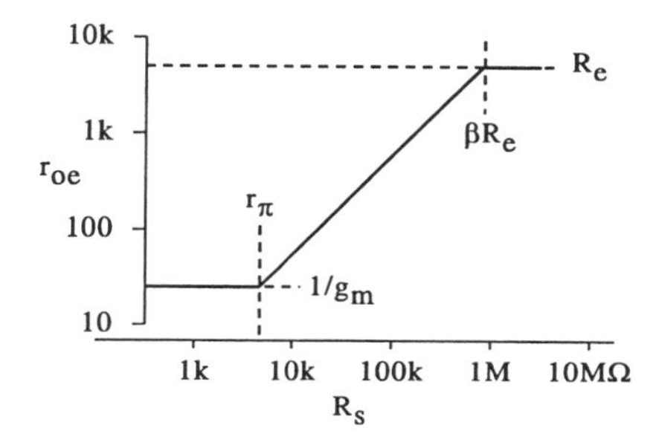

Fig. 7.4 Variation of low-frequency output impedance r_{oe} of emitter-follower of Fig. 7.3 with signal source impedance R_s.

Common-emitter source (Fig. 7.1(a))

With $R_s \approx (R_l = 5k\Omega)$,

from the plot $r_{oe} << (R_e = 4.4k\Omega)$

With $r_{oe} \approx [(R_s + r_x)/\beta + 1/g_m]//R_e$ as in [7.5]

and with $r_x << R_l$

then $r_{oe} \approx R_l/\beta + 1/g_m$

and thus **$r_{oe} \approx 50\Omega$.**

Only ×2 higher than the $1/g_m = 25\Omega$ for $R_s = 0$.

Input impedance

With the emitter-follower represented as in Fig. 7.5 then

$$v_o = e_s \frac{r_{ie}}{R_s + r_{ie}} a_v \quad (7.6)$$

where the voltage gain $a_v = v_o/v_{in}$, and r_{ie} is the follower incremental input impedance.

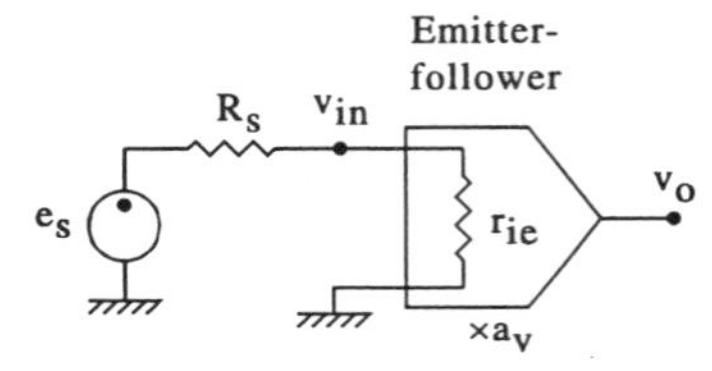

Fig. 7.5 Representation of emitter-follower showing effect of amplifier input impedance r_{ie}.

[7.3] manipulated into this form becomes

$$v_o = e_s \frac{(\beta + 1)R_L}{R_s + (\beta + 1)R_L + r_\pi} \quad (7.7)$$

giving $r_{ie} = (\beta + 1)R_L + r_\pi$.

With r_x included, and with $r_o >> R_e$ still omitted,

then $r_{ie} \approx r_x + (\beta + 1)(R_L//R_e) + r_\pi$ (7.8)

With $r_x << r_\pi$ and $\beta >> 1$

then $r_{ie} \approx \beta(R_L//R_e) + r_\pi$ (7.9)

giving the variation of r_{ie} with R_L as in Fig. 7.6.

Lower limit For $R_L = 0$ in [7.9] then $r_{ie} \approx r_\pi$ giving **$r_{ie} = 5k\Omega$.**

With the transition where

$$R_L//R_e = (r_\pi/\beta = 1/g_m)$$

and with $R_e >> 1/g_m$ then $R_L \approx 1/g_m$

giving **$R_L = 25\Omega$.**

Upper limit For $R_L = \infty$ and with $\beta R_e >> r_\pi$ then $r_{ie} \approx \beta R_e$ giving **$r_{ie} \approx 1M\Omega$**

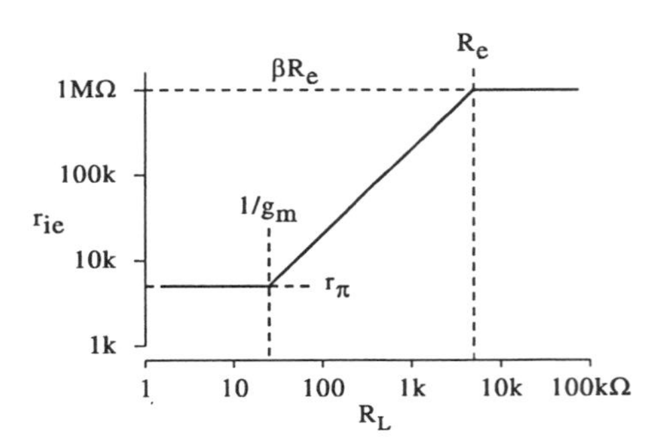

Fig. 7.6 Variation of input impedance r_{ie} of emitter-follower stage in Fig. 7.3 with R_L.

with the transition where $R_L = R_e$, for which **$R_L = 4.4k\Omega$.**

For the central region where $1/g_m < R_L < R_e$

it is confirmed that $r_{ie} \approx \beta R_L$.

7.2 Emitter-follower capacitances

With the capacitances c_π and c_μ in Fig. 7. 2(b), we consider here their effect of the follower input impedance, output impedance, and voltage gain. With the output impedance shown to be inductive for part of the frequency range, the possibility of resonance with a capacitive load is examined in the next section.

Effect of c_π on output impedance

With $r_{oe} = (R_s + r_\pi)/(\beta + 1)$ as in [7.4]

and with $\beta = g_m r_\pi$

then for the added effect of c_π, r_π is replaced by

$$z_\pi = r_\pi // x_\pi$$

to give $z_{oe} = (R_s + z_\pi)/(1 + g_m z_\pi)$ (7.10)

With $z_\pi = r_\pi/(1 + jf/f_\pi)$ (7.11)

where $f_\pi = 1/2\pi c_\pi r_\pi$ (7.12)

then
$$z_{oe} = \frac{R_s(1 + jf / f_\pi) + r_\pi}{\beta + 1 + jf / f_\pi} \quad (7.13)$$

With r_{oe} as in [7.4], this may be written as

$$z_{oe} = r_{oe} \frac{1 + jf / f_{num}}{1 + jf / f_{den}} \quad (7.14)$$

where

$$f_{num} = 1/[2\pi c_\pi (R_s // r_\pi)] \quad (7.15)$$

also with $\beta = g_m r_\pi$ and $\beta >> 1$

$$f_{den} \approx g_m/(2\pi c_\pi) \quad (7.16)$$

Large signal-source impedance ($R_s = 200k\Omega$) $>> r_\pi$

With $$z_{oe} = \frac{R_s(1 + jf / f_\pi) + r_\pi}{\beta + 1 + jf / f_\pi}$$ as in [7.13]

and with $\beta >> 1$ then

$$z_{oe} \approx \frac{R_s(1 + jf / f_\pi)}{\beta + jf / f_\pi} \qquad (7.17)$$

For f=0 then $z_{oe} = R_s/\beta$,

while for $f=\infty$ $z_{oe} = R_s$.

With the transition frequencies f_π and βf_π then

$$\mathbf{f_\pi = 637kHz} \quad \text{and} \quad \mathbf{\beta f_\pi = 127MHz},$$

giving the frequency response of $|z_{oe}|$ as in Fig. 7.7(a).

Signal source CE stage of Fig. 7.1(a) As above, the zero-frequency $\mathbf{r_{oe} = 50\Omega}$, as plotted in Fig. 7.7(b).

With the general $f_{num} = 1/[2\pi c_\pi(R_s//r_\pi)]$ as in [7.15]

and here $R_s \approx R_1$ then

$$f_{num} \approx 1/2\pi c_\pi(R_1//r_\pi)$$

giving $\mathbf{f_{num} \approx 1.27MHz}$.
the first transition in the plot of (b).

With still $f_{den} \approx g_m/(2\pi c_\pi)$ as in [7.16]

then f_{den} remains $\beta f_\pi = 127MHz$ to complete the plot as shown.

Zero signal source impedance ($R_s = 0$) With

$$r_{oe} \approx [(R_s + r_x)/\beta + 1/g_m]//R_e \qquad \text{as in [7.5]}$$

with $r_x/\beta << 1/g_m$ and with $1/g_m << R_e$

then $r_{oe} \approx 1/g_m$
giving, as plotted in (c), $\mathbf{r_{oe} = 25\Omega}$.

With r_x for R_s in [7.15] for f_{num}, and with $r_x << r_\pi$

then $f_{num} \approx 1/(2\pi c_\pi r_x)$

giving $\mathbf{f_{num} = 21.2MHz}$.
and again f_{den} unchanged.

The plots of (d) to (f) for decreasing R_s are of academic interest only, since r_x is the minimum effective R_s possible.

Effect of c_π on voltage gain

With $$a_{ef} \approx (R_L//R_e)/(z_{oe} + R_L//R_e) \qquad (7.18)$$

the voltage gain, we calculate the resulting cut-off frequency f_c for the common-emitter source and for the convenient example $\mathbf{R_L//R_e = 1k\Omega}$.

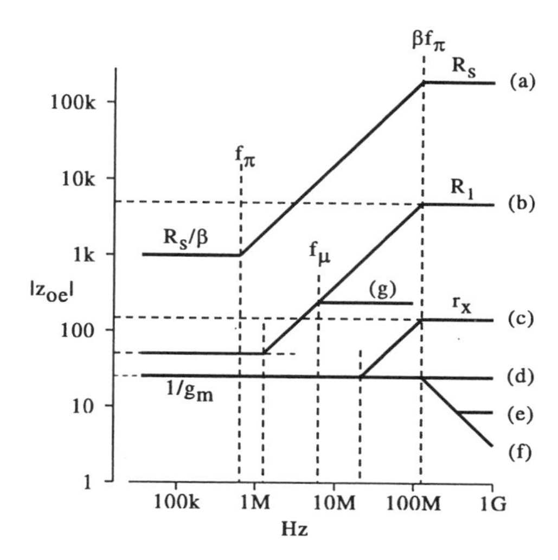

Fig. 7.7(a) to (g) Frequency response output impedance z_{oe} of emitter-follower stage for various values of signal source impedance R_s.
(a) ($R_s = 200k\Omega$) $>> r_\pi$.
(b) $R_s \approx R_1$ in Fig. 7.1.
(c) $R_s = r_x = 150\Omega$ only.
(d) $R_s = r_p/\beta$ ($r_x = 0$).
(e) As (d) but $R_s < r_\pi/\beta$.
(f) As (d) but $R_s = 0$.
(g) As (b) but with c_μ.

With f_c the value of f for which $|z_{oe}| = (R_L//R_e = 1k\Omega)$

and with $|z_{oe}|$ as in Fig. 7.7(b) then f_c is well within the two transition frequencies.

With $$z_{oe} = r_{oe}\frac{1 + jf / f_{num}}{1 + jf / f_{den}}$$ as in [7.14]

then $|z_{oe}| \approx r_{oe}f/f_{num}$ giving $f_c = f_{num}(R_L//R_e)/r_{oe}$

and thus $\mathbf{f_c = 20.7MHz}$.

Effect of c_π upon input impedance

With $r_{ie} \approx r_x + (\beta + 1)(R_L//R_e) + r_\pi$ as in [7.8]

and with r_{ie} replaced by z_π then

$$z_{ie} \approx r_x + (g_m z_\pi + 1)(R_L//R_e) + z_\pi \qquad (7.19)$$

Fig. 7.7(h) shows the resulting variation of $|z_{ie}|$ with frequency for again $R_L//R_e = 1k\Omega$. With here

$$r_x << r_\pi, \quad g_m r_\pi = \beta \quad \text{and} \quad \beta >> 1$$

then, as plotted, the initial and final values are

$$r_{ie} \approx \beta(R_L//R_e) = 200k\Omega \quad \text{and} \quad R_L//R_e = 1k\Omega$$

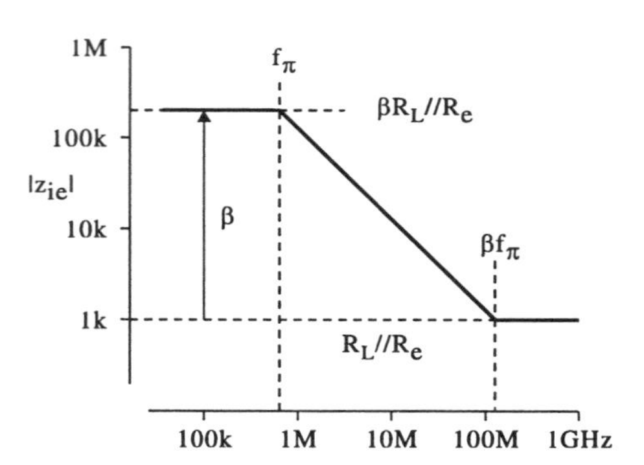

Fig. 7.7(h) Frequency response of emitter-follower input impedance z_{ie} for $R_L//R_e=1k\Omega$.

With $z_{ie} \approx r_x+(g_m z_\pi+1)(R_L//R_e)+z_\pi$ as in [7.19]

the first transition is at f_π, where $g_m z_\pi$ (initially β) begins to decrease. The second is ×β higher, where $|g_m z_\pi|$ has fallen to unity.

Effect of c_μ

From Fig. 7.2 the effect of c_μ is essentially to shunt the signal source with c_μ. For R_s and c_μ alone, this gives a cut-off in voltage gain a_{ef} at the frequency

$$f_\mu=1/(2\pi R_s c_\mu) \qquad (7.20)$$

For $R_s \approx R_1$ then $\mathbf{f_\mu=6.37MHz.}$

Output impedance

The rise in $|z_{oe}|$ in Fig. 7.7(a) to (c) will be checked at this frequency, to alter the response of Fig. 7.7(b) to that in (g).

Input impedance

Regarding r_x as part of R_s then c_μ simply shunts the remaining component of z_{ie}.

7.3 Resonant capacitive loading of emitter-follower

As for the voltage-follower, there is the possibility of resonance when the emitter-follower load is capacitive. For example, where the signal source for the follower is the common-emitter stage of Fig. 7.1(a) then the frequency response of the follower output impedance z_{oe} is as in Fig. 7.7(b), clearly inductive in the central region, and able to resonate with the capacitive load.

The z_{oe} response is replotted in Fig. 7.8(a), together with that in (b) for the reactance X_{line} of a capacitive line that the follower is required to drive. The combined parallel impedance thus resonates as in (c). With the resonant frequency f_{res} that for which

$$|z_{oe}|=|X_c|$$

then $$f_{res}=[\beta f_\pi/(2\pi C_{line}R_1)]^{1/2} \qquad (7.21)$$

With C_{line}=100pF for the plot shown $\mathbf{f_{res}=6.37MHz.}$

Q-factor of resonance

There are two components determining Q. The first associated with the lower transition at f_{num} and the second at the upper transition at $f=\beta f_\pi$. Thus

$$Q_1=f_{res}/f_{num} \quad \text{and} \quad Q_2=\beta f_\pi/f_{res},$$

giving $\mathbf{Q_1=5.0}$ and $\mathbf{Q_2=20.}$

With $Q=Q_1//Q_2$ then $\mathbf{Q=4.}$

Fig. 7.8(e) shows the corresponding oscillatory step response. With the number of oscillations within the decay period equal to $Q/\pi\approx1.5$, the diagram is representative.

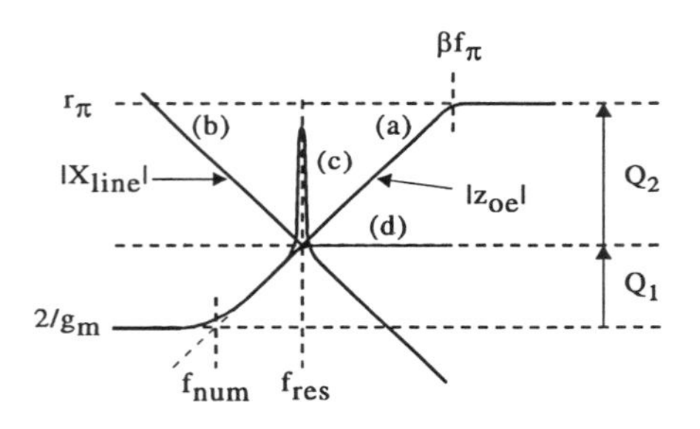

Fig. 7.8 (a) to (d) Resonance due to loading of inductive output of emitter-follower in Fig. 7.1(a,b) by capacitive line.
(a) Frequency response of $|z_{oe}|$.
(b) Frequency response of $|X_{line}|$.
(c) Resonant frequency response of $|z_{oe}//X_{line}|$.
(d) As (a) but with damping due to emitter-follower c_μ.

Damping by c_μ

As for Fig. 7.7(b) and (g), the effect of c_μ on $|z_{oe}|$ in Fig. 7.8 is to alter the response of (a) to that in (d). With the plots as shown the damping is near critical, so there will be no resonance. However, with 100pF the capacitance of only ~60cm of coax, then C_{line} can be larger.

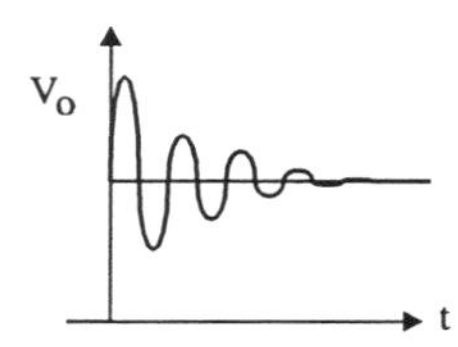

Fig. 7.8(e) Step response without damping, as in (c).

Inductive source impedance

A more likely manifestation of the above effect is where the signal source impedance has an inductive component. This can resonate with c_μ, giving rise to frequency ranges where, not only is the reactive component of z_{oe} inductive, but the resistive component is negative. Connection of the capacitive load can then cause oscillation.

7.4 Common-base amplifier

The essentials of the common-base circuit are shown in Fig. 7.9(a). Many of the properties of the circuit can be inferred directly from the previous analyses of the common-emitter and the emitter-follower stages.

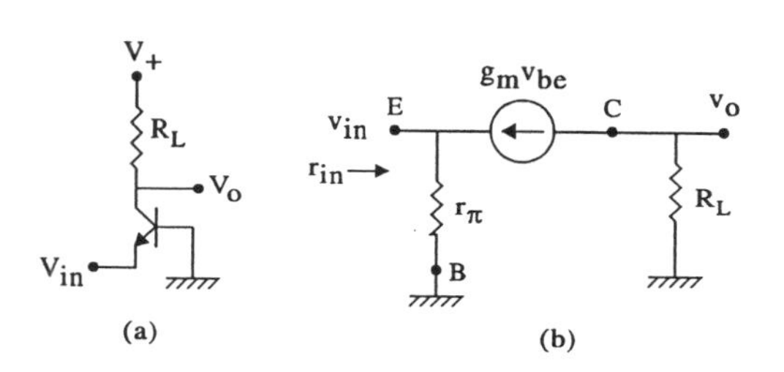

Fig. 7.9 Common-base amplifier.
(a) Essentials of circuit diagram.
(b) Simplified low-frequency incremental circuit model ignoring $r_o >> R_L$ and $r_x << r_\pi$.

Voltage gain

Since the common-base stage is simply the common-emitter with the input terminals reversed then the voltage gain a_{cb} will be essentially the same, except now with no sign reversal. Thus from [6.9] for the common-emitter stage

$$a_{cb} \approx g_m R_L \tag{7.22}$$

giving $\mathbf{a_{cb} \approx 200.}$

Input impedance

Also the input impedance z_{ib} of the common-base stage is simply the output impedance z_{oe} of the emitter-follower.

Thus

$$r_{ib} \approx 1/g_m \tag{7.23}$$

giving $\mathbf{r_{ib} \approx 25\Omega.}$

Current gain

While for the CE stage the current gain is β (typically 200), for the CB it is the much lower $\alpha = I_c/I_e \approx 1$.

7.5 Cascode amplifier

The relatively low current gain, and the associated low input impedance would not seem to commend the common-base stage. However, it has the considerable advantage, when combined with the common-emitter in the cascode of Fig. 7.10(a), of almost completely eliminating the Miller feedback capacitance due to c_μ, which is the dominant mechanism limiting the upper frequency response of the CE stage alone.

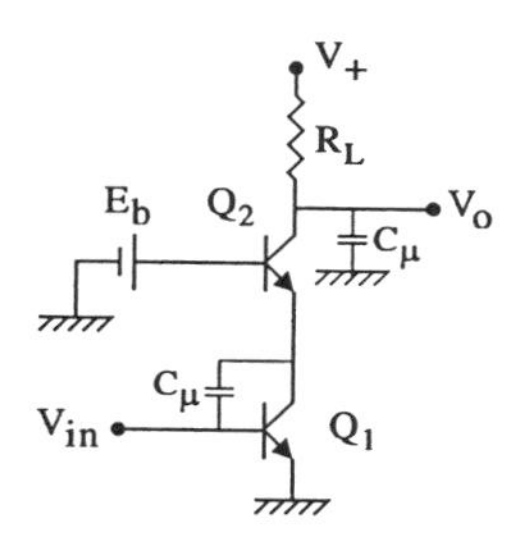

Fig. 7.10(a) Essentials of cascode amplifier.

Voltage gain

With the lower stage voltage gain $a_{v1} \approx -g_m r_{ib}$ where r_{ib} is the input impedance of the upper stage, and with

$$r_{ib} \approx 1/g_m \qquad \text{as in [7.23]}$$

then

$$a_{v1} \approx -1.$$

With the upper stage gain

$$a_{v2} = g_m R_L \qquad \text{as in [7.22]}$$

and with the total cascode voltage gain

$$a_{ca} = a_{v1} \times a_{v2} \quad \text{then} \quad a_{ca} \approx -g_m R_L \tag{7.24}$$

the same as for the CE stage alone. Alternatively, for the first stage

$$i_{c1} = g_m v_{in}$$

and for the second

$$v_o = -i_{c2} R_L$$

With $i_{c2} \approx i_{c1}$ then a_{ca} is as before.

Lower-stage transistor capacitances

With $|a_{v1}| \approx 1$ and the Miller capacitance

$$c_m = (|a_{v1}| + 1)c_\mu \tag{7.25}$$

then the much reduced $\mathbf{c_m = 10pF.}$

With the input cut-off

$$f_{in} = 1/[2\pi(c_m + c_\pi)r_x] \tag{7.26}$$

then $\mathbf{f_{in} = 17.7MHz}$

much higher than the 1MHz for the CE stage.

Upper-stage transistor capacitance

The upper stage c_μ is in parallel with R_L to give the output cut-off

$$f_o = 1/(2\pi c_\mu R_L) \qquad (7.27)$$

for which, with $R_L = 5k\Omega$ as below, $\mathbf{f_o = 6.37\ MHz.}$

This is well below $f_{in} = 17.7MHz$ but can be increased, at the expense of a_{ca}, by reducing R_L. With f_{in} as in [7.26] and f_o as in [7.27], and for $f_{in} = f_o$, then

$$R_L = r_x(c_m + c_\pi)/c_\mu$$

giving $R_L = 1.8k\Omega$.

With $|a_{ca}| = g_m R_L$
then $\mathbf{|a_{ca}| = 72.}$

Compared with the common-emitter stage, the cut-off is ≈ 18 times higher with still a useful amount of gain.

Emitter-follower

The above effect of c_μ is made worse if the cascode has a capacitive load, such as when a line is being driven. Then it is usual to include the emitter-follower stage Q_3 in Fig. 7.10(b). Even here the c_μ for the follower adds to that of Q_2, doubling the total.

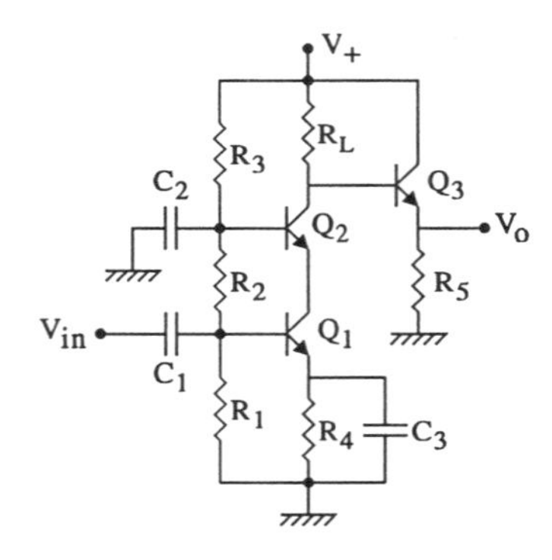

Fig. 7.10(b) Cascode amplifier.
Full circuit including output emitter-follower stage Q_3.

Circuit values

With $\mathbf{V_+ = 15V,\quad V_{e1} = 1V,\quad V_{e2} = 5V,\quad I_c = 1mA}$

V_+ has been increased from the previous 10V for the CE stage, to give V_o the same range as before. Also with V_{e1} remaining 1V the same degree of bias stabilisation is obtained. Thus as before $\mathbf{R_L = 5k\Omega}$ and $\mathbf{R_e = 1k\Omega.}$

With $I_{c3} \approx 1mA$ and $V_{c2} = 10V$ for the emitter-follower then $\mathbf{R_5 = 10k\Omega.}$

Exercise 7.1 gives as suitable for the bias chain

$$\mathbf{R_1 = 5k\Omega,\quad R_2 = 12.5k\Omega,\quad R_3 = 29.4k\Omega.}$$

Miller feedback

With the voltage gain of $Q_1 \approx 1$ the signal amplitude at the collector is relatively low, allowing V_{e2} to be much lower than the present ≈5V, and thus perhaps the original $V_+ = 10V$ to be retained. The reduced V_{cb}, however, would significantly increase $c_{\mu 1}$ and thus increase the Miller feedback.

Decoupling capacitors

C_1, C_2 and C_3 are all decoupling capacitors, typically 10nF to 100nF disc-ceramics, with the relatively low self-inductance of the ceramic type.

The low inductance is particularly important for C_2. Remember the oscillation that can take place for an emitter-follower stage with an inductive source. The inductance of C_2 at the base of Q_2, with the partially capacitive output impedance of the lower stage, gives much the same configuration.

7.6 Long-tailed pair

The circuit of the 'long-tailed pair' (LTP) is shown in Fig. 7.11(a), with the name alluding to the tail resistor R_e. Taking say V_{o+} as the single output V_o, the response of V_o to the differential input $V_{id} = V_{i+} - V_{i-}$ is considerably greater than that to the 'common-mode' V_{ic} applied to both input terminals at once. Also, with the addition of the active tail and current mirror in Fig. 7.13(a), the common-mode response is reduced by a further large factor. With the function thus closely approaching that of the ideal differential amplifier, for which the common-mode response is zero, the LTP is the basis of the input stage of the op-amp of the previous chapters.

Circuit values

For the following calculations

$$\mathbf{V_+ = 10V,\quad V_- = -10V,\quad I_c = 1mA,\quad I_e = 2mA.}$$

With $V_o = V_+/2$ for mid-point bias and with $V_{be} = 600mV$ then $\mathbf{R_L = 5k\Omega,\quad R_e = 4.7k\Omega.}$

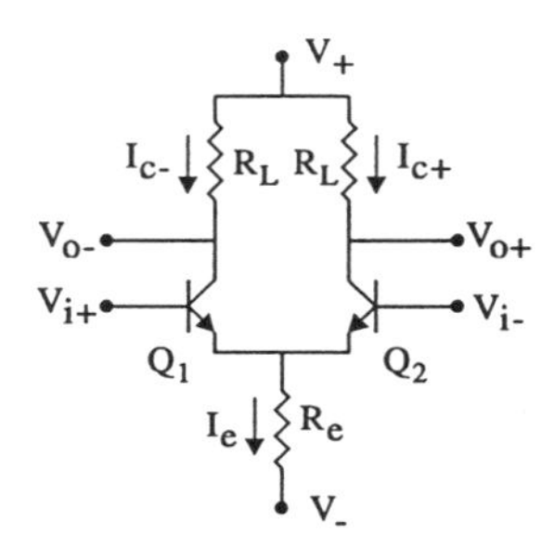

Fig. 7.11(a) Long-tailed pair circuit diagram.

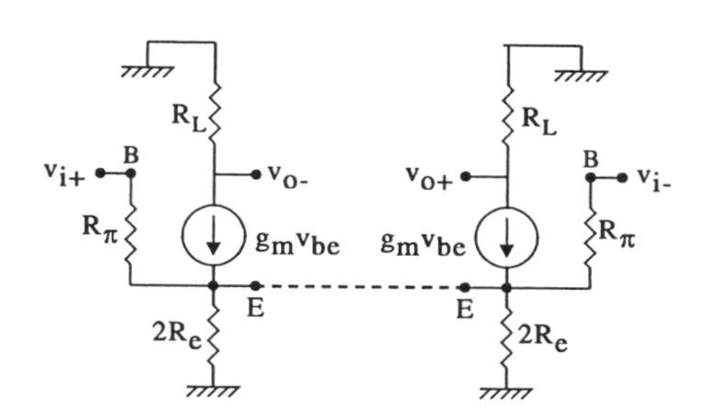

Fig. 7.11(b) Incremental circuit diagram for long-tailed pair of (a).

Common- and difference-mode components

With the incremental circuit diagram of the LTP as in (b), and resolving v_{i+} and v_{i-} into the common- and difference-mode components v_{ic} and v_{id} as in Fig. 7.12, then

$$v_{i+}=v_{ic}+v_{id}/2 \tag{7.28}$$

and

$$v_{i-}=v_{ic}-v_{id}/2 \tag{7.29}$$

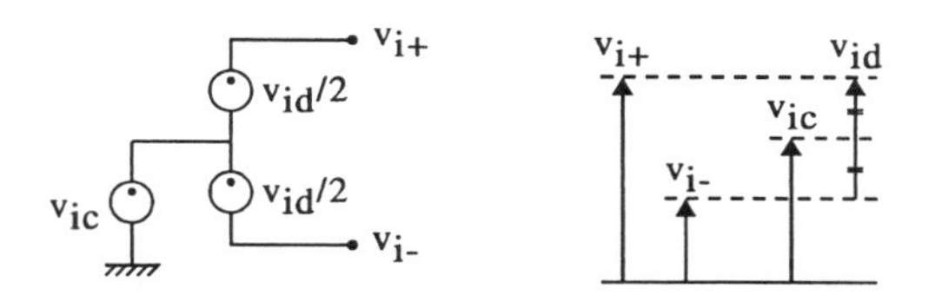

Fig. 7.12 Resolution of differential amplifier inputs v_{i+} and v_{i-} into common- and difference-mode components v_{ic} and v_{id}.

With the general $\quad v_o=a_+v_{i+}+a_-v_{i-} \quad (7.30)$

and for v_{ic} applied alone then

$$v_{i+}=v_{i-}=v_c \quad \text{giving} \quad v_o=v_{ic}a_c$$

where the 'common-mode gain'

$$a_c=a_++a_- \tag{7.31}$$

Conversely, with v_{id} applied alone

$$v_{i+}=v_{id}/2 \quad \text{and} \quad v_{i-}=-v_{id}/2$$

giving $\quad v_o=v_{id}a_d$

where the 'difference-mode gain'

$$a_d=(a_+-a_-)/2 \tag{7.32}$$

With the general $\quad v_o=a_cv_{ic}+a_dv_{id} \quad (7.33)$

then for the ideal $a_c=0$, $a_-=-a_+$.

With the essential feature of the differential amplifier the degree to which a_c is less than a_d, this is expressed by the 'common-mode rejection factor'

$$CMR=|a_d|/|a_c|$$

with typically for the op-amp $\quad$ **CMR=10^5.**

With a_d, a_+ and $|a_-|$ all very closely equal, the values normally specified are a_d (the 'op-amp' a_o in the earlier chapters) and CMR.

With a_d and a_c thus of more direct importance than a_+ and a_-, and with the expressions for a_d and a_c the easier to derive, we derive these first, and then the expressions for a_+ and a_- from the expressions for the other two.

Common-mode voltage gain a_c

For v_{ic} only applied, and with the incremental circuit in Fig. 7.11(b) fully symmetrical, then no current will flow through the emitter link. Thus the analysis can be for the link broken. For the second half of the circuit alone this is the same as for the common-emitter stage with feedback bias in Fig. 6.20(a), for which

$$v_o/v_{in}\approx-R_L/R_e \quad \text{as in [6.76]}$$

For the present half-circuit then

$$a_c\approx-R_L/2R_e \tag{7.34}$$

giving $\quad$ **$a_c=-0.53$.**

Difference-mode voltage gain a_d

With now v_{id} applied alone and the circuit still symmetric then the common v_e becomes zero. Thus the incremental analysis can be for the emitter point grounded. Each half of the circuit then becomes a simple common-emitter single stage, for which the voltage gain

$$a_{ce}=-g_mR_L \quad \text{as in [6.9]}$$

For the second half-circuit then

$$v_{o+}=-g_mR_Lv_{i-}$$

which with $\quad v_{i-}=-v_{id}/2$

gives

$$a_d=g_mR_L/2 \tag{7.35}$$

and thus $\quad$ **$a_d=100$.**

Single-input voltage gain

With a_c and a_d as in [7.31] and [7.32]

$$a_+=a_d+a_c/2 \tag{7.36}$$

and

$$a_-=-a_d+a_c/2 \tag{7.37}$$

With a_c and a_d as in [7.34] and [7.35] then

$$a_+=g_mR_L/2-R_L/(4R_e) \tag{7.38}$$

and

$$a_-=-g_mR_L/2-R_L/(4R_e) \tag{7.39}$$

giving $\quad a_+ \approx 100, \quad a_- \approx -100$

Gain reduction relative to CE stage

With $\quad a_d = g_m R_L/2 \quad$ as in [7.35]

this is one half of $\quad a_{ce} = -g_m R_L$

in [6.9] for the normal CE stage, arising as follows.

- With v_{i+} applied alone to the LTP, Q_1 may be regarded as an emitter-follower driving Q_2 as a common-base stage.
- With the output impedance of one ($\approx 1/g_m$) equal to the input impedance of the other then v_{i+} is divided equally between the v_{be1}.
- With $v_{be2} \approx v_{i+}/2$ rather than v_{i+} then v_o/v_{i+} is halved relative to that for the CE stage.

Were v_o to be taken between the two collectors as

$$v_{od} = v_{o+} - v_{o-}$$

then the loss would be made good.

Common-mode rejection ratio

With the first term in [7.38] and [7.39] largely determining a_+ and a_-, it is the second term that makes the two unequal in magnitude, giving the finite CMR. With

$$\text{CMR} = |a_d|/|a_c|$$

and with a_c and a_d as in [7.34] and [7.35]

then $$\text{CMR} = g_m R_e \tag{7.40}$$

giving $\quad$ **CMR = 188.**

Active tail

With the above figure lamentably below the CMR $\approx 10^5$ typical for the op-amp, the refinements in Fig. 7.13(a) which bring the figure closer to that normally obtained are the current mirror and the active tail.

For the active tail, initially without R_{et}, then R_e in [7.40] is replaced by the considerably larger incremental r_{o5}. With

$$I_e = 2(I_{c1} = I_{c2}) \quad \text{and} \quad r_{o5} \propto 1/I_e$$

then $r_{o5} \approx 35k$ giving $\quad$ **CMR ≈ 1400.**

With r_{tail} the general tail resistance, and with R_{et} added, then r_{tail} is increased further. From the single-node (E) analysis of the incremental equivalent circuit of the active tail, and with $g_m >> g_o$,

$$r_{tail} \approx r_o[1 + (r_\pi // R_{et})g_m] \tag{7.41}$$

With the bias E_b set to give $V_e = V_{i-}/2$ then

$$R_{et} = (|V_{i-}|/2)/I_{tail}$$

giving $\quad$ **$R_{et} = 2.5k\Omega$.**

With $g_{m5} = 80mS$ and $r_{\pi 5} = 2.5k\Omega$ then $\quad$ **$r_{tail} = 3.54M\Omega$.**

With other factors limiting the overall CMR, we term the present component CMR_{tail}. With r_{tail} in place of R_e in [7.40] then the general

$$\text{CMR}_{tail} = g_m r_{tail} \tag{7.42}$$

with now $\quad$ **$\text{CMR}_{tail} = 141k$.**

Current source tail

With here $r_{tail} = \infty$ then [7.42] gives $\text{CMR}_{tail} = \infty$. However, if the LTP r_o, so far ignored, are allowed for then the single-node (E) analysis, as for i_o with the v_o node grounded, gives

$$\text{CMR}_{tail} = \beta g_m r_o$$

Thus the limiting $\quad$ **$\text{CMR}_{tail} = 560k$.**

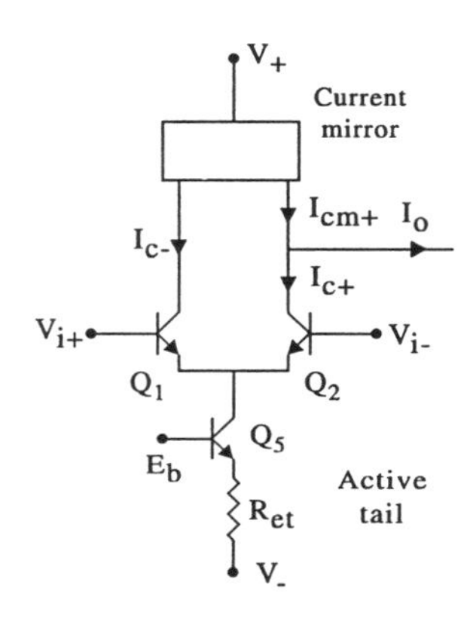

Fig. 7.13(a) Enhanced long-tailed pair. Bias $E_b \approx V_-/2$.

Current mirror

As will now be explained, the current mirror increases both the LTP voltage gain and the CMR. We consider the CMR first.

Ideal mirror

Were the differential output $V_{od} = V_{o+} - V_{o-}$ to be taken for the LTP without mirror in Fig. 7.11(a) then for V_{ic} applied alone, and with the circuit fully symmetric, $V_{od} = 0$ giving CMR $= \infty$.

The mirror in Fig. 7.13(a) parallels this by forming $I_o = I_{c-} - I_{c+}$, doing so by 'reflecting' the previously wasted I_{c-} as the ideally equal I_{cm+}.

With the mirror circuit as in (b), CMR is the same for i_o in the incremental equivalent of (c) as it is for the corresponding v_o with the output node open. With the derivation of CMR thus much simplified, this approach will be taken.

For the ideal mirror the impedance presented to i_{c-} is zero. With the impedance presented to i_{c+} in (c) also zero then the symmetry of the LTP not upset, retaining $i_{cd} = 0$ for v_{ic} applied alone. With $i_o = -i_{cd}$ for the ideal mirror then $i_o = 0$ giving CMR $= \infty$ as before.

Real mirror

The mirror comes close to the above ideal as follows. First, with $r_{o3} >> r_{\pi3}$ and the voltage across r_{o4} zero then the r_o are omitted in (c). With $\beta = r_\pi g_m$ then

$$i_o = i_{c-}\beta/(\beta+2) - i_{c+} \quad (7.43)$$

Resolving i_{c+} and i_{c-} into the common- and difference-mode components i_{cc} and i_{cd}

$$i_{cc} = (i_{c+} + i_{c-})/2 \quad (7.44)$$

and

$$i_{cd} = i_{c+} - i_{c-} \quad (7.45)$$

For i_{cd} alone, and with $\beta >> 1$, then

$$i_o \approx -i_{cd} \quad (7.46)$$

while for i_{cc} alone $\quad i_o \approx -2i_{cc}/\beta \quad (7.47)$

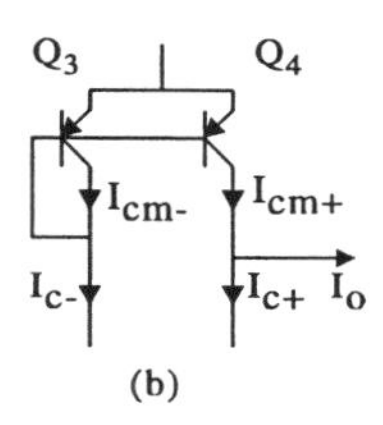

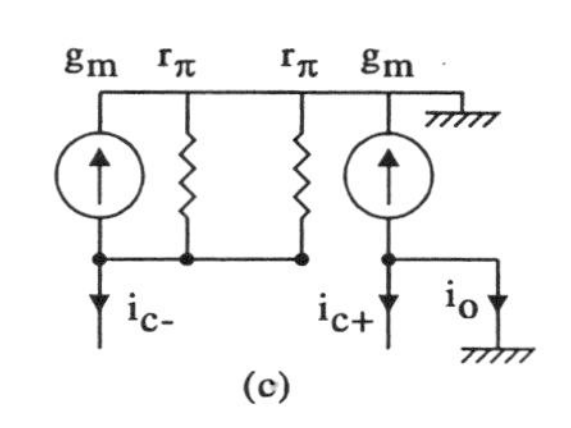

Fig. 7.13(b,c) Current mirror in long-tailed pair of (a).
(a) Circuit diagram.
(b) Incremental equivalent.

Difference-mode LTP transconductance g_d

With $\quad g_d = i_o/v_{id}$

for v_{id} applied alone, and with the E node here able to be grounded in the incremental circuit of Fig. 7.11(b), then

$$i_{c+} = -g_m v_{id}/2 \quad \text{and} \quad i_{c-} = g_m v_{id}/2.$$

With $\quad i_o \approx -i_{cd} \quad$ and $\quad i_{cd} = i_{c+} - i_{c-}$

then $\quad g_d \approx g_m$

Common-mode LTP transconductance g_c

With $\quad g_c = i_o/v_{ic} \quad$ for $v_{id} = 0$

with the E link here able to be broken, and with the negative feedback afforded by the now enhanced $2r_{tail}$, then

$$i_{cc} \approx v_{ic} \times (g_{tail}/2)$$

With i_o as in [7.47] then

$$g_c \approx -g_{tail}/\beta,$$

Common-mode rejection

With $\quad CMR_{tail} = |g_d|/|g_c| \quad (7.48)$

then $\quad CMR_{tail} \approx \beta g_m/g_{tail} \quad (7.49)$

giving $\quad$ $\mathbf{CMR_{tail} = 28.3M}$

With the previous $\quad CMR_{tail} = g_m r_{tail} \quad$ as in [7.42]

the factor β by which the mirror improves upon this is the result of the $\times 2/\beta$ reduction in g_c and the doubling of g_d by the recovery of the previously wasted i_{c-}.

Component mismatch

The inevitable small differences in doping concentration and device area over the integrated circuit cause the device parameters such as g_m to differ slightly for the two halves of the LTP. With the matching typically 1% then this is the other main factor limiting the final CMR.

LTP g_m mismatch

Without the tail, the 1% g_m mismatch alone would give CMR as only ~100, heavily dominating the above CMR_{tail}. But with the $v_{be} = v_{ic}$ for v_{ic} applied alone, addition of the tail considerably reduces v_{be}, proportionately reducing the effect of the mismatch component Δg_m. From the single (E) node analysis

$$v_{be} = v_{ic}(g_o + g_{tail}/2)/(g_m + g_\pi + g_o + g_{tail}/2)$$

$$\approx v_{ic}(g_o + g_{tail}/2)/g_m.$$

With the resulting

$$i_{cd} = \Delta g_m v_{be}, \quad i_o \approx i_{cd}, \quad \text{and } g_c = i_o/v_{ic}$$

then $\quad g_c \approx (\Delta g_m/g_m) \times (g_o + g_{tail}/2).$

With $\quad g_d \approx g_m \quad$ and $\quad CMR = g_d/g_c \quad$ then

$$CMR(LTP\ \Delta g_m) = (g_m/\Delta g_m) \times [g_m/(g_o + g_{tail}/2] \quad (7.50)$$

giving $\quad$ $\mathbf{CMR(LTP\ \Delta g_m) = 277k.}$

LTP g_o mismatch

With Δg_o the mismatch in the two g_o, and for v_{ic} applied alone giving $v_e \approx v_{ic}$, then

$$i_{cd} = v_{ic}\Delta g_o$$

giving $\quad CMR(LTP\ \Delta g_o) = (g_o/\Delta g_o) \times (g_m/g_o) \quad (7.51)$

and thus $\quad$ $\mathbf{CMR(LTP\ \Delta g_o) = 280k.}$

LTP r_π mismatch

For v_{ic} applied alone

$$i_{c-} = i_{c+} = g_m(v_{ic} - v_e) - v_e g_o.$$

With this independent of r_π, a differential change in the r_π has no effect. All a change in one of the r_π does is to change v_e, and this changes the i_c equally.

Mirror g_m mismatch

Here the associated Δg_m gives the component

$$i_o = i_{cc} \times (\Delta g_m/g_m)$$

With $\quad i_{cc} \approx v_{ic}/2r_{tail}$

for v_{ic} applied alone, and with $g_c = i_o/v_{ic}$, then

$$g_c = (g_{tail}/2) \times (\Delta g_m/g_m)$$

giving $\quad \text{CMR(mirror } \Delta g_m) = g_m r_{tail} \times (g_m/\Delta g_m) \quad (7.52)$

and thus $\quad$ **CMR(mirror Δg_m)=14.1M.**

Review of components of CMR

Tables 7.2 and 7.3 summarise the above relations and calculated values for the various components of the common-mode rejection ratio CMR, with the source equation numbers given.

For the worst-case combination of the mismatch components in Rows 3 and 4 then the final CMR(LTP mismatch) ≈ 140kΩ.

With the mismatch resulting in the differential i_{cd} rather than the common-mode i_{cc}, and with the mirror responding fully to i_{cd}, then the mirror cannot reduce the mismatch components.

With CMR(LTP mismatch) thus setting an upper limit to the overall CMR then the ×β increase in CMR_{tail} given by the mirror (Rows 1 and 2) gives only a mere ×2 increase in the overall CMR.

1	LTP alone	CMR_{tail}	$=g_m r_{tail}$	[7.42]	141 $\times 10^3$
2	LTP+ mirror	CMR_{tail}	$=\beta g_m r_{tail}$	[7.49]	28.3 $\times 10^6$

Table 7.2 CMR components for enhanced tail, apart from component mismatch.

3	LTP Δg_m	CMR (LTP Δg_m)	$=(g_m/\Delta g_m)$ $\times(g_m/(g_o)$	[7.50]	277 $\times 10^3$
4	LTP Δg_o	CMR (LTP Δg_o)	$=(g_o/\Delta g_o)$ $\times(g_m/g_o)$	[7.51]	280 $\times 10^3$
5	Mir. Δg_m	CMR (mir. Δg_m)	$=g_m r_{tail}$ $\times(g_m/\Delta g_m)$	[7.52]	14.1 $\times 10^6$

Table 7.3 CMR components for 1% LTP component mismatch, also with enhanced tail.

Enhanced LTP voltage gain

The other advantage of the current mirror is the increase in the voltage gain a_{LTP} of the LTP. This is for two reasons.

- With the mirror added both of the i_c are used.
- The mirror presents the active load r_{o4} to Q_2, instead of the previous R_L.

With $\quad a_{LTP} = g_d r_{o.LTP}$

where $r_{o.LTP}$ is the LTP output impedance,

with $\quad r_{o.LTP} = r_{o4} // r_{o2}$

and with $\quad g_d \approx g_m$

then $\quad a_{LTP} \approx -g_m r_{o4} // r_{o2} \quad (7.53)$

giving $\quad$ **$a_{LTP} = -1400$**

a considerable increase relative to the value of 100 without the mirror. The value is the same as for the CE stage with the active PNP load in Section 6.3.

The relation $r_{o.LTP} = r_{o4} // r_{o2}$ is not quite as obvious as it looks. See Exercise 9.1 for the MOS-based LTP for the comparable discussion.

Thermal stability

For the LTP the I_c are stabilised against temperature change by the tail resistor, much as for the feedback bias circuit of Fig. 6.20 but now with DC coupling to the input.

Also with Q_1 and Q_2 close together on the same chip, the temperature dependence of the differential input offset is considerably reduced from the 2mV/° of the single transistor. Together this gives the typical 5μV/° for the op-amp.

Frequency response

With the differential voltage gain a_d the same as if the emitter was grounded, then the high frequency cut-off for the LTP is virtually the same as for the single-stage common-emitter amplifier.

7.7 Bipolar inverter switch

The emitter-collector path for the bipolar transistor can be used as a switch, with the inverter of Fig. 7.14 a simple example. This is the basis of the earliest form of digital logic, giving the input-output relations of Table 7.4. This, the 'resistor-transistor logic' (RTL) family, has long been outdated, first by the 'transistor-transistor logic' (TTL), and now by the high-speed CMOS (HCMOS) family.

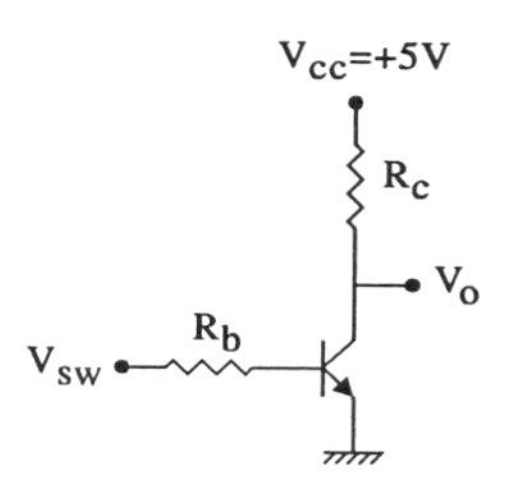

Fig. 7.14 Inverter based on bipolar transistor switch.

However, more generally the circuit forms the analogue shunt switch of Fig. 7.17(a) and (c), with V_{cc} replaced by the analogue input voltage V_{in}. With an understanding of the factors limiting the switching speed needed here, it is convenient to examine this for the case of the inverter.

More recently even the function of the bipolar analogue switch has largely been taken over by the CMOS switch, so the treatment will be brief.

Input		Output	
$V_{in}=0$	Logical 0	$V_o=V_{cc}$	Logical 1
$V_{in}=V_{cc}$	Logical 1	$V_o=V_{sat}$	Logical 0

Table 7.4 Input-output relations for inverter of Fig. 7.14. $V_{cc}=5V$. Saturation voltage $V_{sat} \approx 150mV$.

Experimental result

Fig. 7.15(b) shows some sketched experimental results from Exercise 7.2, where the BC182 is square wave switched as in (a). With this transistor designed for linear operation, the switching speed is much lower than for the switching type. It is used here because the periods T_b, etc., are larger, and thus better able to be displayed experimentally. Also, for clarity, some of the smaller intervals such as T_b and T_m in Fig. 7.15(b) have been sketched larger than actually obtain. The modifications for the normal switching transistor are described at the end of the section.

Transistor capacitances

With c_π and c_μ in Fig. 7.15(a) the incremental values, the extent to which these can be used to predict the large-signal switching transients in (b) is limited.

With $c_{\pi d}$ and $c_{\pi s}$ the depletion layer and 'stored charge' components of c_π, and with c_μ that of the depletion layer, the values of $c_{\pi d}$ and c_μ are approximately valid for the present case. Even here, as discussed later in Section 21.2, these vary with junction voltage, but by no means as much as does $c_{\pi s} \propto I_c$ with I_c.

Allowing for the $\approx$2pF track-to-track capacitance of the Experimentor board, and for the 11pF scope probe capacitance, the approximate effective

$$C_{\pi d}=12pF, \quad C_\mu=7pF, \quad C_p=13pF.$$

Interval T_b With $T_b=R_bC_{\pi d}$
then **$T_b=120ns$.**

Interval T_m With the Miller effect dominating, then

$$dV_o/dt=-I_b/C_\mu.$$

With $dV_o/dt \approx V_{cc}/T_m$

then $T_m=C_\mu V_{cc}/I_b$

giving **$T_m=70ns$.**

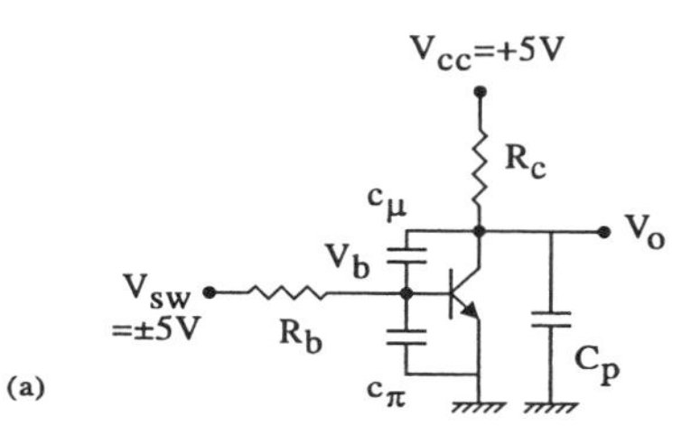

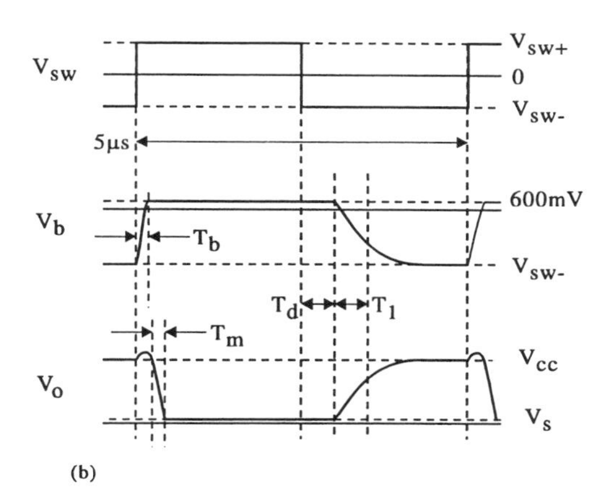

Fig. 7.15 Switching transients for inverter of Fig. 7.14. Sketched experimental result for $R_b=R_c=10k\Omega$.
(a) Circuit diagram showing transistor capacitances.
(b) Waveforms.

Stored charge

The delay interval T_d and the following T_1 are the consequence of the base stored charge Q_s discussed in Section 6.5 for the unsaturated transistor. While $c_{\pi s}$ is the direct consequence of Q_s, for the large-signal equivalent we must reason afresh, starting with Q_s.

With the electron density ρ in the base as in Fig. 6.12(a), where the diffusion controlled I_c is proportional to the concentration gradient, for the saturated transistor this becomes as in Fig. 7.16.

For I_b increasing from zero then (a) represents ρ at the point of saturation. Here V_{bc} has risen to a value close to the $V_{be} \approx 600mV$ needed for significant conduction of the junction. Thus the reverse flow of electrons from collector to emitter has just begun to be comparable with the forward current. Beyond this point the two components of ρ become as in (c) giving the total as in (b). With the gradient remaining constant beyond (a), this is consistent with I_c remaining constant beyond the value of I_b at which saturation occurs.

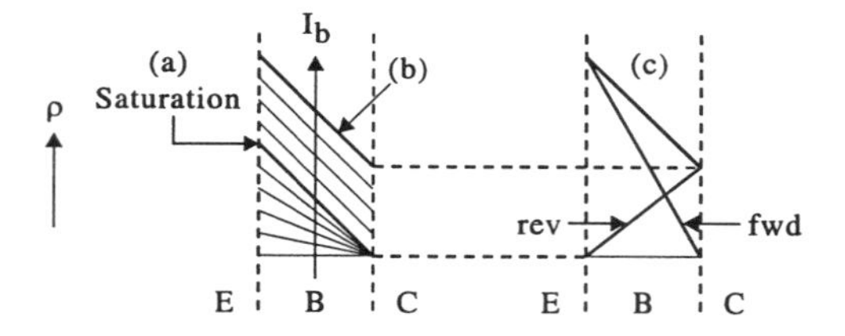

Fig. 7.16 Electron concentration in NPN transistor base
(a) For I_b at the point of saturation.
(b) For I_b beyond saturation.
(c) Components of (b) arising from forward and reverse components of I_c.

Intervals T_d and T_l Recall from Section 6.4 that there are two components of I_b, I_{b1} due to recombination in the base, and I_{b2} due to the flow of holes from base to emitter, with the corresponding β_1 and β_2. With T_r and T_τ the base recombination and transit times then

$$I_b = Q_s/T_r \quad \text{and} \quad \beta_1 = T_r/T_\tau$$

With $T_\tau = 1ns$ and $\beta_1 = 400$ then **$T_r = 400ns$.**

With the magnitude of the negative I_b just after the falling transition in V_{sw} approximately equal to that of the positive I_b just before, then the time $T_d + T_l$ taken to extract Q_s from the base is equal to T_r.

Much as in Fig. 5.25(c,d) for the switched diode, the precise division of T_r into T_d and T_l is a complex matter. Here we shall be content that the calculated sum T_r is consistent with the experimental result.

Switching type of bipolar transistor

With the present BC182 designed for analogue applications then β is made as high as possible. With $T_r \propto \beta_1$ this is counterproductive for switching.

Gold doping With the β required for switching ~10, in contrast to the present 200, this is reduced by gold doping the base. While the exact value of the higher β for the analogue transistor is extremely uncertain, fortunately the lower value for the switching transistor can be controlled much more closely.

Area reduction With T_r thus reduced, the depletion capacitances $C_{\pi d}$ and C_μ become relatively more important. For the switching transistor these are reduced by reducing the device area.

With 1/f noise tending to be associated with the edges of the diffusions on the surface of the planar device, and with area reduction tending to increase the proportion close to the edges, then the i/f noise is increased. This does not matter for the switching transistor, but for the linear transistor is the reason for the area being somewhat larger.

7.8 Bipolar analogue shunt switch

With the general shunt and series switch configurations as in (a) and (b), the bipolar transistor is better suited to the shunt configuration, as in (c), because for the series arrangement I_b has to be returned either through the source or, usually even less desirably, through the load. For the shunt switch I_b is returned via the ground connection

With the switching speed covered above, the remaining points of concern are the 'on'-state switch resistance R_{on}, the small output voltage offset V_{os} when 'on', and the allowable range of V_{in}.

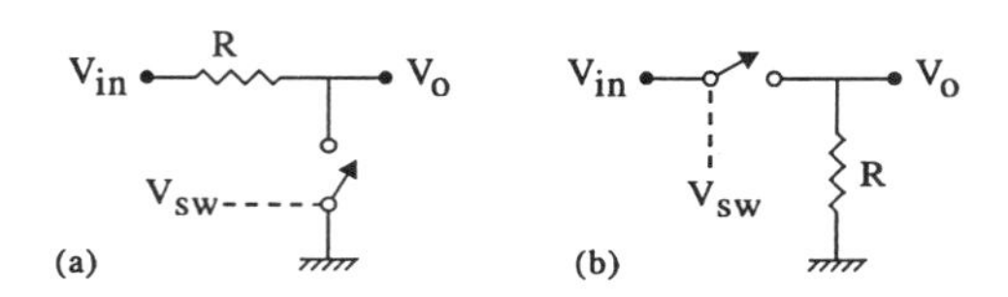

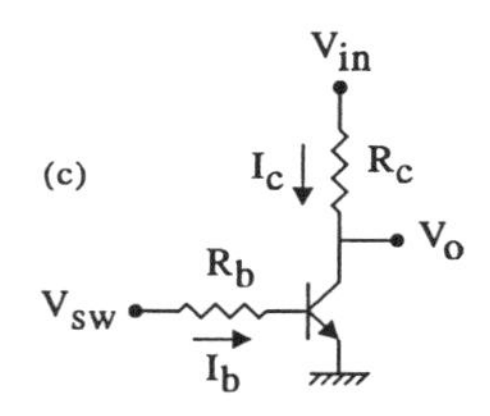

Fig. 7.17 Analogue switch types.
(a) Shunt switch.
(b) Series switch.
(c) Implementation of shunt switch using bipolar transistor.

Forward and reverse β

With the transistor saturated when in the 'on' state then, as for Fig. 7.16(c), both the forward and reverse currents across the base are significant. Here the associated reverse β is termed β_r, with the normal forward β now distinguished as β_f. For the following reasons, apart from a few exceptional transistor types, $\beta_r << \beta_f$. For the unexceptional BC182

$$\beta_f = 200 \quad \text{and} \quad \beta_r = 4$$

Differential doping

As explained in Section 6.4, one of the measures needed to give $\beta_f >> 1$ is to make the emitter doping much greater than that of the base. This is to dominate the flow of holes from base to emitter and thus to prevent β_2 from degrading the overall $\beta = \beta_1 // \beta_2$. For the reverse flow, the method of fabrication involves making the corresponding collector doping even less than that of the base, leaving the associated hole to collector flow far from being dominated.

Base electric field

Also with the method of fabrication, the doping across the base is not constant but somewhat graded. This is normally beneficial, in that it creates an electric field in the base which adds to the effect of diffusion in carrying the electrons across the base in the normal forward direction. Any reverse current, however, is opposed, thus lowering β_r further.

Geometry

Finally, with the geometry of the transistor in Fig. 6.2(a), the area of the collector is larger than that of the emitter. Thus electrons leaving the emitter all face a portion of the collector, but part of the collector has no facing emitter. Thus the electrons leaving the collector from the unfaced region have nowhere to go but to recombine with the base holes, to constitute added base current.

Transistor I_c-V_{ce} relation

With the I_c-V_{ce} relation as in Fig. 6.6 for V_{ce} positive (there for various values of I_b), Fig. 7.18 shows the extension to negative V_{ce}, for just one value of I_b. For simplicity, the slight increase in I_c with V_{ce} above the saturation point V_{sat} is ignored.

Here for $V_{ce} >> V_{sat}$ then $I_c \approx \beta_f I_b$ as usual, but for the remainder the Ebers and Moll relations (Ebers and Moll, 1954) are required giving, as plotted,

$$I_c \approx I_b \beta_r \left\{ \exp\left(\frac{qV_{ce}}{kT} \right) - 1 \right\} \qquad (7.54)$$

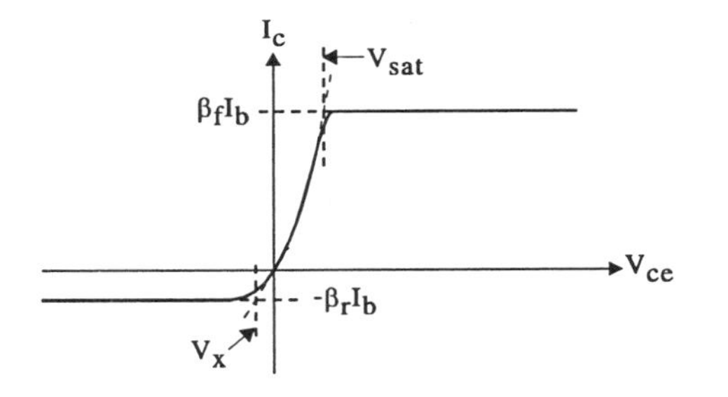

Fig. 7.18 Voltage-current relations for bipolar transistor in Fig. 7.17(c).

With the $I_c \approx \beta_r I_b$ for V_{ce} large and negative, this is as for the circuit an emitter-follower with the transistor C and E connections reversed, and with R_b the signal source impedance. For the following calculations

$$\beta_f = 200, \quad \beta_r = 4, \quad V_{sw} = \pm 5V, \quad R_c = R_b = 10k\Omega.$$

Saturation voltage

With $I_c \approx \beta_f I_b$ for $V_{ce} > V_{sat}$, and otherwise I_c as in [7.54] then V_{sat} is the value of V_{ce} at the intersection giving

$$V_{sat} = (kT/q) \times [\ln(\beta_f/\beta_r) + 1] \qquad (7.55)$$

and thus $\mathbf{V_{sat} = 123mV.}$

On-state resistance

With $R_{on} = (dI_c/dV_{ce})^{-1}$ in [7.54] then for $V_{ce} = 0$

$$R_{on} = (kT/q)/(I_b \beta_r) \qquad (7.56)$$

giving $\mathbf{R_{on} = 12.5\Omega.}$

Transition V_x

With V_x as shown $\quad V_x = -R_{on} I_b \beta_r$

With R_{on} as above then

$$V_x = -kT/q \qquad (7.57)$$

giving $\mathbf{V_x = -25mV.}$

On-state attenuation

For the switch 'on' the transmission factor α_{on} should ideally be zero. With

$$R_{on} << R_c \quad \text{then} \quad \alpha_{on} \approx R_{on}/R_c.$$

With R_{on} as in [7.56] and with $I_b \approx V_{sw+}/R_b$ then

$$R_{on} \approx \frac{(kT/q)}{V_{sw+}} \cdot \frac{R_b}{\beta_r}$$

giving

$$\alpha_{on} \approx \frac{R_b}{R_c} \cdot \frac{(kT/q)}{V_{sw+}} \cdot \frac{1}{\beta_r} \qquad (7.58)$$

and thus $\mathbf{\alpha_{on} \approx 1.25 \times 10^{-3}.}$

Offset voltage

Ebers and Moll (1954) give

$$I_c = -I_e \beta_f / \beta_r \quad \text{for} \quad V_{ce} = 0$$

which with $\beta_f << \beta_r$ gives $I_c \approx I_b$.

With $V_{os} = I_c R_{on}$ then $V_{os} \approx (kT/q)/\beta_r \qquad (7.59)$

giving $\mathbf{V_{os} \approx 6.25mV}$

Inverted connection

If the C and E connections of the transistor are transposed then

$$I_c = -I_e\beta_r/\beta_f \quad \text{giving} \quad I_e \approx I_b$$

and thus $I_c \approx -I_b\beta_r/\beta_f$.

With $V_{os} = I_cR_{on}$ and R_{on} as before

then V_{os} is reduced to

$$V_{os} = (kT/q)/\beta_f \qquad (7.60)$$

giving **V_{os}(inverted) = 125μV.**

Input range

The following values are as calculated in Exercise 7.3.

Switch nominally 'on' This is the range of V_{in} over which the transistor remains saturated. For V_{in} positive it is β_f=200 which is operative, otherwise β_r=4, giving

$-20V < V_{in} < 1000V$.

Switch nominally 'off' This is the range of V_{in} over which the transistor remains non-conducting. Here for V_{in} positive the limit is set by the base-collector reverse breakdown voltage $V_{ce.bk}$, otherwise the much lower base-emitter $V_{be.bk}$, giving

$-5.6V < V_{in} < 55V$.

Inverted transistor connection Here to get any 'off' state range at all, the V_{sw} must be reduced, optimally to ± 2.8V. Even here the range is reduced to

$-2.8V > V_{in} > +2.8V$.

Reference

Ebers, J. J. and Moll, J. L. 1954: *Large-Signal Behaviour of Junction Transistors,* Proc. IRE, **42**.
Gibbons, J. F. 1966: *Semiconductor Electronics*. McGraw-Hill.

7.9 Exercises

Exercise 7.1 Cascode

Calculation. Component values For the circuit of Fig. 7.10(b), calculate suitable values for the bias chain R_1 to R_3 with the given

$$V_+ = 15V, \quad V_{e1} = 1V, \quad V_{e2} = 5V$$

and where the acceptable variation ΔI_c in I_c due to the production spread in β is 5%. For the present BC182

$$\beta_{max} = 480 \quad \text{and} \quad \beta_{min} = 100.$$

With $\Delta I_{b1} = I_c/\beta_{min} - I_c/\beta_{max}$

and $\beta_{max} >> \beta_{min}$

we assume the slightly larger $\Delta I_{b1} = I_c/\beta_{min}$.

With $\Delta V_{e1} \approx \Delta V_{b1}$ and $V_{e1} = I_cR_e$

then the acceptable $\Delta V_{b1} = 0.05 \times V_e$.

With $\Delta V_{b1} = \Delta I_{b1}R_1//(R_2+R_3)$

and with $R_1 << (R_2+R_3)$

from the above voltages, we assume the slightly larger

$$\Delta V_{b1} = \Delta I_{b1}R_1$$

giving $\Delta V_{b1} = I_cR_1/\beta_{min}$.

Thus $R_1 = 0.05 \times V_e/(I_c/\beta_{min})$
giving **$R_1 = 5k\Omega$.**

With $R_2 = R_1(V_{b2} - V_{b1})/V_{b1}$
then **$R_2 = 12.5k\Omega$.**

With $R_3 = R_1(V_+ - V_{b2})/V_{b1}$
then **$R_3 = 29.4k\Omega$.**

Exercise 7.2. Bipolar transistor switching time

Experiment Connect the circuit of Fig. 7.15(a) using the BC182 transistor and $R_b = R_c = 10k\Omega$. Thus confirm that the waveforms are essentially as in (b), with the calculated switching times

$T_b = 120ns$, $T_m = 70ns$, $T_d + T_1 = T_r = 400ns$

and with $C_p = 13pF$ somewhat extending the measured values.

Repeat with R_b and R_c both reduced to 1kΩ. Confirm that this reduces T_b in proportion, makes little difference to T_d, and reduces T_1 somewhat.

Replace the BC182 with a 2N2369, a typical switching transistor, with $R_b = R_c$ restored to 10kΩ. Note how T_d is much reduced, but T_1 only to $C_pR_c = 130ns$.

Reduce $R_b = R_c$ to 1kΩ and note how T_1 is now reduced also.

Exercise 7.3 Analogue switch input range

Calculate the approximate range limits for the 'on' and 'off' switch states for both the normal and inverted transistor connections.

For the BC182 used **$V_{be.bk} = -5V$, $V_{cb.bk} = 60V$.**

Also **$V_{sw} = \pm 5V$, $R_b = R_c = 10k\Omega$.**

'Off' – lower limit The BC junction conducts for $V_{bc} > 600mV$.

Here $V_b = V_{sw-}$ and $V_c = V_{in}$

so conduction is avoided if

$$V_{in} > V_{sw-} - 600mV$$

giving **$V_{in} > -5.6V$.**

With $V_{sw-} = V_{be.bk}$ then no further extension by lowering V_{sw-} is possible.

'Off' – upper limit With

$$V_c = V_{in} \quad \text{and} \quad V_b = V_{sw-},$$

and the limit where $V_{cb} > V_{cb.bk}$

then the limiting $V_{in} = V_{cb.bk} + V_{sw-}$

giving **$V_{in} < 55V$.**

'On' – upper limit With the transistor remaining saturated, for

$$V_{in}/R_c < \beta_f I_b \quad \text{and with} \quad I_b \approx V_{sw+}/R_b$$

then the limiting $V_{in} = \beta_f (R_b/R_c) V_{sw+}$

giving **$V_{in} < 1000V$.**

'On' – lower limit Similarly

$$V_{in} > \beta_r (R_c/R_b) \times (-V_{sw+})$$

giving **$V_{in} > -20V$.**

Inverted connection For the 'on' state the ample limits are much as before. For the 'off' state the range is reduced. With

$$V_{be.bk} = -5V \quad \text{and} \quad V_{sw-} = -5V$$

then the normal BE junction breaks down for $V_{in} > 0V$. With the lower limit at

$$V_{sw-} - 600mV$$

as for the normal connection then V_{sw-} must be increased to –2.2V to obtain a symmetric range of from

$$-2.2V - 600mV \quad \text{to} \quad -2.2V + 5V$$

giving **$-2.8V > V_{in} > +2.8V$.**

8

Basic MOS circuits

Summary

Advances in the development of the 'metal-oxide silicon field-effect transistor' (MOSFET, or MOS for brevity) are leading to its much wider use in analogue integrated circuits, and its exclusive use in almost all digital integrated circuits. This chapter broadly covers the sequence of Chapters 6 and 7 for the bipolar transistor, thus for the single stage

MOS	Bipolar
Common-source	Common-emitter
Source-follower (common-drain)	Emitter-follower (common-collector)
Common-gate	Common-base

While the MOS cascode is covered, the MOS LTP is left to the next chapter, as the first stage in the development of the MOS-based op-amp. MOS-based active filters are covered in Chapter 13.

The final topic is the MOS switch, briefly as used in digital logic, and more fully as for analogue switching.

For the simpler calculations the results are set bold to the right-hand margin, and are cumulative. Details of the less direct calculations are given in the Exercises section, derived from the primary values in Table 8.2.

8.1 MOS operation (descriptive)

Fig. 8.1(a) shows a cross-section of the N-channel MOS before any connection is made to it, with the N-type regions constituting source and drain introduced by impurity diffusion in the normal way.

When a positive gate-to-source voltage V_{gs} is applied as in (b) then the mobile holes under the gate are driven away, leaving the depletion region shown, an extension of the negatively charged half of the source and drain depletion layers. With this balancing the positive charge at the lower surface of the gate, then the gate and substrate constitute the two plates of a capacitor.

As V_{gs} is increased there comes a point where the additional negative charge begins to be in the form of a conducting 'channel' between source and drain, as in (c). With V_{th} the 'threshold voltage' at which the channel begins to form then typically **V_{th}=700mV.**

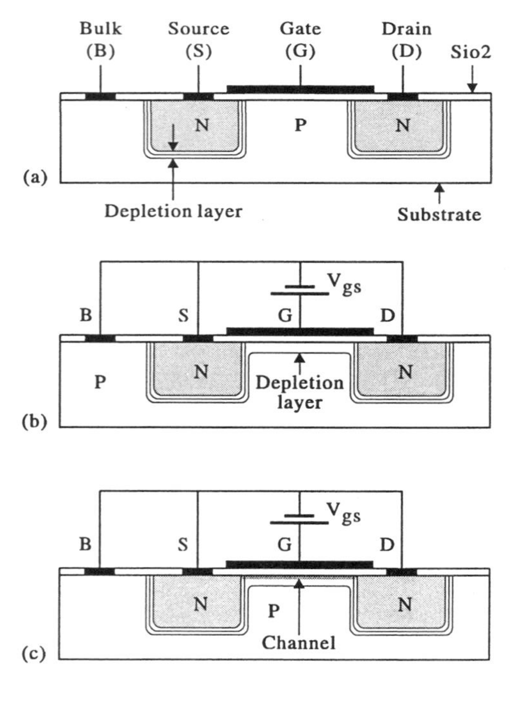

Fig. 8.1 N-channel MOS transistor, formation of channel.
(a) Without external connection.
(b) V_{gs} forms depletion layer under gate.
(c) Increased V_{gs} forms conducting 'channel'.

With G_c the channel conductance, and for $V_{gs} > V_{th}$, then $G_c \propto V_{gs} - V_{th}$, giving a voltage-controlled resistor. Alternatively for V_{gs} switched from below V_{th} to well above, then the device becomes a voltage-controlled switch.

The terms 'bulk' or 'body' are used as synonyms for 'substrate', to be distinct from the 'source' in suffixing.

Pinch-off

The main analogue application of the MOS is as a voltage amplifier. Here the MOS is connected as for the bipolar transistor, with the positive supply V_+ connected to the drain via a usually active load, and the signal input applied to the gate, with the positive bias needed to set $V_{gs} > V_{th}$.

As for the bipolar transistor, the requirement for voltage gain is that $\partial I_d/\partial V_{ds} << \partial I_d/\partial V_{gs}$. At present the two are comparable, but as V_{ds} is increased then the 'pinch-off' shown in Fig. 8.2 has the required effect.

Open channel

With Q_a the charge per unit area viewed from above the channel, and for $V_{ds}=0$, then Q_a is uniform along the length of the channel as in (a), with the value Q_{ao} in (b).

Closing channel

With V_{gd} the gate-to-drain voltage, and as V_{ds} is applied, then V_{gd} is reduced by V_{ds}, causing Q_a to be reduced at the drain end as in (c), ultimately to 'pinch off' the channel completely.

Pinched-off channel

For V_{ds} higher still, the effect is as in (d). With R_c now the resistance of the channel from the source to the tip of the semi-triangular region, then I_d becomes fixed at the value giving I_dR_c equal to that of V_{ds} at pinch-off.

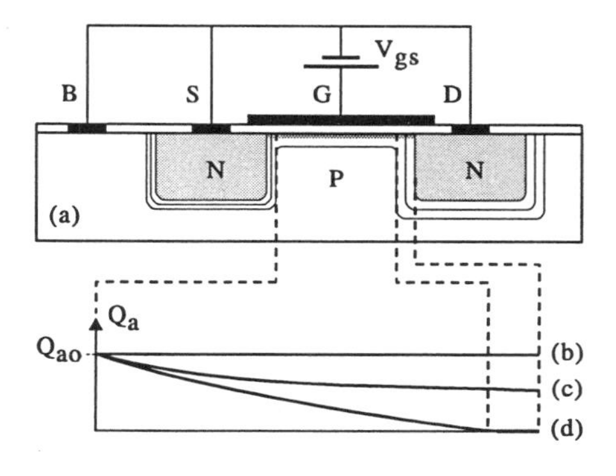

Fig. 8.2(a) to (d) MOS transistor pinch-off.
(a) MOS with V_{ds} supply, but for $V_{ds}=0$.
(b) Uniform charge/area Q_{ao} at surface of substrate for $V_{ds}=0$.
(c) Q_a for $V_{ds} > 0$ but below pinch-off value V_p.
(d) Q_a for $V_{ds} > V_p$.

The remaining component $V_{ds} - I_dR_c$ of V_{ds} is developed across the short region for which $Q_a=0$, somewhat as if it was part of a depletion layer.

Drain voltage-current relation

With initially $I_d=V_{ds}/R_c$ then I_d increases in proportion to V_{ds} as in (e).

As the channel begins to close then R_c increases, causing the slope to decrease. Finally at pinch-off there should be no further increase in I_d at all.

The reason for the small increase that does occur is that, as the pinched-off region extends back into the channel, then the length of the triangular region is reduced. This reduces R_c, requiring a larger I_d to maintain I_dR_c equal to V_p.

The relation differs from the bipolar transistor, in that beyond pinch-off the slope does not become constant but continues to decrease.

The final 'short channel' upswing is the result of the length of the channel having been so much reduced as to no longer be large compared with the depth.

With $\partial I_d/\partial V_{ds}$ and $\partial I_d/\partial V_{gs}$ comparable for V_{ds} below pinch-off, and with the large reduction in $\partial I_d/\partial V_{ds}$ beyond pinch-off, then beyond pinch-off $\partial I_d/\partial V_{ds} << \partial I_d/\partial V_{gs}$ as required for potential voltage gain.

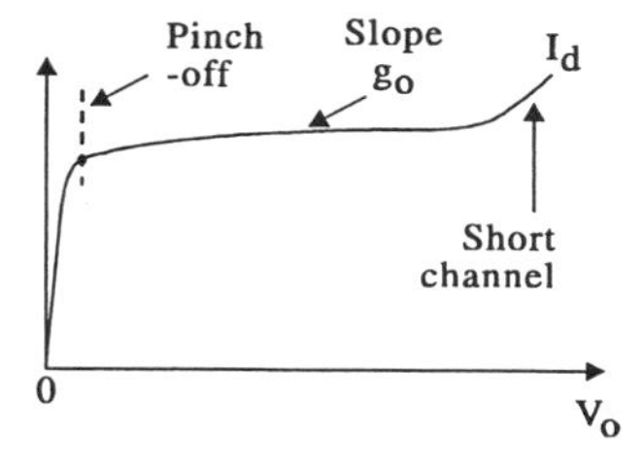

Fig. 8.2(e) Variation of channel current I_d with V_{ds}.

Gate current

Because the gate is insulated from the rest of the device then the gate current I_g is vastly less than the bipolar transistor I_b. This is the major difference between the two types.

Bulk (body) effect

In Fig. 8.1(c), it is the potential between gate and substrate that forms and controls the channel. Thus if B is disconnected from S and a voltage V_{bs} applied, then this will have an effect on I_d comparable with that of V_{gs}.

Mainly due to the relative widths of the layer of SiO_2 under the gate and the channel-bulk depletion layer, V_{gs} has the greater effect, with typically $\partial I_d/\partial V_{bs}=\partial I_d/\partial V_{gs} \times (1/5)$.

With B and S usually connected together as shown, the effect does not occur.

CMOS technology

The P-channel MOS also exists, being the N-channel device with the doping types reversed.

Fig. 8.3(b) shows how both are incorporated into the P-type substrate, with the depletion layers (not shown) Insulating one region from the next.

For the P-channel device the terms 'substrate' and 'bulk' have distinct meanings: 'substrate' refers to the whole slab supporting all devices, while 'bulk' refers to the N-well.

For the N-channel device bulk and substrate are the same, with the suffix 'b' always used, to distinguish substrate from source.

For this technology the P-channel device is a little inferior to the N for two reasons.

(i) The carrier mobility μ, upon which $\partial I_d/\partial V_{gs}$ depends, decreases with increasing channel doping concentration. This needs to be about ×10 higher for the well than for the substrate in order to provide the insulating depletion layer.
(ii) Hole mobility is generally lower than electron mobility by ≈×1/2.5.

For the less often used P-well technology all dopings are reversed. More recently the 'twin-tub' technology has been used. Here the initial substrate doping is reduced, all devices are placed in wells, and the well doping is comparable with the present P_-. Thus the performance of the P-channel device is brought closer to that of the N. Also the bulk effect is eliminated.

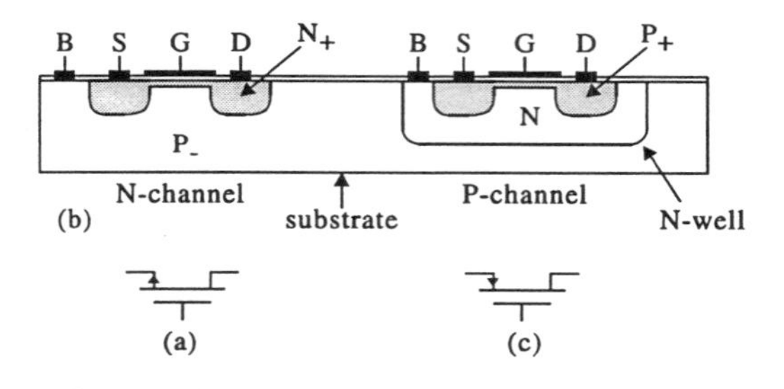

Fig. 8.3(b) CMOS N-well technology.
P_-: light P-channel doping ≈ $2 \times 10^{21}/m^3$.
N: moderate N-Channel doping ≈$1.5 \times 10^{22}/m^3$.
N_+, P_+: heavy doping > $10^{25}/m^3$
(a,c) Circuit symbols.

8.2 MOS operation (quantitative)

In this section the expressions for the MOS drain current I_d, the mutual conductance $g_m=\partial I_d/\partial V_{gs}$ and the output resistance $r_o=1/(\partial I_d/\partial V_{ds})$ are derived. The account is initially for the N-channel MOS.

Channel current I_d

With V_{th} the 'threshold' value of V_{gs} at which the channel begins to form, the 'effective'

$$V_{eff}=V_{gs}-V_{th} \quad (8.1)$$

V_{ds} well below pinch-off With the total channel charge Q_c, and T_c the time taken for Q_c to traverse the length L of the channel,

$$I_d=Q_c/T_c \quad (8.2)$$

With the length L and width W of the channel as seen from above the substrate as in Fig. 8.4(c), and Q_{ao} the channel charge per unit area across the substrate for $V_{ds}=0$ as in (a),

$$Q_c \approx WLQ_{ao} \quad (8.3)$$

With C_{ox} the capacitance per unit surface area of the gate-to-channel capacitor

$$Q_{ao}=C_{ox}V_{eff} \quad (8.4)$$

With μ the channel mobility then the electron velocity

$$dx/dt=\mu V_{ds}/L \quad (8.5)$$

With $dx/dt=L/T_c$ and the above relations

$$I_d=\mu C_{ox}(W/L)V_{ds}V_{eff} \quad (8.6)$$

With g_{co} the channel conductance dI_d/dV_{ds} for $V_{ds}=0$

$$g_{co}=\mu C_{ox}(W/L)V_{eff} \quad (8.7)$$

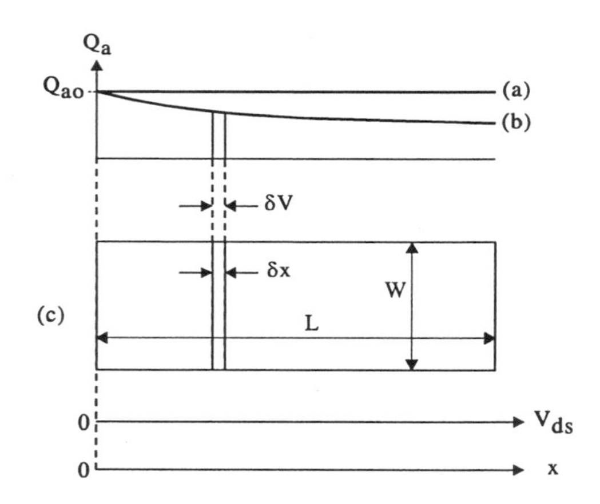

Fig. 8.4 Diagrams for calculation of variation of I_d with V_{ds}.
(a) Uniform charge per unit area Q_a along length L of channel for $V_{ds}=0$.
(b) Variation of Q_a for $V_{ds} > 0$ but below pinch-off.
(c) Channel as seen from above substrate.

V_{ds} rising to pinch-off With δQ the charge for the segment in (c), and with δT_c the time taken to traverse the segment,

$$I_d=\delta Q/\delta T_c \quad (8.8)$$

With V_x the channel voltage at point x

$$Q_a=Q_{ao}(V_{eff}-V_x)/V_{eff} \quad (8.9)$$

With Q_a as in (b)

$$\delta Q = W\delta x Q_a \qquad (8.10)$$

With $\qquad dx/dt = \mu(dV_x/dx)$,

with Q_{ao} as in [8.4], and with [8.8] to [8.10],

$$\int_0^{V_{ds}} \mu C_{ox} W(V_{eff} - V_x).dV_x = I_d \int_0^L dx$$

giving

$$I_d = \mu C_{ox} \frac{W}{L}\left(V_{eff}V_{ds} - \frac{V_{ds}^2}{2}\right) \qquad (8.11)$$

V_{ds} at pinch-off With here $V_{gd} = V_{th}$, and with $V_{eff} = V_{gs} - V_{th}$ as in [8.1],

$$V_{ds} = V_{eff} \qquad (8.12)$$

With I_d as in [8.11]

$$I_d = \mu C_{ox}(W/L)V_{eff}^2/2 \qquad (8.13)$$

Here $\partial I_d/\partial V_{ds}$ for [8.11] is zero, as expected.

V_{ds} beyond pinch-off The small increase in I_d with V_{ds} is allowed for by adding the term λ to [8.13], to give

$$I_d = \mu C_{ox} \frac{W}{L}\frac{V_{eff}^2}{2}[1 + \lambda(V_{ds} - V_{eff})] \qquad (8.14)$$

where

$$\lambda = \frac{k_{ds}}{2L\sqrt{V_{ds} - V_{eff} + \psi_o}} \qquad (8.15)$$

and

$$k_{ds} = \sqrt{2\varepsilon_s\varepsilon_o / (qD)} \qquad (8.16)$$

Here ψ_o is the built-in junction potential, ε_o the permittivity of free space, ε_s the relative permittivity of Silicon, q the electron charge, and D the doping concentration at the channel (Johns and Martin, 1997).

Mutual conductance

With $g_m = \partial I_d/\partial V_{eff}$ at pinch-off and beyond, and with I_d as in [8.13]

$$g_m = \mu C_{ox}(W/L)V_{eff} \qquad (8.17)$$

Alternatively, with V_{eff} as in [8.13],

$$g_m = [2\mu C_{ox}(W/L)I_d]^{1/2} \qquad (8.18)$$

With I_d as in [8.13], and with g_m as in [8.17], these are plotted in Fig. 8.5. While the I_d relation in (a) is parabolic rather than linear, the degree of non-linearity is considerably less than for the exponential variation of I_c with V_{be} for the bipolar transistor.

Output conductance

With $g_o = \partial I_d/\partial V_{ds}$ and I_d as in [8.14] to [8.15] then

$$g_o = \lambda I \qquad (8.19)$$

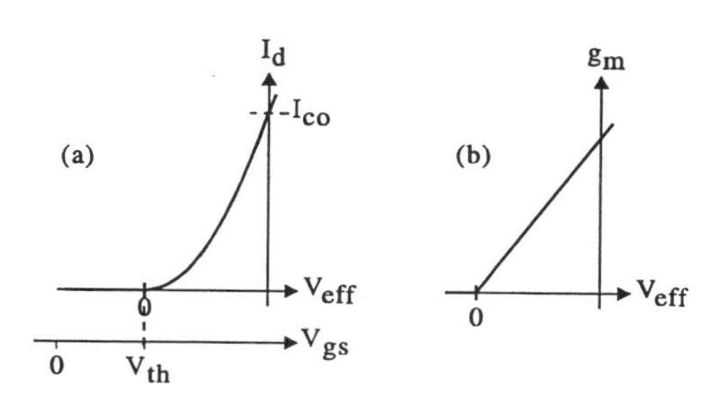

Fig. 8.5 Variation of MOS drain current I_d and $g_m = \partial I_d/\partial V_{gs}$ with $V_{eff} = V_{gs} - V_{th}$.

8.3 Common-source voltage amplifier

Fig. 8.6(a) shows the essentials of the common-source voltage amplifier, here with the load the ideal current source I_L set to the design $I_d = 100\mu A$ for the transistor.

The bias component of V_{in} is then set to give the 'mid-point' $V_o = V_+/2$.

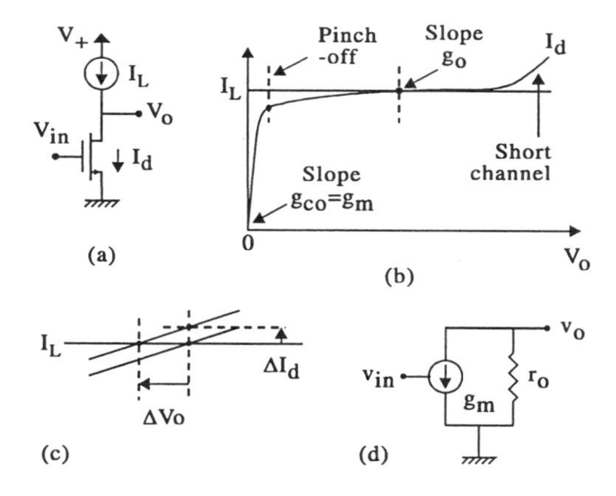

Fig. 8.6 Elementary common-source MOS voltage amplifier.
(a) Circuit diagram.
(b) Output characteristic.
(c) As (b) expanded at operating point, for increment ΔV_{gs}.
(d) Incremental circuit for (a).

Incremental equivalent circuit

With $\qquad i_d = (\partial I_d/\partial V_{gs}) \times v_{gs} + (\partial I_d/\partial V_{ds}) \times v_{ds}$,
and

$$g_m = (\partial I_d/\partial V_{gs}) \quad \text{and} \quad g_o = (\partial I_d/\partial V_{ds})$$

then

$$i_d = g_m v_{gs} + v_{ds} g_o \qquad (8.20)$$

With the incremental $i_d = 0$ for the current-source load, and with $v_{ds} = v_o$, then the incremental equivalent circuit becomes as in Fig. 8.6(d).

Voltage gain

From the incremental circuit, and with the voltage gain of the common-source stage $a_{cs}=v_o/v_{in}$, then

$$a_{cs}=-g_m r_o \quad (8.21)$$

Fig. 8.6(b) and (c) gives an alternative viewpoint. Here the horizontal line in (c) represents the voltage-current relation for the current source, with the V_{in} bias set to place the 'operating point' at the required mid-point $V_o=V_+/2$.

The expanded section in (c) then shows the transistor I_d-V_{ds} relation for two values of V_{gs} differing by the relatively small ΔV_{gs}. With the incremental common-source voltage gain

$$a_{cs}=dV_o/dV_{in},$$

then
$$a_{cs}=-(\partial I_d/\partial V_{gs})/(\partial I_d/\partial V_{ds}) \quad (8.22)$$

With $g_m=\partial I_d/\partial V_{gs}$ and $r_o=1/(\partial I_d/\partial V_{ds})$

then [8.21] is confirmed.

Circuit values and parameters

For **$V_+=10V$ and $I_d=100\mu A$**
with mid-point bias giving **$V_o=5V$**

the resulting values of V_{eff}, g_m and r_o are calculated in Exercise 8.1, and listed in Table 8.1.

	N-channel	P-channel
V_{eff}	220mV	311mV
g_m	910μS	644μS
r_o	118kΩ	320kΩ

Table. 8.1 Incremental values for N- and P-channel MOS for drain current $I_d=100\mu A$.

With $a_{cs}=-g_m r_o$ as in [8.21], then for the present N-channel MOS **$a_{cs}=-108.$**

With the load impedance in reality finite then a_{cs} will be below this value. However, the term $g_m r_o$ is frequently encountered throughout this and the next chapter, so we define

$$a_v=g_m r_o \quad (8.23)$$

with $a_{cs}=a_v$ for the present circuit.

Ratio of output slopes

From [8.7] and [8.17], the slope g_{co} at the origin of the output characteristic in Fig. 8.6(b) is equal to the value of g_m beyond pinch-off, as marked in the diagram.

With $|a_{cs}|=g_m/g_o$ then $|a_{cs}|$ is given by the ratio of the slope at the origin to that at the operating point.

With the two slopes widely different, this confirms the ability of the transistor to give a voltage amplification >>1.

Load resistor

If the ideal current-source load is replaced by the load resistor R_L then

$$a_{cs}=-g_m(r_o//R_L) \quad (8.24)$$

With $R_L=(V_+-V_o)/I_d$ then for mid-point bias $R_L=50k\Omega$ giving **$a_{cs}=-32.0.$**

Active load

$|a_{cs}|$ can be brought closer to a_v by using the active load Q_P in Fig. 8.7(a). With r_{oN} and r_{oP} the r_o for the N- and P-channel transistors then

$$a_{cs}=-(r_{oN}//r_{oP})g_{mN} \quad (8.25)$$

giving **$a_{cs}\approx-78.4.$**

With the amplifier output impedance r_{oa} the combined $r_{oN}//r_{oP}$ then **$r_{oa}=86.1k\Omega.$**

Output characteristics

With the usual load-line for R_L as in Fig. 8.7(c), that for the active load is the output characteristic of Q_P, as in (b). Notice how the slope for Q_P is lower than that for Q_N, corresponding to the higher r_o.

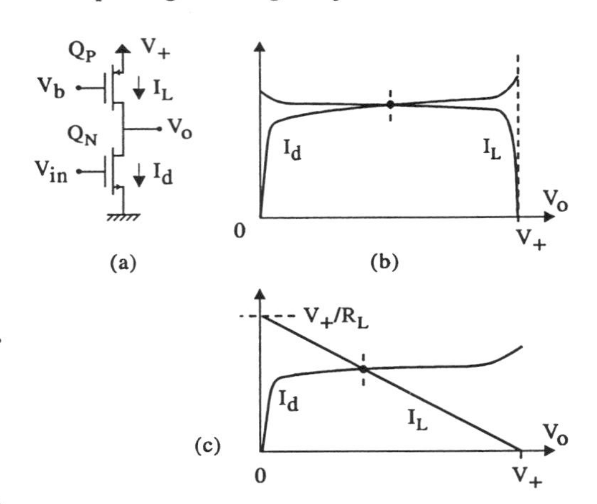

Fig. 8.7 Common-source stage with active load Q_P.
(a) Circuit diagram. V_b=bias.
(b) Output characteristics of Q_N and load Q_P. (L: load, d: drain).
(c) As (a,b) but with load resistor R_L in place of active load.

MOS capacitances

Fig. 8.8 shows the principal capacitances for the MOS at pinch-off, with typically

$c_{gs}=0.2pF$, $c_{gd}=0.05pF$, $c_{sb}=0.2pF$, $c_{db}=0.1pF$.

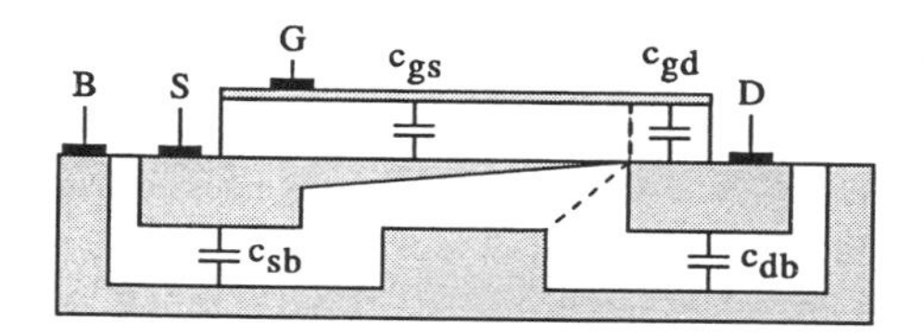

Fig. 8.8 MOS capacitances at pinch-off, dimensions not to scale.

Fig. 8.9(a) shows the incremental circuit of Fig. 8.6(d) with the source impedance

$$\mathbf{R_s=100k\Omega}$$

included, with $r_{oa}=r_{oN}//r_{oP}$ as for the active load, and with the capacitances added. For the active load of Fig. 8.7(a), the total $c_{in}=c_{gs}$, $c_f=c_{gd}$ and $c_o=2c_{db}$ giving

$$\mathbf{c_{in}=0.2pF,\ c_f=0.05pF,\ c_o=0.2pF.}$$

With S and B both grounded, c_{sb} has no effect.

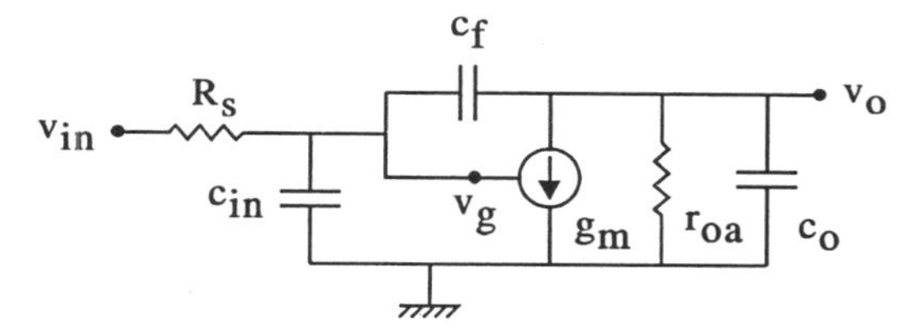

Fig. 8.9(a) Incremental circuit model for common-source stage of Fig. 8.7(a) including source R_s and transistor capactances.

Frequency response

The frequency response for the above circuit is that of a_{cs} in Fig. 8.9(b). Here $|a_o|$ is for the response without the negative Miller feedback through c_f while $|a_{cs}|$ is that for the feedback included. As usual for log plotting, the feedback loop gain is given by the distance between the two, with the transition frequencies f_m and f_{p2} those at which the loop gain is unity, and with the loop gain >1 between the two points.

The sequence is essentially the same as for the bipolar transistor in Section 6.6 where the rather lengthy nodal analysis was carried out, with the result then interpreted. This time we construct the curves by determining the transition points on the curves and the order of the variation with f between the transitions.

Zero frequency gain

With a_{csz} the zero frequency value of a_{cs}, then $a_{csz}=-78.4$ as above for the active load, and as plotted.

Transition at f_m

The Miller enhancement of c_f gives the shunt

$$c_m=(1+|a_{csz}|)c_f \tag{8.26}$$

giving

$$\mathbf{c_m \approx 4pF}$$

With c_m and R_s forming a low-pass filter of cut-off

$$f_m \approx 1/(2\pi c_m R_s) \tag{8.27}$$

giving

$$\mathbf{f_m=401\ kHz.}$$

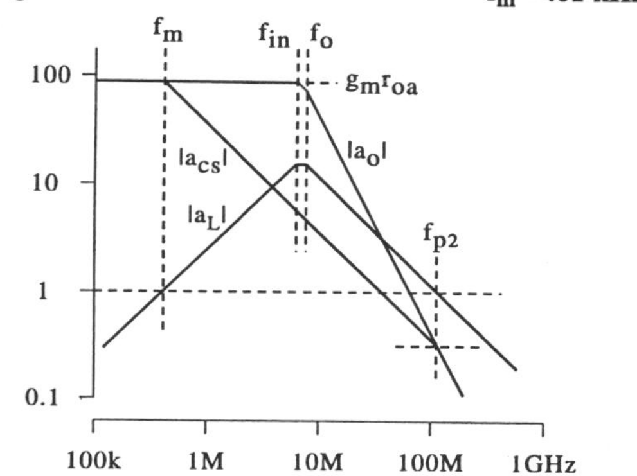

Fig. 8.9(b) Frequency responses derived from incremental model of (a) for common-source stage of Fig. 8.7(a).
$a_{cs}=v_o/v_{in}$ for circuit as shown.
$a_o=v_o/v_{in}$ without feedback through c_f.
a_L=Loop gain for feedback through c_f.

Loop gain

For $f \sim f_m$ the effect of c_{in} and c_o is negligible and $R_s << x_f$. Thus the Miller feedback fraction $\beta_m \approx R_s/|x_f|$.

With $|a_L| \approx |a_{csz}|\beta_m$ then $|a_L| \approx |a_{csz}|2\pi f c_f R_s$.

With $c_m \approx |a_{csz}|c_f$ then

$$|a_L| \approx 2\pi f c_m R_s \tag{8.28}$$

confirming that f_m in [8.27] is the lower limit for $|a_L|>1$.

With $|a_L| \propto f$ and $|a_L|=1$ for $f=f_m$, it is confirmed that $|a_L|$ is as plotted for $f \sim f_m$.

Transition at f_{in}

The above $\beta_m \approx R_s/|x_f|$ only obtains while $(x_{in}//x_f) << R_s$. Beyond this point β_m becomes constant at $\approx x_{in}/(x_f+x_{in})$, causing $|a_L|$ to become constant also. The transition frequency

$$f_{in}=1/[2\pi(c_{in}+c_f)R_s]. \tag{8.29}$$

giving

$$\mathbf{f_{in}=6.37MHz.}$$

Transition at f_o

There is also the frequency f_o at which $|a_L|$ begins to decrease because no longer is $x_o' >> r_{oa}$ where $x_o'=x_o//(x_f+x_{in})$. Here $c_o'=c_o+c_{fin}$ where the series combination

$$c_{fin}=1/(1/c_f+1/c_{in}) \tag{8.30}$$

giving

$$c_o'=c_o+1/(1/c_f+1/c_{in}) \tag{8.31}$$

This too, apart from the transition at f_{in}, would cause $|a_L|$ to level, now at

$$f_o=1/(2\pi c_o' r_{oa}) \tag{8.32}$$

As calculated in Exercise 8.2, **f_o=7.70MHz.**

Thus, as in the plot, the region where $|a_L|$ is constant is brief.

Transition at f_{p2}

This is the 'second pole' frequency at which $|a_L|=1$ and so $|a_{cs}|$ stops being determined by the Miller feedback and instead follows $|a_o|$.

With g_o and G_s both insignificant beyond f_o then the fraction α of $-g_m v_g$ which flows through c_{in} becomes constant at

$$\alpha = c_{fin}/(c_o + c_{fin}) \quad (8.33)$$

Here $|a_L| \approx g_m \alpha x_{in}$ giving

$$f_{p2} = \alpha g_m/(2\pi c_{in}) \quad (8.34)$$

and thus **f_{p2}=121MHz.**

Transition at f_z

There is one further transition (not shown), of which more is said in Chapter 9. This is where the feed-forward transfer through c_f exceeds that via g_m.

Beyond f_z the entire circuit apart from R_s is combined into the impedance $1/g_m//c_{in}//c_o$. Thus $|a_{cs}|$ reverts to being $\propto 1/f$ with f_z thus that of a zero.

This is now the frequency response of a simple low-pass, with the sign inversion of the amplifier reversed. All this happens at far too high a frequency to be of consequence here, but it does become significant later, in the development of the CMOS op-amp. There it is found to be disadvantageous, requiring compensation.

Reduced R_s

For the extreme of $R_s=0$ then neither c_{in} nor the feedback through c_f has any effect. Thus the cut-off becomes f_o=7.70MHz given by $r_{oa}//c_o'$.

Unity gain frequency f_u

This is the frequency for which the gain per stage is unity in a cascade of identical stages. f_u is easier to calculate if the feedback through c_f can be ignored, so we shall do so first.

Starting at the stage v_{gs}, this is multiplied by $g_m(r_o//c_o//c_f//c_{in})$ to equal v_{gs} for the following stage.

For $f \approx f_u$ then r_o can be ignored giving

$$f_u \approx g_m/[2\pi(c_o + c_f + c_{in})] \quad (8.35)$$

and thus **f_u=322 MHz.**

This is comfortably above f_{p2}=119MHz at which the c_f feedback ceases to have effect, so the approximation is valid.

Upper frequency limit

With the above cut-off frequencies typical, and the dramatic recent increase in computing speed, one asks how much more? According to the following, possibly not all that much.

- The unity gain frequency $f_u \approx g_m/2\pi C$ as in [8.35], where C broadly represents the transistor capacitances,
- $g_m=[2\mu C_{ox}(W/L)I_d]^{1/2}$ as in [8.18],
- broadly $C \propto WL$,

giving

$$f_u \propto \left[\frac{I_d}{W}\right]^{1/2} \times \frac{1}{L^{3/2}} \quad (8.36)$$

Here I_d/W is limited by the allowable current density for the device, and L by the limiting potential gradient for breakdown. Thus if L is to be reduced then so also must the supply V_+, with the following consequences.

- MOS circuits able to withstand the normal ±15V supplies for analog circuits based on the bipolar transistor are considerably slower than those like the HC4316 CMOS analogue switch described below, for which the supply is ±5V.
- There is a present move to lower the supply voltage for high-speed CMOS digital logic from the established 5V to now ~3V.

8.4 Source-follower

The function of the source-follower (common-drain) circuit is essentially that of the voltage-follower in Section 2.4, to act as a buffer between signal source and load, without altering the overall voltage gain.

Fig. 8.10(a) shows the circuit, which is as for the common-source stage of Fig. 8.7(a), but with the active load, here Q_2, now in series with the source rather than the drain. The effect of this is to place the output voltage between the signal source and the V_{gs} input, thus giving the 100% negative feedback which is characteristic of the voltage-follower.

Voltage gain

Because the source of Q_1 is no longer grounded then the body effect comes into play, as shown in (b) and (c). Ignoring this for the moment, the incremental circuit becomes as in (c) with r_b omitted. From the single-node analysis then

$$a_{cd} = g_{m1}/(g_{m1} + g_{o1} + g_{o2}) \quad (8.37)$$

giving **a_{cd}=0.99**

very close to the ideal $a_{cd}=1$.

From the form of the expression, this is as for $a_{cd}=1$, with the output impedance $1/g_{m1}$, and with $r_{o1}//r_{o2}$ the follower 'self-load'. Any additional load R_L simply comes in parallel with $r_{o1}//r_{o2}$, to lower the value of a_{cd} further.

Body effect

As noted in Section 8.1, this gives

$$\partial I_d/\partial V_{bs} \sim \partial I_d/\partial V_{gs} \times (1/5)$$

which can therefore be represented by g_{mb} in (b) for Q_1 alone, with $g_{mb} \sim g_m/5$.

With B and D both grounded in (c) then this becomes $r_b = 1/g_{mb}$ as shown, giving **$r_b = 5.50\text{k}\Omega$.**

With the self-load now $r_{o1}//r_{o2}//r_b$ then

$$a_{cd} = g_{m1}/(g_{m1} + g_{mb} + g_{o1} + g_{o2}) \qquad (8.38)$$

giving **$a_{cd} = 0.820$.**
a considerable reduction.

Inverted configuration

The above loss can be avoided by inverting the circuit as in (d), using P-channel transistors. This places Q_1 in the P-channel well, allowing B and S to be connected together.

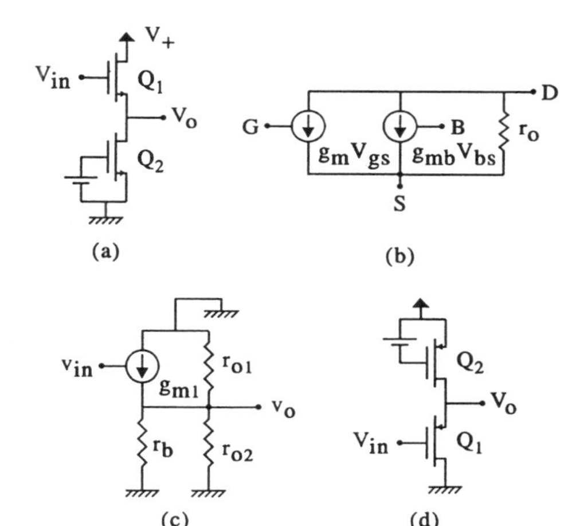

Fig. 8.10 Source-follower.
(a) Circuit diagram for source-follower Q_1 with active load Q_2.
(b) Incremental equivalent circuit of Q_1 with body effect g_{mb}.
(c) Incremental equivalent of (a).
(d) Complementary circuit eliminating r_b.

Effect of capacitances

Fig. 8.11 shows the incremental circuit for the inverted follower of Fig. 8.10(d), with the relevant capacitances included. Here (a) represents the output of a common-source stage which the source-follower is being used to buffer, this being its most common function. With $r_s = r_{oN}//r_{oP}$ **$r_s = 86.1\text{k}\Omega$.**

With r_L in (c) now representing both external and self-load, and initially with no external load, $r_L = r_{o1}//r_{o2}$ and $c_L = c_{sb1} + c_{db2}$. Starting with $c_s = 0$, from the nodal analysis,

$$\frac{v_o}{v_s} = \frac{g_m + b_{gs}}{g_m + b_{gs} + y_L(1 + b_{gs}r_s)} \qquad (8.39)$$

where y_L is the admittance of the load in (c) and $b_{gs} = \omega c_{gs}$.

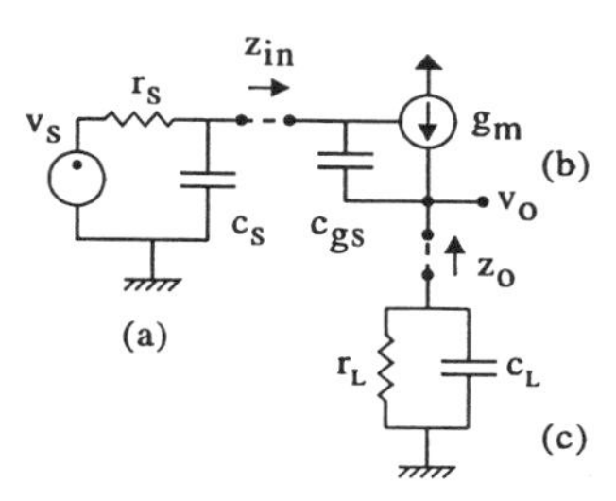

Fig. 8.11 Incremental circuit for inverted source-follower of Fig. 8.10(d) including capacitances.
(a) Output of preceding common-source stage as signal source.
(b) Source-follower apart from self-load.
(c) Self-load.

Output impedance

[8.39] can be written as

$$v_o/v_s = z_L/(z_o + z_L) \qquad (8.40)$$

where

$$z_o = r_s \frac{g_s + b_{gs}}{g_m + b_{gs}} \qquad (8.41)$$

being the output impedance of the follower with signal source.

The transition frequencies for num(z_o) and den(z_o) in [8.41] are respectively

$$f_1 = 1/(2\pi c_{gs} r_s) \qquad (8.42)$$

giving **$f_1 = 9.24\text{MHz}$**

and

$$f_2 = g_m/(2\pi c_{gs}) \qquad (8.43)$$

giving **$f_2 = 512\text{MHz}$.**

Thus the frequency response of $|z_o|$ becomes that emphasised in Fig. 8.12(a) for $c_s = 0$.

For $f \ll f_1$ then the follower reduces the source impedance from $r_s = 86.1\text{k}\Omega$ to the much lower $1/g_m = 1.55\text{k}\Omega$, while for $f \gg f_2$ this reduction has been lost, with z_o approaching r_s.

Resonance

Much as for the voltage-follower and the emitter-follower, z_o is inductive between f_1 and f_2 and so can resonate with a capacitive load. For the present self-load, Exercise 8.3 gives

$r_L = 160\text{k}\Omega$, $c_L = 0.3\text{pF}$, $f_L = 3.32\text{MHz}$.

to give $|z_L|$ as also plotted in (a).

The resonance then occurs at the intersection of the $|z_L|$ and $|z_o|$ lines. With z_o as in [8.41], and $g_m >> b_{gs} >> g_s$ in the region of the intersection then $z_o \approx r_s b_{gs}/g_m = j\omega c_{gs} g_m r_s$. With $z_L \approx x_{cL}$ then $z_L = 1/j\omega c_L$ giving the resonant frequency

$$f_{res} = 1/[2\pi(r_s c_L c_{gs}/g_m)^{1/2}] \qquad (8.44)$$

and thus **f_{res}=56.2MHz**

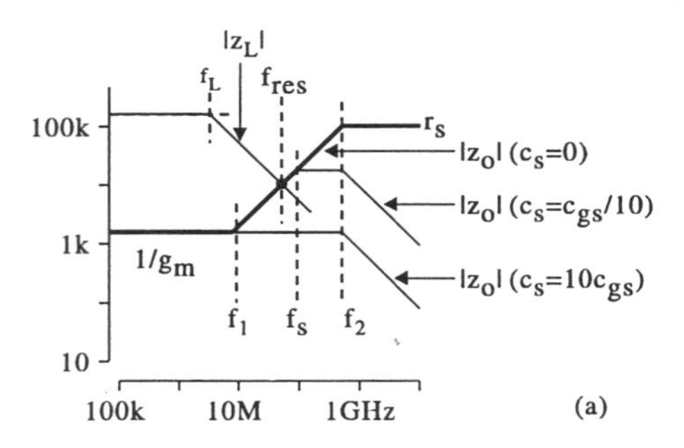

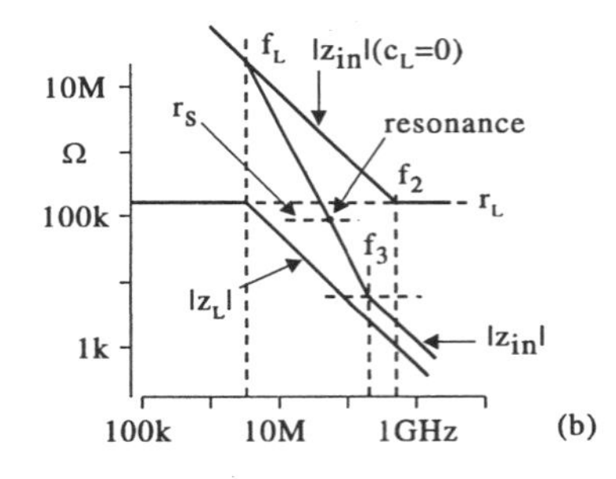

Fig. 8.12 Frequency responses for source-follower of Fig. 8.11. (a) $|z_o|$ for differing c_s showing resonant intersection with $|z_l|$. (b) $|z_{in}|$ showing resonant intersection for c_s=0.

Q-factor

The Q of the resonance is determined by the proximity of the transition frequencies, beyond which the mainly reactive impedances become mainly resistive. Thus

$$Q_1 = f_{res}/f_1, \quad Q_2 = f_2/f_{res}, \quad Q_L = f_{res}/f_L$$

giving **Q_1=6.08, Q_2=9.12, Q_L=16.9.**

With the overall $Q = Q_1 // Q_2 // Q_L$ then **Q=3.00.**

Fortunately this is moderated by c_s to give

$$z_o = \frac{g_s + b_s + b_{gs}}{(g_s + b_s)(g_m + b_{gs})} \qquad (8.45)$$

with the source transition frequency

$$f_s = 1/(2\pi c_s r_s) \qquad (8.46)$$

With $c_s = 2c_{db}$ then c_s=0.2pF giving **f_s=9.24MHz**

Here coincidentally $c_s = c_{gs}$ which with [8.42] and [8.46] gives $f_s = f_1$. This makes it best to consider and plot first the responses for $c_s >> c_{gs}$ and $c_s << c_{gs}$.

$c_s = 10c_{gs}$ Here b_{gs} is negligible in the numerator of [8.45] giving $z_o \approx 1/(g_m + b_{gs})$ and thus the plot shown.

$c_s = c_{gs}/10$ Here it is b_s which is negligible in the numerator. With f_s now increased to **f_s=92.4MHz** the plot becomes as also shown.

$c_s = c_{gs}$. For this, the actual case, the plot is not much different from that for $c_s = 10c_{gs}$. The only difference is a ×2 increase between f_1 and f_2. The main point here is that the strong resonance for c_s=0 is entirely avoided. Beware, however, that for a signal source with a lower effective c_s then the resonance can occur.

Input impedance

Again we start with c_{gs} the only capacitor. Manipulating [8.39] into the form $v_o/v_s = \text{num}/(r_s + z_{in})$, with the numerator independent of r_s, then

$$z_{in} = r_L + x_{gs}(1 + g_m r_L) \qquad (8.47)$$

With $g_m r_L >> 1$ then $z_{in} \approx r_L + x_{gs} a_v$ with the transition frequency f_2 in [8.43] to give the plot in Fig. 8.12(b) for c_L=0. Again above f_2 the follower ceases to have any effect, giving $z_{in} \approx r_L$.

With c_L added then $|z_L|$ plots as in (a), now replicated here in (b), causing the added transition in $|z_{in}|$ at f_L. With [8.47] suitably adapted

$$z_{in} = z_L + x_{gs}(1 + g_m z_L) \qquad (8.48)$$

Here for $f >> f_L$ then $z_{in} \approx x_L + x_{gs} + g_m x_{gs} x_L$ with the transition at

$$f_3 = g_m/[2\pi(c_{gs} + c_L)] \qquad (8.49)$$

giving **f_3=205MHz**

For $f > f_3$ g_m has negligible effect and z_{in} is simply the series combination of c_{gs} and c_L which, as expected, is somewhat greater than x_L alone.

Also from f_L to f_3, $|z_{in}| \approx 1/f^2$ as shown. Over this region z_{in} is close to being a pure negative resistance and thus resonates with r_s for c_s=0. This is another way of looking at the potentially resonant behaviour.

Amplitude frequency response

The cut-off frequency for v_o/v_s is given by the point where $|z_L|$ and $|z_o|$ in Fig. 8.12(a) intersect. For c_s=0 this is at the resonant point of 56.2MHz, with the resonance damped for the actual c_s. Beyond this, with $|z_o| \propto f$ and $|z_L| \propto 1/f$ then $|v_o/v_s| \propto 1/f^2$, a second-order cut-off.

8.5 Common-gate stage

Fig. 8.13(a) shows the essentials for the common-gate stage and (b) the incremental circuit. From (b) the common-gate voltage gain

$$a_{cg} = \frac{g_s(g_m + g_o)}{g_o g_s + g_L(g_s + g_m + g_o)} \qquad (8.50)$$

which for $r_s=0$ reduces to

$$a_{cg} = (a_v + 1)\frac{r_L}{r_L + r_o} \qquad (8.51)$$

where $a_v=g_m r_o$ as in [8.23]. Apart from the sign reversal, this is much as for the common-source a_{cs} in [8.24], which is to be expected, since one stage is essentially the other with the input terminals reversed.

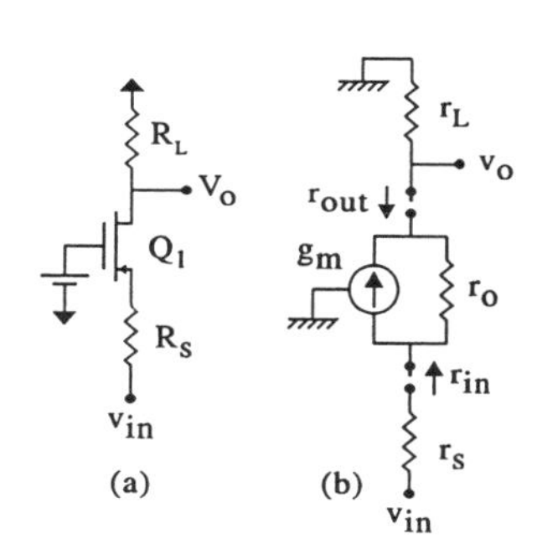

Fig. 8.13 Essentials of common-gate stage.
(a) Basic circuit.
(b) Small-signal model.

Output impedance

Manipulating [8.50] into the form $a_{cg} \propto r_L/(r_{out}+r_L)$ gives r_{out} in (b) as

$$r_{out}=r_o+r_s(a_v+1) \qquad (8.52)$$

which reduces to $r_{out}=r_o$ for $r_s=0$, as expected.

Input impedance

Manipulation of [8.51] to the form $a_{cg} \propto r_{in}/(r_s+r_{in})$ gives r_{in} in (b) as

$$r_{in}=(r_L+r_o)/(a_v+1) \qquad (8.53)$$

With $g_m >> g_o$ and for $r_L=0$ then $r_{in} \approx 1/g_m$ giving

$$r_{in}=1.10\text{k}\Omega.$$

This is a major difference from the common-source stage, for which essentially $r_{in}=\infty$.

Notice too how in [8.52] r_s can considerably increase r_{out} from the r_o for the common-source stage.

These differences will be of repeated significance in the following development, particularly that of the MOS-based op-amp in the next chapter.

8.6 Cascode

The cascode of Fig. 8.14(a) is the common-source stage of Fig. 8.7(a) with the common-gate stage Q_2 interposed between Q_1 and the active load Q_3.

Miller feedback

The function of Q_2 is to reduce the Miller feedback due to the common-source c_{gd}, which otherwise limits the upper cut-off to f_m=400kHz.

With Q_2 and Q_3 constituting a common-gate stage then, following [8.53], and with r_{in2} the input impedance to the stage

$$r_{in2}=(r_{o3}+r_{o2})/(a_{v2}+1).$$

With $a_{v2}=g_{m2}r_{o2} >> 1$ then

$$r_{in2} \approx (1/g_{m2}) \times (1+r_{o3}/r_{o2}) \qquad (8.54)$$

with $r_{o3}/r_{o2}=2.72$ then $r_{in2}=3.72/g_{m2}$.

With r_{in2} the load for Q_2, with the voltage gain of Q_1 $a_{ce1}=-g_{m1}r_{in2}$, and with $g_{m1}=g_{m2}$ then $a_{c1} \approx -3.72$.

Thus $|a_{c1}|$ is much reduced from the $g_{m1}r_{o3}$ for Q_2 absent, with the Miller capacitance

$$c_m=c_{gd1}(|a_{ce1}|+1)$$

thus also reduced.

With c_{gd1}=0.05pF, and with $a_{ce1}=-78.4$ in the absence of Q_2 then $c_m \approx$4pF, but with Q_2 included then c_m is reduced to

$$c_m=0.236\text{pF}.$$

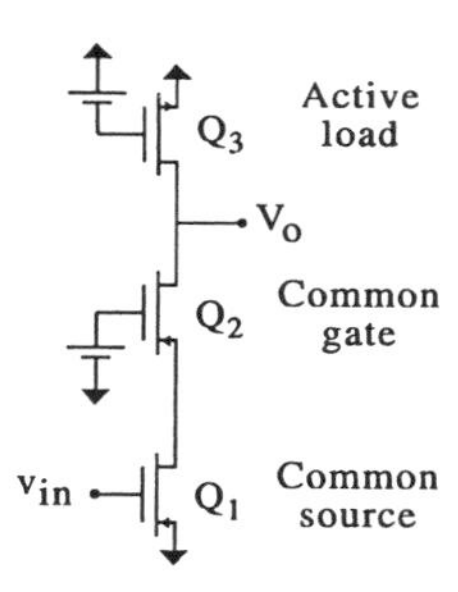

Fig. 8.14(a) Cascode. Essentials of circuit.

Cascode voltage gain

With $r_{in2} \approx 4/g_{m2}$ somewhat greater than the $1/g_{m2}$ that would obtain for Q_3 absent, still $r_{in2} << r_{o1}$ so virtually the whole of i_{d1} passes through Q_2, and hence through Q_3.

Following [8.52], and with r_{o12} the output impedance of Q_2 with Q_1 then

$$r_{o12}=r_{o2}+r_{o1}(a_{v1}+1).$$

With thus $r_{o12} >> r_{o3}$, and with a_{ca} the cascode voltage gain then

$$a_{ca} \approx -g_{m1}r_{o3} \qquad (8.55)$$

giving **$a_{ca}=-291$.**

This is considerably higher than the

$$a_{cs}=g_{m1}r_{o1}//r_{o3}=-78.4$$

for the common-source stage obtaining when Q_2 is absent.

Frequency response

Fig. 8.14(b) shows the incremental circuit for the cascode, with c_m omitted for clarity but with the other transistor capacitances included. With the resulting frequency response as in (c), the features are as follows.

Zero frequency voltage gain This is the above $|a_{ca}|=291$, as plotted.

Cut-off for c_{gs1} With $c_m=0.236$pF no longer dominating $c_{gs1}=0.2$pF then the input cut-off

$$f_{in}\approx 1/[2\pi(c_m+c_{gs})R_s] \tag{8.56}$$

For $R_s=100\text{k}\Omega$ then **$f_{in}=3.65$MHz.**

Cut-off f_1 for c_1 This is where the impedance of c_1 becomes equal to r_{in2}.

With $r_{in2}=4.09\text{k}\Omega$ and $c_1=c_{gs2}=0.2$pF **$f_1\approx 194$MHz** too high to be of concern.

Cut-off f_2 for c_2 From [8.52] $r_{out2}=r_{o2}+r_{o1}(a_{v2}+1)$.

Thus $r_{out2}>>r_{o3}$ so the cut-off for c_2 becomes

$$f_2\approx 1/(2\pi r_{o3}c_2).$$

With $c_2=c_{db2}+c_{gd2}+c_{db3}+c_{gd3}=0.3$pF and $r_{o3}=322\text{k}\Omega$ then **$f_2=1.66$MHz.**

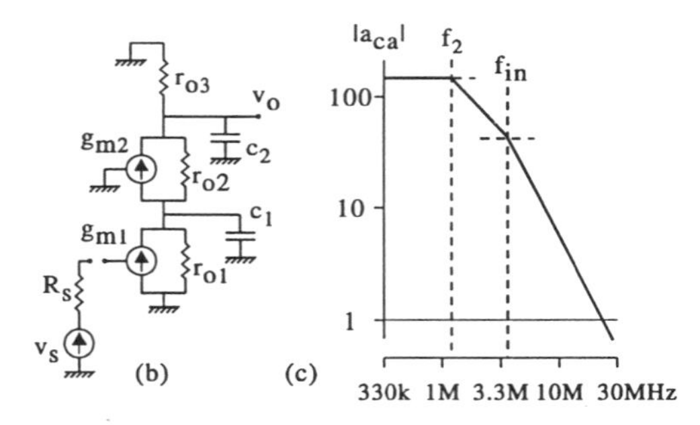

Fig. 8.14(b) and (c). Cascode.
(b) Small-signal equivalent.
(c) Frequency response.

Shunt R_L

With f_2 the dominant pole, and with $|a_{ca}|=291$, then it would probably be acceptable to reduce the effective r_{o3} by a shunt resistor R_L. With $R_L=11\text{k}\Omega$ giving $|a_{cs}|=10$ then **$f_2=48.2$MHz.**

Here r_{in2} is reduced to $\approx 1/g_{m2}$ giving $|a_1|\approx 1$, $c_m\approx 0.1$pF and thus **$f_m=15.9$MHz.**

With $f_{in}=3.65$MHz now dominant, the $f_2=48.2$MHz would be realised for a lower R_s.

Body effect

With the gate of Q_2 grounded, the only influence this effect has, apart from C_{db}, etc., is to reduce g_m by about 20%. This is hardly enough to warrant complementing the whole circuit, enabling Q_2 to be put in a well. There would also then be a small increase in c_1.

Dual-gate MOS

Fig. 8.15 shows a very widely used MOS device known as the 'dual-gate' FET. This is a pair of N-channel devices on a single chip, and is intended for use as shown. The circuit is a cascode, with a parallel resonant load setting the central frequency and bandwidth for the intended narrow-band band-pass frequency response. There will normally be a similar parallel resonator at the input, including the transistor input capacitance as part of the C.

Because of the very low gate currents the biasing resistors can be of very high value, reducing the degree of damping imposed on the input resonator by R_3 and R_4.

As above, the value of the cascode is in the blocking of c_{gd} for the equivalent single-MOS common-source stage. If not causing instability, this would certainly make the tuning of the two resonators interdependent.

The C_d are low-inductance hi-k decoupling capacitors. R_s is sometimes included to stabilise the common I_d.

The arrangement is widely used in communications circuitry, for amplifiers and frequency changers in the range from 1MHz to several GHz.

The higher frequency devices use GaAs in place of Si and are popularly known as 'gasfets'.

As discussed in Section 16.3, the gain of the stage can be controlled by varying V_{g2}.

The circuit is also used as a frequency changer, with the local oscillator output added to V_{g2}.

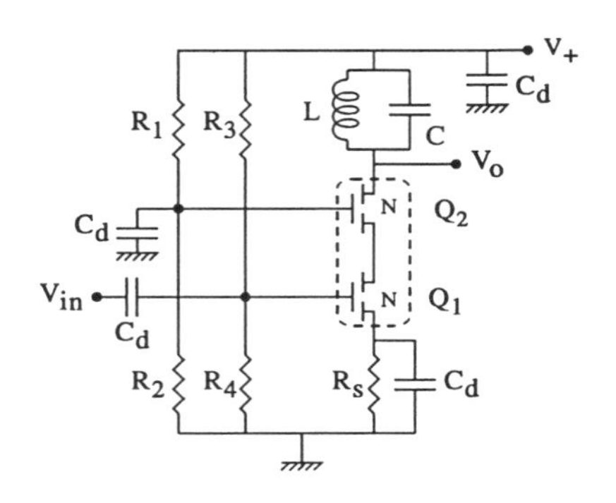

Fig. 8.15 Dual-gate MOSFET cascode tuned RF amplifier.

8.7 CMOS switch

Digital switch

Fig. 8.16 shows the CMOS digital inverter. This and any other such logic gate simply switches V_o between $V_{dd}=5V$ (logic 1) and ground (logic 0).

The load capacitance $C_L=NC_{in}+C_{line}$ where C_{in} is the input capacitance of each of the usually several logic gates that the inverter is driving, N is the number of driven gates, and C_{line} is the line capacitance. With $C_{in}=2C_{gs}$ and $C_{gs}=0.2pF$ then for a fully on-chip design with $N=10$ and C_{line} negligible **$C_L=4pF$.**

With $V_{eff}=V_{gs}-V_{th}$ and $V_{gs}=V_{dd}$ for the downward transition of V_o then $V_{eff}=4.3V$. For most of the switching transient the conducting transistor is beyond pinch-off, for which

$$I_d=\mu C_{ox}(W/L)V_{eff}^2/2$$

as in [8.13].

With the values in Table 8.2 then $I_d=38.3mA$. With

$$I_d=C_L dV_o/dt \quad \text{and} \quad T_{sw}=V_{dd}/(dV_o/dt)$$

then **$T_{sw}=0.522ns$**

comparable with the speed of the present-day PC processor. However, for this order of speed, and for all but the shortest connecting lines, the transmission-line view of Chapter 22 needs to be taken.

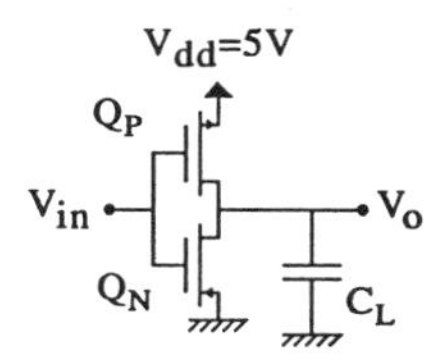

Fig. 8.16 CMOS digital inverter.

Analogue switch

Fig. 8.17(a) and (b) show the general series and shunt analogue switches, with (c) the CMOS switching device.

Choice of configuration

For the bipolar transistor switch, the shunt configuration is most suitable, because of the need to return the base current I_b via ground. With nothing comparable for the MOS transistor the choice is based on the following factors.

To reduce the effects of capacitance, R needs to be as small as possible.

For the series switch when closed there should ideally be no signal loss. With R_{on} the switch 'on' resistance, and $V_o=V_{in}R/(R+R_{on})$ then the requirement is for $R>>R_{on}$.

For the shunt switch when closed there should ideally be no signal transmission. With $V_o=V_{in}R_{on}/(R+R_{on})$ then again the requirement is for $R>>R_{on}$. With a small loss for the transmitting state normally more acceptable than transmission in the nominally non-transmitting state, then the series switch is preferred.

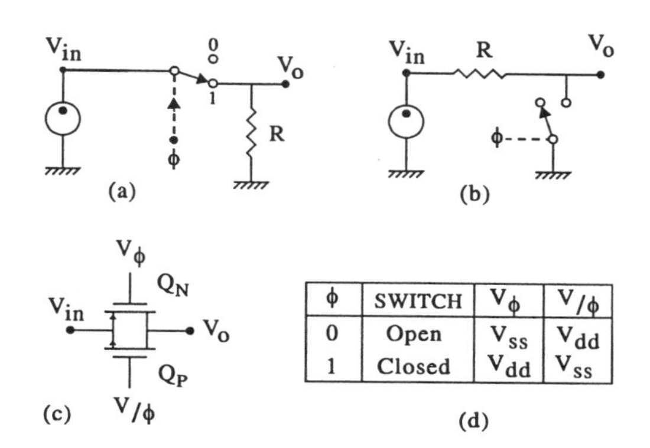

φ	SWITCH	$V_φ$	$V/φ$
0	Open	V_{ss}	V_{dd}
1	Closed	V_{dd}	V_{ss}

Fig. 8.17 Analogue CMOS switch.
(a) Series switch.
(b) Shunt switch.
(c) CMOS switching device.
(d) Switch truth table. $V_{dd}=+5V$, $V_{ss}=-5V$.

Switch resistance

With the voltage-current relation for the above R_{on} non-linear, we consider first the incremental r_{on} for the small-signal v_{in}. Here

$$g_{on}=g_n+g_p$$

where g_n is for Q_N and g_p for Q_P.

N-channel With $\quad g_{co}=\mu C_{ox}(W/L)V_{eff}$

as in [8.7], with $\quad V_{eff}=V_{gs}-V_{th}$

with $\quad V_g=V_{dd}$

as for the device conducting, and with $V_s=V_{in}$, then

$$g_n=\mu_n C_{ox}(W/L)(V_{dd}-V_{in}-V_{thn}) \qquad (8.57)$$

P-channel Similarly

$$g_p=\mu_p C_{ox}(W/L)(-V_{ss}+V_{in}-V_{thp}) \qquad (8.58)$$

With the g_{on} plotted as in Fig. 8.18(a) it is clear why the complementary MOS are needed. With just one there is a range of V_{in} for which the switch when nominally closed is actually open.

For the device dimensions equal, the asymmetry arises from the differing mobility μ_n and μ_p.

Non-linearity Fig. 8.18(c) shows the transfer function for the series switch, which for $R=\infty$ is simply the de-emphasised unity slope line shown. For R finite then the slope dV_o/dV_{in} is reduced by $\approx g_{on}R$. With g_{on} as in (a) then the transfer becomes non-linear.

As always, it is preferable that any non-linearity be anti-symmetric, requiring the variation of g_{on} with V_{in} in (a) to be symmetric. With $g_{co}=\mu_n C_{ox}(W/L)(V_{eff})$ as in [8.7], this can be obtained by altering W/L to compensate for the differing μ_n and μ_p.

Fig. 8.18(b)(i) shows the resulting symmetric g_{on}, and (c)(i) the corresponding anti-symmetric transfer function.

With the distances between the V_{th} points and the edges exaggerated in the diagrams for clarity, then a possible measure is to restrict V_{in} to the relatively wide central linear region. However, g_{on} in (a) is for $R=\infty$. For R finite then as $|V_{in}|$ increases so also does V_{ds}. As V_{ds} approaches pinch-off then g_{on} increases, as shown in (b)(ii) apart from the effect in (i). Fortunately the two effects tend to cancel.

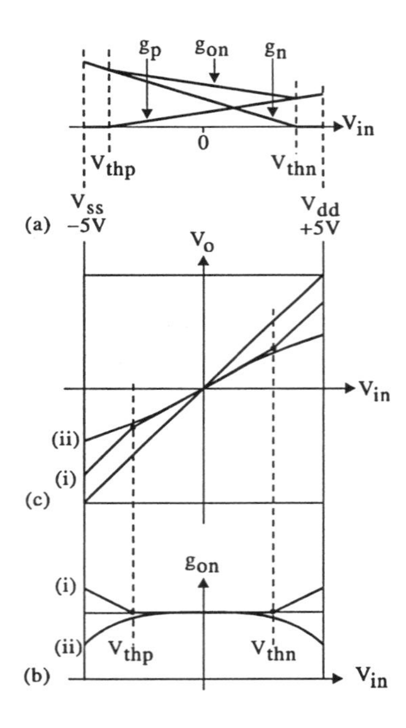

Fig. 8.18 Characteristics of CMOS series switch in Fig. 8.17(a,c).
(a) Variation of switch conductances with V_{in}.
(b)(i) g_{on} in (a) made symmetric by adjusting MOS W/L ratio.
(ii) Reduction in g_{on} as V_{ds} approaches pinch-off.
(c) Effect of (b) on switch transfer function.

Integrated circuits

There are two types of integrated circuit suitable for analogue switching.

HC4316 This is a quad CMOS switch, which is one of the high-speed CMOS (HCMOS) digital logic family. However, with two supplies, the usual $V_{dd}=5V$ and now $V_{ss}=-5V$, bipolar analogue input voltages over the ±5V range can also be switched. For $C_L=50pF$ in Fig. 8.1[6] the switching time is 5ns, which is broadly consistent wit[h] the above 0.5ns for $C_L=4pF$. $R_{on}=30\Omega$ as above. Th[e] maximum allowed output current is 25mA limiting R t[o] 200Ω.

DG-type This is a large class of switching IC, differin[g] only from the HC4316 in that the main supplies are th[e] normal ±15V for analogue circuitry. The attendant disad[-]vantage is that the switching time is increased to 100ns[.] R_{on} remains 30Ω.

For both types the control input is the usual 0V and +5V for logic 0 and 1. The DG equivalent of the HC4316 i[s] the DG212 quad SPST switch. There are many variants[:] quad SPDT, eight-way SPST, etc. The DG3,4,5,6 range[s] offer improving parameters, the DG2.. being the original

CMOS switch capacitances

With increasing frequency, these have the effect of slow[-]ing the response when the switch is nominally 'on', an[d] of giving a degree of transmission when the switch i[s] nominally 'off'. Also as the gates are switched, switchin[g] transients are coupled to the output.

Fig. 8.19(a) shows, in stylised form, the distribute[d] nature of the internal capacitances for the CMOS switc[h]ing device in Fig. 8.17(c). A reasonable approximation i[s] that in Fig. 8.19(b) where, for simplicity, we assume a[ll] of the ΔC to be equal. The full circuit thus becomes as i[n] (c), with $C_o=C_{in}=4\Delta C$.

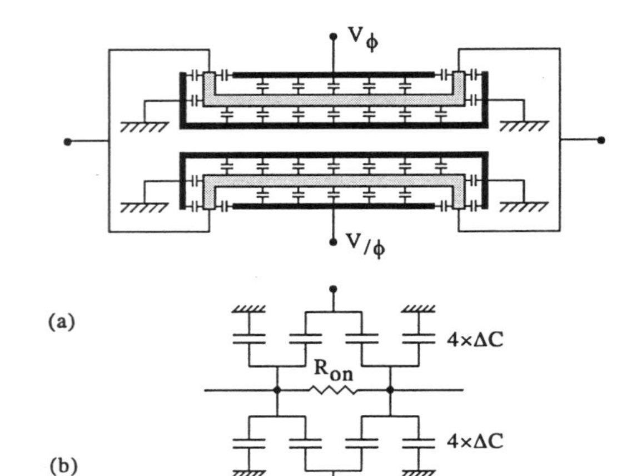

Fig. 8.19(a) and (b) CMOS switch capacitance.
(a) Distributed.
(b) Lumped equivalent.

With $C_{gs}=C_{ox}WL$ and with the values in Table 8.2[,] $C_{gs}=0.276pF$.

With $\Delta C=C_{gs}/2$ then $\quad C_o=C_{in}=\mathbf{0.552pF}$

With C_{ds} effectively zero then C_{st} is the 'stray' capacitance comprising the chip pad-to-pad capacitance plus that of the associated pcb tracks. With the pads usually adjacent, typically **$C_{st} \approx 1pF$.**

Finally we take as typical **$C_L=3pF$ and $R_s=50\Omega$.**

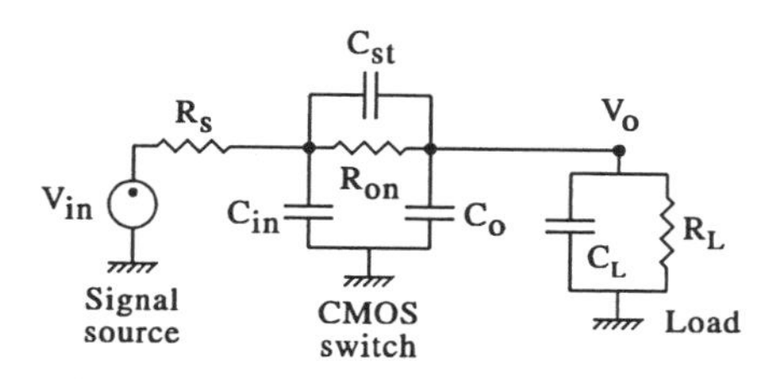

Fig. 8.19(c) CMOS switch capacitance effects. Small-signal equivalent circuit for $V_{in}=0$ and switch 'on'.

'On'-state step response

With typically $R_{on}=30\Omega$ then $R_{on}C_{st}=32ps$ (6GHz) so C_{st} is negligible. For $C_{in}=0$ and $R_L=\infty$ then the circuit resolves to a low-pass filter of response-time

$$T_{on}=(R_s+R_{on})\times(C_o+C_L) \quad (8.59)$$

giving **$T_{on}=0.284ns$ (560 MHz).**

With C_{in} included this introduces a second pole of $T_r \approx$ 28ps which has little additional effect.

'Off'-state transmission For $R_L=\infty$ then C_{st}, C_o and C_L form a potential divider giving the transmission factor

$$\alpha=C_{st}/(C_{st}+C_o+C_{in})$$

and thus **$\alpha=0.220$.**

With R_L included then a high-pass filter is formed of $C_{st}+C_o+C_L$ and R_L, with the cut-off

$$f_{off}=1/[2\pi R_L(C_{st}+C_o+C_L)] \quad (8.60)$$

With R_L the limiting 200Ω then **$f_{off}=175MHz$.**

Below f_{off} the transmission decreases $\propto f/f_{off}$.

Thus the choice of R_L is a compromise between on-state linearity and off-state high-frequency signal rejection.

Switching transients

The switching voltages V_ϕ and $V_{/\phi}$ couple through the associated C_{gs} to give the waveforms shown in Fig. 8.20. Here V_{on} is the component of V_o due to V_ϕ coupling via the C_{gs} of Q_N, and V_{op} is that due to $V_{/\phi}$ via the C_{gs} of Q_P.

On-off transient Assuming the opening of the switch to be immediate then V_ϕ is coupled to V_o via the potential divider ΔC and $3\Delta C+C_L$ to give

$$V_{pk.\phi}=(V_{dd}-V_{ss})\times\Delta C/(4\Delta C+C_L) \quad (8.61)$$

The coupling of $V_{/\phi}$ is the same so for all the ΔC the same then $V_{pk}=0$. With the actual value depending on how well the MOS are matched, and taking the matching factor M=5% as typical, then **$V_{pk}=19.4mV$.**

With $T_{on\text{-}off}$ the decay time for the on-to-off transition

$$T_{on\text{-}off}=R_L(4\Delta C+C_L) \quad (8.62)$$

For $R_L=5k\Omega$ then **$T_{on\text{-}off}=17.8ns$.**

Off-on transient This is more complex, with C_{in} and C_o divided by R_{on}. For $C_{in}=0$ then $|V_{pk}|$ becomes as before. For $R_L>>(R_s+R_{on})$ as for the present $R_L=5k\Omega$ then

$$T_{off\text{-}on}\approx(R_s+R_{on})\times(4\Delta C+C_L) \quad (8.63)$$

giving **$T_{off\text{-}on}=284ps$**

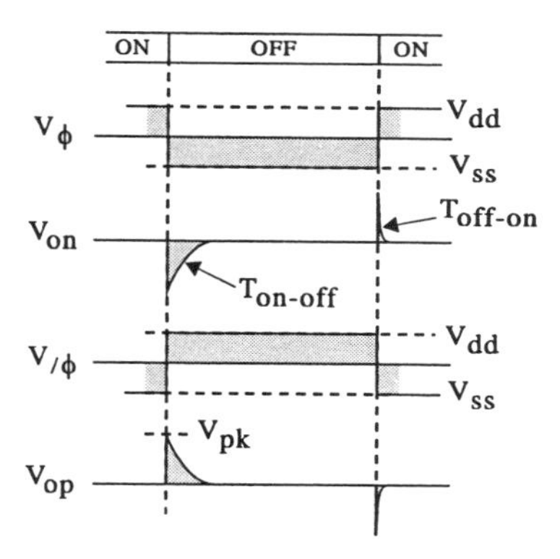

Fig. 8.20 Switching transients with V_{on} and V_{op} near-cancelling components of V_o.

DC output component

For periodic switching at the frequency f_{sw} then the difference in area of the two transients results in a DC mean component $\overline{V_o}$ of V_o. With M=5% the MOS matching factor, and for $f_{sw}=10MHz$, then

$$\overline{V_o}\approx MV_{dd}T_{on\text{-}off}f_{sw} \quad (8.64)$$

giving **$\overline{V_o}=44.5mV$.**

The low-pass filter sometimes used to suppress the switching transients does not remove this component, making it an important consideration for DAC deglitching.

8.8 Exercises

D_n	N-type doping concentration	2×10^{21} atoms/m^3
D_p	P-type doping concentration	1.5×10^{22} atoms/m^3
ε_o	Dielectric constant of space	8.85×10^{-12} F/m
ε_{ox}	Relative permittivity of SiO_2	3.9
ε_s	Relative permittivity of Si	11.8
I_d	Drain current	100μA
k	Boltzmann's constant	1.38×10^{-23} J/K
L	Gate length	2μm
μ_n	Surface electron mobility, Si	0.06 m^2/Vs
μ_p	Surface hole mobility, Si	0.03 m^2/Vs
q	Electron charge	1.6×10^{-19} C.
t_{ox}	Gate SiO_2 layer thickness	250×10^{-10}m
V_{ds}	Drain-source voltage	5V
V_{th}	Threshold voltage	700mV
W	Gate width	100μm
ψ_o	Built-in junction potential	880mV

Table 8.2 Primary values used in calculations.

MOS capacitance values

c_{gs}**=0.2pF,** c_{gd}**=0.05pF,** c_{sb}**=0.2pF,** c_{db}**=0.1pF.**

Exercise 8.1 MOS operation

Calculation Using the values in Table 8.2, calculate those of C_{ox}, V_{eff}, g_m and r_o for the N- and P-channel MOS with I_d=100μA.

Gate capacitance per unit area C_{ox}. With $C_{ox}=\varepsilon_{ox}\varepsilon_o/t_{ox}$ then $\mathbf{C_{ox}=1.38\times10^{-3}F/m^2}$.

Effective gate-source voltage With $I_d=\mu C_{ox}(W/L)V_{eff}^2/2$ as in [8.13] then $V_{eff}=[(2I_d)/(\mu C_{ox}(W/L)]^{1/2}$ giving $\mathbf{V_{effN}=220mV,\ V_{effP}=311mV}$.

Mutual conductance With $g_m=[2\mu C_{ox}(W/L)I_d]^{1/2}$ as in [8.18]

$\mathbf{g_{mN}=910\mu S,\ g_{mp}=644\mu S}$.

Output resistance With $k_{ds}=\sqrt{2\varepsilon_s\varepsilon_o/(qD)}$ as in [8.16] then

$\mathbf{k_{dsN}=8.08\times10^{-7},\quad k_{dsP}=2.95\times10^{-7}}$.

With
$$\lambda=\frac{k_{ds}}{2L\sqrt{V_{ds}-V_{eff}+\psi_o}}$$

as in [8.15] then

$\mathbf{\lambda_N=8.49\times10^{-2},\ \lambda_\Pi=3.12\times10^{-2}}$.

With $g_o=\lambda I_d$ as in [8.19] then

$\mathbf{r_{oN}=118k\Omega,\ r_{oP}=320k\Omega}$.

Exercise 8.2. Frequency response of common-source stage (Fig. 8.9)

Calculation With $a_o(f)$ the voltage gain for no Miller feedback through c_f, calculate the value of f_o in the frequency response of a_o.

With $c_o'=c_o+1/(1/c_f+1/c_{in})$ as in [8.31], c_o'=0.24pF.

With $f_o=1/(2\pi c_o' r_{oa})$ as in [8.32] then $\mathbf{f_o=77.0MHz}$.

Calculation Calculate the value of f_{p2}.

With $c_{fin}=1/(1/c_f+1/c_{in})$ as in [8.30], $c_{f.in}$=0.04pF.

With $\alpha=c_{fin}/(c_o+c_{fin})$ as in [8.33], α=1/6.

With $f_{p2}=\alpha g_m/(2\pi c_{in})$ as in [8.34] then $\mathbf{f_{p2}=121MHz}$.

Exercise 8.3. Source-follower self-load

Calculation Calculate the values of r_L and c_L for the source-follower self-load impedance in Fig. 8.11(c), and the associated transition frequency f_L.

With $r_L=r_{o1}//r_{o2}$ $\mathbf{r_L=160k\Omega}$.

With $c_L=c_{sb1}+c_{db2}$ then c_L=0.2pF+0.1pF giving $\mathbf{C_L=0.3pF}$

With the transition frequency $f_L=1/(2\pi c_L r_L)$ then $\mathbf{f_L=3.32MHz}$.

9

MOS-based operational amplifiers

Summary

There are two main types of MOS-based operational amplifier. The first, as in Fig. 9.1(a), is the conventional type of Chapter 2 and the second, as in (b), is the 'operational transconductance amplifier' (OTA) or 'transductor'. The conventional op-amp is ideally an ideal dependent voltage source, with the output voltage V_o dependent only on the differential input voltage $V_{id}=V_{i+}-V_{i-}$, while the OTA is an ideal dependent current source, with the output current I_o dependent on V_{id}. For the conventional type the output impedance is ideally zero, while for the OTA it is infinite.

This chapter covers the development of first the OTA and then the two-stage Miller MOS op-amp, a typical MOS implementation of the conventional type of op-amp.

The OTA section includes a discussion of the several types of current mirror that are used, as a means of improving both the gain and the common-mode rejection ratio (CMR).

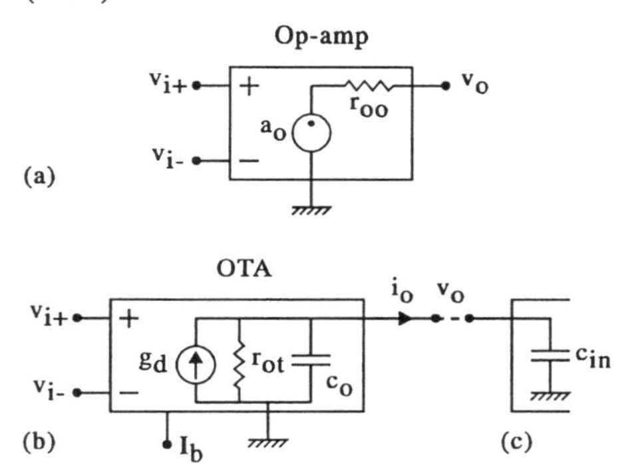

Fig. 9.1 Conventional operational amplifier and 'operational transconductance amplifier' (OTA).
r_{oo} (o: output, o: op-amp). r_{ot} (o: output, t: transductor).
(a) Op-amp, $v_o=a_o(v_{i+}-v_{i-})$.
(b) OTA, $i_o=g_d(v_{o+}-v_{o-})$. $g_d \propto I_b^{1/2}$.
(c) Input circuit of following MOS stage.

The final implementation of the OTA is as the folded-cascode. This circuit gives a much wider range of output voltage than the earlier types.

Towards the end there is a short section illustrating the possible use of the OTA as the voltage-feedback amplifier of Section 2.2. Finally a supply and temperature independent bias supply is described.

With almost all of the numeric calculations merely a matter of using appropriate values in the equation derived then, starting with the values in Table 9.1, the derived values are set bold to the right-hand margin, with the sequence cumulative.

The Exercises section is mainly devoted to further insights into the relations derived in the main text.

Conventional op-amp and OTA

While for the ideal op-amp, the voltage gain $a_o=\infty$ with

$$v_o=a_o(v_{i+}-v_{i-}) \tag{9.1}$$

for the ideal OTA, the difference-mode transconductance $g_d=\infty$ with

$$i_o=g_d(v_{i+}-v_{i-}) \tag{9.2}$$

Implementation

Both types of op-amp are an extension of the long-tailed pair (LTP) differential amplifier.

The conventional op-amp adds a further common-source stage to give the required high a_o, and a source-follower for the low r_{oo}.

The high output impedance r_{ot} of the LTP makes it essentially an OTA. However, a common-gate stage is normally added to increase r_{ot} further.

With the greater simplicity, the cut-off frequency of the OTA is the higher.

Application

A feature of the OTA is the relatively limited number of applications to which it is directly suited.

With the ideal $r_{ot}=\infty$, and with the internal c_o, then the OTA in (b) is inherently an integrator.

When driving the following MOS stage in (c) this simply adds to c_o, with the possible addition of a further C as required for the integrator time constant. This suits the OTA best to the implementation of the low-pass OTA-C filter in Section 13.7.

Feedback amplifiers, high-pass filters, etc., are possible, but with the OTA less directly suited.

Unlike the integrator of Section 2.8 based on the conventional op-amp, there is no direct negative feedback for the OTA-C integrator, making the linearity of the OTA more important. Thus lower-case small-signal symbols are used throughout, signifying the incremental values.

Gain control

Both a_o for the op-amp and g_d for the OTA can be controlled by the LTP tail current I_{tail}, with I_{tail} proportional to the bias current I_b in (b).

Both OTA and conventional op-amp are implemented in bipolar and MOS form. While for the bipolar type, $g_m \propto I_{tail}$ and thus $a_o \propto I_{tail}$, for the MOS $g_m \propto I_{tail}^{1/2}$. This is an inconvenience in the control of say a MOS-based voltage-controlled filter.

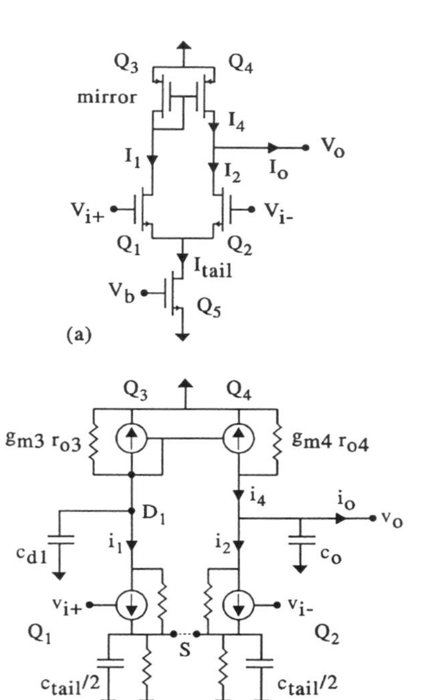

Fig. 9.2(a,b) MOS-based long-tailed pair (LTP) with active tail and current mirror.
(a) Circuit diagram.
(b) Incremental equivalent.

9.1 LTP difference-mode transconductance

Difference-mode transconductance g_d

Fig. 9.2(a) shows the MOS implementation of the LTP, with the current mirror included, and with the incremental circuit as in (b).

With the LTP the basic OTA of Fig. 9.1(b), with i_o as for the v_o point in Fig. 9.2(b) grounded, and with $v_{id}=v_{i+}-v_{i-}$, then the difference-mode transconductance

$$g_d = i_o/v_{id} \tag{9.3}$$

Current mirror

This in (a) replaces the single active load for the common-source stage of the last chapter.

The ideal current mirror 'reflects' I_1 to give $I_4=I_1$ and thus $I_o=I_1-I_2$. Here, not only is the wastage of the signal component of I_1 avoided, but also the common-mode rejection CMR is much improved.

A further property of the ideal mirror is that the impedance presented by Q_3 to Q_1 is zero. With the v_o node in (b) also grounded then the LTP section of (b) becomes fully symmetric.

For v_{id} only applied ($v_{i-}=-v_{i+}$) then the voltage at the S node is zero. This point can therefore be grounded without making any difference to the action of the incremental circuit. With the point thus grounded then

$$i_1 = g_{m1}v_{id}/2 \quad \text{and} \quad i_2 = g_{m2}(-v_{id}/2)$$

With $\quad i_o = i_1 - i_2 \quad$ and $\quad g_{m1}=g_{m2}, \quad$ say g_m,

then $\quad i_o = g_m v_{id}$.

With g_d as in [9.3] and with the real mirror approaching the ideal then

$$g_d \approx g_m \tag{9.4}$$

Voltage gain

With $\quad i_o = g_d(v_{i+}-v_{i-}) \quad$ as in [9.2]

with $\quad v_o = i_o r_{ot} \quad$ from Fig. 9.1(b),

and with the voltage gain

$$a_o = v_o/(v_{i+}-v_{i-}) \tag{9.5}$$

then

$$a_o = -g_d r_{ot} \tag{9.6}$$

Output impedance

The two-node analysis (S, D_1) gives the transductor output impedance

$$r_{ot} \approx r_{o2}//r_{o4} \tag{9.7}$$

Tail

For Q_5 both W and I_d are made twice that of $Q_{1,2}$, making Q_5 as for the other two in parallel giving

$$r_{tail} = r_{o1}/2 \tag{9.8}$$

[9.6] for r_{ot} is not quite as obvious as it might seem. See Exercise 9.1 for discussion of the 'back-door' effect.

Device parameters

With the following calculations based on the same N- and P-channel MOS as in the last chapter, then the incremental parameters in Table 9.1 are much as there. The only difference is that V_{ds} for the P-type is considerably lower. For the current mirror of Fig. 9.2(a)

$$V_{ds3} = V_{th3} + V_{eff3}$$

For the typical values

$$V_{th} = 700\text{mV} \quad \text{and} \quad V_{eff3} \approx 300\text{mV}$$

then $\mathbf{V_{ds3} \approx 1V.}$

Strictly, to maintain the balance needed for low common-mode response V_{ds1} and V_{ds2} should be equal, giving $V_{ds4} \approx 1V$ also.

While the effect of the V_{ds} not being equal is small compared with the likely mismatch in device area, etc., for simplicity we calculate for the two V_{ds} the same. Previously the supply was V_+ with V_o set to the mid-point 5V, while now for the supplies $\mathbf{V_+ = +5V,\ V_- = -5V.}$

With $V_{d1,2} \approx 4V$ and $V_{s1,2} \approx -1V$ then $V_{ds1,2} \approx 5V$ as before. With $\mathbf{I_d = 100\mu A}$

as before, and with the primary values as in Table 8.1, then the device parameters become as below.

	N-channel	P-channel
g_m	910μS	644μS
r_o	118kΩ	289kΩ
$a_v = g_m r_o$	107	186

Table. 9.1 Incremental MOS parameters for $I_d = 100\mu V$, $V_{dsN} = 5V$, $V_{dsP} = 1V$.

With these values in the above relations

$$\mathbf{g_d \approx 910\mu S,\ r_{ot} \approx 83.7k\Omega,\ a_o = -76.2,\ r_{tail} = 58.9k\Omega.}$$

9.2 LTP common-mode transconductance

As for the bipolar LTP in Fig. 7.12, v_{i+} and v_{i-} may be resolved into the difference- and common-mode components

$$v_{id} = v_{i+} - v_{i-} \ldots\text{(a)} \qquad v_{ic} = (v_{i+} + v_{i-})/2 \ldots\text{(b)} \qquad (9.9)$$

giving

$$i_o = g_d v_{id} + g_c v_{ic} \qquad (9.10)$$

where g_d and g_c are the difference- and common-mode transconductances. Also the common-mode rejection ratio

$$CMR = g_d/g_c \qquad (9.11)$$

For the ideal differential amplifier then

$$g_c = 0, \quad \text{giving} \quad CMR = \infty.$$

For the LTP thus g_c needs to be determined. Here with i_o as in [9.10]

$$g_c = i_o/v_{ic} \quad \text{for} \quad v_{id} = 0.$$

Ideal mirror

With v_{id} as in [9.9], and for $v_{id} = 0$, then $v_{i-} = v_{i+}$. With the circuit again symmetrical about S for the ideal mirror then, instead of v_s being zero as for the evaluation of g_d, now it is the current through the S-S link that is zero, thus making no difference if the link is broken. With this condition, and with $i_1 = i_2$ written as the common-mode i_c, then the single node (S) analysis gives

$$i_c = v_{ic}[g_m(g_{tail}/2)/(g_{tail}/2 + g_o + g_m)] \qquad (9.12)$$

With

$$g_m >> g_o + g_{tail}/2$$

then

$$i_c \approx v_{ic}/(2r_{tail}) \qquad (9.13)$$

With $i_o = i_1 - i_2$ for the ideal mirror then $i_o = 0$ giving the ideal $g_c = 0$, whatever the value of i_c.

Real mirror

Here the ideal $g_c = 0$ is compromised as follows.

(i) i_4 is not quite equal to i_1.
(ii) The impedance presented by Q_3 is not quite zero.
(iii) The two halves of the LTP may not be matched.

Defining the difference- and common-mode currents

$$i_d = i_1 - i_2 \ldots\text{(a)} \qquad i_c = (i_1 + i_2)/2 \ldots\text{(b)} \qquad (9.14)$$

then (iii) produces i_d, which the ideal mirror responds to fully, while (i) and (ii) determine the degree to which the real mirror suppresses i_c in [9.12]. For the real mirror

$$i_o = \alpha_d i_d + \alpha_c i_c \qquad (9.15)$$

where α_d and α_c are the mirror difference- and common-mode current transfer factors.

For the ideal mirror $\alpha_d = 1$ and $\alpha_c = 0$. For the real mirror $\alpha_d \approx 1$ and $\alpha_c << 1$.

With the mismatch (iii) largely independent of the tail we define g_c(tail) as the component of g_c due to (i) and (ii). With i_c as in [9.13], then

$$g_c(\text{tail}) \approx \alpha_c g_{tail}/2 \qquad (9.16)$$

Mirror common-mode response α_c

From the two-node (D_1 and S) analysis

$$\alpha_c = -\frac{\Delta g_m + g_{o3} + g_{o1}}{g_{m3} + g_{o3} + g_{o1}/2} \qquad (9.17)$$

With $\qquad g_{m3} >> (g_{o3} + g_{o1}/2)$

then $\qquad \alpha_c \approx -(\Delta g_m + g_{o3} + g_{o1})/g_{m3} \qquad (9.18)$

With $\qquad \Delta g_m = g_{m3} - g_{m4}$

and for a typically 1% match $\qquad \mathbf{\Delta g_m/g_{m3} = 10 \times 10^{-3}}$.

With g_{o3}/g_{m3} representing the leakage of i_1 through g_{o3}
$\mathbf{g_{o3}/g_{m3} = 5.37 \times 10^{-3}}$.

With g_{o1} representing the effect of i_1 being reduced by the non-zero impedance $\approx 1/g_m$ of Q_3, $\quad \mathbf{g_{o1}/g_{m3} = 13.2 \times 10^{-3}}$.

With the components uncorrelated then the most probable
$\mathbf{\alpha_c = 21.1 \times 10^{-3}}$.

The g_{o1} term is not as obvious as may appear. See Exercise 9.2 for further discussion.

LTP mismatch

With $\alpha_c << 1$ and $\alpha_d \approx 1$ then it is the resulting i_d that is of concern. With i_d much as for D_1 grounded then the single S-node analysis gives

$$g_c(\Delta g_{mL}) = -\Delta g_m \times \frac{g_o + g_{tail}/2}{g_m + g_o + g_{tail}/2} \qquad (9.19)$$

and

$$g_c(\Delta g_{oL}) = -\Delta g_o \times \frac{g_m}{g_m + g_o + g_{tail}/2} \qquad (9.20)$$

where

- $g_c(\Delta g_{mL})$ is the component of g_c due to the mismatch Δg_m in the LTP g_m,
- $g_c(\Delta g_{oL})$ is the component of g_c due to the mismatch Δg_o in the LTP g_o.

With $\qquad CMR(\Delta g_{mL}) = g_c(\Delta g_{mL})/g_m$,

with g_m dominant in [9.19], $\quad$ with $g_{tail}/2 = g_o$,

and with $\qquad a_v = g_m/g_o$

then $\qquad CMR(\Delta g_{mL}) \approx (\Delta g_m/g_m)/(a_v/2) \qquad (9.21)$

Similarly $\qquad CMR(\Delta g_{oL}) \approx (\Delta g_o/g_o)/a_v \qquad (9.22)$

With the typically 1% mismatch, the reduction by $\times 2/a_v$ and $\times 1/a_v$ is important, since the mirror gives no further suppression. See Exercise 9.3 for insight into the two suppression factors.

Table 9.2 shows the calculated components of g_c and the final expected value, using the values for g_m, etc., in Table. 9.1, and with the equation used identified. The Active Tail column is that currently relevant. The Ideal Tail column is for $g_{tail} = 0$ and shows the modest potential advantage of increasing r_{tail} by using the cascode tail described in the next section. The final g_c is for the components uncorrelated, as for

$$x = (x_1^2 + x_2^2 + x_3^2 \ldots)^{1/2}$$

Symbol	Active tail	Ideal tail	Eqn.
g_c(tail)	179nS	0	[9.16]
$g_c(\Delta g_{ml})$	166nS	84.1nS	[9.19]
$g_c(\Delta g_{ol})$	83.3nS	84.1nS	[9.20]
g_c	253mS	119nS	

Table 9.2 Components of common-mode conductance g_c.

Common-mode rejection ratio

With $\qquad CMR = g_d/g_c \qquad$ as in [9.11]

and with the above values

then $\qquad$ **CMR(active tail) = 3.59k = 71.1dB**

CMR(ideal tail) = 7.65k = 77.7dB

With the cascode tail the ≈78dB should be realised, but this still falls short of the typical specified CMR = 90dB, so we seek an explanation.

Inverted configuration

With it unlikely that the component matching could be much closer than the assumed 1%, then a more likely contributing factor is the use of the inverted configuration, where the P-channel and N-channel devices are transposed. With

- the mismatch effects suppressed by $\times 1/a_v$,
- the N-channel $a_v = 107$ and the P-channel $a_v = 206$, both for $V_{ds} = 5V$

then the mismatch suppression for the inverted arrangement is twice as high.

W increase

With $\qquad g_m = [2\mu C_{ox}(W/L)I_d]^{1/2} \qquad$ as in [8.18]

and $\qquad g_o = \lambda I_d \qquad$ as in [8.19]

and with $\qquad a_v = g_m/g_o$,

then for a given I_d and L, $\qquad a_v \propto W^{1/2}$.

Thus a ×4 increase in W to 200 × L would account for most of the remainder of the difference.

Common-mode feedback

Fig. 9.3 shows how the operation of the LTP involves common-mode negative feedback as the means whereby the tail resistor reduces the common-mode g_c. With r_o omitted for simplicity, this is a diagrammatic representation of the circuit equations. By the symmetry of the diagram, a difference-mode input v_{id} generates no feedback and so gives

$$i_o = g_m v_{id}.$$

But with the loop gain for the common-mode input

$$A_L = 2g_m r_{tail}$$

then the $i_o = 2g_m v_{ic}$ that would obtain without the feedback is reduced by $\times 1/(A_L+1)$ to give

$$I_o \approx v_{ic}/(2r_{tail}) \quad \text{confirming [9.13].}$$

Here the common-mode feedback is inherent. Sometimes it is applied more deliberately.

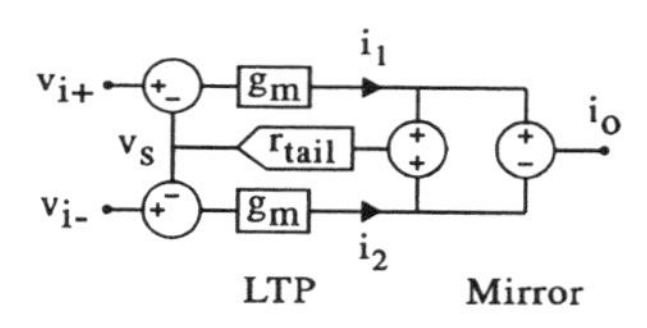

Fig. 9.3 Circuit equations for LTP of Fig. 9.2(b) giving common-mode feedback view of LTP operation.

9.3 LTP frequency response

Common- and difference-mode conductance

Fig. 9.4(a) shows the amplitude-frequency responses of g_d and the three components of g_c in Table 9.2, with the contributing capacitances c_{d1} and c_{tail} in Fig. 9.2(b).

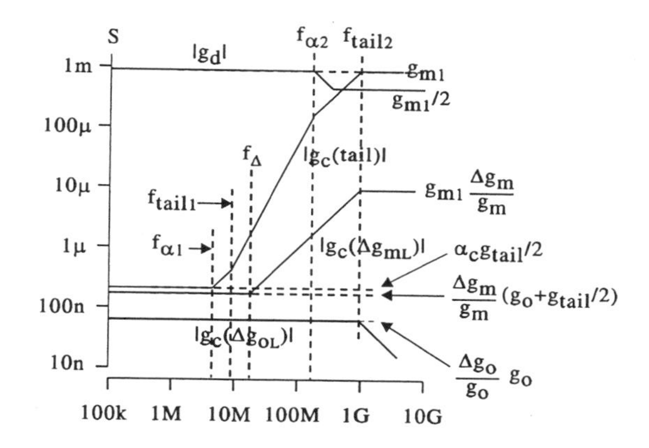

Fig. 9.4(a) Frequency responses for LTP of Fig. 9.2. $g_d = i_o/v_{id}$ and components of $g_c = i_o/v_{ic}$. (c: common-mode, d: difference-mode, i: input).

g_c(tail)

With $g_c(\text{tail}) \approx \alpha_c g_{tail}/2$ as in [9.16], this is subject to changes in α_c and g_{tail}.

Transitions in α_c

With c_{d1} across g_{o3} then $b_{d1} = j\omega c_{d1}$ is added to g_{o3}. With α_c as in [9.17] then

$$\alpha_c = -\frac{\Delta g_m + g_{o3} + g_{o1} + b_{d1}}{g_{m3} + g_{o3} + g_{o1}/2 + b_{d1}} \tag{9.23}$$

Here the numerator transition frequency

$$f_{\alpha 1} = (\Delta g_m + g_{o1} + g_{o3})/(2\pi c_{d1}) \tag{9.24}$$

while for the denominator

$$f_{\alpha 2} \approx g_{m3}/(2\pi c_{d1}) \tag{9.25}$$

$f_{\alpha 1}$ is the frequency at which b_{d1} becomes significant, causing α_c to start to increase from its initially low value.

$f_{\alpha 2}$ is the frequency at which $b_{d1} = g_{m3}$, beyond which i_1 ceases to contribute significantly to i_o. Here $|\alpha_c| = 1$ and so ceases to increase.

Transitions in g_d

With the loss of i_1 beyond $f_{\alpha 2}$, g_d is reduced from g_m to $g_m/2$ as shown.

Transitions in g_{tail}

With $c_{tail}/2$ across $2r_{tail}$ then b_{tail} is added to g_{tail}. With

$$i_c = v_{ic}[g_m(g_{tail}/2)/(g_{tail}/2 + g_o + g_m) \quad \text{as in [9.12]}$$

then
$$i_c = v_{ic}\frac{g_m(g_{tail} + b_{tail})/2}{(g_{tail} + b_{tail})/2 + g_o + g_m} \tag{9.26}$$

The numerator transition frequency

$$f_{tail1} = g_{tail}/(2\pi c_{tail}) \tag{9.27}$$

With
$$g_m >> g_o + g_{tail}/2$$

the denominator transition

$$f_{tail2} \approx g_m/(c_{tail}/2) \tag{9.28}$$

Here f_{tail1} is where $|y_{tail}| = g_{tail}$

and f_{tail2} is where $|y_{tail}|/2 = 1/g_m$

with $1/g_m$ the impedance presented by the MOS to node S, giving no further increase in $|i_c|$.

Beyond all four transitions S is effectively grounded and i_1 of no effect giving, as plotted,

$$i_o = g_m v_{ic} \quad \text{and thus} \quad g_c = g_m$$

Transitions in $g_c(\Delta g_{mL})$

Including b_{tail} into [9.19] then

$$g_c(\Delta g_{mL}) = -\frac{\Delta g_m}{g_m} \times \frac{g_m[g_o + (g_{tail} + b_{tail})/2)]}{g_m + g_o + (g_{tail} + b_{tail})/2} \quad (9.29)$$

With f_Δ in the plot the numerator transition frequency

$$f_\Delta = (g_o + g_{tail}/2)/(2\pi c_{tail}/2) \quad (9.30)$$

This is where $b_{tail}/2$ in the half-circuit becomes comparable with $g_o + g_{tail}/2$, causing the initially small fraction of v_{ic} which constitutes v_{gs} to start to increase. This continues until f_{tail2} where $b_{tail}/2$ exceeds the source output conductance g_m, as for the denominator. Beyond this $v_{gc} \approx v_{ic}$ so there is no further suppression. For the 1% mismatch then $g_c = g_m/100$.

Transition in $g_c(\Delta g_{oL})$

Similarly, including b_{tail} into [9.20]

$$g_c(\Delta g_{oL}) = -\frac{\Delta g_o}{g_o} \times \frac{g_m g_o}{g_m + g_o + (g_{tail} + b_{tail})/2} \quad (9.31)$$

With no numerator transition, the denominator transition is also at f_{tail2}. With v_s driving g_o and initially $v_s \approx v_{ic}$ then at f_{tail2} where v_s begins to decrease, so also does $g_c(\Delta g_{oL})$.

Capacitances

As in the last chapter,

$$\mathbf{c_{sb} = 0.1pF,\ c_{gs} = 0.2pF,\ c_{gd} = 0.05pF,\ c_{db} = 0.1pF.}$$

With $\quad c_{d1} = c_{db1} + c_{db3} + c_{gs3} + c_{gs4} \quad$ $\mathbf{c_{d1} = 0.6pF.}$

With the tail width W twice that of $Q_{1,2}$

$$c_{gd.tail} = 2c_{gd} = 0.1\text{pF}.$$

With $\quad c_{tail} = 2c_{sb} + c_{gd.tail} \quad$ $\mathbf{c_{tail} = 0.3pF.}$

With $\quad c_o = c_{db1} + c_{db4} \quad$ $\mathbf{c_o = 0.2pF.}$

Thus the transition frequencies become as follows.

$f_{\alpha 1} = (\Delta g_m + g_{o1} + g_{o3})/(2\pi c_{d1})$	3.98MHz	[9.24]
$f_{\alpha 2} \approx g_{m3}/(2\pi c_{d1})$	171MHz	[9.25]
$f_{tail1} = g_{tail}/(2\pi c_{tail})$	9.00MHz	[9.27]
$f_{tail2} \approx g_m/(c_{tail}/2)$	966MHz	[9.28]
$f_\Delta = (g_o + g_{tail}/2)/(2\pi c_{tail}/2)$	18.0MHz	[9.30]

Table 9.3 Transitions for frequency responses in Fig. 9.3(a).

Voltage gain a_o

With the main pole $\quad f_o = 1/(2\pi r_{ot} c_o) \quad (9.32)$

then, as in Fig. 9.4(b)(i), $\quad$ $\mathbf{f_o = 9.50MHz.}$

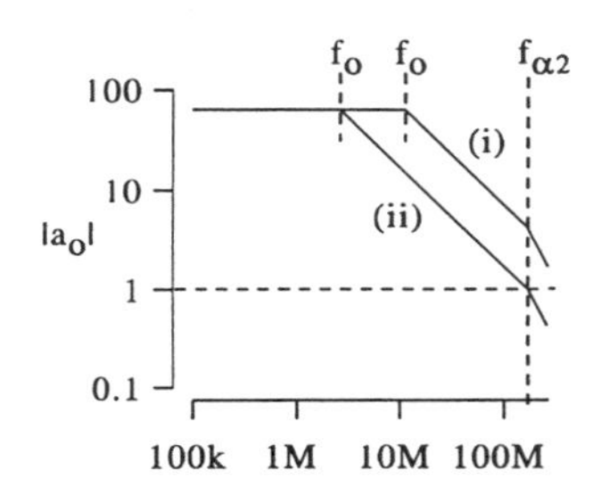

Fig. 9.4(b) Frequency response of voltage gain $a_o = v_o/v_{id}$ for LTP of Fig. 9.2.
(i) For c_o alone.
(ii) With C_o in Fig. 9.1(c) added to give closed-loop stability.

With the remaining transitions as for g_d in (a) then the next is at $f_{\alpha 2} = 171$MHz, as also in (b)(i).

Stability

Were the circuit to be used as an op-amp then $|a_o|$ needs to have fallen below unity before $f_{\alpha 2}$ is reached. This is arranged by increasing c_o, with (b)(ii) as for the limiting $|a_o| = 1$ for $f = f_{\alpha 2}$.

With $\quad f_{\alpha 2} \approx g_{m3}/(2\pi c_{d1}) \quad$ as in [9.25]

and with $\quad a_o \approx g_{m1}(1/j\omega c_o) \quad$ for $f \sim f_\alpha$

then for $\quad |a_o| = 1 \quad$ at $\quad f = f_{\alpha 2}$,

$$c_o = c_{d1}(g_{m1}/g_{m3}) \quad (9.33)$$

giving $\quad$ $\mathbf{c_o = 0.84pF}$

With f_o as in [9.32] then the reduced $\quad$ $\mathbf{f_o = 2.24MHz}$

9.4 Cascode tail

With g_c(tail) in Table 9.2 marginally the largest zero frequency component of the LTP common-mode transconductance g_c, and with

$$g_c(\text{tail}) \approx \alpha_c g_{tail}/2 \quad \text{as in [9.16]}$$

it is advantageous to reduce g_{tail}. This is afforded by the cascode tail shown in Fig. 9.5, with $Q_{4,5}$ the cascode and the remainder the biasing chain.

Tail impedance

With the output impedance of the common-gate stage

$$r_{out} = r_o + r_s(a_v + 1) \quad \text{as in [8.52]}$$

where r_s is the signal source impedance, and with here $r_s = r_{o2}$, then

$$r_{tail} = r_{o4} + r_{o2}(1 + a_{v4}) \approx r_{o2} a_v \quad (9.34)$$

With $a_{v4} >> 1$ and $r_{o4} \sim r_{o2}$ then $r_{tail} \approx r_{o2} a_{v4}$.

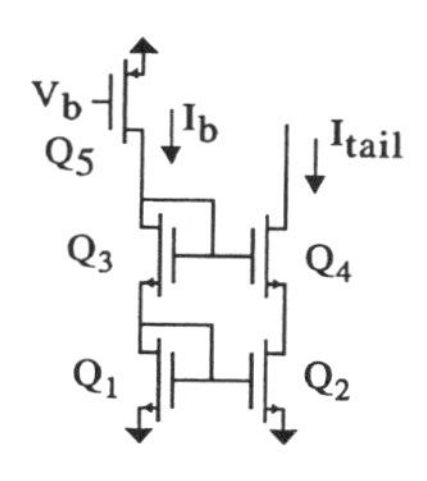

Fig. 9.5 Cascode LTP tail.

With r_{o2} the previous r_{tail}, this is increased $\times a_{v4}$.

With the tail transistor effectively two of the upper pair in parallel, a_{v4} remains 107 as in Table 9.1.

With $r_{o2}=58.8k\Omega$ as before then $\mathbf{r_{tail}=6.43M\Omega}$.

However, the component of c_{tail} due to the tail remains $c_{dg}+c_{db}$ so there is little change in g_c(tail) from the previous values beyond f_{tail1} in the plot.

Biasing

Q_1 and Q_3 are said to be 'diode connected', because the bipolar equivalent behaves as a diode. V_b from the bias supply sets the bias I_b to the required I_{tail}. The diode connections then develop the appropriate V_g for this current, which are then coupled to Q_2 and Q_4 to give the required tail current. For the diode connection

$$V_{ds}=V_{gs}=V_{th}+V_{eff}.$$

With $V_{th}=700mV$ and $V_{eff}=220mV$

then $V_{ds}=920mV$ is well in excess of $V_{eff}=220mV$, the value at which the channel is pinched-off, so the devices are set well into the active region. The circuit is also an example of the cascode current mirror of the next section, 'reflecting' the bias I_b as I_{tail}.

9.5 Cascode current mirrors

Fig. 9.6 shows the LTP with the basic current mirror of Fig. 9.2 replaced by the cascode current mirror used as the cascode tail above. With there

$$r_{tail}=r_{o4}+r_{o2}(1+a_{v4}) \quad \text{as in [9.34]}$$

then here the mirror output impedance

$$r_{om}=r_{o4}+r_{o6}(1+a_{v4}) \quad (9.35)$$

With $a_{v4} >> 1$ then $r_{om} \approx a_{v4}r_{o6}$
giving $\mathbf{r_{om}=54.5M\Omega}$.

With the previous $a_o=-g_d r_{ot}$ as in [9.6]

where $r_{ot} \approx r_{o2}//r_{o4}$ as in [9.7]

giving $a_o=-76.2$, now r_{o4} is replaced by the much higher r_{om}, giving $r_{ot} \approx r_{o2}=118$ and thus increasing $|a_o|$ to $\approx a_{v2}=107$.

We shall shortly be considering the cascode OTA of Fig. 9.8, where r_{o2} is effectively increased $\times a_v$ also, giving a much larger increase in a_o.

Cascode mirror α_c

With i_c the common-mode component of i_1 and i_2, then for the incremental circuit

$$v_{gs5}=-i_c/(g_{m5}+g_{o5}) \quad \text{and} \quad v_{gs3}=v_{gs5}.$$

With v_o grounded and with $\alpha_c=i_o/i_c$ then the nodal analysis gives

$$\alpha_c=-2/(a_v^2+3a_v+2) \approx 2/a_v^2 \quad (9.36)$$

Basic mirror α_c

With $\alpha_c \approx -(\Delta g_m + g_{o3} + g_{o1})/g_{m3}$ as in [9.18]

for the basic mirror, the first component represents the parameter mismatch, and the last the effect of the mirror impedance on i_1.

With neither of these allowed for in [9.36] then only the term $-g_{o3}/g_{m3}$ in [9.18] can be compared with the RHS of [9.36]. With

$$-g_{o3}/g_{m3}=-1/a_v$$

for the basic mirror, then in this respect the cascode mirror is superior by $\times 2/a_v$.

However, with the mirror mismatch $\Delta g_m/g_{m3}$ typically 1%, with $a_v \sim 100$, and with the effect obtaining equally for the cascode mirror, then the previous advantage is largely overridden.

Moreover, with the impedance presented by the mirror to Q_1 doubled for the cascode mirror, then the component corresponding to $-g_{o1}/g_{m3}$ for the basic mirror is also doubled, approximately cancelling the effect of the suppression of the $-g_{o3}/g_{m3}$ component.

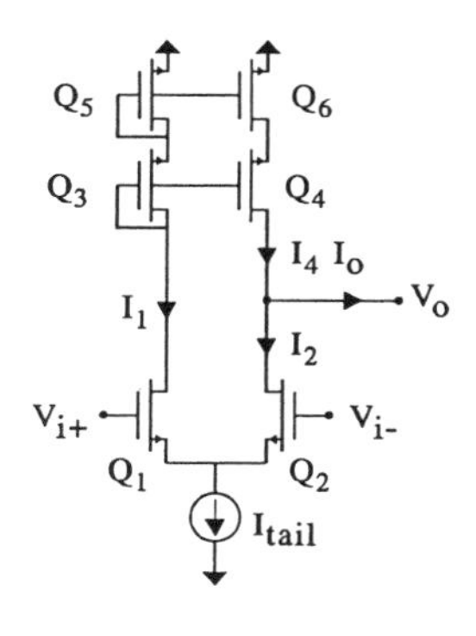

Fig. 9.6 LTP of Fig. 9.2 with cascode current mirror load.

Wide-swing cascode current mirror

This, shown in Fig. 9.7(b), avoids the above increase in impedance presented to Q_1. Also, as the name implies, it gives a larger range of output voltage V_o. With the two types compared in (a) and (b), it is the N-channel versions which are shown, for ease of understanding.

V_o limit

Standard In (a) the diode-connected Q_1 and Q_3 give

$$V_y = V_{th} + V_{eff} \quad \text{and} \quad V_z = 2(V_{th} + V_{eff}).$$

With the limit to V_o for Q_4 to remain pinched-off $V_z - V_{th}$ then we require

$$V_o > (V_{th} + 2V_{eff})$$

Wide-swing In (b) Q_2 pinches off at V_{eff} so V_y can be lowered by V_{th} from the present $V_{th}+V_{eff}$. With this adjustment, Q_4 pinches off at $2V_{eff}$, requiring

$$V_o > 2V_{eff}$$

an extension of the range by V_{th}.

Bias

With $V_y = V_{eff}$ for the wide-swing circuit then

$$V_z = V_{th} + 2V_{eff}$$

With $I_d \propto V_{eff}^2$ then V_z has the required value for

$$I_b = 4(I_1 = I_2)$$

Alternatively, since $I_d \propto (W/L)V_{eff}^2$, all the I_d can be made equal and W/L for the bias transistor reduced by × 1/4.

Common-mode current gain α_c

For the LHS and RHS single-node analyses

$$v_{d3} = \frac{i_1}{g_m} \times \frac{a_v(a_v+2)}{a_v^2 + a_v + 1} \tag{9.37}$$

and

$$i_2 = g_m v_{d3} \frac{a_v + 1}{a_v + 2} \tag{9.38}$$

With $\quad i_o = i_1 - i_2 \quad$ and $\quad i_1 = i_2 = i_c$

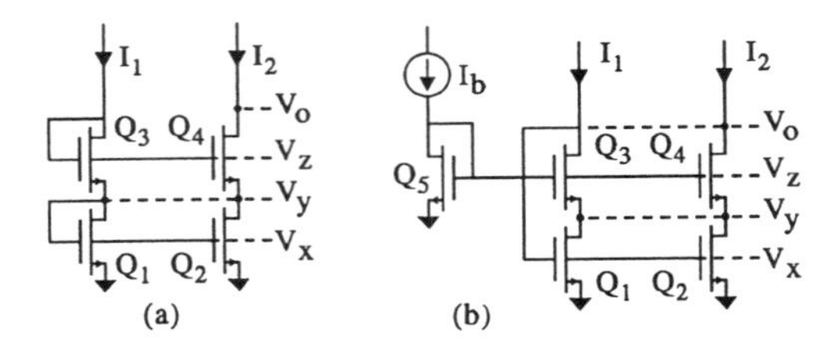

Fig. 9.7 Cascode current mirrors.
(a) Standard. (b) Wide-swing.

then $$i_o/i_c = 1/(a_v^2 + a_v + 1) \tag{9.39}$$

and thus $i_o/i_c \approx 1/a_v^2$, half of the $2/a_v^2$ for the standard cascode mirror.

Also with v_{d3} as in [9.37] then $i_i \approx v_{d3}g_m$, confirming that the impedance presented to the LTP Q_1 is restored to $\approx 1/g_m$. For insight into the results of the nodal analysis see Exercise 9.4.

9.6 Cascode OTA

Fig. 9.8 shows the arrangement of Fig. 9.6 with the added cascode pair Q_{34}. While the folded form of Fig. 9.10 is that normally used, the present arrangement gives an introduction.

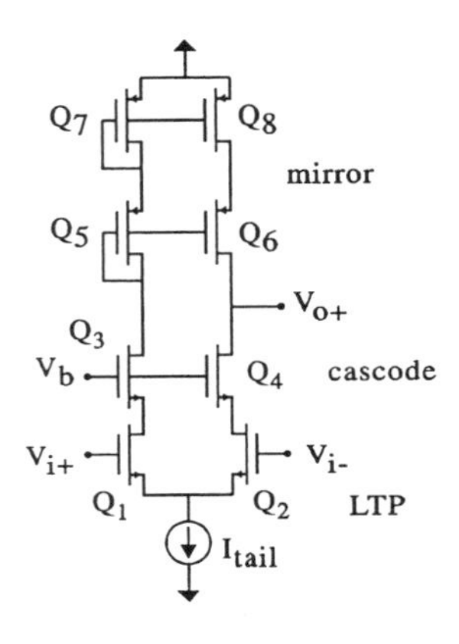

Fig. 9.8 LTP of Fig. 9.6 with added cascode stage.

Output impedance

The main advantage of the circuit is that it gives a much higher transductor output impedance r_{ot}, leading to a correspondingly higher voltage gain a_o where significant. For the LTP with cascode mirror as in Fig. 9.6, $r_{ot} = r_{om}//r_{o2}$. Here r_{om} is increased by $\sim\times a_v$ relative to the $\approx r_o$ for the standard mirror, but r_{o2} remains as normal. Thus the increase in r_{ot} relative to where the basic mirror is used is only from $\approx r_o/2$ to $\approx r_o$. With the cascode Q_4 added then r_{o2} is effectively also increased $\sim\times a_v$ to give r_{ot} now $\approx a_v r_o/2$.

More precisely, with

$$r_{tail} = r_{o4} + r_{o2}(1 + a_{v4})$$

as in [9.34] for the cascode tail then for the present Q_{24} cascode, including the back-door effect, the effective

$$r_{o24} = r_{o4} + r_{o2}(1 + a_{v4}) \approx r_{o2} \times a_{v4} \tag{9.40}$$

With $r_{ot} = r_{om}//r_{o24}$ then

$$r_{ot} \approx (r_{o8} \times a_{v6})//r_{o2} \times a_{v4}) \approx r_o a_v/2 \tag{9.41}$$

With this $\quad (r_{oP} \times a_{vP})//r_{oN} \times a_{vN})$

then $\mathbf{r_{ot}=10.4M\Omega}$ and $\mathbf{a_{oz}=9.47\times 10^3}$.

Frequency response

For the basic LTP of Fig. 9.2, c_o needed to be increased to 0.84pF for stability with full negative feedback. Here we choose the slightly larger c_o=1pF to give the frequency response of Fig. 9.9. With

$$f_o=1/(2\pi r_{ot}c_o) \quad \text{as in [9.32]}$$

for the basic LTP still valid then $\mathbf{f_o=15.3kHz.}$

With the unity gain frequency

$$f_u=g_m/(2\pi c_o)$$

then $\mathbf{f_u=145MHz}$

giving f_u is a little below f_α=171MHz as required.

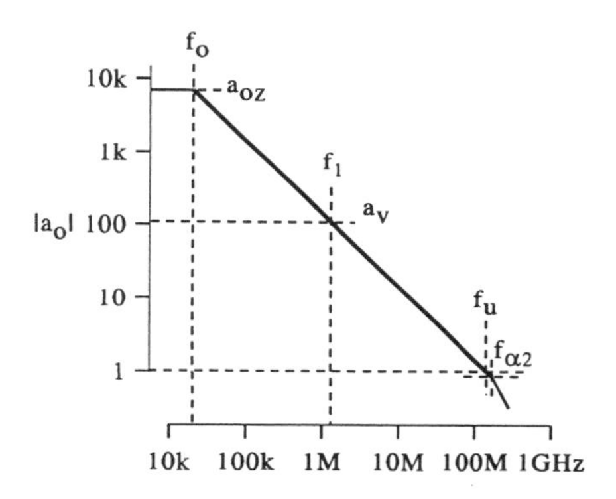

Fig. 9.9 Frequency response of voltage gain a_o for cascode OTA of Fig. 9.8.

Miller effect

For v_{i+} the Miller feedback via c_{gd1} is negligible, because of the low impedance presented by Q_3, and the same is so for v_{i-} with the v_o node grounded.

But with the node released and with the impedance presented by $Q_6\approx r_o a_v$, then that presented by Q_4 becomes $\approx r_o$ giving the normal Miller feedback as for the common-source stage. Exercise 9.5 explores this point further.

9.7 Folded cascode

With the large number of stacked MOS, the output range of the cascode OTA of Fig. 9.8 is so much reduced as to make it impractical. Also V_o still cannot cross zero into the negative region.

The folded cascode of Fig. 9.10 resolves these limitations. Here the circuit folds around the active loads $Q_{3,4}$, which provide the required I_d. With the cascode transistors $Q_{5,6}$ biased close to the upper supply and the wide-swing cascode current mirror used then the limits to V_o are little short of either supply voltage.

Gain reduction

The addition of Q_{34} gives a small reduction in output impedance r_{ot}, and therefore in a_o, but fortunately not in g_d.

Differential transconductance

With $g_d=i_o/v_{id}$ for the v_o node grounded then say $r_{in5}\approx 1/g_{m5}$. With $1/g_{m5}<<r_{o3}$ then the fraction of i_{d1} leaked through r_{o3} is negligible. Thus g_d remains $\approx g_{m12}$.

Output impedance

Without Q_4 in the incremental circuit and with r_{out6} the component of r_{ot} presented by Q_6 then r_{out6} would be $\approx a_{v6}r_{o2}$. With Q_6 present then the term r_{o2} is reduced to $r_{o2}//r_{o4}$. With $I_{d4}=I_{d2}+I_{d6}$ and for $I_{d2}=I_{d6}$ then $r_{o4}=r_{o2}/2$ giving a reduction of ×1/3 in r_{out6} causing r_{ot} to be halved.

Bias chain

Fig. 9.10(a) shows the bias chain. If all of the other transistors are identical then W/L for the MOS providing V_{b3} will need to be one quarter of that for all the others, to bias the current mirror correctly. With its increased r_{ot}, retained g_d, high CMR, and wide bipolar output range, the folded cascode is at last a circuit worthy of use.

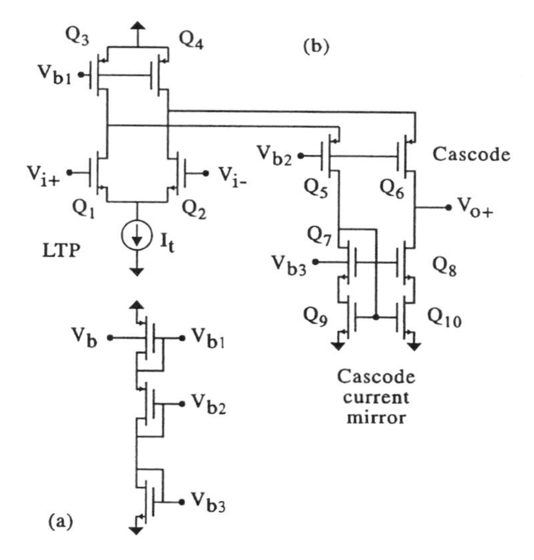

Fig. 9.10 Folded cascode adaptation of Fig. 9.8.
(a) Bias chain.
(b) Cascode.

Pad capacitances

One, two, or perhaps four of such circuits might be fabricated on a single chip, with the relevant leads brought out to the IC connections. With the pad capacitances each typically $\mathbf{C_{pad}=3pF.}$

For one OTA driving another, with the driving stage $c_o = 2c_{gd} + 2c_{db}$, and with the driven stage $c_{in} = c_{gs} + c_{gd}$, then the total

$$c_{mos} = 3c_{gd} + 2c_{db} + c_{gs}$$

giving $\mathbf{c_{mos} = 0.55pF.}$

With the $2C_{pad} = 6pF$ for the two pad capacitances then these are dominant, much reducing the unity gain frequency from the otherwise 263MHz to

$$f_u \approx g_m/(2\pi C) \tag{9.42}$$

giving $\mathbf{f_u = 22.1MHz.}$

With $g_m = [2\mu C_{ox}(W/L)I_d]^{1/2}$

as in [8.18] then g_m can be somewhat increased by increasing I_d.

For the current density already at the maximum allowable, then W will need to be increased in proportion, giving the more rapid effective $g_m \propto I_d$.

However with $c_{mos} \propto W$ then c_{mos} increases in proportion, making the I_d increase only of value up to the point where $c_{mos} = 2C_{pad}$.

For a ×10 increase in I_d and W then c_{mos} is increased to 5.5pF which is comparable with $2C_{pad} = 6pF$ giving

$$\mathbf{I_d = 1mA, \quad C = 11.5pF, \quad g_m = 9.10mS,}$$

and thus $\mathbf{f_u = 126MHz.}$

Negative feedback

While the OTA is normally used without feedback, where feedback is used then the value of the second cut-off at $f_{\alpha2}$ needs to be kept in mind. With previously $f_{\alpha2} = 171MHz$, and with the associated capacitance increasing $\propto I_d$ together with the g_m then $f_{\alpha2}$ remains unchanged. Here a somewhat smaller increase in I_d would be more appropriate.

9.8 Two-stage Miller op-amp

With the folded cascode a suitable MOS implementation of the OTA in Fig. 9.1(b), Fig. 9.11 now shows a corresponding implementation for the more conventional op-amp in Fig. 9.1(a).

The first stage in (a) is the LTP with mirror of Fig. 9.2. This drives the second stage Q_8, a conventional inverted common-source circuit with the active load Q_7. Then Q_{10} is the source-follower with active load Q_9, as needed to give the low output impedance r_{oo} required of the op-amp.

While the main pole for the OTA is set by the output capacitor c_o, for the conventional op-amp it is set internally, by the Miller feedback over Q_8, initially by c_{gd8} but increased by the added C_f.

Q_6 is operated below pinch-off as a voltage-controlled resistor, providing frequency compensation.

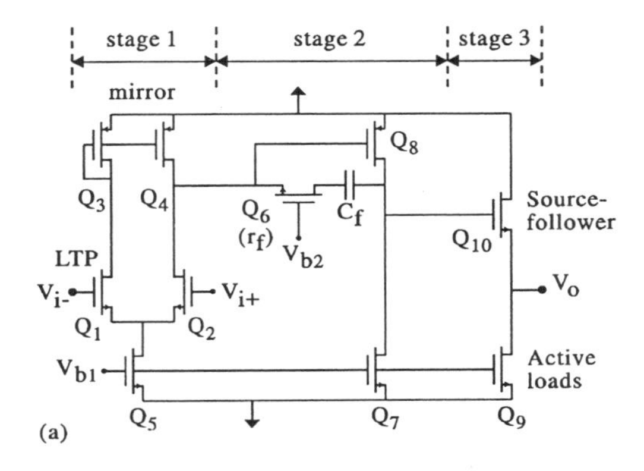

Fig. 9.11(a) Two-stage CMOS op-amp. Circuit diagram.

It is equally possible to have the overall complement, with the $Q_{1,2}$ P-channel MOS, etc. There are advantages and disadvantages (Johns and Martin, 1997) to either arrangement, and the complement tends to be the more popular. However, the present arrangement is a little easier to follow.

W/L value

If W/L is made our usual 50 for Q_{123478} and 100 for the tail Q_5 then for symmetric supplies V_{d8} will be zero for $V_{i+} = V_{i-}$. This is as required for zero input offset, apart from the small V_{gs10}.

Normally W for $Q_{9,10}$ is increased further, to give the higher $I_d \times (W/L)$ needed for a higher g_m, and thus a lower final output impedance.

Zero-frequency voltage gain a_{oz}

Fig. 9.11(b) shows the incremental equivalent of (b). Here the zero-frequency voltage gain

$$a_{oz} = a_1 \times a_2 \times a_3 \tag{9.43}$$

where

$$a_1 = -g_{m1}r_{o24} \quad \ldots(a)$$
$$a_2 = -g_{m8}r_{o78} \quad \ldots(b)$$
$$a_3 \approx 1 \quad \ldots(c) \tag{9.44}$$

giving

$$\mathbf{a_1 = -76.2, \quad a_2 = -53.9, \quad a_{oz} = 4.11k.}$$

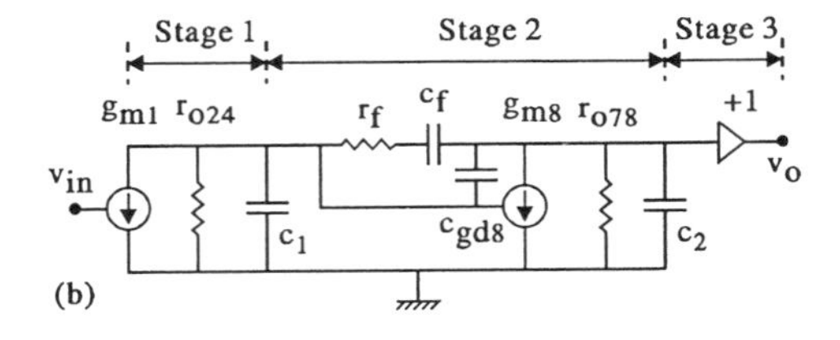

Fig. 9.11(b) Incremental circuit for (a).

Frequency response

Apart from Q_6, Q_8 is the common-source stage of Section 8.3 with the frequency response in Fig. 8.9(b). Here the Miller capacitance $c_m \approx c_{gd8}|a_2|$ giving

$$c_m \approx c_{gd8}|a_2| \tag{9.45}$$

With the Miller cut-off

$$f_m = 1/(2\pi c_m r_{o24})$$

then **$c_m = 2.69$pF and $f_m = 706$Hz.**

The other transition there is at **$f_{p2} = 121$MHz**

which will be much the same here. With these values then $|a_o|$ plots as in Fig. 9.13(a).

Increased c_f

While for the single-stage $|a_o|<1$ at the second pole f_{p2} now, due to the additional stage, $|a_o|>>1$.

To avoid closed-loop instability then c_{gd8} must be effectively increased by the addition of c_f. For the plot shown $c_f=3$pF giving **$c_m = 162$pF and $f_m = 11.8$kHz**

and thus a potentially stable response.

Unwanted zero

With c_f added, instead of the expected (c), the frequency response becomes as in (b).

With the focus of interest on the transition at f_z, and with f_{p2} determined by c_1 and c_2 in Fig. 9.11(c), then initially we ignore these, to give the much simplified overall incremental circuit of Fig. 9.13(a). With

$$v_o(s) = v_x - g_m(1/sc_f)v_x,$$

and with the impedance presented to v_x equal to $1/g_m$, then $v_x = -v_{in}$ giving

$$a_o(s) = (s_z - s)/s \tag{9.46}$$

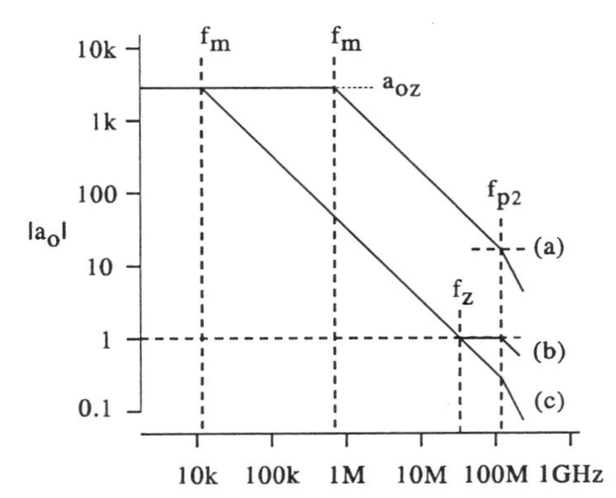

Fig. 9.12 Frequency responses for two-stage MOS op-amp of Fig. 9.11.
(a) For $c_f=0$. (b) With c_f increased ×100.
(c) As (a) but with r_f included.

where $$s_z = g_m/c_f \tag{9.47}$$

The corresponding pole-zero diagram is as for the pole and right-hand zero in (e), giving the frequency response in (c), and thus confirming that at the f_z transition in Fig. 9.12(b).

From the pole-zero diagram the phase shift ϕ starts at $-\pi/2$ for $\omega=0$ and then moves towards $-\pi$ for $\omega=\infty$, as marked in Fig. 9.13(c).

This is in sharp contrast to the response for the circuit in Fig. 9.13(b). For $1/g_m=0$ this is the straight integrating response of Fig. 9.12(b) before the f_z transition. With $1/g_m$ added then the amplitude-frequency response becomes again as in Fig. 9.12(b) and Fig. 9.13(c).

However, the pole-zero diagram becomes as in Fig. 9.13(e) but now with the more normal left-hand zero.

With the amplitude-frequency response the same for both the zero placements, for the left-hand zero the phase swings back to zero after f_z, improving the stability, while for the right-hand zero the stability is further degraded.

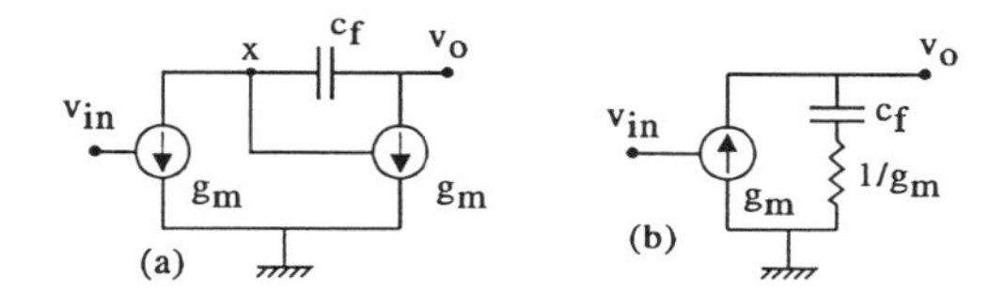

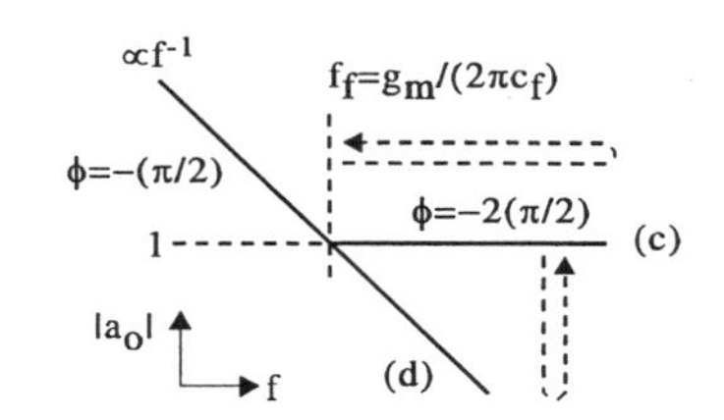

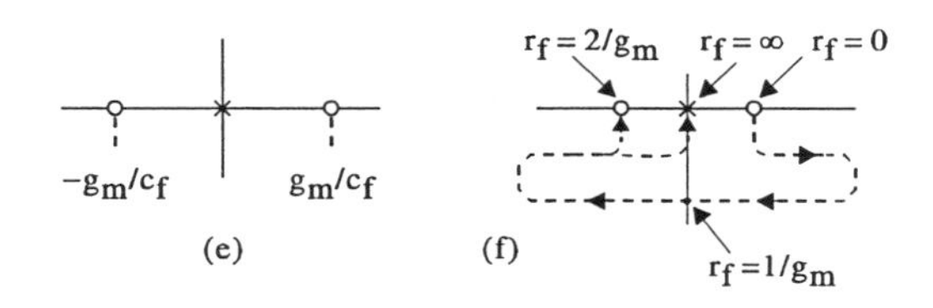

Fig. 9.13 Effect of c_f and r_f in Fig. 9.11 on frequency response of Fig. 9.12.
(a) Simplified small-signal circuit diagram of Fig. 9.11(c). $g_m = g_{m1} = g_{m2}$. $r_f = 0$.
(b) Circuit giving same amplitude-frequency response (b) as for (a) but with LH zero in (e).
(c) Frequency response for (a).
(d) Frequency response with $r_f = 1/g_m$.
(e) Pole-zero diagrams. LH-zero for (d). RH-zero for (a).
(f) Locus of zero in (e) as r_f is varied from zero to ∞, also traced in (c,d).

Zero correction

Addition of r_f (Q_6) in Fig. 9.11 gives, for c_{gd8} negligible,

$$a_o \approx 1 - g_m(r_f + 1/sc_f) \qquad (9.48)$$

For $r_f = 1/g_m$ this becomes the ideal integrating response g_m/sc_m of Fig. 9.13(d). Thus the overall frequency response becomes as in Fig. 9.12(c), leaving only f_{p2} to set the limit to the overall open-loop gain. With r_f included, the zero

$$s_z = g_m/[c_f(1 - g_m r_f)] \qquad (9.49)$$

As r_f increases from zero, then s_z traces the path

$$g_m/c_f \rightarrow \infty \rightarrow -\infty \rightarrow -g_m/c_f \rightarrow 0$$

shown in (f). The horizontal section of (c), and f_z, then follow as also shown.

Normally it is the integrating response of (d) that is required, for which $r_f = 1/g_m$. However, it is usual to set r_f a little higher, to allow for error and production spread, and also for c_{gd8}.

Bias

Fig. 9.14 shows a suitable bias chain for the Miller op-amp of Fig. 9.11(a). The zero correction requires $r_f = 1/g_{m8}$. With g_{co} the channel conductance for $V_{ds} = 0$

$$g_{co} = \mu C_{ox}(W/L)V_{eff} \qquad \text{as in [8.7]}$$

and with

$$g_m = \mu C_{ox}(W/L)V_{eff} \qquad \text{as in [8.17]}$$

then

$$g_{co} = g_m.$$

Thus if Q_6 and Q_8 are identical, and $V_{gs6} = V_{gs8}$ then r_f will equal $1/g_{m8}$ as required.

The V_{gs} will be equal if the bias chain Q_{b1} and Q_{b2} are identical. Where r_f needs to be a little higher than this, to allow for manufacturing errors, then $(W/L)_{b2}$ is slightly increased.

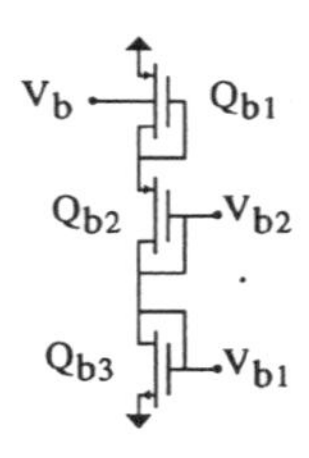

Fig. 9.14 Bias chain for two-stage CMOS op-amp of (a).

9.9 OTA-based feedback amplifier

While the function to which the MOS-based OTA is best suited is the integrator used in the OTA-C filter, the device can be used more widely. Fig. 9.15(a) shows its use in the normal non-inverting feedback amplifier of Section 2.2, with the frequency responses as in (b).

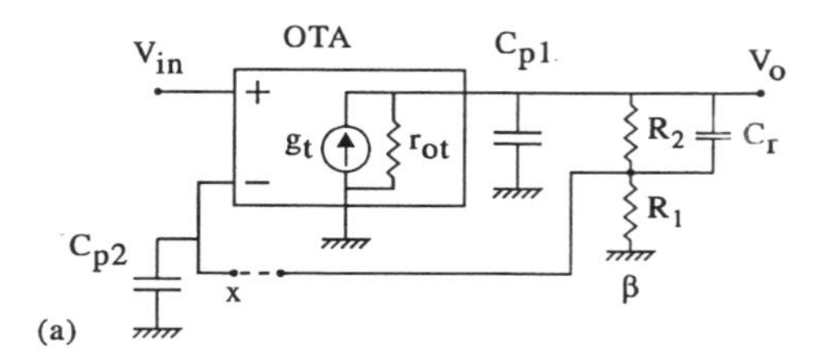

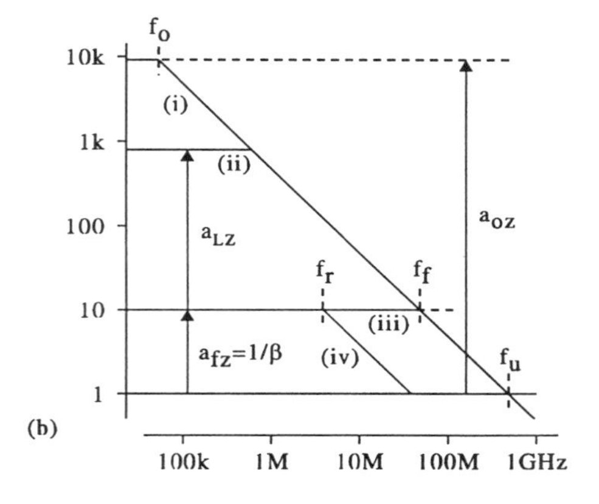

Fig. 9.15 OTA based non-inverting feedback amplifier.
(a) Circuit diagram. (b) Frequency responses.
(i) OTA and C_1 alone with point x grounded.
(ii) As (i) but with resistors added, x still grounded.
(iii) As (ii) but with feedback applied (x released and linked, C_p omitted).
(iv) As (iii) but with C_r included (suitable C_p restores to (iii)).
Symbols. a: voltage gain, β: feedback fraction, f: frequency.
Subscripts: L: loop, f: feedback amp, o: ota, p: pad, u: unity gain, r: C_r, z: zero-frequency.

Parasitic C_r

This is the parasitic capacitance of R_2, giving a transition in β such that

$$f_r = 1/(2\pi C_r R_2) \qquad (9.50)$$

Thus the closed-loop frequency response is changed from (iii) to (iv).

This is without the pad capacitance C_{p2}. If C_{p2} is slightly increased to give $C_r R_2 = C_{p2} R_1$ then (iii) is restored.

Exercise 9.6 confirms the values of the transition frequencies, etc., as in (b). These are for the increased $I_d = 1\text{mA}$ as in Section 9.7, to moderate the otherwise dominating effect of the pad capacitances C_{p1} and C_{p2}.

As usual the loop gain a_L is given by the difference between the open- and closed-loop responses. Here a_{LZ} is only ≈80 which is inferior to the typical value for a normal bi-fet op-amp. Note, however, the considerably higher cut-off frequencies.

Other configurations

Applying V_{in} to the tail of R_1, as for the equivalent inverting configuration, gives an input impedance equal to R_1, here 10kΩ. This is entirely at variance with the principle of maintaining the very high MOS-type input impedance.

However, a further OTA can be connected as a voltage follower, to buffer the input of the inverting circuit. Here the two C_p combine to give just one pole. The follower output impedance is $1/g_m$, here ≈100Ω.

The same considerations apply for the feedback integrator, differentiator, adder and subtractor configurations, albeit with the number of required voltage followers rapidly multiplying.

Also the voltage-follower can be used in the Sallen–Key filter configurations of Section 13.1.

9.10 Supply-independent bias source

Integrated circuits are normally designed to operate over a wide range of supply voltages, e.g. ±2.5V to ±8V for the CA3130 op-amp.

Over this range the device parameters should not change unduly, so suitable biasing circuitry is needed. Fig. 9.16(a) shows the essentials of such a supply. With

$$I_d = \mu C_{ox}(W/L)V_{eff}^2/2 \qquad \text{as in [8.13]}$$

then

$$I_1 = \mu C_{ox}(W/L)_1 V_{eff1}^2/2 \quad \ldots(a)$$

$$I_2 = \mu C_{ox}(W/L)_2 V_{eff2}^2/2 \quad \ldots(b) \qquad (9.51)$$

From the circuit

$$V_{eff1} = V_b - V_{th1} \quad \ldots(a)$$

$$V_{eff2} = V_b - V_{th2} - I_2 R \quad \ldots(b) \qquad (9.52)$$

With the current mirror maintaining $I_1 = I_2$ which we term I_d, and for $V_{th1} = V_{th2}$ then

$$R\sqrt{I_d \mu_n C_{ox}/2} = 1/\sqrt{(W/L)_1} - 1/\sqrt{(W/L)_2} \qquad (9.53)$$

Fig. 9.16(b) shows the variation of the I with V_b.

$(W/L)_2$ is made larger than $(W/L)_1$, so I_2 is larger to start with but is later linearised by R, causing the curves to cross. With the current mirror maintaining $I_1 = I_2$ then it is the crossing point that determines the I, and not the supply values.

V_b then becomes the bias voltage for the supply chain, maintaining the I_d for all the other MOS equally supply-independent.

The feedback for the loop is positive, but with loop-gain < 1 the loop is stable. There are, however, two stable states, the one shown and where the I are both zero. A 'start-up' trigger is therefore needed to place the circuit in the required state.

Temperature θ independence

In [9.53] the θ-dependent V_{th1} and V_{th2} do not appear. Being equal they have cancelled. This leaves the mobility μ_n as the only θ-dependent term. However, with

$$g_m = [2\mu C_{ox}(W/L)I_d]^{1/2}$$

as in [8.18], and with the circuit maintaining $\mu_n I_d$ independent of θ, then g_m becomes independent of θ also.

The same will apply for all of the other MOS. With the variation of μ_n for θ from 0 to 70°C in the region of 2:1, so also is I_d. However, with $g_m \propto \mu I_d$ constant, this is acceptable.

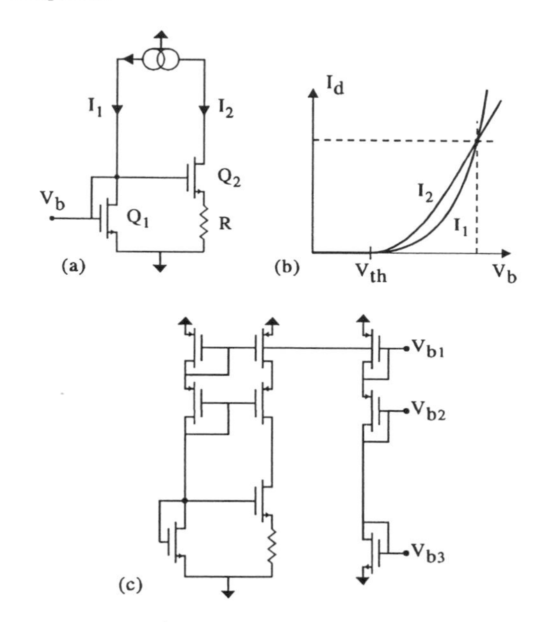

Fig. 9.16 Supply- and temperature-independent bias source.
(a) Circuit essentials.
(b) I/V relations.
(c) With cascode current mirror in (a) providing V_{b1} for Miller op-amp bias chain of Fig. 9.14 (redrawn).

Bias chain Fig. 9.16(c) shows the more complete arrangement. A cascode current mirror is used, to reduce the remaining small supply dependence of the I_d due to the finite r_o for the MOS. It is more convenient to take V_b there as the V_{b1} shown here.

References

Gray, P. R. and Meyer, R. G. 1993: *Analysis and Design of Analogue Integrated Circuits*. John Wiley.
Johns, D. A. and Martin, K. 1997: *Analog Integrated Circuit Design*. John Wiley.
Laker, K. R. and Sansen, W. M. C. 1994: *Design of Analog Integrated Circuits and Systems*. McGraw Hill.
Massobrio, G. and Antognetti, P. 1993: *Semiconductor Device Modelling with SPICE*. McGraw Hill.

Much of the material of the present chapter is based on the first three references, particularly Johns and Martin.

9.11 Exercises

Calculations here are based on the values in Table 9.1 unless stated otherwise.

Exercise 9.1 LTP output impedance

> **Explanation** Explain why $r_{ot} \approx r_{o2}//r_{o4}$ as in [9.7].

Impedance presented by Q_2. With the LTP of Fig. 9.2(a) redrawn as Fig. 9.17(a), it will first be shown that the impedance r_{out2} presented by Q_2 at the output node is not r_{o2} as the expression suggests but $\approx 2r_{o2}$. With the output impedance of the common-gate stage

$$r_{out}=r_o+r_s(a_v+1) \qquad \text{as in [8.52]}$$

and with Q_2 regarded as such, then

$$r_{out2}=r_{o2}+r_{in1}(a_{v2}+1)$$

where r_{in1} is the impedance presented by Q_1 to Q_2. With the input impedance of the common-gate stage

$$r_{in}=(r_L+r_o)/(a_v+1) \qquad \text{as in [8.53]}$$

with Q_1 regarded as such, with the load impedance r_L presented by $Q_3 << r_{o1}$, and with $a_{v1}=g_{m1}r_{o1}$ then

$$r_{in1} \approx 1/g_{m1}$$

confirming, for Q_1 and Q_2 identical, that

$$r_{out2} \approx 2r_{o2}$$

Fig. 9.17(a) MOS-based long-tailed pair (LTP).

Back-door effect Why then is not

$$r_{ot}=r_{o4}//2r_{o2}?$$

As shown in Fig. 9.17(b), for v_o externally applied the resulting $i_2=v_o/2r_{o2}$ flows down through Q_2 and, with very little reduction, up through Q_1, to be reflected as a reduction in i_4 and thus as an addition by $v_o/2r_o$ to i_o. Thus the effect of i_2 is doubled, to give $r_{ot}=r_{o2}//r_{o4}$ after all.

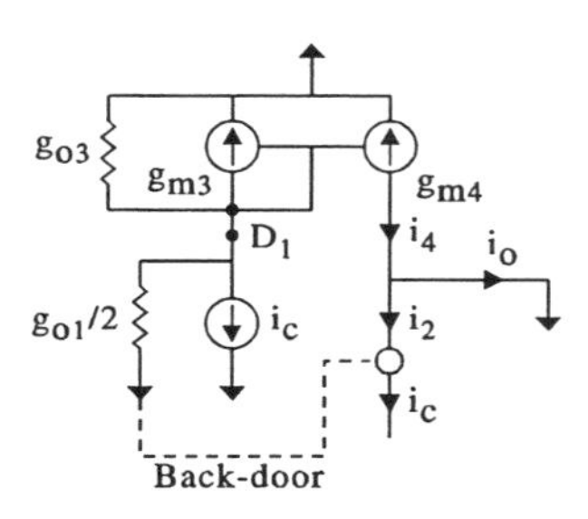

Fig. 9.17(b) Back-door route for MOS-based long-tailed pair in (a).

Exercise 9.2. Current-mirror common-mode response

> **Explanation** With the common-mode mirror current gain α_c as in [9.17], account for the term g_{o1}/g_{m3}.

The term accounts for the extent to which the impedance $\approx 1/g_{m3}$ presented to Q_1 by Q_3 reduces i_1 and so upsets the balance between i_1 and i_2. The expression suggests that the impedance r_{out1} presented by Q_1 to Q_3 is r_{o1}, while, as for r_{out2} in the last exercise, $r_{out1} \approx 2r_{o1}$.

Once more it is the back-door effect that accounts for the difference, with the reduction in i_1 coupled round as an equal increase in i_2, thus doubling the effect upon i_o.

Exercise 9.3 LTP parameter mismatch

> **Explanation** Explain why the LTP parameter mismatch terms Δg_m and Δg_o are suppressed by $\times 2/a_v$ in [9.19] and $\times 1/a_v$ in [9.20] for the common-mode transconductance g_c.

With the mirror responding fully to the result of the imbalance, the effect of the non-zero impedance of Q_3 is secondary and can be ignored.

For the mirror impedance zero then the LTP circuit becomes symmetric. With v_{ic} only applied then the current through the S-node is zero, so the circuit can be split at the node without making any difference to the operation.

Δg_m suppression The response to v_{ic} is then as for a source-follower with self-load $r_o//2r_{tail}$, reducing v_{gs} from v_{ic} by $\times(g_o+g_{tail}/2)/g_m$. With $g_{tail}/2=g_o$ and $a_v=g_m/g_o$ then the suppression is by $\times 2/a_v$.

Δg_o suppression With $g_m >> g_o+g_{tail}/2$ in [9.20] then Δg_o is not really suppressed at all. For the same 1% mismatch Δg_o is simply less than Δg_m because g_o is smaller.

Note, however, the differing effects as r_{tail} is reduced to zero. For Δg_m the suppression is lost as $r_{tail}/2$ falls below the output impedance $1/g_m$ of the source-follower, increasing the fraction of v_{ic} reaching v_{gs}, finally to unity. In contrast, as r_{tail} is decreased then v_s falls to zero. With v_s the voltage driving g_o then $g_c(\Delta g_{oL})$ falls to zero also.

Exercise 9.4 Wide-swing cascode current mirror (Fig. 97(b))

Explanation Explain the nodal analysis result of [9.39] that $|i_o/i_c| \approx 1/a_v^2$.

For the incremental circuit diagram corresponding to Fig. 9.7(b) for the mirror, ground the v_o node to derive the output current $i_o = i_2 - i_4$, and inject $i_1 = i_2 = i_c$.

Using superposition for the common v_{d3} and v_{g1} we first set v_{d3} to zero. This effectively grounds the point and renders the circuit fully symmetric. Then, whatever the value of v_{g1}, i_2 will equal i_1 giving $i_o = 0$.

Thus it is only v_{d3} which makes α_c non-zero. With v_{d3} applied and $v_{g1} = 0$, this makes no difference to i_2 but adds the component $\Delta i_1 = v_{d3}/r_{oc}$ to i_1, where r_{oc} is the output impedance of the LHS cascode. With $r_{oc} \approx a_{v3} r_{o1}$ and with $i_1 \approx v_{d3} g_m$ then $\Delta i_1/i_1 = 1/a_v^2$.

Exercise 9.5 Cascode OTA Miller effect (Fig. 9.8)

Problem Explain

- why the Miller effect for v_{i+} is negligible,
- why the effect is also negligible for v_{i-} when the V_o node is grounded,
- why for v_{i-} with the node released,

(i) for $f < f_o$ the Miller input capacitance is equal to $c_{gd2}(a_v/2 + 1.5)$,
(ii) for $f_o < f < f_1$ the impedance changes to the resistance $r_m \approx (1/g_m)(c_o/c_{gd})a_v = 2.35\text{M}\Omega$,
(iii) for $f_1 < f$ the impedance becomes that of c_{gd2}.

Here f_1 is the frequency at which $|a_o| = a_v$ as in Fig. 9.9.

Exercise 9.6 OTA-based feedback amplifier

Confirm the values shown for the frequency responses of Fig. 9.15(b) for the OTA-based feedback amplifier of (a).

Here I_d has been increased to 1mA to give

$$g_d = 9.10\text{mS},\quad r_{ot} = 1\text{M}\Omega,\quad C_{p1} = C_{p2} = 3\text{pF},$$
$$R_1 = 10\text{k}\Omega,\quad R_2 = 90\text{k}\Omega,\quad C_r = 0.5\text{pF}.$$

With C_{p1} dominant, ignore the MOS capacitance.

Unity gain frequency From the circuit $f_u = g_d/(2\pi C_{p1})$ giving **$f_u = 484\text{MHz}$.**

OTA alone (i) Here $a_{oz} = g_d r_{ot}$ giving **$a_{oz} = 9.14\text{k}\Omega$.**

With $f_o = 1/2\pi r_{ot} C_{p1}$ then **$f_o = 53.1\text{kHz}$.**

Resistors added (ii) Here a_{oz} is reduced to

$$a_{oz} = g_d r_{ot}//(R_1 + R_2)$$

With $r_{ot}//(R_1 + R_2) = 90.9\text{k}\Omega$ then **$a_{oz} = 9.14\text{k}$.**

Feedback applied (iii) With $a_{fz} = 1/\beta$

where $\beta = R_1/(R_1 + R_2$
then **$a_{fz} = 10$.**

With $f_f = f_u/a_{fz}$
then **$f_f = 48.4\text{MHz}$.**

Parasitic capacitance With $f_r = 1/(2\pi C_r R_2)$ as in [9.50] then **$f_r = 3.54\text{MHz}$.**

10

Junction field effect transistor

Summary

The JFET preceded the MOSFET by a few years and there are now few applications where either the bipolar or MOS device does not give a better performance. However, there remain some, notably as the LTP at the input of the TL081 bi-fet op-amp used in most of our experimental work, and as the JFET audio power amplifier, as used in the popular 500W car blaster.

Because the operation so closely parallels that of the MOS the treatment will be brief. Calculations are all within the main text, with the cumulative calculation results set bold to the right-hand margin as usual.

10.1 JFET operation

Fig. 10.1 shows the essential structures of the N- and P-channel JFET, with the corresponding circuit symbols. The present account will be for the N-channel device.

As for the MOS, the JFET works by the control of the conducting channel by the potential of the non-conducting gate. With the structure in Fig. 10.1(a) that of the semiconductor diode, reverse bias applied to the gate increases the width of the depletion layer, thus decreasing the width of the channel, and eventually closing the channel. Thus, while for the MOS the channel is normally absent, requiring the gate-to-channel voltage V_{gs} to be greater than the threshold voltage V_{th} for its creation, for the JFET the channel is normally present, requiring $V_{gs} < V_{th}$ for its closure.

Voltage-controlled resistor

With V_{gs} controlling the width and therefore the conductance of the channel, the JFET becomes a voltage-controlled resistor.

With the channel width needing to be somewhat less than that of the depletion layer for effective operation, the variation of channel conductance g_c with V_{gc} is approximately linear, giving

$$g_c \approx g_{co} \times (V_{gc} - V_{th})/(-V_{th}) \qquad (10.1)$$

where g_{co} is the value of g_c for $V_{gs}=0$.

With [10.1] only obtaining for V_{ds} sufficiently small, g_c and g_{co} are incremental values.

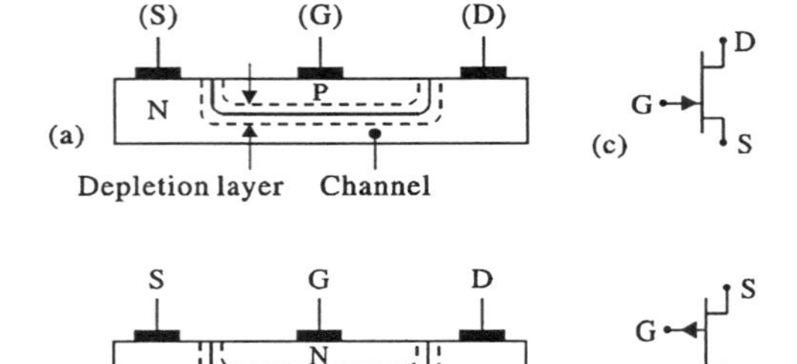

Fig. 10.1 JFET cross-sections and symbols.
(a) N-channel JFET with symbol in (c).
(b) P-channel JFET with symbol in (d).

10.2 Common-source voltage amplifier

Fig. 10.2(a) shows the essentials of the N-channel JFET common-source amplifier circuit. This is much as for the bipolar transistor or MOS equivalents, but with E_b now negative and approximately mid-way between V_{th} and 0V.

Pinch-off

As for the MOS, the ability of the device to function as a voltage amplifier is dependent upon it being pinched-off.

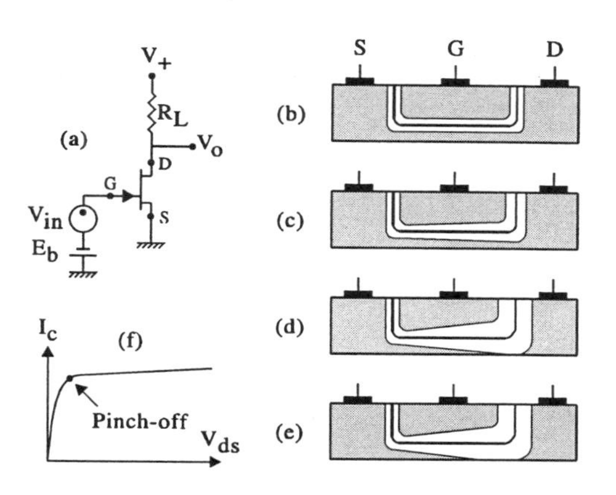

Fig. 10.2 JFET amplifier showing development of pinch-off as V_{ds} is increased from zero.
(a) Circuit diagram.
(b) $V_{ds}=0$.
(c) V_{ds} rising to pinch-off.
(d) V_{ds} at pinch-off.
(e) V_{ds} beyond pinch-off.
(f) Variation of I_c with V_{ds} for above sequence.

Fig. 10.2(b) to (e) shows the development, with (f) the resulting variation of channel current I_c with drain-to-source voltage V_{ds}.

Pinch-off channel current With the variation of JFET channel width much as for the variation of MOS channel charge density in Fig. 8.4, at pinch-off

$$I_c = g_{co}(V_{gs} - V_{th})^2/(-2V_{th}) \qquad (10.2)$$

With I_{co} the value of I_c for $V_{gs}=0$ then

$$I_c = I_{co}(1 - V_{gs}/V_{th})^2 \qquad (10.3)$$

giving the parabolic plot of Fig. 10.3(a).

Mutual conductance

With the mutual conductance $g_m = \partial I_c/\partial V_{gs}$ then

$$g_m = 2I_{co}[(V_{gs} - V_{th})/V_{th}^2] \qquad (10.4)$$

or

$$g_m = 2\sqrt{I_c I_{co}}\ /\ (-V_{th}) \qquad (10.5)$$

With g_m as in [10.4] and g_{mo} the value of g_m for $V_{gs}=0$,

$$g_m = g_{mo}(1 - V_{gs}/V_{th}) \qquad (10.6)$$

giving the linear plot of Fig. 10.3(b).

As for the MOS, the concept of an effective gate voltage

$$V_{eff} = V_{gs} - V_{th} \qquad (10.7)$$

gives the simplified relations

$$I_c \propto V_{eff}^2 \quad \text{and} \quad g_m \propto V_{eff},$$

with the additional scales as shown.

For the typical **$V_{th}=-2V$ and $I_{co}=4mA$**

and with g_m as in [10.4] then

$$g_{mo} = 2I_{co}/(-V_{th})$$

giving **$g_{mo}=4mS$.**

With I_c as in [10.2] $I_{co} = -g_{co}V_{th}/2$,

giving $g_{co}=g_{mo}$ and thus also **$g_{co}=4mS$.**

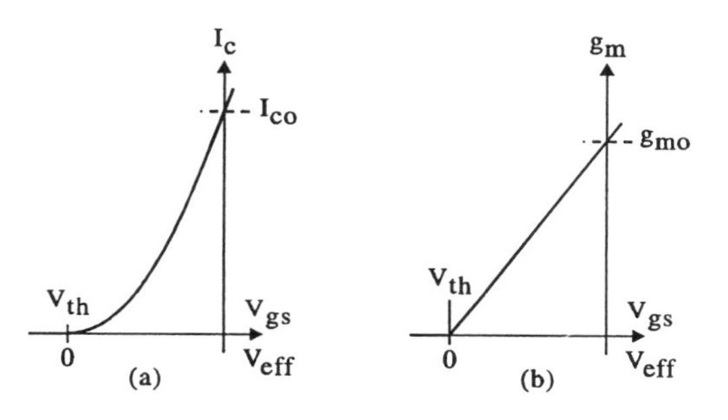

Fig. 10.3 Variation of pinch-off I_c and g_m with V_{gs} and V_{eff}.
(a) I_c.
(b) $g_m = \partial I_c/\partial V_{gs}$.

Output resistance

The incremental JFET output resistance $r_o = \partial I_c/\partial V_{ds}$ is the slope of the output characteristic in Fig. 10.2(f) beyond pinch-off. As for the bipolar transistor in Fig. 6.6, the family of characteristics for differing V_{gs} converge at a point equivalent to the Early voltage, with the JFET voltage typically −100V.

With $I_c \propto V_{eff}^2$, $I_{co}=4mA$ and for $V_{eff}=|V_{th}|/2$

then **$I_c=1mA$** giving **$r_o \approx 100k\Omega$.**

Pinch-off

At pinch-off $V_{gs} - V_{ds} = V_{th}$ giving the pinch-off voltage

$$V_p = V_{gs} - V_{th} \qquad (10.8)$$

With V_{eff} as in [10.7] then $V_p = V_{eff}$.

Output range

For V_{gs} at the mid-point $V_{th}/2$ then $V_p=|V_{th}|/2$ giving **$V_p=1V$**

which is considerably higher than the equivalent saturation voltage $V_{sat} \sim 150mV$ for the bipolar transistor. Here not only is the range of V_o accordingly limited, but also the efficiency of the JFET power amplifier is proportionally reduced. This is an advantage that the bipolar power amplifier has over both types of FET power amplifier.

Potential voltage gain

With $g_m \propto V_{eff}$ as in Fig. 10.3(b), and with $V_{eff}=|V_{th}|/2$, as above then $g_m=g_{mo}/2$ giving **g_m=2mA/V.**

With the common-source voltage gain $a_{cs}=-g_m r_o$ then **a_{cs}=−200**

much inferior to the equivalent −2800 for the bipolar.

Resistive load

With $R_L=(V_+ - V_p)/2I_c$ then **R_L=4.5kΩ.**

With $R_L \ll r_o$ then $a_{cs} \approx -g_m R_L$ giving the even more meagre **a_{cs}=−9.**

This is why the JFET is used only for the differential input stage of the bi-fet op-amp.

Problem

For $V_{gs}=0$ then the slope of the output characteristic in Fig. 10.2(f) at the origin is g_{co}. Show that the slope of the line joining the origin to the pinch-off point is $g_{co}/2$.

10.3 JFET capacitances

For the simplified structures of Fig. 10.1(a,b) with no applied voltage there is just the one depletion layer capacitance. At pinch-off, however, this is split into c_{gs} and c_{gd}, with c_{gs} the larger. Apart from parasitics c_{ds} is essentially zero.

Detailed device structure

With the channel width W_c comparable with that of the depletion layer, this would make the structures of Fig. 10.1 far too thin for mechanical stability. Also the device may need to be combined with others in a single integrated circuit.

N-channel For the N-channel device this calls for the structure of Fig. 10.4(a). First the N-type substrate is made as thick as is convenient. Then the insulating P^+-type diffusion is followed by the operative N- and P-type diffusions in (a). These can be made as thin as required.

The insulation afforded by the P^+ layer is by virtue of the depletion layer between this and the channel. As well as defining the lower edge of the channel, the layer also serves to isolate the FET from other devices on the chip.

However, we now have the relatively large associated c_{cb} (c: channel, b: base/substrate). Beyond pinch-off this too is split, into c_{sb} and c_{db}, with c_{sb} the larger. Thus the small-signal equivalent circuit of the device is as in Fig. 10.5(a), with c_{gb} essentially zero.

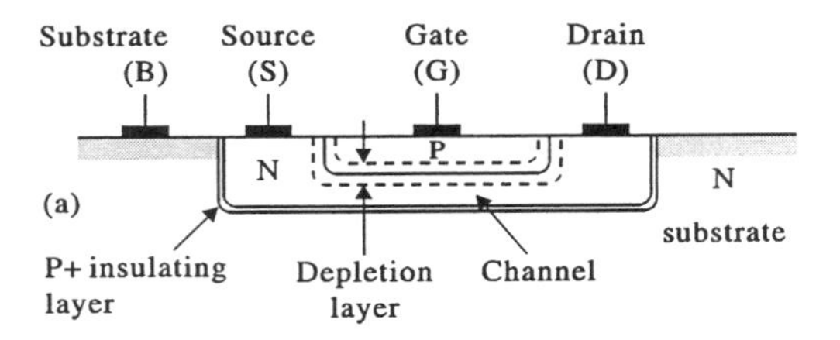

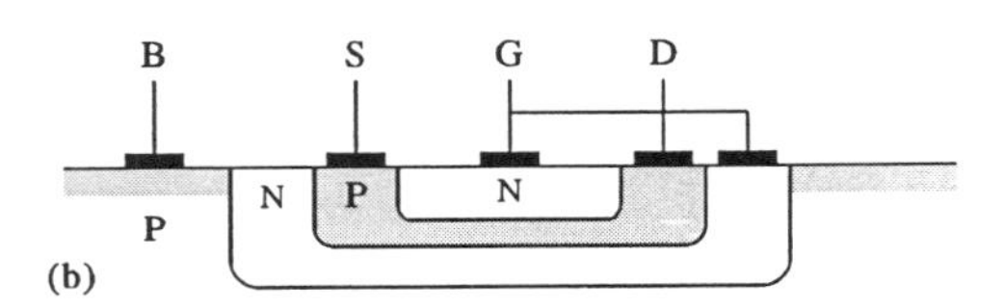

Fig. 10.4
(a) Incorporation of N-channel JFET into an N-type substrate for mechanical stability.
(b) Essentials for P-channel type in bi-fet op-amp.

P-channel The structure shown in Fig. 10.4(b) is as for the JFET LTPs at the input of a bi-fet op-amp. Here the FET needs to be of the P-channel type, with a P-type substrate. With this structure the gate extends to both sides of the channel, pinching off from both sides. Now c_{gb} is the largest capacitance, while c_{sb} and c_{db} are both zero. The equivalent circuit is thus as in Fig. 10.5(b).

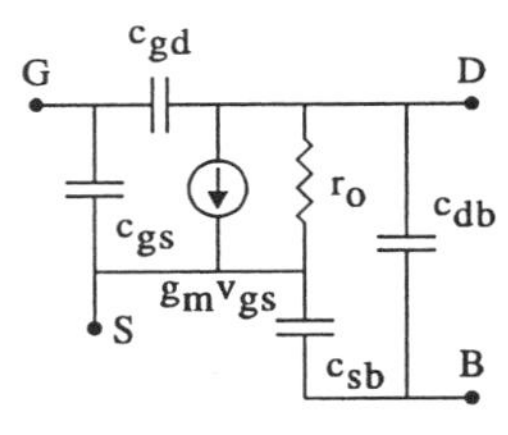

Fig. 10.5(a) JFET small-signal capacitances for N-channel JFET of Fig. 10.4(a).
Typ. c_{gs}=2pF, c_{gd}=0.5pF, c_{db}=1pF.

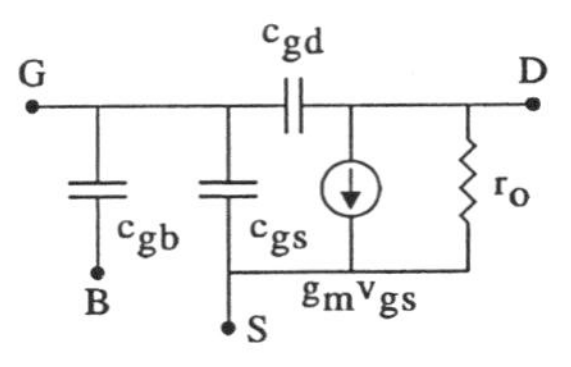

Fig. 10.5(b) JFET small-signal capacitances for P-channel JFET of Fig. 10.4(b).
Typ. c_{gs}=2pF, c_{gd}=0.5pF, c_{gb}=4pF.

10.4 Frequency response

For the common-source amplifier, c_{gd} has the same effect as c_μ in the bipolar common-emitter amplifier, and has a comparable value, each being depletion layer capacitances. Thus the Miller effect presents the input capacitance $c_m \approx |a_{cs}c_{gd}|$ in parallel with c_{gs}, tending to dominate c_{gs}.

With a_{cs} as for the above R_L then $c_m=4.5pF$, while for the active load the ideal a_{cs} is approached giving $c_m=100pF$.

Unity current gain frequency

For the common-source amplifier S and B are both grounded.

For the incremental circuit, the incremental current gain $\beta=i_o/i_{in}$, where i_{in} is the input current and i_o the current flowing to ground with the output node D grounded. With $\beta=-g_m/\omega c_{in}$ where c_{in} is the total input capacitance, then the unity gain frequency

$$f_u=g_m/(2\pi c_{in}) \tag{10.9}$$

N-channel With $c_{in}=c_{gs}+c_{gd}$ giving $c_{in}=2.5pF$ then **$f_u=127MHz$.**

P-channel With $c_{in}=c_{gb}+c_{gs}+c_{gd}$ giving $c_{in}=6.5pF$ then **$f_u=49MHz$.**

With the values of the JFET capacitances proportional to the area of the device, these will be much higher for the power JFET, but so also is g_m, leaving f_u much the same.

10.5 Temperature coefficient

With the channel current I_c and the gate-to-source voltage V_{gs}, then the effect of temperature on the variation of I_c with V_{gs} is as shown in Fig. 10.6.

Here the first effect of increasing temperature is to reduce the carrier mobility, giving the general reduction in I_c shown.

The other is to increase the magnitude of V_{th}. This occurs as follows.

- The increase in temperature causes an increase in the minority carrier density, causing an increase in the minority carrier flow across the depletion layer.
- The depletion layer width W_d is reduced, reducing the junction potential ψ_o, increasing the majority flow to balance the increase in minority flow.
- The decrease in W_d gives an equal increase in the channel width, increasing the magnitude of V_{th} needed to close the channel.

While the crossover point varies somewhat with temperature, it is still advantageous to set the bias E_b at the intersection as shown.

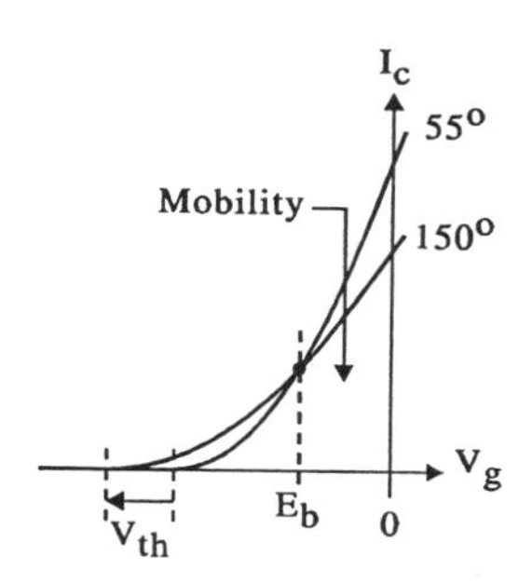

Fig. 10.6. Diagram showing the effect of temperature increase on the JFET mutual characteristic.

10.6 JFET applications

Bi-fet op-amp

While JFET implementations of the source-follower, common-gate, cascode, etc., are all realisable, the bipolar or MOS equivalents are usually superior in performance. The main exception is the LTP at the input of the bi-fet op-amp.

Bias current and input impedance

With these about three orders of magnitude less than for the bipolar transistor, as in Table 10.1, this is the main reason for preferring the FET input stage.

	Bipolar	JFET	MOS
I_b	5μA	30pA	1pA
r_{in}	5kΩ	1GΩ	25GΩ

Table 10.1 Signal input current and impedance for bipolar, JFET, and MOS input op-amp types.

With the voltage gain of the JFET so much inferior to that of the bipolar transistor, this is the reason why the remainder of the circuit is implemented using bipolar transistors.

The reason that the JFET is preferred to the MOS for the input LTP is that for the MOS the noise is somewhat higher, particularly the 1/f-noise.

A disadvantage of the bi-fet op-amp relative to the bipolar is that it has a larger voltage offset temperature coefficient. With the basic sensitivities comparable, this is probably due to the better thermal matching that can be obtained for the bipolar transistors.

Power JFET

As for the MOS, the JFET transfer function is parabolic, rather than exponential as for the bipolar transistor. For the audio power amplifier this improvement in linearity is somewhat offset by the much higher voltage gain for the bipolar transistor, allowing a higher degree of negative-feedback. However, the power FET still gives the better overall linearity.

While the vertical VMOS structure (Mohan et al, 1995) allows the MOS to be implemented as a power device, the power JFET appears to be the more robust, and so is the more usual choice.

For the I_{co} of several amps required by the power JFET, many of the smaller devices are effectively connected in parallel by increasing the device width W across the surface of the substrate. With channel length L fixed by other constraints then W becomes many times L, with the resulting ribbon-like structure snaked to better fit the normally square shape of the substrate.

Constant-current source

As in Fig. 10.7(a) the JFET provides a passable and very simple implementation of the ideal current source, since above pinch-off the variation of I_c with V_{ds} is small.

The thermal stability is improved by setting V_{gs} to E_b in Fig. 10.6, as in (b), with $R_s = E_b/I_c$. Sometimes the circuit is manufactured as a two-terminal device for this purpose.

Voltage reference

For the higher voltage Zener reference it is avalanche breakdown which dominates. From [17.3] the shot-noise $\overline{I_n^2} \propto q$ where q is the single electron charge. This is the area of each component impulse that makes up the noisy current.

But for avalanche breakdown each component impulse is multiplied by the number N of electrons that are generated as the result of each random 'seed' electron. Thus q in the shot-noise expression becomes Nq, increasing $\overline{I_n^2}$ by this factor.

Connected as in (c) the JFET can give a more noise-free voltage reference. Here R_L is set to make $V_o = I_c R_L$ where V_o is the required reference voltage.

For V_o positive as here the P-channel JFET is needed, with the N-channel for V_o negative,

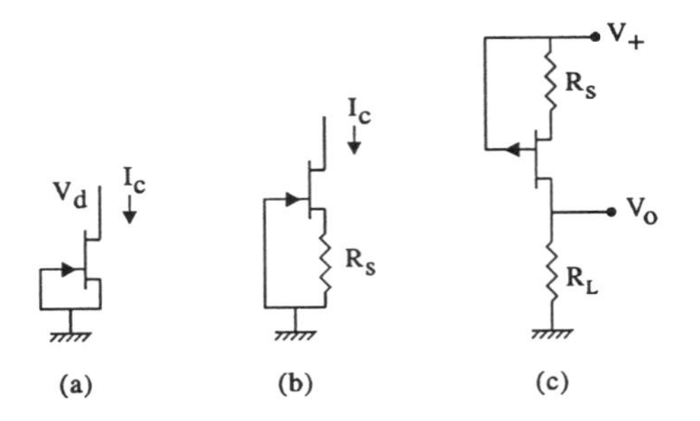

Fig. 10.7 JFET used as a constant-current source.
(a) Basic circuit.
(b) Addition of R_s to give improved temperature independence.
(c) Low-noise voltage reference using P-channel JFET.

JFET switch

Fig. 10.8(a) shows the JFET analog series switch, largely to show its disadvantage relative to the CMOS equivalent.

With the bipolar transistor saturation voltage V_{sat} and the FET threshold V_{th}, and for the switch nominally open, conduction begins for $V_{in} < V_{sat} + |V_{th}|$. There is no practical way in which two complementary JFETs can be combined to avoid this limitation.

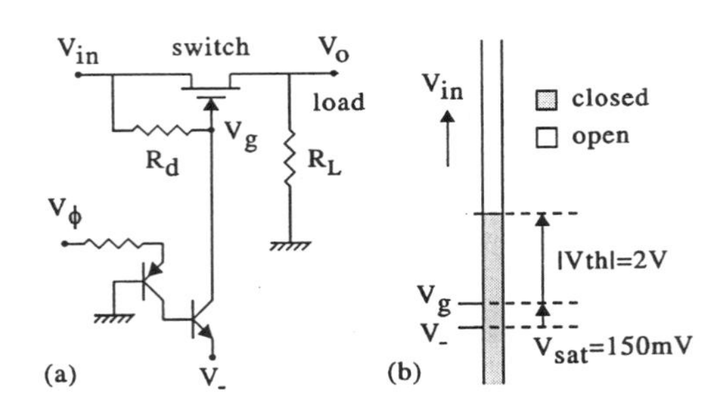

Fig. 10.8 JFET series switch.
(a) Circuit diagram.
(b) Diagram showing range of V_{in} for which the nominally open switch is actually closed.

Reference

Mohan, N., Undeland, T. M. and Robbins, W. P. 1995, *Power Electronics, Applications and Design*, John Wiley.

11

Single-section passive RC filters

Summary

The topic for this and the next two chapters is the electronic filter. This is a circuit that will transfer a sine wave input of frequency f to the output for some ranges of f but not for others. Fig. 11.1 shows the ideal responses for four of the main types.

A passive filter is one which uses only R, C and L components, while the active filter also uses the transistor, usually in the form of the op-amp, with the associated power supplies.

The first two sections of this chapter cover the simple single-section passive low- and high-pass filters, composed of one resistor and one capacitor. For each the step response, the frequency response, the impulse response, and the time-averaging properties are considered.

The same functions can be implemented using a single L and R but this is not favoured, because the inductor is relatively large, costly, usually cannot be implemented in microcircuit form, and tends to radiate and pick up magnetic interference.

Band-pass and band-stop filters can also be implemented in passive RC form but then the relative bandwidth is large, typically with $f_u/f_L > 3$.

For the 'narrow band' types, where the bandwidth $\Delta f \ll f_o$, the passive implementation requires the LCR resonator of the next chapter, or the active RC implementation of Chapter 13. Again the RC implementation is used where possible.

Other types of filter are the 'comb' and 'all-pass'. Comb filters have a series of alternating band-pass and band-stop regions, usually regularly spaced. The 'all-pass' transmit for all frequencies but introduce a phase shift that varies with frequency in a chosen manner. These types are left for further reading (Van Valkenburg, 1982).

The remaining sections cover the pole-zero diagram and the Laplace transform. These methods are needed for the more advanced types of filter covered in the next two chapters.

A relatively large amount of experimental work with numeric examples is given for the low- and high-pass filters of the first two sections. These are then used as the examples for the pole-zero and Laplace methods of the other two.

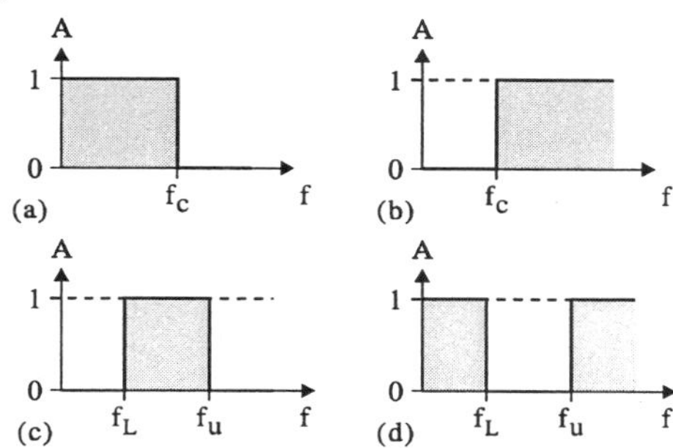

Fig. 11.1 Gain-frequency responses for ideal (abrupt cut-off) filters of pass-band gain A = 1.
(a) Low-pass.
(b) High-pass.
(c) Band-pass.
(d) Band-stop.

11.1 Simple RC low-pass

The single-section low-pass filter of Fig. 11.3(a) is very widely used. The effect also occurs without invitation, as the result of device capacitances such as the bipolar transistor c_π and c_μ. With X_c the capacitor impedance, and with

$$|X_c| = 1/\omega C,$$

then for frequencies below that for which $|X_c| = R$ the filter transmits, while for higher frequencies the signal is attenuated, hence the low-pass action.

When a voltage step is applied to the filter, the capacitor takes the time T_r to charge to the input value, also to discharge when the input is returned to zero. Here T_r is the 'response time' of the filter.

If an impulse is applied to the filter then this gives the capacitor C an initial charge, the following discharge of C back through R constitutes the 'impulse response' of the filter.

For any input waveform the 'smearing' effect of the filter is much as for the forming of a 'running average' of the input, with the averaging time T_r.

In this section we consider the step response, the frequency response, the impulse response, and the averaging function of the filter, in that order. But, prior to any, we consider the simple case of the discharge of C through R, as for the impulse response.

Capacitor discharge

With $I=-CdV_o/dt$ and $I=V_o/R$ in (a)

then $$dV_o/dt=-V_o/CR \quad (11.1)$$

With $\Delta V_o \approx (dV_o/dt) \times \Delta t$

in (b) then for any value of V_o the next can be calculated, allowing the V_o waveform to be constructed.

Problem Try writing and running the two-line computer program needed to carry out this repeated operation.

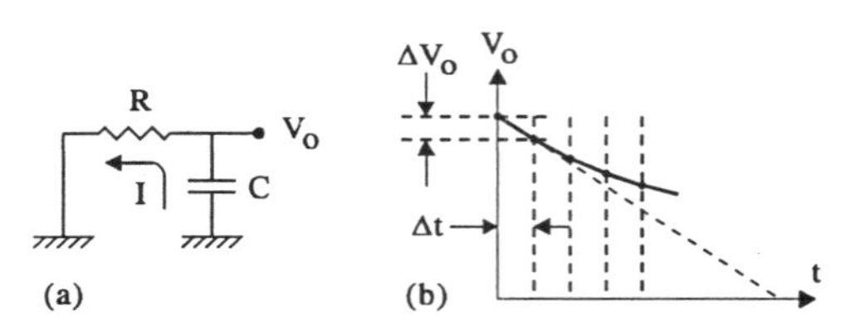

Fig. 11.2 Capacitor discharge.
(a) Circuit diagram.
(b) Development of V_o waveform for $dV_o/dt \propto -V_o$.

Analytic expression

While the above approach does not directly give the analytic expression for V_o, it does give the appropriate tests to check that a proposed expression is the correct one. These are

- dV_o/dt must be as in [11.1],
- the starting value of V_o must be correct.

With $dV_o/dt \propto V_o$ in [11.1], this is the known property of the exponential function, so we propose

$$V_o=k_a.\exp(k_b t) \quad (11.2)$$

and see if values of the constants k_a and k_b can be found for which the expression satisfies the two tests. With

$$dV_o/dt=k_b V_o$$

then this accords for

$$k_b=-1/CR.$$

With k_a equal to the starting value of V_o then

$$V_o=V_o(t=0) \times \exp(-t/CR) \quad (11.3)$$

satisfies the tests, and so correctly describes the V_o waveform.

Step response

The above method can now be extended to give the step response in Fig. 11.3(b). With

$I=CdV_o/dt$ for C, $I=(V_{in}-V_o)/R$ for R,

and $V_{in}=E$,

then $$dV_o/dt=(E-V_o)/CR \quad (11.4)$$

With this expression the waveform for V_o can be derived much as in Fig. 11.2(b), but now with dV_o/dt positive causing V_o to rise towards V_{in}, with the rate decreasing as V_o is approached. We propose

$$V_o=k_a+k_b \exp(k_c t) \quad (11.5)$$

With $V_o=0$ for $t=0$ then $k_b=-k_a$ giving

$$V_o=k_a(1-\exp(k_c t)) \quad (11.6)$$

With $$dV_o/dt=-k_a k_c \exp(k_c t)$$

then this must equal

$$(E-V_o)/CR=[E-k_a(1-\exp(k_c t))]/CR$$

for all values of t. Thus

$$k_a=E \quad \text{and} \quad k_c=-1/CR,$$

giving

$$V_o=E[1-\exp(-t/CR)] \quad (11.7)$$

with the filter response-time

$$T_r=CR \quad (11.8)$$

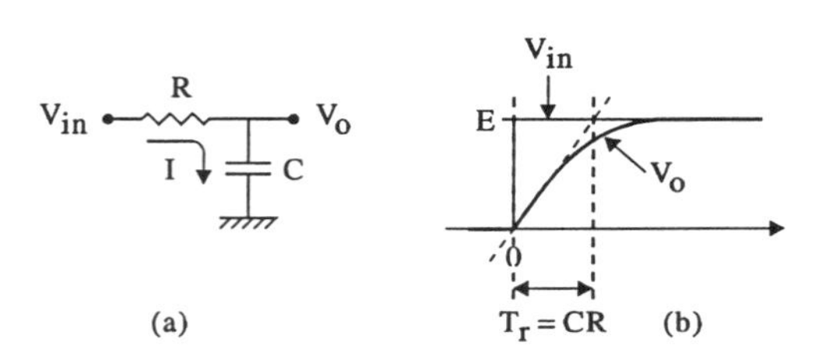

Fig. 11.3 Simple single-section RC low-pass filter.
(a) Circuit diagram.
(b) Step response.

Near-integration

For V_{in} periodic and of period $T_{in} << T_r$, then $V_o << V_{in}$ giving $I \approx V_{in}/R$. With the capacitor voltage V_o proportional to the integral of I then V_o is approximately the integral of V_{in}.

For example, the integral of a square wave is a triangular wave, as is evident from the step response when such a square wave is applied to the present filter. Exercise 11.1 demonstrates the above topics experimentally, with appropriate numeric calculations.

Frequency response

With $X_c = 1/j\omega C$ as in Section 1.7 then the filter gain

$$A(\omega) = \frac{1/j\omega C}{R + 1/j\omega C} \tag{11.9}$$

giving

$$A(\omega) = \omega_c/(\omega_c + j\omega) \tag{11.10}$$

where the filter cut-off

$$\omega_c = 1/CR \tag{11.11}$$

From [11.10]

$$|A(\omega)| = \frac{\omega_c}{\sqrt{\omega_c^2 + \omega^2}} \tag{11.12}$$

and the phase shift

$$\phi(\omega) = -\tan^{-1}(\omega/\omega_c) \tag{11.13}$$

giving the plots of Fig. 11.4. While the transition in $|A|$ is relatively abrupt, that of ϕ is much more gradual.

Near-integration

For $\omega >> \omega_o$

$$A(\omega) \approx \omega_c/j\omega \tag{11.14}$$

With V_o lagging V_{in} by nearly 90° this is as for the integral of the sine wave.

Exercise 11.2 demonstrates the above frequency response experimentally, with appropriate numeric calculations.

Decibel scale

The frequency and amplitude scales in Fig. 11.4(a) are both made logarithmic to adequately indicate the nature of the decrease in the stop-band. The corresponding linear increment is the decibel (dB), as shown on the amplitude axis in Fig. 11.14. Here the 'Bel' (Graham Alexander Bell) is given by

$$1 \text{ Bel} = \log(P_2/P_1) \quad \text{for} \quad P_2/P_1 = 10$$

where P is signal power. With $P \propto V^2$ and V_o here $\propto |A|$ then $|A|$ in Bels is $2\log(|A|)$. The decibel (dB) is one tenth of a Bel giving

$$|A|(\text{dB}) = 20\log(|A|) \tag{11.15}$$

The decibel is said to be the smallest increment in sound level that the ear can detect.

An advantage of working in dBs is that for two cascaded amplifiers of voltage gain A_1 and A_2 the resulting

$$A(\text{dB}) = A_1(\text{dB}) + A_2(\text{dB}),$$

corresponding to

$$A = A_1 \times A_2.$$

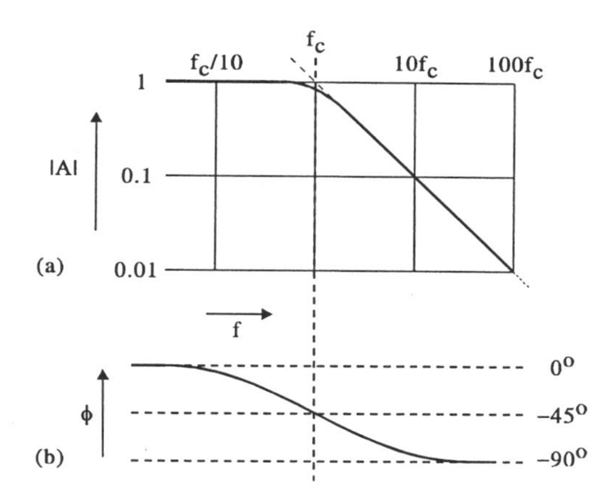

Fig. 11.4 Frequency response curves for low-pass filter of Fig. 11.3(a).
(a) Amplitude-frequency response.
(b) Phase-frequency response.

Problem Confirm that for A_1=10dB and A_2=20dB then A=31.6.

Roll-off

With $|A(\omega)|$ approximately $\propto 1/\omega$ in the above low-pass filter stop-band, and with $|A|(\text{dB}) = 20\log(|A|)$, then this is a 'roll-off' of 20dB/decade, where a 'decade' is a ×10 change in frequency.

With ×2(dB) ≈ 6dB, this is sometimes also referred to as a '6dB/octave' roll-off, with the octave a ×2 change in frequency as for the piano.

Impulse response

An 'impulse' is what a pulse of height h and width Δt becomes as $\Delta t \rightarrow 0$ while keeping the area $h\Delta t$ under the pulse constant, thus giving $h \rightarrow \infty$.

With the 'unit impulse' $\delta(t)$ an impulse of area = 1 occurring at time t, then the 'impulse response' $g_i(t)$ is the output of the filter following $\delta(t=0)$ applied to the input.

When $\delta(t=0)$ is applied to the low-pass filter of Fig. 11.2(a) then during the impulse V_{in} is infinitely more than V_o giving $I = \delta(0)/R$.

With the capacitor charge thus 1/R then the resulting $V_o = 1/CR$. With this the starting value for the capacitor discharge in [11.3] then, as in Fig. 11.5,

$$g_i(t) = (1/CR)\exp(-t/CR) \tag{11.16}$$

Differentiation of unit step

Another method of obtaining $g_i(t)$ is to note that, with the area under the unit impulse unity, the integral of the unit impulse is the unit step, so the differential of the unit step is the unit impulse. Thus $g_i(t)$ is the differential of the unit step response $g_s(t)$.

With the step response

$$V_o = E[1 - \exp(-t/CR)]$$

as in [11.7] then for the unit step

$$g_s = 1 - \exp(-t/CR)$$

giving $\quad dg_s/dt = (1/CR)\exp(-t/CR)$

thus confirming [11.16]. This kind of mental exercise is needed a good deal in what follows.

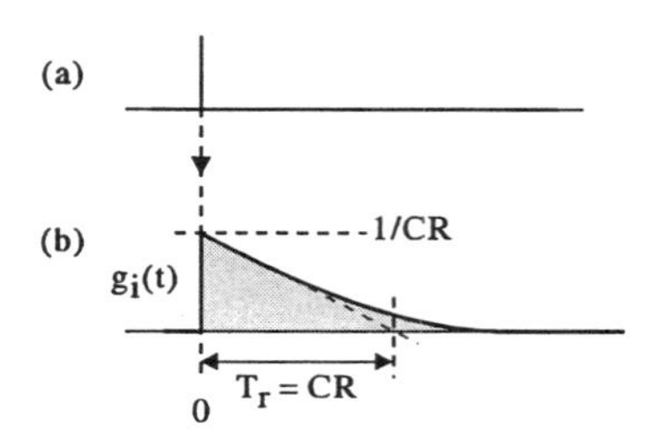

Fig. 11.5 Impulse response of low-pass filter of Fig. 11.2(a).
(a) Unit input impulse.
(b) Impulse response $g_i(t)$.

Averaging

One of the features of any low-pass filter is that the output V_o is a form of running average of V_{in}, with the averaging time the filter response time T_r.

Basic running average

For $V_o(t)$ the basic running average of V_{in} over T_r

$$V_o(t) = \frac{1}{T_r} \int_{t-T_r}^{t} V_{in}(t')dt' \qquad (11.17)$$

For V_{in} the input step in Fig. 11.6, this is as if the 'memory function' M were to run along the step, tracing the average of V_{in} within the box as V_o.

This is much as for the step response of the simple low-pass in Fig. 11.3, with the same T_r, except that there the response is exponential rather than the present ramp.

Tapered running average

Fig. 11.7 shows the corresponding process for the simple low-pass. Here the small segment of width Δt in (a) approximates to an impulse of area $\Delta t \times V_{in}(t')$, producing the transient shown in (b). The value ΔV_o in (d) is then the resulting component of V_o at time t.

For segments of V_{in} occurring earlier, the transient decay at t is greater, soon to make the contribution insignificant, while those after t make no contribution at t. Thus the contributions to $V_o(t)$ are subject to the exponential memory function M in (c), which is $g_i(t)$ reversed in time, and placed to become $g_i(t - t')$. This takes the place of the memory box in Fig. 11.6 to give the step response its exponential curve. With ΔV_o in (b) the component of V_o due to the segment of V_{in} of width Δt

$$\Delta V_o = V_{in}(t')g_i(t - t')\Delta t \qquad (11.18)$$

Summing for all ΔV_o

$$V_o(t) = \int_{-\infty}^{t} V_{in}(t')g_i(t - t')dt' \qquad (11.19)$$

Compared with the simple running average of [11.17], $V_o(t)$ in [11.19] is a weighted running average, with the weighting by $g_i(t - t')$.

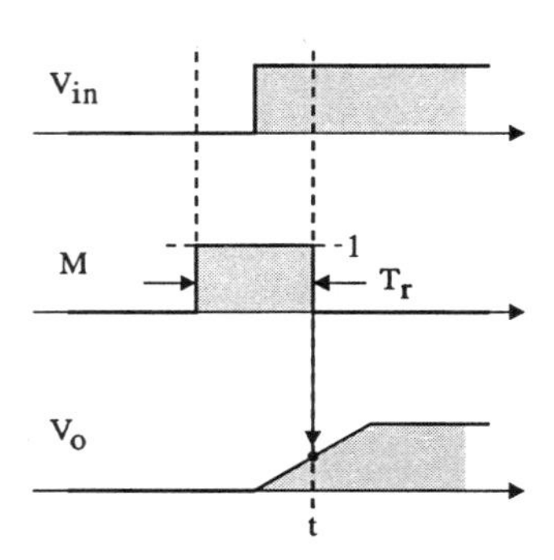

Fig. 11.6 Formation of running average of voltage step.

Normalised average

The weighted average of any variable y is said to be properly 'normalised' if the average is equal to y for all the y equal. This is clearly so for the present filter, since for V_{in} constant then the long-term average of V_o is equal to V_{in}.

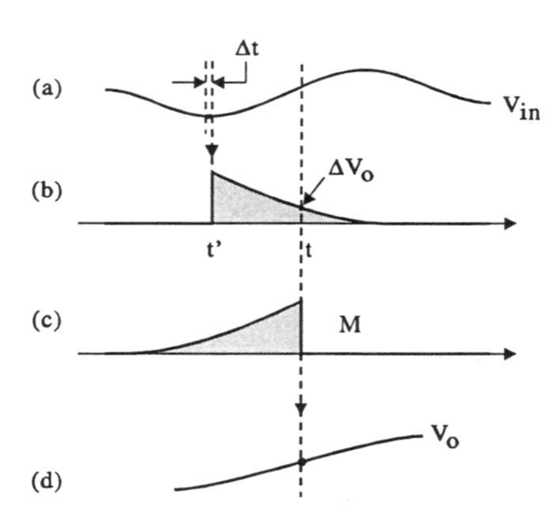

Fig. 11.7 Averaging function of low-pass filter of Fig. 11.2(a).
(a) Filter input V_{in}.
(b) Component of V_o arising from segment of V_{in} at time $t - t'$.
(c) Filter memory function $g_i(t - t')$.
(d) Filter output V_o.

Fading memory

For the low-pass the memory fades, while the true running average remembers completely for the period of length T_r prior to the current time t, and forgets completely anything earlier.

Convolution

The process of [11.19] is termed the 'convolution' of V_{in} with $g(t-t')$, signified by

$$V_o = V_{in} \otimes g_i(t-t').$$

Problem

Show that for the present low-pass filter the area under $g_i(t)$ is unity, as would be expected from the charge and discharge of C.

With $g_i(t)$ as in [11.16] for the present low-pass, thus confirm that the weighting by $g_i(t-t')$ in [11.19] is such that the average is equal to V_{in} for V_{in} constant.

11.2 Simple RC high-pass

Fig. 11.8(a) shows the simple single-section RC high-pass filter corresponding to the low-pass of Fig. 11.3(a). This is simply the low-pass with the R and C transposed.

Step response

Thus the high-pass step response of Fig. 11.8(b) is the input step with the low-pass step response in Fig. 11.3(b) subtracted. With $V_o=E[1-\exp(-t/CR)]$ as in [11.7] for the low-pass then for the high-pass

$$V_o = E.\exp(-t/CR) \qquad (11.20)$$

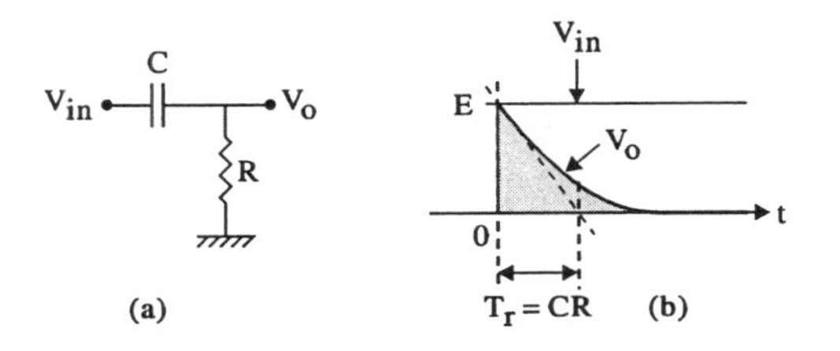

Fig. 11.8 Simple single-section high-pass RC filter.
(a) Circuit diagram.
(b) Step response.

Near-differentiation

For V_{in} a square wave of period T_{in}, and for $CR << T_{in}$, then V_o approaches the series of alternating impulses that would be the output of a true differentiator.

Frequency response

From the circuit of (a)

$$A(\omega) = \frac{R}{R + 1/j\omega C} \qquad (11.21)$$

giving

$$A(\omega) = j\omega/(j\omega + \omega_c) \qquad (11.22)$$

where the filter cut-off $\omega_c = 1/CR$ (11.23)
as for the low-pass in [11.11]. Thus

$$|A(\omega)| = \frac{\omega}{\sqrt{\omega^2 + \omega_c^2}} \qquad (11.24)$$

and

$$\phi = 90^o - \tan^{-1}(\omega/\omega_c) \qquad (11.25)$$

which plot as in Fig. 11.9. The phase shift ϕ is now a lead, rather than a lag as for the low-pass.

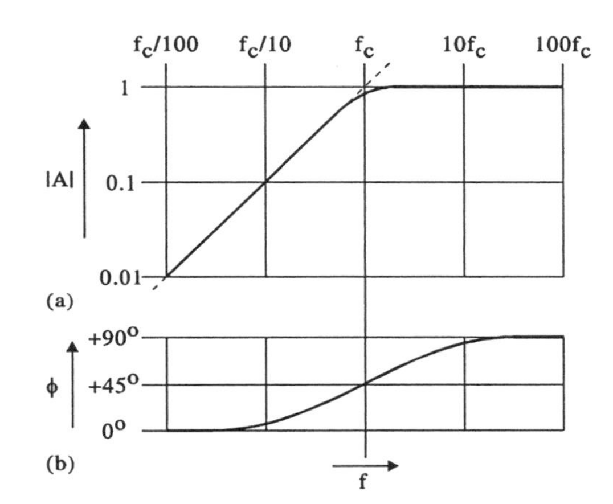

Fig. 11.9 Frequency response of high-pass filter of Fig. 11.8(a).
(a) Gain-frequency response.
(b) Phase-frequency response.

Impulse response

Since the unit impulse is the differential of the unit step then the impulse response of a filter is the differential of its step response. With the unit step response of the present high-pass filter as in Fig. 11.10(a) then the impulse response is as in (b).

In detail, with the response of the high-pass filter to the step of height E

$$V_o = E.\exp(-t/CR)$$

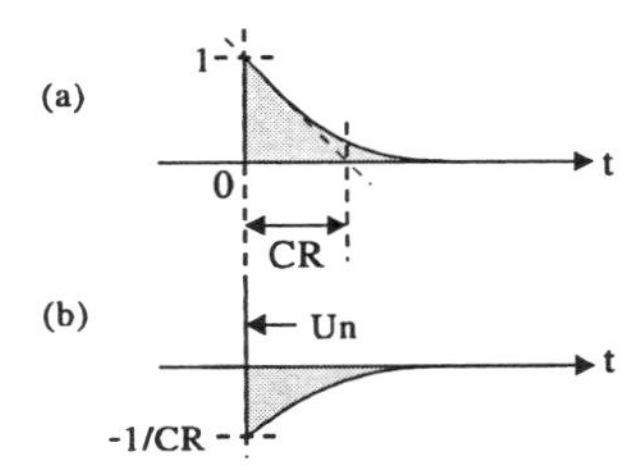

Fig. 11.10 Derivation of impulse response of high-pass filter of Fig. 11.8.
(a) Filter unit step response $g_s(t)$.
(b) Filter unit impulse response $g_i(t)=d[g_s(t)]/dt$.

as in [11.20] then for the unit step

$$g_s(t)=\exp(-t/CR) \qquad (11.26)$$

as in Fig. 11.10(a). With the unit impulse response

$$g_i(t)=dg_s(t)/dt$$

then

$$g_i(t)=(-1/CR)\exp(-t/CR) \qquad (11.27)$$

as in (b).

Exercise 11.3 gives experimental work, problems and calculations relating to the high-pass filter.

11.3 Pole-zero diagrams

Fig. 11.11 shows the 'pole-zero' diagrams for the present low- and high-pass filters. From these the frequency, impulse and step responses can be derived, as will now be shown for the above low- and high-pass filters.

Exponential waves

The pole-zero diagram is based on a wider interpretation of the exponential function exp(st). Previously s has only been real, for example

$$g_i(t)=(1/CR)\exp(-t/CR)$$

as in [11.16] for the impulse response of the simple low-pass. In principle s can be

- negative as here, giving the exponential decay,
- positive giving the corresponding exponential rise,
- zero giving $\exp(0\times t)=1$ which is constant.

s can also be complex, with $s=\sigma+j\omega$, and with both σ and ω real.

Level corkscrew

For the rotating phasor V of Section 1.7, after an interval δt the increment δV is 90° ahead of V and of magnitude $V\omega\delta t$ (draw to confirm). Thus

$$dV/dt=j\omega V.$$

With the exponential the function proportional to its own differential then V may be written as

$$V=|V|\exp(j\omega t) \qquad (11.28)$$

If the rotating V is viewed with time as the third axis the result is an anti-clockwise corkscrew. Since the amplitude does not change as time progresses we call this a 'level corkscrew'.

Damped and rising corkscrews

With $\qquad \exp(\sigma+j\omega t)=\exp(\sigma t)\times\exp(j\omega t)$

then σ negative gives an exponentially decaying corkscrew, while for σ positive the corkscrew expands with time.

Problem

Sketch these three-dimensional functions noting also how as t increases in the negative direction then the decaying corkscrew expands, while the expanding corkscrew decays.

None of the above waveforms, complex or real, is changed in shape by differentiation or integration, so if any were to be applied to a filter as the input then the output would be of the same shape. There would only be a difference in the phase on the complex plane, and in the overall amplitude.

For example, for $\quad V(s)=|V|\exp(st)$

applied to the capacitor C, and with $I=CdV/dt$, then

$$I(s)=sCV(s)$$

Thus the susceptance of C becomes sC and the reactance 1/sC. This is consistent with the previous $j\omega C$ for the level corkscrew, since there $\sigma=0$.

Simple low-pass filter

For any such input applied to the low-pass of Fig. 11.3(a) then the complex gain

$$A(s)=\frac{1/(sC)}{R+1/(sC)} \qquad (11.29)$$

giving

$$A(s)=\frac{-s_p}{s-s_p} \qquad (11.30)$$

where

$$s_p=-1/CR \qquad (11.31)$$

Pole-zero diagram

For a function such as A(s) in [11.30], a 'pole' is a value of s for which $A(s)=\infty$, while a 'zero' is a value of s for which $A(s)=0$.

The 'pole-zero diagram' is then the poles and zeros plotted on the complex s-plane, with the poles marked by a cross and the zeros by a circle.

The present A(s) has no zeros and only one pole s_p giving the diagram of Fig. 11.11(a). As shown shortly the corresponding high-pass also has one zero, and this at the origin as in (b).

Low-pass frequency response

This is derived from the pole-zero diagram as follows. With A(s) as in [11.30] and for $\sigma=0$ then

$$A(\omega)=\frac{-s_p}{j\omega-s_p} \qquad (11.32)$$

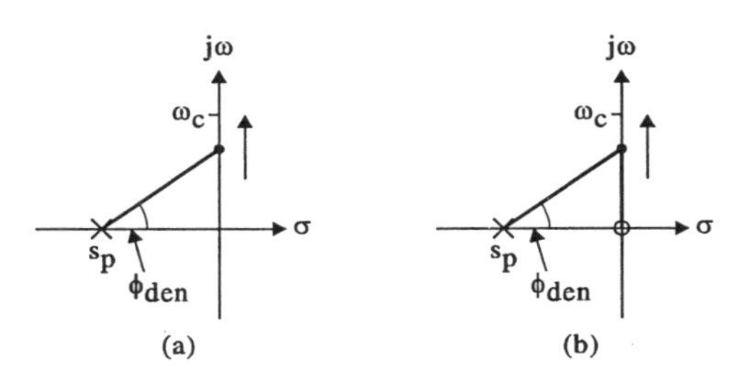

Fig. 11.11 Filter pole-zero diagrams.
(a) For low-pass filter of Fig. 11.3(a).
(b) For high-pass filter of Fig. 11.8(a).

With 'num' and 'den' the numerator and denominator of a fraction, here of [11.32], then

$$|A(\omega)| = |\text{num}|/|\text{den}|$$

With |num| here constant,

then $$|A(\omega)| \propto 1/|\text{den}|$$

where |den| is given by the length of the x-to-dot line shown. With

$$s_p = -1/CR \qquad \text{as in [11.31]}$$

and $$\omega_c = 1/CR \qquad \text{as in [11.11]}$$

then $$\omega_c = -s_p \qquad (11.33)$$

which is marked on the jω axis in the diagram.

Thus as jω varies from 0 to j∞ along the jω axis then |A(ω)| is given by the reciprocal of the length of the line joining the fixed pole × and the moving jω •.

Also $\phi(\omega)$ is given by $-\phi_{den}$ shown, confirming |A(ω)| as in the plots of [11.12] and $\phi(\omega)$ as in [11.13].

Low-pass impulse response

With $A(s) = \infty$ at the pole s_p then there can be an output of the form $\exp(s_p t)$ for zero input. But the output that occurs for zero input after the initial stimulus is the impulse response. Thus each pole represents a component of the impulse response.

For the present case, with $s_p = -1/CR$ then the function given by the pole is $\exp(-t/CR)$, which is consistent with

$$g_i(t) = (1/CR)\exp(-t/CR) \qquad (11.34)$$

in [11.16] for the impulse response.

Basic filter responses

The more advanced filters of the next two chapters simply have larger numbers of poles and zeros, with the frequency response derived much as above, and with a component of the impulse response for each pole. What is then needed is a starting library of impulse responses for the various single pole placements. Table 11.1 gives four basic entries, derived as follows.

Low-pass unit impulse response (Row 4)

With $$A(s) = -s_p/(s - s_p) \qquad \text{as in [11.30]}$$

and $$g_i(t) = -s_p \exp(s_p t) \qquad \text{as in [11.34]}$$

then for a filter for which

$$A(s) = 1/(s - s_p) \qquad (11.35)$$

the impulse response

$$g_i(t) = \exp(s_p t) \qquad (11.36)$$

giving Row 4 of the table. Also Row 1 is obvious.

Unit step (Row 2)

For the unit integrator (one of unit time constant) with

$$V_{in} = \exp(st) \quad \text{and} \quad V_o(t=0) = 0$$

then $$V_o = V_{in}/s$$

giving $$A(s) = 1/s.$$

Also the impulse response of the unit integrator is the unit step. This is also Row 4 with $s_p = 0$.

Unit ramp (Row 3)

The unit ramp is the result of applying the unit impulse to the input of the unit integrator to produce the unit step, and then applying this to the input of a second unit integrator. Thus

$$A(s) = 1/s^2$$

		A(s)	$g_i(t)$
1	Direct transfer	1	Unit impulse
2	Single integrator	$1/s$	Unit step
3	Two cascaded integrators	$1/s^2$	Unit ramp
4	Single-section low-pass	$1/(s - s_p)$	$\exp(s_p t)$

Table 11.1 Impulse response $g_i(t)$ for filters of various A(s).

Simple high-pass filter

From the circuit diagram of Fig. 11.8(a)

$$\frac{V_o}{V_{in}} = \frac{R}{R + 1/sC} \qquad (11.37)$$

giving

$$A(s) = \frac{s}{s - s_p} \qquad (11.38)$$

where $s_p=-1/CR$ as for the low-pass. Hence the zero at $s=0$ and the previous pole at $s=s_p$, as in Fig. 11.11(b).

High-pass frequency response

As before, to determine the frequency response s is replaced by $j\omega$, now in [11.38] to give

$$A(\omega)=\frac{j\omega}{j\omega - s_p} \qquad (11.39)$$

which with $\omega_c=-s_p$ as in [11.33] confirms [11.22] for $A(\omega)$.

With again $\quad |A\omega)| = |num|/|den|$

$|den|$ is taken from the diagram as before, but now we also have $|num|$ as the length of the vertical circle-to-dot line.

For $\omega << \omega_c$ then $|den|$ is close to being constant while $|num| \propto \omega$, giving $|A(\omega)|$ nearly $\propto\omega$, as for the frequency response of Fig. 11.9(a).

Conversely for $\omega >> \omega_c$ then $|num|/|den| \approx 1$ which is also as for the plotted response.

The phase-frequency response of Fig. 11.9(b) is confirmed in the same way, with

$$\phi=\phi_{num}-\phi_{den} \quad \text{and} \quad \phi=90^o-45^o$$

for $\omega=\omega_c$ as there.

High-pass impulse response

This is the first example of how Table 11.1 can be used to determine the impulse response of another circuit. First [11.38] for the high-pass is manipulated into the form

$$A(s)=1+\frac{s_p}{s-s_p} \qquad (11.40)$$

Thus $g_i(t)$ has two components, with the unity term in [11.40] giving the unit impulse as in Row 1, and the other term giving $s_p \exp(s_p t)$ as in Row 4.

With $\quad s_p= -1/CR$

as in [11.31] then the two components combine to confirm the result of Fig. 11.10(b).

Low-pass step response

With the unit step the impulse response of the unit integrator then the step response of a filter is given by the impulse response of that filter preceded by the unit integrator. This corresponds to multiplying the normal A(s) of the filter by 1/s. With

$$A(s)=-s_p/(s-s_p)$$

as in [11.30] for the low-pass impulse response then with the added integrator

$$A(s)=\frac{-s_p}{s(s-s_p)} \qquad (11.41)$$

which can be resolved into

$$\frac{-s_p}{s(s-s_p)}=\frac{A}{s}+\frac{B}{s-s_p} \qquad (11.42)$$

To obtain B we multiply both sides of the expression by $(s-s_p)$ and set $s=s_p$, giving $B=-1$.

Similarly for A we multiply by (s–0) and set $s=0$, giving $A=1$. Thus

$$A(s)=\frac{1}{s}-\frac{1}{(s-s_p)} \qquad (11.43)$$

This process is known as 'removing the pole'.

From the table, the first term 1/s corresponds to the unit step and the second to $-\exp(s_p t)$.

With $\quad s_p=-1/CR$

as in [11.31], this confirms the result of [11.7] and Fig. 11.3(a).

High-pass step response

Here the factor 1/s is added to the RHS of [11.38] to give

$$A(s)=\frac{1}{s-s_p} \qquad (11.44)$$

From the table then

$$g_i(t)=\exp(-t/CR)$$

confirming [11.20] and Fig. 11.8(b) for the input step height E.

Cascaded low-pass sections

Where two simple low-pass sections are cascaded, for simplicity with a voltage-follower between them, then the above method of obtaining the impulse response can be applied directly, provided the two sections are not identical. Here

$$A(s)=\frac{s_a s_b}{(s-s_a)(s-s_b)}=\frac{A}{(s-s_a)}+\frac{B}{(s-s_b)} \qquad (11.45)$$

from which

$$A=\frac{s_a s_b}{(s_a-s_b)} \quad \text{and} \quad B=\frac{s_a s_b}{(s_b-s_a)}$$

giving

$$g_i(t)=\frac{s_a s_b}{(s_a-s_b)}[\exp(s_a t)-\exp(s_b t)] \qquad (11.46)$$

and thus the impulse response as in Fig. 11.12(a) for $|s_a|>>|s_b|$.

However, for $s_a = s_b$ the decomposition of [11.45] fails. Here $g_i(t)$ is best obtained from [11.46] in the limit where the two poles are equal. For

$$\Delta s = s_b - s_a$$

then

$$\exp(s_a t) - \exp(s_b t) = \exp(s_a t)[1 - \exp(\Delta s \times t)]$$
$$\approx \exp(s_a t)(-\Delta s \times t) = (s_a - s_b).t.\exp(s_a t)$$

With this in [11.46]

$$g_i(t) = s_a^2.t.\exp(s_a t) \qquad (11.47)$$

as in Fig. 11.12(b).

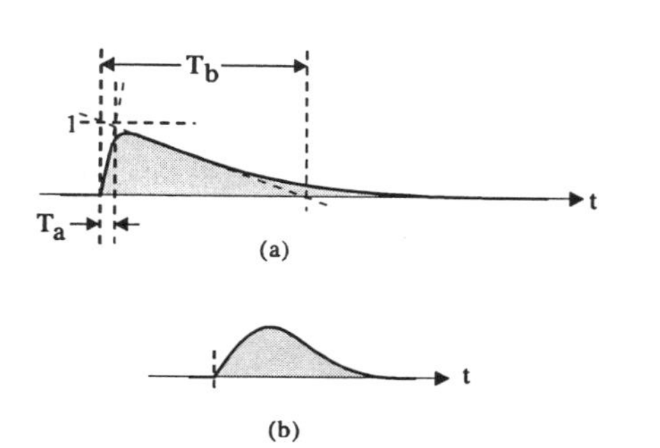

Fig. 11.12 Impulse response for two cascaded low-pass filters
(a) Response times differing.
(b) Response times the same.

11.4 Laplace transform

The Laplace transform (Holbrook, 1966) is essentially an extension of the Fourier analysis of Section 1.8. There say a square wave of frequency f_1 is resolved into a series of sine wave components of frequency nf_1 where n is integer. It is also possible the resolve a single transient, such as the impulse or step, into Fourier components. For the square wave this is a matter of letting the fundamental frequency f_1 decrease to zero with the mark period remaining constant.

A periodic train of impulses can be reduced to one impulse in the same way. Here the spacing of the components is reduced to zero, giving a continuum of appropriate spectral density.

However, not all transient signals can be so derived. The ramp $V \propto t$ is one such. But the ramp multiplied by $\exp(\sigma t)$ with σ negative is made to decay to zero for $t = \infty$, and so can be resolved into Fourier components, as level corkscrew waves.

If the ramp damped by $\exp(\sigma t)$ can thus be represented by a continuum of level corkscrew waves then the undamped ramp can be represented by the corresponding series of rising corkscrew waves, the level corkscrews multiplied by $\exp(-\sigma t)$ with σ still negative.

The operation which gives F(s) for a given function F(t) is the Laplace transform, given by

$$F(s) = \int_0^\infty F(t)\exp(-st).dt \qquad (11.48)$$

This is somewhat similar to the Fourier transform of Section 1.8 and, as there, we shall not present the detailed argument leading to the transform. Instead a series of exercises in the method confirm the results of Table 11.1, previously obtained using the pole-zero method.

The Laplace transform has been used to establish a large library of transient signals and their associated transforms (Holbrook, 1966; Spiegel, 1968). Thus when the complex transfer function of a given circuit is split into its various component responses it is usually possible to find these in the library and thus determine the associated impulse response. Exercise 11.4 confirms the transforms in Table 11.1.

References

Holbrook, J. G. 1966: *Laplace Transforms for Electronic Engineers*. Pergamon.

Spiegel, M. R. 1968: *Mathematical Handbook of Formulas and Tables,* McGraw-Hill, Schaum's Outline Series.

Van Valkenburg, M. E. 1982: *Analog Filter Design*. Holt, Rinehart and Winston.

11.5 Exercises and experiments

Exercise 11.1 Low-pass filter step response

Calculation. Response time For the low-pass filter of Fig. 11.3, calculate T_r for $R = 10k\Omega$, $C = 10nF$.

With $T_r = CR$ as in [11.8] then **$T_r = 100\mu s$.**

Experiment. Step response Connect the filter and apply a 10V p-p square wave of $f_{in} = 400Hz$ and observe the V_o waveform.

Measure T_r by projecting the initial slope as in Fig. 11.3(b), and also as the time at which the decay is by

$$\times \exp(-1) = 36.8\%$$

Thus confirm the calculated value.

Increase f_{in} until $T_{in} = 1/f_{in} << T_r$ and note how the then much attenuated V_o approximates to a triangular wave.

Calculation. Triangular output wave Calculate the peak-to-peak value $V_{o.p\text{-}p}$ of V_o for V_{in} the present 10V p-p square wave and $T_{in} = T_r/10$.

With $\qquad dV_o/dt \approx V_{in}/CR$

as here then $\Delta V_o \approx V_{in}(T_{in}/2)/CR$

where ΔV_o is the change in V_o over one half cycle of V_{in}.

With $T_{in}=T_r/10$ and $V_{o.p\text{-}p}=\Delta V_o$ then **$V_{o.p\text{-}p} \approx 250mV$.**

> **Experiment. Output waveforms** Set T_{in} as above and confirm $V_{o.p\text{-}p}=250mV$.
>
> Reduce f_{in} to see the transition from the low-amplitude near-triangular wave to the full-amplitude square wave of rise time T_r.
>
> Set $T_{in}=T_r$ and explore the effect of varying R and C.

Oscilloscope-probe

The scope-probe is provided to prevent the loading that would otherwise occur by the scope input capacitance C_{in}, and C_L that of the connecting line. The standard scope input circuit is

$$(C_{in}=30pF)//(R_{in}=1M\Omega)$$

With C_L typically 100pF then this is $\approx 1M\Omega//130pF$.

With the present filter $R=10k\Omega$ and $C=10nF$ this is not a major source of error, but for say $C=100pF$ for the low-pass then the observed T_r will be more than twice the unloaded value.

Fig. 11.13 shows the details of the ×10 probe. The basic principle is to add $R_p=9M\Omega$ and C_p which is equal to the shunt capacitance ×1/9 This transmits components of all frequencies equally, with the constant reduction of × 1/10. At the same time it reduces both resistive and capacative loading by 1/10.

The variable component of C_{in} is included so that C_{in} can be set to the standard 30pF. This is to allow any correctly adjusted ×10 probe to be used.

With C_p largely the natural capacitance of R_p, C_v is included to bring the total shunt capacitance to nine times C_p.

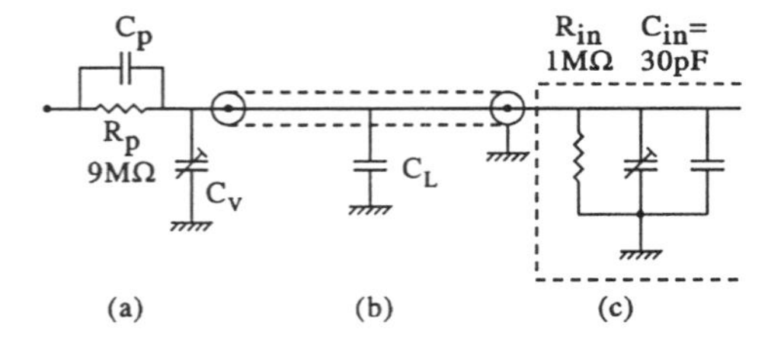

Fig. 11.13 Scope probe.
(a) ×10 probe.
(b) Connecting line.
(c) Scope input.

The probe has a switch (not shown) which eliminates th components in (a) to give the ×1 setting.

The ×10 is thus not a gain but actually a loss (of ×1/10 It is that by which a scope reading has to be multiplied b to give the true value of the measured voltage.

For example, 100mV displayed corresponds to 1V be ing probed.

> **Experiment. Oscilloscope-probe capacitance** For the present low-pass filter, remove C and measure T_r for the two settings. Thus confirm the above values of C_p.

Exercise 11.2 Low-pass filter frequency response

> **Calculation. Cut-off frequency** Calculate the cut-off frequency f_c for the low-pass filter of Fig. 11.3.

With $\omega_c=1/CR$ as in [11.11] then **$f_c=1.59kH$**

> **Calculation. Output amplitude** Calculate $|A(\omega)|$ for the input frequency $f_{in}=10f_c$.

With $$|A(\omega)| = \frac{\omega_c}{\sqrt{\omega_c^2+\omega^2}}$$

as in [11.12] then **$|A(\omega)| \approx 0.$**

> **Calculation. Phase lag** Calculate ϕ for $f_{in}=f_c/10$.

With $\phi(\omega)=-\tan^{-1}(\omega/\omega_c)$ as in [11.1

then **$\phi \approx 0.1$ rad$=-5.71$**

Log scales

The axes for the amplitude-frequency response of Fi 11.4(a) are both logarithmic, where the plotted distances and y are proportional to log(f) and log(|A|) respectivel This kind of plotting is needed when the ratio of max mum to minimum for the variable is large.

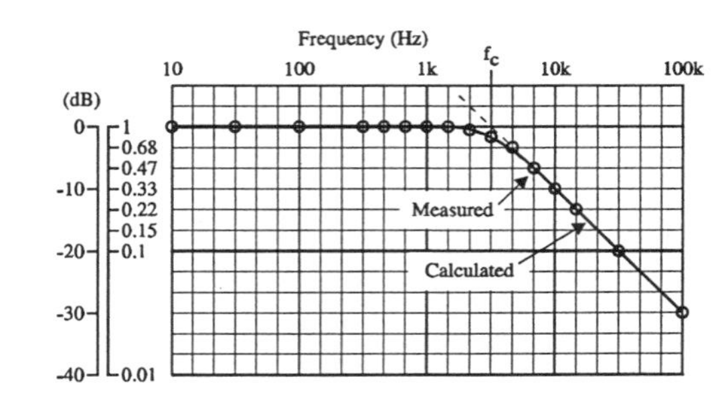

Fig. 11.14 Example of experimental frequency response plottir using preferred values of Fig. 11.15.

Preferred values

For such log plotting it is quite unnecessary to go to the expense of using a sheet of log-log graph paper for every plot. The practising electronic engineer has the preferred component values firmly in memory. These are approximately equally divided on the log scale from one to ten, to appear as to the left of Fig. 11.14. Thus ordinary squared paper can be used.

Fig. 11.15 shows the preferred values for varying precision. Note that Fig. 11.14 is an example, with f_c there not necessarily the same as for the present circuit.

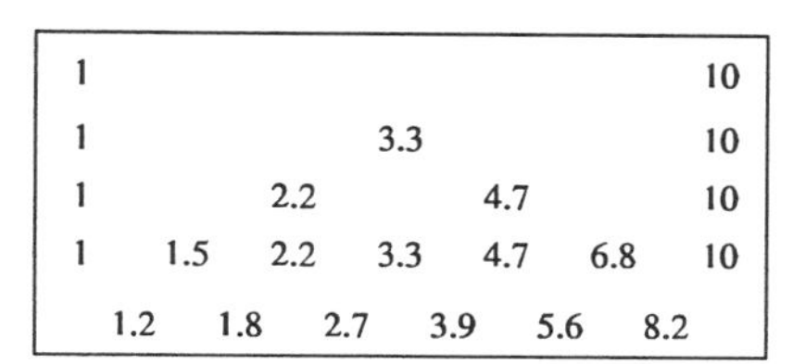

1												10
1						3.3						10
1				2.2				4.7				10
1		1.5		2.2		3.3		4.7		6.8		10
	1.2		1.8		2.7		3.9		5.6		8.2	

Fig. 11.15 Preferred component values for varying precision.

Experiment. Amplitude-frequency response Switch to a sinusoidal input and plot the frequency response of $|A(\omega)|$ in [11.12] for a range of f_{in} large enough to identify clearly the band-pass and band-stop regions. Thus confirm the above calculated value of $|A(\omega)|$ for $f_{in}=10f_o$. Use log axes as in Fig. 11.14, setting f to the preferred value at each step.

Experiment. Phase-frequency response Plot the phase frequency response of Fig. 11.4(b), using the linear ϕ axis. Thus confirm the above calculated value of ϕ for $f_{in}=f_c/10$. Note how, for this far into the filter pass-band, ϕ is still significant, while the gain has become very close to unity.

Experiment. Near-integration For $f_{in}=10f_c$ re-view the waveforms confirming the near 90° phase-lag. Then switch to the scope x-y mode to view the resulting near-right ellipse.

Vary f_{in}, noting in particular how the ellipse becomes skewed as f_{in} is decreased, finally to become a straight line of unit slope.

Problem Derive the expressions needed for calculation of ϕ from the dimensions of the ellipse.

Exercise 11.3 High-pass filter

Experiment. Step response Transpose C and R as above and repeat the series of step response measurements carried out for the low-pass. Thus check the waveform of Fig. 11.10(b) and the calculated value of T_r.

Experiment. Near-differentiation Reduce C to note the series of alternating narrow pulses approximating the alternating impulse train for the step response of an ideal differentiator.

Experiment. Scope AC-input To resolve a low amplitude alternating signal from a much larger DC component the scope has an 'AC' input setting. Here a capacitor C is connected in series with the input, to form a high-pass filter with the standard scope $R_{in}=1M\Omega$. With the resulting T_r usually 100ms, determine the value experimentally from the step response.

Problem. Waveform sag Remembering the semi-integrating function of the low-pass filter for an input square wave of period $T_{in} \ll CR$, derive an expression for the approximate value of the % sag for $T_{in}=CR/30$.

Discuss the effect from the Fourier viewpoint.

With
$$\Delta V_o \approx V_{in}(T_{in}/2)/CR$$
as the amplitude of the near-triangular wave output of the low-pass filter driven by a square wave of frequency well above the filter cut-off, then it is this which constitutes the sag for the same filter connected as the high-pass.

Fourier view

With
$$|A(\omega)| = \frac{\omega}{\sqrt{\omega^2 + \omega_c^2}}$$
as in [11.24], then for ω only a few times ω_c the attenuation is negligible. But with
$$\phi = 90° - \tan^{-1}(\omega/\omega_c)$$
as in [11.25] for the phase advance then this persists much further into the high-pass pass-band. Thus for the square wave the fundamental is advanced relative to the harmonics.

For any sine wave a small phase advance is equivalent to adding a proportional small component 90° ahead of the original, to the original. Sketch the result to show that this is beginning to have the features of the near-linear sag.

Experiment. Waveform sag Check the above calculated value experimentally.

Experiment. Scope-probe adjustment With the probe as in Fig. 11.13 then C_p and R_{in} alone gives a high-pass filter, while R_p and the shunt capacitances alone give a low-pass. Thus incorrect adjustment of C_v gives a transition from one kind of response to the other. Calculate the expected response time, set f_{in} accordingly, and vary C_v to confirm.

Exercise 11.4 Laplace transforms

Unit impulse Derive the Laplace transform for the unit impulse $\delta(t=0)$.

With the transform as above

$$F(s) = \int_0^\infty \delta(t=0)\exp(-st).dt = \int_0^\infty \delta(t=0).dt = 1$$

confirming Row 1 of Table 11.1.

Unit step Derive the Laplace transform for the unit step.

With

$$F(s) = \int_0^\infty F(t)\exp(-st).dt$$

for the transform, as in [11.48], and with $F(t)=1$ for $t>0$ for the unit step, then

$$F(s) = \int_0^\infty 1.\exp(-st).dt = 1/s$$

confirming Row 2 of the Table 11.1.

Unit ramp Derive the Laplace transform for the unit ramp $R=t$.

Here

$$F(s) = \int_0^\infty t.\exp(-st).dt$$

By parts

$$\int_0^\infty t.\exp(-st).dt = \int_0^\infty \frac{\exp(-st)}{s} - \left[\frac{t.\exp(-st)}{s}\right]_0^\infty$$

giving

$$F(s) = \frac{1}{s^2}$$

confirming Row 3 of the Table 11.1.

Exponential Derive the Laplace transform for $\exp(s_p t)$.

Here

$$F(s) = \int_0^\infty \exp(s_p t)\exp(-st)dt$$

$$= \int_0^\infty \exp((s_p - s)t)dt = \left[\frac{\exp((s_p - s)t)}{s_p - s}\right]_0^\infty$$

$$= \frac{1}{s - s_p}$$

confirming Row 4 of the Table. 11.1.

12

Simple LC resonator-based filters

Summary

In this chapter it is first shown how the LC series or parallel resonator can be used as a simple band-pass or band-stop filter, normally of the narrow-band type where the width Δf of the pass- or stop-band is small relative to the central frequency f_o.

The LC circuit can also be used to form a second-order low- or high-pass filter. Here the order refers to the variation with f in the stop-band. For example, for the first-order low-pass the stop-band gain $\propto 1/f$, while for the second-order the variation is with $1/f^2$. Similarly for the high-pass the variation is $\propto f$ and $\propto f^2$.

The LC network can also be used for impedance transformation. This is as for the transformer but now for use at a single frequency only, usually with a resonant peak at that frequency.

For this chapter all numeric calculations are reserved for the last section, which also includes a larger than usual amount of supporting experimental work.

12.1 Series resonator narrow-band band-pass filter

Frequency response

For the circuit of Fig. 12.1 the transfer function

$$A(\omega) = \frac{R}{R + j\omega L + 1/j\omega C} \qquad (12.1)$$

where $A(\omega) = V_o/V_{in}$. With the reactive term

$$X = j\omega L + 1/j\omega C = j(\omega L - 1/\omega C)$$

then $X = 0$ at the resonant frequency

$$\omega_o = 1/\sqrt{LC} \qquad (12.2)$$

giving a resonant peak at ω_o, at which $A(\omega) = 1$.

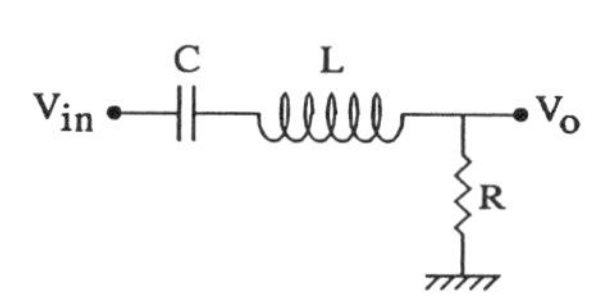

Fig. 12.1 Narrow-band band-pass filter based on series resonator.

With [12.1] written as

$$A(\omega) = 1\Big/\left[1 + j\left(\frac{\omega L}{R} - \frac{1}{\omega CR}\right)\right] \qquad (12.3)$$

then Fig. 12.2(a,b) shows the frequency response plotted on log axes.

First (a) shows den [12.3] (den: denominator) with the component terms also plotted. The reactive component

$$|X| = \omega L/R - 1/\omega CR$$

dips to zero at resonance ($\log(0) = -\infty$), to be checked by the unity term as shown.

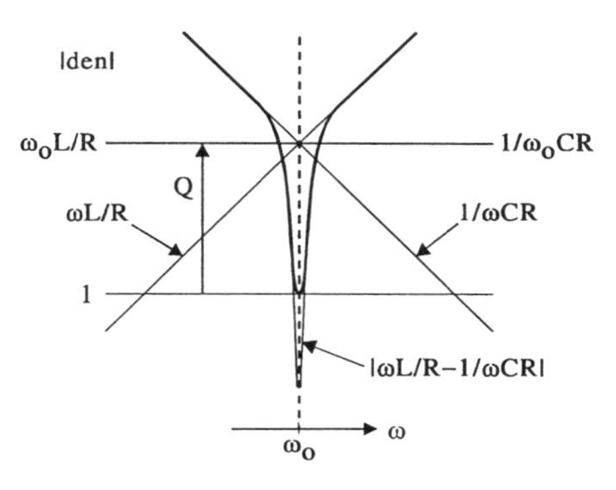

Fig. 12.2(a) Frequency response of band-pass filter of Fig. 12.1. Denominator of [12.3]. Log axes.

With $\text{num}[12.3] = 1$

then, as in [1.69], the plot for

$$|A(\omega)| = |\text{num}|/|\text{den}|$$

becomes as in (b).

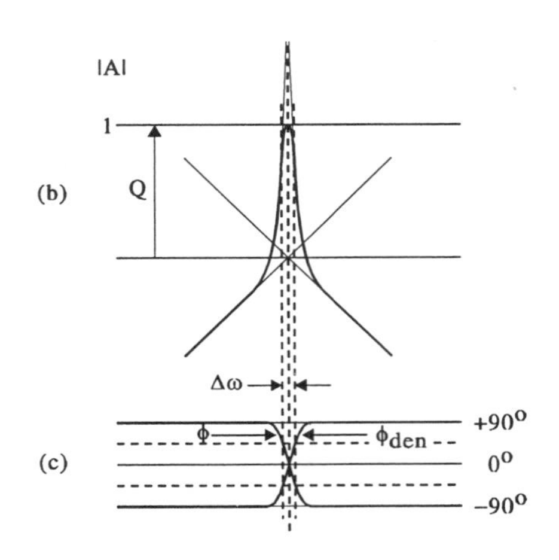

Fig. 12.2(b,c) Frequency response of band-pass filter of Fig. 12.1.
(b) Amplitude-frequency response.
(c) Phase-frequency responses for (a) and (b). Lin-log axes.

Quality factor

From Fig. 12.2(a) the quality factor of the resonance

$$Q = \omega_o L/R = 1/\omega_o CR \qquad (12.4)$$

giving

$$A(\omega) = 1\bigg/\left[1 + jQ\left(\frac{\omega}{\omega_o} - \frac{\omega_o}{\omega}\right)\right] \qquad (12.5)$$

Bandwidth

Defining the bandwidth Δf as in Fig. 12.3(a) then for $Q >> 1$

$$\Delta\omega \approx \omega_o/Q \qquad (12.6)$$

Thus a narrow-band response requires a high Q.

Conversely for $Q << 1$ the lower and upper limits

$$\omega_L \approx \omega_o/Q \ldots(a) \qquad \omega_u \approx Q\omega_o \ldots(b) \qquad (12.7)$$

The form of [12.5], where the circuit component values of [12.3] are replaced by the more general Q and ω_o, is applicable to all sorts of resonators, mechanical as well as electrical. We shall use it frequently.

Phase shift

With the phase-shift $\phi = \phi_{num} - \phi_{den}$ as in [1.69]

and with ϕ_{num} of $[12.5] = 0$ then $\phi = -\phi_{den}$.

With
$$\phi_{den} = \tan^{-1}\left[Q\left(\frac{\omega}{\omega_o} - \frac{\omega_o}{\omega}\right)\right] \qquad (12.8)$$

then for $Q >> 1$
$$\phi_{den} \approx \tan^{-1}[(\omega - \omega_o)/(\Delta\omega/2)]$$
giving
$$\phi \approx \tan^{-1}[(\omega_o - \omega)/(\Delta\omega/2)] \qquad (12.9)$$

Thus ϕ_{den} and ϕ plot as in Fig. 12.2(c), with $\Delta\omega$ spanning the points where $\phi = +45^o$ and -45°.

Complex plane plot

Fig. 12.3(b) shows the complex plane plot of $A(\omega)$ in [12.5]. This confirms the above $\phi = +45^o$ and -45° at the limits of $\Delta\omega$, also that $|A(\omega)| = 1$ at resonance.

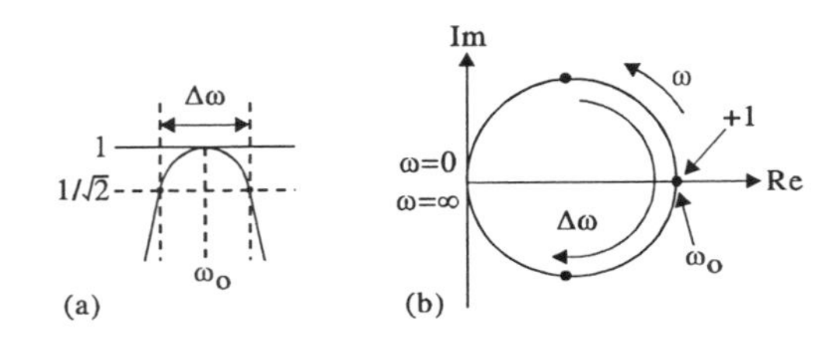

Fig. 12.3 Tip of resonance curve in Fig. 12.2(b).
(a) Bandwidth $\Delta\omega$ of band-pass filter of Fig. 12.1.
(b) Complex plane plot of A in [12.2] showing limits of $\Delta\omega$.

Problem Use normalisation to resolve $A(\omega)$ in [12.5] into the real and imaginary parts $A(\omega)_r$ and $jA(\omega)_i$. Thus confirm that $A(\omega)$ traces the circle shown.

Pole-zero method

This can also be applied to the resonator, first to confirm the above frequency response, and then to give the step and impulse responses. Replacing $j\omega$ by s in [12.5] gives the s-plane transfer function as

$$A(s) = \frac{s(\omega_o/Q)}{s^2 + s\omega_o/Q + \omega_o^2} \qquad (12.10)$$

Factoring the denominator

$$A(s) = \frac{\omega_o}{Q}\frac{s}{(s - s_1)(s - s_2)} \qquad (12.11)$$

giving the poles

$$s_1 = -\omega_o/2Q + \omega_o\sqrt{(1/2Q)^2 - 1} \ldots(a)$$

$$s_2 = -\omega_o/2Q - \omega_o\sqrt{(1/2Q)^2 - 1} \ldots(b) \qquad (12.12)$$

For increasing Q the pole-zero diagram follows the sequence in Fig. 12.4(a) to (c).

- For $Q=0$ then $s_1=0$ and $s_2=-\infty$.
- For $0<Q<1/2$ the poles move as in (a).
- For $Q=1/2$ the poles coincide as in (b), distinguished as a double pole by the number 2.
- For $1/2<Q<\infty$ the poles move as in (c).
- For $Q=\infty$ then $s_1=\omega_o$ and $s_2=-\omega_o$.

For $Q>1/2$ the poles are conjugate, giving

$$s_1=\sigma_p+j\omega_p \quad \ldots(a)$$
$$s_2=\sigma_p-j\omega_p \quad \ldots(b) \qquad (12.13)$$

where

$$\sigma_p=-\omega_o/2Q \quad \ldots(a)$$

and

$$\omega_p=\omega_o\left(1/(2Q)^2-1\right)^{1/2} \quad \ldots(b) \qquad (12.14)$$

Frequency response

Here as usual s in [12.11] is replaced by $j\omega$. With

$$|A(\omega)| \propto \frac{|j\omega|}{|j\omega-s_1| \times |j\omega-s_2|}$$

then the three magnitudes are given by the lengths of the three lines in Fig. 12.4(d).

With the poles usually closer to the vertical axis than shown, it is $|j\omega-s_1|$ which gives the sharp dip in $|den|$ and thus the sharp peak in $|A|$.

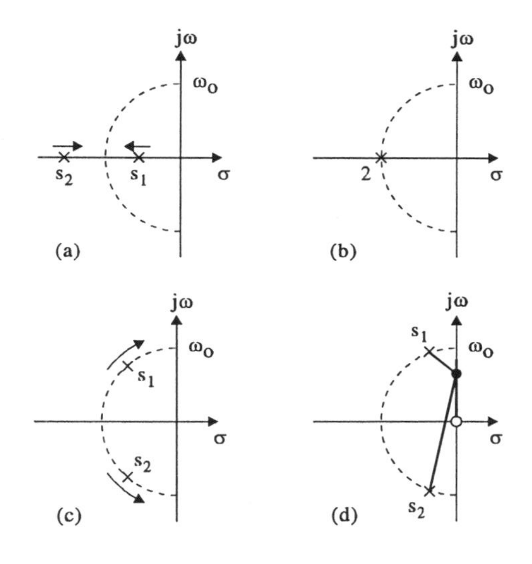

Fig. 12.4 Pole-zero diagram for band-pass filter of Fig. 12.1 with Q increasing.
(a) $Q<1/2$.
(b) $Q=1/2$.
(c) $Q>1/2$.
(d) As (c), showing lines for deriving frequency response.

Problem

For $Q>>1$ show from Fig. 12.4(d) that
- for $\omega<<\omega_o$ then $|A| \propto f$ and $\phi \approx +90°$
- for $\omega>>\omega_o$ then $|A| \propto 1/f$ and $\phi \approx -90°$

both as in Fig. 12.2(b,c).

Impulse response

Resolving A(s) as in [11.42], etc.,

and with
$$A(s)=\frac{\omega_o}{Q}\frac{s}{(s-s_1)(s-s_2)} \quad \text{as in [12.11]}$$

then

$$A(s)=\frac{A_1}{(s-s_1)}+\frac{A_2}{(s-s_2)} \qquad (12.15)$$

For A_1 we multiply by $(s-s_1)$ and set $s=s_1$, and similarly for A_2, giving

$$A_1=\frac{\omega_o}{Q}\frac{s_1}{(s_1-s_2)} \quad \ldots(a)$$

$$A_2=\frac{\omega_o}{Q}\frac{s_2}{(s_2-s_1)} \quad \ldots(b) \qquad (12.16)$$

From Table 11.1, Row 4 the impulse response

$$g_i(t)=\frac{\omega_o}{Q}\left[\frac{s_1}{s_1-s_2}\exp(s_1t)+\frac{s_2}{s_2-s_1}\exp(s_2t)\right] \qquad (12.17)$$

For $Q>>1$ and with [12.13], etc., then

$$g_i(t)\approx(\omega_o/2Q)[\exp(\sigma_pt).\exp(j\omega_ot)$$
$$+\exp(\sigma_pt).\exp(-j\omega_ot)] \qquad (12.18)$$

giving

$$g_i(t)\approx\frac{\omega_o}{Q}\cos\omega_ot.\exp\left(\frac{-\omega_o}{2Q}t\right) \qquad (12.19)$$

as plotted in Fig. 12.5.

With σ_p negative the two components in [12.17] are decaying corkscrew waves. These start at the common real value $\omega_o/2Q$ and then twist in opposite directions at the same rate ω_o, decaying at the same rate $\omega_o/2Q$, and resulting in the real $g_i(t)$ shown.

This 'ring' at the resonant frequency ω_o is characteristic of the transient response of a high-Q resonator type of filter, whether band-pass or band-stop, impulse response or step response. For the transient response it is usually unwelcome. With the decay time constant $2Q/\omega_o$

$$Q=\pi N_d \qquad (12.20)$$

where N_d is the number of cycles of oscillation over the decay period. This is a convenient way of estimating Q experimentally.

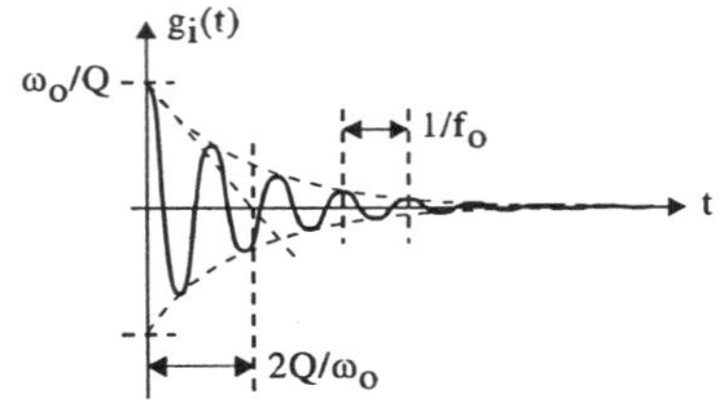

Fig. 12.5 Impulse response of band-pass filter of Fig. 12.1 showing $Q=\pi N_d$ where N_d=no of cycles over decay time.

Problem

With the step response of the band-pass filter obtained by adding the factor 1/s to the RHS of [12.11] for A(s), show that the oscillatory component is as in Fig. 12.5 but with $\sin(\omega_o t)$ in place of $\cos(\omega_o t)$.

Working Q-factor

Fig. 12.1 ignores both the ohmic resistance R_L of the inductor and the output impedance R_s of the signal source. With these included then

$$A(\omega_o)=R/(R_s+R_L+R) \qquad (12.21)$$

and the 'working' Q-factor

$$Q_w=\omega_o L/(R_s+R_L+R) \qquad (12.22)$$

Defining

$$Q_s=\omega_o L/R_s \quad \ldots(a)$$

$$Q_L=\omega_o L/R_L \quad \ldots(b)$$

$$Q_r=\omega_o L/R \quad \ldots(c) \qquad (12.23)$$

then

$$Q_w=1/(1/Q_s+1/Q_L+1/Q_r) \qquad (12.24)$$

or more concisely

$$Q_w=Q_s//Q_L//Q_r$$

Incorporation into long-tailed pair

For the RF circuitry in which the LC filter is used the signal source will usually be a previous common-emitter, common-source, or most likely a dual-gate MOS cascode stage. Both the output impedance of these, and the corresponding input impedance of the following stage, will all be far too high for the present purpose, making the parallel resonator of the next section the more suitable choice.

Fig. 12.6 shows an exception, where the inductor is connected between the two halves of a long-tailed pair. The first half is an emitter-follower of relatively low output impedance $1/g_m$ giving $R_s=1/g_m \sim 25\Omega$, while the second is a common-base stage with the input impedance also $1/g_m$, giving $R=1/g_m$.

Exercise 12.1 gives example calculations for this circuit.

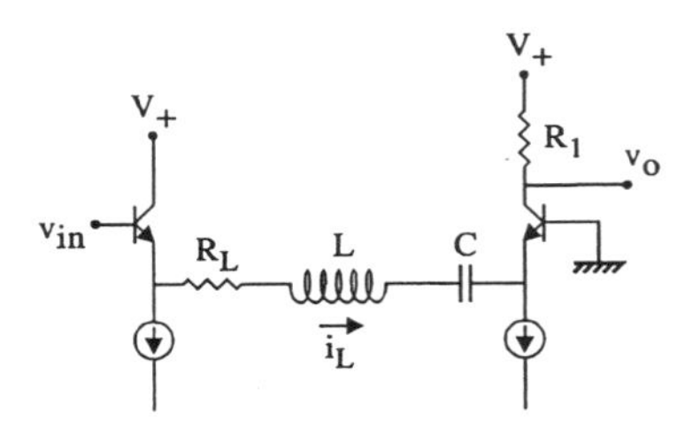

Fig. 12.6 Incorporation of series resonator into long-tailed pair.

12.2 Parallel resonator

Fig. 12.7(a) shows the essentials of the parallel resonator. Here $V_o=I_{in}Z$ where Z is the impedance of the resonator. Thus $|Z|$ is a direct indication of the variation of $|V_o|$ with frequency. From the diagram the corresponding admittance

$$Y(\omega)=1/R+1/j\omega L+j\omega C \qquad (12.25)$$

Fig. 12.7(b) shows the variation of $|Y(\omega)|$ with ω for $Q>>1$. As for the series resonator, this is built from the frequency dependencies of the separate components. Also as before, each of the three components dominates over a different part of the frequency range. However, now the component that dominates is that with the highest admittance, rather than that of the highest impedance.

As for the series resonator, the resonant frequency ω_o is that at the crossover, giving

$$\omega_o=1/\sqrt{LC}$$

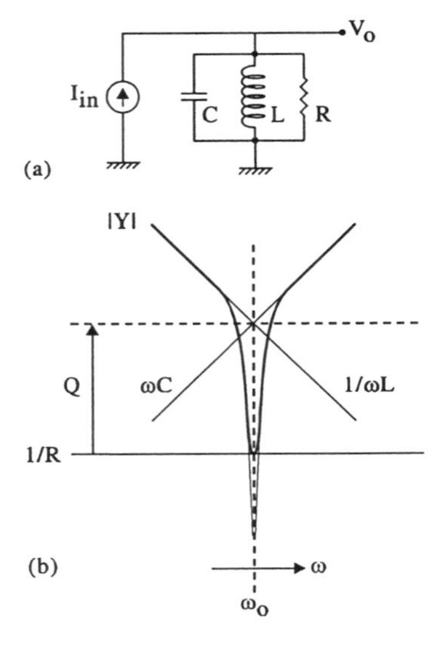

Fig. 12.7 Parallel resonator.
(a) Circuit diagram.
(b) Frequency response $|Y|$, Y=resonator admittance.

However, now $Q = R/(\omega_o L) = \omega_o CR$ (12.26)

requiring $R >> \omega_o L$ for $Q >> 1$, rather than $R << \omega_o L$ as before. For both types the bandwidth for $Q >> 1$ remains

$$\Delta f \approx f_o/Q \qquad (12.27)$$

Much as before, $|Z(\omega)|$ is obtained by inverting $|Y(\omega)|$ in (b) to give the resonant peak. Expressed in terms of Q and ω_o, Z(s) has the same form as [12.10] for the series resonator, and thus the same pole-zero diagram and the same kind of impulse and step responses.

Tuned RF amplifier

Fig. 12.8(a) shows the parallel resonator as used with the dual-gate MOS amplifier, with (b) the small-signal equivalent circuit. With r_o the output impedance of the first stage, the input impedance of the second is only capacitive, constituting part of C.

Both R_L and r_o contribute to the final parallel impedance and the working Q-factor Q_w. With Q as in [12.26] and with Q_o the component due to r_o then

$$Q_o = r_o/\omega_o L \qquad (12.28)$$

Series to parallel transformation

For the effect of R_L upon Q_w we transform R_L to the equivalent parallel R_p. For L and R_L in series

$$Z(\omega) = R_L + j\omega L$$

giving

$$Y(\omega) = 1/(R_L + j\omega L)$$

Normalising

$$Y(\omega) = (R_L - j\omega_o L)/(R_L{}^2 + (\omega_o L)^2)$$

giving $Y(\omega) = G_p(\omega) + B_p(\omega)$

and thus at resonance the equivalent parallel

$$R_p = (R_L{}^2 + (\omega_o L)^2)/R_L \quad ...(a)$$
$$X_p = (R_L{}^2 + (\omega_o L)^2)(-j\omega_o L) \quad ...(b) \qquad (12.29)$$

With $Q_L = \omega_o L/R_L$ then for $Q_L >> 1$

$$R_p \approx Q_L{}^2 R_L ...(a) \qquad L_p \approx L ...(b) \qquad (12.30)$$

With $Q_w = (r_o // R_p)/\omega_o L$

and with Q_o as in [12.28]

then $Q_w = 1/[1/Q_o + 1/Q_L]$ (12.31)

i.e., $Q_w = Q_o // Q_L$

Problem

Derive Q_w by arguing the other way, with r_o transforming to $r_o/Q_o{}^2$ in series with R_L. Exercise 12.2 gives example calculations for the circuit.

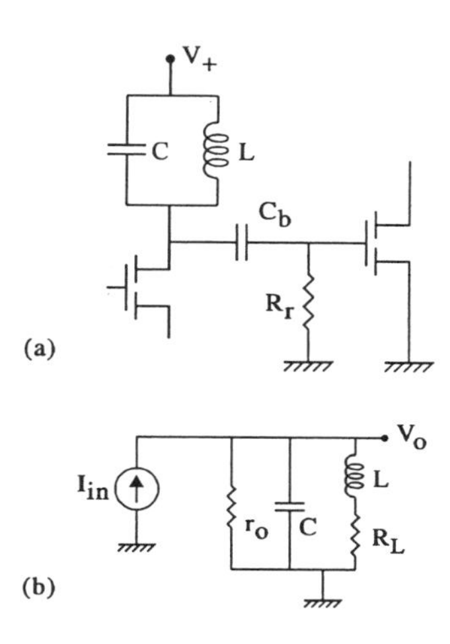

Fig. 12.8 Use of parallel resonator in coupling from one dual-gate MOS stage to the next.
(a) Circuit diagram.
(b) Incremental equivalent circuit.

12.3 Narrow-band band-stop filter

While the series resonator does not lend itself well to narrow-band band-pass filter applications, it is somewhat better suited to the corresponding band-stop function.

As in Fig. 12.9(a), this is simply the band-pass filter of Fig. 12.1 with the components reversed. Thus what was transmitted for the band-pass is now prevented for the band-stop, and vice versa.

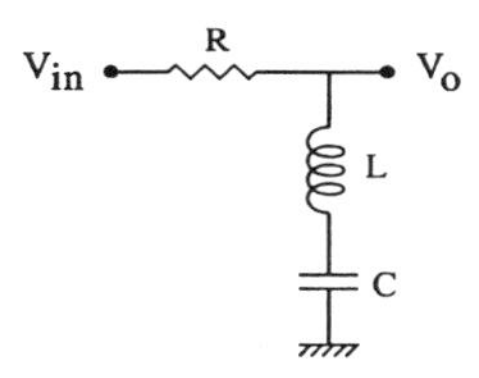

Fig. 12.9(a) Narrow-band band-stop filter based on series resonator. Circuit diagram.

Frequency response

From the circuit, with $A(\omega) = V_o/V_{in}$ then

$$A(\omega) = \frac{j\omega L + 1/(j\omega C)}{R + j\omega L + 1/(j\omega C)} \qquad (12.32)$$

or in terms of s, ω_o and Q

$$A(s) = \frac{\omega_o^2 + s^2}{s^2 + s\omega_o/Q + \omega_o^2} \qquad (12.33)$$

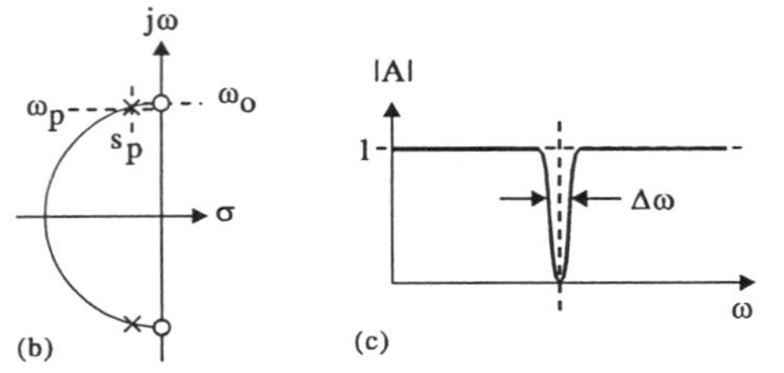

Fig. 12.9(b,c) Narrow-band band-stop filter based on series resonator.
(b) Pole-zero diagram.
(c) Frequency response of $|A\omega| = |V_o|/|V_{in}|$.

Pole-zero diagram

This in Fig. 12.9(b) is derived as follows.

- For the band-pass filters of Fig. 12.1, and Fig. 12.7

$$A(s) = \frac{s(\omega_o / Q)}{s^2 + s\omega_o / Q + \omega_o^2} \quad \text{as in [12.10]}$$

- With the denominator here the same as in [12.33] for the present band-stop filter, the poles are the same for both, as in Fig. 12.4(c) for the band-pass, and above for the band-stop.
- From num[12.33] for the band-stop there are zeros at $\pm \omega_o$, also plotted in the present (b).

With the frequency response derived as in Fig. 12.4(d) for the band-pass, now it is the upper zero that gives the notch, with the poles and zeros combining for

$$|\omega - \omega_o| >> \sigma_p \quad \text{to give} \quad |A(\omega)| \approx 1.$$

Inductor resistance

With the inductor resistance R_L included

$$A(\omega_o) = R_L/(R_L + R) \tag{12.34}$$

increasing $A(\omega_o)$ from the required zero, and requiring $R >> R_L$. But the working

$$Q_w = \omega_o L/(R + R_L) \tag{12.35}$$

requiring $R << R_L$. Thus we have a conflict of interests.

Exercise 12.3 gives associated calculations with experimental verification for the above circuit.

12.4 Notch filters

The circuit of Fig. 12.10 (Van Valkenburg, 1982) gives the required $A(\omega) = 0$ at resonance without unduly compromising the circuit Q. With the total effective capacitance now C/2 then

$$\omega_o = 1/(LC/2) \tag{12.36}$$

while for $A(\omega_o) = 0$

$$R_p R_L = 1/(\omega_o C)^2 \tag{12.37}$$

One application of the notch filter is in a radio receiver, where it is required to eliminate an interfering carrier within the bandwidth of say a required speech signal. To minimise the loss of required signal, the width of the notch needs to be as small as possible consistent with the spectral purity of the unwanted signal.

A lower frequency example is the need to eliminate 50Hz or 100Hz interference induced by the mains supply. This would be implemented using the active notch filter of the next chapter.

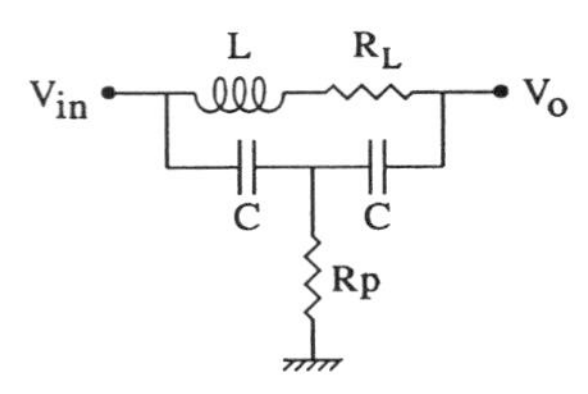

Fig. 12.10 Bridged-T notch filter giving $|A(\omega)| = 0$ at notch.

Q-enhancement by negative feedback

With the required notch width ~30Hz for the above communications receiver then Exercise 12.4 gives the required $Q_w > 15k$. With this figure well above that for the LC resonator, enhancement is provided by the negative-feedback circuit of Fig. 12.11. Here the feedback keeps the closed-loop gain $A_f \approx 1$ until almost the bottom of the notch, where the loop gain falls to below unity and the feedback has no effect. Thus $\Delta\omega$ is reduced by the factor $1/A_o$ which, in principle, can be made as small as is wished.

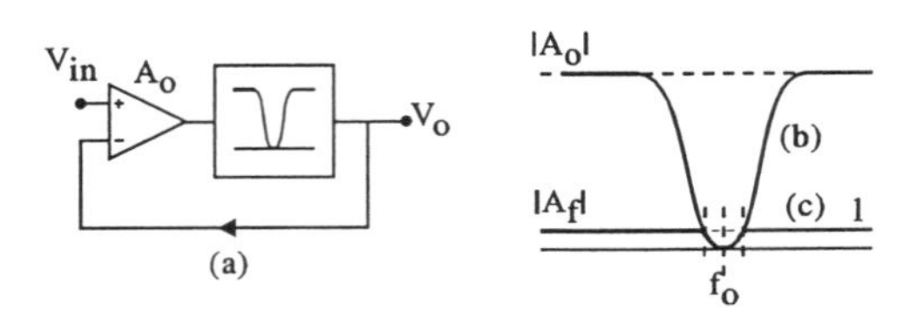

Fig. 12.11 Narrowing of notch by negative feedback.
(a) Block diagram.
(b) Response of op-amp and notch filter without feedback.
(c) Response with feedback

Band-pass incorporating notch

Another commonly used way of inserting a notch is as in Fig. 12.12(a). Here C and L_2 with the associated R_L form a series resonator tuned to the notch frequency ω_s in (b).

With ω_p the parallel resonant frequency for which the narrow-band band-pass transmission is required, and for $\omega_p << \omega_s$, then at ω_p the impedance of the series resonator is close to being just the reactance of C. Thus L_1 is simply set to parallel resonate with C.

With the two frequencies closer the relations are a little more complex, but nevertheless the two resonances remain, with that at ω_s unaltered.

Problem Show that if L_1 is replaced by C_1 then $\omega_p > \omega_s$.

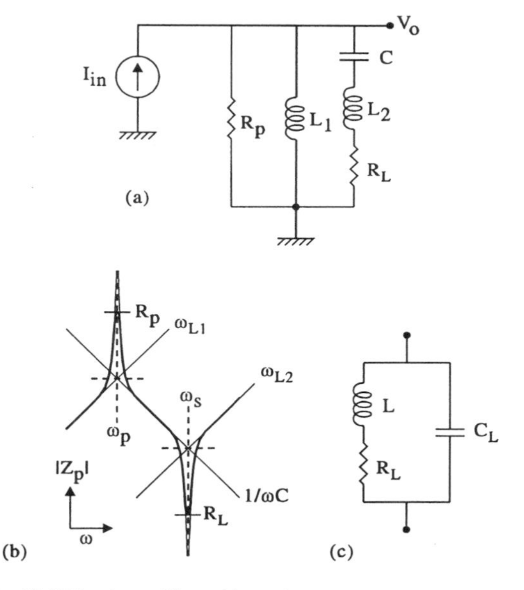

Fig. 12.12 Band-pass filter with notch.
(a) Circuit diagram.
(b) Frequency response.
(c) Inductor equivalent circuit showing winding capacitance C_L.

Inductor self-resonance

Fig. 12.12(c) shows the self-capacitance C_L of the inductor, causing the inductor to become 'self-resonant' at the frequency

$$f_L = 1/[2\pi(LC_L)^{1/2}] \qquad (12.38)$$

The only effect C_L has on a parallel resonance is to shift the resonant frequency, which can be allowed for. For the series resonator, however, this adds a spurious parallel resonance. For the present circuit this will give a further peak, this time above ω_s, usually a highly undesirable result.

Narrow-band band-pass filter based on parallel resonator

It is thus preferable to use the parallel resonator for narrow-band band-stop applications, for example as in Fig. 12.13, in the absence of C_{in}. With C_L thus combined with C then there is only the one required parallel resonance.

Even here, with the circuit as for the experimental measurement of Exercise 12.5, and with C_{in} the unavoidable scope-probe capacitance, there is a secondary resonance giving an unwanted peak.

With ω_2 the frequency of the secondary resonance, and ignoring R_{in},

$$1/\omega_2 C_{in} = j\omega_2 L // (1/j\omega_2 C_s)$$

giving

$$\omega_2 = 1/[L(C_L + C_{in})]^{1/2} \qquad (12.39)$$

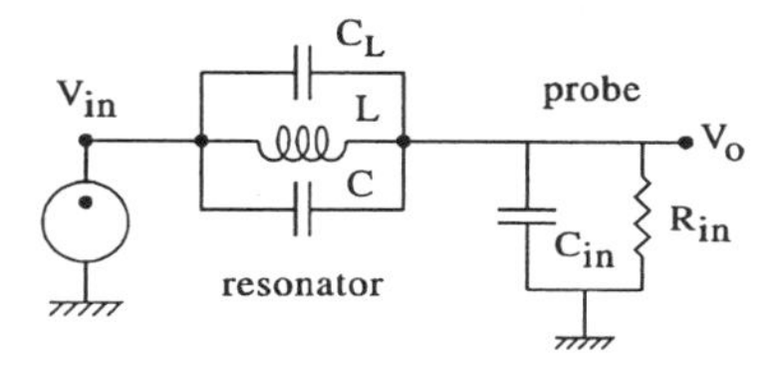

Fig. 12.13 Narrow-band band-stop filter based on parallel resonance, as for experimental work.

12.5 Second-order low- and high-pass filters

With

$$A(\omega) = \frac{1/j\omega C}{R + j(\omega L - 1/\omega C)} \qquad (12.40)$$

and

$$A(\omega) = \frac{j\omega L}{R + j(\omega L - 1/\omega C)} \qquad (12.41)$$

for the low- and high-pass filters of Fig. 12.14 then the corresponding

$$A(s) = \frac{\omega_o^2}{\omega_o^2 + s\omega_o/Q + s^2} \qquad (12.42)$$

and

$$A(s) = \frac{s^2}{\omega_o^2 + s\omega_o/Q + s^2} \qquad (12.43)$$

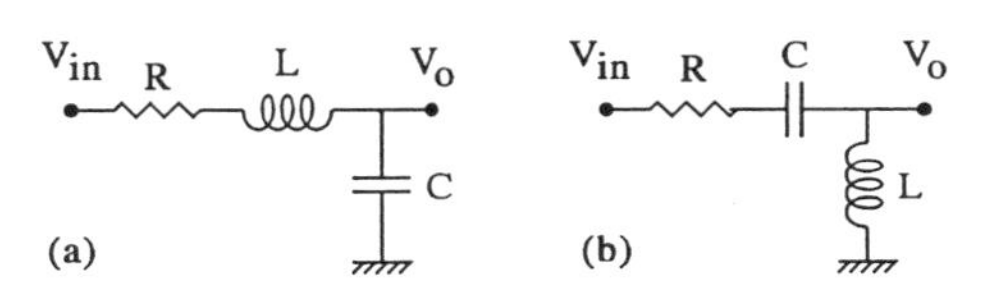

Fig. 12.14 Second-order low- and high-pass filters.
(a) Low-pass.
(b) High-pass.

giving the frequency responses of Fig. 12.15(a) to (f) for $Q >> 1$.

Here, apart from the resonant peak, we have the second-order low- and high-pass filters, with $|A(\omega)| \propto 1/f^2$ in the stop-band of the low-pass, and $|A(\omega)| \propto f^2$ in the stop-band of the high.

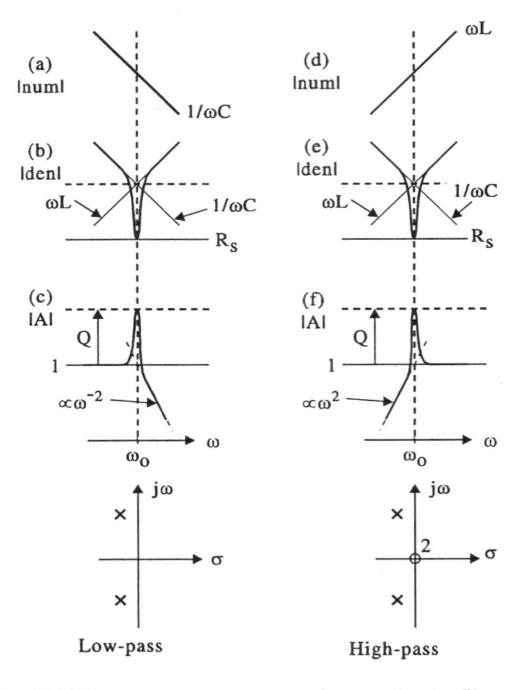

Fig. 12.15 Frequency responses curves for second-order filters of Fig. 12.14 with resonant peak.

Low-pass	High-pass
(a) |num[12.40]|	(d) |num[12.41]|
(b) |den[12.40]|	(e) |den[12.41]|
(c) |A| = |num|/|den|	(f) |A| = |num|/|den|
(g) Pole-zero diagram	Pole-zero diagram

num: numerator, den: denominator

Pole-zero diagrams

With A(s) as in [12.10] for the band-pass filter of Fig. 12.1 then the denominator of this and the above two equations are all the same. Thus the poles in Fig. 12.15(g) and (h) are as in Fig. 12.4(d) for the band-pass. With the zeros all at the origin we have the sequence

- no zeros for the low-pass as in the present (g),
- one zero for the band-pass as in Fig. 12.4(d),
- two zeros for the high-pass in the present (h).

The frequency responses of (c) and (f) can also be derived from the pole-zero diagrams as usual.

Damping

Where the resonant peak is not required, this can be removed by setting the value of R needed for critical damping ($Q = 1/2$). Fig. 12.16 shows the resulting frequency and step responses for the previous low-pass, for under, critical, and overdamping.

While $Q = 1/2$ is the value for which the poles are about to leave the real axis, to begin to give overshoot in the step response, Q may be advanced to $1/\sqrt{2}$ before overshoot occurs in the frequency response. This is the Butterworth response of the next chapter, and places the poles at the 45° points on the circle

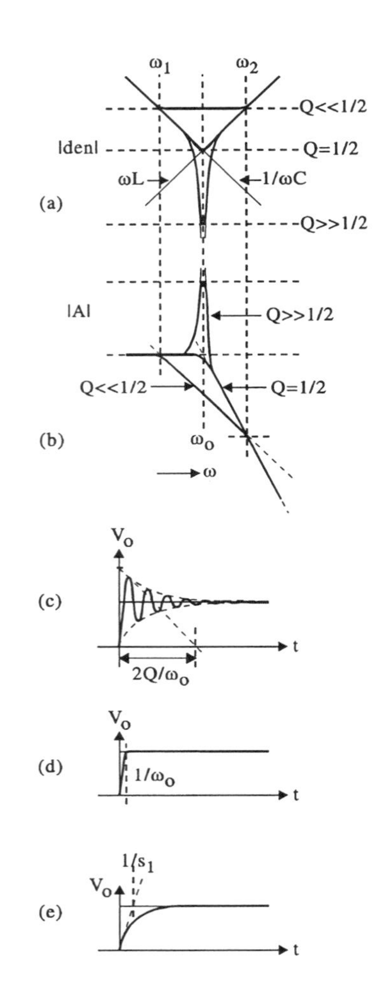

Fig. 12.16 Damping of second-order low-pass filter of Fig. 12.14(a) to remove resonant peak.
(a) |den[12.40]|.
(b) |A| = |num[12.40]|/|den[12.40]|.
(c) Step response for under damping.
(d) Critical damping.
(e) Over damping.

12.6 LC impedance matching

As in [12.30], for a high-Q parallel resonator at resonance the series resistance R_L of the inductor appears as the parallel $R_p \approx Q^2R_L$. Thus, taking R_L as zero for the moment, the impedance of a load R_o can be increased in this way to match the source impedance R_s as in Fig. 12.17.

However, with Q not necessarily >>1, the more general design equations are required. For R_o and L in series

$$Z(\omega)=R_o+j\omega L$$

giving $$Y(\omega)=1/(R_o+j\omega L)=G_p+B_p \quad (12.44)$$

where $$G_p(\omega)=R_o/[R_o{}^2+(\omega L)^2] \quad \ldots(a)$$

and $$B_p(\omega)=-j\omega L/[R_o{}^2+(\omega L)^2] \quad \ldots(b) \quad (12.45)$$

giving $$R_p(\omega)=R_o(1+Q^2) \quad \ldots(a)$$

$$X_p(\omega)=j\omega L(1+1/Q^2) \quad \ldots(b) \quad (12.46)$$

Thus the value of L is set to give the Q needed for $R_p=R_s$, and the parallel C is set to resonate with X_p, leaving the impedance presented to the source purely resistive and equal to R_s.

Design equations

Allowing for $R_s < R_o$ by designating the higher and lower of the two as R_{high} and R_{low} then the design equations become

$$Q=(R_{high}/R_{low}-1)^{1/2} \quad \ldots(a)$$

$$|X_{ser}|=QR_{low} \quad \ldots(b)$$

$$|X_p|=R_{low}(Q+1/Q) \quad \ldots(c) \quad (12.47)$$

For $R_s > R_L$ the LC network is laterally transposed. For a high- rather than a low-pass response the L and C are exchanged, both prior to the value calculations.

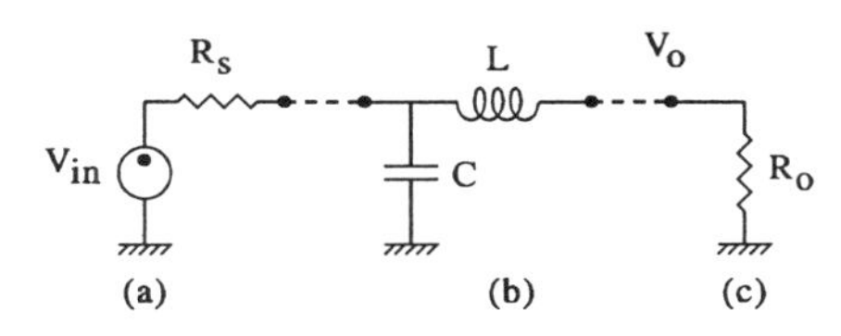

Fig. 12.17 L-C network giving matching of R_o and R_s. (a) Signal source. (b) Low-pass L-C network. (c) Load R_o.

Exercise 12.6 gives example calculations for the matching of a receiver input and a transmitter output.

Link coupling

The above impedance transforming filters give a resonance with either a second-order low-pass or high-pass characteristic. It is possible that a true band-pass response might be preferred to either of these, where the response is $\propto$f below the resonance and $\propto$1/f above it.

Fig. 12.18 shows a suitable circuit. Here the mutual inductance

$$M=k\sqrt{L_1L_2} \quad \text{as in [1.38]}$$

where the coupling factor k has the maximum value of unity when all the flux is generated by one coil links with all the turns of the other. The mutual inductance couples the effective impedance

$$R_{s2}=\omega^2M^2/R_s$$

in series with L_2, which then appears as

$$R_{p2}=Q^2R_{s2}$$

in parallel with L_2. With

$$Q=\omega_oL/R_{s2}$$

and M as above, this gives

$$R_{p2}=R_sL_2/(k^2L_1) \quad (12.48)$$

and for k=1 $$R_{p2}/R_s=L_2/L_1$$

With $$L_2/L_1=(N_2/N_1)^2$$

then $$R_{p2}/R_s=(N_2/N_1)^2$$

as in [1.29], for the normal transformer. The circuit also produces no spurious resonances.

Problem

Show that for k=1 then the true band-pass response is obtained, while for lower values there is a further transition above ω_o at which the $\propto 1/\omega$ changes to $\propto 1/\omega^2$, usually even more desirable.

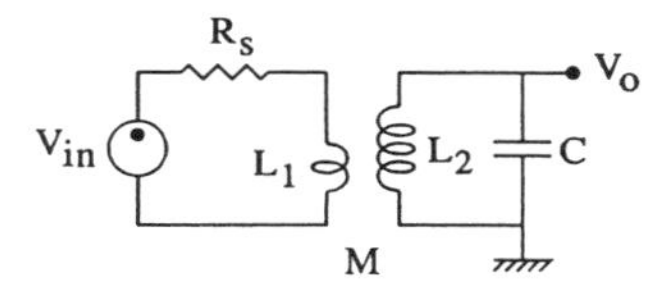

Fig. 12.18 Link-coupling giving impedance step-up with true band-pass response.

12.7 Broad-band band-pass and band-stop filters

The only band-pass and band-stop filters that have so far been discussed have been of the narrow-band type, based on a simple, usually high-Q resonator. The frequency response of these falls a good deal short of the idealised abrupt cut-off band-pass and band-stop filters of Fig. 11.1(c) and (d). Here, as for the low- and high-pass, the ideal can be approximated to better by a higher order filter, i.e. one with more sections. For the band-pass and band-stop these will commonly consist of a sequence of cascaded active or passive resonator type circuits, with the resonant frequencies and Q-factors suitably chosen.

Coupled LCR resonators

Fig. 12.19 shows a well-established means of improving on the response of a single resonator. This is simply two such resonators coupled by mutual inductance. The two will normally be tuned to the same frequency. Then as the coupling is increased the single second-order resonance curve initially flattens at the top and then breaks into two separate peaks.

The coupling is usually set at the critical point or a little beyond it. Here the slight double-hump is an acceptable price to pay for the more abrupt cut-off obtained at the edges of the pass-band.

A normal AM radio tends to have two such double resonators in the intermediate frequency (IF) section. Here tuning is usually by adjustable powdered iron cores as shown, to adjust L to the usual resonant condition $\omega_0 = 1/LC$.

With the circuit as shown, the Q is determined by the inductor resistances R_L. Should this Q be too high, additional shunt resistors would be used. Here the output impedance of the real source and the input impedance of the destination (load) must be allowed for.

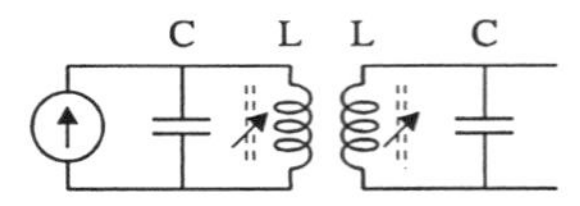

Fig. 12.19 Mutually coupled resonators for band-pass filtering.

Quartz crystal filters

For narrow-band band-pass filters where the requirement is for a high degree of flatness within the band-pass, and a very abrupt and deep cut-off outside it, the filter will probably be in the form of a series of cascaded quartz crystal resonators. This approach is widely used in modern communications receivers and transmitters.

A considerably cheaper alternative, where the requirements are less demanding, is the range of filters based on ceramic resonators. For the intermediate frequency of 7MHz used for FM radio these are available for about 50p. A good multi-section quartz filter, in contrast, will run to £50+.

12.8 Inductor specifications

For the inductor used as the example for the calculations and experimental work in the Exercises section (Maplin order code WH47) **L=1mH.**

As is usual for a distributor's catalogue, several other specifications are given, as follows.

Self-resonant frequency

This is the frequency f_L at which the wiring capacitance C_L in Fig. 12.12(c) forms a parallel resonance with L. For the chosen device $\mathbf{f_L \approx 1MHz.}$

DC resistance

Sometimes the inductor is used for power line decoupling, with the inductor in series with the line and low-inductance capacitors from each terminal to the zero line or ground, thus forming a low-pass filter in either direction. For a load current I_L drawn from the supply, and with R_{dc} the DC resistance, then a drop of $I_L R_{dc}$ in the supply voltage results. Here $\mathbf{R_{dc} = 14\Omega.}$

AC resistance

This is R_L in Fig. 12.12(c), higher than the DC value, and somewhat frequency dependent, though always a resistance.

One reason for the increase is the skin effect, which confines the current in a conductor to the regions close to the edge.

Added to this is the proximity effect which, for adjacent turns on the coil, tends to confine the current to the parts of the wire closest to its neighbours.

Finally inductors with L in the region of the 1mH, as for that chosen, will usually have a ferrite core. Here the core losses increase with frequency.

Q-factor

Because of its frequency-dependent nature, the AC resistance is not specified directly, but indirectly as the Q-factor. This too is frequency dependent, so together with the Q-factor is quoted a 'test frequency' f_o.

With $Q = \omega_o L/R_L$ then apparently $Q \propto \omega_o$, but with R_L also increasing there comes a point where the increase in R_L has the greater effect. This is where the Q has its highest value, with the quoted 'test frequency' no doubt comparable. Also one may need to stay clear of the self-resonance at f_L. For the chosen L=1mH

Q>45 for $\mathbf{f_o = 800kHz.}$

Not far from $f_L = 1MHz$.

With the present device having the highest value of L for the range, the frequency for maximum Q increases with decreasing L, being 300MHz for L=0.22μH, for which Q>50. Thus the most suitable device for the operating frequency can be chosen.

With Q=45 the minimum value for production spread for the chosen L=1mH, the typical value will be higher. However, this is the value that will be used for calculations.

12.9 LC oscillator

The oscillator circuit of Fig. 12.20(a) is popular because of its simplicity. The 100% DC feedback is negative, setting the gate input initially at its threshold. With the network ideally giving a 180° phase lag at the resonant frequency of the split LC resonator, then the feedback becomes positive and the circuit oscillates.

Unfortunately for the gate output impedance $R_s=0$ the phase lag at resonance is not 180° but 90°, causing the oscillation to be above the resonant frequency, to give the extra phase lag needed.

With off-resonance oscillation undesirable, in that it gives a greater sensitivity to changes in the gate parameters, the oscillation can be brought closer to ω_o by adding to R_s. The limit to R_s is determined by the need for the loop gain at resonance to exceed unity.

The quartz crystal version of (c) is possibly even more popular, being the normal circuit for the digital clock oscillator. Here the negative DC feedback is via R, usually 1MΩ or above. Exercise 12.8 gives experimental work.

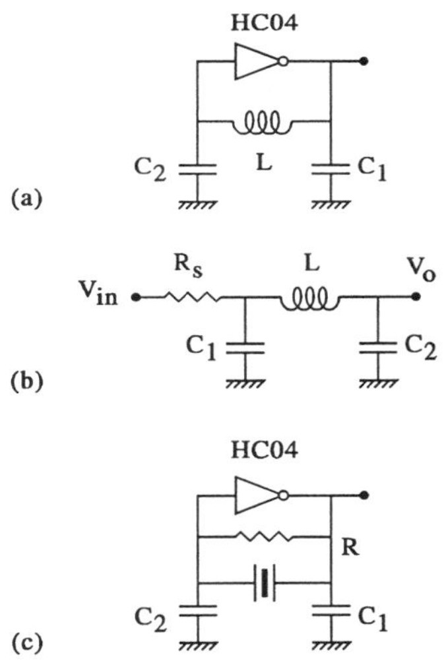

Fig. 12.20 LC oscillator based on HCMOS gate.
(a) Circuit diagram.
(b) Network including gate output impedance $R_s \approx 50\Omega$.
(c) As (a) but for quartz crystal. $R \sim 1M\Omega$.

Problem

Confirm the above statement that for $R_s=0$ then at resonance V_o in (b) lags V_{in} by 90°.

With the source $Q_s=R_s/|X_{c1}|$ and with normally $C_1=C_2$, show that for $Q_s << 1$ the phase lag at resonance approaches 180°.

Reference

Van Valkenburg, M. E. 1982: *Analog Filter Design*. Holt, Rinehart and Winston.

12.10 Exercises and experiments

Starting values for cumulative calculations As above,

$$L=1mH, \quad Q_L>45, \quad f_o=800kHz, \quad f_L \approx 1MHz.$$

Transistor collector current $I_c=1mA$.

Thermal energy $kT/q=25mV$.

Exercise 12.1 Incorporation of series-resonator into long-tailed pair (Fig. 12.6)

Calculations Calculate
- C for the circuit to resonate at 800kHz,
- the inductive reactance ω_oL,
- the effective inductor series resistance R_L,
- the transistor mutual conductance $1/g_m$,
- the incremental voltage gain a_{LTP} at resonance, given $R_l=5k\Omega$'
- the working Q-factor Q_w.

With $\omega_o=1/\sqrt{LC}$ as in [12.2] then **$C=39.6pF$.**

With $\omega_oL=2\pi\times800kHz\times1mH$ then **$\omega_oL=5.03k\Omega$.**

With $Q=\omega_oL/R$ as in [12.4]

then here $R_L=\omega_oL/Q_L$

giving **$R_L=112\Omega$.**

With $g_m=I_c/(kT/q)$ as in [6.6]

then **$g_m=40mS$.**

With $R_s=R=1/g_m$

then the total resistance

$$R_T=R_s+R_L+R$$

giving $R_T=162\Omega$.

With the inductor current

$$i_L=v_{in}/R_T \quad \text{and} \quad v_o=i_LR$$

then $a_{LTP}=R/R_T$

giving **$a_{LTP}=30.9$.**

With $Q_w=\omega_oL/(R_s+R_L+R)$ as in [12.22]

then **$Q_w=31.1$.**

Exercise 12.2 Tuned RF cascode amplifier

Cascode of Fig. 12.8(b) with chosen inductor.

Calculate the resonator impedance R_p at parallel resonance.

With $R_p \approx Q_L{}^2 R_L$

as in [12.30] then **$R_p = 226k\Omega$.**

> **Calculate** the component of the Q-factor due to the MOS output impedance $r_o = 150k\Omega$ and the overall Q.

With $Q_o = r_o/\omega_o L$ as in [12.27]

then **$Q_o = 29.8$.**

With $Q = 1/[1/Q_o + 1/Q_L]$ as in [12.31]

then **$Q = 17.9$.**

Exercise 12.3 Narrow-band band-stop filter based on series resonance (Fig. 12.9(a))

> **Calculate** the working Q-factor Q_w and the gain $A(\omega_o)$ at resonance for $R = 10\Omega$ to $100k\Omega$ in decade steps.

With $A(\omega_o) = R_L/(R_L + R)$ as in [12.34]

and $Q_w = \omega_o L/(R + R_L)$ as in [12.35]

then

R	10Ω	100Ω	1kΩ	10kΩ	100kΩ
Q_w	41.2	23.7	4.5	0.5	0.05
$A(\omega_o)$	0.92	0.53	0.1	0.01	0.001

> **Experimental verification** Connect the chosen inductor as in Fig. 12.9(a) with $C = 39.6pF$ as above for $f_o = 800kHz$.
>
> Observe the notch for the above values of R to confirm the calculated values of $A(\omega_o)$ and Q.
>
> For $R = 100k\Omega$ note the peak at $f_L \approx 1MHz$ due to the inductor self-resonance.
>
> Note the distorted low-level waveform then seen at resonance. This is the sum of the harmonics arising from the inevitable slight distortion of the signal generator output, made visible by the suppression of the fundamental. The method is a sensitive way of assessing such distortion. But don't forget the possible non-linearity of the inductor core.

Exercise 12.4 Notch filter (Fig. 12.11)

> **Calculate** the required Q-factor for communications receiver notch. The filtering is to be done at the intermediate frequency $f_{if} = 455kHz$. Speech ranges from $\approx$ 300Hz to 3kHz.

With the frequency of the interfering signal possibly corresponding to the low end of the speech spectrum then the width Δf of the notch needs to be below 10% of 300Hz giving $\Delta f < 30Hz$. With

$$Q > f_{if}/\Delta f$$

then **$Q > 15.2k$.**

Exercise 12.5 Narrow-band band-stop filter based on parallel resonator (Fig. 12.13)

The circuit uses the chosen inductor with the usual $f_L = 800kHz$.

> **Calculations** Calculate
> - the inductor winding capacitance C_L,
> - the value of C needed for the parallel resonance to be at 800kHz,
> - the approximate value of f_s, the frequency of the spurious series resonance, for the ×10 scope-probe capacitance $C_{in} = 10pF$.

With $\omega_L = 1/(LC_L)^{1/2}$

then **$C_L = 25.3pF$.**

With the previous $C = 39.6pF$ for resonance at 800kHz ignoring C_L, now

$$C = 39.6pF - 25.3pF$$

giving **$C = 14.3pF$.**

With $\omega_s = 1/[L(C + C_L + C_{in})]^{1/2}$ as in [12.39]

then **$f_s = 715kHz$.**

> **Experiment** Connect the circuit and observe the deep notch at f_p and the resonant peak at f_s. Thus confirm the above calculated values. Note that at the peak $|A(\omega_s)| >> 1$.

Exercise 12.6 LC impedance matching

> **Receiver input** With the source impedance R_s for a receiver normally the standard 50Ω co-ax cable impedance, this is vastly below the input impedance R_{in} of the dual-gate MOSFET that is likely to be the first RF stage. Develop the expressions for design of the matching network.

The matching network will be as in Fig. 12.17(b) but reversed. With $R_{in} \sim 1G\Omega$ and the matching

$$Q = (R_{high}/R_{low} - 1)^{1/2}$$

as in [12.47] then $Q \approx 4.5k$ which is unobtainable. Thus the point of interest becomes the voltage gain $A(\omega_o)$ that the input resonator provides at resonance. We calculate for the usual 800kHz resonance, ignoring C_L for simplicity.

With $|A(\omega_o)| = Q_w$ where Q_w is the working Q-factor, and with

$$Q_w = \omega_o L/(R_s + R_L)$$

where R_L is the inductor resistance then

$$|A(\omega_o)| = \omega_o L/(R_s + R_L)$$

giving **$|A(\omega_o)| = 31.1$.**

> **Transmitter output** Design a suitable LC section for matching the output of the final stage of a 10MHz, 25W transmitter to $R_L = 50\Omega$ coax. The DC supply V_+ is a 12V car battery.

This is not matching in the sense of presenting an impedance to the power transistor output equal to its own output impedance. This is a criterion valid only for a linear circuit, and the output transistor is operating essentially as a switch. Instead the required impedance R'_L is that which will draw the rated power from the circuit.

With the V_o swing at full output almost from 0V to $2\times V_+$ then the peak output power

$$P_{pk} \approx V_+^2/R'_L.$$

With $P_{mean} = P_{pk}/2$ then $R'_L = V_+^2/2P_m$

giving $R'_L = 2.9\Omega$.

With $R'_L < R_L$ then the L-C section is again reversed.

With the low-pass response giving the better suppression of harmonics, we use the series L and parallel C of Fig. 12.14(a), rather than (b). With

$$Q = (R_{high}/R_{low} - 1)^{1/2} \quad \text{as in [12.45]}$$

then $Q = 4.04$.

With $|X_{ser}| = QR_{low}$ as in [12.47]

and here $|X_{ser}| = \omega L$ then $L = R_{low}Q/\omega_o$

giving **L=185nH.**

With $|X_p| = R_{low}(Q + 1/Q)$ as in [12.45]

and here $|X_p| 1/\omega C$

then $C = 1/[\omega R_{low}(Q + 1/Q)]$

giving **C=1.29nF.**

Exercise 12.7 Parallel-resonator narrow-band band-pass filter

With the self-capacitance C_L of the inductor tending to give a secondary resonance of the opposite type then, unless this is specifically required, as in Fig. 12.12 to add a notch to the band-pass response, it is advisable to use the resonator in such a way that C_L becomes part of the main resonator C.

For the present exercise we take this to the extreme of considering the chosen 1mH inductor of Section 12.8 as a resonator, with the resonance that of C_L with L. The signal is then coupled in and out of the resonator as in Fig. 12.21 via the small coupling capacitances $C_c \approx 2$pF, C_c being capacitance between adjacent tracks on the Experimentor board.

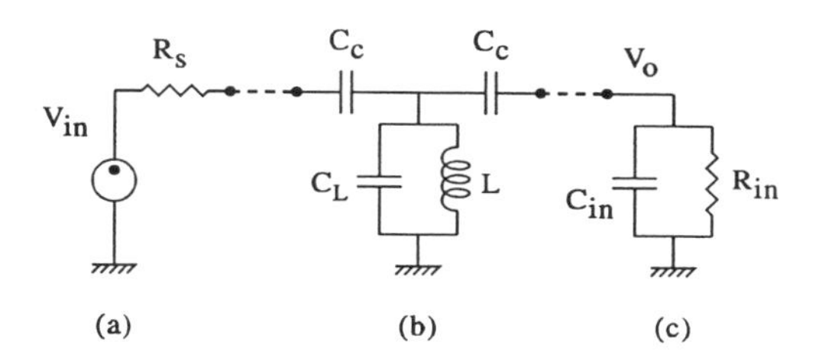

Fig. 12.21 Narrow-band band-pass filter used for experimental work.
(a) Signal generator.
(b) Resonator. L=1mH, C_L=25.3pF (Ex. 12.5).
(c) ×10 oscilloscope probe. R_{in}=10MΩ, C_{in}=10pF.

> **Calculation** Calculate the approximate change in the resonant frequency ω_o from that of the self-resonance due to the coupling via the C_c.

Approximately $2C_c$ is added to the self-capacitance C_L.

With $\omega_o = 1/\sqrt{LC}$ as in [12.2]

then here $\Delta\omega_L/\omega_L = -(2C_c/C_L)/2$

giving **$\Delta\omega_L/\omega_L = -10\%$**

> **Calculation. Component of Q due to scope-probe** Show that the reduction in the resonator Q-factor due to the scope-probe R_{in} is negligible.

With $|X_{in}| = 1/\omega_L C_{in}$ $|X_{IN}| = 1/\Omega_L C_{IN}$

then $|X_{in}| = 15.9\text{k}\Omega$.

With $Q_{in} = R_{in}/|X_{in}|$

then $Q_{in} = 628$.

With $Q_{in} >> 1$ then $R_{in}//C_{in}$ transforms to the series C'_{in} and R'_{in} where

$$R'in \approx R_{in}/Q_{in}^2 \quad \text{and} \quad C'_{in} \approx C_{in},$$

giving $R'_{in} = 25.3\Omega$.

With C_c also in series and then the effective $C' = 1.67$pF.

With $|X'| = 1/\omega_L C'$

then $|X_{cc}| = 95.5\text{k}\Omega$.

With the resulting $Q'_{in} = |X_{cc}|/R'_{in}$

then $Q'_{in}=3.77k$.

With $Q'_{in}>>1$, this transforms to the parallel

$$R''_{in}=R'_{in}\times Q'^{2}_{in}$$

giving $R''_{in}=360M\Omega$.

With $\omega_L L=6.28k\Omega$ and with $Q''_{in}=R''_{in}/\omega_L L$

then we have the negligible **$Q''_{in}=57.3k$.**

> **Problem. Component of Q due to signal source coupling** Show that $Q'_s=20k$ where Q'_s is that imposed by the source and C_c, and thus that Q'_s is also negligible.

> **Experiment. Resonance** Couple the signal generator and ×10 scope-probe to the inductor as in Fig. 12.21. From the observed resonance and bandwidth $\Delta\omega$ confirm the above calculations, remembering the approximate nature of the specified f_L and Q_L.
>
> Switch to a square wave input to view the oscillatory transient of Fig. 12.5. Count the cycles for the period $2Q/\omega_o$ shown and thus confirm the specified $Q_L>45$.

Direct connection of resonator to scope

The second C_c is now discarded and the probe connected directly to the resonator, giving a realistic representation of the resonator coupling of a low-impedance source to a high-impedance load. With C_{in} now in parallel with C_L the effect is to lower the resonant frequency f_o somewhat, but still with only one resonance. There are no unsuspected notches or secondary peaks.

> **Calculations** Calculate
>
> - the resulting resonant frequency ω_o,
> - the associated Q-factor,
> - the voltage gain at resonance $|A(\omega_o)|$.

With $\omega_o=1/\sqrt{LC}$ as in [12.2]

then $\omega_o=[L(C_L+C_{in})]^{-1/2}$

giving **$f_o=847kHz$.**

With $Q_{in}=R_{in}/\omega_o L$

then $Q_{in}=1.88k$.

With $Q=1/(1/Q_{in}+1/Q_L)$

then **$Q=43.9$.**

For the source, C_c and $(C_{in}+C_L)$ the Thevenin voltage

$$V_T\approx C_c/(C_{in}+C_L)\times V_{in}.$$

With $\alpha_{in}=V_T/V_{in}$

then $\alpha_{in}=56.6\times10^{-3}$.

Having established that the effect of R_s as coupled via C_c is negligible then

$$V_o\approx V_T\times Q$$

giving

$$|A(\omega_o)|=\alpha_{in}Q$$

and thus **$|A(\omega_o)|=2.49$.**

> **Problem** Confirm that apart from the resonance the response is essentially of the second-order high-pass type.

> **Experiment** Repeat the previous experiment with the probe connected directly to the resonator as above.
>
> Observe the resonance of somewhat reduced frequency, confirming the above calculations.
>
> Confirm the otherwise high-pass response giving $|A|\propto 1/f^2$ below the resonance and constant above.

Power match

The above gives a voltage gain $A>1$ despite the weak coupling. As explained in Section 1.4, the coupling giving the greatest output is that which presents the source with an impedance equal to R_s.

> **Calculate** the value of coupling capacitor for matching.

With the parallel impedance of the resonator at resonance

$$R_p\approx Q^2R_L$$

then **$R_p=216k\Omega$,**

compared with which $R_{in}=10M\Omega$ may be ignored.

It is simplest to calculate the value of C_c needed for R'_s, the transformed R_s, to equal R_p. With

$$Q_s=(1/\omega_o C_c)/R_s,$$

$$R'_s=Q_s^2R_s \text{ and } R'_s=R_p$$

then

$$C_c=1/\omega_o(R_sR_{Lp})^{1/2}$$

giving **$C_c=57.2pF$.**

> **Calculate** the voltage gain $|A(\omega_o)|$ at resonance when matched.

With the available power from the source

$$P_{av}=V_{in}^2/4R_s$$

For P_{av} equal to the resonator power V_o^2/R_p then

$$A=[R_p/(4R_s)]^{1/2}$$

giving $|A(\omega_o)|=32.8$.

Experiment. Power match Use a potential divider of two resistors to give $R_s \approx 50\Omega$ regardless of the signal generator, and increase C_c to the above value.

Observe the resonance to confirm the calculated $|A(\omega_o)|$ remembering that V_{in} is the open-circuit value.

The above calculation is a considerable approximation because the increased C_c nearly halves ω_o. Consider whether this calls for a higher or lower C_c and confirm experimentally. The original value appears to be close to optimum.

Confirm that the effective source voltage dips to be halved at resonance, as for the presented impedance equal to R_s.

From the decay time for the step response confirm that Q has been reduced. This will be halved by the power matching, and possibly reduced a little further due to the reduction in frequency.

Experiment. Link coupling Twist two turns of wire around the inductor to give the coupling of Fig. 12.18. Observe the resonance, and the true band-pass response of $|A|\propto f$ below the resonance and $\propto 1/f$ above it.

Experiment. Dual-resonator band-pass filter Set a further one of the chosen 1mH inductors about 3cm to the left of the first, so that the wires can be bent to bring the two side by side, but keep them apart to start with.

Connect the scope-probe to the RH inductor as before, and by loosely coupling the signal generator output to this find the resonance and note the value of ω_o.

Connect a 30pF adjustable capacitor C_a across the LH inductor and couple the signal generator to this via C_c. Find the resonance by loosely coupling the probe to the circuit. Adjust C_a to give the same ω_o.

Reconnect the probe to the RH circuit and observe the resonance, now with $|A|\propto 1/\omega^2$ above the resonance and $\propto \omega^2$ below.

Carefully bring the two inductors together to see the way that the resonance first widens to a fairly flat top, and then splits into two.

Exercise 12.8 Experiment: LC oscillator

Experiment Connect the circuit of Fig. 12.20(a) with $C_1=C_2=100$pF and the usual 1mH inductor. Observe the waveforms at the gate input and output.

Disconnect the network from the gate and drive with the signal generator as in (b), with the generator adapted as before to ensure $R_s=50\Omega$. Thus confirm the $\approx 90^o$ phase shift at resonance. Increase R_s to 10kΩ and confirm the increase to nearly 180°.

Change C_1 to 470pF and remove C_2, relying on the probe $C_{in}=10$pF. Apply a 5V p-p square wave as normally output by the gate. Thus confirm the near phase inversion at resonance, with the output a sine wave of ≈3V p-p. This is of adequate amplitude to drive the gate input.

Connect to the gate and check the waveforms. The kink in the otherwise good sine wave at the gate input is due to the 2pF track-to-track capacitance of the Experimentor board, and would be much reduced on a normal printed circuit board.

13

Active RC filters

Summary

Today the inductor is only used where necessary, for a number of reasons. Firstly, compared with the capacitor and resistor, it is expensive to manufacture, and particularly difficult to realise in microcircuit form. Next it tends to pick up interference from stray alternating magnetic fields. Finally its parasitic elements are more evident than for R and C.

Thus for values of $f < 5MHz$ for which the op-amp is effective, the inductor-based filters of the last chapter are usually replaced by 'active' op-amp-based filters, using only R and C components.

First considered are the principal low- and high-pass types of active filter. Then come the corresponding narrow-band band-pass and band-stop designs.

More recent developments are the switched-capacitor filter, the OTA-C filter and the MOSFET-C filter. These are suited to whole on-chip designs, although the switched-capacitor type is also available at the level of two units per chip, for use in an otherwise off-chip design.

The last section gives example calculations, design examples, and experimental work for the topics covered.

13.1 Low- and high-pass Butterworth filters

The Butterworth response for any order of filter is that where the transition from pass-band to stop-band is as abrupt as possible without a resonant peak at the transition. Fig. 13.1(a) shows the Butterworth responses for the first few orders of low- and high-pass.

Butterworth second-order low-pass response

With the second-order low-pass voltage gain A(s) as in [12.42] then

$$A(\omega) = \frac{\omega_o^2}{\omega_o^2 + j\omega\omega_o / Q - \omega^2}$$

giving

$$|A(\omega)| = \frac{\omega_o^2}{\left[\left(\omega_o^2 - \omega^2\right)^2 + \left(\omega\omega_o / Q\right)^2\right]^{1/2}} \quad (13.1)$$

and thus

$$|A(\omega)| = \frac{\omega_o^2}{\left[\omega_o^4 + \omega^2(\omega_o^2 / Q^2 - 2\omega_o^2) + \omega^4\right]^{1/2}} \quad (13.2)$$

For $Q = 1/\sqrt{2}$, then the terms in ω^2 cancel to give

$$|A(\omega)| = \frac{\omega_o^2}{\left(\omega_o^4 + \omega^4\right)^{1/2}} \quad (13.3)$$

the most abrupt possible transition without a resonance.

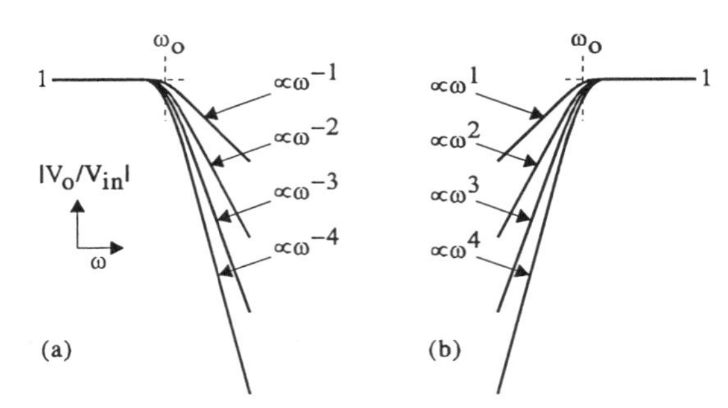

Fig. 13.1 Butterworth filter gain-frequency response for first four orders.
(a) Low-pass.
(b) High-pass.

Pole-zero diagram

For the pole-zero diagram of Fig. 12.4, the value of Q for which the poles leave the real axis to begin to travel around the circle is 1/2. This is the value for which the impulse response and therefore the step response begins to be oscillatory, giving 'overshoot'. With $Q=1/\sqrt{2}$ for the Butterworth response slightly higher, while there is no overshoot in the Butterworth amplitude-frequency response, there is therefore a little in the step response. From [12.13] and [12.14] the conjugate pole pair

$$s_p=\sigma_p+j\omega_p \quad \text{and} \quad s_p^*=\sigma_p-j\omega_p$$

for the second-order low-pass with Q>1/2 are given by

$$\sigma_p=-\omega_o/2Q \qquad \ldots(a)$$

$$A(\omega)=\frac{\omega_o^2}{\omega_o^2+j\omega\omega_o/Q-\omega^2} \qquad \ldots(b) \qquad (13.4)$$

With $Q=1/\sqrt{2}$ for the Butterworth response, then

$$\sigma_p=-\omega_o/\sqrt{2} \text{ and } \omega_p=j\omega_o/\sqrt{2}$$

giving the pole positions shown in Fig. 13.2(a). The amplitude-frequency response, derived as usual from (a), then becomes as in (b).

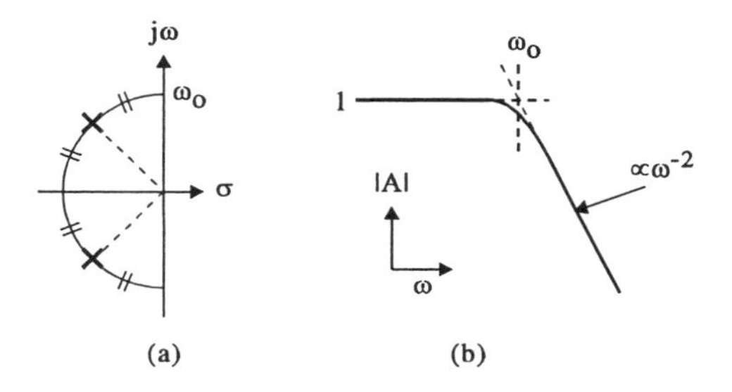

Fig. 13.2 Second-order Butterworth low-pass.
(a) Pole-zero diagram.
(b) Gain-frequency response.

Sallen-Key

This popular implementation of the active low-pass second-order filter is shown in Fig. 13.3. The passive alternative is the direct cascading of two identical first-order sections, with or without an interposing voltage-follower. Here, without the follower the Q-factor is 1/3 and with the follower 1/2, both short of the required $Q=1/\sqrt{2}$. With the positive feedback for the Sallen-Key circuit the Q can, in principle, be made as high as is wished.

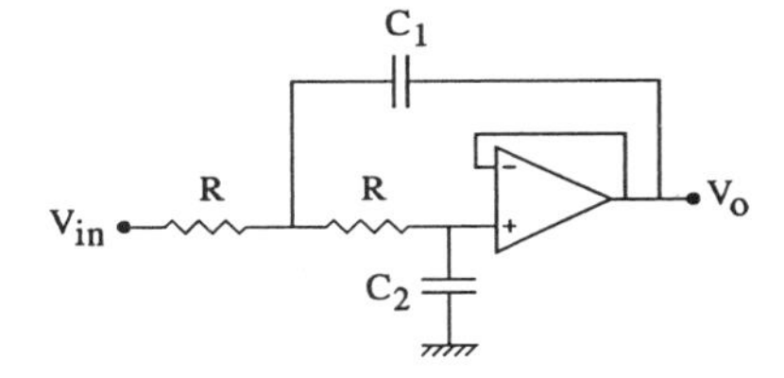

Fig. 13.3 Sallen-Key configuration for second-order low-pass.

Design equations

As for the single-section low-pass, the cut-off (Van Valkenburg, 1982)

$$\omega_o=1/CR \qquad (13.5)$$

With $\quad C_1=\gamma C$ and $C_2=C/\gamma$

then $\quad Q=\gamma/2 \qquad (13.6)$

For $\quad Q=1/\sqrt{2}$ then $\gamma=\sqrt{2}$ giving

$$C=1/\omega_oR \ldots(a) \quad C_1=C\times\sqrt{2} \ldots(b) \quad C_2=C/\sqrt{2} \ldots(c) \quad (13.7)$$

Higher order Butterworth filters

With $|A(\omega)|$ as in [13.3] for the second-order Butterworth low-pass, then for the nth order, with the now multiple terms in the denominator cancelling,

$$|A(\omega)|=\frac{\omega_o^n}{\left(\omega_o^{2n}+\omega^{2n}\right)^{1/2}} \qquad (13.8)$$

while for the high-pass

$$|A(\omega)|=\frac{\omega^n}{\left(\omega_o^{2n}+\omega^{2n}\right)^{1/2}} \qquad (13.9)$$

This gives the sequence of frequency responses in Fig. 13.1 and the pole-zero diagrams as in Fig. 13.4.

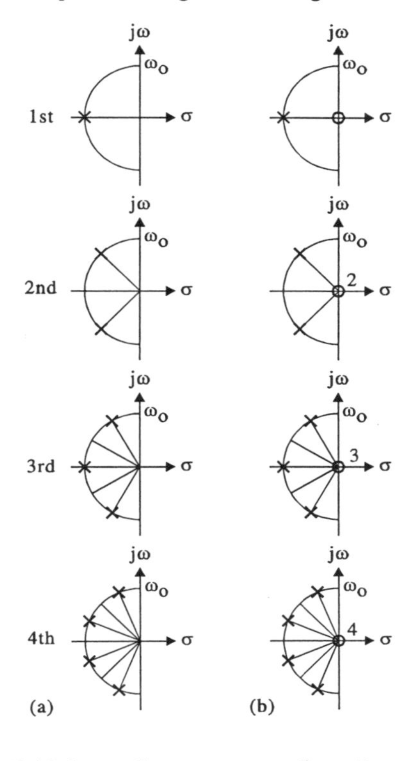

Fig. 13.4 Pole-zero diagrams corresponding to Butterworth frequency responses in Fig. 13.5(a) Low-pass. (b) High-pass.

Sallen-Key

Fig. 13.5(a) shows the Sallen Key implementations for the first three orders of low-pass filter. The first-order is the familiar single-section RC filter of Chapter 11, while the second-order is as in Fig. 13.3 with the values of [13.7].

For the even-order filters, Fig. 13.4(a) shows the poles to be all in conjugate pairs. Each pair can be implemented as a single Sallen-Key circuit with the appropriate value of Q.

For the odd-order filters the real-axis pole can be provided by a single passive RC section, as for the third-order circuit in Fig. 13.5(a).

Here, however, if the above simple design equations are to remain then the voltage-follower shown needs to follow the single passive section, to avoid it being loaded by the following stage.

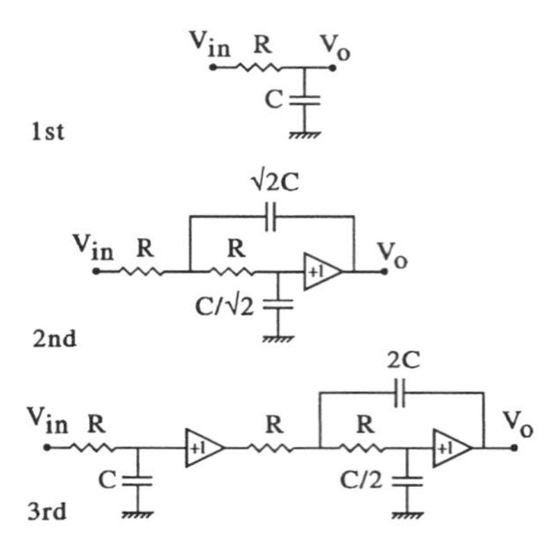

Fig. 13.5(a) Circuit configurations for active low-pass Butterworth filters of the first three orders.

High-pass

For the corresponding high-pass responses the C and R are simply transposed with the appropriate scaling, as in Fig. 13.5(b). The resulting pole-zero diagrams in Fig. 13.4(b) are then those for the low-pass in (a), but with the appropriate number of zeros added at the origin, one zero for each pole.

Holbrook third-order

If the voltage-follower is omitted then the Butterworth response can still be obtained, but the design equations must be modified and the resulting values of C become highly non-integral.

Holbrook (private communication) has devised the alternative third-order low-pass filter of Fig. 13.6(a). Here by increasing the gain of the second voltage-follower to +4 the voltage-follower is no longer required while the C values remain integral.

It might be countered that the values of C for the Butterworth second-order filter are not integrally related either. However, for a filter of as low an order as this, the values do not have to be very precise, so the nearest preferred value will do.

Exercise 13.1 gives experimental work and example calculations for the above Butterworth filters

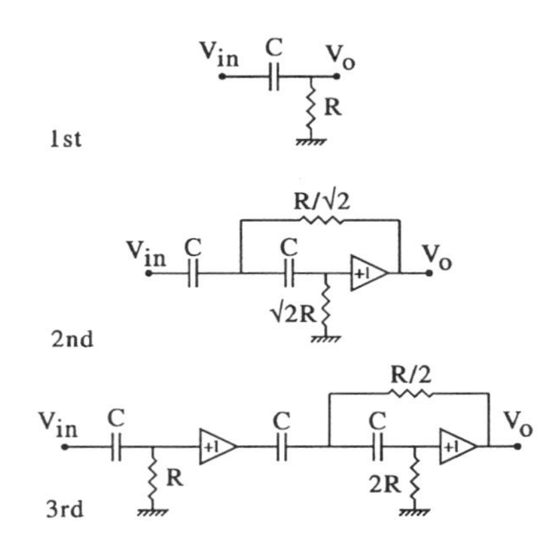

Fig. 13.5(b) Circuit configurations for active high-pass Butterworth filters of the first three orders.

Tchebyschev and other filter classes

There are a number of other widely used classes of filter. These offer a more abrupt cut-off than the Butterworth but at the expense of a calculated degree of ripple in the pass-band (Van Valkenburg, 1982).

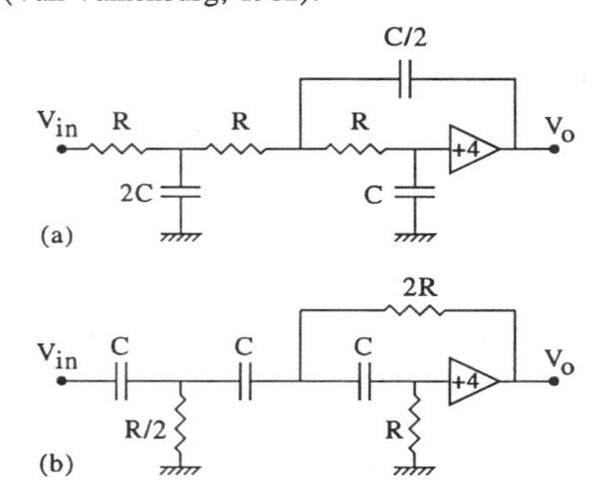

Fig. 13.6 Holbrook third-order filters.
(a) Low-pass.
(b) High-pass.

13.2 Linear-phase filters

While the Butterworth response gives the most abrupt transition in the amplitude-frequency response without a resonant peak, for the step response the variation of phase with frequency is also important.

As noted in Chapter 11, the sag imposed by a high-pass filter upon an input square wave of frequency only moderately above the filter cut-off is mainly attributable to the phase advance of the fundamental relative to the harmonics.

A frequency-dependent phase shift does not necessarily distort the waveform. Indeed a phase shift that is independent of frequency will cause distortion.

With the phase-shift ϕ, the signal frequency f_s and the harmonic frequency nf_s, then

$$T_d=-\phi/2\pi nf_s$$

where T_d is the time delay for the harmonic. Thus for all ϕ the same then T_d is different for each value of n, causing each of the Fourier components to be shifted differently.

With T_d as above then for this to be the same for each component requires $\phi \propto f$ where $f=nf_s$ is the frequency of the component. This is the response of the 'linear-phase filter'. Giving, for a sine wave signal of frequency f, the frequency-independent

$$T_d=\phi/\omega \tag{13.10}$$

The closest approximation to the linear-phase response for a given number of poles requires some change from the Butterworth positions. With the third-order low-pass pole-zero diagram as in Fig. 13.7, and with the diagram used as usual to derive the amplitude- and phase-frequency responses, then it is possible to see that a small shift of the real pole towards the origin gives a more linear phase-frequency variation.

For this and higher orders, computer simulation can find the optimum placements, with the implementation then probably the Sallen-Key.

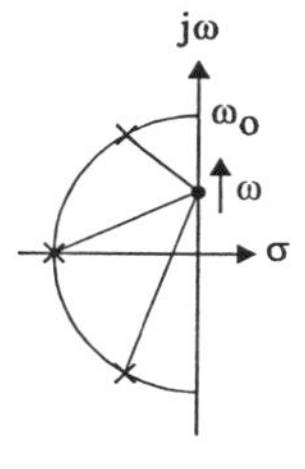

Fig. 13.7 Pole-zero diagram for third-order low-pass Butterworth.

13.3 Narrow-band band-pass filter

Wein network

The Wein network of Fig. 13.8(a) is the single-section low-pass filter followed immediately by the corresponding high-pass. With the two cut-off frequencies comparable this gives a band-pass response of very low Q. Analysis of the network gives the gain $A_w(\omega)=V_o/V_{in}$ as

$$A_w(\omega)=\frac{j\gamma\omega\omega_o}{\omega_o^2+j\omega\omega_o/Q_w-\omega^2} \tag{13.11}$$

where $$\omega_o=1/CR \tag{13.12}$$

Also the Wein Q-factor $$Q_w=1/(\gamma+2/\gamma) \tag{13.13}$$

for $$R_1=R/\gamma\ \ldots(a) \qquad R_2=\gamma R\ \ldots(b) \tag{13.14}$$

Thus $$A_w(\omega_o)=Q_w\gamma \tag{13.15}$$

With A(s) as in [12.10] for the LC band-pass filter of Fig. 12.1, then [13.11] is essentially of the same form but now with the gain at resonance $A(\omega_o)<1$ and with $Q<1$.

With $\gamma=1$ for $R_1=R_2=R$

then $Q_w=1/3$ and $A_w=1/3$,

the optimum $Q=1/(2\sqrt{2})$ being only marginally higher.

The Wein network exists in two other forms, consider what they may be.

Positive feedback

The Q-factor of any resonator-type network can be increased by positive feedback, with Fig. 13.8(b) showing this applied to the Wein network.

Wein oscillator

With A_w at resonance equal to 1/3, and for the gain of the non-inverting feedback amplifier in the loop greater than three, then the circuit will oscillate at ω_o. This is often the way in which the laboratory sine wave signal generator is implemented, with a suitable arrangement to stabilise the amplitude at a value below that at which saturation of the amplifier would otherwise do so, causing distortion.

For the arrangement as shown, and with R_a increased slightly from the value of $R_b/2$ needed for oscillation, the circuit becomes an amplifier of much increased gain and proportionally enhanced Q.

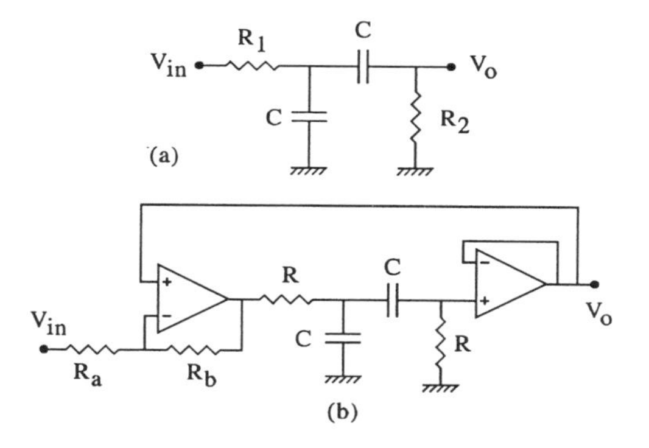

Fig. 13.8
(a) Wein network.
(b) Q-enhancement of Wein network by positive feedback.

Sensitivity

A severe disadvantage of the arrangement is that the closer one gets to the threshold of oscillation the finer the adjustment of R_a becomes and, more seriously, the greater the effect of say a small thermal change in the value of R_a.

Oscillation hysteresis

This is an annoying effect whereby once the system is oscillating, then to stop the oscillation it is necessary to reduce the amplifier gain to a value giving only a rather poor Q-enhancement once the oscillation has stopped.

Exercise 13.2 gives experimental demonstration of the Wein oscillator and narrow-band band-pass filter, with associated example calculations.

Delyiannis-Friend circuit

This much preferred way of enhancing the Q-factor of the Wein network is shown in Fig. 13.9(a). With the feedback now negative, it is surprising that enhancement rather than reduction of Q_w is obtained. The major advantage is that the enhancement is now without the large increase in parameter sensitivity for positive feedback.

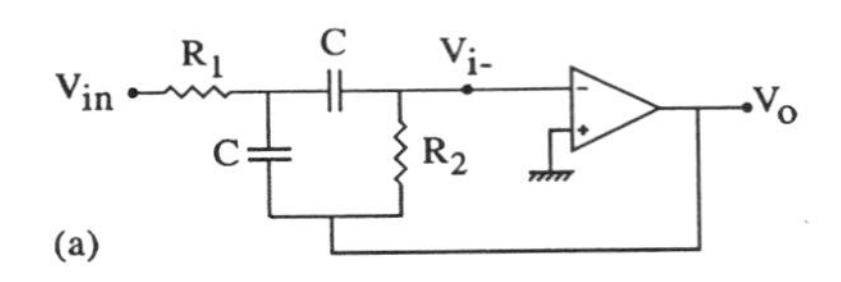

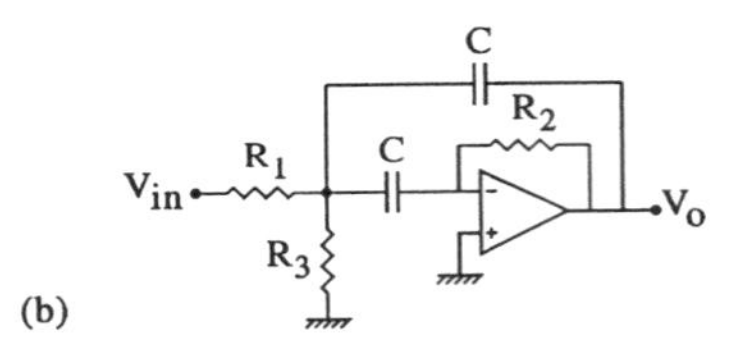

Fig. 13.9 Delyiannis-Friend band-pass filter.
(a) Basic form.
(b) With gain reduction.

Fig. 13.10(a) shows the complex plane plot of A_w. With the form of [13.11] for $A(\omega)$ the same as that for the LC resonator, then the diagram is as in Fig. 12.3 for the resonator, but with the much lower Q.

Even more paradoxically, successful operation of the circuit requires Q_w to be further reduced from the already low optimum $Q_w \approx 1/3$.

With $\quad Q_w = 1/(\gamma+2/\gamma) \quad$ as in [13.13]

this requires $\gamma >> 1$.

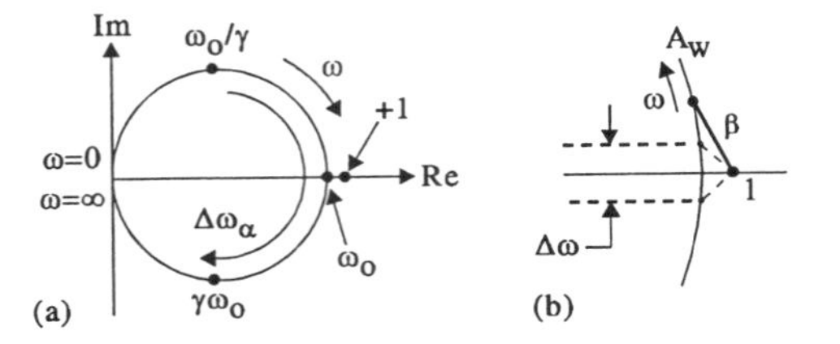

Fig. 13.10 Complex plane plot of Wein transfer function $A_w = V_o/V_{in}$ in Fig. 13.8(a).
(a) Normal size.
(b) Expansion of (a) close to resonance.

For say $\gamma = 10$ then we have a two-decade region from $\omega_o/10$ to $10\omega_o$ for which $|A(\omega)| \approx 1$, with a high-pass cut-off at the lower frequency limit, and a low-pass at the upper frequency limit, all very unpromising as a starting point for a high-Q band-pass response.

Q-enhancement

With $\quad A_w(\omega_o) = Q_w\gamma \quad$ as in [13.15]

and $\quad Q_w = 1/(\gamma+2/\gamma) \quad$ as in [13.13]

then $\quad A_w(\omega_o) = \gamma/(\gamma+2/\gamma)$,

giving $A_w(\omega_o)$ only a little less than unity for $\gamma >> 1$, as shown. Next with

$$V_{i-} = \alpha V_{in} + \beta V_o$$

where α and β are the usual feed-forward and feed-back fractions for the inverting feedback amplifier, and with

$$V_{i-} = V_{in} \quad \text{for} \quad V_o = V_{in} \quad \text{then} \quad \alpha+\beta = 1.$$

With $\quad \alpha = A_w \quad$ then $\quad \beta = 1 - A_w$

as in the expanded section of Fig. 13.10(a) shown in (b).

With the feedback amplifier gain

$$A_f = -\alpha/\beta \quad \text{and} \quad \alpha \approx 1$$

then the sharp dip in β gives the corresponding sharp peak in $|A_f|$, and thus the high-Q resonance. See Exercise 13.3 for completion of this argument.

Expression for $A_f(\omega)$

With $\quad \alpha(\omega) = A_w(\omega) \quad$ and $\quad A_w(\omega)$ as in [13.11]

then
$$\alpha(\omega) = \frac{j\gamma\omega\omega_o}{\omega_o^2 + j\omega\omega_o / Q_w - \omega^2} \qquad (13.16)$$

With $\quad \beta = 1 - \alpha$

and $\quad Q_w = 1/(\gamma+2/\gamma) \quad$ as in [13.13]

then
$$\beta(\omega) = \frac{\omega_o^2 + j\omega\omega_o(2/\gamma) - \omega^2}{\omega_o^2 + j\omega\omega_o / Q_w - \omega^2} \qquad (13.17)$$

With $\quad A_f = -\alpha/\beta$

then
$$A_f(\omega) = \frac{-j\gamma\omega\omega_o}{\omega_o^2 + j\omega\omega_o(2/\gamma) - \omega^2} \qquad (13.18)$$

Comparing with Q_w in [13.16] for the Wein network alone, for the active filter

$$Q = \gamma/2 \qquad (13.19)$$

With $\quad Q_w = 1/(\gamma+2/\gamma) \quad$ as in [13.13]

giving $\quad Q_w \approx 1/\gamma \quad$ for $\quad \gamma >> 1$

then $\quad Q \approx 1/2Q_w$,

confirming the initially surprising result that a high Q requires a low Q_w.

Finally with $Q=\gamma/2$ then the sensitivity of Q to the circuit values has no enhancement.

Gain

With A_f as in [13.18] then at resonance $A_f=-\gamma^2/2$. With $Q=\gamma/2$ then

$$A_f=-2Q^2. \quad (13.20)$$

This is usually much higher than is required. For example, the modest $Q=30$ gives $A \approx 2000$. Thus the circuit is usually modified as in Fig. 13.9(b). Here R_1 and R_3 form a potentiometer which reduces the gain to the required

$$A=A_fR_3/(R_1+R_3).$$

With A_f as in [13.20] then

$$R_3/(R_1+R_3)=-A/2Q^2.$$

Also $R_1//R_3$ is made equal to the previous value of $R_1=R/\gamma$ in [13.14]. With $Q=\gamma/2$ as in [13.19] then

$$R_1//R_3=R/2Q$$

giving
$$R_1=QR/|A| \quad (13.21)$$

$$R_3=QR/(2Q^2-|A|) \quad (13.22)$$

With or without R_3, the circuit is usually drawn as in (b) rather than (a), (a) being better for explaining the operation.

Design equations

With R_1 and R_3 as above, $\omega_o=1/CR$ as in [13.12], $R_2=\gamma R$ as in [13.14], and $Q=\gamma/2$ as in [13.19] then the design equations can be grouped as

$$R=1/(2\pi f_oC)\ldots(a) \quad R_1=QR/|A| \quad \ldots(b)$$
$$R_2=2QR \quad \ldots(c) \quad R_3=QR/(2Q^2-|A|)\ldots(d) \quad (13.23)$$

Exercise 13.3 gives a design example with experimental verification.

13.4 Notch filter

Fig. 13.11(a) shows the modification needed to convert the basic Delyiannis-Friend band-pass circuit of Fig. 13.9(a) into a notch type band-stop filter. For the op-amp ideal, the output component corresponding to the upper part of the circuit is equal to

$$V_{in}(-\alpha/\beta)$$

while that for the lower part is

$$V_{in}\alpha_{ab}/\beta$$

where
$$\alpha_{ab}=R_b/(R_a+R_b)$$

Thus overall
$$A(\omega)=\frac{\alpha_{ab}-\alpha}{\beta} \quad (13.24)$$

With α_{ab} set equal to α at resonance then $A(\omega_o)$ becomes zero to give the notch.

Fig. 13.11(c) shows α and β as in Fig. 13.10(a) for the band-pass filter, but with the above frequency-independent α_{ab} added. With the lines shown representing $(\alpha-\alpha_{ab})$ and β then it is seen how $A\approx1$ off resonance.

Also with $(\alpha-\alpha_{ab})=0$ at resonance giving the notch, then the notch bandwidth $\Delta\omega$ becomes the same as that for the equivalent band-pass version. With $\alpha=\gamma Q_w$ at resonance as in [13.16],

$$Q_w=1/(\gamma+2/\gamma)$$

as in [13.13], $Q=\gamma/2$ as in [13.19], and $\alpha_{\alpha\beta}=\alpha$ at resonance then

$$\alpha_{ab}=2Q^2/(1+2Q^2) \quad (13.25)$$

requiring

$$R_a=R_b/(2Q^2) \quad (13.26)$$

Design equations

With R_3 in Fig. 13.9(b) no longer required then $R_1=R/\gamma$ and $R_2=\gamma R$ as in [13.14]. With $\gamma=2Q$ as in [13.19] then the design equations become

$$R=1/(2\pi f_oC) \ldots(a) \quad R_1=R/(2Q) \ldots(b)$$
$$R_2=2QR \ldots(c) \quad R_a=R_b/2Q^2 \ldots(d) \quad (13.27)$$

Exercise 13.4 gives a design example with experimental verification.

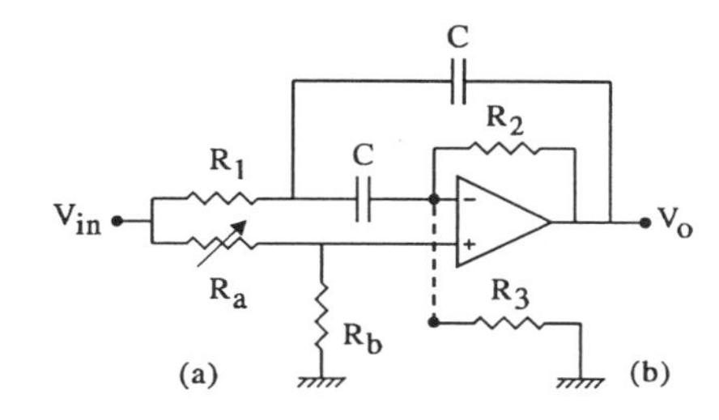

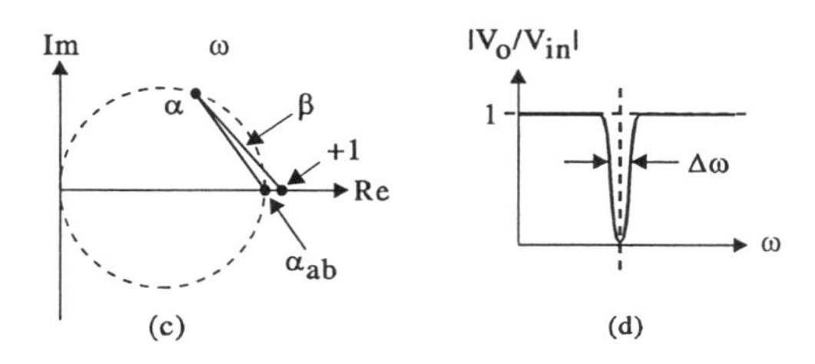

Fig. 13.11 Notch filter based on Delyiannis-Friend circuit.
(a) Basic circuit.
(b) Additional component needed for zero shift in Fig. 13.12.
(c) Complex plane plot for α as in Fig. 13.10(a) showing variation of β and $\alpha-\alpha_{ab}$.
(d) Frequency response.

13.5 Notch-enhanced low-pass

By adding the resistor R_3 to the Delyiannis-Friend notch filter circuit as in Fig. 13.11(b), the frequency response can be converted to that in Fig. 13.12(b). This is essentially the second-order low-pass but with the added notch giving a considerably more abrupt cut-off, still without overshoot in the frequency response.

This is at the expense of a non-zero gain for $\omega >> \omega_z$, but this can be made good by adding a further simple RC section.

Without R_3, as for the original notch filter, then the pole-zero diagram becomes as for the LC notch filter in Fig. 12.9(b). With R_3 added then the two zeros are moved out along the $j\omega$ axis from their previous positions at $\pm\omega_o$, to become the $\pm\omega_z$ in Fig. 13.12(c). Also, to give the otherwise second-order Butterworth response, the poles are backed off from their previous positions close to the ends of the half-circle, to become as for the second-order pole-zero diagram in Fig. 13.4(a).

With A as in [12.33] for the LC notch filter then to shift the zeros as above we simply replace the ω_o in the numerator by ω_z to give

$$A = \frac{\omega_z^2 + s^2}{s^2 + s\omega_o / Q + \omega_o^2} \qquad (13.28)$$

With $Q = 1/\sqrt{2}$ as for the second-order Butterworth response then the frequency response of |num(A)| and |den(A)| become as in Fig. 13.12(a). These combine in the usual way to give the response of |A| in (b). Were Q higher then a resonant peak would be added at ω_o, usually not required.

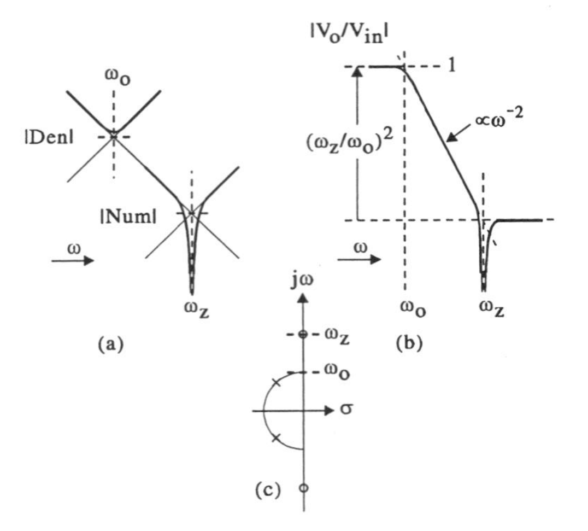

Fig. 13.12 Frequency response of notch circuit of Fig. 13.11(a) with zero shift as in (b).
(a) Frequency response of numerator and denominator of transfer function A.
(b) Frequency response of A.
(c) Pole-zero diagram.

Design equations

Van Valkenburg (1982) gives these as

$$R = 1/(2\pi f_o C) \quad \ldots(a)$$
$$R_1 = R/(2Q) \quad \ldots(b)$$
$$R_2 = 2QR \quad \ldots(c) \qquad (13.29)$$

as before, but now

$$R_3 = 2QR/[(\omega_z/\omega_o)^2 - 1] \qquad (13.30)$$

and

$$R_a = (R_b/2Q^2)(\omega_z/\omega_o)^2 \qquad (13.31)$$

Exercise 13.5 gives a design example with experimental verification.

13.6 Switched-capacitor filters

A considerable limitation of the active filters so far described is the complexity involved when they are to be tuned. For example, for a mere third-order Butterworth to be tuned over say four decades will require three ganged variable resistors and four sets of three switched capacitors. Here the switched-capacitor (SC) filter gives much simplification.

Sometimes too a filter needs to be adaptive, for example where the band-pass filter ω_o needs to 'track' an input signal of varying frequency. The SC filter allows this kind of control.

State-variable filter

There is a wide variety of active filter types based on the feedback integrator of Fig. 2.16. Any of these can be adapted to the SC method and we discuss just one, the state-variable filter.

For the single-section RC low-pass filter the output $V_o(t)$ depends not only on $V_{in}(t)$ but also on the capacitor voltage $V_c(t=0)$. With V_c defining the initial state of the circuit then V_c is the 'state variable'.

Similarly for the equivalent RL filter $V_o(t)$ is dependent on the inductor current $I_L(t=0)$. Both V_c and I_L represent stored energy and both are, in effect, the output of integrators. For the capacitor, Q is the integral of I and $Q = CV_c$ giving

$$V_c = \frac{1}{C}\int I\,dt \qquad (13.32)$$

while for the inductor

$$I = \frac{1}{L}\int V_L\,dt \qquad (13.33)$$

Thus for the second-order LC low-pass of Fig. 13.13, V_c and I are the state variables. Also

$$V_L = V_{in} - (IR + V_c) \qquad (13.34)$$

so the active implementation must be of these three relations.

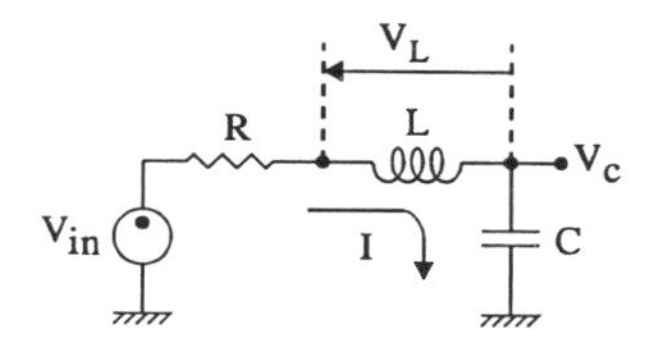

Fig. 13.13 Resonator-based low-pass filter.

Active implementation

Fig. 3.14(a) shows the direct active implementation, where the equation number for each block is given. It is then convenient to make all of the variables voltages as in (b).

Low-pass

With

$$V_2(s)=V_1(s)/(sT_o)$$

and

$$V_3(s)=V_2(s)/(sT_o)$$

in (b) then

$$\frac{V_3}{V_{in}}=\frac{\omega_o^2}{s^2+ks\omega_o+\omega_o^2} \qquad (13.35)$$

where

$$\omega_o=1/T_o \qquad (13.36)$$

With

$$A(s)=\frac{\omega_o^2}{s^2+s\omega_o/Q+\omega_o^2}$$

as in [12.42] for the second-order low-pass then [13.35] is of the same form, with $Q=1/k$, ω_o the cut-off frequency, and with a peak at resonance for $Q>1$.

Band-pass

Since $V_3=(\omega_o/s)V_2$, then V_2/V_{in} is as in [13.35] but with the $\omega_o s$ replacing ω_o^2 in the numerator. This is the expression for the resonator band-pass.

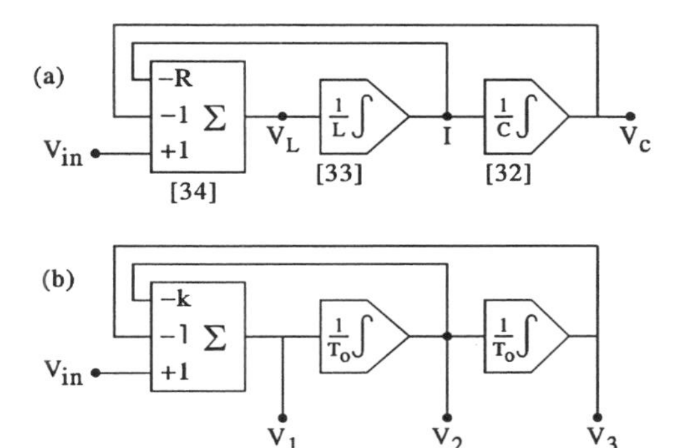

Fig. 13.14 Active simulation of filter in Fig. 13.13.
(a) Direct implementation (equation numbers as in text).
(b) Adaptation for all variables voltages.

High-pass

Similarly with $V_2=(\omega_o/s)V_1$ then V_1/V_{in} is given by [13.35] with the numerator now s^2. This is the normal expression for the second-order high-pass. Thus the same unit provides any or all of the classic second-order low- band- and high-pass characteristics as required.

Switched-capacitor integrator

Adaptation of the above circuit to SC operation is merely a matter of using SC integrators in place of the above conventional op-amp-based integrators. Fig. 13.15 shows the SC-based integrator. First (a) shows the conventional integrator which gives

$$V_o=-(1/T_o)\int V_{in}dt \quad \text{with} \quad T_o=CR,$$

and thus a smooth descending ramp for constant positive V_{in}. Then (b) shows the simplest SC equivalent. For each cycle of the switch, C_1 is charged to V_{in} and then discharged into C_2. This gives the downward step

$$\Delta V_o=-V_{in}C_1/C_2 \qquad (13.37)$$

in V_o. Thus V_o becomes the descending staircase of (c).

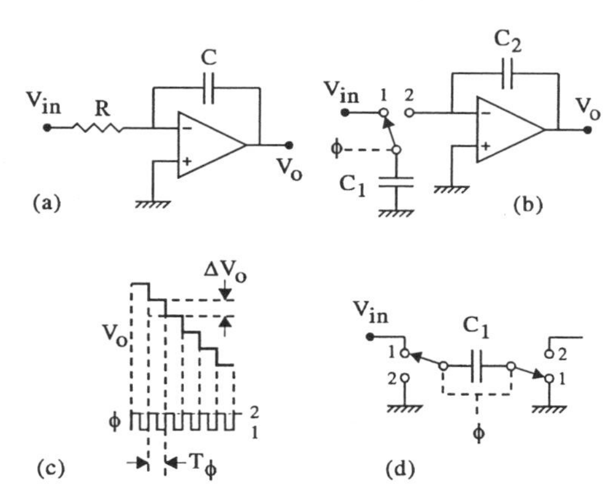

Fig. 13.15 Development of switched-capacitor integrator.
(a) Conventional integrator.
(b) Basic SC integrator.
(c) Waveforms for (b) with V_{in} constant.
(d) Adaptation of (b) to eliminate sign inversion.

The staircase is only an approximation to the smooth ramp for a true integrator. However, if the switch clock period T_ϕ is made small enough relative to the signal period this will be adequate.

Both Fig. 13.15(a) and (b) include sign inversion, which is not required for the simulation of Fig. 13.14. Fig. 13.15(d) shows how the inversion can be cancelled, using a slightly more complex switching system.

With ΔV_o as in [13.37] then for the non-inverting SC integrator

$$dV_o/dt=V_{in}(C_1/C_2)/T_\phi$$

giving the integrator time constant

$$T_o = T_\phi C_2/C_1 \qquad (13.38)$$

where T_ϕ is the clock period.

Capacitance ratio

With the cut-off $f_o = 1/2\pi T_o$

as in [13.36], $T_\phi = 1/f_\phi$

and T_o as above then

$$f_\phi = 2\pi f_o(C_2/C_1).$$

Here $f_\phi = 100f_o$ is normally considered adequate, giving $C_2 \approx 16C_1$. This ratio can be implemented in microchip technology, with C_1 comfortably in excess of stray capacitances and C_2 not too large to be implementable.

Microcircuit SC filter

Fig. 13.16 shows how the SC filter is implemented using a suitable chip such as the MF10. This includes the two SC integrators and the op-amp for the current-follower. All that is then needed are the four external resistors.

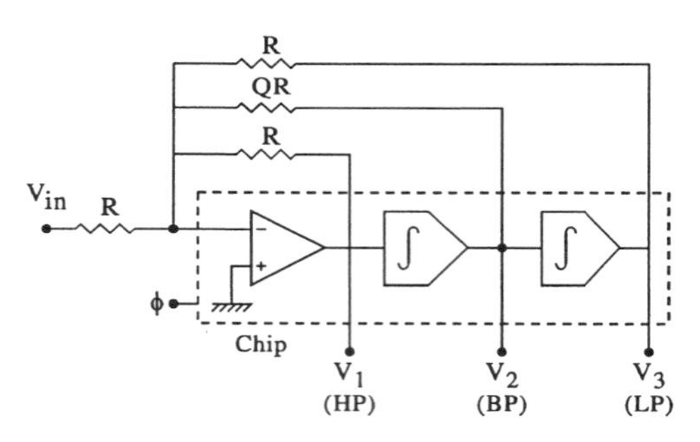

Fig. 13.16 Second-order switched-capacitor filter based on microcircuit chip.
HP: High-pass, BP: Band-pass, LP: Low-pass.

Tuning

The main point of the SC adaptation is that, in order to tune the filter, all that is needed is to vary the switch clock frequency $f_\phi = 1/T_\phi$. Consider, for example, a ninth-order Butterworth low-pass. This comprises four of the present second-order sections giving the four required complex pole pairs, and a further section giving the single real-axis pole. The last is provided by one of the present sections but with the second integrator not used.

Voltage-tuneable filter

With $f_\phi >> f_o$ for the SC filter, much of the staircase ripple can be removed by a following low-pass of cut-off f_c between f_o and f_ϕ. For varying f_ϕ then f_c must vary in proportion, with the low-pass non-SC.

Fig. 13.17 shows an arrangement the author has used. With k_m the constant for the multiplier in the voltage-controlled low-pass filter then

$$V_2 = (V_1 - V_2)k_m V_c T_o/s \qquad (13.39)$$

giving

$$V_2(s)/V_1(s) = \omega_c/(s + \omega_c)$$

as for the low-pass, with

$$\omega_c = k_m V_r/T_o$$

and thus $f_c \propto V_r$ as required. With the ratemeter output $V_r \propto f_\phi$ then $f_c \propto f_\phi$ as also required.

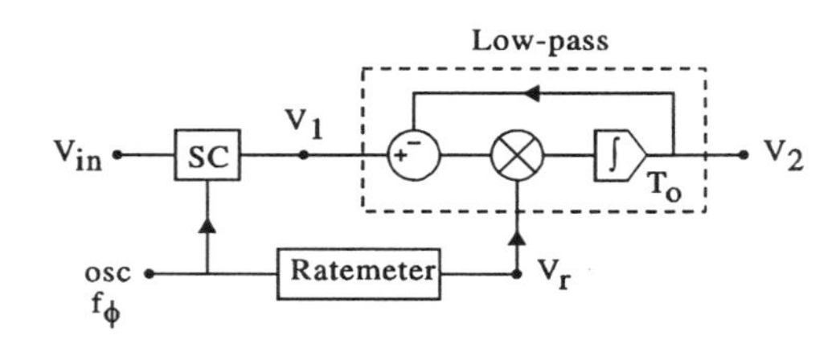

Fig. 13.17 Voltage-tuneable low-pass filter used to remove stepping from output of switched-capacitor filter.

13.7 OTA-C filters

This is another class of filter for which the cut-off frequency is controllable by an external signal. Fig. 13.18 shows the function of the 'operational transconductance amplifier' (OTA), as covered in Chapter 9. With

$$i_o = g_m(v_{i+} - v_{i-})$$

in (a) and with the output impedance ideally infinite in (c)

then $$v_o = \frac{g_m}{C}\int (v_{i+} - v_{i-}).dt$$

With g_m controllable by the bias current I_b in (a) then we have an ideal current-controllable integrator, the basis of the OTA-C filter.

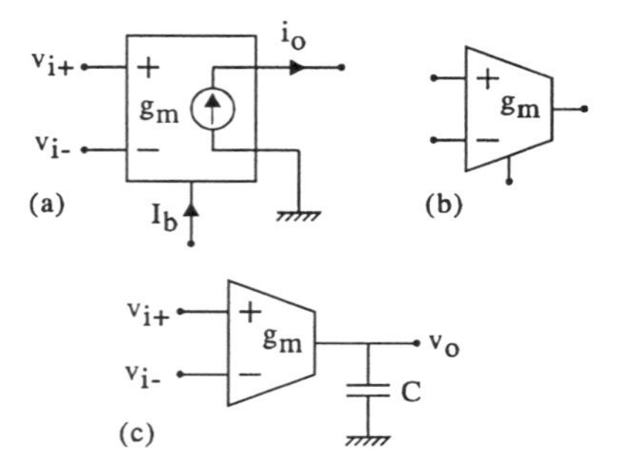

Fig. 13.18 Operational transductance amplifier (OTA).
(a) Ideal OTA with g_m controlled by I_b.
(b) Conventional symbol. (c) OTA-C integrator.

Order of control

With the OTA able to be implemented in either the bipolar or the CMOS form, for both the control of g_m is essentially that of the first-stage LTP tail current. For the bipolar $g_m \propto I_c$ as in [6.6] while for the CMOS $g_m \propto I_d^{1/2}$ as in [8.18]. With the CMOS the type normally used for large on-chip designs, it is unfortunate that it is here that the control is non-linear.

First-order low-pass

Fig. 13.19(a) shows the conventional single-section low-pass filter, and (b) its OTA-C equivalent. With the OTA thus connected the output impedance is $1/g_m$.

With $\quad v_o/v_{in}=s_o/(s+s_o) \quad$ (13.40)

where $\quad s_o=-1/CR \quad$ as in [11.31]

for (a) then for (b) $\quad s_o=g_m/C \quad$ (13.41)

giving $s_o \propto I_b$ for the bipolar and $s_o \propto I_b^{1/2}$ for the CMOS.

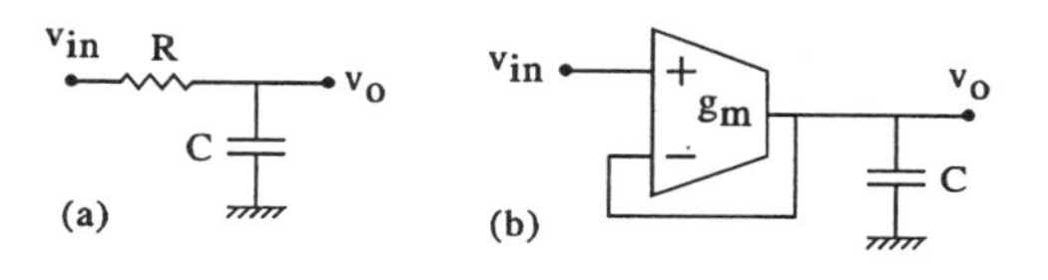

Fig. 13.19 First-order low-pass filter.
(a) Conventional RC type.
(b) OTA type.

First-order high-pass

The OTA does not lend itself so naturally to implementation of the high-pass function. Fig. 13.20 shows an example. Here

$$r_{in2}=1/g_{m2} \quad \text{and} \quad r_{out1}=1/g_{m1}$$

giving

$$\frac{v_o}{v_{in}}=\frac{1/g_{m2}}{1/g_{m1}+1/sC+1/g_{m2}} \qquad (13.42)$$

with the cut-off

$$f_o=1/[2\pi(1/g_{m1}+1/g_{m2})C] \qquad (13.43)$$

Unlike the low-pass, this shows a pass-band loss, of ×1/2 for the two g_m equal. This is of no great consequence, being easily made up elsewhere.

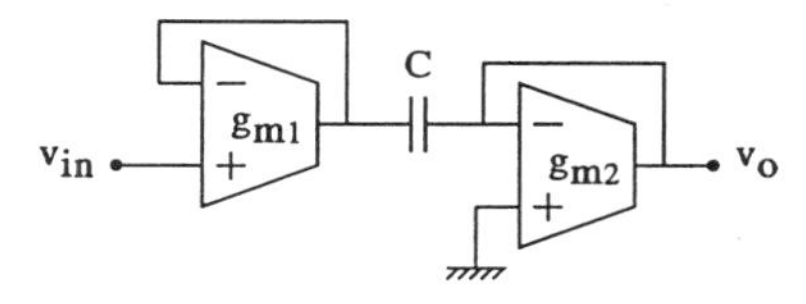

Fig. 13.20 First-order high-pass OTA-C filter.

What is more important is that part of C in the low-pass of Fig. 13.19(b) is the output capacitance of the OTA and part of the input capacitance of the following OTA. For the high-pass these components remain, adding an upper cut-off possibly not much higher than the required cut-off f_o of the high-pass.

OTA linearity

Unlike the op-amp-based integrator of Fig. 2.16, the OTA-based integrator does not use negative feedback. Thus such non-linearity as the OTA exhibits is not suppressed, increasing the importance of the OTA linearity.

Second-order low-pass

With this as in Fig. 13.21, for the 'band-pass' terminal grounded,

then $\quad v_o=X_{c2}(-g_{m3}v_o-g_{m2}v_x)$

and $\quad v_x=X_{c1}g_{m1}(v_o-v_{in})$.

For the two integrators identical we write

$$C_1=C_2=C, \quad \text{and} \quad g_{m1}=g_{m2}=g_m.$$

Then

$$\frac{v_o}{v_{in}}=\frac{s_o^2}{s^2+ss_o/Q+s_o^2} \qquad (13.44)$$

where $\quad \omega_o=s_o=g_m/C \quad$ (13.45)

and $\quad Q=g_m/g_{m3} \quad$ (13.46)

This is the normal second-order low-pass response with the common bias current I_b controlling the cut-off ω_o, while leaving Q unaltered.

The summation of the currents $-g_{m3}v_o$ and $-g_{m2}v_x$ is achieved by adding the currents in C_2. This is the normal way of forming the sum in an OTA-C filter.

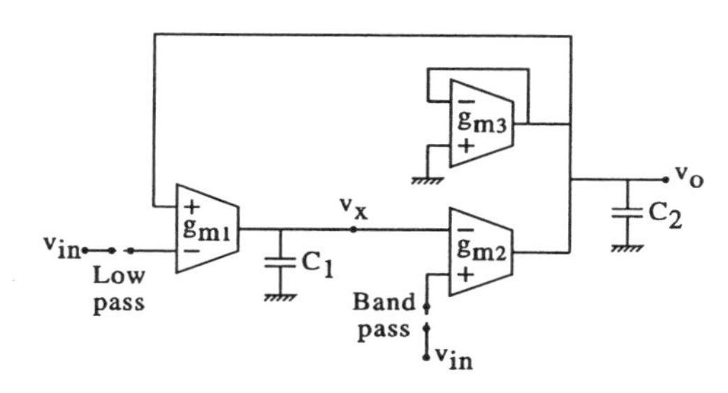

Fig. 13.21 Second-order OTA-C filter giving low-pass and band-pass responses according to input point.

Second-order band-pass

The corresponding band-pass filter uses the same circuit but with the 'band-pass' input used and the 'low-pass' input grounded. With the same s_o, f_o and Q then

$$\frac{v_o}{v_{in}}=\frac{ss_o}{s^2+ss_o/Q+s_o^2} \qquad (13.47)$$

The normal band-pass response. With the above building blocks, a wide variety of such filter responses can be synthesised.

Tuning

The chip-to-chip production spread for the OTA g_m is about 20% which is generally unacceptable for OTA-C filter implementation, and particularly so for a band-pass filter of even modest Q. Thus each chip has to be separately 'tuned' to give the required cut-off or resonant frequency ω_o. This could be done by measuring ω_o and adjusting the bias current I_b accordingly. However, a less labour-intensive method is normally required.

g_m adjustment

For all of the above filters it is $\omega_o \propto g_m/C$ that must have the required value. With the production spread of C lower than that of g_m, sometimes the setting of g_m to a defined value is sufficient.

Fig. 13.22 shows such an arrangement. Here the negative feedback adjusts I_b to the value needed to make the error current $I_e = E(R - 1/g_m)$ zero, thus setting $1/g_m$ to equal the well-defined off-chip R.

With g_m correct for this OTA then the others on the chip will be close, typically within 1%.

As an exercise confirm that the above feedback is negative.

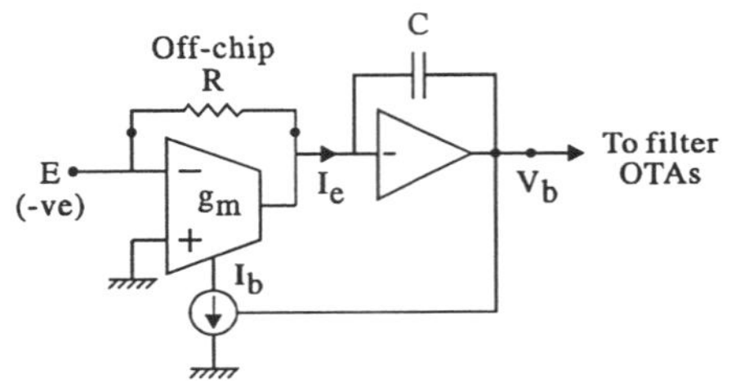

Fig. 13.22 Automatic adjustment of OTA g_m to equal 1/R.

Filter frequency adjustment

Where reliance on the value of C being correct is inadequate then it is $\omega_o \propto g_m/C$ which must be set to the correct value. Fig. 13.23 shows one approach. Here V_{in} is a reference square wave of constant frequency f_{in} and constant amplitude. This is applied to an OTA-C low-pass filter of the type of Fig. 13.19(b).

With the filter cut-off $f_o \approx f_{in}$ then the waveforms become as shown, with the mean value of the full-wave rectifier output V_d increasing with increasing f_o. V_d is then compared with the reference V_r to give the necessary negative feedback to maintain f_o at the value determined by f_{in}.

The arrangement will work equally well with a sine wave input. The reason a square-wave is shown is that frequently the filter will be part of a system with digital components, where various clock signals of highly stable frequency are available.

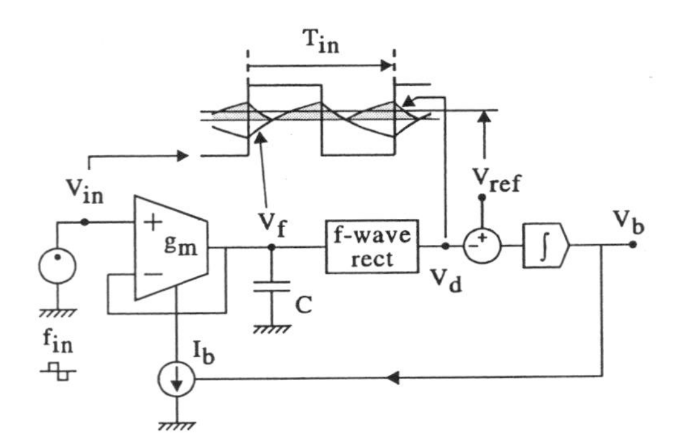

Fig. 13.23 Automatic adjustment of OTA-C filter $\omega_o = g_m/C$ by adjustment of g_m.

Phase-locked loop method

Many variants exist. Another method is to replicate the loop of Fig. 13.21 without the damping g_{m3}. The circuit then becomes an oscillator of frequency equal to the filter f_o. V_o is then phase-locked to f_{in} using a PLL with the phase-detector output controlling g_m.

A complication here is that the level of oscillation must be kept at the normal signal level. Otherwise slewing can change the frequency of oscillation from the ω_o for the other OTAs on the chip, which are not saturating.

Phase-detection method

Fig. 13.24 shows an arrangement where the test filter is of the band-pass type, e.g. as in Fig. 13.21. With the amplitude-frequency and phase-frequency responses of the filter as in (a) and (b), it is (b) that is the most effective indicator of the degree to which the resonant frequency of the filter f_o is equal to f_{in}. Thus the controller becomes as in (c), probably with the phase detector the simple XOR-gate type of Fig. 19.3.

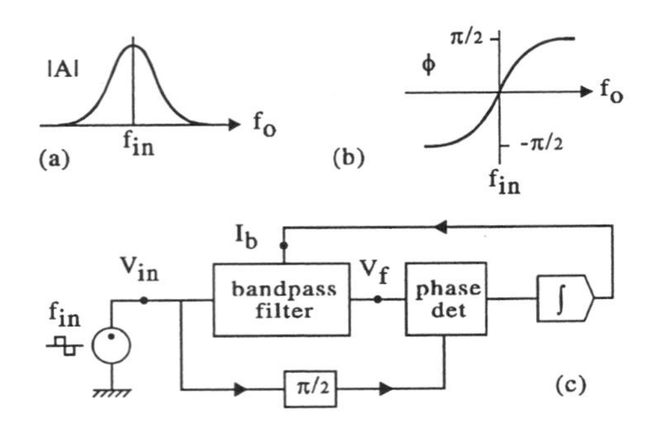

Fig. 13.24 Phase-detector tuning of narrow-band band-pass filter.
(a) Variation of |A| with filter f_o where $A = V_f/V_{in}$.
(b) Variation of filter phase shift ϕ with f_o.
(c) Controller based on detection of phase ϕ.

Direct tuning

All of the above methods are known as 'indirect tuning', in that what is 'tuned' is not the actual filter but one well matched to it, on the same chip. For more demanding circuits, such as the band-pass filter of Fig. 13.21 with $Q >> 1$, 'direct tuning' is needed. This is as above except that, instead of having the added filter, the one to be used is switched from its point of application to the control circuit. The required value of I_b is then noted and maintained while the filter is restored to its working position.

Fully differential configuration

Most OTA-C filters are implemented in fully differential form, where the OTA also has complementary outputs such that one is the inverse of the other. Fig. 13.25 shows the fully differential equivalent of the first-order low-pass filter of Fig. 13.19(b).

The extra OTA is needed because the differential input of the first becomes that of the filter. No longer is half of it available for subtraction of v_o from v_{in} as for the single-ended implementation. Instead the subtraction is implemented by the currents into the C.

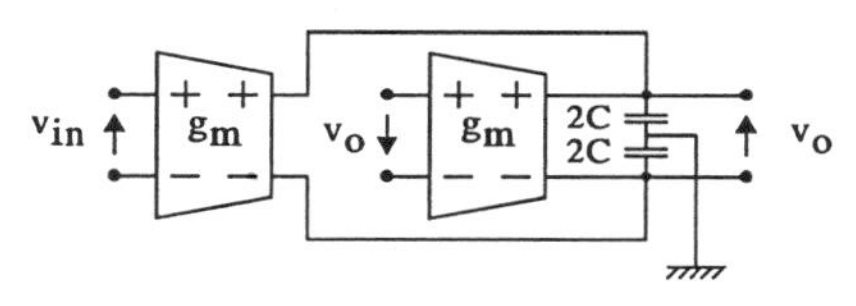

Fig. 13.25 Fully differential equivalent of low-pass filter of Fig. 13.19(b).

One reason put forward for the fully differential arrangement is the need to improve the OTA linearity. This is less than convincing because the principal non-linearity in the transfer function of the OTA is anti-symmetric. Adapting to the fully differential configuration will only cancel symmetric non-linearity.

A more convincing point is the need to reject interference from digital logic on the same chip. This tends to make the substrate 'noisy' and fully balancing the OTA will reduce such pick-up.

Common-mode offset

The input offset voltage $V_{i.off}$ of the single-output OTA will normally be such as to drive to output to one of the saturation points for both of the inputs zero. However, when say the 100% feedback for the OTA-C low-pass filter of Fig. 13.19(b) is applied, to give the zero-frequency gain of unity, then all that happens is that the small $V_{i.off}$ is transferred directly to the output.

But the fully differential OTA has both common- and difference-mode input voltage offsets. With the feedback difference-mode only then the common-mode component is not reduced, and may still drive the two outputs to a common saturation point. Here either a suitably low common-mode gain, or some form of common-mode feedback, is needed.

13.8 MOSFET-C filter

This is yet another type of externally controllable filter that is well suited to microcircuit fabrication. The circuits are somewhat as for the active RC filters in Section 13.1, but with the resistors replaced by MOS transistors operating below pinch-off. From [8.7] the channel conductance for the drain-to-source voltage $V_{ds}=0$ is given by

$$g_{co}=\beta V_{eff} \qquad (13.48)$$

where the 'effective'

$$V_{eff}=V_{gs}-V_{th},$$

V_{th} is the threshold voltage, V_{gs} is the gate-to-source voltage, and

$$\beta=\mu C_{ox}(W/L)$$

which is independent of V_{eff}.

For the RC single-section low-pass filter of Fig. 13.19(a) the cut-off $\omega_o=1/RC$ as in [13.5]. Thus if a CMOS pair is used in place of the resistor then $R=1/g_{co}$ allowing ω_o to be controlled by V_{gs}.

Here it is most convenient if the source S is either connected to the real or a virtual ground. This tends to favour the op-amp-based integrator of Section 2.8 and hence the type of active filter built around such integrators, as in Fig. 13.14 for the switched-capacitor implementation.

Non-linearity

A limitation of the MOS resistor is that its voltage-current relation is not completely linear. Restating [8.11] for the more general V_{ds}, the channel (drain) current

$$I_d=\beta(V_{eff}V_{ds}-V_{ds}^2/2) \qquad (13.49)$$

which only gives $I_d \propto V_{ds}$ for $V_{ds} << V_{eff}$.

Fully differential operation

In contrast to the OTA, fully differential operation completely eliminates this component of non-linearity. Consider two MOS transistors with both sources grounded, and V_{in} applied to one drain and $-V_{in}$ to the other. Here

$$I_{d1}=\beta(V_{eff}V_{in}-V_{in}^2/2) \quad \ldots(a)$$

$$I_{d2}=\beta(V_{eff}(-V_{in})-V_{in}^2/2) \quad \ldots(b) \qquad (13.50)$$

giving

$$I_{d1}-I_{d2}=2\beta V_{eff}V_{in} \qquad (13.51)$$

the required fully linear relation.

To make use of this concept the fully differential op-amp (not OTA) of Fig. 13.26 is needed.

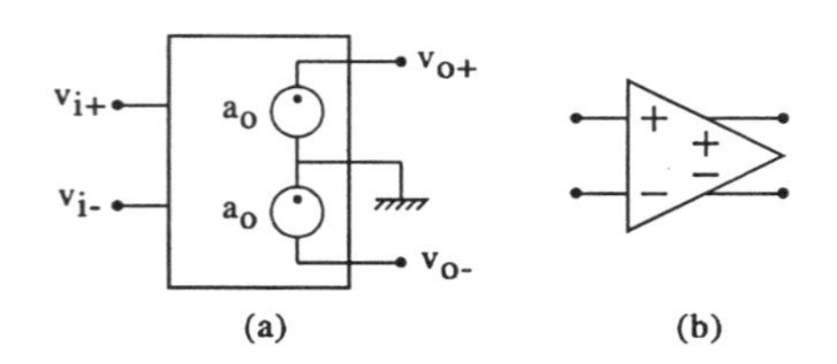

Fig. 13.26 Fully differential op-amp.
(a) Functional diagram.
(b) Conventional symbol.

Fully differential feedback amplifier

For the circuit of Fig. 13.27(a) with

$$V_{in+} = V_{in} \quad \text{and} \quad V_{i-} = -V_{in}$$

then both of the op-amp input terminals remain at virtual ground. For the R_1 the above CMOS devices then

$$I_{d1} - I_{d2} = 2\beta V_{eff} V_{in} \qquad \text{as in [13.51]}$$

For the R_2 normal resistors then

$$(V_{o+} - V_{o-}) \propto V_{eff}(I_{d1} - I_{d2})$$

giving $$(V_{o+} - V_{o-}) \propto V_{eff} V_{in}.$$

Thus we have a fully linear negative-feedback amplifier with the gain proportional to the control V_{eff}.

Fully differential feedback integrator

More important for the MOSFET-C filter is the integrating version of the above feedback amplifier, shown in (b).

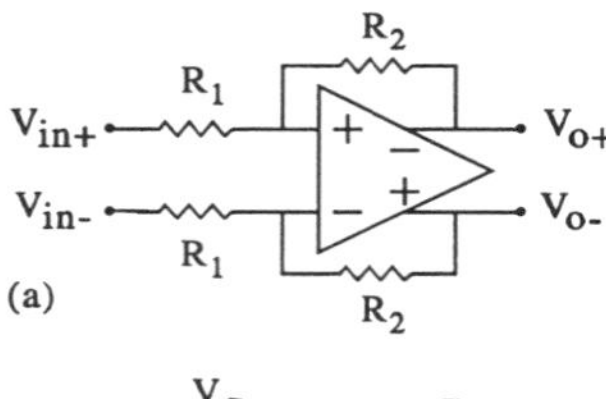

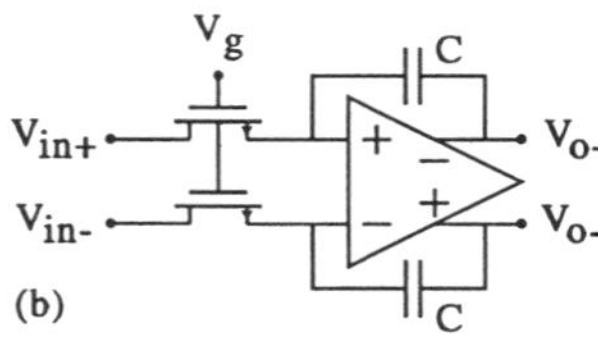

Fig. 13.27 MOSFET-C circuits based on fully differential op-amp.
(a) Feedback amplifier.
(b) Feedback integrator.

Higher order distortion

Unfortunately the strictly parabolic non-linearity of the I_d-V_{ds} relation in [13.49] is an approximation. In reality there are higher order terms, and some of them of odd-order, which the fully differential arrangement does nothing to reduce. Returning to

$$I_d = \beta(V_{eff} V_{ds} - V_{ds}^2/2) \qquad \text{as in [13.49]}$$

another approach is possible. For the two grounded source transistors of [13.50], we now apply V_{in} to both and adjust V_{eff} to $V_{eff} + \Delta V_{eff}$ and $V_{eff} - \Delta V_{eff}$

Here
$$I_{d1} = \beta(V_{eff} + \Delta V_{eff})V_{in} + V_{ds}^2/2) \qquad \ldots(a)$$
$$I_{d2} = \beta(V_{eff} - \Delta V_{eff})V_{in} + V_{ds}^2/2) \ldots(b) \qquad (13.52)$$

giving
$$I_{d1} - I_{d2} = 2\beta \Delta V_{eff} V_{in} \qquad (13.53)$$

again fully linear. The reason this works is that the distortion term $V_{ds}^2/2$ is independent of V_{eff}. This also tends to be so for the higher-order distortion components, so they are removed as well.

Fig. 13.28 shows the corresponding integrator, which incorporates both methods of cancellation, also retaining the two zero virtual-ground voltages.

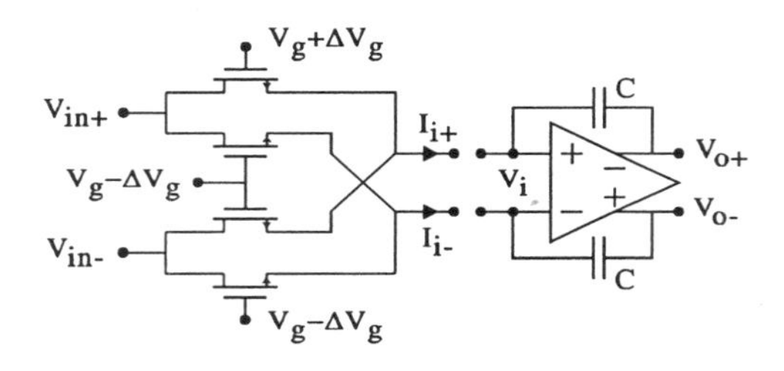

Fig. 13.28 MOSFET-C integrator with additional cancellation of higher-order distortion terms.

MOSFET CR filter

Fig. 13.29(a) and (b) shows the first-order low and high-pass types, with the MOS simply replacing the R in the normal RC implementation. For each the pass-band gain is unity with the cut-off. These do not incorporate the higher-order distortion cancellation of Fig. 13.27.

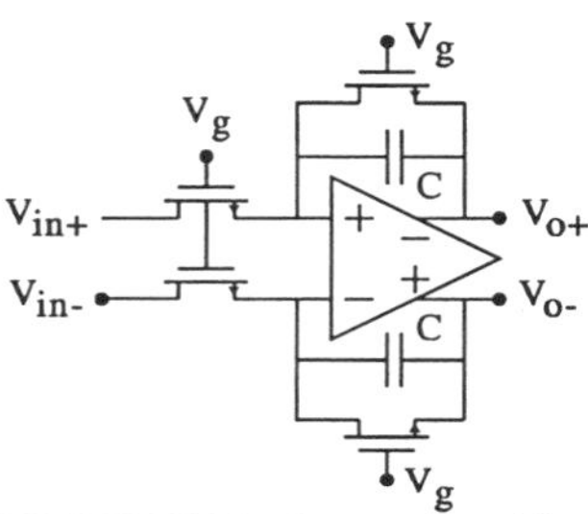

Fig. 13.29(a) MOSFET-CR implementation of first-order low-pass filter.

This kind of arrangement is termed a MOSFET-CR filter, while the term MOSFET-C filter tends to be reserved for

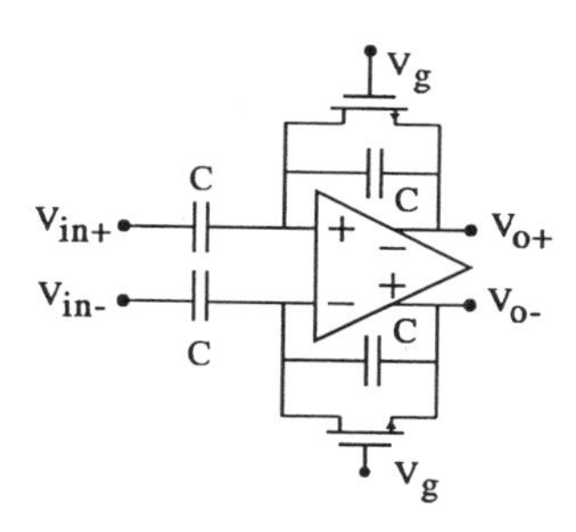

Fig. 13.29(b) MOSFET-CR implementation of first-order high-pass filter.

the kind of arrangement using the MOSFET-C integrator in the 'integrators and adder' type of configuration of Fig. 13.14.

References

Coughlin, R. F. and Driscoll, F. F. 1982: *Operational Amplifiers and Linear Integrated Circuits*. Prentice Hall.

Gray, P. R. and Meyer, R. G. 1993: *Analysis and Design of Analogue Integrated Circuits*. John Wiley.

Holbrook, J. G. 1966: *Laplace Transforms for Electronic Engineers*. Pergamon.

Johns, D. A. and Martin, K. 1997: *Analog Integrated Circuit Design*. John Wiley.

Laker, K. R. and Sansen, W. M. C. 1994: *Design of Analog Integrated Circuits and Systems*. McGraw Hill.

Van Valkenburg, M. E. 1982: *Analog Filter Design*. Holt, Rinehart and Winston.

13.9 Exercises and experiments

Calculated values set bold are cumulative, being used from one example to the next. Use the TL081 bi-fet op-amp for all experimental work.

Exercise 13.1. Butterworth low- and high-pass filters

Calculation. Single-section low-pass Calculate the cut-off frequency f_o for the simple single-section low-pass filter for which **C=10nF, R=10kΩ.**

With $\omega_o=1/CR$ as in [13.5] then **f_o=1.59MHz.**

Calculation. Sallen-Key Butterworth low-pass Calculate the values of C_1 and C_2.

With $C_1=C\times\sqrt{2}$ and $C_2=C/\sqrt{2}$ as in [13.7] then **C_1=14.1nF, C_2=7.07nF.**

Experiment. Second-order low-pass Cascade two single-section low-pass RC filters with R=10kΩ and C=10nF as above. Confirm $|A|\propto 1/f^2$ in the stop-band with a gradual transition from $|A|\approx 1$ in the pass-band, also no overshoot in the step response (Q=1/3, poles on real axis).

Interpose the voltage follower and confirm the slight improvement in frequency response with still no overshoot in the step response (Q=1/2, poles still on real axis).

Assemble the Sallen-Key circuit of Fig. 13.3 with $C_1=C_2=C$ giving again Q=1/2. Confirm no change from above.

With C_1=15nF and C_2=6.8nF suitable preferred-value approximations to the above, confirm the further improvement in the frequency response, now with slight overshoot in the step response.

With C_1=10C and C_2=C/10 giving Q=5, confirm the resonant peak in the frequency response and the oscillatory transient in the step response.

Experiment. Third-order low-pass Alter to the Sallen-Key to the third-order low-pass Holbrook circuit of Fig. 13.6(a) and confirm the $\propto 1/f^3$ stop-band frequency response, with still no overshoot in the frequency response.

Calculation. Second-order high-pass Calculate the values for $R_1=R/\sqrt{2}$ and $R_2=R\sqrt{2}$ for the second-order Butterworth high-pass in Fig. 13.5, with C=10nF as above.

With still R=10kΩ then **R_1=7.07kΩ, R_2=14.1kΩ.**

Experiment. Second-order high-pass Connect the second-order high-pass in Fig. 13.5(b) with the preferred-values for R_1 and R_2. Confirm

- the expected value of the cut-off frequency f_o,
- that the stop-band response $\propto f^2$,
- the absence of overshoot in the frequency response,
- the slight overshoot in the step response.

Exercise 13.2. Wein oscillator/narrow-band band-pass filter

> **Calculation** Calculate suitable component values for the circuit of Fig. 13.8(b).

With R and C as above, and with the required feedback fraction $\beta=1/3$. For say $R_a=10k\Omega$ then

$$\mathbf{R_a=10k\Omega,\ R_b=20k\Omega.}$$

> **Experiment** For the Wein network alone, plot the frequency response to confirm the low Q_w and $A_w=1/3$ at resonance.
>
> Complete the circuit using $R_a=6.8k\Omega$ in series with a $47k\Omega$ variable resistor.
>
> Confirm that the circuit oscillates at f_o for $R_a=6.8k\Omega$, while the narrow-band band-pass response is obtained a little below the threshold of oscillation.

Exercise 13.3 Delyiannis-Friend narrow-band band-pass filter

Q-factor

The argument based on Fig. 13.10(b) requires completion. This was that, as the ω point moves around the circle shown, the distance $|\beta|$ between the ω and the $+1$ points undergoes a sharp dip. With the filter gain $A_f\approx 1/\beta$ and β as shown then $|A_f|$ exhibits the corresponding high-Q resonant peak.

With this requiring the circle to approach the $+1$ point closely, for which the Wein network $Q_w << 1$, then $Q >> 1$ requires $Q_w << 1$.

But as Q_w decreases so does the rate at which the ω point moves round the circle, in direct proportion to Q_w. Thus for the argument to be valid, the circle must approach the $+1$ point at a rate more rapidly than directly with Q_w.

With $\quad Q_w=1/(\gamma+2/\gamma) \quad$ as in [13.13]

and $\quad A_w(\omega_o)=Q_w\gamma \quad$ as in [13.15]

then $\quad A_w(\omega_o)=\gamma/(\gamma+2/\gamma)$

With $\quad Q_w\approx 1/\gamma \quad$ for $\gamma>>1$

then $\quad A_w(\omega_o)\approx 1-2{Q_w}^2$

Thus the distance between the A_w circle and the $+1$ point $\propto {Q_w}^2$, while the rate at which the ω point moves around the circle is only $\propto Q_w$.

Note too that at resonance both the low- and high-pass sections of the Wien network are operating well within their pass-bands. As in say Fig. 11.4 with [11.12] and [11.13] for the low-pass, within the pass-band the decrease from unity gain decreases much more rapidly with the distance from the cut-off than does the phase shift ϕ.

> **Calculation. Sensitivity** Calculate the % change in Q for a 1% change in R_1.

With $\quad Q=\gamma/2 \quad$ as in [13.19]

and $\quad R_1=R/\gamma$ and $R_2=\gamma R \quad$ as in [13.14]

then $\quad Q=(R_2/R_1)^{1/2}/2$

giving $\quad dQ/dR_1=-Q/2R_1.$

With $\quad \Delta Q=(dQ/dR_1)\times\Delta R_1$

then $\quad \Delta Q/Q=(-1/2)\times\Delta R_1/R_1$

giving $\quad \mathbf{\Delta Q/Q=0.5\%}$

> **Calculation. Realisable Q-factor** For the TL081 op-amp the unity-gain frequency $f_u=3MHz$. For the above narrow-band band-pass filter with $f_o=1kHz$, calculate the maximum Q that can be realised.

With $\quad A_f=-2Q^2 \quad$ as in [13.20]

and the op-amp gain $\quad A_o\approx f_u/f_o$

then $A_o=1.88\times10^3$.

With $\quad Q=(|A_f|/2)^{1/2}$

then $\quad \mathbf{Q=30.7}$

Any higher design Q will not be realised because $A_f\approx 1/$ only for $A_f << A_o$.

> **Design example** Design a Delyiannis narrow-band band-pass filter with f_o and C as above, the marginally realisable $Q=30$, and $|A_f|=1$ at resonance.

$R=1/(2\pi f_oC)$ ...(a) $\quad R_1=QR/|A|$ ...(b)

$R_2=2QR$ ...(c) $\quad R_3=QR/(2Q^2-|A|)$...(d)

as in [13.22] and with $R=10k\Omega$ as above,

$$\mathbf{R_1=300k\Omega,\ R_2=600k\Omega,\ R_3=167\Omega.}$$

> **Experiment** Using components of the nearest preferred value, connect the above circuit and confirm the design resonant frequency, bandwidth and gain.

Exercise 13.4 Delyiannis-Friend notch filter

Calculation Calculate the component values needed to convert the band-pass filter of the previous exercise to the above notch filter, recalling $R \approx 10k\Omega$ and $Q=30$.

With

$$R_1=R/(2Q),\ R_2=2QR,\ R_a=R_b/2Q^2,$$

as in [13.27], and with R as before,

$$\mathbf{R_1=167\Omega,\ R_2=600k\Omega.}$$

With $R_b=2Q^2R_a$ and say $R_a=1k\Omega$ then

$$\mathbf{R_a=1k\Omega,\ R_b=1.8M\Omega.}$$

Experiment Adapt the circuit of the last experiment to the above notch filter using the nearest preferred values to those calculated above. Make a suitable portion of R_a adjustable to allow for the difference between the preferred and the calculated values. Thus set $A(\omega)$ at the centre of the notch to zero and confirm that the frequency response and bandwidth is then as expected.

Exercise 13.5 Notch-enhanced second-order low-pass

Calculation. High-Q In the following experiment we convert the notch filter of Example 13.7 to the notch-enhanced low-pass, initially with the original $Q=30$ to give a peak at f_o in addition to the notch at f_z. With still $f_o=1.59kHz$, $Q=30$ and $R=10k\Omega$, calculate the component values for $f_z=3.2kHz$.

With R_1 and R_2 and R_a as before and with

$$R_3=2QR/[(\omega_z/\omega_o)^2-1]$$

as in [13.30], $\mathbf{R_3=197k\Omega.}$

With

$$R_b=2Q^2R_a/(\omega_z/\omega_o)^2$$

as in [13.31], $\mathbf{R_b=445k\Omega.}$

Calculation. Low-Q Repeat the calculation for the $Q=1/\sqrt{2}$ needed to remove the peak from the otherwise second-order Butterworth response.

With R_1 and R_2 as R_a as before,

$$\mathbf{R_3=4.65k\Omega,\ R_b=247k\Omega.}$$

Experiment Add R_3 as in Fig. 13.11(b), making R_3 adjustable to allow the notch gain to be set to zero.

For the first set of values calculated above, confirm that the frequency response is as in Fig. 13.12(b), but with the resonant peak at f_o.

Change to the second set of values to confirm that the peak has been removed.

14

Modulator and demodulator applications

Summary

Modulation is the process of imposing an analogue signal such as speech on a higher frequency sine wave 'carrier', originally in order to be more efficiently radiated from an antenna. Demodulation is the process of recovering the modulating signal from the received signal. The circuits for these processes are the topic of the next two chapters. In this chapter the subject is introduced by considering the main applications.

Amplitude modulation (AM) and frequency modulation (FM) are the processes of imposing the modulating voltage V_m on the amplitude and the frequency of the carrier. These three chapters are mainly concerned with AM, with FM covered later, in Chapter 19.

Amplitude modulation involves multiplication of the carrier voltage V_c by the modulating voltage V_m, basically requiring the analogue multiplier.

While this circuit is covered in the next chapter, it can not be fully implemented for the higher values of the carrier frequency f_c. Here the normal method is to subject V_m to a reversing switch, operated at f_c. This gives the corresponding amplitude-modulated square wave, which is then converted to the sinusoidal form by a narrow-band band-pass filter tuned to f_c.

The simplest AM demodulator is the envelope-detector of Fig. 4.19, but sometimes a form of synchronous demodulation is needed. Here the analogue-multiplier or the reversing-switch again serve.

Another application of the modulator circuit is that of 'frequency changing', where f_c is changed to a more convenient value for filtering and amplification, with V_m still imposed. This is the subject of Chapter 16. The present chapter ends with some worked examples.

14.1 AM radio link

Fig. 14.1 shows the essentials of the transmitter and receiver in an AM radio link. With the modulator in (a) shown as an analogue multiplier, then the low-frequency speech waveform V_m is imposed on the amplitude of the high-frequency carrier V_c in (d) to give the output waveform in (e). With V_o' in (e) the input to the detector in the receiver, this is proportional to the transmitter output V_o.

With the detector assumed ideal then the detector output V_d becomes as also shown in (e). With the following power-amplifier AC-coupled then the unwanted DC component of V_d is removed, prior to application to the loudspeaker LS.

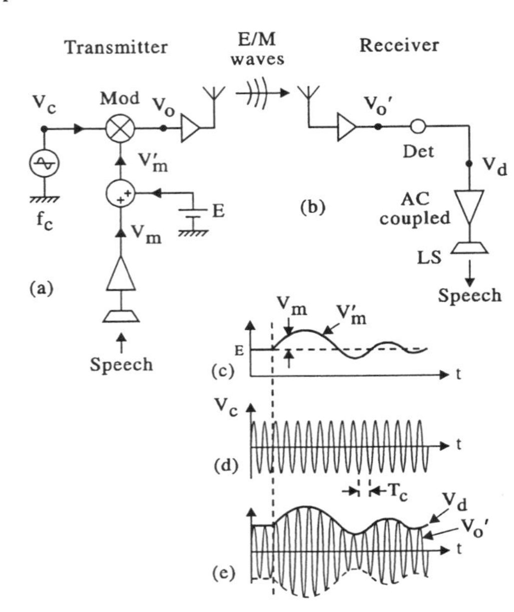

Fig. 14.1 AM radio link.
(a) Transmitter.
(b) Receiver.
(c) Voltage V'_m applied to modulator in (b).
(d) Carrier voltage V_c in (a).
(e) Demodulator waveforms for ideal envelope-detector.

Problem

To show why E in (a) is needed, draw the distorted V_d waveform that would result in its absence.

The processes of modulation and demodulation are needed here for two reasons.

- For efficient radiation (radiated power >> antenna ohmic losses) the antenna length needs to be comparable with one wavelength.
- Direct radiation of V_m would only allow one channel, in contrast to the many that are currently in use over the electromagnetic spectrum.

Problem

With the velocity of light $c=3\times10^8$m/s, frequency f, and wavelength λ, then $\lambda=c/f$. With the speech spectrum extending from 300Hz to 3kHz, show that for direct transmission of the speech via the antenna then for the lowest frequency speech component the antenna length would need to be comparable with the likely distance between transmitter and receiver.

Spectra

Fig. 14.2 shows the spectra for the various waveforms, first in (a) for a single sine wave speech component of frequency f_m, and then in (b) for the composite speech spectrum. With the multiplier constant unity then, from the transmitter diagram,

$$V_o=V_c(E+V_m) \quad (14.1)$$

For $$V_m=\hat{V}_m\cos(\omega_m t)$$

and with the 'modulation index' $M=\hat{V}_m/E$

$$V'_m=E[1+M\cos(\omega_m t)] \quad (14.2)$$

For $V_c=\hat{V}_c\cos\omega_c t$ and $E=1$ then

$$V_o=\hat{V}_c\cos\omega_c t(1+M\cos\omega_m t) \quad (14.3)$$

giving

$$V_o=\hat{V}_c\cos\omega_c t+M(\hat{V}_c/2)[\cos(\omega_u t)+\cos(\omega_L t)] \quad (14.4)$$

with the upper and lower 'sideband' frequencies

$$\omega_u=\omega_c+\omega_m\ \ldots(a) \quad \omega_L=\omega_c-\omega_m\ \ldots(b) \quad (14.5)$$

Thus the spectra become as in Fig. 14.2(a).

With the spectrum of V'_m as in (b) for the speech, then that for the output V_o becomes as also shown.

Frequency-domain multiplexing

By making f_c different for each transmitter, and including the corresponding band-pass filter at the input of each receiver, then many channels become available, with the channel separation twice the frequency of the highest frequency component of f_m. The same method can be used for transmitting a large number of different signals down a common line.

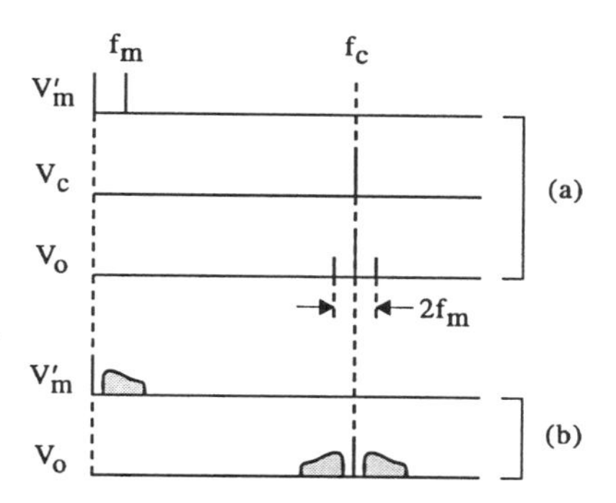

Fig. 14.2 Spectra for amplitude-modulator in Fig. 14.1.
(a) Modulating voltage V_m a single sine wave of frequency f_m.
(b) Speech modulation.

14.2 Reversing-switch modulator

Fig. 14.3 shows how the reversing-switch modulator gives the modulated square-wave in (c) having the same envelope as for the modulated sine-wave in Fig. 14.1(e).

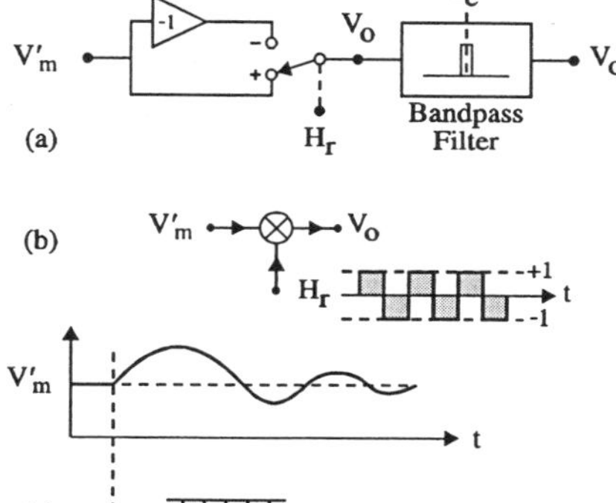

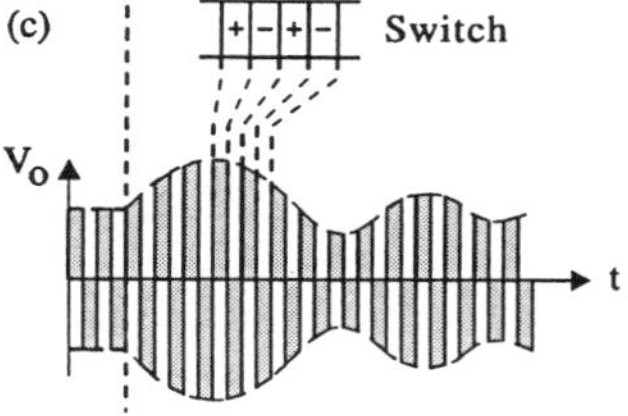

Fig. 14.3 Use of reversing switch as amplitude modulator in Fig. 14.1.
(a) Block diagram.
(b) Diagram showing reversing switch function.
(c) Waveforms.

Spectra

To see how this is restored to the sinusoidal form the spectra of Fig. 14.4 are needed. With the function of the reversing switch shown diagramatically in Fig. 14.3(b), and with the switching function H_r the unity amplitude square wave, from the Fourier analysis of [1.80]

$$H_r = \frac{4}{\pi}\left(\sin\omega_c t + \frac{1}{3}\sin 3\omega_c t + \frac{1}{5}\sin 5\omega_c t \ldots\right) \quad (14.6)$$

giving the H_r spectrum in Fig. 14.4(a).

With sideband pairs formed for each component of H_r then the spectrum of V_o becomes as also shown.

With speech giving the V_o spectrum in (b) then the final output spectrum becomes the same as for the analog-multiplier if the band-pass filter having the frequency response shown is added, restoring the output waveform to the sinusoidal form. Exercise 14.1 gives an example calculation.

Problem Explain the disadvantages in E/M spectrum usage if the band-pass filter were to be omitted.

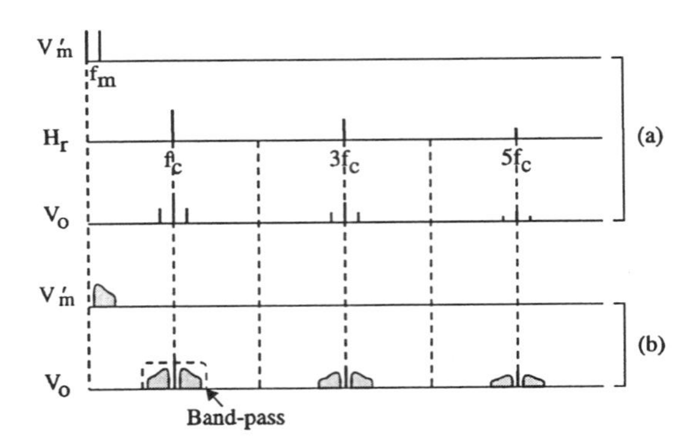

Fig. 14.4 Spectra for reversing switch modulator of Fig. 14.3.
(a) Sine wave modulation.
(b) Speech modulation.

14.3. Reversing-switch demodulator

Fig. 14.5 shows the reversing-switch used as the demodulator in a resistor-bridge strain gauge. The gauge resistor R_g is attached to a mechanical component which, when stressed, stretches the resistor to increase R_g. This alters the bridge balance to produce the amplified voltage $V_g \propto S$, the component strain.

The arrangement would work in principle with a DC voltage applied to the bridge. However, the output then includes the sensing amplifier offset, thermal drift, and 1/f-noise. With the bridge AC driven these effects are avoided, essentially as a property of the demodulator, usually backed by AC-coupling the signal amplifier.

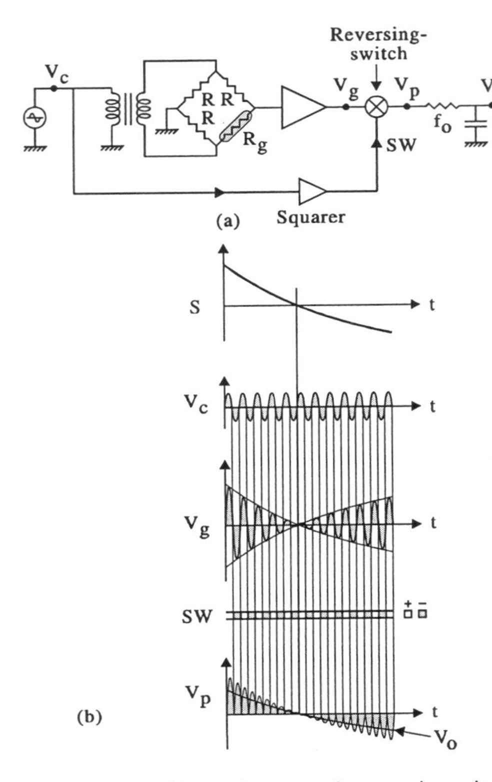

Fig. 14.5 Resistor-bridge strain gauge using reversing-switch demodulator.
(a) Circuit diagram.
(b) Waveforms.

Used in this way the reversing-switch is sometimes called a 'phase-sensitive detector'; the envelope-detector of Fig. 14.1 would produce only $V_o \approx |S|$.

As for the AM detector, the cut-off frequency f_o of the following low-pass needs to be well below the frequency f_c of the applied V_c for V_o to be the smoothed V_p.

Spectra

For the demodulator, these in Fig. 14.6 are somewhat as in Fig. 14.4 for the corresponding modulator. However, the component of frequency f_{in} shown for the amplified gauge output V_g is one of any frequency, possibly a noise component, with $f_{in}=f_c$ only for the required signal.

For f_{in} increasing as shown then the lower sidebands move downwards as also shown, eventually to place the first within the band-pass of the output filter, giving the first 'acceptance' shown. The other acceptances follow as the other lower sidebands come within the output filter band-pass.

With $f_{in}=f_c$ for the required signal then this is placed in the centre of the main acceptance. The lower sideband is then of zero frequency, the DC $V_o \propto S$.

With there being no acceptance at zero frequency and thereabouts, the demodulator rejects the amplifier offset and low-frequency noise, even if the gauge amplifier is not AC coupled.

The higher-order acceptances serve no useful purpose and add extra noise. This is avoided if the previous band-pass filter is now added at the demodulator input, thus suppressing all but the required acceptance.

However, it is equally acceptable to drive the bridge with a square wave. In this case there is a signal component at the centre of each acceptance so the band-pass filter is no longer appropriate. Exercise 14.2 gives some associated points to think about.

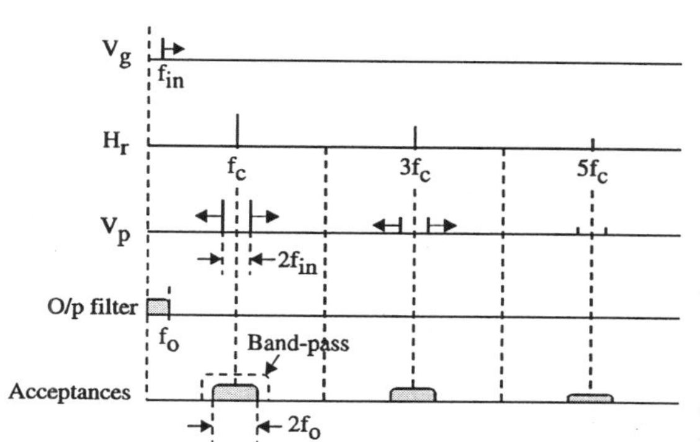

Fig. 14.6 Spectra for reversing-switch demodulator in Fig. 14.5.

14.4. Single-sideband suppressed-carrier radio link

The AM system of Fig. 14.1 is extremely wasteful of power. As shown in Exercise 14.1 the carrier power is twice the sum for the sidebands even for 100% modulation (modulation index M=1).

Since the carrier gives no information, this can be suppressed at the transmitter and reinserted locally at the receiver. Also, with both sidebands 'saying the same thing', then more efficient use of the available frequency space is obtained by suppressing one of the sidebands.

Fig. 14.7 shows the usual arrangement, with the waveforms as in Fig. 14.8. With the modulating signal V_m' as in Fig. 14.8(a) for normal AM modulation then all that is needed for suppression of the carrier is to remove the component E, to give the V_m' waveform as in (b).

With the demodulator in Fig. 14.7(b) the reversing switch then the associated waveforms become as in Fig. 14.8(d) and (e), giving the final output V_d proportional to the modulating V_m as required.

(f) is added to show the gross distortion that would result if the simple 'asynchronous' detector for the AM system were to be used, with the modulating speech waveform then effectively full-wave rectified.

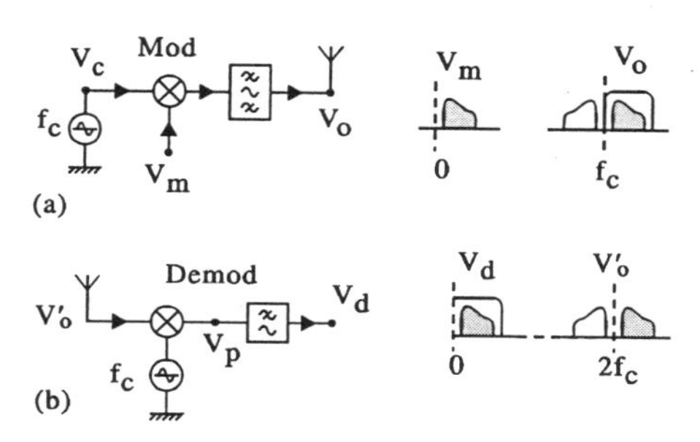

Fig. 14.7 Block diagram for suppressed-carrier radio link.

For single-sideband operation the only essential change is the removal of the unwanted sideband by the band-pass filter following the modulator, as shown in Fig. 14.7(a).

Carrier synchronisation

The waveforms of Fig. 14.8 for double-sideband suppressed-carrier operation assume exact synchronisation and alignment of the outputs from the two sources of frequency f_c in Fig. 14.7.

With this normally impossible there will be a small frequency difference (error) f_e causing the phase error ϕ_e to drift at a constant rate.

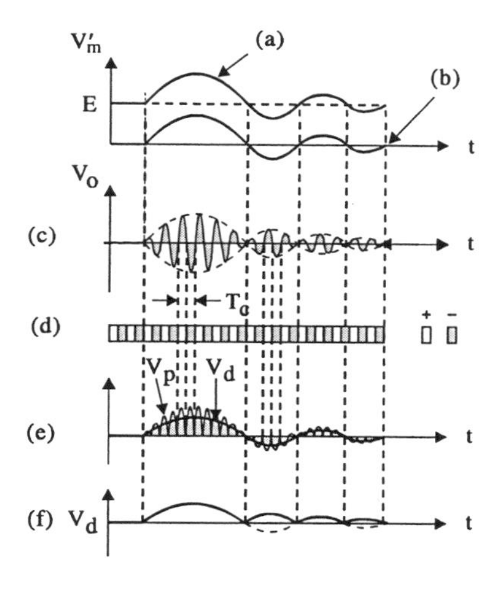

Fig. 14.8 Waveforms for suppressed-carrier operation of radio link of Fig. 14.1.
(a) Modulator input V_m' with E included as for AM.
(b) As (a) but without E as for suppressed carrier.
(c) Modulator output V_o for (b).
(d) Switching control for reversing-switch demodulator; as for modulator in Fig. 14.3.
(e) Demodulator output waveforms.
(f) Waveform for output V_d of envelope detector of Fig. 14.1 used in place of reversing switch.

It is left as an exercise to show that for a given ϕ_e the value of the demodulator output V_d is reduced by the factor $\cos(\phi_e)$. For $f_e \sim 0.1$Hz, then the amplitude of the received speech becomes subject to a periodic slow fade, becoming zero twice for each period $T_e = 1/f_e$.

For $f_e > 1$Hz the ear loses track of the fade, perceiving the speech as garbled to the point of unintelligibility. Here each speech component of frequency f_m is split into two, of frequency $f_m \pm f_e$. To hear it is to believe it!

This limitation is much relieved for the single-sideband system. Here the demodulator acts as a frequency-changer, restoring the V_d spectra to the original 'base-band' values of V_m. With the effect of f_e thus being to shift each speech component by f_e, then this needs to approach $\approx$10Hz for the effect even to be noticeable, and can be up to $\approx$100Hz before intelligibility is significantly impaired. How does this effect differ from a tape recorder played at the wrong speed?

14.5 Frequency-changer

One of the most costly components in a single-sideband transmitter is the band-pass filter following the modulator. With the required speech components extending roughly from 300Hz to 3kHz then a filter is required that will transmit fully over the corresponding range at the modulator output, but which will cut off sufficiently abruptly to reduce all components of the unwanted sideband by typically 60dB (×1/1000). Any remaining carrier component that may be present due to modulator imperfection will also need to be reduced to a comparable level. The filter will therefore usually be a multi-section quartz crystal type, at a cost of typically £50.

With the quartz filter of fixed frequency, and with the transmitter output frequency f_o normally needing to be variable, then the 'frequency-changer' of Fig. 14.9(a) is required.

With this just another modulator, and with f_{LO} the local-oscillator frequency, then the two output components are at $|f_{LO} \pm f_{if}|$. With the component of frequency $f_o = f_{LO} - f_{if}$ selected, then f_o is varied over the range shown by varying f_{LO}.

The 8MHz chosen for the intermediate frequency is one of the standard frequencies for which the quartz filter is available; 455kHz is another.

Receiver

The receiver in (c) is the reverse arrangement, the component blocks in (a) being suitably transposed. With most of the filtering and amplification done at the intermediate frequency, the only place where LC filtering is used is the 2MHz low-pass at the antenna interface.

Double conversion

For a wider range of f_o the double-conversion arrangement of (b) is used. Here the 70MHz 'roofing filter' is of more modest specification, needing only to reject the widely separated component at f_{LO2}+70MHz. For the receive configuration again the blocks are suitably transposed. Exercise 14.4 gives example calculations.

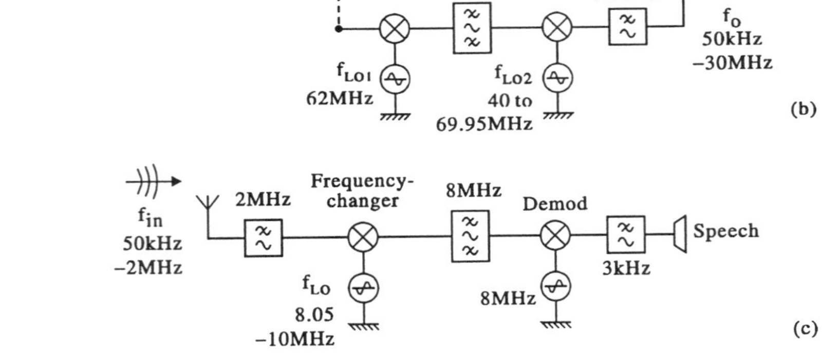

Fig. 14.9 Single-sideband transceiver.
(a) Transmitter configuration. (b) Extension giving higher output frequency range. (c) Receiver configuration for (a).

Terminology

While the term 'frequency-changer' tends to be reserved for this kind of function, both the modulator and demodulator for the single-sideband mode of Fig. 14.7 are frequency-changers. The modulator shifts the base-band speech spectrum up by f_c, and the demodulator shifts it down again.

14.6 Exclusive-OR

The last class of multiplier/modulator to be considered is the exclusive-OR (XOR) gate of Fig. 14.10(c), a member of any digital logic family. (a) shows an analogue multiplier driven by two saturating op-amps, giving the output table in (b).

With this corresponding to the XOR truth table in (d) then the XOR preceded by two comparators constitutes a simple replacement of the analogue multiplier, where saturation of the two input signals is acceptable. The PLL phase-detector in Fig. 19.3 is an example. The reversing switch can be regarded as the analogue multiplier of (a) with only one of the inputs saturated.

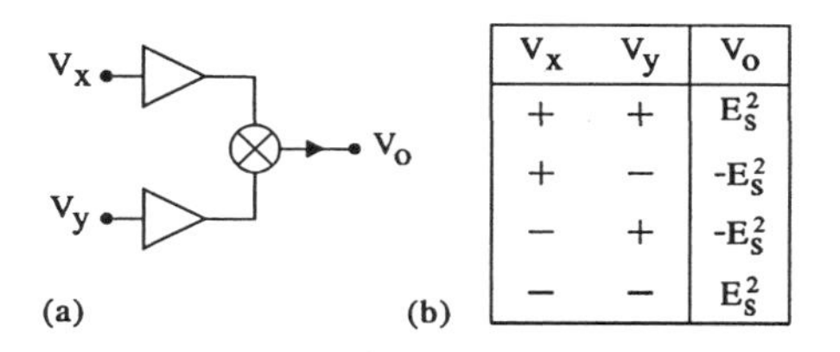

V_x	V_y	V_o
+	+	E_s^2
+	−	$-E_s^2$
−	+	$-E_s^2$
−	−	E_s^2

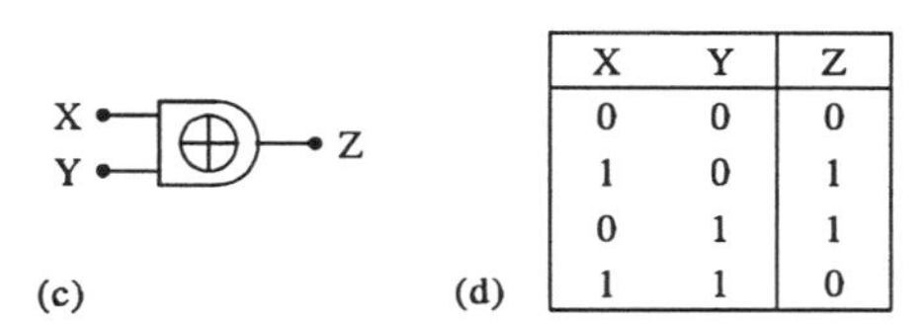

X	Y	Z
0	0	0
1	0	1
0	1	1
1	1	0

Fig. 14.10 Comparison of input-output relation of XOR gate with that for analogue multiplier preceded by saturating amplifiers of infinite gain.
(a) Analogue multiplier circuit.
(b) I/O table for (a).
(c) XOR gate.
(d) Truth table for (c).

References

Smith, J. 1986: *Modern Communication Circuits*. McGraw Hill.
Wilmshurst, T. H. 1990: *Recovery of signals from noise in electronic instrumentation*. Adam Hilger.

14.7 Exercises

Exercise 14.1. AM modulator

> **Calculation. Efficiency** Show that the maximum $\eta = 33.3\%$.

With V_o as in [14.4] and for $M=1$ (100% modulation) then $\hat{V}_u = \hat{V}_c/2$ for the upper sideband V_u.

With the power $\quad P_u \propto \hat{V}_u^2$,

etc., then $\quad P_u = P_c/4$.

Thus $\quad (P_u+P_L)/(P_u+P_L+P_c) = 1/3$

giving the $\eta = 33.3\%$.

> **Waveforms** Draw the waveforms for sine wave modulation with the modulation index $M>1$, thus showing the gross distortion for V_d.

Exercise 14.2 Reversing-switch modulator (Fig. 14.3)

> **Calculation** The modulator is used to produce a 50% amplitude sine wave modulated carrier with $E=5V$. Calculate the peak value $\hat{V}_o$ of the output of the band-pass filter in Fig. 14.4(b).

With $\quad V_m' = E[1+M\cos(\omega_m t)]$

as in [14.2] and with $M=0.5$ then the maximum value of V_m' is $1.5 \times E$.

With $\quad V_o = V_m' H_r$

and H_r as in [14.6] then $\quad$ **$\hat{V}_o = 9.55V$.**

Exercise 14.3 Problems: resistor-bridge strain-gauge (Fig. 14.5)

> **Analogue-multiplier waveforms** Draw the waveforms for the analogue multiplier used as the demodulator.

> **Analogue-multiplier acceptances** Show that only the main acceptance in Fig. 14.6 results when the analogue multiplier is used as the demodulator, and thus that the band-pass filter is no longer needed, even for a sine wave signal.

Square wave drive Show that the signal loss incurred by the band-pass filter outweighs the reduction in noise.

Chopping-switch When this is used in place of the reversing-switch then $H_r=\pm1$ is replaced by $H_c=1$ or 0. Derive the input acceptance pattern for the chopping-switch demodulator, showing that AC coupling of the signal amplifier is now essential in order to remove amplifier drift, etc.

Confirm by drawing the demodulator waveforms for a drifting DC voltage at the signal amplifier output.

Switching waveform For the reversing-switch demodulator, show that if the mark-space of H_r ratio differs from unity, or if the two magnitudes are not quite equal, then a range of minor acceptances at even multiples of the switching frequency is added, again requiring the AC-coupled amplifier.

Exercise 14.4 Transceiver (Fig. 14.9)

Determine whether the arrangement shown gives upper or lower sideband.

With f_m the frequency of the modulating speech component, and f'_{if} that of the resulting selected component at the modulator output, then

$$f'_{if}=f_c+f_m.$$

With f'_o the frequency of the resulting final output component then

$$f'_o = f_{LO} - f'_{if} = f_{LO} - f_c - f_m$$

giving **LSB.**

Calculate the value of f_c needed for upper sideband operation.

With f_1 and f_2 the lower and upper limits of the quartz filter pass-band, and with f_{cL} and f_{cu} the values of f_c for lower and upper sideband operation, then for the initial lower sideband operation

$$f_{cL}=8\text{MHz},\quad f_1=8\text{MHz}+300\text{Hz}$$

and

$$f_2=8\text{MHz}+3\text{kHz}.$$

With

$$f_{cu}=f_2+300\text{Hz}$$

then

$$f_{cu}=8\text{MHz}+3\text{kHz}+300\text{Hz}$$

giving $\mathbf{f_{cu}=8.0033MHz.}$

Calculate the corresponding change Δf_{LO} in f_{LO}.

With f_o the frequency of the suppressed carrier at the final output then

$$f_o=f_{LO}-f_c.$$

With f_o needing to be the same for both modes then

$$f_{LOL}-f_{cL}=f_{LOU}-f_{cu}.$$

With

$$f_{cu}=f_{cL}+3.3\text{kHz}$$

and

$$\Delta f_{LO}=f_{LOU}-f_{LOL}$$

then $\mathbf{\Delta f_{LO}=3.3kHz.}$

15

Long-tailed pair modulator circuits

Summary

Both the analogue-multiplier and the reversing-switch modulators of the last chapter can be implemented by suitable development of the long-tailed pair (LTP). In this chapter the development of the basic bipolar transistor LTP modulator is traced, through the reversing-switch modulator, and finally to the full analogue multiplier. The last section gives experimental work with associated example calculations.

15.1 Long-tailed pair modulator

The LTP modulator in Fig. 15.1 is the LTP of Fig. 7.11 with the added transistor in the tail. With the negative feedback imposed by R_e then

$$I_e \approx (V_m - 600\text{mV} - V_-)/R_e \qquad (15.1)$$

With $I_{c+}=I_{c-}=I_e/2$ and $g_m=I_c/(kT/q)$ as in [6.6] then

$$g_m = \frac{V_m - 600\text{mV} - V_-}{2(kT/q)R_e} \qquad (15.2)$$

With the voltage gain $a_{LTP} \propto g_m$ then a_{LTP} varies linearly with the modulating voltage V_m as required.

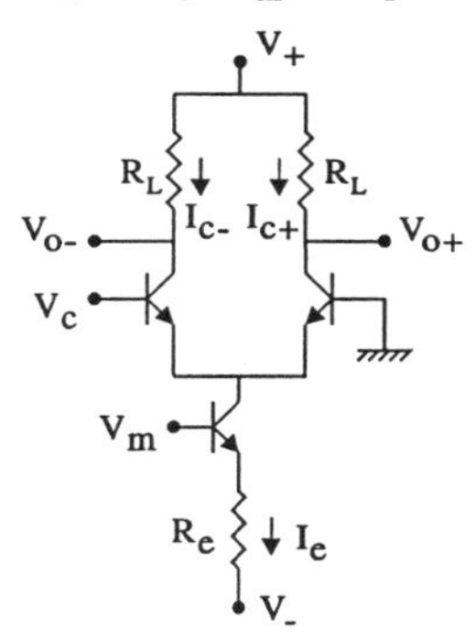

Fig. 15.1 Long-tailed pair modulator.

However with $g_m \propto I_e$ and I_e unable to be negative then the factor by which the incremental input v_c is multiplied can only be positive. This makes the circuit only suitable for the basic AM modulator of Fig. 14.1(a), not for the suppressed-carrier modulator of Fig. 14.7. With the incremental differential output

$$v_{od} = v_{o+} - v_{o-},$$

and with

$$a_{LTP} \approx g_m R_L \qquad (15.3)$$

as in Section 7.6, then

$$v_{od} = k_m v_c V'_m \qquad (15.4)$$

where

$$V'_m = V_m - 600\text{mV} - V_- \qquad (15.5)$$

and the multiplier constant

$$k_m = R_L/[2R_e(kT/q)].$$

Reversing-switch

While the negative feedback due to R_e linearises the variation of a_{LTP} with V_m, the response to V_c remains far from linear. With

$$I_c \approx I_{co}\exp(qV_{be}/kT)$$

as in [6.1] for the single transistor, it is shown below that the corresponding function for the LTP is

$$\tanh(V_{be}/(2kT/q)).$$

With $kT/q=25$mV then linearity requires $V_c << 50$mV. With 50mV comparable with the thermal drift, the usual approach is to make no attempt at a linear response to V_c, but instead to increase the level to 'hard-switch' I_e from one LTP transistor to the other, giving

$$V_{od} \approx \pm V'_m R_L/R_e \qquad (15.6)$$

essentially the reversing-switch operation. Fig. 15.2(a) and (b) compares the waveforms for V_c well below and well above the amplitudes for hard switching.

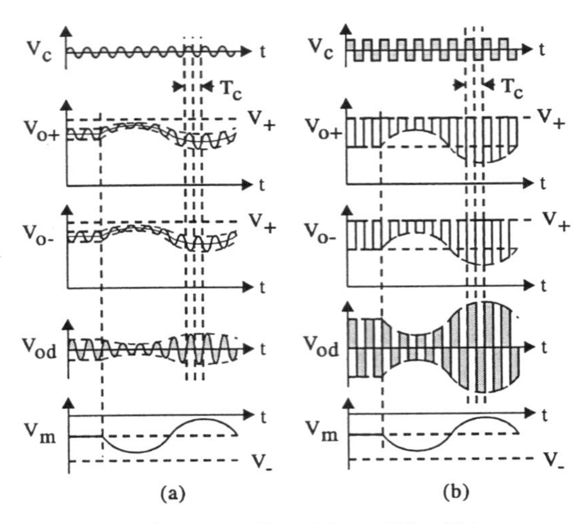

Fig. 15.2 Waveforms for LTP modulator of Fig. 15.1.
(a) Linear modulation, $|V_c| << kT/q$.
(b) Reversing switch modulation, $|V_c| >> kT/q$.

Bow-tie diagram

For the common-emitter amplifier, Fig. 6.1 shows how the amplifier transfer function converts the input waveform to that of the output, with it possible to display the transfer function using the scope X-Y mode.

The bow-tie diagram of Fig. 15.3 is the modulator equivalent of the amplifier transfer function, now serving the function of checking the linearity of the variation of output carrier amplitude with modulating voltage V_m. Thus V_m is connected to the scope X-plates and V_{od} to the Y.

For the V_c input hard switched, only the outer bounds are seen, but for a sinusoidal output the waveforms and the bow-tie are filled, as shown. Exercise 15.1 gives experimental demonstration of the LTP modulator with example calculations.

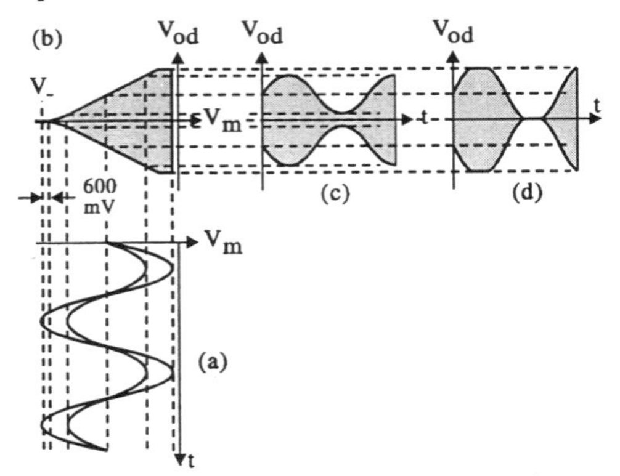

Fig. 15.3 Bow-tie diagrams for LTP modulator of Fig. 15.1.
(a) Waveforms for V_m within and beyond linear range of LTP.
(b) Bow-tie diagram.
(c) V_{od} waveform for V_m within linear range.
(d) V_{od} waveform for V_m beyond linear range.

15.2. LTP large signal analysis

Fig. 15.4(a) shows the LTP now with the tail represented by the ideal current source I_e. For the two transistors

$$I_{c-} = I_{co} \exp\left(\frac{qV_{be+}}{kT}\right) \quad \ldots(a)$$

$$I_{c+} = I_{co} \exp\left(\frac{qV_{be-}}{kT}\right) \quad \ldots(b) \qquad (15.7)$$

With $I_e = I_{c+} + I_{c-}$ then

$$\frac{I_{c-}}{I_{c+}} = \exp\left(\frac{qV_{bd}}{kT}\right) \qquad (15.8)$$

where the differential $V_{bd} = V_{b+} - V_{b-}$. Hence

$$I_{c-} = \frac{I_e}{2}\left[1 + \tanh\left(\frac{qV_{bd}}{2kT}\right)\right] \quad \ldots(a)$$

$$I_{c+} = \frac{I_e}{2}\left[1 - \tanh\left(\frac{qV_{bd}}{2kT}\right)\right] \quad \ldots(b) \qquad (15.9)$$

With $V_{o+} = V_+ - R_L I_{c+}$, $V_{o-} = V_+ - R_L I_{c-}$

and $V_{od} = V_{o+} - V_{o-}$ then $V_{od} = R_L(I_{c-} - I_{c+})$

giving
$$V_{od} = I_e R_L \tanh\left(\frac{qV_{bd}}{2kT}\right) \qquad (15.10)$$

Thus the variation of V_{o+}, V_{o-} and V_{od} with V_{bd} becomes as in Fig. 15.4(b).

Differentiation of [15.9] (a) and (b) gives the sloping dashed lines shown, and thus that

$$V_{bd} << (2kT/q \approx 50\text{mV})$$

is the criterion for linearity.

Also differentiation of [15.10] with $g_m = I_c/(kT/q)$ and $I_c = I_e/2$ confirms $a_{LTP} \approx g_m R_L$ as in [15.3].

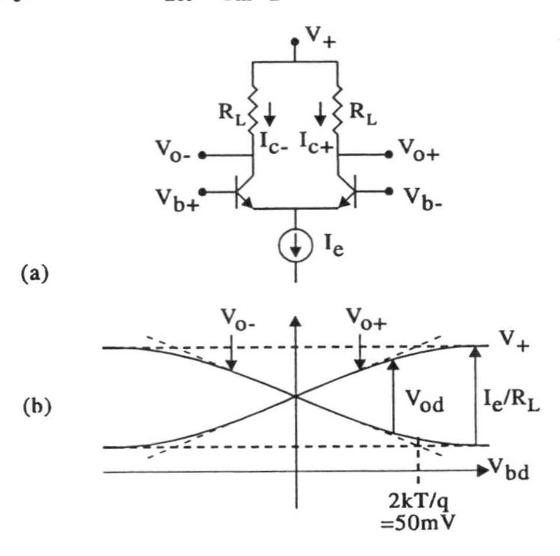

Fig. 15.4 Large-signal response of long-tailed pair.
(a) Circuit diagram.
(b) Input-output relations for input $V_{bd} = V_{b+} - V_{b-}$.

15.3 Transposed input LTP

For a modulator giving the output $z \propto x \times y$, and if this is only so for both x and y positive, then this is a 'single-quadrant' modulator.

For the LTP connected as above $v_{od} \propto v_c g_m$ with g_m varying linearly with the modulating V_m. Here v_c can be both positive and negative but g_m can only be positive. Thus the modulator is of the two-quadrant type.

If, however, the inputs are transposed then Fig. 15.5 shows four-quadrant operation. This is largely academic because of the small range of values for which the response to V_m remains linear. However, we do now have a bow-tie worthy of the name. Exercise 15.2 gives experimental demonstration.

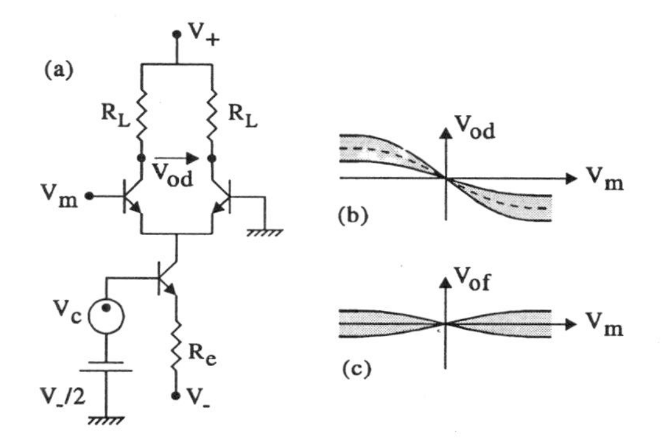

Fig. 15.5 Long-tailed pair modulator with modulation and carrier inputs transposed from the normal.
(a) Circuit diagram.
(b) Bow-tie.
(c) Bow-tie with high-pass filter at output. V_{of} = filter output.

15.4 Four-quadrant modulator

Fig. 15.6 shows a better way of operating the LTP as a four-quadrant modulator. Here two two-quadrant LTP modulators are combined in such a way that, although V'_m is never negative, the differential output $V_{od} \propto V_m$ for V_m both positive and negative.

U in the diagram refers to the 'upper' LTP inputs, previously labelled V_c, and L to the 'lower' input, previously labelled V_m.

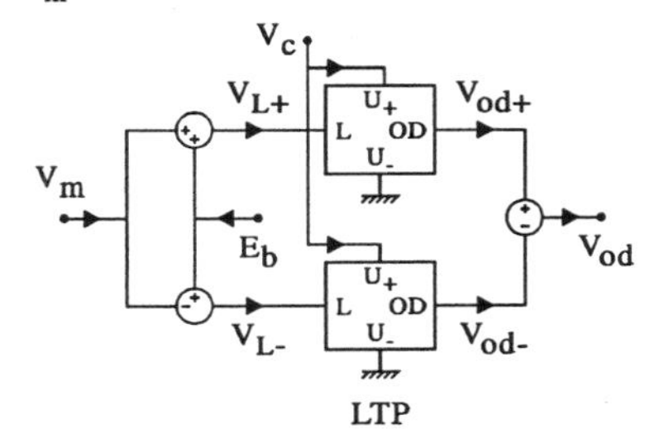

Fig. 15.6(a) Diagram showing how two LTP modulators can be combined to form a four-quadrant modulator.

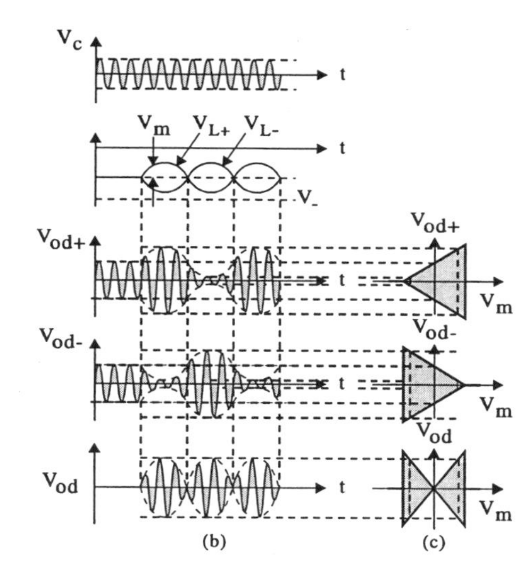

Fig. 15.6(b,c) Waveforms and bow-ties for circuit of (a)
(b) Waveforms.
(c) Bow-ties.

Problem Draw the quadrants for the four possible types of single-quadrant response. How many different types of two-, three- and four-quadrant response are there?

15.5 Gilbert cell

Fig. 15.7(a) shows the above principle as implemented in the Gilbert cell (1968). Here the phase splitting of V_m at the left of Fig. 15.6 is implemented by the two tail transistors. The waveforms when the upper input is 'hard switched' for reversing switch operation are shown in Fig. 15.7(b). From the above large signal LTP analysis

$$I_{c1-} = I_{e1}/2[1 + \tanh(qV_c/2kT)] \quad \ldots(a)$$
$$I_{c1+} = I_{e1}/2[1 - \tanh(qV_c/2kT)] \quad \ldots(b) \qquad (15.11)$$

$$I_{c2-} = I_{e2}/2[1 + \tanh(qV_c/2kT)] \quad \ldots(a)$$
$$I_{c2+} = I_{e2}/2[1 - \tanh(qV_c/2kT)] \quad \ldots(b) \qquad (15.12)$$

Also

$$I_{e1} = I_{eo} + V_m/R_e \quad \ldots(a)$$
$$I_{e2} = I_{eo} - V_m/R_e \quad \ldots(b) \qquad (15.13)$$

$$I_{c1} = I_{c1-} + I_{c2+} \quad \ldots(a)$$
$$I_{c2} = I_{c1+} + I_{c2-} \quad \ldots(b) \qquad (15.14)$$

Hence

$$I_{c1} = I_{eo} + (V_m/R_e).\tanh(qV_c/2kT) \quad \ldots(a)$$
$$I_{c2} = I_{eo} - (V_m/R_e).\tanh(qV_c/2kT) \quad \ldots(b) \qquad (15.15)$$

For the output voltages

$$V_{o+} = V_+ - I_{c1}R_L \quad \ldots(a)$$
$$V_{o-} = V_+ - I_{c+}R_L \quad \ldots(b) \qquad (15.16)$$

giving

$$V_{o+} = V_+ - I_{eo}R_L - (R_L/R_e)V_m \tanh(qV_c/2kT) \quad (15.17)$$

$$V_{o-} = V_+ - I_{eo}R_L + (R_L/R_e)V_m \tanh(qV_c/2kT) \quad (15.18)$$

With $V_{od} = V_{o+} - V_{o-}$

then
$$V_{od} = -\frac{R_L}{R_e} 2V_m \tanh\left(\frac{qV_c}{2kT}\right) \quad (15.19)$$

Finally for $V_c << 2kT/q$

$$v_{od} \approx -\frac{R_L}{R_e} V_m v_c \frac{q}{kT} \quad (15.20)$$

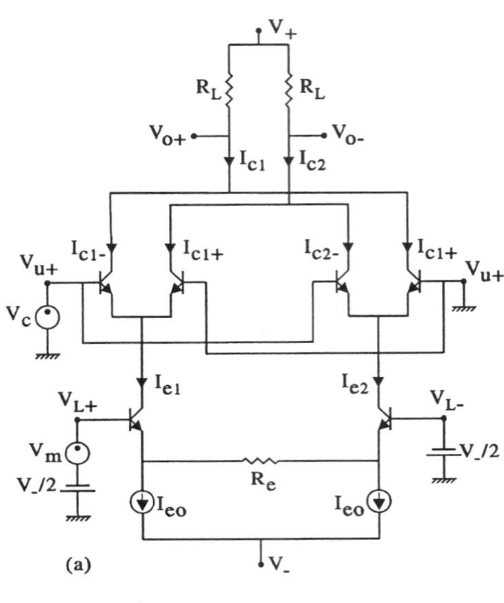

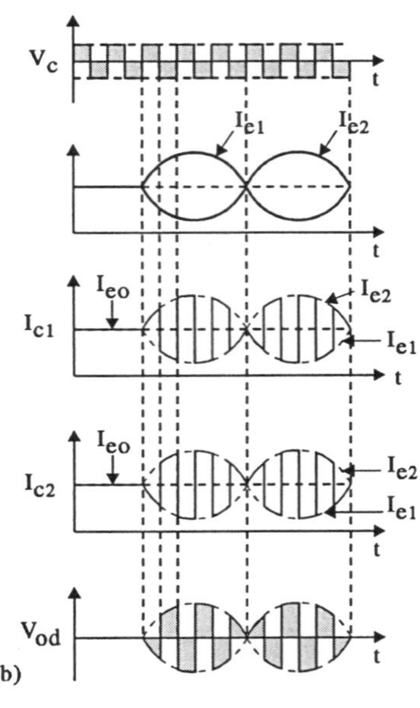

Fig. 15.7 Gilbert cell implementation of four-quadrant modulator of Fig. 15.6.
(a) Circuit diagram.
(b) Waveforms. $V_{od} = V_{o+} - V_{o-}$

15.6 Full range analogue multiplier

The remaining limitation of the above circuit is that still $|V_c| << 2kT/q = 50\text{mV}$ is the criterion for linear response to V_c, making the circuit only really suitable as a switching modulator, albeit now with a bipolar dependence on V_m.

The circuit of Fig. 15.8 provides the required linearisation. With

$$V_{od} \propto \tanh(qV_c/2kT)$$

as in [15.19] then if the V_c input is preceded by a circuit with the appropriate $\tanh^{-1}$ transfer function then the non-linearity will be corrected. For two diodes D_+ and D_-

$$I_{d-} = I_{do} \exp(-qV_{d-}/kT) \quad \ldots(a)$$
$$I_{d+} = I_{do} \exp(-qV_{d+}/kT) \quad \ldots(b) \quad (15.21)$$

Thus
$$\frac{I_{d+}}{I_{d-}} = \exp\left(\frac{qV_{dd}}{kT}\right) \quad (15.22)$$

where $V_{dd} = V_{d+} - V_{d-}$

With $I_{dd} = I_{d+} - I_{d-}$

and from the circuit $I_{d+} + I_{d-} = 2I_z \quad (15.23)$

then
$$I_{dd} = 2I_z \tanh\left(\frac{qV_{dd}}{2kT}\right) \quad (15.24)$$

Also from the circuit $I_{dd} = V_{id}/R$

where $V_{id} = V_{in+} - V_{in-}$

Thus
$$V_{id} = 2I_z R \tanh\left(\frac{qV_{dd}}{2kT}\right) \quad (15.25)$$

With the output V_{dd} applied to the differential upper input to the double-balanced modulator, to effectively constitute V_c in [15.19] then

$$V_{od} = V_{id}V_m(-R_L/(I_zR_eR)) \quad (15.26)$$

and so the tanh non-linearity is corrected.

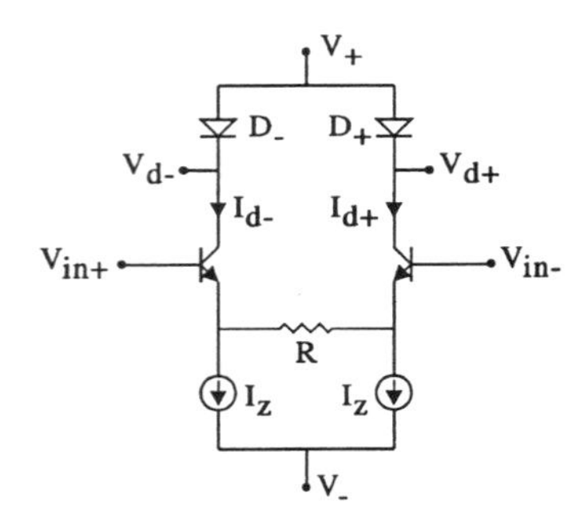

Fig. 15.8 Tanh^{-1} converter used to linearise upper input of Gilbert cell of Fig. 15.7.

The converter is an ingenious method of causing the two diodes to parallel the behaviour of the LTP transistors in the modulator, but in reverse with the diode currents the input and the diode voltages the output.

Fig. 15.9 shows how the linearity is maintained. The $\tanh^{-1}$ transfer function is set on its side in (b), for the output to match the modulator input for the modulator bow-tie in (c). It is thus seen how the increase in the slope of the $\tanh^{-1}$ function in (b) compensates for the decrease in the slope of the bow-tie in (c) give the linear bow-tie of (d).

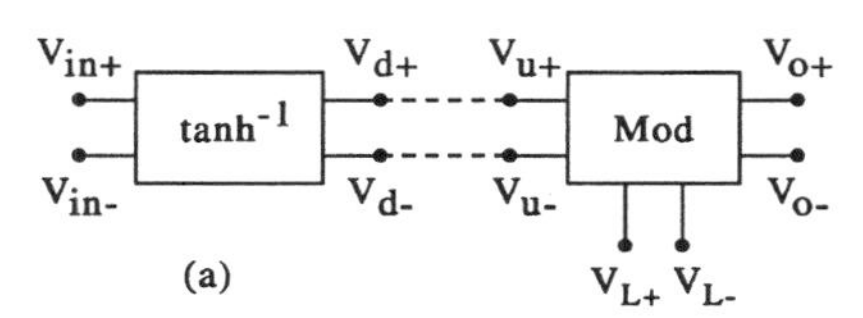

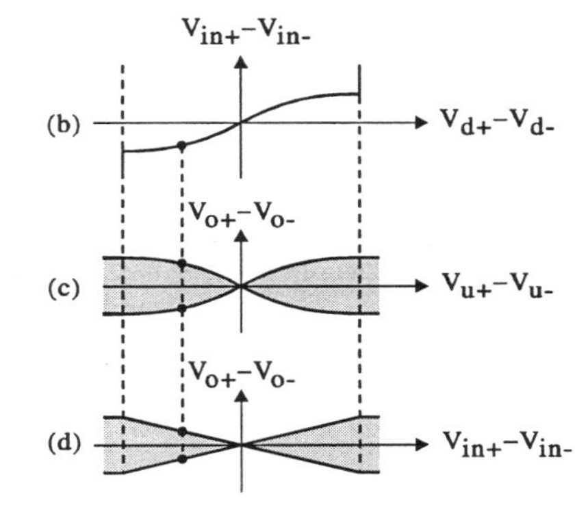

Fig. 15.9 Connection of $\tanh^{-1}$ converter of Fig. 15.8 to Gilbert cell modulator of Fig. 15.7.
(a) Block diagram.
(b) Tanh^{-1} converter transfer function.
(c) Non-linear bow-tie for modulator.
(d) Bow-tie for linearised modulator.

15.7 Detector circuits

Fig. 15.10(a) shows the analogue multiplier used as a mean-square detector, as for measuring the value for noise, etc.

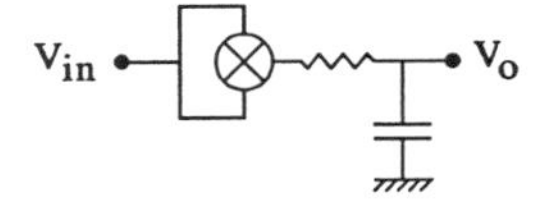

Fig. 15.10(a) Mean-square detector using analogue multiplier.

For frequencies beyond the range of the op-amp the circuit of (b) is an effective alternative to the precision rectifier of Fig. 4.17. Here the comparator output, suitably reduced in amplitude, drives the upper input to hard-switch the low-level signal at the lower input according to its sign, as would the ideal diode rectifier.

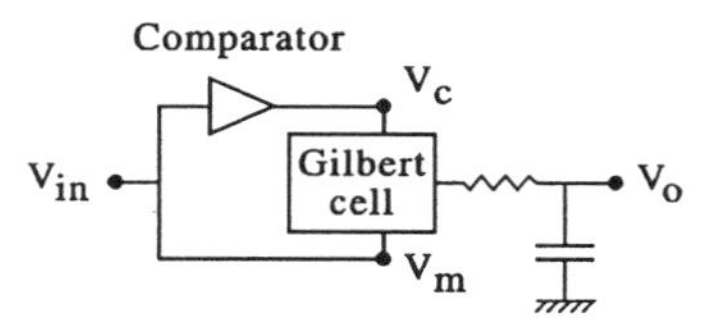

Fig. 15.10(b) Full-wave rectifier using Gilbert cell as reversing-switch modulator.

Reference

Gilbert, B. 1968: *A Precise Four-Quadrant Multiplier with Subnanosecond Response*. IEEE Journal of Solid State Circuits, **SC3**.

15.8 Exercises and experiments

Exercise 15.1 LTP modulator

All of the experimental work in this section is based on the LTP modulator of Fig. 15.11.

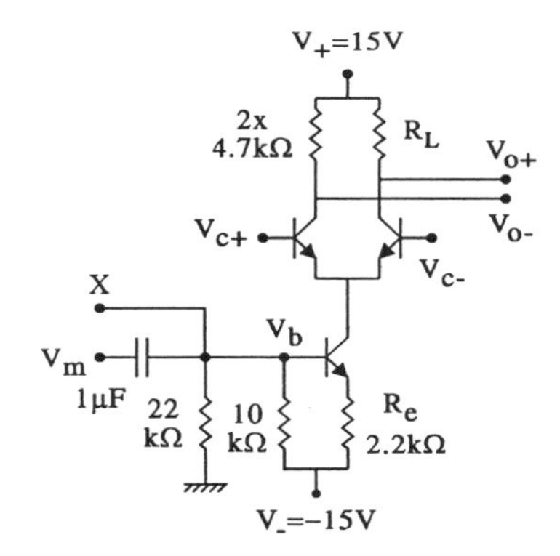

Fig. 15.11(a) LTP modulator used in experimental work.

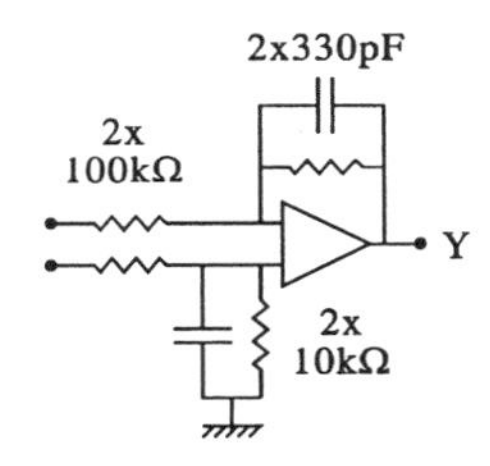

Fig. 15.11(b) Differential amplifier used to give $Y \approx V_{od} = V_{o+} - V_{o-}$ in (b).

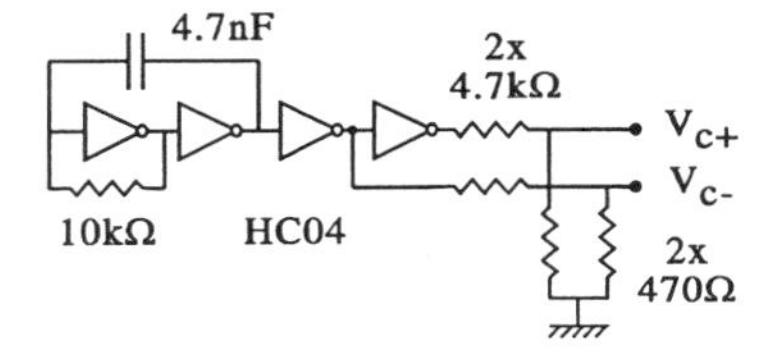

Fig. 15.11(c) 10kHz, 0.5V source for hard switching (a).

Calculation. Operating voltages and currents Calculate the approximate values for these for the LTP in (a), with the subtractor and the 5V supply to the switching circuit in (c) disconnected.

For $I_b=0$ $V_b=(-15V)\times 22k\Omega/(22k\Omega+10k\Omega)$

giving $\mathbf{V_b \approx -10.3V.}$

With $V_e \approx V_b-600mV$ then $\mathbf{V_e=-10.9V.}$

With $I_e=(V_e-V_-)/R_e$ then $\mathbf{I_e=1.86mA.}$

For $\beta=200$ then $I_b=1.86mA/200$ giving $I_b=9.29\mu A$.

With the resulting drop in

$$V_b=I_b\times 1/(1/10k\Omega+1/22k\Omega)=63.9mV$$

this can be neglected.

With the $V_o=V_+-(I_e/2)\times R_L$

then $V_o=15V-(1.86mA/2)\times 4.7k\Omega$ giving $\mathbf{V_o=10.6V.}$

Experiment Confirm the above values experimentally.

Calculation. Peak output voltage values Calculate these for the V_{o+} and V_{o-} waveforms for $V_m=0$ and the 10kHz oscillator operating.

For the LTP transistor conducting $I_c=I_e$ giving $V_o=6.26V$. Thus V_c is switched between $V_+=15V$ and 6.26V.

Experiment With $V_m=0$ confirm that V_e is as before while V_{o+} and V_{o-} are anti-phase 10kHz square waves with the above limits.

Apply V_m as a 100Hz sine wave and use the scope dual y-t mode to compare the V_m waveform with each of V_+ and V_{o-}. These are the 'hard-switched' waveforms of Fig. 15.2(b).

The upper limit of the envelope is where I_e is cut off. The other is saturation. Which will occur first? If saturation, will this be of the tail transistor or the LTP?

Subtractor The most direct way to monitor the differential $V_{od}=V_{o+}-V_{o-}$ would be to use the subtractor of Fig. 2.13 with all the resistors equal, to give the subtractor output equal to V_{od}. Here a voltage-follower would be needed at each input to avoid the subtractor loading the modulator. However,

(i) with V_+ the upper limit of the V_o then the op-amp will not follow to this limit,
(ii) the amplitude of the subtractor output would be such as to exceed the op-amp output range.

Thus the subtractor gain is reduced to $\approx 1/10$ and the input resistors are made $>>R_L$ to make the loading small.

The 330pF capacitors are added to give something of the 'filled in' effect for the output envelope when the modulated waveform is a sine wave, rather than the present square wave.

Experiment. Differential output Connect the subtractor and thus observe the scaled down V_{od} waveform, as in Fig. 15.3(c).

Switch the scope to the X-Y mode, and connect to points X and Y in the diagram to show the bow-tie of Fig. 15.3(b).

How should the bias chain be altered to increase the linear range?

Exercise 15.2 LTP modulator with transposed inputs

Experiment Apply the output from the final gate in Fig. 15.11(c) to the lower LTP input, and V_m to one of the upper inputs, grounding the other. Thus observe the waveform, and bow-tie of Fig. 15.5(b).

A 10kHz Delyiannis-Friend band-pass filter as in Fig. 13.9 would allow (c) to be viewed.

16

RF amplifiers and frequency-changers

Summary

This chapter completes the discussion of the modulator and amplifier circuits to be found in the transceiver circuits of Fig. 14.9 and elsewhere.

RF and VHF ranges

For brevity, the term 'RF' in the title is taken loosely to include both the more normally defined RF (radio frequency) range from ~100kHz to ~30MHz, and the VHF (very high frequency) range from ~30MHz to ~1000MHz. Within the chapter the more normal definitions will be followed.

For amplifiers and modulators of this class the input and output circuits may be in the form of an LC resonator, a wideband transmission-line transformer, or a ceramic or quartz filter.

Frequency-changer

The term 'frequency-changer' is traditionally reserved for the conversion of the output of the RF or VHF stages to the intermediate frequency (IF) and back. Remember, however, that the conversion from the audio 'base-band' to the first IF in the single-sideband transmitter, and the corresponding demodulation, are both a form of frequency conversion. We shall follow the traditional interpretation

RF and VHF amplifiers

In addition to the modulators, demodulators, frequency-changers and filters shown in the transceivers of Fig. 14.9, the circuits will also include RF, IF and VHF amplifiers. The first section describes the low-noise RF and VHF amplifiers used at the input of the related receivers, and the RF and VHF power output stages for the corresponding transmitters.

While the VHF circuits are resonator coupled, giving operation essentially at one frequency only, the RF circuits use the transmission-line transformer, giving coverage of most of the RF range for any one transformer.

Envelope distortion

For an amplifier or frequency-changer with a narrow-band band-pass filter at the output, tuned to the frequency of the required output component, as the amplitude of the input sine wave is increased beyond the point of distortion, so the amplitude of the output sine wave ceases to increase in proportion to that of the input. For the input amplitude modulated then the envelope of the output becomes symmetrically distorted, for example with the blunting of the lower peaks the same as that of the upper. Here some of the factors causing distortion in a wideband amplifier do not contribute to the envelope distortion.

Automatic gain control

For the transceiver the overall gain needs to be automatically controlled. For the receive mode this is 'automatic gain control' (AGC) where the gain is adjusted to suit the widely differing input signal levels.

For the transmit mode the equivalent is 'automatic level control' (ALC). Usually the full transmitter output power is needed for the speech peaks, and ALC ensures that this does not overdrive the final power amplifier, giving envelope distortion and unwanted harmonics of the carrier frequency. In each case the output level is sensed by a detector circuit, compared with the reference, and the error signal amplified to control the gain.

Alternatives to LTP modulator

As finally described, these comprise

- the dual-gate MOS,
- the switched JFET amplifier,
- the JFET or CMOS switch,
- the diode ring,
- the single-diode microwave circuit.

16.1 RF and VHF amplifiers

VHF receive-mode first stage

Fig. 16.1(a) shows the typical dual-gate tuned amplifier between the antenna and the frequency-changer in the VHF (very high frequency) receiver, with the input and output coupling by LC resonators.

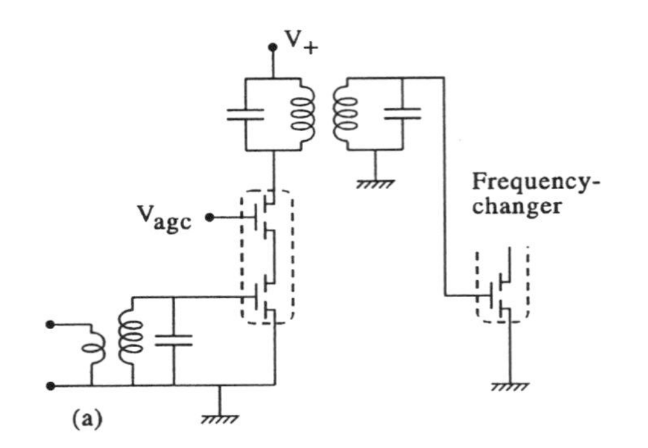

Fig. 16.1(a) VHF receiver first stage and frequency-changer.

Image response

For VHF operation it is normally not possible for the intermediate frequency f_{if} to be above the input signal frequency f_{in}. With f_{LO} the local oscillator frequency and with

$$f_{if} = |f_{LO} - f_{in}|$$

then for a given f_{if} and f_{LO} there are potential input responses at

$$f_{in} = f_{LO} \pm f_{if},$$

For $f_{in} >> f_{if}$, as may well be the case, then the two responses, that required and the 'image', are relatively close, requiring narrow-band band-pass filtering at the input of the frequency-changer. The circuit shown has three such resonators. Sometimes more are required.

Example For f_{if} the standard 455kHz, f_{in}=30MHz and

$$f_{LO} = 30\text{MHz} + 455\text{kHz}$$

then the image response will be at

$$f_{LO} + 455\text{kHz} = 30\text{MHz} + 2 \times 455\text{kHz}$$

Up-conversion

With the image separation for the above example only 3% of f_{in}, and also with the need for the filter to be tuneable, the implementation becomes complex.

This is why where possible, as for the RF transceiver, f_{if} is made larger than f_{in}. For f_{in}=30MHz as before and the usual f_{if}=70MHz then f_{LO} can be $f_{if}+f_{in}$=100MHz giving the image at $f_{LO}+f_{if}$=130MHz. All that is then needed to eliminate the image response is a fixed cut-off low-pass filter at the frequency-changer input.

Noise figure

Without the input tuned circuit, the amplifier input capacitance would give considerable signal loss. With the impedance and voltage step-up afforded by the resonator the signal level is increased to give a normally excellent noise figure.

RF receive-mode first stage

With up-conversion normally following, and with the MOS capacitances of relatively little effect, this stage can be relatively wideband, using the transmission-line transformers shown. The input transformer will give some step-up, in order to improve the noise figure (see Exercise 16.3).

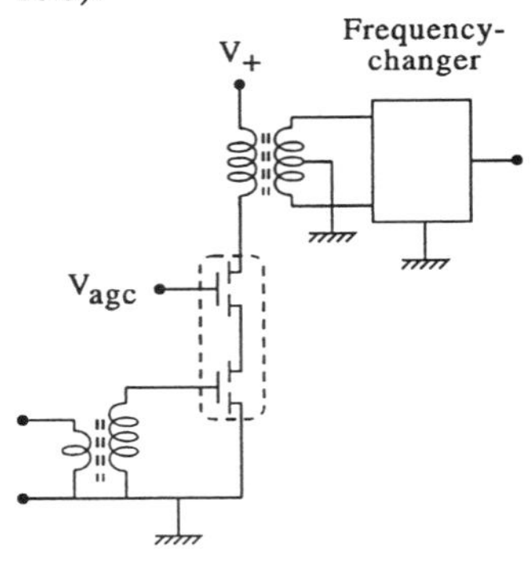

Fig. 16.1(c) First stage for RF receiver of Fig. 14.9(c).

Second intermediate frequency

With it not possible to implement the high-grade band-pass filter needed for single-sideband operation at 70MHz, then a further stage of frequency conversion is needed, usually down to 8MHz where the quartz filter can be implemented.

Bipolar transistor

Fig. 16.1(c) gives a somewhat extreme example showing why the bipolar transistor is less suitable than the MOS for an amplifier for which the input is tuned. With c_π here constituting the whole of the C for the resonator, the resonator Q is heavily damped by the transistor spreading resistance r_x, with also a contribution from r_π (see Exercise 16.2).

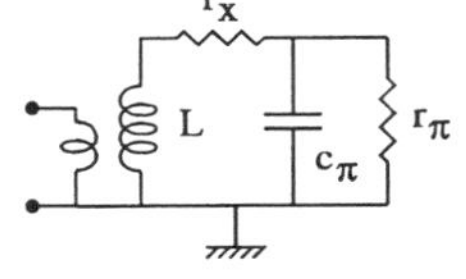

Fig. 16.1(c) Input capacitance c_π of bipolar transistor forming first-stage resonator C.

AGC control

The AGC voltage is applied as in Fig. 16.1(a) and (b) for the dual-gate MOS amplifiers. For full gain $V_{agc} \sim V_+/2$, with the gain reduced as V_{agc} is lowered. Supply and AGC decoupling capacitors are omitted for simplicity.

Local oscillator rejection

A limitation of the up-converter is that for $f_{in} << f_{if}$ then $f_{LO} \approx f_{if}$. For example for

$$f_{in}=50\text{kHz} \quad \text{and} \quad f_{if}=70\text{MHz}$$

then $$f_{LO}=70\text{MHz}+50\text{kHz}.$$

With the 70MHz 'roofing' filter of only the relatively low grade needed to remove the major unwanted components then this value of f_{LO} would be well within the filter band-pass. Thus to avoid overload of later stages it is necessary that the frequency-changer be of the 'balanced' type, ideally giving no output component at f_{LO}. This will usually need the first-stage output transformer to be split phase as shown.

Input transformer

For the nearly three-decade f_{in} range (50kHz to 30MHz) for the double-conversion system, the transformers will need to be of the transmission-line type in Section 22.6. However, with the antenna noise temperature increasing from thermal at 100MHz at the rate of ≈30dB/decade, then the input transformer is not really needed below ~ 10MHz. Thus a transformer that is conventionally wound on the ferrite ring, and switched out above the upper transformer limit of ≈1MHz, will suffice.

Transmit-mode final stage (VHF)

With the typical circuit as in Fig. 6.2(a), L and C provide the resonant impedance transformation needed to convert the output 50Ω load to the lower impedance for the design output power.

L_1 and L_2 are only needed for the DC connection. They therefore are of relatively high impedance. Similarly the DC blocking capacitors C_b are of relatively low impedance.

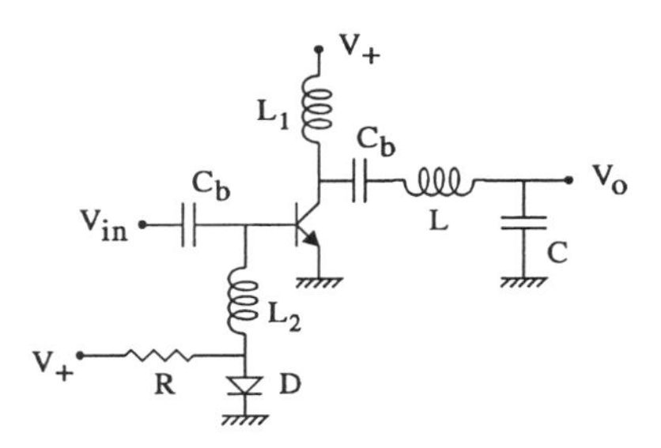

Fig. 16.2(a) Typical VHF final power amplifier circuit.

R is adjusted to bias the transistor to a modest 'quiescent' current for $V_{in}=0$.

As the sine wave V_{in} is applied then the transistor conducts for each half-cycle of V_{in}. The resonant LC network resolves the fundamental from this highly distorted waveform to some degree, but not nearly enough to reduce the harmonic radiation to an acceptable level. Thus further filtering is needed, probably with the second-order resonant low-pass response of Fig. 12.15(c).

Transmit-mode final stage (RF)

With operation possible here over the typical range of 100kHz to 30MHz without tuning, the transmission-line transformers of Fig. 16.2(b) can be used to provide the required input and output power matching.

Here, unlike for the narrow-band output response of the VHF amplifier output, the harmonics of the output frequency are no longer suppressed.

The normal method for harmonic suppression over the nearly three-decade range of f_o is to have a set of switched low-pass filters, each of abrupt cut-off f_L and with the cut-off increment ×1.5. This places the second harmonic well into the filter stop-band even for the lowest value of f_o used for the filter.

The push-pull arrangement of (b) ideally fills in the missing half-cycles of V_{in}, thus reducing the harmonic radiation. Also such non-linearity as remains is anti-symmetric. As in Section 1.8 this produces only odd-order harmonics, placing the first harmonic of any consequence even further into the filter stop-band (see Exercise 16.4).

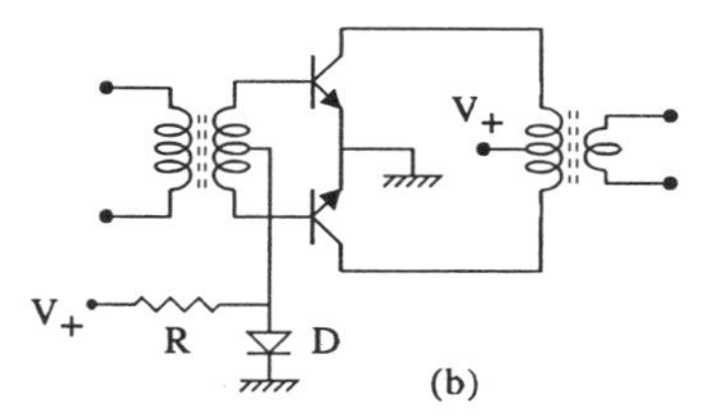

Fig. 16.2(b) Typical RF final power amplifier circuits.

16.2 Envelope distortion

Envelope distortion is where the amplitude of the output of a tuned amplifier or frequency-changer ceases to be proportional to that of the input. Consider the effect of applying a sine wave of frequency f_o to the input of an amplifier, with the amplifier output subject to an ideal narrow-band band-pass filter tuned to f_o. As the input amplitude is increased to the point of distortion of the amplifier then components of frequency nf_o are produced, with n integer and including zero. Of these only the component of frequency f_o will emerge, with the amplitude now no longer proportional to that of the input, usually eventually symmetrically clipping the peaks of the output envelope. This is envelope distortion.

With the general transfer function for the amplifier

$$V_o = k_o + k_1 V_{in} + k_2 V_{in}^2 + k_3 V_{in}^3 \ldots \quad (16.1)$$

the odd-order terms represent the anti-symmetric component of the transfer function and the even-order terms the symmetric component. Since the symmetric component produces only even-order harmonics of f_o, including that at zero frequency, none of these is accepted by the filter. Thus it is only the odd-order terms that give the envelope distortion.

Field-effect transistor

The above point is of particular consequence for the FET. With the channel current

$$I_c = I_{co}(1 - V_{gs}/V_{th})^2$$

as in [10.3], with the gate-source voltage V_{gs} biased to the mid-point $V_{th}/2$, and with V_{in} added, then

$$I_c = k_o + k_1 V_{in} + k_2 V_{in}^2 \quad (16.2)$$

where $k_o = I_{co}/4$, $k_1 = -I_{co}/V_{th}$ and $k_2 = I_{co}/V_{th}^2$.

Here the only distortion term is of even-order (symmetric) so we expect no envelope distortion.

For $$V_{in} = \hat{V}_{in}\cos(\omega_{in}t) \quad (16.3)$$
then

$$I_c = k_o + k_1 \hat{V}_{in}\cos(\omega_{in}t) + k_2 \hat{V}_{in}^2[(1+\cos(2\omega t_{in})]/2 \quad (16.4)$$

With only the component $k_1 \hat{V}_{in}\cos(\omega_{in}t)$

selected by the output filter, then the amplitude of the output signal is strictly proportional to that of the input, confirming the absence of envelope distortion.

Power transistor

Regardless of this rather remarkable result, it is the bipolar transistor that is used most frequently in the power amplifiers of Fig. 16.2, for the following reasons.

RF operation

While for the bipolar transistor the variation of collector current I_c with base-emitter voltage V_{be} is exponential, and therefore highly non-linear, for the power transistor the current levels are such that for low-frequency operation I_b is largely controlled by the spreading resistance r_x. With $I_c \propto I_b$ then the non-linearity is much reduced.

Also negative feedback may be applied, by AC-coupled resistors from collector to base.

Finally, each of the many component transistors that make up the single power transistor has its own small emitter resistor. This is partly to equalise the collector currents as described in Section 21.3, but also gives additional negative feedback.

VHF operation

Here the effect of the base 'stored charge' Q_s is to make the input impedance apart from r_x even lower, thus making the variation of I_b with V_{in} even more linear. With

$$I_c = T_\tau^{-1}\int I_b dt$$

as in [6.46] for the charge-control model then the linearity obtains equally for I_c.

Efficiency

Because of the higher FET saturation voltage, giving greater transistor power dissipation at current peaks, the efficiency of the bipolar circuit is higher, particularly for mobile operation where the supply is the relatively low 12V of the car battery.

General non-linearity

With the general transfer function again

$$V_o = k_o + k_1 V_{in} + k_2 V_{in}^2 + k_3 V_{in}^3 \ldots \quad (16.5)$$

and with $V_{in} = \hat{V}_{in}\cos(\omega_{in}t)$ as in [16.3] then Fig. 16.3(a) shows how the output components develop.

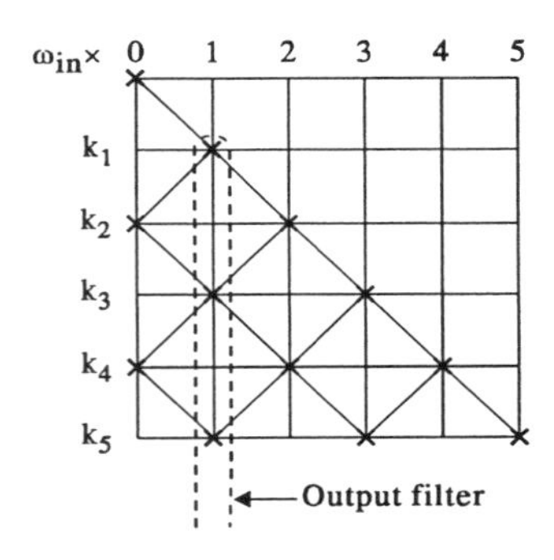

Fig. 16.3(a) Envelope distortion. Output components for $V_{in} = \hat{V}_{in}\cos(\omega_{in}t)$ applied to general non-linear transfer function.

Row k_o. This is the zero-frequency component.

Row k_1. This is the required component

$$V_1 = k_1 \hat{V}_{in}\cos(\omega_{in}t)$$

Row k_2. With

$$V_2 = k_2 V_{in}^2 = k_2(\hat{V}_{in}\cos(\omega_{in}t))^2$$

then as above

$$V_2 = k_2 \hat{V}_{in}^2(1+\cos 2\omega_{in}t)/2 \quad (16.6)$$

Row k_3. With

$$V_3 = k_3 V_{in}^3 = k_3(\hat{V}_{in}\cos(\omega_{in}t))^3$$

then $$V_3 = k_3 \hat{V}_{in}^3[(3/4)\cos\omega_{in}t + (1/4)\cos 3\omega_{in}t)] \quad (16.7)$$

Here the components for any Row r are obtained by multiplying each of the components of Row r−1 by $\cos(\omega_{in}t)$, thus splitting each component into two. With only the components of the ω_{in} column selected by the output filter, the amplitudes A_r are as follows.

A_0	A_1	A_2	A_3	A_4	A_5	$A_6 \ldots$
0	$k_1\hat{V}_{in}$	0	$(3/4)k_3\hat{V}_{in}{}^3$	0	$\approx k_5\hat{V}_{in}{}^5$	0

Table 16.1 Amplitudes A_r of components of input frequency ω_{in} due to k_r terms in [16.5], also in r_{th} row of Fig. 16.3(a).

Thus while $A_1 \propto \hat{V}_{in}$ as required, A_3, A_5, etc., are not. It is these which give the envelope distortion. For the ideal FET only the first three rows of (a) are populated, hence again the absence of envelope distortion.

Spectral spread

With the envelope distortion as if the modulating V_m had been subject to anti-symmetric distortion, then the corresponding V_m spectrum would be as in Fig. 16.3(b), with only odd-order harmonics.

With the undistorted output spectrum for the suppressed carrier modulation of Fig. 14.7 as in Fig. 16.3(c) then the effect of the envelope distortion is to give the added sidebands of (d).

An undesirable feature here is that for normal speech modulation, not only is the received speech distorted, but the spectrum of the transmitted signal spreads to the neighbouring channels.

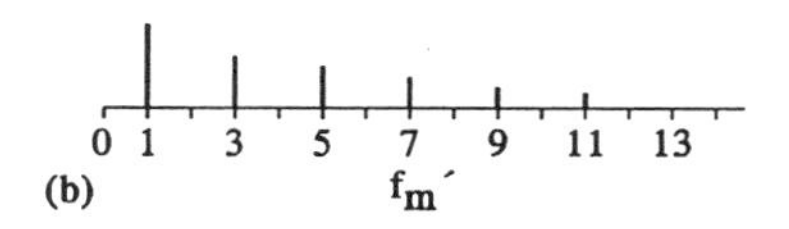

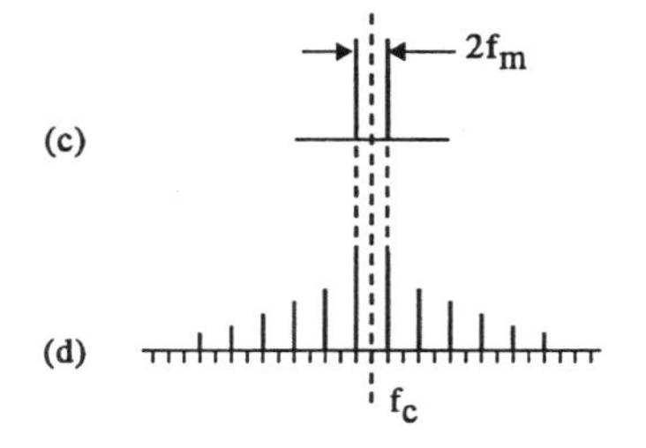

Fig. 16.3(b) to (d) Envelope distortion.
(b) V_m spectrum for envelope distortion attributed to anti-symmetric distortion of V_m waveform.
(c) Output spectrum for suppressed-carrier sine wave modulation using analogue multiplier.
(d) As (c) but with envelope distortion.

Two-tone test

The above sideband spread becomes significant long before the envelope distortion can be perceived on the waveform. Thus the normal test procedure is to apply the above suppressed carrier amplitude modulation and to measure the resulting sidebands using a spectrum analyser. Here it is important that the test modulator be highly non-linear and in practice this is side-stepped by generating the test waveform as the sum of two pure sine waves of slightly differing frequency and the same amplitude. Thus, as in Fig. 16.4

$$V_{test} = \hat{V}_{test}\left[\cos(\omega_{m1}t) + \cos(\omega_{m2}t)\right] \qquad (16.8)$$

gives the required $V_{test} =$

$$2\hat{V}_{test} \times \cos\left(\frac{\omega_{m1} + \omega_{m2}}{2}t\right) \times \cos\left(\frac{\omega_{m1} - \omega_{m2}}{2}t\right) \qquad (16.9)$$

For testing a transmitter, the 'two tones' can be of differing audio frequencies, sufficiently far apart to suit the resolution of the spectrum analyser, but within the 3kHz bandwidth of the transmitter.

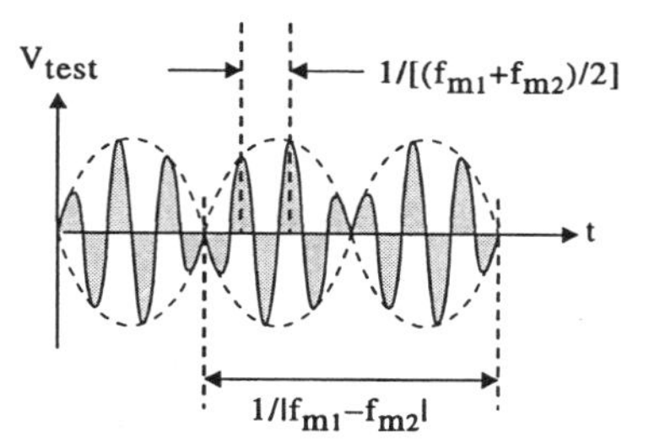

Fig. 16.4 Waveform for two-tone test.

Third-order intercept

This is the value indicating the extent to which a tuned amplifier, or the whole transceiver, will give envelope distortion. With the A_r in Table 16.1 the amplitudes of the components of frequency ω_{in} selected by the output filter as in Fig. 16.3(a) then Fig. 16.5 shows the variation of each of these with the amplitude $\hat{V}_{in}$ of V_{in}, together with the amplitude $\hat{V}_o$ of the total component of frequency ω_{in}.

Here for $\hat{V}_{in} \ll V_{int}$ the required component of amplitude A_1 dominates, to give $\hat{V}_o \propto \hat{V}_{in}$ as required.

As $\hat{V}_{in}$ increases then the envelope distortion causes curvature in the $\hat{V}_o$ response shown. There is considerable interest in the degree of distortion for $\hat{V}_{in}$ significantly below V_{int} and there it is the 'third-order' component A_3 that dominates the distortion, there being no A_2 component.

Thus for any value of $\hat{V}_{in} \ll V_{int}$ the degree of envelope distortion can be calculated from the values of $\hat{V}_{in}$

and the 'third-order intercept' point V_{int}. With A_1 and A_3 as in the table, and for $A_1 = A_3$,

$$V_{int} = [4k_1/(3k_3)]^{1/2} \qquad (16.10)$$

For the whole receiver the value is more usually expressed as P_{int}, the available power from a 50Ω signal source giving the intercept.

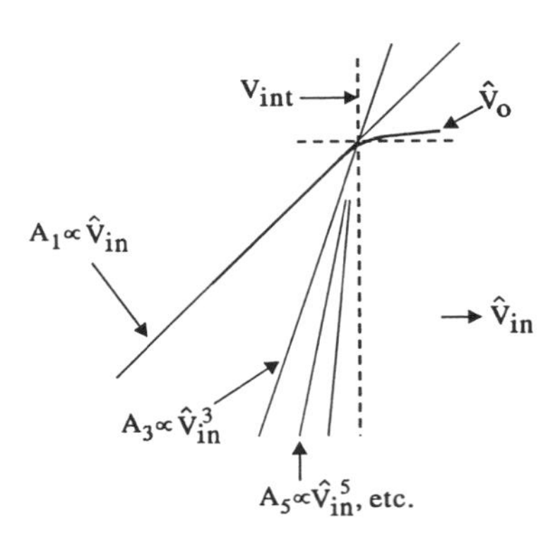

Fig. 16.5 Third-order intercept V_{int}.

Measurement of third-order intercept

With the principal unwanted effect of the envelope distortion the spectral sideband spread that it incurs, this is a suitable way of measuring V_{int}. With

$$I_c = k_o + k_1 V_{in} + k_2 V_{in}^2 + k_3 V_{in}^3 \ldots$$

as in [16.5] and expressing the two-tone test voltage V_{test} in the more familiar form

$$V_{test} = \hat{V}_{test} \cos(\omega_c t).\cos(\omega_m t) \qquad (16.11)$$

• the k_1 term simply reproduces the input sidebands at $\omega_c \pm \omega_m$, each of amplitude

$$B_1 = k_1 \hat{V}_{test}/2 \qquad (16.12)$$

• the k_2 term produces no terms in the region of ω_c, only in the region of zero and $2\omega_c$;
• the k_3 term produces distortion sidebands at $\omega_c \pm 3\omega_m$ of amplitude

$$B_3 = k_3 \hat{V}_{test}^3/4 \qquad (16.13)$$

and also at $\omega_c \pm \omega_m$ of amplitude

$$B_{13} = k_3 \hat{V}_{test}^3/2 \qquad (16.14)$$

Here the B_1 and B_{13} components are of the same frequency preventing the value of either to be assessed separately. However, $\hat{V}_{test}$ can be reduced to the point where $B_{13} << B_1$ and then B_1 can be measured; the assessment of B_3 presents no such problem. With V_{int}, B_1 and B_3 as above,

$$V_{int} = \hat{V}_{test}\left(\frac{2B_1}{3B_3}\right)^{1/2} \qquad (16.15)$$

For measurement the receiver is used as its own spectrum analyser. The two tones are separated by ~50kHz, which is well beyond the 3kHz bandwidth of the 8MHz filter, but well within the bandwidth of the part of the receiver preceding the filter. Thus the receiver can be tuned to the sidebands one at a time.

Sideband spread

In use the effect is much as if the transmitter output envelope was distorted, with the increased sideband spread. Thus a strong signal in a channel close to that in use can have its sidebands spread into the channel in use, this time due to the receiver overload.

Exercise 16.5 gives example calculations relating the third-order intercept and also to noise.

16.3 Dual-gate MOS

Gain-controlled amplifier

For the required automatic gain control, the gain of the dual-gate MOS is controlled by varying the upper gate potential, as in Fig. 16.1(a) for the receiver first stage. With the device used for most of the other stages, the AGC voltage is applied to all in the same way.

Fig. 16.6(a) shows the essentials of the device, with (b) and (c) showing how $\partial I_c/\partial V_{g1}$ is reduced when V_{d1} is below the pinch-off value. Here Q_2 acts as a source-follower controlling V_{d1}.

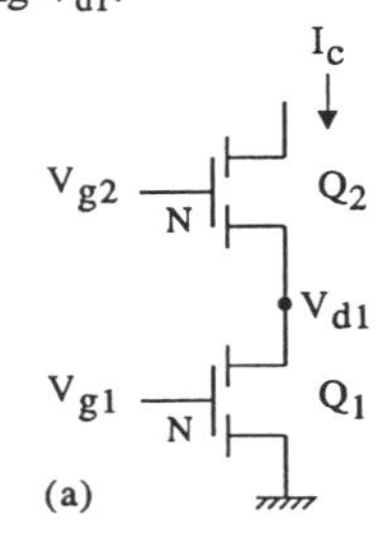

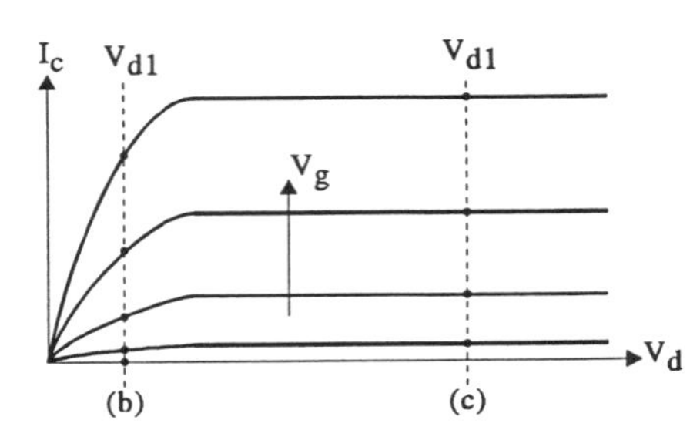

Fig. 16.6 Dual-gate MOS cascode.
(a) Essentials.
(b) Output characteristics for Q_1 below pinch-off.
(c) Above pinch-off.

Combined output characteristics

Fig. 16.7 shows the situation rather more accurately. With Q_2 pinched-off the variation of I_{d2} with V_{gs2} is parabolic. The load-line presented to Q_1 by Q_2 thus becomes parabolic as shown.

As V_{g2} is lowered from (c) to (b) then the mutual characteristic is reduced much as in Fig. 16.6, but now being traced by the intersections in Fig. 16.7(b).

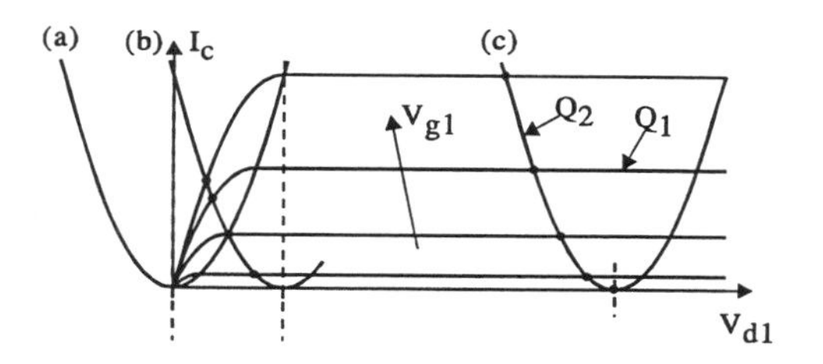

Fig. 16.7 Output characteristics of Q_1 in Fig. 16.5 with the load-lines imposed by Q_2 superimposed.
(a) $V_{gc}=V_{th}$,
(b) $V_{gc} > V_{th}$ but Q_1 below pinch-off.
(c) $V_{gc} >> V_{th}$, Q_1 above pinch-off.

Here for part of the resulting mutual characteristic the FET is above pinch-off and part below. Thus the previously parabolic nature of the characteristic is compromised, also compromising the immunity to envelope distortion obtained above pinch-off.

Note too that the gain-control function is far from linear, being more nearly parabolic. This is of little consequence for use as a gain-controlled amplifier in an AGC system, but renders the device unsuitable as a linear modulator.

Frequency-changer

The device is also widely used as a frequency-changer, with the local oscillator output superimposed on V_{g2}, and with the amplitude such as to virtually switch the device on and off. With the parabolic nature of the normal mutual characteristic retained, envelope distortion is largely avoided.

16.4 Balanced frequency-changers

A modulator is said to be 'singly-balanced' if, in addition to the required sum and difference frequency components, there is an output component at only one of the input frequencies, and 'doubly balanced' if neither is present.

The dual-gate frequency-changer is in no sense balanced. With f_s and f_{LO} the signal and local-oscillator frequencies then there are output components of both of these, to say nothing of the DC component.

As stated above, and with f_{if} the intermediate frequency, an example of where at least single balancing is needed is where for the up-converting receiver

$$f_s=50\text{kHz},\quad f_{if}=70\text{MHz},\quad \text{and}\quad f_{LO}=70.050\text{kHz}$$

normally giving f_{LO} well within the intermediate frequency filter band-pass.

Fig. 16.8 is a notional diagram showing how two unbalanced modulators can be combined to give singly balanced operation, and then how two of the singly balanced pairs can be combined to give doubly balanced operation.

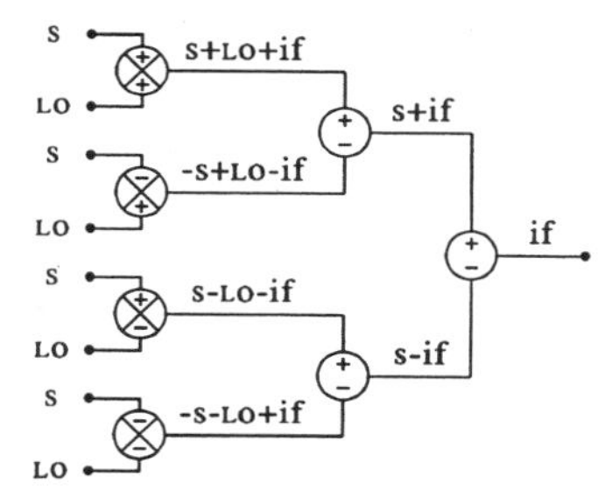

Fig. 16.8 Combination of four unbalanced frequency-changing modulators can be combined to give doubly balanced operation. s: signal, LO: local oscillator, if: intermediate frequency.

Problem Devise singly and doubly balanced types using two and four dual-gate transistors.

16.5 JFET frequency-changers

Fig. 16.9(a) shows two JFETs combined as in Fig. 16.8 to give a singly balanced frequency-changer, while (b) shows four combined to give the doubly balanced circuit (see Exercise 16.6).

For (a) the transistors operate as switched common-source amplifiers, with both switched on and off at the same time by the local oscillator. Here the balancing is such as to eliminate the local oscillator component at the output.

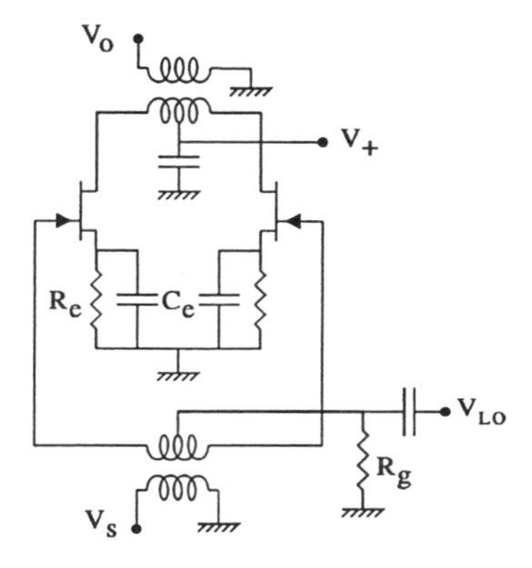

Fig. 16.9(a) Singly balanced JFET frequency-changer.

For (b) the transistors are simply operated as switches, giving no gain at all. This is acceptable for the transmitter but less so for the receiver, where the noise figure is more important.

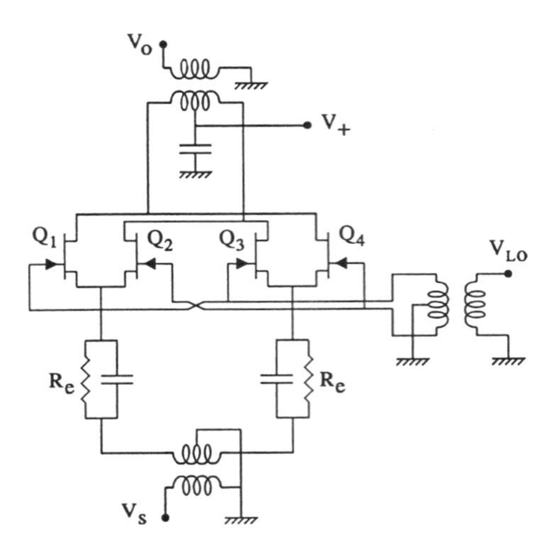

Fig. 16.9(b) Doubly balanced JFET frequency-changer.

Problem

How would the Gilbert cell of Fig. 15.7 compare with the two JFET frequency-changers of Fig. 16.9 from the point of view of noise figure and third-order intercept?

16.6. CMOS switching modulator

Fig. 16.10 shows the HC4316 integrated circuit used as a reversing-switch demodulator. The strength of this arrangement is the very high input overload levels, close to the ±5V supply range. Its weakness lies in the typically 10ns time delay of the switch inverter. This restricts the operation to carrier frequencies of below about 10MHz.

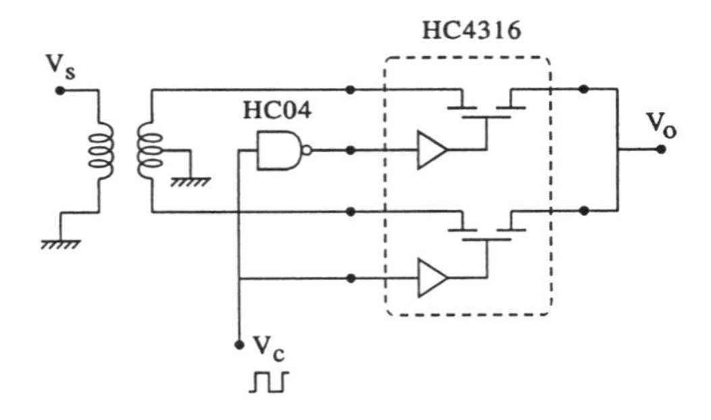

Fig. 16.10 Reversing-switch demodulator based on HC4316 (s: signal, c: carrier).

16.7 Diode ring modulator

This, in Fig. 16.11, is another form of reversing-switch modulator, this time with the diodes the switches. With the carrier V_c 'hard switching' the diodes, the connection between the modulating V_{in} and the output V_o is periodically reversed at the carrier frequency. For normal AM the waveforms are thus as in Fig. 14.3(c).

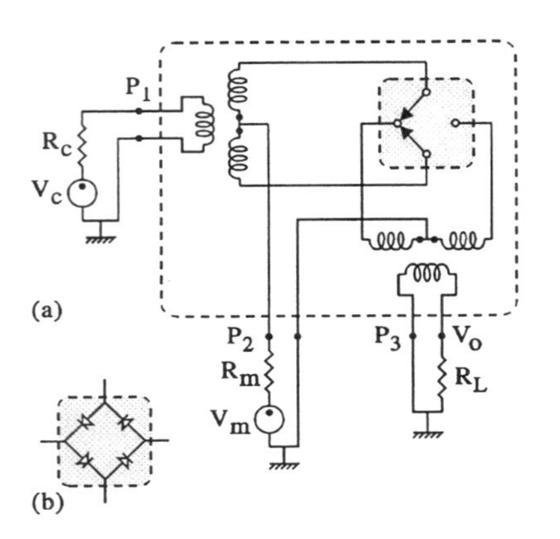

Fig. 16.11 Diode-ring double balanced modulator. (a) Circuit diagram. (b) Diode ring.

Because of the symmetry, the carrier input transformer presents no impedance in the path from the modulating source to the ring. With the usual 600mV across each of the two left-hand diodes, these cancel at the ring output.

As for any reversing-switch, the device can equally be used as a demodulator or frequency-changer. Here ports P_2 and P_3 are transposable, but the signal of lowest frequency is normally at P_2. For example, when used as a demodulator then P_2 is the output port.

Conversion gain (ideal diodes)

When used as a frequency-changer, the ring modulator shares the low conversion gain of the CMOS type. For the switch resistance $R_{sw}=0$ and the source and load matched ($R_s=R_L$) then the conversion gain α of the otherwise ideal switching modulator is –4dB (see Exercise 16.7).

Upper frequency limit

To avoid further loss requires $R_{sw} << (R_s+R_L)$. For the CMOS HC4316 $R_{sw} \approx 30\Omega$ requiring $R_s+R_L \sim 300\Omega$. With the switch capacitances ~5pF this gives a time-constant of 1.5ns, corresponding to ≈10MHz. This is comparable with the limit imposed by the switch inverter propagation time.

For the diode ring the typical quoted maximum diode current $I_d \approx 14$mA giving $R_d=(kT/q)/I_d \approx 1.8\Omega$. Also there are two 'on' diodes in parallel making 0.9Ω, much lower than the 30Ω for the CMOS.

Schottky diodes are invariably used, to avoid the stored charge of a PN diode, and for these the junction capacitance is typically 1pF. With $R_s = R_L = 50\Omega$ the time constant is 100ps corresponding to 150MHz, 15 times better than for the CMOS switch.

Linearity

A weakness of the ring modulator is that, unlike the CMOS, the signal amplitude must remain small compared with that of the local oscillator. Otherwise the signal will begin to alter the times at which the diodes are switched. Also some further non-linearity arises from R_d changing with diode current. To reduce this effect, the minimum effective value of R_d is sometimes limited by adding a small resistor in series with each diode.

Conversion gain (typical)

Typical values of conversion gain obtained in practice for the ring modulator are −8dB. A further contributor here is the loss in the transformers, ohmic in the wire, and also in the ferrite core.

Bandwidth

The transformers in the ring modulator give a low lower frequency limit f_L as well as the above upper limit f_u. Transmission-line transformers are used as in Section 22.6 to give a typical three-decade range of from $f_L = 200\text{kHz}$ to $f_u = 200\text{MHz}$. Broadly, the ring modulator is best suited to high-frequency applications and the CMOS switch to low.

16.8 Microwave modulator

The higher the signal frequency f_s the simpler the modulator circuit needs to become. Fig. 16.12 shows a waveguide circuit suited to $f_s \sim 10\text{GHz}$. This is really just the envelope detector of (b) and can also be used as such.

The circuit is shown operating as a down-converting frequency-changer. The phasor diagram in (c) shows how the amplitude of the combined signal incident on the diode is amplitude modulated at f_{if}, to give the corresponding output from the detector.

The device can also be used as a modulator. If the carrier is injected as a single wave incident on the diode, and a modulating signal of much lower frequency f_m applied via the coax, then the reflected wave is suitably modulated. For balanced (suppressed-carrier) modulation the diode is matched to the waveguide, giving no reflection of the incident carrier. This is a matter of suitable tuning stubs etc.

Directional microwave side couplers are available, allowing the carrier or local oscillator signal to be injected with little interruption of the waveguide path of the required signal.

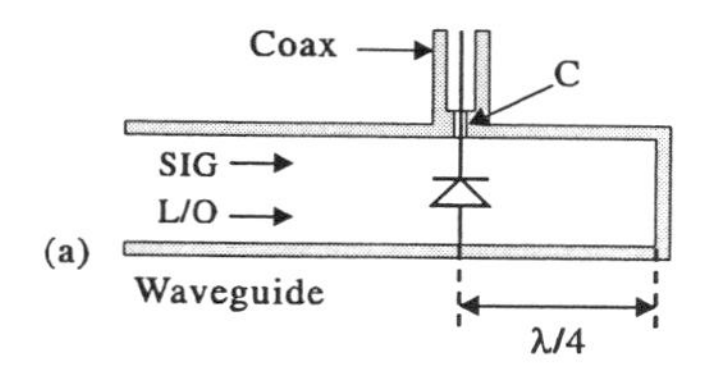

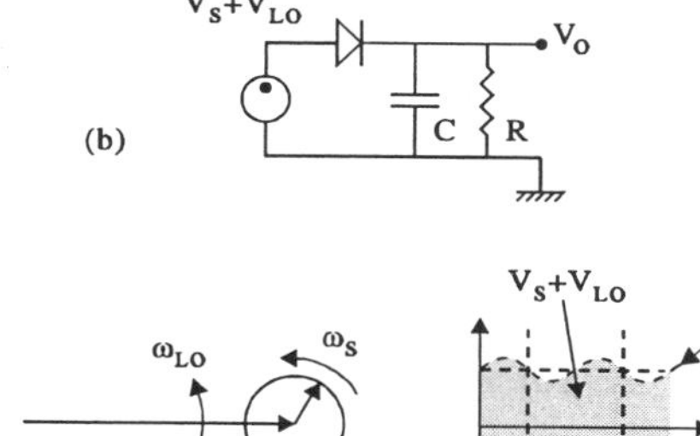

Fig. 16.12 Microwave frequency-changer.
(a) Waveguide mount.
(b) Effective circuit diagram.
(c) Rotating phasors for V_{lo} and V_s.
(d) Waveforms.

Reference

Smith, J. 1986: *Modern Communication Circuits*. McGraw Hill.

16.9 Exercises

Exercise 16.1 Receiver image response

Calculation The frequency-changer and first-stage amplifier of Fig. 16.1(a) is used in a VHF receiver for which $f_{in} = 150\text{MHz}$, $f_{if} = 8\text{MHz}$ and $f_{LO} = 158\text{MHz}$. Calculate the frequency of the image response and the image rejection factor α if $Q_1 = 30$ for the first resonator and $Q_2 = 100$ for the other two.

With the input responses at $f_{LO} \pm f_{if}$ then $f_{image} = f_{LO} + f_{if}$ giving **$f_{image} = 166\text{MHz}$.**

With the rejection factor per resonator $(\Delta f/2)/(2f_{if})$ where $\Delta f = f_{in}/Q$,

$\alpha = [f_{in}/(2Q_1 f_{if})] \times [f_{in}/(2Q_2 f_{if})]^2$ giving **$\alpha = 1/364 = 51.2\text{dB}$.**

This is below the generally required 60dB so a fourth resonator would be required.

Exercise 16.2 Resonator damping by bipolar transistor input

> **Calculate** the Q-factor for the bipolar transistor used in the circuit of Fig. 16.1(b) for f_{in}=30MHz. c_π=40pF, r_π=5kΩ, r_x=150Ω.

With $X_c=1/\omega_0 c_\pi$ then X_c=133Ω.

With $Q_{r\pi}$ that for r_π alone then $Q_{r\pi}$=37.7.

With Q_{rx} that for r_x alone, Q_{rx}=0.88 giving **Q ≈ 0.88.**

Exercise 16.3. Receiver first-stage input transformer

For the transformer-coupled RF receiver first-stage of Fig. 16.1(b) the FET input capacitance C_{gs}=5pF and the FET input noise resistance R_{nv}=1kΩ.

> **Calculation** For the source impedance R_s=50Ω, and the maximum value of the signal input frequency f_{in}=30MHz, calculate the maximum permissible transformer step-up ratio N, and the noise figure F with and without the transformer.

Transformer step-up ratio With the transformed impedance $R_s'=N^2R_s$ and the upper cut-off $f_u=1/2\pi R_s'C_{in}$,

$N^2=1/2\pi f_u C_{gs}R_s$ giving N=4.6. Thus the suitable **N=4.**

Noise figure With $R_s'=N^2\times R_s$ then R_s'=800Ω.

With $F=(R_s'+R_{nv})/R_s'$ then **F=2.25=3.5dB.**

Without the transformer $F=(R_s+R_{nv})/R_s$ giving **F=21=13.2dB.**

Exercise 16.4 RF transmitter output filters

> **Calculation** The cut-off frequencies f_L of the set of low-pass switched filters at the transmitter output are in steps of ×1.5. For the filter of fourth-order, calculate the filter attenuation α for the worst-case second and third harmonics.

The worst case is where $f_o=f_L/1.5$. Below this the next filter in the sequence can be used.

Second harmonic Here $f_2=2f_L/15$ giving $\alpha_2=(1.5/2)^4$ and thus **α_2=10dB.**

Third harmonic Here $f_3=3f_L/15$ giving $\alpha_2=(1.5/3)^4$ and thus **α_3=24dB.**

These values are inadequate, requiring the filter to have notches (zeros) shortly beyond f_L.

Exercise 16.5 Third-order intercept and noise

> **Calculation** For one of the frequency-changers in Fig. 16.9, with the following stages but without the first-stage amplifier, the resulting noise figure F_2=12dB, and the third-order-intercept power P_{int2}=30dBm (1W).
>
> The first stage F_1=2dB, and when this circuit is added the overall F becomes 3dB. Calculate the resulting reduction in P_{int}. The first stage may be considered distortion free.

With P_{th} the source thermal noise power, and P_1 the first stage noise power referred to the source, then

$$F_1=(P_{th}+P_1)/P_{th} \quad (16.80)$$

Similarly for the remaining stages used without the first stage

$$F_2=(P_{th}+P_2)/P_{th} \quad (16.81)$$

With P_{tot} the total noise power referred to the source and G_1 the first-stage power gain

$$P_{tot}=P_{th}+P_1+P_2/G_1 \quad (16.82)$$

With $F=P_{tot}/P_{th}$ then $F=F_1+(F_2-1)/G_1$ giving

$$G_1=(F_2-1)/(F-F_1) \quad (16.83)$$

With F(dB) = 10 × log(F)

F=2.00, F_1=1.58 and F_2=15.5,

giving G_i=(15.5–1)/(2–1.58) and thus G_1=34.5=15.4dB.

With $P_{int}=P_{int.1}/G_1$ then **P_{int}=14.6dBm=29.0mW.**

Exercise 16.6 JFET frequency-changer component values

> **Calculation** The frequency-changer in Fig. 16.9(a) is to be used in the receiver of Fig. 14.9(c). For the JFET V_{th}=–2V and I_{co}=4mA. Calculate suitable component values.

For full range of V_{gs} then $V_{gs}=V_{th}/2$ giving V_{gs}=–1V.

Resistor R_e. With $I_c=I_{co}(1-V_{gs}/V_{th})^2$ as in [10.3], I_c=1mA.

With $R_e=V_s/I_c$ and $V_{gs}=-V_s$, then **R_e=1kΩ.**

Capacitor C_e. We require X_{ce} << $1/g_m$.

With $g_m=2I_{co}[(V_{gs}-V_{th})/V_{th}^2]$

as in [10.4] then g_m=2mS.

With $X_{ce}=1/2\pi fC_e$, we require $C_e >> g_m/2\pi f_{min}$.

With $f_{min}=50kHz$ then $C_e >>=6.37nF$,
with **$C_e=100nF$**

both fully adequate and readily available in the low-inductance high-k ceramic form.

With the FET bias current $I_b \approx 50pA$ then R_g could be very large. Choose **$R_g=100k\Omega$**

giving I_bR_g the negligible $5\mu V$.

Local-oscillator coupling capacitor C For $X_c << R_g$ at f_{min} we require $1/2\pi f_{min}C << R_g$ giving $C >> 31.8pF$.

Choose **$C_{LO}=1nF$.**

Exercise 16.7 Diode ring modulator

> **Demonstration** Fig. 16.11. This is used as a frequency-changer with the signal input and output load impedances R_s and R_L matched. Show that for zero diode resistance the conversion loss $\alpha \approx -4dB$.

The load voltage that would obtain for direct connection of R_L to R_s is modulated by the reversing function $H_r(t)$. This is a square wave of unit peak value for which [14.6] gives the fundamental as $4/\pi$. Each output component is of half this amplitude giving **$\alpha=2/\pi=0.637 \approx -4dB$.**

17

Noise

Summary

Some definitions are clear: 'noise' is that which I do not wish to hear; 'junk' is that which I do not wish to possess; 'afternoon' is when I have had my lunch; and so on. So 'electronic noise' is that which I had rather not find at the output of my amplifier. But offset and drift and distortion have already been covered, and interference is the topic of the next chapter. Thus the 'noise' of the title is the random electronic noise generated within the circuit.

Random electronic noise is that caused by thermal agitation of current carriers (thermal noise), or the quantised nature of carrier flow across a junction (shot noise). These are fully random and, having a spectral density which is independent of frequency, are termed 'white'.

1/f noise has the spectral distribution that the name implies, and is thus somewhat less random. This type is usually associated with imperfections in the device structure, particularly at the surface.

Two types of noise circuit model are used, one generalised and the other specific to the amplifying device, for example the bipolar transistor or the fet.

The remaining subjects are noise in the long-tailed pair, the effect of negative feedback on noise, and the noise performance of the photodetector.

All numeric calculations are within the main body of the text, with the values cumulative and set bold to the right-hand margin. The final Exercises section consists mainly of detailed derivations.

17.1 Thermal noise

Thermal noise arises from thermal agitation of the otherwise neutral charge distribution in a resistor. The mobile negatively charged electrons move at random relative to the positively charged lattice. The resulting fluctuating charge imbalance then causes a corresponding fluctuation in potential across the ends of the device. The noise waveform observed depends upon the frequency response of the system between the resistor and the point of observation. For a simple low-pass response with the cut-off f_c, the waveform becomes as in Fig. 17.1(a).

Noise amplitude

This may be expressed as the 'mean-square' value described in Section 1.1 for the sine wave. For the noise voltage V_n this is $\overline{V_n^2}$ and for the noise current I_n, $\overline{I_n^2}$. Alternatively the root-mean-square (rms) value, e.g. $\tilde{V}_n = \sqrt{\overline{V_n^2}}$, may be given. The mean-square value is sometimes loosely referred to as the noise 'power'. For a resistor R

$$\overline{V_n^2} = 4kTRB \qquad (17.1)$$

where B is the bandwidth of the filter through which V_n is observed, k is Boltzmann's constant, and T is the absolute temperature. Correspondingly the rms

$$\tilde{V}_n = \sqrt{4kTRB} \qquad (17.2)$$

These expressions are valid whether for a low-pass or band-pass filter, or indeed for a filter of several separated pass-bands, with B the total bandwidth. With

$k=1.38 \times 10^{-23} J/^{o}$,

and the standard **$T_o=290^{o}$,**

then the frequently used value **$4KT \approx 1.6 \times 10^{-20}$.**

Example For a 10kΩ resistor, and with the bandwidth B=30kHz **$\tilde{V}_n = 2.19\mu V$.**

Verification

In Exercise 17.1 the basic structure of Nyquist's derivation of [17.1] is outlined, with the completion of the argument left as an exercise. There the example is a transmission line carrying thermal noise. Here we shall give and even simpler argument, albeit with the result approximate, based on the noise voltage at the output of the simple single-section low-pass RC filter.

Here

- the mean energy per degree of freedom of a physical system is kT/2,
- the energy stored in a capacitor C is $CV^2/2$,
- a resistor R in parallel with C forms a low-pass filter of cut-off frequency

$$f_c = 1/(2\pi CR).$$

With these relations

$$\overline{V_n^2} = 2\pi kTRf_c.$$

With $f_c = B$ in [17.1] there is fair agreement. The $\times\pi/2$ difference is due to the cut-off of the present RC filter not being abrupt as for [17.1].

Spectral components

When noise is applied to the input of a high-Q band-pass filter, one for which the bandwidth Δf_o is small compared with the central frequency f_o, then the waveform of the filter output voltage is as in Fig. 17.1(b). This is a sine wave, essentially of frequency f_o, which fluctuates in amplitude at the approximate frequency Δf_o. There is also a less obvious fluctuation in phase, also at the frequency Δf_o.

White noise

The distinguishing feature of white noise is that the long-term mean-square value of each spectral component is the same. The filtered white noise of (a) is thus the sum of a series of the components in (b), with the component frequency ranging from zero to the low-pass cut-off f_c, and with the long-term mean-square value of each the same.

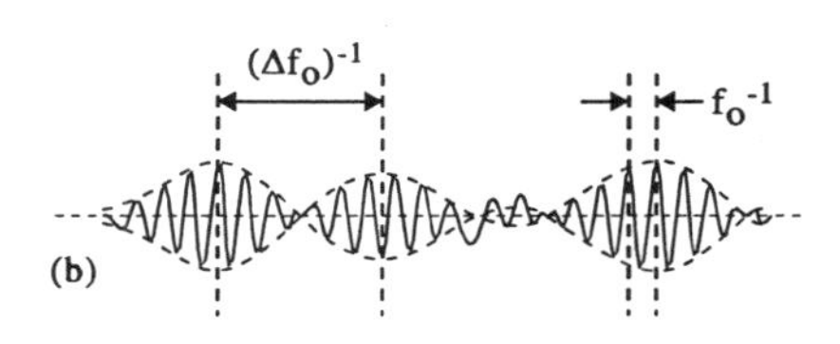

Fig. 17.1 Waveforms for white noise.
(a) Low-pass filtered with filter cut-off f_c.
(b) Narrow-band band-pass filtered, bandwidth $\Delta f_o \ll$ central frequency f_o.

Fully unfiltered white noise can never exist, because the number of components would be infinite. For thermal noise the limit is described by quantum theory and lies in the infra-red, beyond the limits of even today's electronic circuitry.

17.2 Shot noise

Shot noise is the result of the random timing of carriers as they pass across a semiconductor junction, for example the current I_c due to electrons passing through an NPN transistor. As each electron passes from emitter to collector there is an associated pulse of current flowing in the path external to the transistor, with the area under the pulse equal to the electron charge q, and the width of the pulse equal to the emitter-to-collector transit time T_τ. With the number of carriers in transit at any one time extremely large, then the pulses overlap to a high degree. Thus I_c is largely constant at the mean value $\overline{I_c}$, but superimposed is the small shot noise component I_{cn}, with the fluctuation time T_τ.

Mean-square value

With q the electron charge then, for a current I subject to shot noise as seen through an ideal filter of bandwidth B,

$$\overline{I_n^2} = 2q\bar{I}B \qquad (17.3)$$

Example For a transistor for which $I_c = 1\text{mA}$, and with $B = 30\text{kHz}$, and with

$q = 1.6 \times 10^{-19}$ Coulomb then $\tilde{I}_n = 3.38\text{nA}$.

Memory-box model

This is a helpful time-domain way of looking at the transistor shot-noise. With the transistor stored charge capacitance

$$c_{\pi s} = T_\tau I_c/(kT/q) \qquad \text{as in [6.38]}$$

then the passage of one electron across the transistor can be thought of as the effect of applying a current impulse of area q to a low-pass filter of response time T_τ. The impulse is effectively broadened to a pulse of width T_τ.

But the transistor has other capacitances, which smear the pulse further, so in (a) we represent the total low-pass filtering effects by the RC section shown, thinking of i_c prior to this as the randomly timed series of impulses shown in (b).

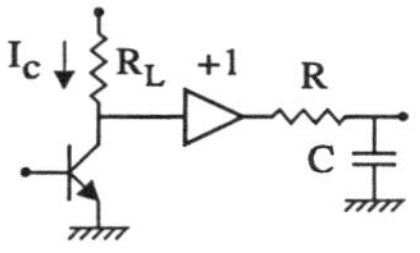

Fig. 17.2(a) Memory-box model of shot-noise in transistor amplifier. Circuit model.

Fig. 17.2(c,d) then shows the filtering of the impulse train in (b) as in Fig. 11.6 and 7. Here the exponential memory function in (c) runs along the impulse train to give I_c as in (d), with the noise component I_{cn} superimposed.

With the 'memory-box' approximation to the exponential function also in (c), then I_{cn} can be thought of as the variation in the number of impulses spanned by the box.

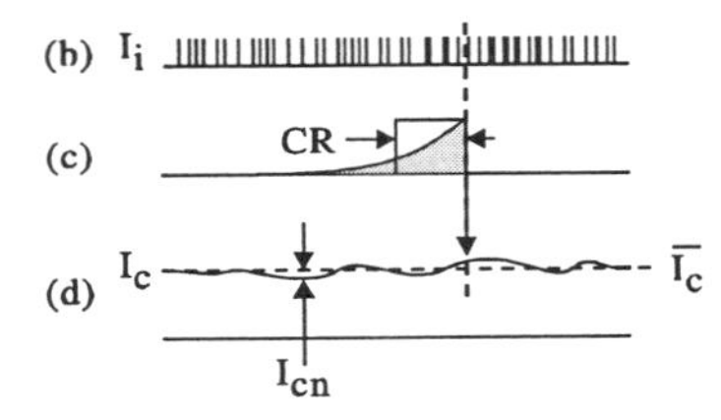

Fig. 17.2(b-d) Memory-box model of shot-noise in transistor amplifier.
(b) Random impulse train for I_c apart from low-pass filter action.
(c) Memory function of filter with box approximation.
(d) I_c as seen through filter.

Standard deviation

Often the result of a measurement is just one value of the system output, rather than a displayed trace. Here the rms noise value remains an indicator of the probable noise error in the single result. However, in such cases it is more usual to refer to the 'standard deviation'.

Where a large number of such single values are recorded then the results form a bell-like 'Gaussian' distribution, the 'probability density function'. It is the approximate half-width of this which is σ_n. For low-pass filtered white noise the distinction is academic, since the values of $\tilde{V}_n$ and σ_n are the same.

Generalised time averaging

With $\tilde{V}_n \propto B^{1/2}$ as in [17.2]

where B is the low-pass filter bandwidth,

with $B \propto 1/T_r$

where T_r is the filter response time,

and with $\tilde{V}_n = \sigma_n$

then for a single observation of the filter output

$$\sigma_n \propto T_r^{-1/2}.$$

With this the effect of the filter forming a noise average over the period T_r, the same is so for any kind of averaging of white noise, for example where the averaging time T_{av} is extended from the filter T_r by visual observation of an oscilloscope display. Thus for white noise generally

$$\sigma_n \propto T_{av}^{-1/2} \quad (17.4)$$

Exercise 17.2 gives an approximate time-domain verification of [17.3] based on the above memory-box model, followed by the more traditional and more exact frequency-domain argument.

17.3 1/f noise and drift

With the noise for a real amplifier normally a mix of white and 1/f noise, the combined spectrum is typically as in Fig. 17.3(a). The 1/f noise waveform of (b) differs from that of the white noise in Fig. 17.1(a) in that it has no clearly defined long-term mean. The greater prominence of the lower frequency components causes a 'walk away' from any apparent initial mean, which increases as time proceeds.

Power spectral density

The variable used to express the noise amplitude in (a) is the 'power spectral density' D. While

$$\overline{V_n^2} \propto \text{filter bandwidth B}$$

only for white noise, for non-white noise subject to a band-pass filter of sufficiently narrow bandwidth Δf

$$\overline{V_n^2} \propto \Delta f$$

giving

$$D = d\left(\overline{V_n^2}\right) / df \quad (17.5)$$

Noise corner

With the noise corner f_k as in (a) then typically for a bipolar or bi-fet op-amp f_k=300Hz, while for MOS f_k=10kHz. The higher value for the MOS is due to it being a surface device where the imperfections are greater, giving more 1/f noise.

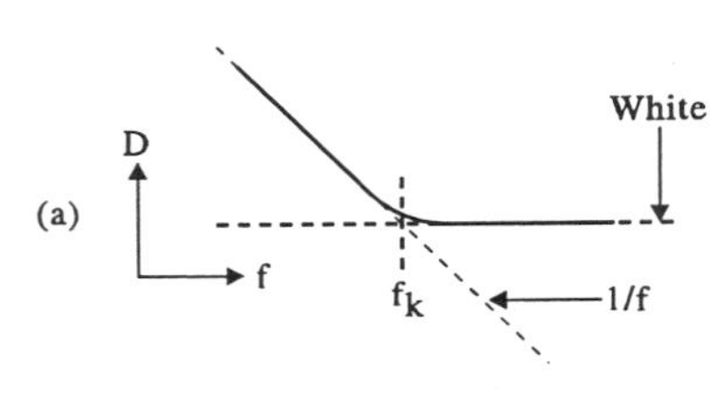

Fig. 17.3 1/f noise.
(a) Spectrum for white and 1/f noise. D: power spectral density (log axes).
(b) Waveform for 1/f noise.

Time-averaged noise errors

While for white noise the noise error $\sigma_n \propto T_{av}^{-1/2}$ is decreased by increasing the averaging time T_{av}, for 1/f noise the emphasis of the lower frequency components causes this advantage to be lost, in fact giving σ_n independent of T_{av}.

Fig. 17.4 shows the variation of σ_n with T_{av} for both types of noise, with T_k corresponding to f_k in Fig. 17.3(a). Also shown is the typical drift error, which increases $\propto T_{av}$. Typically $T_k = 1$ms and the drift transition $T_d = 1$s, so the range over which the 1/f noise dominates is large.

Various techniques are employed to eliminate the effects of drift and 1/f noise, which otherwise frustrate attempts to reduce the white noise error by increasing T_{av}. These comprise various forms of base-line subtraction, multiple time averaging and phase-sensitive detector methods (Wilmshurst 1990).

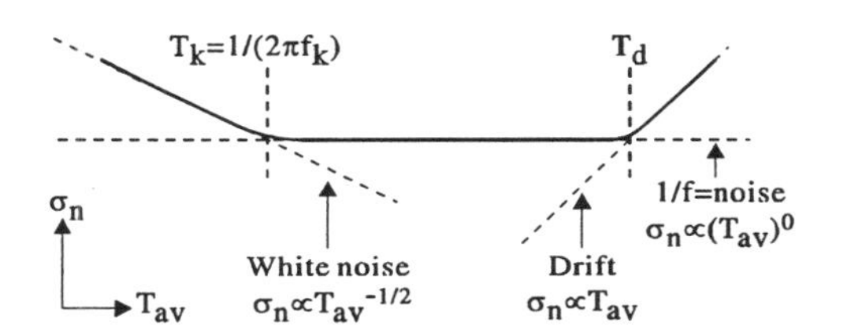

Fig. 17.4 Variation of noise error (standard deviation) σ_n with averaging time T_{av} for various types of noise (log axes).

Problem

For the monitoring of a square-wave of period T_s in the presence of 1/f noise, it is acceptable to use a low-pass filter of response time $T_s/30$ and a high-pass of response time $30T_s$. Show that the noise amplitude at the combined filter output is independent of T_s.

17.4. Dual source noise model

Fig. 17.5 shows a simple general method of representing amplifier noise. V_{ns} in (a) represents the thermal noise of the signal source impedance R_s, while V_n and I_n in (b) represent the amplifier noise.

Noise factor/figure

With V_{nos} the noise component of the amplifier output V_o originating from the source, and V_{noa} that from the amplifier then the 'noise factor'

$$F = \overline{\left(V_{nos}^2 + V_{noa}^2\right)} \Big/ \overline{V_{nos}^2} \tag{17.6}$$

This is the figure of merit for the noise performance, with F=1 for a noiseless amplifier.

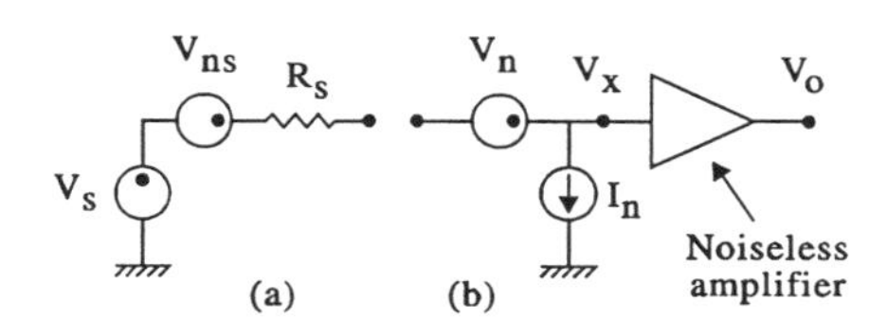

Fig. 17.5 Dual source model for amplifier noise.
(a) Source.
(b) Amplifier model.

Once the point has been reached where $F \approx 1$ there is clearly little value in trying to reduce the amplifier noise further.

The 'noise figure' is simply F expressed in decibels. Thus

$$F(dB) = 10 \log(F)$$

giving F(dB)=0 for the noiseless amplifier.

For the circuit of Fig. 17.5 the value of F will be the same for V_x with the noiseless amplifier absent as for V_o with the amplifier present. With V_{nx} the noise component of V_x then

$$V_{nx} = V_{ns} + V_n - I_n R_s \tag{17.7}$$

giving

$$\overline{V_{nx}^2} = \overline{\left(V_{ns} + V_n - I_n R_s\right)^2} \tag{17.8}$$

Correlation

Expansion of the average in [17.8] gives many cross-terms

$$\overline{V_n I_n}, \quad \overline{V_{ns} V_n}, \text{ etc.}$$

With V_{ns} and V_n from different sources

$$\overline{V_{ns} V_n} = 0.$$

Such components are said to be 'uncorrelated'. Also, as will be shown, for both the bipolar transistor and the FET, V_n and I_n originate from independent sources. Thus these too are uncorrelated.

Unfortunately, for frequencies above the amplifier cut-off, the circuit capacitances do introduce cross-correlation, making the method too complex to be useful. Assuming zero correlation for the present circuit then all cross-terms cancel to give

$$\overline{V_{nx}^2} = \overline{V_{ns}^2} + \overline{V_n^2} + \overline{I_n^2} R_s^2 \tag{17.9}$$

and thus

$$F = \left(\overline{V_{ns}^2} + \overline{V_n^2} + \overline{I_n^2} R_s^2\right) \Big/ \overline{V_{ns}^2} \tag{17.10}$$

Equivalent noise resistance

It is often useful to think of a noise voltage in terms of the resistance that would generate a thermal noise voltage of magnitude equal to the actual voltage.

With the general $\overline{V_n^2} = 4kTRB$

as in [17.1] then for V_n here the 'equivalent noise resistance' R_{nv} is given by

$$\overline{V_n^2} = 4kTR_{nv}B \quad (17.11)$$

Similarly for a noise current I_n the equivalent resistance R_{ni} is that which, when shorted, will pass the current I_n through the short. Thus

$$\overline{I_n^2} = 4kTG_{ni}B \quad (17.12)$$

Neither R_{nv} nor R_{ni} are real resistances, they are merely convenient expressions of the noise spectral densities, being independent of bandwidth B.

Finally for the signal source

$$\overline{V_{ns}^2} = 4kTR_sB \quad (17.13)$$

With F as in [17.10] then

$$F = \frac{R_s + R_{nv} + R_s^2 / R_{ni}}{R_s} \quad (17.14)$$

which is plotted against R_s in Fig. 17.6. Relative to the numerator of [17.10]

- the downward sloping line represents the $\overline{V_n^2}$ term,
- the horizontal unity line represents the $\overline{V_{ns}^2}$ term,
- the upward sloping line represents the $\overline{I_n^2}R_s^2$ term.

As shown, there is normally a wide range of values of R_s for which F is near ideal. The optimum value F_{opt} of F is that at the intersection of the two sloping lines in the diagram, at which

$$R_s/R_{ni} = R_{nv}/R_s$$

Thus $$R_{opt} = \sqrt{R_{nv}R_{ni}} \quad (17.15)$$

giving $$F_{opt} = 1 + 2\sqrt{R_{nv} / R_{ni}} \quad (17.16)$$

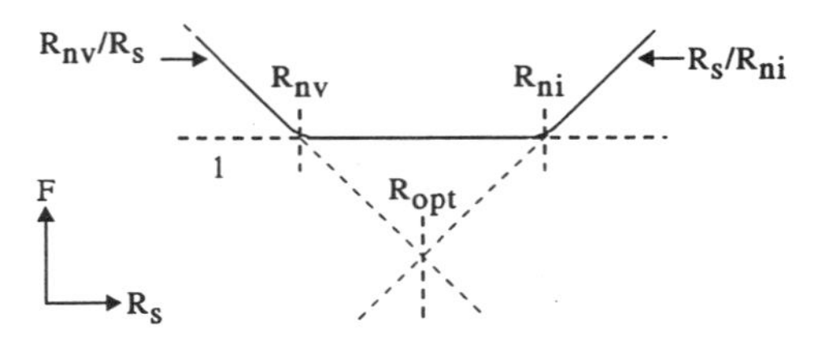

Fig. 17.6 Variation of components of noise figure F with source resistance R_s for dual-source amplifier noise model of Fig. 17.5 (log axes).

Example For $R_{nv}=1k\Omega$ and $R_{ni}=100k\Omega$ then $R_{opt}=10k\Omega$ giving $F_{opt}=1.2=0.79dB$.

Variation of V_{nx} with R_s

Further insight into the form of Fig. 17.6 can be obtained by plotting the variation of $\overline{V_{nx}^2}$ with R_s. With

$$\overline{V_{nx}^2} = \overline{V_{ns}^2} + \overline{V_n^2} + \overline{I_n^2}R_s^2 \quad \text{as in [17.9]}$$

$$\overline{V_n^2} = 4kTR_{nv}B \quad \text{as in [17.11]}$$

$$\overline{I_n^2} = 4kTG_{ni}B \quad \text{as in [17.12]}$$

$$\overline{V_{ns}^2} = 4kTR_sB \quad \text{as in [17.13]}$$

then $$\overline{V_{nx}^2} = 4kTB\left(R_{nv} + R_s + R_s^2 / R_{ni}\right) \quad (17.17)$$

The three stages in the plot are then as follows.

• $R_s < R_{nv}$. Amplifier voltage noise R_{nv} dominant, $\overline{V_{nx}^2}$ independent of R_s, F high but decreasing.
• $R_{nv} < R_s < R_{ni}$. Source noise R_s dominant, $\overline{V_{nx}^2} \propto R_s$, $F \approx 1$.
• $R_{ni} < R_s$. Amplifier current noise R_s^2/R_{ni} dominant, $\overline{V_{nx}^2} \propto R_s^2$, F high and increasing.

High-frequency amplifiers

These normally have to be designed for the signal source impedance R_s the standard 50Ω of the connecting coaxial cable. With the principal value of the dual-source model that it allows calculation of F for varying R_s, this is of no particular advantage here. Thus the unsuitability of the dual-source model on account of the cross-correlation of the voltage and current noise sources is of little importance. All that needs to be specified is the single value of F, or perhaps the variation of F with operating frequency.

Amplifier noise temperature

From [17.13] the source noise amplitude varies with temperature T. Thus when F is specified it must be for a stated source temperature. This is normally the 'standard' temperature T_o of 290K.

Some sources differ in temperature widely from T_o. For example, a microwave dish antenna pointing upwards can have a temperature of only a few tens of degrees. In such cases the closeness of F to unity ceases to be a useful design parameter. Then a more helpful value is that of the 'amplifier noise temperature' T_a. This is simply the increase in source temperature needed to account for the amplifier noise. Then the point at which there is no value in reducing the amplifier noise further is when T_a becomes small compared with the source temperature T_s.

17.5 Bipolar transistor amplifier

The two-source noise model of Fig. 17.5 is a generalised one, while the method of this section is specific to the single-stage common-emitter bipolar transistor amplifier.

Fig. 17.7 shows the low-frequency noise equivalent circuit, which is that of Fig. 6.11 with the noise sources added. R_x is initially omitted, being added to R_s later. Here V_{ns} is the normal source thermal noise, for which

$$\overline{V_{ns}^2} = 4kTR_sB \tag{17.18}$$

Collector noise current

For the NPN transistor, when electrons leave the emitter they can either reach the collector or recombine in the base. Those reaching the collector cause, in the process of moving from emitter to collector, a current pulse of area q and width T_τ in the external C-E circuit, with T_τ the base transit time. These pulses combine to constitute I_c which will then have the shot component I_{nc}. Here

$$\overline{I_{nc}^2} = 2qI_cB \tag{17.19}$$

Base noise current

For the base current I_b, an electron leaving the emitter and recombining in the base requires the injection of a compensating hole into the base, provided by the base connection. Thus the pulse of area q now flows in the external B-E circuit. Also a hole passing from the base to the emitter to recombine there causes a similar pulse in the circuit. It is the sum of these two pulse trains that constitute I_b, with the noise component $I_{n\pi}$. Here

$$\overline{I_{n\pi}^2} = 2qI_bB \tag{17.20}$$

Load noise current

I_{nl} (not shown) is the thermal noise due to R_L, for which

$$\overline{I_{nL}^2} = 4kTG_LB.$$

With I_{nc} as in [17.19] and

$$g_m = I_c/(kT/q)$$

as in [6.6] then

$$\overline{I_{nc}^2}/\overline{I_{nL}^2} = g_mR_L/2.$$

With the stage voltage gain

$$a_{ce} = -g_mR_L$$

as in [6.9] then I_{nl} is normally negligible.

Noise figure

If the output node C is grounded then F for the resulting noise current I_{on} flowing to ground will be the same as for V_o, giving

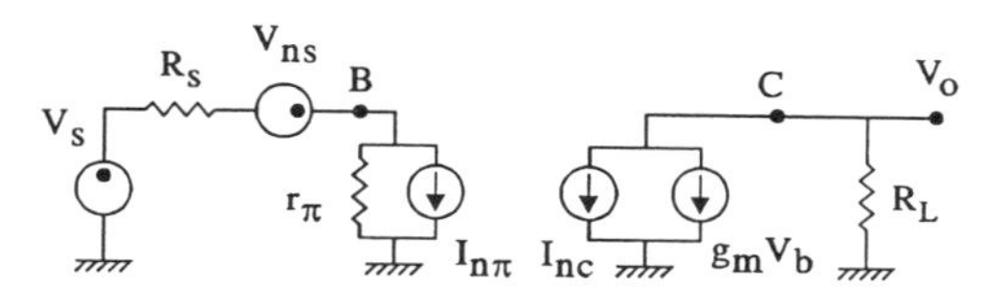

Fig. 17.7 Low-frequency small-signal circuit diagram for bipolar transistor amplifier including noise.

$$I_{on} = I_{nc} + g_m\left(\frac{V_{ns}R_\pi}{R_\pi + R_s} + I_{n\pi}\frac{R_\pi R_s}{R_\pi + R_s}\right) \tag{17.21}$$

Here I_{nc} and $I_{n\pi}$ are uncorrelated, since the time at which one electron leaves the emitter to reach the collector is totally unrelated to that at which another leaves the emitter to recombine in the base, etc. Thus none of the three noise sources is correlated with any of the others. With

$r_\pi = (kT/q)/I_b$ as in [6.27]

$g_m = qI_c/(kT)$ as in [6.6]

and $\beta = I_c/I_b$ as in [6.21]

then

$$F = 1 + \frac{R_s}{r_\pi}\frac{(\beta+1)}{2\beta} + \frac{r_\pi}{2\beta R_s} + \frac{1}{\beta} \tag{17.22}$$

which is plotted against R_s in Fig. 17.8. From the diagram, the R_s for optimum F

$$R_{opt} = r_\pi / \sqrt{\beta+1} \tag{17.23}$$

giving

$$F_{opt} = 1 + \frac{\sqrt{\beta+1}}{\beta} + \frac{1}{\beta} \approx 1 + \frac{1}{\sqrt{\beta}} \tag{17.24}$$

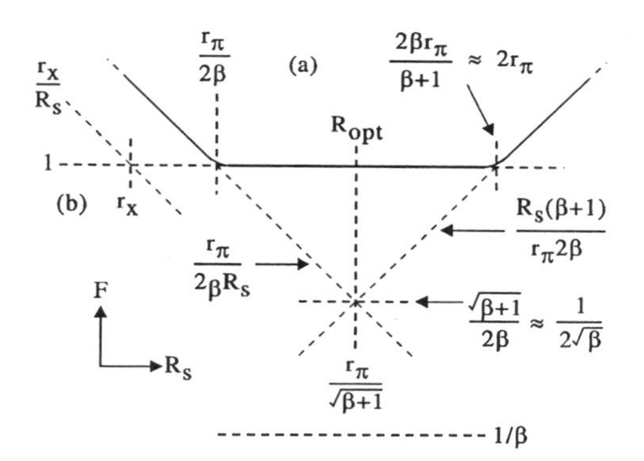

Fig. 17.8 Variation of noise figure F with source resistance R_s for bipolar transistor amplifier of Fig. 17.7 (log axes).
(a) Without r_x.
(b) Contribution of r_x.

With β typically 200 then F_{opt} is close to the ideal of unity. The 'bathtub' form of Fig. 17.8 confirms that of Fig. 17.6 for the general dual-source noise model. It also confirms that the crossover point is well below the F=1 line, allowing R_s to vary considerably from R_{opt} before F increases significantly from unity.

Example

For the typical $r_\pi=5k\Omega$ and $\beta=200$ then $R_{opt}=353\Omega$ giving $F_{opt}=1.07=0.3dB$.

Transformer noise-matching

If R_s does lie beyond either of the two transition values in Fig. 17.8, then R_s can be restored to within the required range using an impedance matching transformer.

This 'noise-matching' is not the same as the normal 'power matching' criterion of Section 1.4. There the objective is to transfer as much power as possible from the source to the load, and for this the load impedance should equal that of the source. In contrast, the noise-matching R_{opt} in [17.23] is well below the load impedance r_π.

Noise matching by I_c variation

Transformers cease to be practical for frequencies of much below 100Hz, and also tend to pick up magnetically coupled interference. Suppose in Fig. 17.8(a) that R_s is out of the bathtub to the left. With both of the transition values $\propto r_\pi$, and with $r_\pi \propto 1/I_c$, then by increasing I_c the tub can be moved to the left to encompass R_s.

Optimum collector current

With $I_{c.opt}$ the value of I_c giving optimum noise figure for a given R_s, with

$$R_{opt} = r_\pi/\sqrt{\beta+1} \qquad \text{as in [17.23]}$$

and with $\quad r_\pi=\beta(kT/q)/I_c \quad$ as in [6.28]

then

$$I_{c.opt} = \frac{\beta(kT/q)}{R_s\sqrt{\beta+1}} \qquad (17.25)$$

Example

For $\quad R_s=10k\Omega$ and $\beta=200$

then $\quad$ **$I_{c.opt}=35.2\mu A$.**

With $\quad r_\pi=\beta(kT/q)/I_c \quad$ as in [6.28]

the resulting $\quad$ **$r_\pi=142k\Omega$,**

much greater than the $r_\pi=10k\Omega$ for power matching.

Effect of r_x

The above noise-matching by adjusting I_c is effective so long as the base spreading resistance r_x in Fig. 6.11 ceases to be significant. With r_x in series with R_s then V_{nx} adds to V_{ns}, thus adding the component

$$\overline{V_{nx}^2}/\overline{V_{ns}^2} = r_x / R_s$$

to F, as plotted in Fig. 17.8(b). For the situation as shown, the added component is insignificant, but as I_c is increased to move the tub left, then the fixed line (b) takes over the function of the left-hand edge of the then narrowing tub. With r_x also increasing the attenuation of V_n relative to the amplifier noise, why is the effect of this on F insignificant here?

Limiting collector current

With $\quad r_\pi/2\beta=r_x$ at the above limit

and with $\quad r_\pi=\beta(kT/q)/I_c \quad$ as in [6.28]

then $\quad I_c=(kT/q)/2r_x$.

For the typical $r_x=150\Omega$ then $\quad$ **$I_c=83.3\mu A$**

At the other extreme, β begins to fall at $I_c \approx 1\mu A$ for which [17.25] gives $\quad$ **$R_{opt}=353k\Omega$**

Overall, with I_c suitably set, we have a wide range of values of R_s for which $F \approx 1$,

from $\quad R_s \approx (r_x=150\Omega)$

to $\quad R_s=350k\Omega \quad$ for declining β

Paralleling of transistors

A way of moving the bathtub left which moves the r_x limit with it is to replace the single input transistor with a number N in parallel. This reduces all of the resistances, both noise and normal, by the factor N, and so moves the tub, together with the r_x/R_s line, left by this factor.

17.6 Long-tailed pair noise

The LTP is the basis of most op-amps and, due to the thermal matching of the two LTP transistors, has much lower drift than the equivalent single-transistor common-emitter stage. However, it sometimes gives a little more random noise.

Fig. 17.9(a) shows the LTP with each transistor represented by the equivalent noiseless device and its pair of generalised noise sources.

If the differential output is taken from the LTP (as it is in any op-amp, using the current mirror) then a current injected into the common-emitter node has no effect on the output. Thus the output is the same as if the noise current sources were returned to ground. With this assumption the noise equivalent circuit becomes as in (b).

Single-ended operation

For the inverting feedback amplifier the V_{i+} terminal is grounded, allowing the generalised noise model of Fig. 7.5 to be used, as in (c).

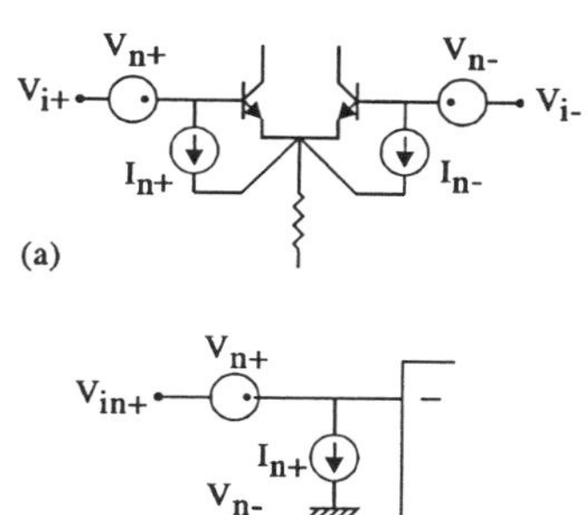

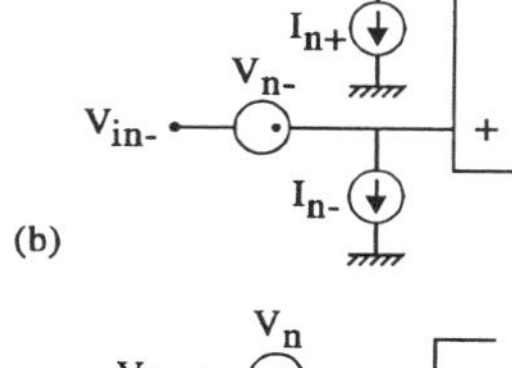

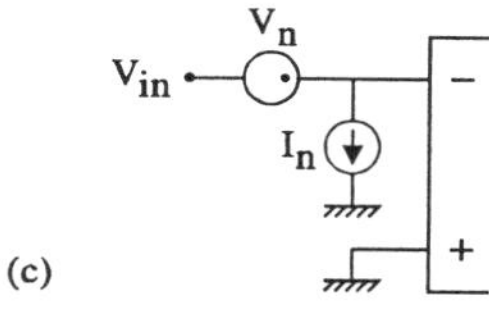

Fig. 17.9 Dual source noise model applied to long-tailed pair.
(a) Circuit diagram.
(b) Noise equivalent for emitter point grounded.
(c) Dual-source equivalent for non-differential use.

Here $\qquad V_n = -V_{n+} + V_{n-}$

with the components uncorrelated giving

$$\overline{V_n^{\,2}} = \overline{V_{n+}^{\,2}} + \overline{V_{n-}^{\,2}} \qquad (17.26)$$

Also with I_{n+} shorted $\qquad \overline{I_n^{\,2}} = \overline{I_{n-}^{\,2}}$

Thus the LTP 'noise power' due to the voltage noise sources is twice that for the equivalent CE stage, while that due to I_n remains unchanged.

The non-inverting feedback amplifier where I_{n+} is not shorted is considered in Section 17.8.

Where the LTP is the input stage for a single-chip op-amp, there is the further disadvantage that it is usually not possible to vary the transistor current I_c, and so optimise F for a given R_s.

17.7 FET noise

The FET is capable of an excellent noise figure, better than the bipolar, except at low source impedances. Fig. 17.10(b) shows a simple noise model for the FET.

FET noise voltage

V_n arises from the channel thermal noise. With g_{co} the channel conductance for $V_{ds}=0$, and with [8.7] and [8.17] giving $g_m = g_{co}$ where g_m is the value at pinch-off, then $R_{nv} \sim 1/g_m$. More exactly, Van der Ziel (1962) gives

$$R_{nv} = (2/3)/g_m \qquad (17.27)$$

where R_{nv} is the equivalent noise resistance for V_n,

Here $\qquad \overline{V_n^{\,2}} = 4kTR_{nv}B \qquad (17.28)$

FET noise current

I_n in (b) is the shot noise component of the bias current I_b

for which $\qquad \overline{I_n^{\,2}} = 2qI_bB \qquad (17.29)$

With $\qquad \overline{I_n^2} = 4kTG_{ni}B \qquad$ as in [17.12]

then $\qquad R_{ni} = 2(kT/q)/I_b \qquad (17.30)$

Example With typically $g_m = 1mS$ and $I_b = 50pA$ then **$R_{nv} = 667\Omega$ and $R_{ni} = 1G\Omega$.**

Optimum noise figure

For the operating frequency high enough for 1/f noise to be insignificant, and low-enough for c_{in} also to be insignificant, and with the physical mechanisms for R_{nv} and R_{ni} uncorrelated, then the present FET noise model resolves to the dual-source model of Fig. 17.5. With

$$R_{opt} = \sqrt{R_{nv}R_{ni}} \qquad \text{as in [17.15]}$$

and $\qquad F_{opt} = 1 + 2\sqrt{R_{nv} / R_{ni}} \qquad$ as in [17.16]

For the present example

$R_{opt} = 817k\Omega$ and $F_{opt} = 1.002 = 0.007dB$.

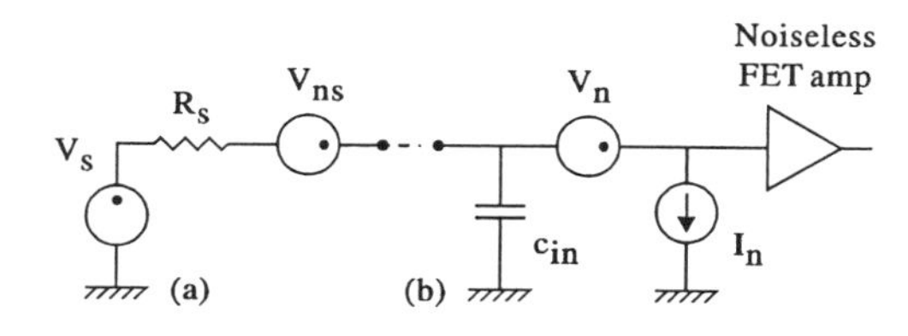

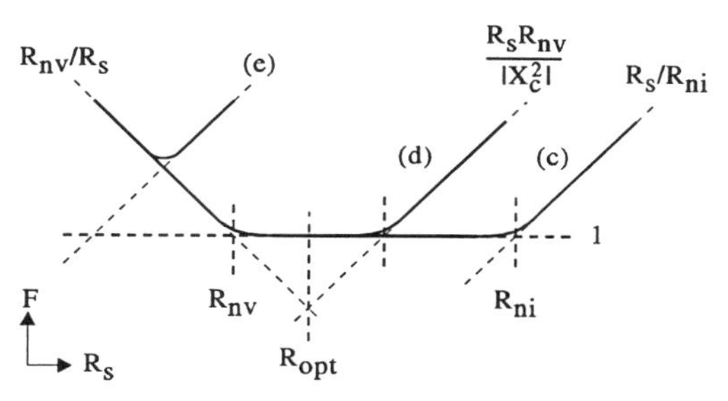

Fig. 17.10 Noise model for FET.
(a) Signal source. (b) FET.
(c) Variation of noise figure F with R_s with $c_{gs}=0$ (log axes).
(d) F increase due to c_{gs}.
(e) As (d) but for higher frequency.

While for the Bi-fet TL081 op-amp

$$\mathbf{R_{nv} \approx 10k\Omega,\ R_{ni} \approx 1G\Omega,\ c_{in} \approx 15pF.}$$

still giving F_{opt} very close to unity.

Effect of gate capacitance

From Fig. 17.10(a,b), following the usual procedure

$$F = 1 + \frac{R_{nv}}{R_s} \times \frac{(R_s^2 + |X_c|^2)}{|X_c|^2} + \frac{R_s}{R_{ni}} \qquad (17.31)$$

where X_c is the impedance of c_{in}. The only addition here to [17.14] for the two-source model is the factor

$$(R_s^2 + X_c^2)/X_c^2$$

This is due to the attenuation of V_{ns} at the FET gate, relative to the other noise components, and adds the term

$$R_{nv}R_s/X_c^2$$

to F, as plotted in (d). Why is there no change in the final term?

Spot noise figure

With $X_c \propto 1/f$ then (d) is for one value of f only, moving say to (e) for f higher.

It is interesting that R_{opt}, the value of R_s for optimum spot noise figure, is also the value for which $|X_c| = R_s$ and thus the cut-off frequency for the low-pass filter formed by c_{gs} and R_s.

17.8 Effect of feedback on noise

Negative feedback improves many features of an amplifier but noise figure F is not one of them. Indeed F can thus be indirectly degraded.

Fig. 17.11 shows the inverting feedback amplifier with the noisy op-amp represented by its noiseless equivalent and the usual dual noise sources V_n and I_n. With the system linear then

$$V_x = \alpha V_{in} + \beta V_o + \gamma V_n + \delta I_n \qquad (17.32)$$

where α to δ are constants determined by the network and the amplifier input impedance. With $V_o = -A_oV_x$ for the noiseless amplifier, where A_o is the op-amp gain, then

$$V_o = \left(\alpha V_{in} + \gamma V_n + \delta I_n\right)\frac{A_o}{1 + A_o\beta} \qquad (17.33)$$

giving the signal-to-noise ratio

$$\frac{S}{N} = \frac{\alpha V_{in}}{\gamma V_n + \delta I_n} \qquad (17.34)$$

Here the feedback fraction β is absent, so in this sense S/N is independent of feedback.

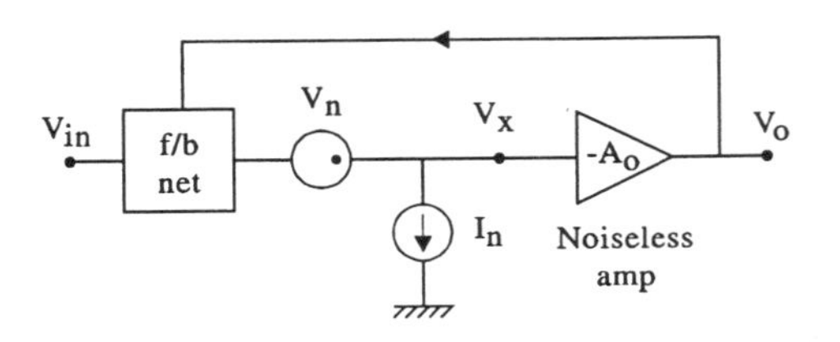

Fig. 17.11 Feedback amplifier noise.

However, for the op-amp output impedance zero, setting β to zero is equivalent to disconnecting the feedback network from V_o and grounding the connection point to the feedback network. With the feedback network still between the signal source and the op-amp input,

- the thermal noise from the feedback resistors will add to the total,
- the feedback network will attenuate the source output by the feed-forward factor $\times(\alpha = R_2/(R_1+R_2)$ thus increasing the relative effect of the amplifier noise.

These are the ways in which the negative feedback can indirectly degrade the noise figure.

Inverting negative-feedback amplifier

Fig. 17.12 shows the circuit with the noise components added. The first and most obvious step here is to remove R_1 with its thermal noise, reducing R_2 to maintain the feedback amplifier gain, now $A_f = -R_2/R_s$, at the required value.

Next some simplifications are made for the calculation of F.

- R_2 is disconnected from V_o and the free end grounded. As above, this makes no difference to F.
- With F the same for V_x as for V_o, and also independent of the loading of the V_x point by the amplifier, then the amplifier is removed altogether and F calculated for the resulting unloaded V_x.

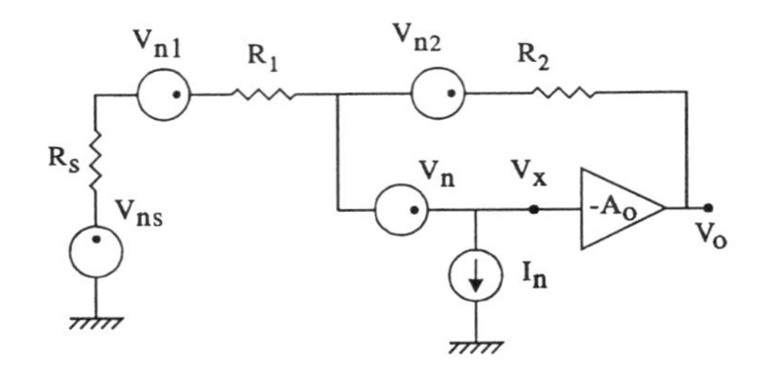

Fig. 17.12 Noise model for inverting feedback amplifier.

Thus

$$F = 1 + \frac{R_{nv}}{R_s}\left(\frac{R_s + R_2}{R_2}\right)^2 + \frac{R_s}{R_{ni}} + \frac{R_s}{R_2} \quad (17.35)$$

Comparing this with

$$F = \frac{R_s + R_{nv} + R_s^2 / R_{ni}}{R_s} \quad \text{as in [17.14]}$$

for the op-amp alone, the added component R_s/R_2 represents the thermal noise from R_s and the multiplier $[(R_s+R_2)/R_2]^2$ represents the feed-forward attenuation α, remembering that F deals in noise 'power' not voltage.

Both effects become negligible for $R_s >> R_s$, as is required anyway with $|A_f| = R_2/R_s$ and with the need for $|A_f|>>1$ in order to make the noise from following stages insignificant.

With $\overline{V_{n2}^2} \propto R_2$ as in [17.1]

it might appear that increasing R_2 to give the required A_f as above would increase, rather than decrease, the effect of the R_2 thermal noise. Why does it not?

Example For R_s=3kΩ connected directly to the TL081 op-amp, with no feedback, then from [17.14]

F=4.33=6.36dB.

With R_2=30kΩ added to give A_f=−10, and with F as in [17.35] then **F=5.13=7.01dB,**

a negligible increase as expected.

Non-inverting feedback ampifier

With the noise model for this circuit as in Fig. 17.13, the factors determining the noise figure F are as follows.

- There is now no feed-forward attenuation.
- R_1 and R_2 will generate thermal noise.
- The effects of V_{n+} and V_{n-} are much as before.
- Both I_{n+} and I_{n-} are now potentially significant.

As above, R_2 is disconnected from V_o and grounded for the comparison.

Noise from feedback resistors

Writing R_p for $R_1//R_2$ then the noise from R_p is insignificant if

$$R_p << (R_{nv}+R_s).$$

For R_s=0 and with R_{nv}=10kΩ for the TL081 then we require **R_p << 10kΩ.**

Noise from op-amp noise current

With I_{n-} developing the noise voltage $V_{ni-}=I_{n-}R_p$ at the lower op-amp input, then for this to be small compared with the remainder requires

$$R_p << (R_sR_{ni}+R_s^2+2R_{nv}R_{ni})^{1/2} \quad (17.36)$$

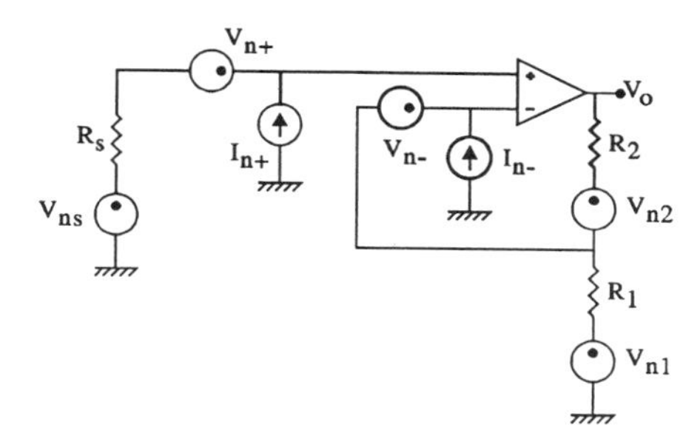

Fig. 17.13 Noise model for non-inverting feedback amplifier.

Again for R_s=0 and with the TL081 op-amp

R_p << 4.47MΩ.

clearly satisfied for the previous R_p<<10kΩ.

17.9. Photodiode amplifier

Fig. 17.14(a) shows the essentials of a photodiode amplifier. With the diode reverse-biased then initially no current flows. But when light is shone on the diode then carriers are generated in the depletion layer and the current I_d results. The rate at which the carriers are generated is proportional to the light intensity and so therefore is I_d. Finally the current-follower gives $V_o \propto I_d$.

With the virtual ground, the reverse bias from V_- is not strictly necessary. Its only purpose is to reduce the junction capacitance c_d, and thus improve the response time.

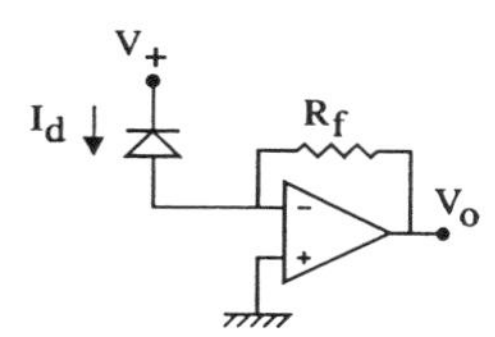

Fig. 17.14(a) Photodetector amplifier

In the noise equivalent circuit of (b) c_{in} is the op-amp input capacitance, V_n and I_n are the uncorrelated noise sources representing the op-amp noise, and I_{nf} is the thermal noise generated by R_f.

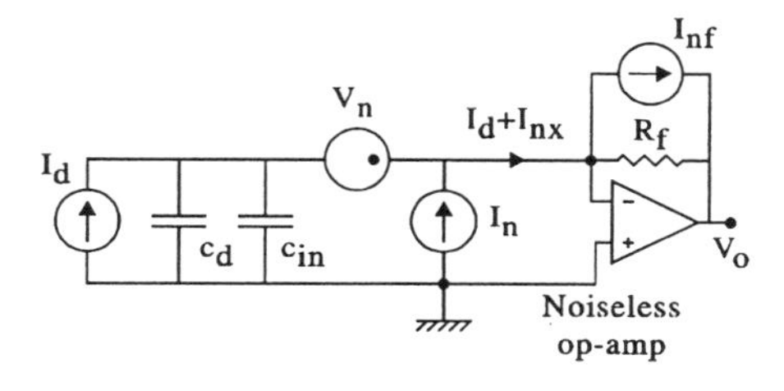

Fig. 17.14(b) Noise equivalent circuit for photodetector amplifier of (a).

With the impedance of the diode simply that of c_d, there is no source thermal noise, making the concept of noise figure inappropriate. Instead we consider the effect of the remaining noise as referred to I_d. With I_d flowing directly into the virtual ground then the comparable noise current is I_{nx}, also flowing into the virtual ground. Here

$$I_{nx}=I_n+V_n/X_c-I_{nf} \qquad (17.37)$$

With all of the noise components uncorrelated then

$$\overline{I_{nx}^2}=\overline{I_n^2}+\overline{V_n^2}/|X_c|^2+\overline{I_{nf}^2} \qquad (17.38)$$

With $\overline{V_n^2}=4kTR_{nv}B$ as in [17.11]

and $\overline{I_n^2}=4kTG_{ni}B$ as in [17.12]

then the equivalent noise conductance

$$G_{nx}=G_{ni}+R_{nv}/|X_c|^2+G_{nf} \qquad (17.39)$$

which, with

$$X_c=1/[2\pi f(c_d+c_{in})] \qquad (17.40)$$

plots as in (c).

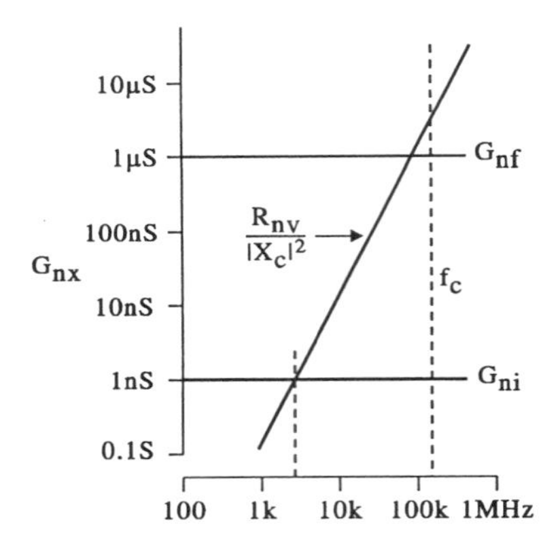

Fig. 17.14(c). Frequency dependence of components of I_{nx} in (b) expressed as equivalent noise conductances (log axes).

First intercept

With f_1 the frequency of the intercept, and this where

$$G_{ni}=R_{nv}/|X_c|^2$$

then $$f_c=1/[2\pi(c_d+c_{in})\times(R_{nv}R_{ni})^{1/2} \qquad (17.41)$$

With the values as above for the TL081 op-amp, and with the diode c_d=5pF then, as plotted, **f_c=2.52kHz.**

Stability

The stability of the circuit is discussed in Section 3.13, where it is shown that usually the small additional damping capacitance C_{da} is needed across R_f.

Here we simply note that a response time T_r=1μs is obtainable, with the corresponding marked cut-off

$$f_c=1/2\pi T_r=160\text{kHz}.$$

Feedback resistor noise

With the present R_{nf}=1MΩ giving G_{nf}=1μS as plotted then R_f is high enough not to add significantly to the noise for $f \approx f_c$.

Cut-off frequency dependence

The strong ($\propto f^2$) frequency dependence of the main noise component

$$R_{nv}/|X_c|^2$$

makes the total output noise V_{on} of the photodetector much more strongly dependent on f_c than for white noise. There the power spectral density D is independent of frequency giving

$$\overline{V_{no}^2}\propto f_c.$$

With $D\propto f^2$ for the photodetector, in contrast,

$$\int_0^{f_c} D.df \propto f_c^3$$

giving $$\overline{V_{no}^2}\propto f_c^3.$$

Thus, while for the white noise

$$\tilde{V}_{no}\propto f_c^{1/2}$$

for the photodetector

$$\tilde{V}_{no}\propto f_c^{3/2}$$

FET area

With the term $R_{nv}/|X_c^2|$

dominant for the present T_r=1μs, and with

$$R_{nv}=(2/3)/g_m \qquad \text{as in [17.27]}$$

then a high g_m is sought. From Table 8.3, increasing both $W\times n$ and $I_d\times n$ for the MOS increases $g_m\times n$. However, this increases $c_{in}\times n$ also, so the measure is only advantageous for $c_{in}<c_d$.

Photodiode area

With photodiodes available with a variety of areas, it is important to choose a device for which the area is no greater than that of the light beam. A larger area simply adds to c_d with no signal advantage. If still $c_d>c_{in}$ then it is advantageous to focus the light beam to a smaller area, allowing the use of a smaller device.

References

Faulkner, E. A. 1968: The design of low-noise audio-frequency amplifiers. *The Radio and Electronic Engineer* **36**, July, 17-30.
Motchenbatcher, C. D. and Fitchen, F. C. 1973: *Low-noise electronic design*. John Wiley.
Van der Ziel, A. 1962. *Thermal noise in field-effect transistors*. Proc. I.R.E. **50**, 1808-12.
Wilmshurst, T. H. 1990: *Recovery of signals from noise in electronic instrumentation*. Adam Hilger.

17.10 Exercises

Exercise 17.1 Thermal noise

Problem By integrating the power flow as a capacitor C is charged to the voltage V, verify that the stored energy is $CV^2/2$.

Problem The verification given for [17.1] for thermal noise power is a simplification of Nyquist's derivation. Here the model is a long transmission line of impedance R with a resistor of value R at each end. Once the system has come into thermal equilibrium the line ends are shorted and the resistors removed, making a multi-mode transmission-line resonator with the energy kT/2 stored in each of the many resonances. Retrace his argument to confirm the exact relation.

Exercise 17.2 Shot noise

Verify [17.3], first approximately using a time-domain argument, and then exactly using a frequency-domain argument.

Time domain With

$$\overline{I_n^2} = 2q\bar{I}B \quad \text{as in [17.3]}$$

p the number of pulses in the memory box of Fig. 17.2(c),
$\bar{p}$ the long-term mean of p,
and with σ_i and σ_p the standard deviations for I and p,

then $$\sigma_i / \bar{I} = \sigma_p / \bar{p} \quad (17.42)$$

For $\bar{p} >> 1$ $$\sigma_p \approx (\bar{p})^{1/2} \quad (17.43)$$

With the width of the box $T_r = CR$

$$\bar{I} = q\bar{p}/T_r \quad (17.44)$$

giving $$\sigma_i = (q\bar{I}/T_r)^{1/2} \quad (17.45)$$

With $$T_r = 1/2\pi f_c$$

for the simple low-pass of Section 11.1, where f_c is the filter cut-off,

$$\overline{I_n^2} = 2\pi q\bar{I}f_c \quad (17.46)$$

The factor $\times\pi$ by which this differs from [17.3] is due to the differences in the filter type assumed.

(i) For [17.3] the frequency-domain cut-off is abrupt.

(ii) For the memory-box it is the time-domain cut-off that is abrupt.

(iii) For the single-section RC filter neither is abrupt.

Frequency domain Fourier analysis of a periodic train of impulses of frequency f_F and area q gives the amplitude of each Fourier component as $2qf_F$ (verify). The component at zero frequency, the average, is the exception, for which

$$\bar{I} = nqf_F .$$

For the random superposition of n such impulse trains the mean-square value of each component becomes

$$2qf_F \times n^{1/2}$$

and thus $$2(qf_F\bar{I})^{1/2}.$$

For $B >> f_F$ there are approximately B/f_L such random components over B, giving a further increase of $\times(B/f_F)^{1/2}$ and thus the verification of [17.3] as $f_F \rightarrow 0$.

Problem With the spectrum for the above periodic impulse train also white, why is the waveform so very different from that of white noise? (hint: phasing)

18

Interference and its prevention

Summary

The topic of this chapter is the various mechanisms whereby interference is generated and picked up by an electronic circuit, and how this may be prevented.

The first is by capacitive coupling from a body of varying voltage to the input terminal of an amplifier. The dual of this is the magnetic coupling from a line carrying current to the loop connecting the amplifier to its signal source.

Conducting electrostatic screening is the preventive measure for the capacitively coupled interference, while for the magnetically coupled interference the equivalent is direct screening by a low-permeability magnetic material such as 'mu-metal'.

More usually the magnetic coupling is reduced by the effect of eddy currents, in a thicker conducting screen than is needed for electrostatic screening.

Ground loop coupling is another topic covered, also the need for low-inductance decoupling capacitors across the power supply lines.

It is appropriate to present the first section in the context of a series of experimental observations. For the remaining sections experimental work and associated calculations are given in the final section.

18.1 Electrostatic screening

Body coupling

Put your finger on the end of the ×10 scope-probe and observe the waveform. The author obtained a 5V p-p somewhat distorted 50Hz sine wave.

Wire coupling

Next remove the probe and poke a 30cm length of wire into the scope BNC socket (being careful not to strain the fingers of the socket outwards making it useless for its normal purpose!). For the author the result was the same distorted 50Hz wave but much reduced in amplitude, to 10mV p-p. Obscuring this was 20mV of high-frequency hash which disappeared when the PC computer monitor was switched off.

Coupling mechanism

Fig. 18.1(a) shows some unshielded two-wire mains cabling. This comprises the 'line' (L) which is at nominally 240V rms relative to the 'neutral' (N), which should be close to ground potential. L generates the alternating electric flux shown, which causes the initially floating wire to have a proportional alternating potential.

In (b) the mechanism is expressed in terms of the capacitance C_{lw} between L and the wire, together with the scope input impedance $R_{in}//C_{in}$.

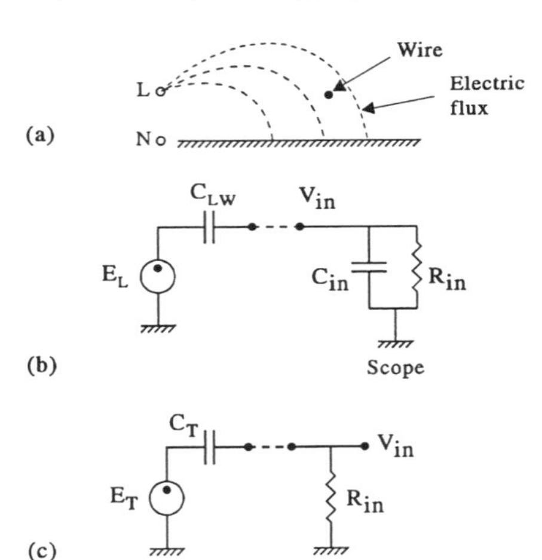

Fig. 18.1 Capacitively coupled interference.
(a) Coupling of mains-induced interference to scope input wire.
(b) Equivalent circuit for (a).
(c) Thévenin division of (b).

Next (c) shows the Thévenin equivalent of E_L and the two capacitors. With $C_{lw} << C_{in}$ then the Thévenin voltage $E_T << E_L$ and the Thevenin capacitance $C_T \approx C_{in}$.

With the circuit of (c) a simple high-pass filter of cut-off

$$f_{in} \approx 1/(2\pi C_{in} R_{in}) \qquad (18.1)$$

and with the standard scope input values

$$C_{in}=30\text{pF} \quad \text{and} \quad R_{in}=1\text{M}\Omega$$

then $\mathbf{f_{in} \approx 5kHz.}$

Coupling differentiation

With f_{in} well above the 50Hz of the mains interference, the coupling approximates closely to that of a differentiator. This explains the distorted nature of the displayed waveform. The original 240V mains waveform is a reasonably good sine wave, but when differentiated the irregularities are exaggerated.

Looking at it another way. The slight distortion of the mains waveform means that there are harmonics present. The differentiation enhances these relative to the fundamental, by ×N where N is the order of the harmonic.

This also explains the relatively strong coupling of the hash from the PC monitor. The hash originates from the line time-base, for which the frequency ~10kHz. This is comfortably within the pass-band of the high-pass filter, while 50Hz is well into the stop-band.

Signal source

More realistically, the above wire might be carrying the output of a signal source to the input of the amplifier. For the interference coupling, this places the source impedance R_s in parallel with R_{in}, usually considerably lowering the effective value of R_{in} in (c).

The overall effect is to raise the filter cut-off frequency f_{in}. For say

$$R_s=50\Omega \quad \text{and} \quad C_{in}=30\text{pF},$$

then $\mathbf{f_{in}=100MHz.}$

With this even further above 50Hz then the amplitude of the 50Hz interference is reduced by the factor 50Ω/1MΩ to a level most unlikely to be significant.

Higher frequency interference, however, can still be significant. So also can the capacitive coupling from amplifier output to input, with its potential for instability.

Op-amp feedback amplifiers

For the inverting feedback amplifier of Fig. 2.7, R_1 becomes the equivalent of R_s above. The fact that the junction of R_1 and C_{lw} is taken to the virtual ground makes no difference to the relative transfer of signal and interference; both flow directly into the virtual ground.

With R_1 typically 10kΩ, this is much larger than the above $R_s=50\Omega$, so the low-frequency coupling is increased. This is another reason for not making R_1 any higher than necessary.

The above is most likely to cause difficulties for an amplifier incorporating low- or high-pass filtering, for cases where the time-constant has to be large. Here leak-free non-electrolytic capacitors are usually obligatory, and these, for reasons of physical size, cannot be much greater than 1μF. This places a lower limit on R_1.

Screening

Screening against capacitively coupled interference is a relatively simple matter. Either the signal amplifier, or the interfering source, or preferably both, are surrounded by an electrically conducting enclosure.

Fig. 18.2(a) shows how this is done using screening boxes and coaxial cable, with the outer of the cable completing the screen. The effect is to split C_{lw} into C_{lw1} and C_{lw2} as in (b), with the centre point grounded.

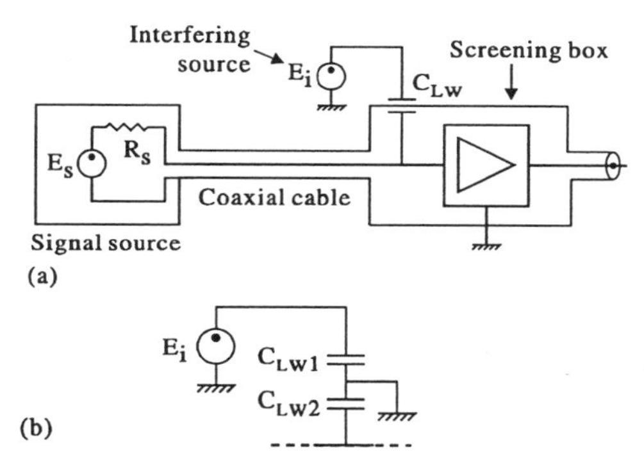

Fig. 18.2 Electrostatic screening.
(a) Conducting screening boxes and coaxial cable.
(b) Equivalent circuit.

Experiment. Coupling transients

Connect the original 30cm of wire to the tip of the ×10 scope-probe and another such length to the output of the signal generator. Set the signal generator to give a 20V p-p 200Hz square wave and observe the series of transients resulting as the two wires are brought together.

Calculation

With the change to the use of the scope-probe now the response time

$$T_r=R_{in}(C_{in}+C_{wg}).$$

With $C_{in}=10\text{pF}$, $C_{wg}=3\text{pF}$ and $R_{in}=10\text{M}\Omega$

then $\mathbf{T_r=130\mu s}$ and $\mathbf{f_r=1.2kHz.}$

From the observed step response, confirm the value of T_r. Then change to a sine wave and confirm that the coupling increases with signal frequency f_s until $f_s=f_r$.

Experiment. Screening

With the signal frequency set to 1kHz, check how a grounded metal sheet (usually aluminium) interposed between source wire and probe has the required effect of removing the interference, or nearly so depending on the size of the sheet relative to the wires.

Check that thin aluminium cooking foil is equally effective, also a reasonably fine wire mesh, or a conducting comb.

The last is most simply made by taking a sheet of circuit board of the type covered with conducting strips, and connecting the strips together at one end.

Wide-mesh coax

Some grades of coaxial cable have a very wide mesh for the outer conductor, particularly that supplied for TV antenna connection. This is perfectly adequate for the intended purpose, attention there being given rather to the grade of the insulating medium, but it does not make for a good low-frequency shield.

Try a length in place of the scope probe. It will be found that gripping the cable by hand well away from the open end will still show a significant 50Hz pick-up. Then confirm that a length of the normal coax exhibits no such effect.

18.2 Magnetic screening

Fig. 18.3 shows the mechanism for magnetically coupled interference. Here the load current I_L for the interfering source sets up the magnetic flux Φ which links into the signal amplifier input. This induces the interfering voltage E_i shown into the signal amplifier input circuit.

Here $E_i = Nd\Phi/dt$, where $N=1$ is the number of turns linking with Φ. With $\Phi \propto I_L$ then once more the process involves differentiation. Thus if the interfering source is the mains supply then the distortion will again be enhanced as it was for the capacitively coupled interference.

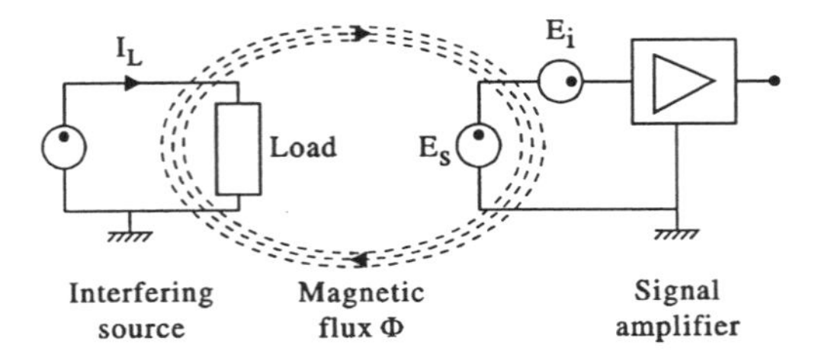

Fig. 18.3 Magnetically coupled of interference.

High-mu screen

Fig. 18.4(a) shows the direct way of screening an amplifier from magnetically coupled interference, which is to encase the circuit within a thick layer of high-permeability ferromagnetic material. This offers a low magnetic impedance path for the flux, directing it past the circuit and thus not linking with the loop at the amplifier input.

High-mu materials have the disadvantages of being costly, difficult to machine, and particularly difficult to do so without degrading the permeability. Thus the alternative of eddy-current screening is used where possible.

Toroids and pot-cores

Where a current carrying inductor or transformer has to be used, its capacity for generating interference can be much reduced by using a toroidal core or pot-core, as in Fig. 18.4(b) to (d). Here the low reluctance path of the core tends to confine the flux to where it is required.

The toroid is only suitable where the number of turns is fairly low, because of the difficulty in winding a toroid for a larger number. Where a higher inductance is needed the pot-core is more suitable, with the winding on a normal bobbin and the two halves of the pot-core then clipped around it. In either case the flux tends to be confined within the closed magnetic loop of the ferrite core material.

For reasons not so obvious, the method also works in reverse. That is, if the input circuit of an amplifier needs to include an inductor, then the tendency to develop the interfering E_i in response to an incident Φ is also much reduced by the use of a toroid or pot-core.

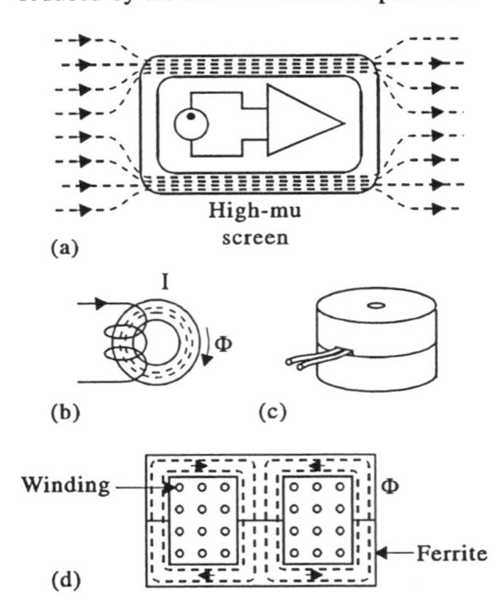

Fig. 18.4 Direct methods for reducing magnetically coupled interference.
(a) High-mu screening.
(b) Use of toroid-cored inductor to confine flux.
(c) Ferrite pot-core.
(d) Cross-section of (d) showing how flux Φ is confined.

This is in preference to an open magnetic circuit where the inductance is made variable by a simple threaded ferrite or dust-iron slug screwed down the axis of the inductor.

For the toroid, the resulting inductor is unfortunately no longer variable. However the pot-core is manufactured with the small hole shown only in (d). This extends down the whole axis of the pot, allowing the insertion of a threaded slug.

18.3 Eddy-current screening

For magnetically coupled interference of frequency higher than a few hundred Hz, eddy-current screening is a cheaper and more effective preventative measure than the above high-mu screen.

In Fig. 18.5(a) the interfering flux Φ_1 is generated as shown. This induces an emf in the second coil which, with the coil shorted, drives the 'eddy current' shown around the coil. This generates the secondary flux Φ_2 which is in opposition to Φ_1, thus reducing the total flux beyond the second coil. If the second coil is reduced to a single turn, and that filled to make a conducting plate, then the same effect obtains. Moreover, the thicker the plate the more effective the reduction.

Fig. 18.5(b) shows what happens at the surface of the conducting plate. Here with increasing distance x into the conductor the opposing flux due to the eddy-current density J reduces the overall flux density B in the exponential manner shown. J then varies in the same manner, so the eddy-current flows only in the thin layer of width δx at the surface of the conductor. With

$$B(x) = B(0) \exp(-x/\delta x) \quad (18.2)$$

and similarly for J, then δx is termed the 'skin-depth'. Because of the exponential variation, a plate of thickness of just a few times δx is thoroughly effective. With

$$\delta x = kf^{-1/2} \quad (18.3)$$

then as f decreases the required plate thickness increases. The constant k varies with the conductivity of the plate material as in Fig. 18.5(c). For aluminium at 1MHz then $\delta x = 0.08$mm.

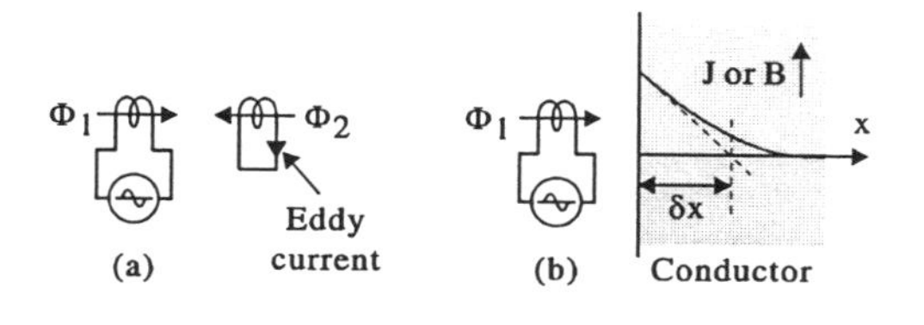

(c)

	k
Silver	0.064
Copper	0.066
Aluminium	0.083
Brass	0.13
Solder	0.18

Fig. 18.5 Eddy-current screening.
(a) Generation of opposing flux Φ_2 by shorted coil.
(b) Exponential reduction of flux density B due to proportional eddy-current density J in conductor.
(c) Value of k in [18.3] in m/sec$^{1/2}$ for various conductors.

Loop reduction

For an arrangement such as that in Fig. 18.3 without screening, the degree of interference can be limited by reducing the area A of the loop at the amplifier input. With the coupled flux $\Phi = BA$, and with the resulting $E_i = Nd\Phi/dt$, then $E_i \propto A$.

The same applies to the loop associated with the interfering source. For the simple case of a single straight current-carrying wire of infinite length, $B = K/x$ where x is the distance from the wire and $K = \mu_0\mu_r/2\pi$ (Compton, 1986). For two such wires separated by the small distance Δx, and carrying equal and opposite currents

$$B = k[1/(x - \Delta x/2) - 1/(x + \Delta x/2)] \approx 2\Delta x/x^2.$$

Thus closing the loop to reduce Δx reduces the interfering B. Note too how B varies with x^{-2}, rather than with x^{-1} for proximity to just one of the wires of a widely open loop.

For printed circuit board designs the loop is kept well closed by making the lower face of the board a conducting 'ground plane'. For a track on the upper face carrying a current from one point to another, the ground plane provides the return.

Cabling

For cabling from one board or enclosure to the next, coaxial cable is preferred. Because of the geometry, the external B, or the coupling to such, is zero, even apart from the effects of eddy-current screening.

Ribbon cables consisting of ten or more parallel wires bonded together in a flat row are now widely used. Here, to reduce the loop areas, each transferred signal needs to have its own adjacent return wire.

Where a non-coaxial wire pair is used, twisting is helpful. Then with L_t the pitch of the twist, and at a distance $r >> L_t$ there is a further reduction by the approximate factor $(L_t/2)/r$.

Exercise 18.1 gives experimental work on magnetic coupling and eddy-current screening, with associated calculations.

18.4 Ground loops

The mechanism for ground-loop interference is shown in Fig. 18.6(a). This shows a signal source E_s connected by the usual two lines to a normal non-differential amplifier. Then, for safety reasons if for no other, both source and amplifier need to be connected to the laboratory and mains ground.

Normally both source and amplifier will be enclosed in conducting screening boxes with these also connected to ground, as in Fig. 18.2, but this makes no difference to the present effect, so the boxes are omitted in (a) and (b), for clarity.

Thus the ground loop shown in (a) is formed. This can be quite extensive. For example, suppose that the source is a signal generator, as is the case for our experiments. This will have its internal mains driven power supply, and the ground return is then via the green and yellow striped line in the three-core mains lead to the unit. Also when the amplifier is that in the oscilloscope, this will be returned to ground in the same way, the loop being completed by the ground strap in the 13 amp mains distributor board.

Here any, say 50Hz, flux Φ linking with the closed loop induces an emf which drives the current I_L shown around the loop. This has the effect of developing a corresponding potential between any two points on the loop. Thus with the amplifier common taken as reference, and so marked 0, the point marked E_g (g: ground) has that potential relative to the reference. The voltage between the two amplifier input terminals is thus not just the required E_s, but E_s+E_g.

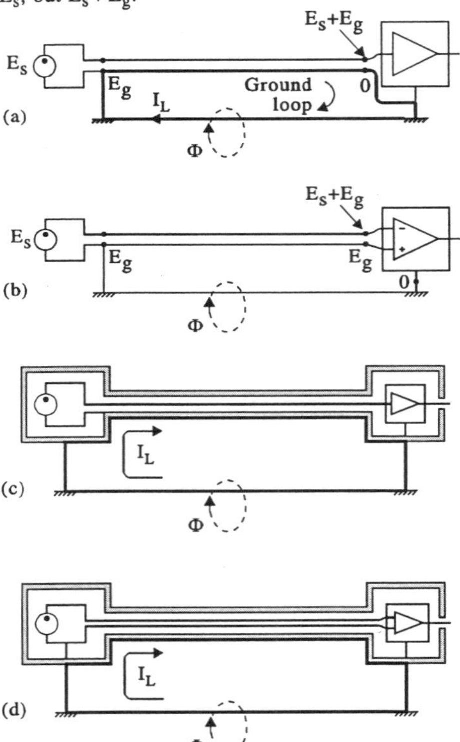

Fig. 18.6 Ground-loop interference.
(a) Mechanism for ground-loop interference.
(b) Rejection of low-frequency components using differential amplifier.
(c) Rejection of high-frequency components by eddy-current screening.
(d) Combination of (b) and (c).

Eddy-current screening

There are two ways of eliminating the above ground-loop interference, the one applicable for low-frequency interference and the other for high.

The high-frequency case in Fig. 18.6(c) is the simplest to follow. Here the frequency of the interference is sufficiently high for the skin depth at the interference frequency to be small compared with the thickness of the conducting shield, which now must be coaxial. Then I_L is confined to the outer skin of the enclosure and so does not couple into the signal circuit.

Differential amplifier

With the skin-depth for aluminium about 12mm at 50Hz, then for this class of ground-loop interference the eddy-current approach ceases to be practical. Instead the differential amplifier of Fig. 18.6(b) is required.

With the ground loop here essentially broken, completely so if the amplifier input impedance is infinite, E_s is somewhat increased. But with E_g now applied equally to the two differential amplifier inputs this is cancelled, leaving only the required E_s.

Wide-band applications

With rejection of both high- and low-frequency interference required here then the arrangement of (d) is used, combining both approaches. The cable used here is a shielded pair, with a suitably thick outer conductor.

Ground strapping

Even for (a), considerable reduction in low-frequency ground-loop interference may be obtained by connecting a heavy conducting strap between the source and amplifier ground connections. This reduces the proportion of the open-circuit E_g in (b) that appears as E_g in (a). Exercise 18.2 gives experimental demonstration of ground-loop coupling.

DAC test

Fig. 18.7(a) shows the usual arrangement for testing a digital-to-analogue converter (DAC). Here interference arising from the previous digital circuitry will be seen at the DAC output, often with a magnitude higher than that of the DAC resolution ΔV_o.

This is usually more a feature of the mode of testing than a real component of the DAC output. The rapid charging and discharging of gate input capacitances, associated with the fast gate switching in the digital circuitry, generates significant transient magnetic fields. Here two effects occur.

(i) The above magnetic fields couple directly into the loop in (a), to generate the interference voltage V_i.
(ii) The fields couple into the external ground loop, causing the current I to flow. This generates the flux Φ, thus adding to V_i.

That V_i is due to these effects may be confirmed by disconnecting the probe tip from the DAC output and connecting it instead to the same ground connection point G as used for the probe return. The observed V_i will still be present.

The relative magnitudes of the two components of V_i can be determined by disconnecting the probe from the point G, with the probe return still clipped to its tip. This will remove the usually dominant ground-loop component associated with I. If the loop is rotated the remaining V_i can then be minimised.

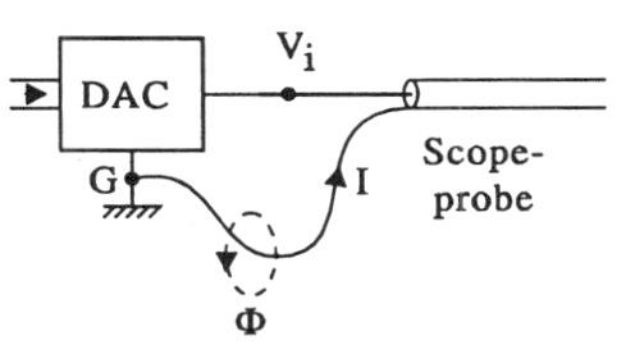

Fig. 18.7(a) DAC testing. Normal use of probe, coupling interference into probe return lead.

Adapted BNC socket

A good way of reducing the above interference, and so observing the true DAC output, is that shown in Fig. 18.7(b). Here use is made of the coaxial adapter provided with most scope probes, which mates with a BNC socket. To the socket is then connected the shortest possible length of twisted wires, which are then connected to the DAC as shown. Here both the area and the length of the remaining loop are much less than that in (a), which is drawn to a different scale.

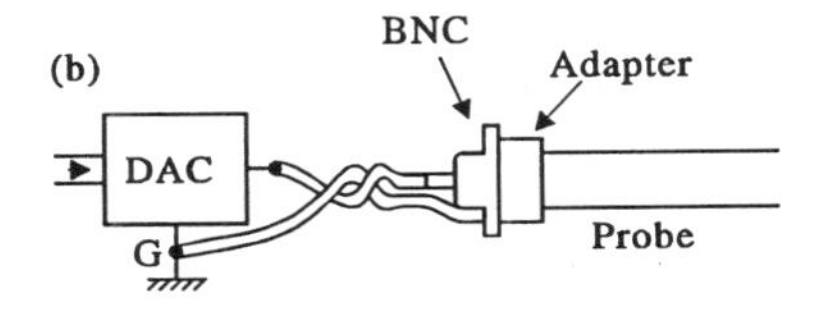

Fig. 18.7(b) DAC testing.
Use of BNC probe adapter to reduce interference.

18.5 Decoupling

An important measure in interference prevention is the use of supply decoupling capacitors. Fig. 18.8(a) shows what happens when they are omitted in a digital logic circuit.

When the first gate output switches from 0 to 1, to take up the position shown, then C_{in}, the input capacitance of the second gate must be charged to 5V from its previous value of zero. (b) shows the typical waveform for the step in V_c. With

$$I_c = C_{in} dV_c/dt$$

then the I_c waveform becomes the pulse in (c). With typically

$$dV_c/dt \approx 5V/10ns$$

and for C_{in}=30pF then $\mathbf{I_c = 15mA}$.

Without decoupling, this current has to be provided by the 5V power supply, to which the leads may be both long and in the form of an open loop, as shown. The transient flux $\Phi \propto I_c$ is thus generated, and this may then link into other parts of the circuit, or elsewhere. Also, the line resistance and, probably more importantly its inductance, will develop significant voltage transients that are imposed on the 5V supply line, and thus to adjacent logic gates.

The way in which the decoupling capacitor C_d resolves this problem is shown in Fig. 18.8(d). Here C_d is usually a high-k disc ceramic capacitor of 100nF. These are just two conducting plates deposited on either side of a disc of very high permittivity. The important feature here, in addition to the adequate capacitance, is the low parasitic series inductance. Thus the charging current path now becomes as in (d). This is highly localised and so the generation of potentially interfering flux is much reduced.

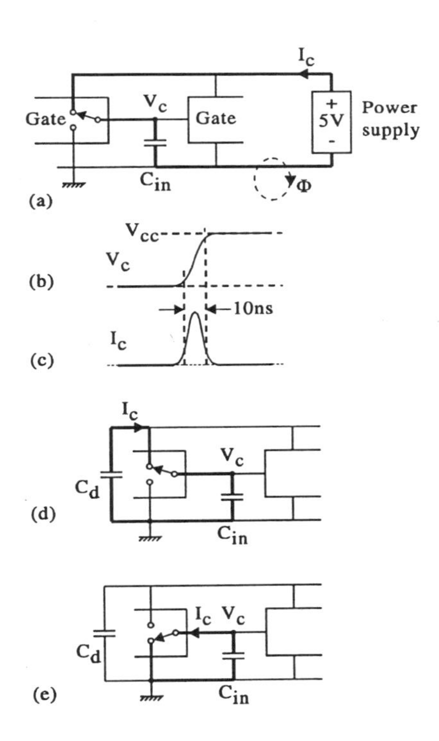

Fig. 18.8 Capacitive supply decoupling of a potentially interference-generating digital logic circuit.
(a) 0 to 1 switching without decoupling, driving I_c around the supply loop and generating interfering flux Φ.
(b) Transient waveforms for V_c during 0 to 1 transition in (a).
(c) I_c waveform for (b).
(d) Localisation of I_c in (a) by decoupling capacitor C_d, thus reducing flux generated.
(e) Current path for 1 to 0 transition in (d) showing local nature of current path regardless of C_d.

With C_d initially providing the charge for C_{in}, this gives a small initial downward step of

$$\Delta V_{cc}=V_{cc}\times C_{in}/C_d$$

in the local supply. For C_d=100nF then **ΔV_{cc}=-1.5mV.**

Even this is subsequently made good by the mains supply. However, to avoid this replenishment from being too rapid, and so once more generating an interfering flux in the supply loop, the impedance of the 5V supply line is sometimes deliberately increased. Usually it is the inductance, rather than the resistance, that is increased, to avoid an overall drop in the mean.

To avoid the added inductance resonating with the C_d the inductance is usually in the form of one or two turns round a small ferrite bead of somewhat lossy material. Also the high-k capacitors are somewhat lossy, for transient currents.

There is a corresponding reversed transient I_c when the gate output switches from 1 to 0, to become as in Fig. 18.8(e). Here, however, the I_c is already localised, as shown, whether C_d is included or not.

Decoupling of signal amplifier

The above is concerned with reduction of interference generated by a piece of switching circuitry. The other approach is the matter of reducing the sensitivity of a signal amplifier to such interfering transients as may be imposed on its supply lines.

Fig. 18.9(a) shows the circuit arrangement. The position of the decoupling capacitor C_d is important, being at the entry point of the shielding box. Here R_d and C_d form a simple low-pass filter cutting off the higher frequency components carried by V_+.

For this purpose, the value of R_d needs to be as high as possible, but this does diminish the supply voltage. Thus 100Ω is typical which, with C_d the usual 100nF, gives a cut-off of 15kHz. This may be much reduced by the addition of say a 10μF in parallel. The position of this is far less critical, and will usually be on the amplifier card within the box.

Feed-through capacitor

For yet higher frequency applications (100MHz and over) the feed-through capacitor of (b) is preferred. This eliminates even the low lead inductance for the previous disc ceramic type.

The feed-through is relatively expensive and usually only of 1nF or so. This is adequate for the higher operating frequencies but for wide-band applications the disc-ceramic will also be needed, soldered on in more or less the same position as before, above or below the feed-through. Similarly the 10μF may also be needed. For the feed-through, the box will need to be of copper or brass, so that the component can be soldered.

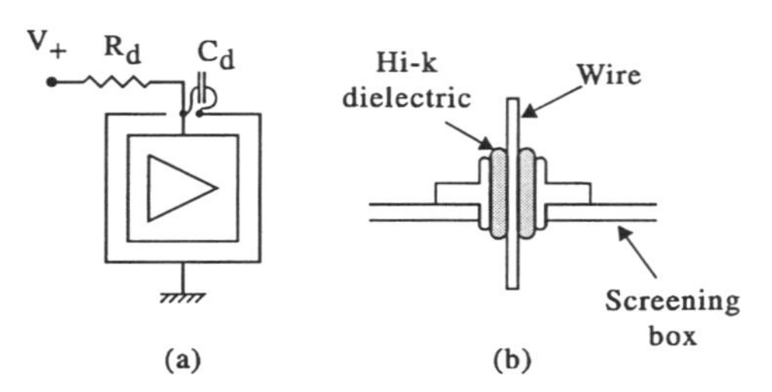

Fig. 18.9 Decoupling of amplifier power line to prevent entry of interference.
(a) Using normal disc ceramic (usually 100nF).
(b) Using soldered feed-through capacitor (usually 1nF).

Where the amplifier current is high the ferrite bead inductor can be used in place of R_r, but this does not give the lower frequency decoupling.

References

Compton, A. J. 1986: *Basic Electromagnetism and its Applications*. Chapman and Hall.

Ott, H. W. 1988: *Noise Reduction Techniques in Electronic Systems*. New York: Wiley and Sons.

Catt, I. Walton, D. and Davidson, M. 1979: *Digital Hardware Design,* MacMillan.

Catt, I. Walton, D. and Davidson, M. 1979: *Digital Electronic Design-Vol 2.* C.A.M.

18.6 Exercises and experiments

Exercise 18.1 Magnetic coupling

Experiment. Mutual inductive coupling. Obtain a pair of coreless flat-wound inductors of L=1mH and connect one to the signal generator through a 470Ω resistor and the other directly to the scope.

For a sine wave of frequency f_{in}=1kHz note the transfer from one to the other when the two are brought together.

Note also that the frequency response is as for a high-pass filter, increasing ∝frequency initially and then becoming level.

A strong resonance at about 50kHz will normally also be observed. This is the 'self-resonance' of the inductor due to the interwiring capacitance, which is effectively in parallel with L.

Calculation Calculate the value of the above high-pass cut-off.

For the signal source E_s the primary current

$$I_1 = E_s/(R_1 + j2\pi fL)$$

giving the resulting output voltage $V_o = j2\pi fMI_1$

where M is the mutual inductance. Thus

$$V_o \propto f/(f_L + jf)$$

where $f_L = R_1/(2\pi L)$ (18.4)

With $R_1 = 470\Omega$ and $L = 1mH$ then **$f_L = 75.8kHz$.**

Experiment. Eddy-current screening Interpose a sheet of aluminium of thickness $d \approx 1mm$ and note the eddy-current screening effect for varying f_{in}, i.e. by how much the observed signal drops when the screen is inserted for each frequency.

Although the variation of skin-depth δx with frequency in [18.3] is fairly weak, the variation of B with x in [18.2] is exponential. Thus once f has increased enough for δx to have fallen below the thickness of the sheet the increase in attenuation with frequency becomes fairly rapid.

Check that the wire mesh is roughly as effective but that the conducting comb has virtually no effect, giving only capacity screening.

Calculation. Limiting frequency for effective screening Calculate the frequency f_{sk} at which the skin-depth δx is equal to the thickness d of the above aluminium sheet.

With $\delta x = kf^{-1/2}$

as in [18.3], and with $\delta x = d$, then

$$f_{sk} = (k/d)^2.$$

With $k = 0.083$ and $d = 1mm$ then **$f_{sk} = 6.9kHz$.**

Screening attenuation Calculate the attenuation α_5 for $f = 5f_{sk}$.

With δx_5 the skin depth for $f = 5f_{sk}$ then

$$\delta x_5 = k(5f_{sk})^{-1/2}.$$

With $d = kf_{sk}^{-1/2}$ then $\delta x_5 = d/5^{1/2}$.

With $\alpha_5 = \exp(-d/\delta x_5)$ then **$\alpha_5 = 0.106$.**

Calculation. Cooking foil as eddy-current screen Calculate the attenuation α at 100kHz for aluminium cooking foil of the typical $d = 12.5\mu m$.

With $\delta x = kf^{-1/2}$
as in [18.3] then

$$\delta x = 0.083/(100kHz)^{1/2} = 263 \times 10^{-6} m.$$

With $\alpha = \exp(-d/\delta x)$
then **$\alpha = 0.95$.**

Experiment Insert the cooking foil and confirm that for $f = 100kHz$ as above there is little reduction. Such as is observed is more likely to be due to the elimination of the capacitive component of the coupling than to the small reduction in magnetic coupling.

Exercise 18.2 Ground loops

Experiment Connect the signal generator to the scope through a few metres of coaxial cable. Significant largely 50Hz ground loop interference should now be visible.

To avoid the ground-loop coupling, some signal generators have the facility of being able to isolate the generator section (as distinct from its power supply) by a removable strap between the lower generator output terminal and ground. For such cases the link should be closed for the above observation. Then remove the link and confirm that the interference is removed. Finally replace the link for the following tests.

Construct a differential amplifier with a gain of ×10 using the normal subtractor circuit of Fig. 2.13. Use 10kΩ and 100kΩ resistors with a TL081 op-amp. Connecting as in Fig. 18.6(b), with the coax braid to the lower amplifier input, confirm that the 50Hz interference is absent. Then ground the point marked E_g to see the interference return. There may still be some higher frequency hash present. This is capacitively coupled to the braid and would be eliminated if screened twin cable were to be used as in Fig. 18.6(d).

19

Phase-locked loop and oscillators

Summary

The phase-locked loop (PLL) is a system for locking the phase of an oscillator to that of a periodic input signal.

The basic function of the PLL is first described and the 'system diagram' derived. This is a graphic representation of the inter-relation of the equations representing the various components of the loop. It allows the response time, the frequency response, and the stability of the loop to be evaluated. The response time T_r is the time taken for the oscillator frequency f_o to respond to a step change in the signal frequency f_s. The frequency response is that of f_o to a sinusoidal modulation of f_s.

Sometimes a loop containing components of significant individual response time, which can make the loop unstable unless suitable precautions are taken. The example discussed is the phase-locking of the rotation of a motor to the output of a quartz-crystal-based oscillator. Here the significant response time is that due to the motor inertia.

The PLL 'lock range' is the range over which f_s can vary without the loop becoming unlocked; the 'capture range' is that which f_s must enter for the unlocked loop to lock.

The PLL may also be used for noise-optimised measurements of the amplitude, the timing, and thus the frequency, of a periodic signal.

Various types of voltage-controlled oscillator (VCO) are used, with the topic of oscillators discussed more generally.

A common application of the PLL is the frequency synthesiser, where a variable frequency oscillator acquires the very high frequency stability of a fixed-frequency quartz-crystal oscillator.

Direct digital synthesis (DDS) is a more recent alternative to the PLL type, not in fact involving phase-locking but conveniently included here.

The final section gives example calculations and some problems to solve. The chapter has no experimental work but SPICE simulation should be possible.

19.1 PLL function

Recall trying to make an observed oscilloscope waveform stand still without using the time-base 'sync' facility. What is being attempted here is a form of phase-locking, with the human operator trying to align the time-base output with the displayed signal. The reason for the difficulty is that it is the time-base *frequency* that is proportional to the setting of the control knob, while what is observed on the screen is the difference in time-base and signal *phase*. With the phase difference essentially the integral of the frequency difference, imagine trying to drive a car using a knob that only controls the speed at which the steering wheel is turning! For true phase-locking the scope waveform stands still - no systematic sideways slip at all, however slow.

Actually the scope sync system operates in a different manner, with a single time-base sweep triggered as the signal crosses a threshold. However, the arrangement for locking a TV picture to the transmitted sync pulses is a PLL.

Essentials of PLL

V_{in} in Fig. 19.1(a) is the periodic signal of frequency f_s to which the voltage-controlled oscillator (VCO) of frequency f_o is to be locked. The phase-detector develops the output V_p proportional to the phase difference ϕ_e (e: error) between the signal and oscillator output. With V_p applied to the VCO control then the loop achieves what the human operator could not. The external control V_{co} allows adjustment of ϕ_e, once locked.

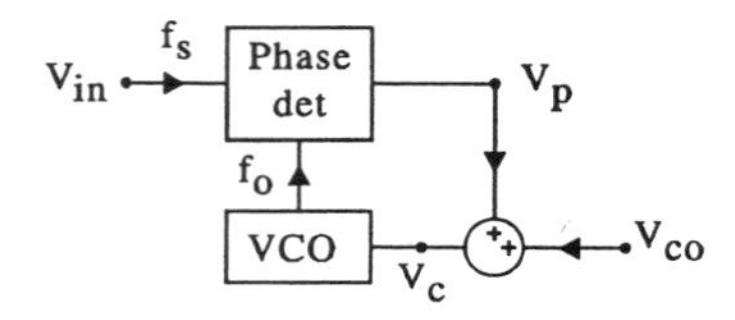

Fig. 19.1(a) Essentials of phase-locked loop.

19.2 System diagram

The system diagram in Fig. 19.1(b) is simply a diagrammatic representation of the equations describing the operation of the loop in (a). With f_e the frequency error

$$f_e = f_s - f_o \qquad (19.1)$$

This is the first component of the system diagram, with the equation number also shown.

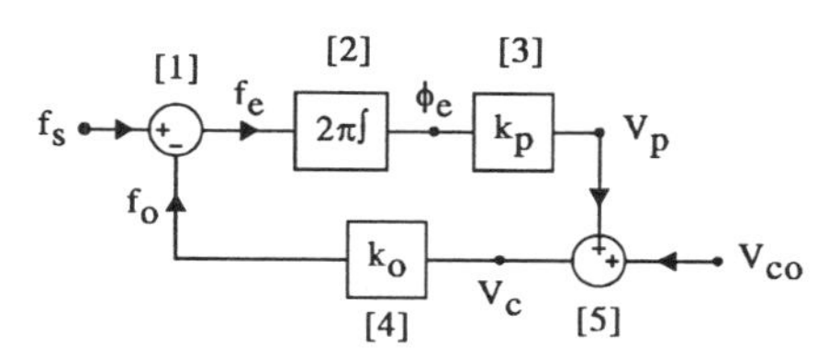

Fig. 19.1(b) Phase-locked loop.
System diagram showing equation numbers.

Phase error

With $\omega_s = 2\pi f_s$ and $\omega_o = 2\pi f_o$ for the phasors in Fig. 19.2 then the phase error

$$\phi_e = 2\pi\int(f_s - f_o)dt$$

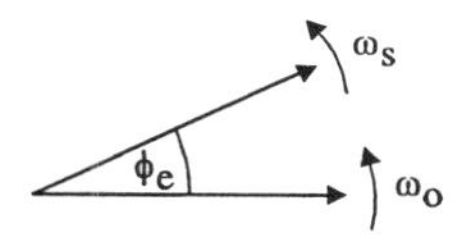

Fig. 19.2 Phasors for VCO and signal.

with f_e as in [19.1], $\quad \phi_e = 2\pi\int f_e dt \qquad (19.2)$

the next block in the system diagram.

Phase detector

With the circuit as in Fig. 19.3(a), the signal SIG and the oscillator output VCO are both logic compatible square waves. For the exclusive-OR (XOR) gate the output is logical 1 only when the two inputs differ. Thus the gate output V_x becomes as in (b).

With the following low-pass filter deriving the detector output V_p as the mean of V_x then the variation of V_p with ϕ_e becomes as in (c). For $0 < \phi_e < \pi$ then

$$V_p = k_p\phi_e \qquad (19.3)$$

as in the system diagram.

For the HC08 XOR-gate with the $V_{cc} = 5V$ supply $k_p = 5V/\pi$.

With f_x the frequency of V_x, $f_x = 2f_s$, and with f_p the cut-off frequency of low-pass filter, then adequate ripple reduction requires $f_p << 2f_s$.

VCO control

With the external control voltage V_{co} as in Fig. 19.1(a)

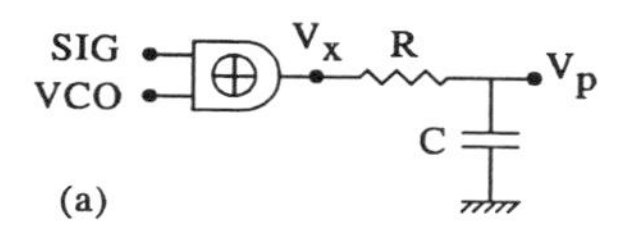

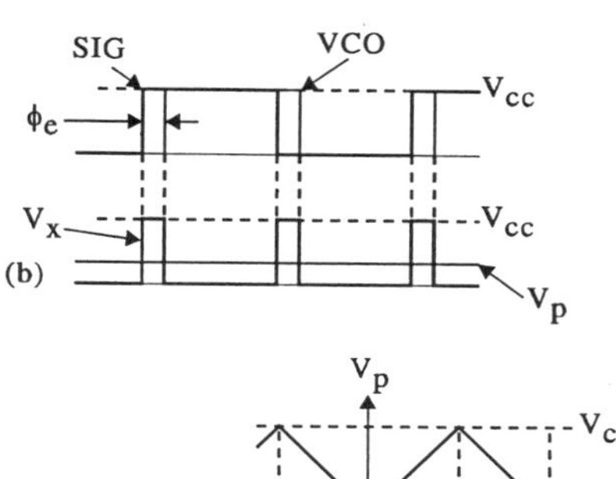

Fig. 19.3 Phase-detector based on exclusive-OR logic gate.
(a) Circuit diagram. (b) Waveforms.
(c) Variation of output V_p with phase error ϕ_e.

$$V_c = V_p + V_{co} \qquad (19.4)$$

With V_c the VCO control voltage, and for $f_o \propto V_c$

$$f_o = k_o V_c \qquad (19.5)$$

Thus the system diagram is completed.

Open- and closed-loop response

Breaking the loop at f_o, and with the open-loop gain $A_L = f_o/f_s$

$$A_L(s) = 2\pi k_p k_o/s \qquad (19.6)$$

With the feedback from f_o applied then

$$f_o = f_s s_L/(s + s_L) \ldots(a) \quad f_e = f_s s/(s + s_L) \ldots(b) \quad (19.7)$$

where $\quad s_L = 2\pi k_p k_o \qquad (19.8)$

These are the familiar unity-gain low- and high-pass responses of Chapter 11, with the closed-loop cut-off frequency

$$f_L = s_L/(2\pi) = k_o k_p \qquad (19.9)$$

The corresponding step responses then become as in Fig. 19.4 where

$$T_L = 1/2\pi f_L \qquad (19.10)$$

with the final values $f_o = f_s$ and $f_e = 0$.

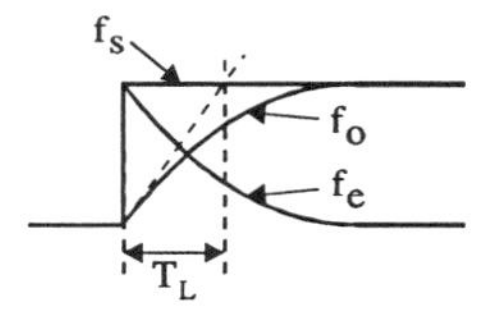

Fig. 19.4 Step response for PLL of Fig. 19.1.

Phase error

For $V_{co}=0$ in the system diagram,

$$\phi_e = f_o/k_o k_p.$$

With f_o as in [19.7], then

$$\phi_e = f_s(1/k_p k_o) \times s_L/(s+s_L) \quad (19.11)$$

again a low-pass response, and with the final value

$$\phi_e = f_s/(k_p k_o) \quad (19.12)$$

With V_{co} added the final value becomes

$$\phi_e = (f_s - k_o V_{co})/(k_o k_p) \quad (19.13)$$

allowing ϕ_e to be adjusted. From Fig. 19.3(c) the most suitable value is the mid-range $\phi_e = \pi/2$, rather than $\phi_e = 0$.

Feedback sense

Fig. 19.5 shows how, for ϕ_e initially in the range where the feedback is positive rather than negative, ϕ_e is quickly moved to the range of negative feedback, to give the same final value.

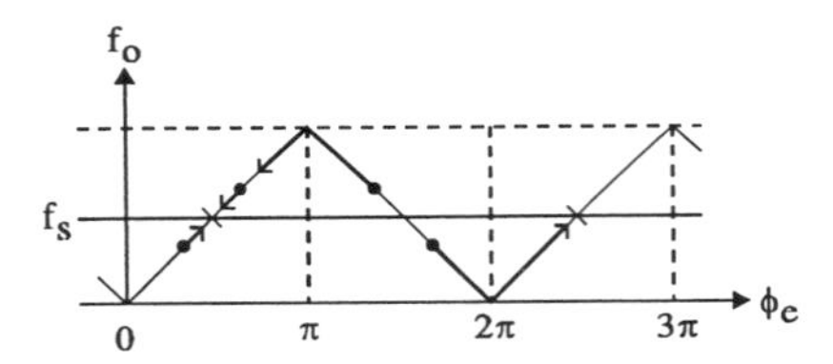

Fig. 19.5 Locking for various initial values of ϕ_e.

19.3. Loop stability, lock range and capture range

With the open-loop response $A_L(s) = 2\pi k_p k_o/s$ as in [19.6], and for sinusoidal variation of f_s at the frequency f, then the resulting sinusoidal variation if f_o is given by $f_o = A_L(f) \times f_s$ where

$$A_L(f) = k_p k_o/jf \quad (19.14)$$

Thus $|A_L(f)| \propto 1/f$ with a 90° phase lag, both due to the integration.

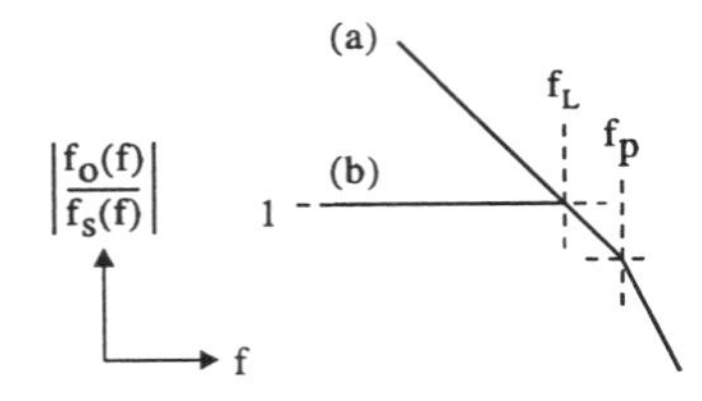

Fig. 19.6 PLL open- and closed-loop frequency responses. f_p= cut-off frequency of phase-detector low-pass filter.
(a) Open loop.
(b) Closed loop.

With the effect of the low-pass filter at the output of the phase-detector included, and with f_p the filter cut-off, then the amplitude-frequency response of $A_L(f)$ becomes as in Fig. 19.6(a).

With the closed-loop response as in (b), then the stability limit to $k_p k_o$ is that for $f_L = f_p$; f_L can be no higher.

Lock-range

For the present phase-detector, the limits to the lock-range are the values of f_s for which $\phi_e = 0$ and $\phi_e = \pi$. With

$$\phi_e = (f_s - k_o V_{co})/(k_o k_p)$$

as in [19.13] then

$$f_{s.max} - f_{s.min} = k_o k_p \pi \quad (19.15)$$

where $f_{s.max}$ and $f_{s.min}$ are the limits of the lock-range.

With $\quad f_L = k_o k_p$

as in [19.9], and $f_L = f_p$ at the stability limit

$$f_{s.max} - f_{s.min} = f_p \pi \quad (19.16)$$

With the frequency of the ripple at the phase-detector output $2f_s$, low ripple requires $f_p << 2f_{s.min}$, thus requiring

$$f_{s.max} - f_{s.min} << 2\pi f_{s.min} \quad (19.17)$$

For say $\quad f_p = 2f_{s.min}/6$

for adequate ripple reduction then

$$f_{s.max}/f_{s.min} \sim 2.$$

RC network

With the above $f_{s.max}/f_{s.min}$ inadequate for many applications, inclusion of the RC network of Fig. 19.7(a) can give the required increase. For C initially absent (short-circuit) then the reduction factor

$$\alpha = R_2/(R_1 + R_2)$$

could be set to below the stability limit.

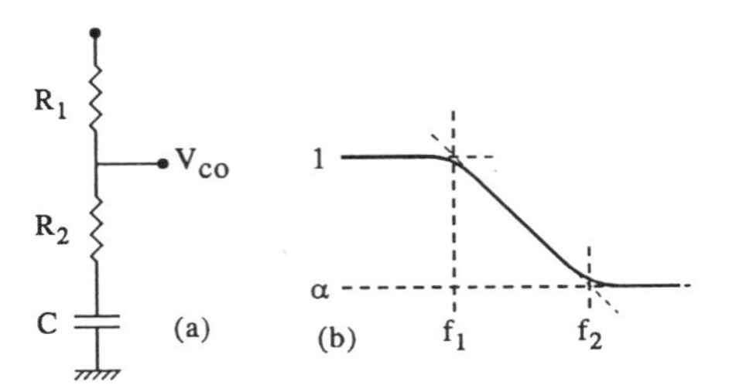

Fig. 19.7 Incorporation of RC network into PLL in order to increase lock-range.
(a) Network.
(b) Network frequency response (log axes).

With C included then the network frequency response becomes as in (b), with the transition frequencies

$$f_1 = 1/[2\pi(R_1 + R_2)] \ldots(a) \qquad f_2 = 1/(2\pi R_2) \ldots(b) \qquad (19.18)$$

Thus if C is chosen to set f_2 well below the open-loop unity-gain frequency f_L in Fig. 19.6 then in the region of f_L the network responds much as if C was short-circuit, thus retaining stability.

For any frequency below f_1, in contrast, the attenuation α is absent, to give the same lock range as for the network absent.

Example. HC4046 PLL

This integrated circuit includes the VCO of Fig. 19.14 and the XOR-gate phase-detector of Fig. 19.3. With f_{om} the upper limit of f_o for the VCO, and with V_{cm} the value of V_c for $f_o = f_{om}$, then f_o varies from zero to f_{om} as V_c is varied from 0V to V_{cm}, with the variation approximately linear.

With V_{cm} close to the supply V_{cc}, and with the phase-detector output range from 0V to V_{cc}, then the potential lock-range is the full working range of the VCO.

Capture-range

Assuming the cut-off f_p for the phase-detector output filter to be abrupt, and with Δf_{ca} the width of the capture range, then, as follows,

$$f_{ca} = 2f_p$$

With $|f_e| > f_p$ for the error frequency f_e then the phase error ϕ_e describes a full-range sawtooth at the frequency f_e. But, with f_e beyond the filter cut-off, the fiter output remains constant at the mid-range value for $\phi_e = 180^o$.

But as soon as f_s is moved to within f_p of f_o then the phase-detector gives the full sawtooth sweep output, soon finding and locking to the signal.

Lock-capture hysteresis

In the absence of the RC network or equivalent, [19.16] gives the width of the lock-range $\Delta f_{LK} = \pi f_p$, so here $\Delta f_{ca} \approx \Delta f_{LK}$. But with the network included then Δf_{LK} is increased $\times 1/\alpha$ while Δf_{ca} remains the same. Once unlocking has occurred then the difference between f_s and f_p has to be reduced $\times\alpha$ for the lock to be re-established.

Range reduction

Sometimes Δf_{ca} is deliberately reduced, by reducing f_p, with the loop gain reduced in proportion to maintain stability. This is as for the FM radio receiver, to prevent 'grabbing' of the tuning by a signal of larger amplitude than that required, at an adjacent frequency. This, however, is more usually a 'frequency-locked' rather than phase-locked loop. In the same way Δf_{LK} may also need to reduced. Otherwise as the tuning knob is turned to find another station it would be difficult to 'let go' of the one currently being locked to.

19.4. Motor phase-locking

Fig. 19.8 shows an arrangement for phase-locking the rotation of a motor to a highly stable quartz-crystal oscillator. This is part of an ellipsometer, an instrument for measuring the polarisation of reflected light, and includes various optical components which must be rotated at a highly constant rate. One type of ellipsometer includes two such assemblies. With these needing be electronically 'geared' to each other, the phase-locking is required.

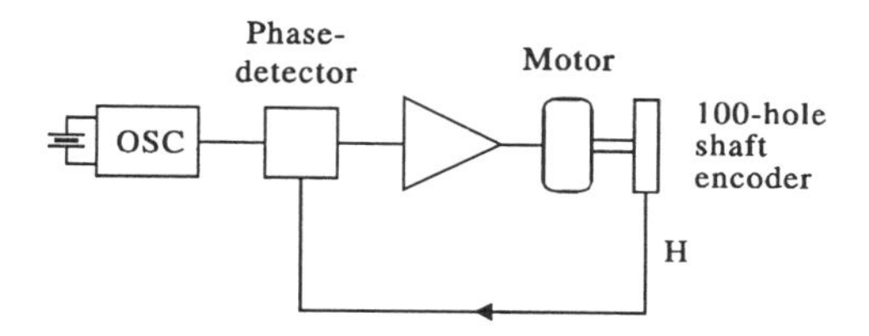

Fig. 19.8 Arrangement for phase-locking motor rotation to quartz-crystal oscillator.

Motor inertia

As noted at the start of the chapter, the problem here is the motor inertia. For the PLL previously described the range of frequencies over which correction could be provided was that below f_p, the cut-off of the low-pass filter at the output of the phase-detector. With f_p the main low-pass component in the loop, and now with the motor inertia setting the very much lower f_i as the dominant cut-off, then the correction range is proportionally reduced. With a flywheel added to reduce the speed fluctuations prior to the PLL control, f_i can be very low indeed.

Ratemeter loop

The solution here is to add the local ratemeter loop in Fig. 19.9. The ratemeter gives the voltage $V_r \propto f_h$, the frequency of H. With the local feedback the closed-loop cut-off can be increased from the open-loop f_i to the much higher cut-off f_r of the ratemeter response. Much as for the phase-detector in Fig. 19.3 for the earlier system, f_r need only be a few times f_h.

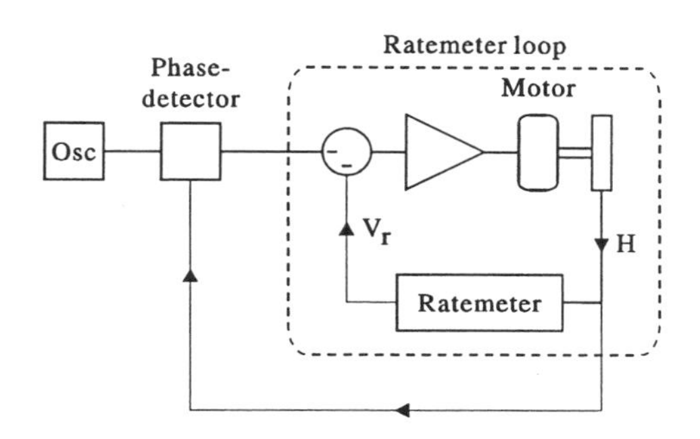

Fig. 19.9 Inclusion of ratemeter loop into PLL of Fig. 19.8.

The arrangement has a double advantage. Firstly the ratemeter loop gives correction for speed fluctuations up to $f_r >> f_i$. Secondly, with f_i effectively increased to f_r, the phase-locked loop is able to give further correction, again up to $f_r >> f_i$.

With the motor speed $\Omega_m = 180$ rpm for the present system ($f_m = 3$Hz) then $f_h = 300$Hz. With the inertial time-constant $T_i = 1$s then $f_i = 160$mHz. With thus $f_h/f_i = 1875$ then the upper limit to the overall correction range is increased by approaching this factor.

Phase-detector

The circuit and waveforms for this highly accurate circuit are shown in Fig. 19.10. With the sawtooth labelled Count representing the value of the 12-bit DAC output were the latch to be transparent, and with the latch is operated by the rising edge of the encoder output H, then the phase-detector output V_p represents the phase difference (error) ϕ_e between H and Count.

With the DAC output range is centred on zero and with ϕ_e is taken as zero for the rising edge of H midway between the falling edges of Count then

$$V_p = k_p \phi_e \tag{19.19}$$

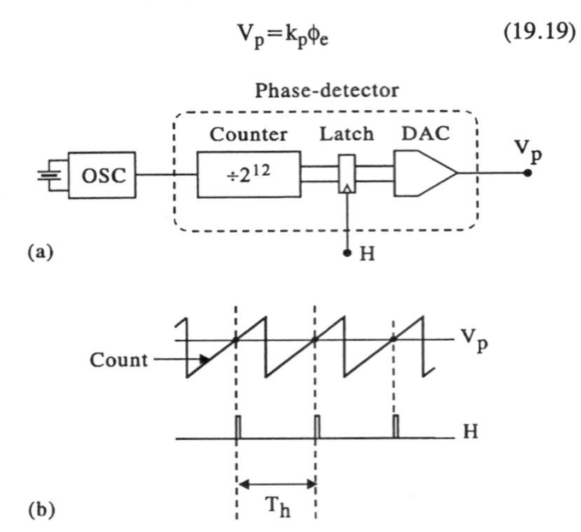

Fig. 19.10 Phase-detector in motor phase-locked loop of Fig. 19.8. (a) Block diagram. (b) Waveforms.

The arrangement improves upon the XOR-gate phase-detector of Fig. 19.3 in two respects. Firstly the input range is $-\pi < \phi_e < \pi$, twice that of the previous arrangement. Secondly V_p is ripple free, with no need for a following low-pass filter.

Ratemeter detector

With this as in Fig. 19.11, the operation is much as for the phase-detector but with the added ÷2 stage and with the counter also reset by the rising edge of H. With $T_h = 1/f_h$, and with T_{ho} the required value of T_h then the ratemeter output

$$V_r = k_r(T_h - T_{ho}) \tag{19.20}$$

where k is the appropriate constant. This is essentially the response of a 'period-meter' rather than a ratemeter. However, with variation of T_h from T_{ho} small, all that is needed is a sign reversal to obtain the required

$$V_r \propto (f_h - f_{ho}).$$

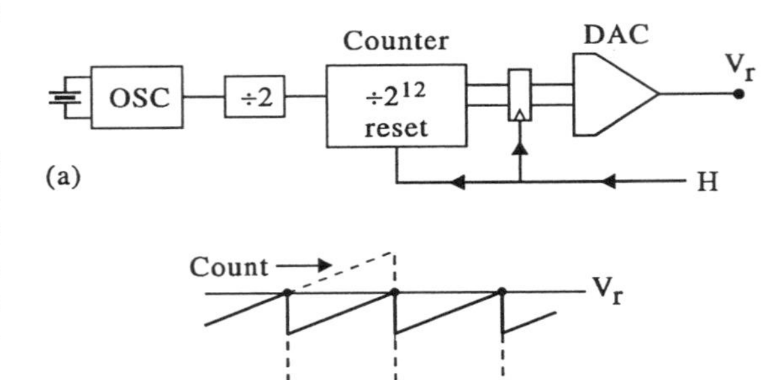

Fig. 19.11 Ratemeter in motor phase-locked loop of Fig. 19.9. (a) Block diagram. (b) Waveforms.

Conventional ratemeter

In sequence, this comprises a comparator, monostable, and low-pass filter. With f_{in} the frequency of the comparator input, T_m the width of the monostable output pulse, the supply V_{cc} the height of the pulse, and with the filter deriving V_r as the average of the monostable output, then $V_r = V_{cc}T_m f_{in}$ giving $V_r \propto f_{in}$ as for the true ratemeter.

As for the phase-detector of Fig. 19.10, the advantage of the ratemeter of Fig. 19.11 over the conventional type is that the low-pass filter is not required. Thus again the response time is able to approach f_h more closely.

Ratemeter loop system diagram

As shown in Fig. 19.12, this is constructed as follows. With V_m the motor voltage, V_{mo} the value of V_m needed to drive the motor at the required speed, with k_m the motor constant, with the motor inertia represented by the low-pass filter of cut-off f_i shown, and with f_m the motor rotation frequency, then the section of the diagram between V_{mo} and f_m becomes as shown. With f_{mo} the required value of f_m, and with

$$V_r \approx k_r(f_m - f_{mo}) \tag{19.21}$$

for the ratemeter, then this too becomes as shown.

With the added scaling factor k_a then the loop is completed, with the negative feedback tending to maintain $f_m = f_{mo}$, and with the phase-detector output V_p giving the final phase-lock.

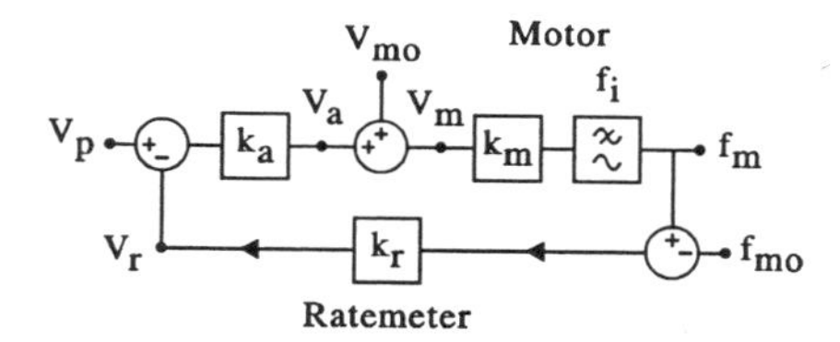

Fig. 19.12 System diagram for ratemeter loop in Fig. 19.9.

Drag perturbation

The effects that the PLL is required to correct for are as follows.

- Long-term changes due to wear, heating, etc.
- Periodic changes at the motor rotation frequency f_m due to flywheel imbalance, bearing eccentricity, etc.
- Shorter term changes due to ball-bearing 'rumble', armature 'cogging', etc.

With each of these constituting a variation in the motor drag coefficient k_m then a given variation Δk_m can conveniently be represented by the equivalent change ΔV_{mo} in V_{mo}, where

$$\Delta V_{mo}/V_{mo}=\Delta k_m/k_m \qquad (19.22)$$

Step response

With the output absent for the present ratemeter, it will now be shown how the time-stepped response of the ratemeter imposes the limit on the closed-loop response of the ratemeter. Here we consider the response of the ratemeter output V_r to a step change in V_{mo}, as simulating a step change in the drag.

With the closed-loop response time small compared with T_i then the response of the inertial low-pass approximates to that of the equivalent integrator. Thus over the ensuing period T_h the resulting

$$\Delta f_m \approx \Delta V_{mo} k_m T_h/T_i$$

With the overall loop gain constant

$$A=k_m k_r k_a \qquad (19.23)$$

then the corresponding

$$\Delta V_a=-\Delta V_{mo} A T_h/T_i \qquad (19.24)$$

For $\quad A=T_i/T_h \quad$ then $\quad \Delta V_a=-\Delta V_{mo}$

restoring the integrator input to zero, and thus giving no further change in any variable. Thus, as in Fig. 19.13(a), in just one step of the encoder output, V_r is adjusted to the value required to correct for the step change in simulated drag. Other values of A give the responses shown.

As an exercise, write the very short computer program needed to verify the above responses.

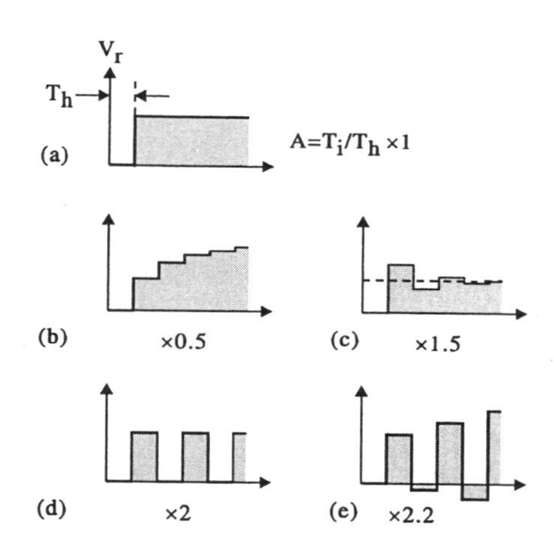

Fig. 19.13 Response of ratemeter output V_r to applied V_{mo} step, for ratemeter loop of Fig. 19.12.
(a) Loop gain factor A of optimum value.
(b,c) A below and above optimium.
(d) A twice optimum.
(d) A over twice optimum giving instability.

Frequency response

With A the optimum T_i/T_h in Fig. 19.13(a) then the frequency response of the ratemeter open-loop gain becomes as in Fig. 19.14(a). With the closed-loop response thus as in (b), then the closed-loop cut-off is increased from f_i to f_h as previously asserted.

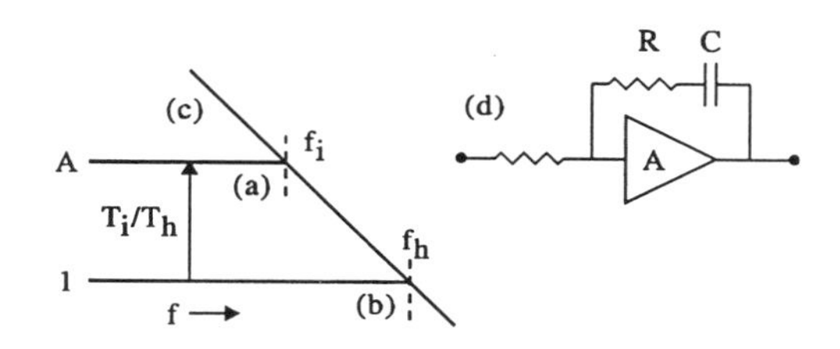

Fig. 19.14 Ratemeter frequency responses.
(a) Ratemeter open-loop.
(b) Ratemeter closed-loop for optimum loop gain constant.
(c) Modification to (a) for pseudo-integrator of (d) included.

Response to drag variation

For the closed ratemeter loop alone (apart from the phase-locking) the change Δf_m due to a change Δk_m in the motor drag can now be calculated.

With $\qquad \Delta V_{mo}/V_{mo}=\Delta k_m/k_m \qquad$ as in [19.22]

with $\qquad f_m(s)=k_m V_m(s)\times s_i/(s+s_i)$

for the motor in the system diagram, with the remaining relations as in the diagram, and for $V_p=0$, then

$$\frac{\Delta f_m(\omega)}{f_{mo}} \approx \frac{\Delta k_m(\omega)}{k_m} \times \frac{T_h}{T_i} \times \frac{\omega_h}{\omega + \omega_h} \qquad (19.25)$$

Thus the effect of changes in drag is reduced $\times T_h/T_i$ for all sinusoidal components of drag up to f_h. Beyond f_h the expression becomes invalid because the time-stepped nature of the ratemeter response has not been allowed for. The reduction remains constant beyond f_i because, although the loop gain begins to decrease here, so also does the inertial filter response.

Phase-locked loop

With the ratemeter loop system diagram as shown then an applied V_p from the phase-detector gives the proportional change $\Delta f_m=V_p/k_r$ making the response of the corresponding change in motor phase proportional to the integral of V_p. Thus the phase-locked loop response becomes the classic one of Fig. 19.1, with the limiting cut-off those at f_h, both for the ratemeter and the phase-detector.

With the open-loop response of the PLL thus as in Fig. 19.6 with the limit at f_h then the closing of the PLL gives a further reduction in the response to drag changes, by the factor f/f_h for $f < f_h$. Thus, ignoring the invalid final term in [19.25]

$$\frac{\Delta f_m(\omega)}{f_{mo}} \approx \frac{\Delta k_m(\omega)}{k_m} \times \frac{T_h}{T_i} \times \frac{f}{f_h} \qquad (19.26)$$

Phase fluctuation

With ϕ_e as in the system diagram of Fig. 19.1(b), f_s constant, and here $f_o=f_m$ then

$$\Delta\phi_e=2\pi\Delta f_m/f \qquad (19.27)$$

giving

$$\Delta\phi_e = 2\pi \times \frac{\Delta k_m}{k_m} \times \frac{T_h}{T_i} \times \frac{f_m}{f_h} \qquad (19.28)$$

Example For a 10% variation in drag and the above values of T_h, etc., then **$\Delta\phi_e=7.54m^o$.**

With the limiting cut-off frequencies the same for both the ratemeter loop alone, and for the phase-locked loop with the closed-loop response of the ratemeter loop assumed immediate, both being equal to f_h, then one or both loop gain scaling factors will have to be relaxed a little to retain stability. However, the above value of $\Delta\phi_e$ is comparable with the $6m^o$ observed experimentally.

Pseudo-integrator

With $\Delta\phi_e \propto \Delta k_m$ as in [19.28], it may be desirable for a constant change Δk_m in drag that the final value of the phase error $\Delta\phi_e$ be zero. This can be achieved by adding the pseudo-integrator of Fig. 19.14(d) to the forward section of the ratemeter loop. With $RC=T_i$ this converts the low-pass open-loop response of (a) to the ideal integrating response of (c).

19.5 Noise-optimised measurements

The PLL can be used in the process of measuring the timing and amplitude of a noisy periodic signal with minimum noise error.

Amplitude measurement

The 'amplitude' A of an observed sine wave V_{in} of frequency f is the number by which the 'reference' sine wave $V_r=\sin(2\pi ft)$ must be multiplied in order to equal V_{in}. For the two in phase then 'equal' means

$$V_{in} - AV_r=0$$

at all times. With noise added to V_{in} then A is set for the integral of $(V_{in}-AV_r)=0$. For minimum white-noise error then the integral is weighted by V_r, to give

$$\int_{-\infty}^{+\infty}(V_{in} - AV_r)V_r dt = 0 \qquad (19.29)$$

Thus

$$A = I_a \Big/ \int_{-\infty}^{+\infty} V_r^2 dt \qquad (19.30)$$

where

$$I_a = \int_{-\infty}^{+\infty} V_r V_{in} dt \qquad (19.31)$$

For a periodic signal such as the sine wave the integrals need to be over a whole number of cycles, while for a single noisy pulse such as that in Fig. 19.15(a) the integrals are over the pulse.

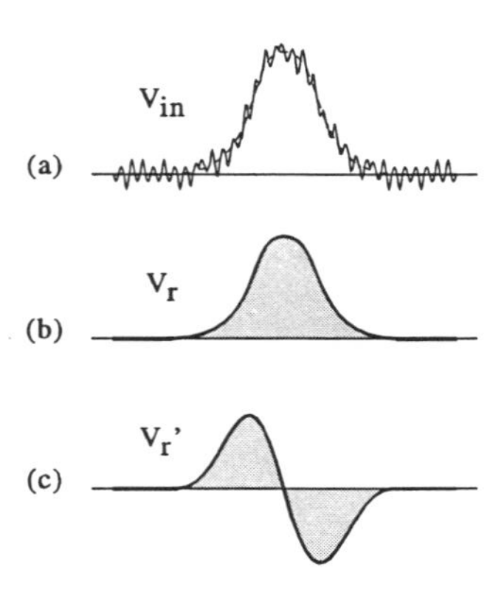

Fig. 19.15 Pulse amplitude measurement.
(a) Noisy signal peak.
(b) Reference waveform V_r.
(c) Reference waveform $V'_r=dV_r/dt$.

Measurement of time of occurrence

By a similar argument (Wilmshurst, 1990) the noise-optimum measurement of the time of occurrence of the noisy pulse in (a) is given by the placing of V_r in (b) such that $I_t=0$ for

$$I_t = \int_{-\infty}^{+\infty} V'_r \, V_{in} dt \qquad (19.32)$$

where, as in (c), $V'_r=dV_r/dt$.

PLL implementation

Fig. 19.16 shows how the above principle may be implemented using a PLL. With the signal input V_{in} the periodic train of noisy pulses in (b) then the required output from the VCO is no longer a square wave as previously, nor even a sine wave, but now the reference waveform V_r, possibly as generated in (c).

With V'_r developed from V_r as in (a) then the analogue multiplier forms the product $V_{in}V'_r$ which is averaged in the following low-pass to effectively form the integral I_t as the phase-detector output V_p.

With the external control V_{co} set to give the VCO frequency $f_o=f_s$ for $V_p=0$, and with V_p then added, the resulting phase-locked loop maintains $(V_p \propto I_t)=0$ and thus V_r' and V_r aligned with V_{in} as required for V_o to correctly indicate the amplitude of V_{in}.

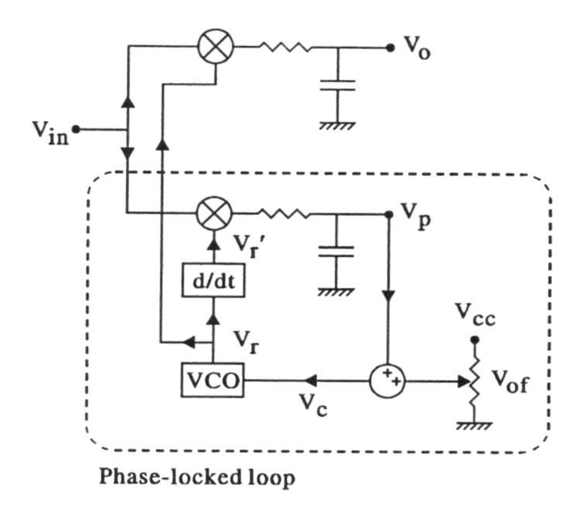

Fig. 19.16(a) Noise-optimised PLL measurement. Circuit diagram.

Frequency measurement

The most common example of the above is where the known shape of the signal is a sine wave. Here the generation of V_r and V'_r is simply a matter of providing a sine wave oscillator with quadrature outputs (see Exercise 19.1).

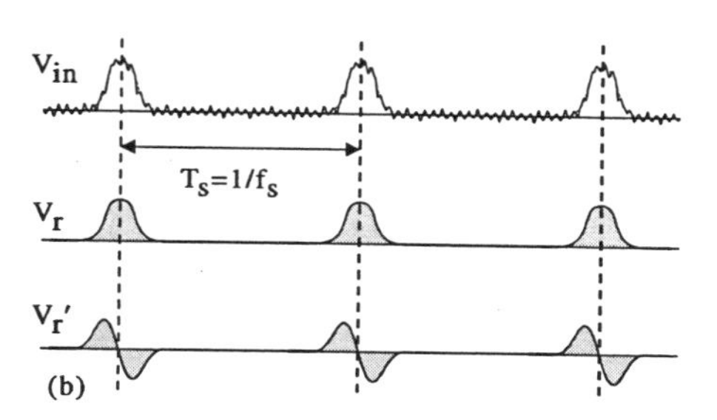

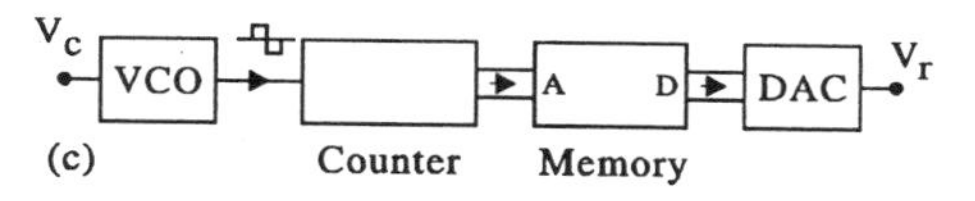

Fig. 19.16(b,c) Noise-optimised PLL measurement.
(b) Waveforms for circuit in (a).
(c) Method for deriving waveform V_r in (a) and (b).
A: memory address, D: memory data.

However, provided the frequency of the signal does not vary too widely, the same degree of noise immunity is given by preceding the PLL with a suitable relatively narrow-band band-pass filter, followed by the basic PLL of Fig. 19.1. Here the zero-crossing of the filter output occurs at just the time when I_t in [19.32] is zero.

If the relation between the control voltage V_c and oscillator frequency is linear then V_c also constitutes the output of the noise-optimum frequency-to-voltage converter. The PLL chip is designed with this in mind.

Gated integrator phase-detector

With the I_t of [19.32] formed in the lower low-pass filter of Fig. 19.16(a), then the filter response time needs to be $\gg T_s$ if there is not to be ripple at the filter output modulating the oscillator frequency. Where the required resolution-time for the measurement is equal to T_s (one value per cycle) then a gated analogue integrator is more suitable. Here the integrator is reset prior to the occurrence of the V_r pulse, and its output sampled after the pulse has finished. In this way there is no ripple at all.

Phase-detector for laser speckle signal

Fig. 19.17(b) shows yet another approach to phase-detection. Here the requirement is to monitor the progress of the rotating shaft in (a). The standard way of doing this is to paint a white line on the shaft which is then sensed using a photodetector. If, however, a finer time resolution than that of one rotation period is required then some sort of shaft encoder must be attached. This is expensive and often unacceptably invasive of the system.

Thus a simple alternative is to attach a band of retro-reflective tape to the shaft and shine a laser beam on it, sensing the reflected speckle pattern with a photodetector. Here the speckle pattern is a noise-like signal with typically 1000 fluctuations per shaft rotation, which repeats at each rotation. First a 'snapshot' of the speckle pattern is taken over one rotation. This is then written into the VCO-driven memory of Fig. 19.16(c) to produce the reference waveform V_r.

With V_r phase locked, largely as in Fig. 19.16(a), to the current signal V_{in} from the photo-detector then the counter addressing the memory works as a binary type shaft encoder, giving moment by moment shaft position.

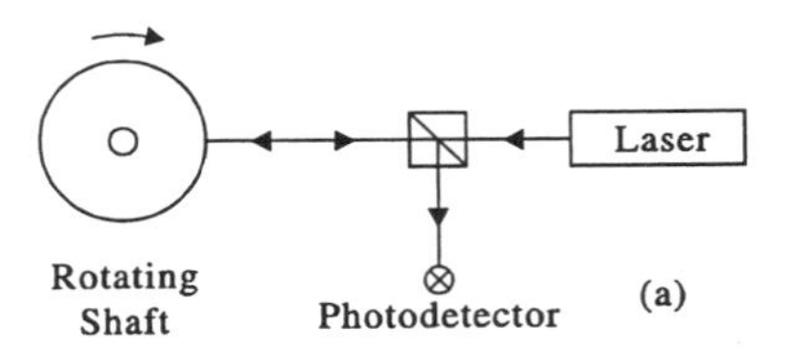

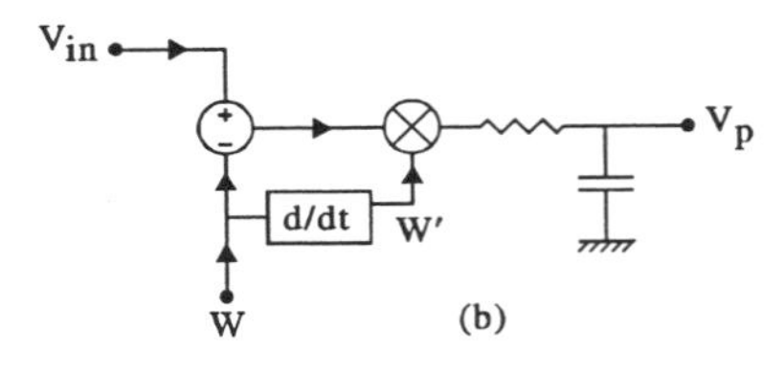

Fig. 19.17 Laser speckle shaft rotation sensor.
(a) Optical system with shaft coated with retro-reflective tape to generate speckle pattern.
(b) Suitable noise-optimised phase-detector.

General timing criterion

While the timing criterion of [19.32] is valid for the signal pulse shown, it is only so for a transient where the initial and final values are the same, as for the pulse. With I_t the integral that the timing of V_r is adjusted to bring to zero, then more generally

$$I_t = \int_{-\infty}^{+\infty} (V_{in} - V_r)V'_r \, dt \qquad (19.33)$$

with now the amplitude of V_r needing to be equal to that of the signal component of V_{in}. As shown in Exercise 19.2, [19.33] reduces to [19.32] for the initial and final values equal.

Where the resolution time of the present measurement is a fraction of the shaft rotation period then the initial and final values will usually not be the same. Fortunately here the recorded V_r is equal to the current signal component. Thus the I_t of [19.33] can be formed as in Fig. 19.17(b).

19.6 Voltage-controlled oscillator

A voltage-controlled oscillator (VCO) is one for which the frequency f_o of the output varies with the control voltage V_c. There are two main classes of oscillator, both potentially voltage controllable. For the first f_o is determined by the time taken to charge and discharge a capacitor C, sometimes through a resistor R. For the second f_o is the resonant frequency of an LC resonator. A sub-class is where the resonator is a quartz crystal. This too can be voltage controlled over a very small range, say for phase-locking to a primary standard.

The RC type is better suited to lower frequencies and gives a larger ratio of maximum to minimum f_o in response to the voltage control, but is more prone to thermal drift and frequency modulated noise (jitter).

Switched-capacitor VCO

Fig. 19.18 shows the essentials of the VCO in the 4046 type PLL chip. This is basically of the RC type, with R the resistor and C the capacitor. The augmented source-follower feedback circuit gives $I_{c1}=V_c/R$, which is reflected by the current mirror to be alternately switched into C as shown by the output bistable.

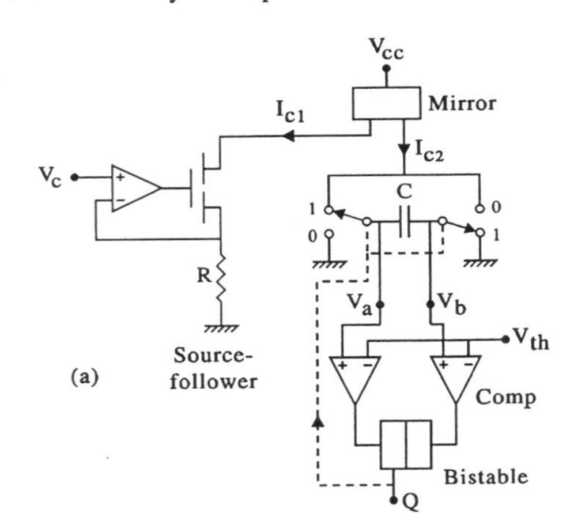

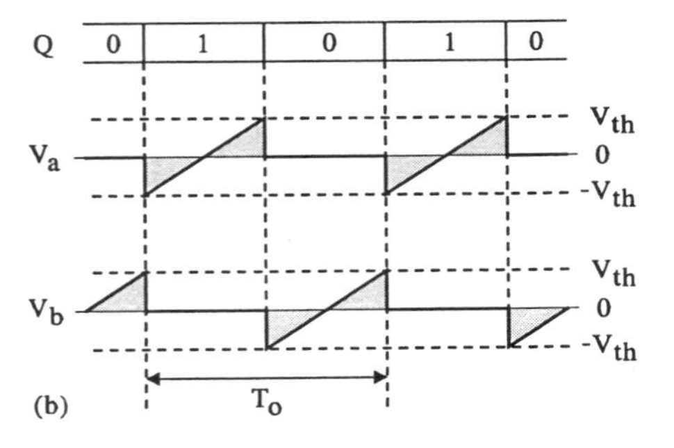

Fig. 19.18 VCO in 4046 phase-locked loop integrated circuit.
(a) Circuit diagram. (b) Waveforms.

Output frequency

For the switch position shown $dV_a/dt = I_{c2}/C$. With $dV_a/dt = 2V_{th}/(T_o/2)$ as in (b), and with $I_{c1} = I_{c2} = V_c/R$, then

$$f_o = V_c/(4V_{th}CR) \qquad (19.34)$$

See also the RC circuit of Exercise 19.3, based on the logic inverter gate.

Thermal drift

The measured f_o for the above VCO changed by about 1% for a 10°C temperature rise. This is not due to changes in C, because thermally isolating C made no difference. It is also unlikely to be due to changes in the comparator input capacitances because these, being of SiO_2 dielectric, are extremely temperature stable. Differential changes in the V_{th} for the two comparators are a possible cause.

Jitter

With f_o as in [19.34] then f_o is subject to the noise fluctuations in the comparator threshold V_{th}. With the half-period $T_o/2 = 2V_{th}CR/V_c$, with one noise perturbation ΔV_{th} per half-cycle, and with Δt the resulting perturbation in the half-cycle period, then

$$\Delta t = (2CR/V_c) \times V_n$$

giving $\qquad \Delta t/(T_o/2) = \Delta V_{th}/V_{th}$.

With V_n the noise voltage then

$$\Delta f_o/f_o = V_n/V_{th} \qquad (19.35)$$

Exercise 19.4 gives an example calculation.

LC-based oscillator

Fig. 19.19(a) is a simplified circuit model showing the essentials of an LC-based oscillator.

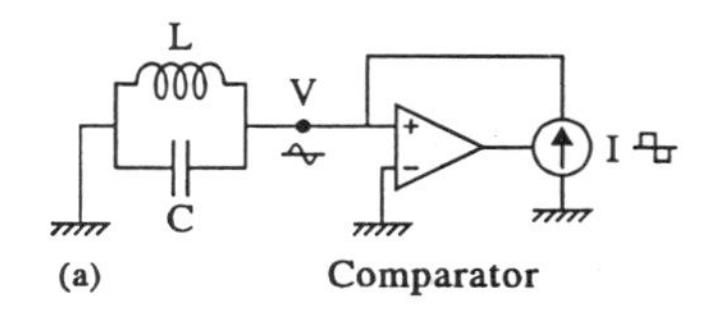

Fig. 19.19(a) Essentials of LC-based oscillator.

This is roughly the LC-based equivalent of the above switched-capacitor circuit. The comparator output is a switched current source of either ±I with I fixed. The feedback is positive and the result is sustained oscillation at the resonant frequency

$$f_o = 1/[2\pi\surd(LC)]$$

of the LC circuit. The comparator output is a square wave of amplitude I, and the LC circuit, which is arranged to have as high a Q-factor as possible, responds significantly only the fundamental component, making the resonator voltage V a nearly pure sine wave.

Thermal drift

With only one comparator, there is no equivalent to the differential thermal drift in f_o for the above RC type, since this originates from a differential change in the two thresholds. Only changes in the L and C values will cause drift, including those of the comparator input and output capacitances, and changes in the comparator propagation time.

Frequency jitter

With V_n the input noise voltage for the present comparator, then the time at which V crosses the comparator threshold, and thus at which I is switched, is altered by

$$\Delta t_{sw} = V_n/(dV/dt).$$

With $\qquad dV/dt = \omega_o \hat{V}$

where $\hat{V}$ is the peak value of V, then

$$\Delta t_{sw} = V_n/(\omega_o \hat{V}) . \qquad (19.36)$$

The change in switching time occurs when the capacitor voltage is crossing zero, and injects a current pulse of width $I\Delta t$ into C, causing a voltage step

$$\Delta V_c = I\Delta t_{in}/C,$$

which is equivalent to a timing jump

$$\Delta t_o = \Delta V_c/(dV/dt)$$

in the resonator sine wave. With again

$$dV/dt = \omega_o \hat{V}$$

then $\qquad \Delta t_o = I\Delta t_{in}/(\omega_o C \hat{V}) \qquad (19.37)$

With $Z(\omega_o)$ the impedance of the resonator at resonance, and with

$$\hat{V} = \hat{I}\, Z(\omega_o), \quad \text{then} \quad Z(\omega_o) = Q/(\omega_o C).$$

With $\hat{I}$ the peak value of the fundamental component of I then $\hat{I} = 4I/\pi$ giving

$$\hat{V} = 4QI/(\pi\omega_o C) \qquad (19.38)$$

With the above relations, and with two uncorrelated transitions per cycle then

$$\Delta f_o/f_o = (V_n/\hat{V})/(Q \times 2^{2.5}) \qquad (19.39)$$

With $\qquad \Delta f_o/f_o = V_n/V_{th} \qquad$ as in [19.35]

for the switched-capacitor VCO, and with the peak output $V_{th}/2$, then jitter for the LC type is less by $\times 1/(Q \times 2^{1.5})$.

Colpitts oscillator

Fig. 19.19(b) shows this widely used LC-based oscillator circuit, with the biasing and decoupling components de-emphasised to show the coupling of resonator and transistor more clearly.

With the resonator split at the centre of the two capacitors, and the centre point grounded instead of as shown, then the circuit becomes a common-emitter stage, with the I_c output fed into one end of the resonator, and the other end providing the input to the base. When operated in this way the resonator gives a phase inversion which, together with the inverting property of the common-emitter, gives the overall positive feedback needed for oscillation. This is unchanged by the choice of ground point.

See also Fig. 12.21(a) for the popular LC oscillator based on the HCMOS logic inverter gate, and the quartz-crystal version in (c).

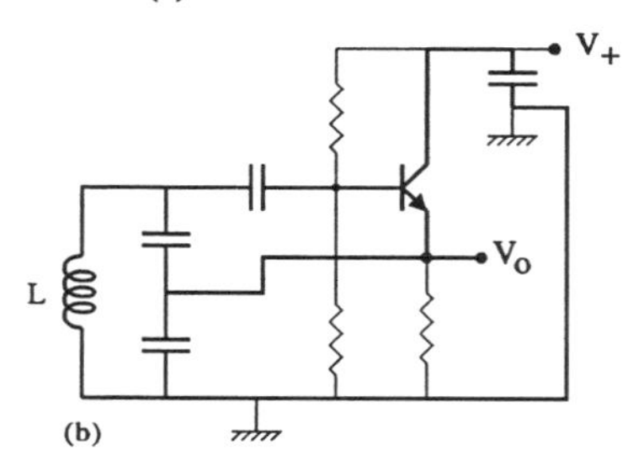

Fig. 19.19(b) Colpitts LC-based oscillator.

Varactor tuning

The 'varactor' is a semiconductor diode reverse biased by the control voltage V_c. With the diode then a capacitor c_d formed of the semi-conducting regions separated by the non conducting depletion layer, and with the width of the layer varying with the bias, then c_d is voltage controllable. Fig. 19.20(a) shows how the frequency f_o of the Colpitts oscillator can thus be voltage controlled. With C_1 a decoupling capacitor, the voltage-dependent c_d is effectively connected across L, and so controls f_o.

Extension of the analysis associated with Fig. 5.14 for the depletion layer gives

$$c_d = c_{do}(1 - V_d/\psi_o)^{-1/2} \qquad (19.40)$$

where c_{do} is the value of c_d for $V_d=0$ and ψ_o is the internal junction potential ~1V. This is plotted in (b).

Tuning range

The normal semiconductor diode is usually inadequate for the above purpose, because of the low range over which f_o can be varied.

With the amplitude of V_L comparable with V_c, it is difficult to make an exact prediction of the tuning range. For the varactor not to conduct, the range of V_c might be from $V_+/2$ to V_+. With the mean of V_L zero, and otherwise ignoring the effect of V_L, then $V_d \approx -V_c$ giving

$$c_{d.max}/c_{d.min} = [(V_+ + \psi_o)/(V_+/2 + \psi_o)]^{1/2} \qquad (19.41)$$

With
$$f_o = 1/[2\pi\sqrt{(LC)}]$$
then
$$f_{max}/f_{min} = [(V_+ + \psi_o)/(V_+/2 + \psi_o)]^{1/4} \qquad (19.42)$$

For the typical $\psi_o = 900mV$ and V_+ the usual 15V then

$$f_{max}/f_{min} = (15V + 900mV)/(15V/2 + 900mV)^{1/4}$$

giving the usually inadequate $\mathbf{f_{max}/f_{min} = 1.17.}$

Hyper-abrupt junction varactor

This has a modified doping profile giving a much stronger dependence of c_d upon V_c. Here c_d can vary in the ratio of 16:1 as V_c varies from 1V to 15V. The device is relatively costly.

Decoupling components

The function of R is to form a low-pass filter with C_1, to prevent high-frequency interference on the control line from frequency modulating the oscillator. In a PLL loop V_c would be derived from the phase-detector, with care as to the effect of RC_1 on the loop stability. For the logic-gate-based oscillators of Fig. 12.21 the varactor is connected across C_2.

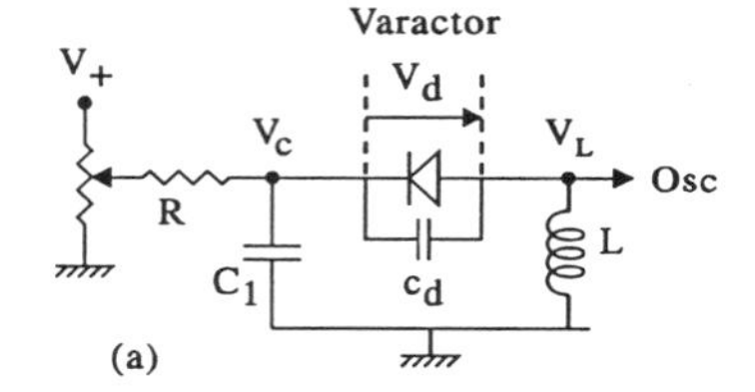

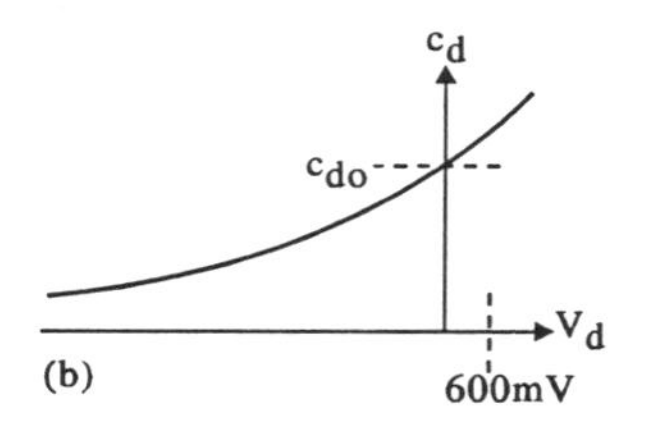

Fig. 19.20 Varactor tuning of Colpitts oscillator of Fig. 19.19(b).
(a) Connection of varactor across oscillator inductor L.
(b) Variation of varactor capacitance c_d with voltage V_d.

19.7 PLL frequency synthesisers

A frequency synthesiser is a means of giving a variable-frequency oscillator the frequency stability of a quartz-crystal oscillator. Apart from the 'direct-digital synthesis' method of the next section, the synthesiser involves one or more phase-locked loops.

Fig. 19.21(a) shows the basic principle. Here the output of the programmable divider is phase-locked to that of the relatively low-frequency quartz-controlled oscillator of output frequency Δf giving

$$f_o = N\Delta f \qquad (19.43)$$

Thus f_o can be set to any multiple of Δf within the range of the VCO. With Δf normally well below the normal operating range of a quartz oscillator, then the block shown includes a suitable fixed frequency divider.

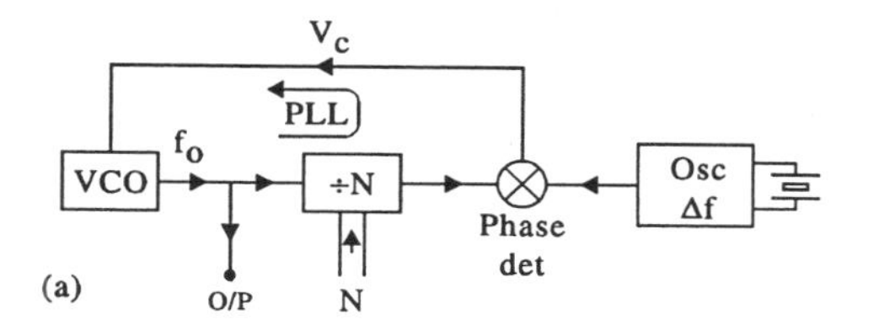

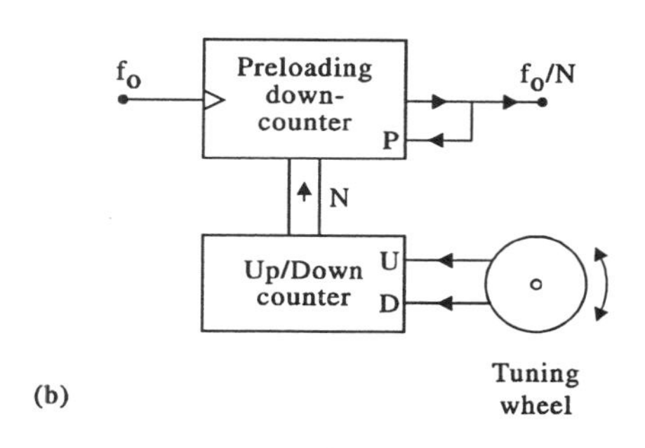

Fig. 19.21 Basic frequency synthesiser.
(a) Circuit diagram giving $f_o = N\Delta f$ where N is integer.
(b) Arrangement for deriving N from a manually operated tuning wheel.

Tuning wheel

The synthesiser is now commonly used as the local-oscillator in a transceiver. Here the 'tuning wheel' of (b) takes the place of the knob that previously controlled the oscillator C. The wheel outputs 'up' pulses for the wheel rotated in one direction, and 'down' pulses for the other, causing N to increase or decrease accordingly. Exercise 19.5 gives a suitable overall design.

Refinements

With the transceiver local-oscillator frequency normally ~ 70MHz as in Fig. 14.9, and the required tuning increment now 1Hz, then the above arrangement requires a degree of refinement beyond our scope (Smith, 1986). There may be up to three PLLs involved, with the final increment probably provided by the 'direct digital synthesis' method.

MSF primary standard

For the highest precision the Δf quartz oscillator will be phase-locked to the 60kHz broadcast from MSF Rugby (or the equivalent from WWV-Boulder, Colorado, etc.). For MSF the 60kHz carrier is interrupted for 100ms in each second to give the clock ticks, and in a more complex fashion each time the time information is encoded. Exercise 19.6 considers the refinements needed when the reference signal is interrupted in this way.

VHF synthesiser

Fig. 19.22 shows the more modest extension of the basic synthesiser for the local oscillator of a channelised VHF transceiver. For this example the range of the signal input frequency f_{in} is from 150MHz to 152MHz, the intermediate frequency f_{if} is the standard 8MHz for this range of f_{in}, and the channel separation $\Delta f = 10$kHz.

With the local-oscillator frequency $f_{Lo} = f_{in} - f_{if}$ then $f_o = 142$ to 144MHz as shown. The main extension is that the local-oscillator output needs to be converted to a frequency low enough for the programmable divider to operate. With the width of the required tuning range 2MHz, then the converted value is comparable, at 2-4MHz.

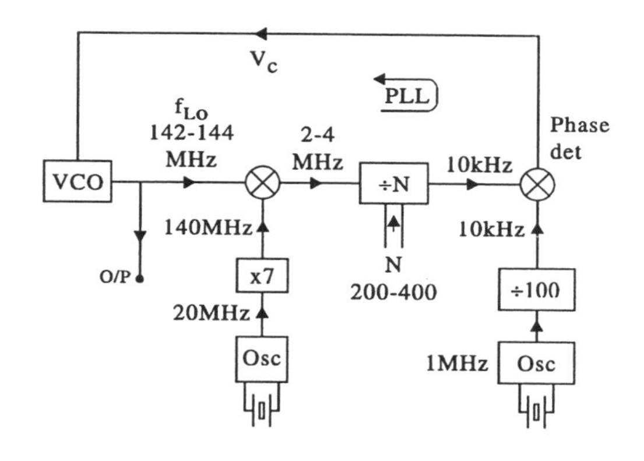

Fig. 19.22 Refinements needed to synthesiser of Fig. 19.21 for f_{Lo} in the VHF frequency range.

140MHz oscillator

This also needs to be of quartz stability. Here 140MHz borders on the limit obtainable for a crystal oscillator, even then probably operating in the fifth overtone mode of the crystal. Thus the output is derived as shown. The ×7 multiplier is merely an amplifier driven heavily into saturation. This produces an I_c waveform rich in odd-order harmonics (being near to a square wave) and the seventh is selected by a suitable narrow-band LC resonator. There may also be a following amplifier also tuned to the seventh harmonic. Apart from this, the operation of the PLL is as before.

19.8 Direct digital synthesis

Fig. 19.23(a) shows the arrangement used for the 'direct digital synthesis' (DDS) of a variable frequency sine wave, with the frequency of the analogue output V_o proportional to the value of the digital input D_{in}. As for the above PLL-based synthesisers, the frequency stability is that of the quartz-crystal oscillator providing the clock input ϕ. The method is 'direct' in that no phase-locking is involved.

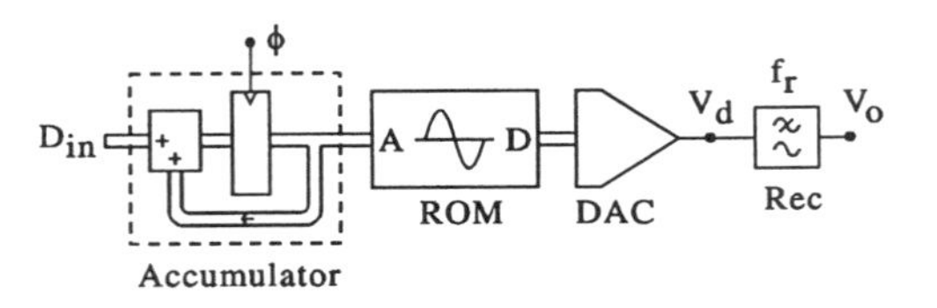

(a)

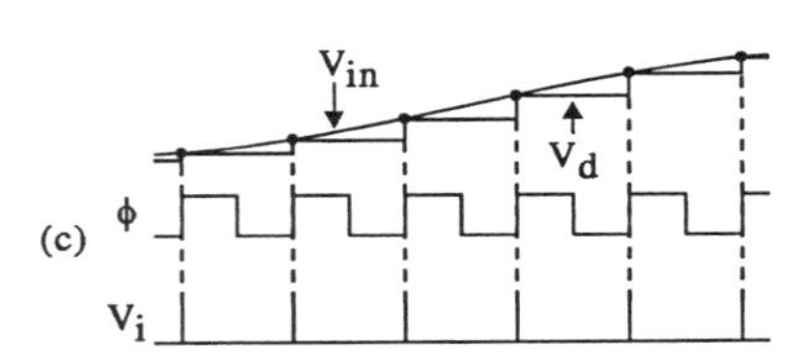

(b)

(c)

Fig. 19.23 Direct digital synthesis of variable frequency sine wave.
(a) Circuit diagram. Rec: reconstruction filter.
(b) Circuit model for derivation of spectra.
(c) Waveforms.

Circuit operation

With the read-only memory (ROM) programmed with a full sine wave cycle, and if the memory address A was to be provided by the output of a binary counter driven by the clock ϕ, then f_o would equal f_ϕ/N_{ad}, where

$$N_{ad} = 2^{B_{ad}}$$

with B_{ad} the number of address bits. But with the accumulator shown then, instead of A advancing only once per clock pulse, the increase is by D_{in}, giving

$$f_o = f_\phi \times D_{in}/N_{ad} \qquad (19.44)$$

Thus f_o can be advanced in N_{ad} equal increments of

$$\Delta f_o = f_\phi/N_{ad} \text{ to } f_\phi.$$

With V_d a stepped version of the required V_o, the function of the following low-pass reconstruction filter is to remove the stepping. As will now be shown, this is only possible for f_o up to $f_\phi/2$.

Circuit model

Fig. 19.23(b) shows a circuit model for the stepping in the synthesiser output V_d, as shown in the waveforms of (c).

In (b) the value of V_d that would obtain in the absence of the stepping is shown as V_{in}, with the stepped V_d the output of the process. With V_i a train of unit impulses then V_{in}' becomes a train of impulses of magnitude V_{in}. These are then spread by the running-average filter, to give the stepped waveform V_d.

Spectra

Fig. 19.24 shows the corresponding spectra, with V_{in}, now generalised to a multi-component signal such as speech. With the spectrum for V_i as shown, then that for V_{in}' is as also shown, comprising the required V_{in} with a series of unwanted upper and lower sideband components.

Running average

With $H_{rav}(f)$ the frequency response of the running-average filter, this is derived as follows. For an input component exp(st) and with the average over the clock period T_ϕ then

$$V_o(t) = \frac{1}{T_\phi}\int_{t-T_\phi}^{t} \exp(st').dt' \qquad (19.45)$$

giving $\quad V_o(t) = (1/s)\exp(st)[1 - \exp(-sT_\phi)]$.

For a sine wave input of frequency f then, as plotted,

$$|H_{rav}(f)| = \sin(\pi f/f_\phi)/(\pi f/f_\phi) \qquad (19.46)$$

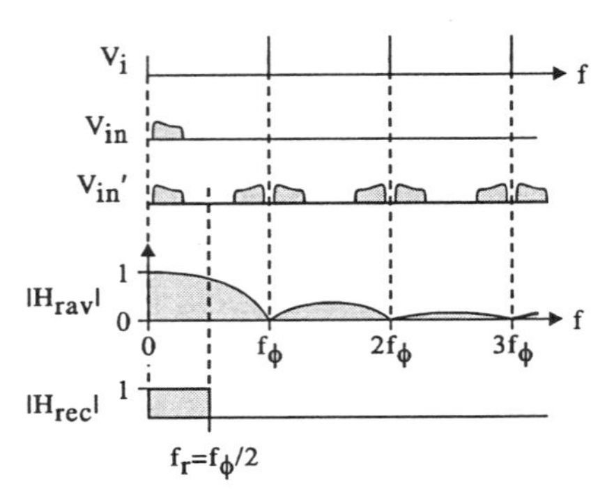

Fig. 19.24 Spectra and frequency responses for stepping of Fig. 19.23(c).
V_{in}: as above but generalised to multi-component signal to be sampled.
H_{rav}: frequency response of running-average filter.
H_{rec}: frequency response of ideal reconstruction filter.

Reconstruction filter

With $H_{rec}(f)$ the frequency response of the ideal reconstruction filter then this alone would remove all of the unwanted components of V_{in}'. The added effect of $H_{rav}(f)$ is disadvantageous in that it imposes a modest attenuation on the wanted components of V_{in}', but advantageous in that it adds to the attenuation of the unwanted components partially admitted by the real reconstruction filter.

Nyquist limit

The above reconstruction is only possible if the unwanted and wanted components of the V_{in}' spectrum do not overlap, requiring that f_ϕ be higher than twice the highest frequency component of V_{in}. This is the Nyquist limit.

Problem

What would the V_d waveform be (i) for $D_{in}=N_{ad}$, as for $f_o=f_\phi$, (ii) for $D_{in}=N_{ad}-1$? Which component of V_{in}' is being seen for (ii)?

Increased frequency resolution

For direct digital synthesis, a realisable maximum value for f_o is 1MHz, with the typical required increment $\Delta f=1$Hz. With the corresponding $N_{ad}=2\times 10^6$ inconveniently large, a resolution higher than that imposed by N_{ad} can be obtained by making the number of accumulator bits $B_{ac}>B_{ad}$, and connecting the memory address lines to the upper B_{ad} bits of the accumulator output.

Unwanted output components

These are as follows.

- Quantisation noise V_{qd} due to the DAC resolution.
- Quantisation noise V_{qa} due to the memory address resolution.
- Sampling step noise V_{os}.

DAC output quantisation

With B_d the number of DAC bits, and with $N_d=2^{B_d}$, then

$$V_{qd}=V_{o.max}/N_d \qquad (19.47)$$

Memory address quantisation

With B_{ac} the number of accumulator bits, B_{ad} the number of memory address bits, and with $B_{ac}>B_{ad}$ as usual, then there is an effective variation of up to one memory address step. With V_{qa} the resulting output noise then

$$V_{qa}=dV_o/dA\times\Delta A,$$

where $\Delta A=1$ is the address increment. With

$$(dV_o/dA)_{max}=2\pi V_{o.max}/N_{ad}$$

then at the point of maximum slope

$$V_{qa}=2\pi\times V_{o.max}/N_{ad} \qquad (19.48)$$

Sampling steps

With

- the frequencies of the unwanted components $nf_\phi\pm f_{in}$,
- the step corresponding to each point sample being subject to the frequency response $H_{rav}(f)$ in [19.46],
- here $f_{in}<<f_\phi$ giving, in [19.46]

$$|H_{rav}|\approx f_o/(nf_\phi\pm f_o) \qquad (19.49)$$

- each unwanted output component subject to both H_{rav} and the response of the reconstruction filter H_{rec}, then the amplitudes of the resulting output components can be calculated as in Exercise 19.7.

Reference

Smith, J. 1986: *Modern Communication Circuits.* McGraw Hill.

19.9 Exercises

Exercise 19.1 Quadrature output oscillator

Design A variable frequency oscillator with quadrature outputs is required for the frequency range from 1kHz to 100kHz.

Show how this can be implemented using two Gilbert-cell modulators with following low-pass filters, and a pair of variable-frequency quadrature square waves of frequency ~1MHz, derived using

- a variable-frequency clock oscillator operating at ~ 4MHz,
- a ÷4 counter,
- a clocked bistable,
- a fixed frequency 1MHz sine wave oscillator.

Exercise 19.2. Noise-optimised measurements

Demonstration With $V'_r=dV_r/dt$ then [19.33] gives

$$I_t=\int_{-\infty}^{+\infty}(V_{in}-V_r)V'_r\,dt$$

as the integral that must be set to zero for the noisy signal V_{in} to be considered correctly aligned with the noise-free reference function V_r.

> Show that for the initial and final values of V_r the same, as for the single pulse in Fig. 19.15(b, then [19.33] reduces to
>
> $$I_t = \int_{-\infty}^{+\infty} V'_r \, V_{in} dt$$
>
> as given by [19.32] for that case.

With

$$I_t = \int_{-\infty}^{+\infty} (V_{in} - V_r)V'_r \, dt = \int_{-\infty}^{+\infty} V_{in}V'_r \, dt - \int_{-\infty}^{+\infty} V_r V'_r \, dt$$

and with $V'_r = dV_r/dt$, then

$$\int_{-\infty}^{+\infty} V_r V'_r \, .dt = \int_{-\infty}^{+\infty} V_r .dV_r = \left[\frac{V_r^2}{2}\right]_{-\infty}^{+\infty}$$

For $V_r(-\infty) = V_r(+\infty)$ this is zero

confirming $$I_t = \int_{-\infty}^{+\infty} V'_r \, V_{in} dt$$ as required.

Exercise 19.3 RC-oscillator based on inverter gate

> **Calculation** For the circuit of Fig. 19.25, first sketch the waveforms. Then derive suitable values of R and C for an output frequency $f_o = 1kHz$.
>
> There are protection diodes at the gate input that prevent the input voltage from going beyond the supply $V_{cc} = 5V$ and 0V by more than 600mV.
>
> The gate threshold $V_{th} = V_{cc}/2$.

From the waveforms, with $T_o = 1/f_o$,

$$V_{th}/(V_{cc} + 600mV) = \exp[-(T_o/2)/CR]$$

giving

$$T_o = 2CR \ln[V_{th}/(V_{cc} + 600mV)]$$

and thus

$$C = T_o/(1.61 \times R).$$

For say $R = 10k\Omega$ then **C=62.1nF.**

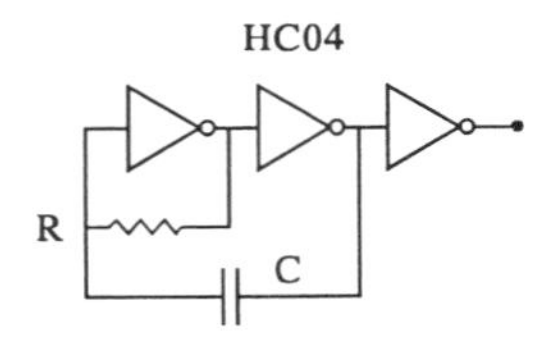

Fig. 19.25 RC-coupled oscillator based on logic inverter gates.

For

- $R = 10k\Omega$,
- the gate input impedance $R_{in} \sim 1G\Omega$,
- the gate output impedance $R_o \sim 30\Omega$.

Thus $R_{in} >> R >> R_o$, making the choice of R suitable.

Also $C >> C_{in}$ where the gate input $C_{in} \sim 3pF$.

Exercise 19.4 Switched-capacitor VCO (Fig. 19.18)

> **Calculation** For the output frequency $f_o = 50kHz$, calculate the frequency jitter Δf_o for the comparator noise resistance $R_n = 20k\Omega$. The comparator response time $T_r = 100ns$ and the amplitude $2V_{th}$ of the V_a waveform is 3V.
>
> Boltzmann's constant $k = 1.38 \times 10^{-23} J/^o$.
>
> The standard temperature $T = 290^o$.

With the bandwidth $B = 1/2\pi T_r$ then $B = 1.59MHz$.

With $\tilde{V}_n = (4kTR_nB)^{1/2}$ then $\tilde{V}_n = 1.3\mu V$.

With the frequency jitter

$$\Delta f_o/f_o = V_n/V_{th} \qquad \text{as in [19.35]}$$

then $$\Delta f_o = f_o \times (\tilde{V}_n/V_{th})$$

giving **Δf_o=43.3mHz.**

Exercise 19.5 Basic frequency synthesiser (Fig. 19.21)

> **Design** a system giving f_o from 10kHz to 1MHz in 10kHz steps.

Δf output With $f_q \sim 8MHz$ the most suitable value for quartz-crystal operation then for $\Delta f = 10kHz$ requires division by $8MHz/10kHz = 800$.

The HCMOS HC40103 is an eight-bit programmable divider. With the output pulse from this lasting for only one input clock pulse then this is not suitable to drive the simple XOR gate phase-detector of Fig. 19.3. Instead the divider is set to divide by 200, with the remaining ÷4 provided by a normal binary counter, giving the required square wave output.

VCO It would not be possible for an LC type to cover this range. The switched-capacitor HC4046 (max $f_o = 20MHz$) is suitable.

Programmable divider With N from 1 to 100 then the eight-bit 40103 would be suitable. Again the final stage needs to be a ÷2 to give the required square wave. Thu nother ÷2 is needed following the Δf_o output, making the hase detection at $\Delta f/2 = 5$kHz.

Jp-down counter Two four-bit HC193 up-down counters ould provide the maximum $N=100$ for divider. Decod-rs to inhibit the up-count for $N=100$ and the down-count or $N=0$ would normally be added.

Open-loop frequency response With the added ÷2 stages ien the loop gain

$$A_L(f) = (k_o/2N)k_p/(jf) \qquad (19.50)$$

Vith A_L thus largest for $N=1$, and $|A_L(f_p)| = 1$ at the tability limit then the limiting

$$k_o k_p = 2f_p.$$

Vith $\quad f_{o.max} - f_{o.min} = k_p k_o \pi \quad$ as in [19.16]

ien $\quad f_{o.max} - f_{o.min} = 2f_p \pi.$

Vith the added ÷2 stage then ripple reduction requires $_{\text{}} << f_{o.min}$ giving

$$f_{o.max}/f_{o.min} << \text{the required} \times 100.$$

'hus the pseudo-integrator or the RC network of Fig. 9.7(a) is needed, with the transition frequency f_2 some-vhat below f_p, say 1kHz.

xercise 19.6. MSF-controlled frequency standard

Design The 60kHz standard from MSF Rugby is normally interrupted for 100ms in each second to give the 1s clock ticks, and in a more complex fashion when the time is encoded. Devise a suitable system for phase-locking a quartz-crystal oscillator to the standard.

he phase-detector output is followed by a sample-and-old circuit controlled by the received signal amplitude.

xercise 19.7 Direct digital synthesis

Calculation For the system of Fig. 19.23(a) with the following values, calculate

- the number of bits B_{ac} needed for the accumulator,
- the exact value of f_ϕ,
- the approximate amplitude of the unwanted components relative to the required output.

ystem values DAC: 12-bit ($B_d=12$). ROM: 12-bit data, 6-bit address ($B_{ad}=16$). Clock frequency $f_\phi \approx 8$MHz. Maximum output frequency $f_{o.max}=1$MHz. Frequency ncrement $\Delta f = 1$Hz. Third-order low-pass reconstruction ilter.

Accumulator and clock With $f_\phi = N_{ac}\Delta f$ where $N_{ac} = 2^{B_{ac}}$, and for $\quad$ **$B_{ac}=23$ then $f_\phi = 8.39 \times 10^6$.**

DAC quantisation noise With $V_{od} = V_{o.max}/N_d$ as in [19.47] then $\quad$ **$V_{qd}/V_{o.max} = 244 \times 10^{-6} = -72.2$dB.**

Address quantisation noise With $V_{qa} = 2\pi \times V_{o.max}/N_a$ as in [19.48] at the point of maximum dV_o/dt, then

$$V_{qa}/V_{o.max} = 2\pi/2^{16} = 95.9 \times 10^{-6}.$$

With the mean approximately one half of this then

$V_{qa}/V_{o.max} \approx 50 \times 10^{-6} = -86$dB.

First sampling noise frequency component With f_o the frequency of the required output component, this is the component at $f_\phi - f_o$.

With $\quad |H_{rav}| \approx f_o/(nf_\phi \pm f_o) \quad$ as in [19.49]

for the effective running-average filter, then here

$$|H_{rav}| \approx f_o/(f_\phi - f_o).$$

For the maximum $f_o = 1$MHz then $|H_{rav}| = 0.135$.

With H_r the response of the third-order reconstruction filter, with f_r set to $f_{o.max}$, and with $f_{o.max} << f_\phi$, then for the unwanted components

$$|H_r| \approx [f_r/(nf_\phi \pm f_o)]^3.$$

For the present component then

$$|H_r| \approx [f_r/(f_\phi - f_o)]^3$$

giving $|H_r| \approx 2.48 \times 10^{-3}$.

With the total $\quad H = H_{rav} \times H_{rec}$

then $\quad |H| = 335 \times 10^{-6}$.

Second sampling noise frequency component With this the component at $f_\phi + f_o$ then, as for the first component,

$$|H_{rav}| = 0.107, \; |H_r| \approx 1.21 \times 10^{-3} \text{ giving } |H| = 129 \times 10^{-6}.$$

With V_{qs} the total sampling noise, with the components of higher n negligible by comparison, and with the first two components taken as uncorrelated, then

$V_{qs}/V_{o.max} = 359 \times 10^{-6} = -68.9$dB.

With the unwanted components uncorrelated and V_{ou} the total of the unwanted components then

$V_{ou}/V_{o.max} \approx 417 \times 10^{-6} = -67.6$dB.

With V_{qs} the largest component, a higher-order reconstruction filter would be advantageous.

20

Analogue-digital data conversion

Summary

Apart from the comparator of Section 4.3, which is a one-bit analogue-to-digital converter (ADC), this chapter covers the circuits which interface between analogue and digital electronic circuitry.

The digital-to-analogue converter (DAC) gives the analogue output V_o proportional to the normally parallel digital input D_{in}. Sometimes the operation is 'unipolar', with D_{in} 'unsigned', where the codes are interpreted as being always positive, with V_o positive in proportion. Otherwise the operation is 'bipolar', with D_{in} 'signed' (ranging from negative to positive) with V_o again in proportion. With the modifications described, the unsigned DAC can be converted to signed operation, sometimes requiring two of the unsigned devices.

The multiplying DAC (MDAC) has one analogue input V_{in} and one digital D_{in}, with $V_o \propto V_{in} \times D_{in}$. An MDAC with D_{in} unsigned and V_{in} only positive is said to be 'single quadrant'. This becomes 'two quadrant' for either D_{in} signed or V_{in} bipolar, and 'four quadrant' for both bipolar. Methods of conversion from single- to two-quadrant operation, etc., are again described.

One of the limitations of a DAC is the brief, but often quite large transient, or 'glitch', which occurs at the point when D_{in} changes from one value to the next. The associated 'settling time' and methods of avoiding the glitch are explained.

For the ADC the digital output D_o is proportional to the analogue input V_{in}. The ADC too can be bipolar or unipolar.

The types of DAC covered are the switched-output and switched-input ladder DACs, the higher speed 0800 type, and the oversampling DAC.

The ADC types are the successive-approximation, flash, dual-ramp, and oversampling types. The now very popular oversampling DAC and ADC are together at the end of the chapter. With experimental work on these and some other types.

20.1 DAC function

Fig. 20.1 shows a unipolar four-bit DAC giving

$$V_o = k_{dac} D_{in} \qquad (20.1)$$

For the DAC constant $k_{dac} = 100\text{mV/bit}$ then the input and output values are as in Table 20.1.

D_{in} (Binary)	D_{in} (Decimal)	V_o (Volts)
0000	0	0
0001	1	0.1
0010	2	0.2
0011	3	0.3
.	.	.
.	.	.
1110	14	1.4
1111	15	1.5

Table 20.1 Input-output relation for DAC of Fig. 20.1 ($k_{dac} = 100\text{mV/bit}$).

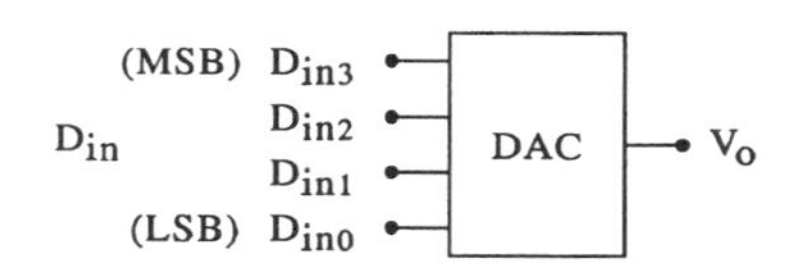

Fig. 20.1 Four-bit DAC (LSB: least significant bit. MSB: most significant bit).

With N_r the number of resolved levels and N the number of bits then

$$N_r = 2^n \qquad (20.2)$$

With normally N=8, 12 or 16, the value N=4 used here and below is for brevity. From the table

$$D_{in.max} = N_r - 1 \qquad (20.3)$$

xample

alculate the number of bits N needed to adequately rep-
esent the number 2536. With $2^{11}=2048$ and $2^{12}=4096$,
e required **N=12.**

oltage levels

s for the comparator, $V_{in}=0$ represents logical 0, and
$_{in}=5V$ logical 1. However, there is a specified toler-
nce, for the high-speed CMOS family (HCMOS) as in
able 20.2.

Output	Lower limit	Upper limit	D	
V_o	$V_{cc}-$ 0.8V	V_{cc}	1	Gate output guaranteed to be within this range
V_{in}	$V_{cc}-$ 1.4V	V_{cc}	1	Gate will respond as to logical 1
V_{in}	0	900mV	0	Gate will respond as to logical 0
V_o	0	400mV	0	Gate output guaranteed to be within this range

able 20.2 Voltage levels for HCMOS logic (supply $V_{cc}=5V$).

he interval of say 400mV to 900mV for D=0 is termed
e 'noise margin'. This is the allowance for interference
mponents, usually capacitively coupled from other parts
the digital circuit.

0.2 Switched output ladder DAC

ost DACs are based on the ladder network of Fig.
.2(a). Working from the right, the load impedance pre-
nted to each horizontal resistor is R, so the voltage drop
r section is ×1/2, with the same for $I_{0,1,2}$. With
$=E_r/2R$ in (a) and I_+ as in (b) then here

$$I_+=(E_r/R)\times(D_{in}/2^4).$$

ith $V_o=-I_+R_f$ for the current follower in (c), for the N-
t converter

$$V_o=-E_r\times(R_f/R)\times(D_{in}/2^n) \quad (20.4)$$

$_f$ is placed on the chip together with the ladder resistors.
hus any thermal change is the same for all, and so can-
ls in V_o. The op-amp is usually external. The reference
$_r$ may be external, or from an on-chip Zener diode pow-
ed by the 5V V_{cc} supply.

esponse time

he DAC response time T_{dac} is largely determined by the
ternal op-amp. With the chip response time $T_c=RC_s$
here C_s is the capacitance of the switch, etc., then R
eds to be low enough to avoid extending T_{dac} further.

Conversely, accuracy requires $R>>R_{on}$, the 'on' resistance of the CMOS switch.

For the present circuit it is not possible to dispense with
the current-follower, and so obtain an improved T_{dac},
because the Thévenin output voltage V_+ is only proportional to the Norton I_+ into the current-follower if the
output impedance R_o is independent of D_{in}. With $R_o=\infty$
for $D_{in}=0$, this is clearly not so.

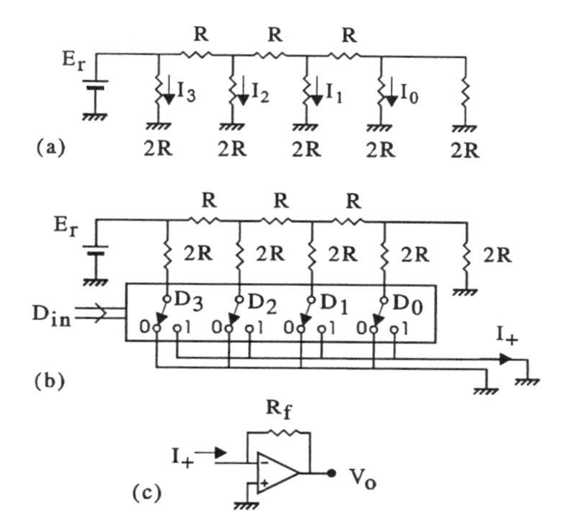

Fig. 20.2 Switched-output ladder DAC.
(a) Ladder network.
(b) Ladder with switches.
(c) Current follower.

20.3 Switched-input ladder DAC

For the switched-input DAC of Fig. 20.3 the output impedance is independent of D_{in}, allowing the current-
follower to be omitted, and thus the chip response time T_c
to be realised. With the V_o terminal grounded the impedance presented by the V-node to each 2R is R. With the
same ×1/2 voltage drop per stage as above then

$$V_3=\frac{1}{3}\left(\frac{E_3}{1}+\frac{E_2}{2}+\frac{E_1}{4}+\frac{E_0}{8}\right) \quad (20.5)$$

With $I_o=V_3/(2R)$ then $I_o=(E_r/3R)\times(D_{in}/2^4)$.

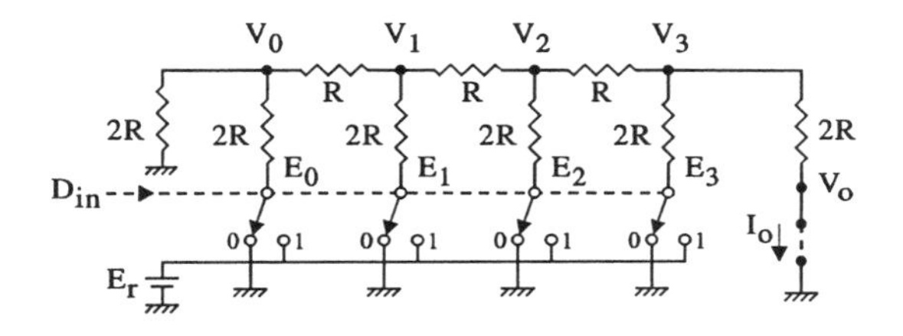

Fig. 20.3 Switched-input ladder DAC.

With the output impedance 3R then the open-circuit

$$V_o = E_r \times (D_{in}/2^4).$$

giving, for the N-bit converter,

$$V_o = E_r \times (D_{in}/2^n) \qquad (20.6)$$

The right-hand 2R is normally omitted, being merely an aid to understanding.

20.4 High-speed DAC (0800 type)

Fig. 20.4 shows a ladder type DAC with the switches non-saturating bipolar transistors. Despite the recent advantages in MOS technology, this still gives a slightly better response time than the above circuits when CMOS switched.

The circuit maintains the emitter voltage V_e of all the transistors in (c) at the same value, a few volts above the negative supply V_-. Then, apart from V_e and V_- replacing the previous E_r and 0V, the ladder network is driven much as before.

The ladder currents are passed through the bank of transistors to the current steering switches detailed in (e), with only the small reduction of $\times\beta/(1+\beta)$.

The value R_L of load resistors is made sufficiently low to avoid saturation of the steering transistors, thus giving the high-speed switching.

With β typically 200 and with D_{max} the maximum value of D_{in} then

$$I_+ \approx D_{in}I_0 \quad \ldots(a)$$

$$I \approx (D_{max} - D_{in})I_0 \quad \ldots(b) \qquad (20.7)$$

For the load resistors

$$V_{o+} = V_+ - I_+R_L \quad \ldots(a)$$

$$V_{o-} = V_+ - I_-R_L \quad \ldots(b) \qquad (20.8)$$

giving

$$V_{o+} = V_+ - R_LD_{in}I_0 \quad \ldots(a)$$

$$V_{o-} = V_+ - R_L(D_{max} - D_{in})I_0 \quad \ldots(b) \qquad (20.9)$$

Positive supply immunity

With V_{o+} and V_{o-} both including the supply voltage V_+, this is not precisely determined. The resulting supply dependence is avoided by adding the circuit of (f). Together with the R_L in (b) this forms a conventional subtractor to give

$$V_o = I_oR_L(2D_{in} - D_{max}) \qquad (20.10)$$

in which the term V_+ is cancelled. V_{o+}, V_{o-} and V_o are plotted against D_{in} in Fig. 20.5.

Negative supply immunity

With $V_{th} \approx +1.5$V in Fig. 20.4(e) then the current-steering switches would operate with the V_b point grounded. But then the ladder currents would be largely determined by the equally imprecisely determined V_-. This is avoided by using the reference supply of Fig. 20.4(a), in which t negative feedback gives

$$I_r = E_r/R_1.$$

With

- the area of Q_3 equal to that in the reference supply,
- I_{c3} the collector current of Q_3,
- the current gain β the same for both transistors,

then

$$I_{c3} = I_r = E_r/R_1$$

regardless of the value of V_{i-}.

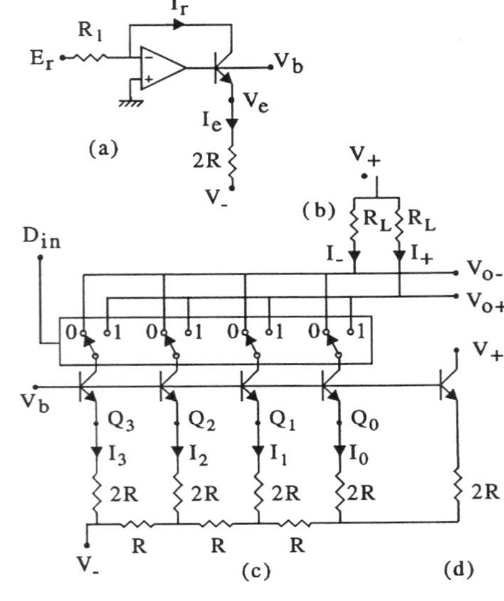

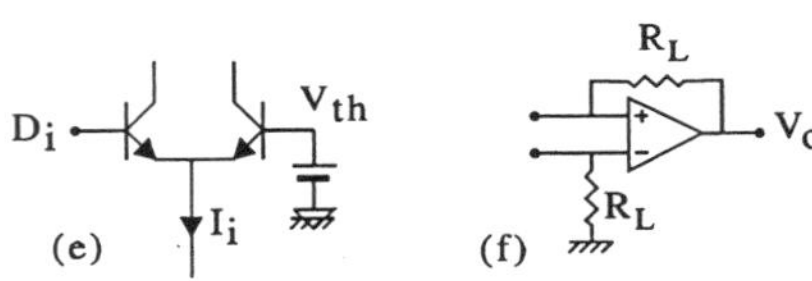

Fig. 20.4 High-speed 0800 type DAC.
(a) Reference supply. (d) Ladder termination.
(b) Load resistors. (e) i_{th} current-steering switch.
(c) Switched ladder network. (f) Output subtractor.

Thermal immunity

For the V_b point grounded the ladder currents would subject to the thermal variation of V_{be} for the lower tr sistors. For the present arrangement this is compensa for by the transistor in the reference supply.

Working from the right, Q_1 consists of two of the t used for Q_o in parallel, and so on, up to eight in para for Q_3 and for the reference supply transistor. In this w the thermal sensitivities and the β are all made the same

This circuit is a good example of the use of microcir techniques to reduce errors and produce a precision dev very cheaply.

put-output relation

'ith $I_3 = E_r/R_1, \quad I_o = I_3/2^3$

d $V_o = I_o R_L (2D_{in} - D_{max})$

in [20.10] then

$$V_o = E_r \times (R_L/R_1) \times (4D_{in} - 2D_{max})/2^4$$

ius for the N-bit converter

$$V_o = E_r \times \frac{R_L}{R_1} \times \frac{4D_{in} - 2D_{max}}{2^N} \qquad (20.11)$$

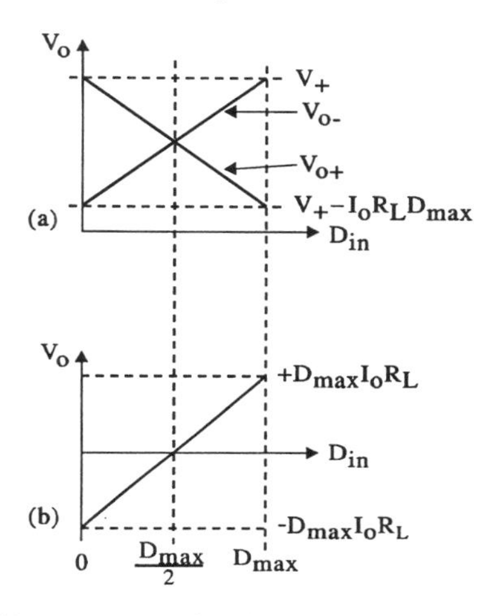

g. **20.5** Input-output relations for 0800 type DAC.
Outputs for Fig. 20.4(a) to (c).
Output for subtractor in (f) added.

oblem

r the 12-bit version it ceases to be either practical or cessary to keep halving the area of the lower transistors th the progression to the right. Determine at which ge the halving ceases to be necessary.

polar operation

g. 20.6(a) shows how a unipolar DAC can be converted bipolar operation, with the input-output relations as in to (d).

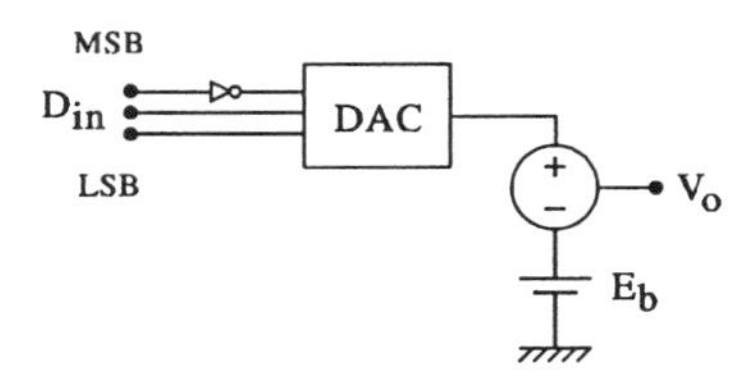

g. **20.6(a)** Conversion of three-bit unipolar DAC to bipolar eration. Circuit diagram.

First the usual input-output relation for the unconverted device is as in (b).

Next (e) shows the change when the output bias E_b in (a) is subtracted, with E_b equal to the previous mid-range value of V_o.

Next (c) and (f) show the result of inverting the MSB as in (a).

Finally (g) shows the result of interpreting the 'unsigned' (unipolar) values in (b) as the 'signed' (bipolar) values in (d) (see Exercise 20.1).

For the 0800 type of DAC with the subtractor of (f) then only the MSB reversal is needed for full bipolar operation.

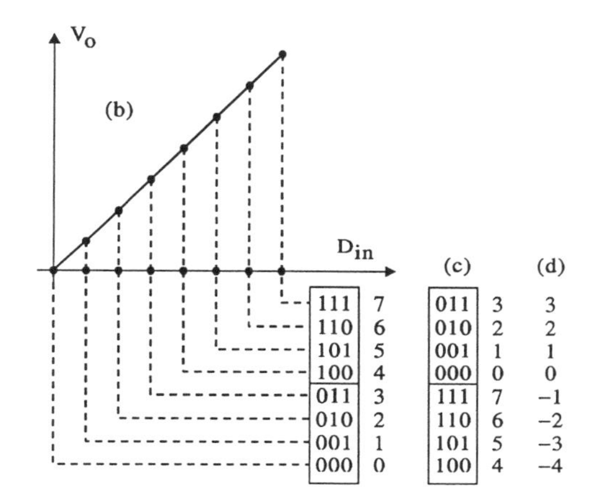

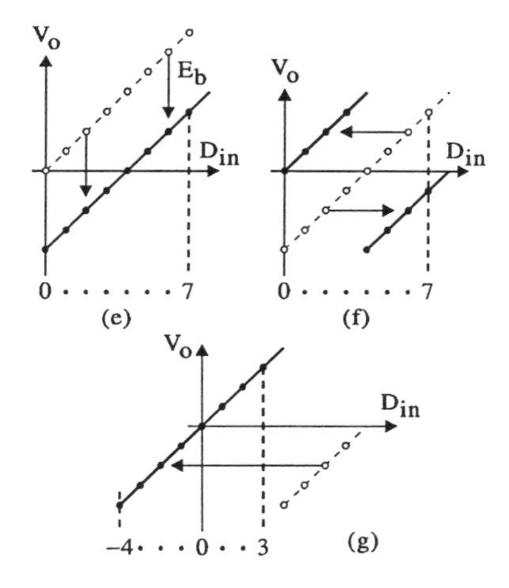

Fig. 20.6(b) to (g) Conversion of three-bit unipolar DAC to bipolar operation.
(b) Input-output relation for unmodified DAC.
(c) D_{in} for MSB inverted as in (a).
(d) Signed interpretation of binary numbers in (b).
(e) Change in input-output relation for bias E_b in (a) added.
(f) Further change in i/o relation for MSB reversal in (a).
(g) Yet further change in i/o relation for D_{in} taken as signed.

20.5 Multiplying DAC

With $V_o \propto E_r \times D_{in}$ for all of the above types of DAC then all are basically capable of MDAC operation. The only distinction is as to whether the operating range is over one, two, or four quadrants.

For the one-time work-horse ZN428 E_r could only be positive, but with the fast CMOS switches now used the range of E_r is bipolar.

For D_{in} to be bipolar for MDAC operation, the circuit of Fig. 20.6(a) must be modified as in Fig. 20.7.

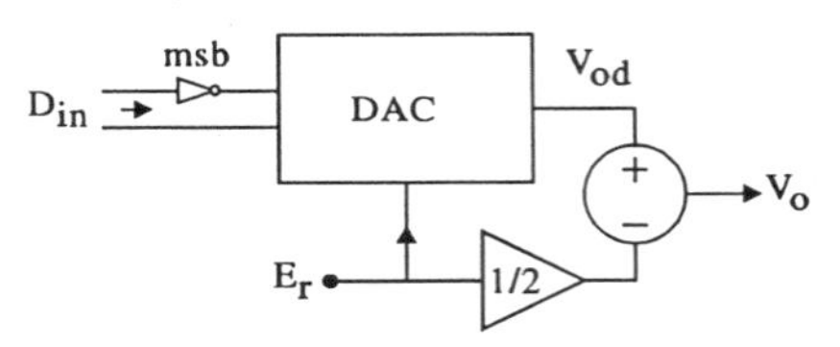

Fig. 20.7 Modification of single-quadrant DAC for two-quadrant MDAC operation.

0800 DAC

For this device the ladder currents in Fig. 20.4 can only be positive. Thus, while D_{in} can always be made bipolar, by reversing the MSB, the factor by which D_{in} is multiplied remains unipolar. For a device with this limitation, two can be combined as in Fig. 20.8 to give full four-quadrant operation (see Exercise 20.2).

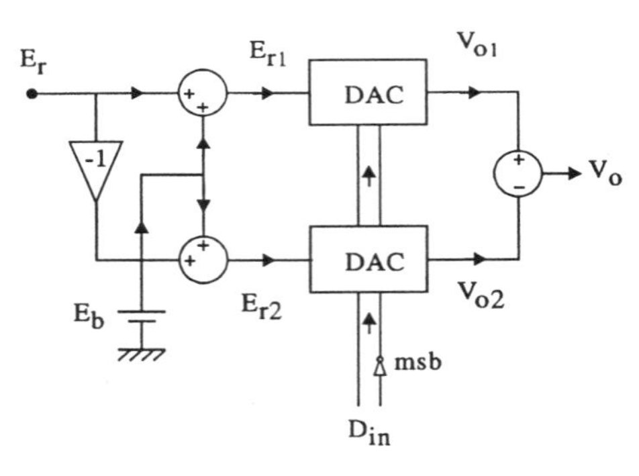

Fig. 20.8 Combination of two 0800 type DACs to give four-quadrant MDAC operation.

20.6 DAC glitches

The DAC 'glitch' is a narrow transient that occurs at the DAC output when the input D_{in} changes. The classic example is where the DAC is used to generate a near-sawtooth waveform by driving it with a continuously incremented ripple-type binary counter. Minor glitches are seen at each transition, but the largest is where the MSB changes at the centre of the count, as D_{in} increments from 011...111 to 100...000. Here the transition is actually one bit at a time, starting with the LSB, and with the time for each step the propagation time for the stage. Thus imm diately before the MSB is about the change, D=0 f every bit. Thus V_o steps briefly down to zero from th mid-range value, for the duration of the glitch.

To confirm this point, sketch the glitch pattern for four-bit DAC driven by counter and clock oscillator.

The obvious measure here is to use a synchrono counter in place of the ripple counter, so that all of th bits change at virtually the same time.

But even then some glitches are seen, as also for th more general case where the DAC is driven by a parall clocked data latch. These are due the switch capacitance as in Fig. 20.9(a). There is thus a component glitch f each bit that changes, with again the MSB change havi the largest effect.

Settling time

The important parameter for the glitch is the length time T_{set} needed for the glitch voltage V_g to 'settle' to value less than the DAC resolution increment ΔV_o. No that T_{set} is *not* the glitch decay time constant T_r in Fi 20.9(b) but typically a few times greater. With the glit amplitude V_g then

$$V_g \exp(-T_{set}/T_r) = \Delta V_o$$

giving

$$T_{set} = T_r \ln(V_g/\Delta V) \qquad (20.1$$

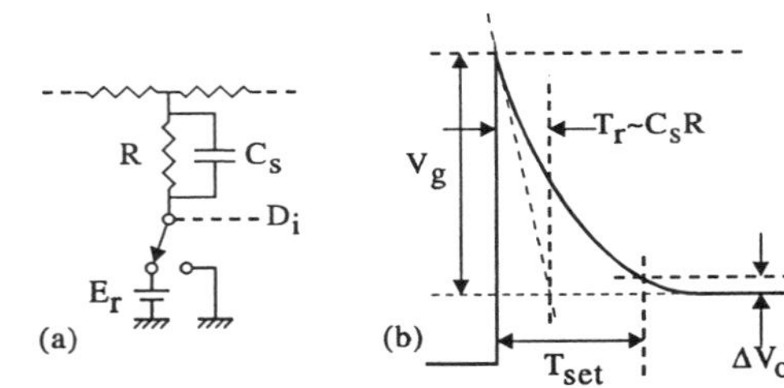

Fig. 20.9 Generation DAC glitch by ladder network stray capacitance C_s.
(a) Circuit diagram.
(b) Comparison of glitch time constant T_r and associated settli time T_{set}.

Example

Calculate T_r and T_{set} for a 12-bit converter with C_s=5p $R=10k\Omega$, and the glitch height V_g one tenth of the DA output range V_{or}.

With $T_r = C_sR$ then **T_r=50n**

For the 12-bit converter $N_r = 2^{12} = 4096$.

With T_{set} as in [20.12] then

$$T_{set} = T_r \ln[(V_{or}/10)/(V_{or}/N_r)$$

giving **T_{set}=300n**

Sampling gate

The glitches are avoided by using the gate of Fig. 20.10 to sample the DAC output V_d away from the glitches.

Not shown in the diagram is the effect of the sampling-gate switch capacitance. This causes a small charge to be transferred to the gate C as the switch is opened, to be reversed as the switch is closed. The result is a relatively low-amplitude square wave at the sampling frequency f_s. This component is periodic at f_s and so can be removed using the low-pass filter shown, with $f_r<f_s$. The filter alone cannot remove the glitches since these are random and thus have a spectrum that extends down to zero frequency.

Unfortunately, as discussed in Exercise 20.3, the square-wave includes a DC component of comparable amplitude which the following low-pass cannot remove. With the amplitudes of both $\propto 1/C$ then C needs to be as large as possible.

With the same so for reduction of the drift due to the bias current of the voltage-follower (not shown) which must follow C, the upper limit to C is set by boxcar effect, determined by the time taken for C to charge through the switch resistance, etc.

For experimental work on DAC glitches and the sampling gate see Exercise 20.4.

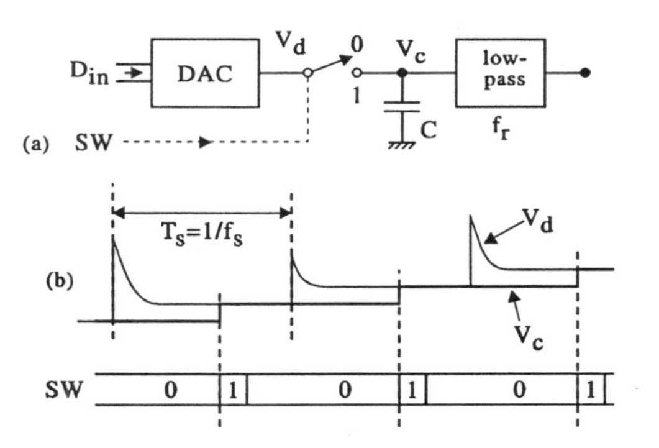

Fig. 20.10 Use of track-and-hold circuit to eliminate ADC glitches.
(a) Block diagram.
(b) Waveforms for D_{in} incrementing at each clock interval T_s.

20.7 Successive-approximation ADC

We now turn to the analogue-to-digital converter (ADC), with Fig. 20.11 showing the widely used successive-approximation type. Suppose initially that the successive-approximation (SAR) register is replaced by a binary counter. This is reset and then allowed to count until V_d reaches V_{in}. With $D_o \propto V_{in}$, then the ADC function is realised.

But this is like weighing a 73g object using one hundred 1g weights, added one at a time until the scale tips. The successive-approximation register (SAR) works more normally. First the 50g weight (the most significant bit) is put on. If the scale tips (comparator registers HI) the 'weight' is taken off, then the '25g weight' is put on - and so on.

For an eight-bit register with a 1MHz clock the conversion is complete in 8µs, rather than a possible 256µs for the counting register.

/CONV resets the SAR and then goes high to allow the count to proceed. BUSY indicates when the conversion is complete.

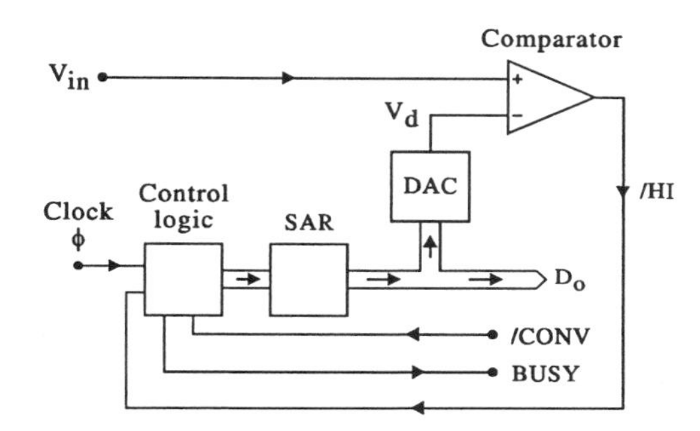

Fig. 20.11 Successive-approximation ADC (SAR: successive-approximation register).

20.8 Flash ADC

This is the type giving the most rapid conversion. With N_r the ADC resolution, the reference E_r in Fig. 20.12(a) is divided into N_r equal increments as shown. The encoder logic then converts the comparator outputs to the appropriate N-bit binary code for D_o. Here a large number of decoding gates is needed.

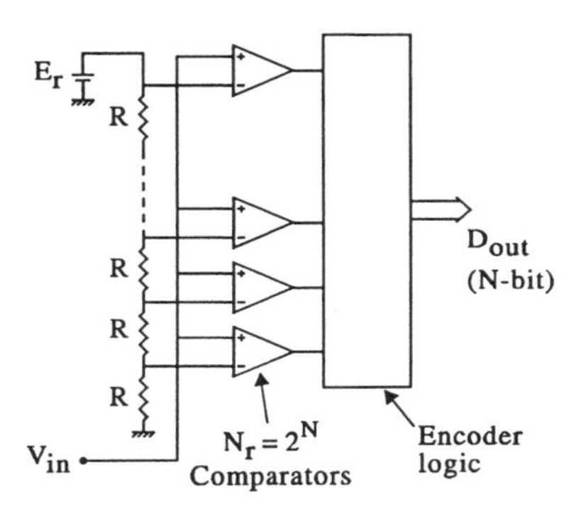

Fig. 20.12(a) Flash ADC. Essentials of circuit diagram.

Switched-capacitor comparator

For the arrangement as shown, the scatter in the comparator offset voltages V_{of} would greatly limit the accuracy. The switched-capacitor comparator of (b) overcomes this limitation as follows.

- With the switch in the Z (zero) position as shown, $V_{i-} = V_{of}$ so C is charged to $E_{ri} - V_{of}$.
- For the R (run) position then $V_{i-} = V_{of}$ for $V_{in} = E_{ri}$ as required.

Conversion time

This is merely the response time of the comparator and following logic, with the clock only needed for the switched-capacitor comparator. 20MHz is typical for the clock, and thus the conversion rate.

An appreciably longer time than this is needed for the comparator capacitors to charge to the above values (see Exercise 20.5).

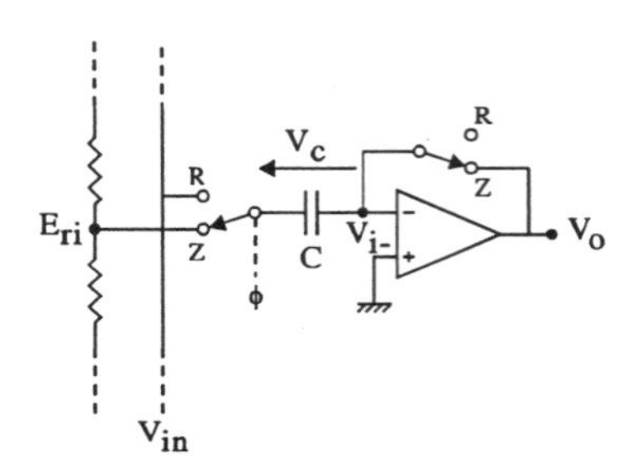

Fig. 20.12(b) Switched-capacitor comparator used in flash ADC of (a).

20.9 Dual-ramp ADC

Both of the above types of ADC rely on the accuracy of a set of resistor values, either for the DAC ladder in the successive-approximation type, or the potential divider in the flash type. The dual-ramp ADC avoids this kind of network, to give greater accuracy, albeit with a much increased conversion time $T_{conv} \sim 100$ms. It was once the basis of most digital test-meters, but the successive-approximation type is now providing a viable alternative.

Fig. 20.13(a) and (b) show the main components. There is also the digital controller (not shown) which implements the following sequence.

- The integrator in (a) and the counter in (b) are reset as in (c) to make both V_x and D_o zero.
- The counter, clocked by the crystal controlled oscillator output ϕ, is allowed to cycle once through its full range and back to zero. Here V_x represents $\int V_{in}$ over the accurately determined interval T_0.
- SWITCH is set to REF. This causes V_x to ramp up at a constant rate until V_x becomes zero, at which point the count is stopped.

At the end of T_o $\quad V_x = -\frac{1}{CR}\int_0^{T_o} V_{in}dt$,

while for the interval T_1

$$V_x = -\frac{1}{CR}E_r T_1 .$$

With the two equal then

$$T_1 = \frac{1}{E_r}\int_0^{T_o} V_{in}dt \qquad (20.13$$

or for V_{in} constant $\quad T_1 = T_o V_{in}/E_r \qquad$ (20.14

Thus $D_o \propto T_1 \propto V_{in}$ as required, more specifically pro portional to the average of V_{in} over T_o.

With the values of both C and R cancelling in the ex pressions then these do not have to be accurately deter mined.

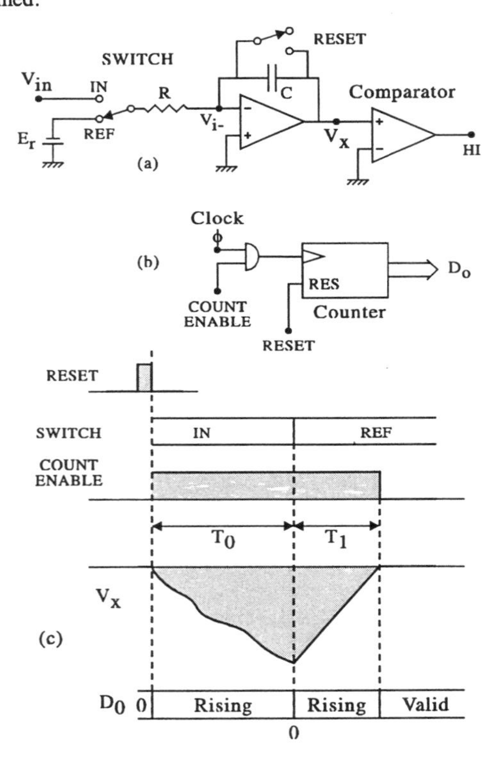

Fig. 20.13 Dual-ramp ADC.
(a) Gated integrator.
(b) Gated counter.
(c) Waveforms.

Example

Calculate the total conversion time T_{conv} for the dual ramp converter for 16-bit accuracy with a 1MHz clock.

$2^{16} \approx 64{,}000$ clock pulses (64ms) are required for T_o and up to another 64 000 for T_1, giving $\quad$ **T_{conv}=128ms**

:ontinuous operation

Iere the counter runs continuously with one extra bit dded. This does not constitute part of D_o but provides the ;WITCH waveform.

For test-meter operation, at some time during the 'ALID period D_o is latched to the visual display, usually t about once per second to maintain the display reasona-ly stable for reading.

0Hz interference

f the period T_o over which V_{in} is effectively averaged is nade an integral multiple of the period T_m of the 50Hz nains then the average of any interfering mains omponent becomes zero. This is normal practice for a est-meter.

xample

:alculate the minimum clock frequency f_ϕ for the above est meter-type operation, and for 16-bit precision.

With T_m=20ms, the conversion period T_{conv}=40ms. Vith

$$T_{conv}=2\times2^{16}\times T_\phi$$

ıen **f_ϕ=3.2768MHz**

:0.10 Oversampling (sigma-delta) ADC

'he oversampling ADC is another arrangement not lim-ed in its accuracy by that of a resistor chain in the feed-ack loop, but with the effective conversion rate much igher than for the dual-ramp converter. With at least 16-it resolution obtainable for signal frequencies of 25kHz r more, the device is now widely used in digital audio ystems. It is also now frequently taking the place of the ual-ramp converter for test meter use.

Fig. 20.14 shows a first-order implementation. In broad utline, with the negative feedback loop containing an ntegrator, this maintains the average of V_d equal to V_{in}. 'hus the average of the one-bit data stream D_b is propor-onal to V_{in}. The low-pass digital filter then extracts the verage of D_b, smoothing out the full-range alternations etween one and zero and giving the mean as the multi-bit arallel output D_o.

While the conversion rate is relatively high, this is still good deal short of the Nyquist rate, hence the term oversampling'. The significance of 'sigma-delta' will be xplained later.

Vaveforms

'he comparator and clocked bistable in Fig. 20.14 con-titute a one-bit clocked ADC. This performs a conversion f the integrator output V_i for each sampling period T_s of ne clock ϕ, to give the one-bit digital data stream D_b (b: istable output).

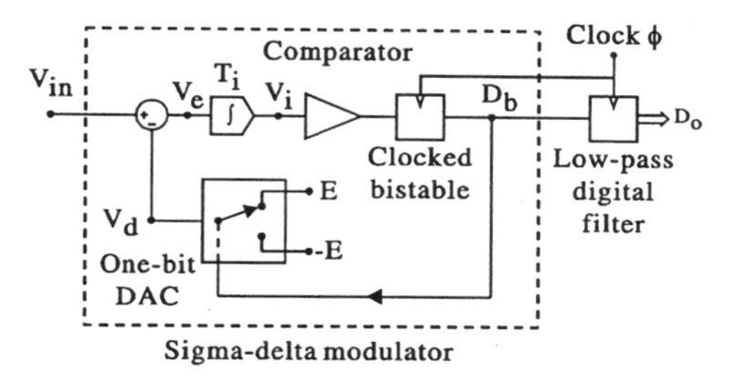

Fig. 20.14 First-order oversampling (sigma-delta) ADC.

With the switch also shown constituting the corresponding one-bit DAC, then the two together constitute a clocked 'quantiser', giving $V_d=\pm E$ according to the sign of V_i.

$V_{in}=0$

Here D_b alternates between 1 and 0 at the frequency $f_s/2$ where $f_s=1/T_s$, giving the mean of $D_b=1/2$, and thus D_o the mid-range value. V_i is a triangular wave, also of frequency $f_s/2$, and of peak-to-peak value ET_s/T_i centred at zero.

V_{in} close to upper limit

For other values of V_{in} within the range $\pm E$, V_i still cycles about zero, but in a more complex way. For V_{in} slightly less than the upper limit E then V_i becomes a sawtooth of period $>>T_s$, as follows.

- For $D_b=1$ then $V_d=E$ giving the error voltage V_e slightly negative.
- V_i falls slowly towards 0V, with D_b maintaining an extended series of ones until it gets there.
- D_b switches to 0 giving $V_d=-E$, and thus $V_e\approx 2E$, much larger than before.
- During this clock period V_i rises to nearly $2ET_s/T_i$, well above 0V, restoring D_b to 1 at the end of the period.

Thereafter the entire cycle repeats, giving V_i the slow sawtooth with $D_b=1$ at all stages apart from one clock period at the upward transition of the sawtooth. Accordingly the mean of D_b is close to one and D_o close to its upper limit.

Periodic sequences

These two waveforms are relatively easy to understand and can be sketched. Those for other values of V_{in} are less simple, usually not repeating at all. Table 20.3 shows the D_b sequences for several values of V_{in} for which the sequence does repeat. With D_o the average of D_b

$$D_o=(V_{in}/E+1)/2 \qquad (20.15)$$

V_{in}/E	D_o
1	11111111
3/4	11111110
1/2	11101110
0	10101010
−1/2	10001000
−3/4	10000000
−1	00000000

Table 20.3 D_b sequences for oversampling ADC of Fig. 20.14.

Counter

A very simple implementation of the low-pass digital filter at the ADC output is a synchronous counter, clocked at the ADC clock rate f_s, and with D_b applied to the counter 'enable' input. With the counter reset at the start of the conversion period T_c, then at the end of the period the counter output is the sum of D_b over T_c, constituting the required D_o. With N the number of counter bits

$$T_c = T_s \times 2^N$$

giving
$$D_o = 2^{N-1} \times (V_{in}/E + 1) \qquad (20.16)$$

Circuit model

With say 16-bit accuracy requiring the conversion time

$$T_c = 2^{16} \approx 64k$$

clock periods for the counter, this is scarcely better than for the dual-ramp converter. Fortunately the 'trade-off' can be much improved

- by using a higher-order digital filter,
- by using a higher-order feedback loop.

Explanation of the improvement requires the circuit model of Fig. 20.15 for the sigma-delta converter in the ADC. The model is first justified as follows.

- The value of the integrator period T_i makes no difference to the function of the integrator and comparator in the sigma-delta loop. Thus it is permissible to assume $T_i = T_s$.
- The quantiser (comparator, clocked-bistable and switch) sample V_i at the clock rate f_s. For $T_i = T_s$ then V_e is added to V_i at each clock pulse. This is the function of the accumulator in the model.
- Without the quantiser V_d would equal V_i. With it then $V_d = V_i + V_q$ where V_q is the 'quantisation error', or 'quantisation noise' component. Thus the model is completed.

The value of the model is that what was previously a highly non-linear sampling system can now be viewed as a sampled linear system with V_q added. All that is previously known about linear sampling feedback systems can thus be applied.

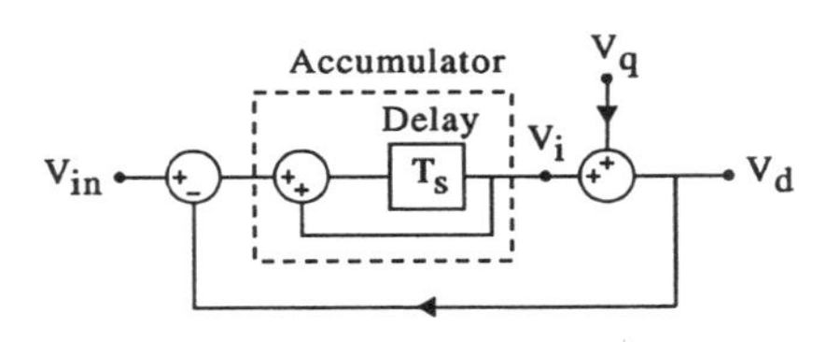

Fig. 20.15 Model of oversampling ADC of Fig. 20.14 showing quantisation noise V_q.

Quantisation noise reduction

First, the model shows the total V_i fed back to the acc mulator input to be zero. Thus the signal component of V is V_{in} delayed by one sampling interval.

Also the quantisation noise component V_{oq} of V_d is th two-point differential of V_q, the current value of V_q wi the previous subtracted from it. The effect of such tw point differentiation upon a sine wave of frequency f is for a filter of frequency response

$$A_d = 2.\sin(\pi f/f_s) \qquad (20.1$$

which is plotted in Fig. 20.16(a).

With V_q a series of pulses of width T_s then the spe trum extends from zero frequency to $\approx f_s/2$, with the pul heights random but within the range $\pm 2E$. Taking th spectral density G_q as uniform over this interval, then (in the diagram shows the considerable reduction obtai able for the low-pass digital filter of ideally abrupt cut-o and with the cut-off frequency $f_r << f_s/2$.

With $f_s/2$ the Nyquist limit for the signal frequency f_i it is thus clear why f_{in} needs to be $<<f_s/2$, to constitu 'oversampling'.

The diagram also shows, in (c), the comparable effe of the counter when used as the digital filter. With th counter essentially a first-order filter of frequenc response $A_c(f)$, then over the filter stop-band $A_c(f) \propto 1/$ With $A_d \propto f$ then the combined response becomes close being uniform as shown, giving considerably less quan sation noise reduction. At least a second-order filter needed to approach the ideal of (b).

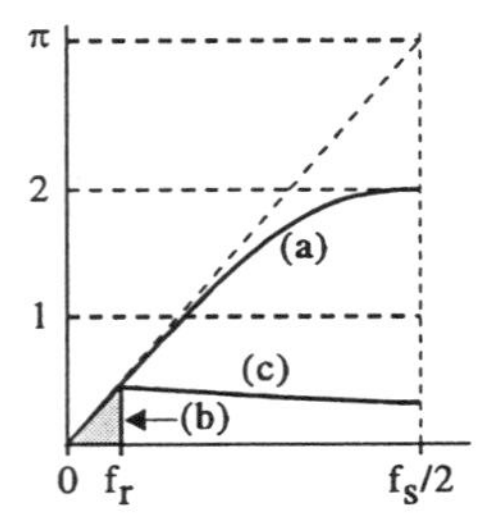

Fig. 20.16 Frequency responses for oversampling ADC
(a) Response of V_d to V_q in Fig. 20.15(a) (two-point differentiator over T_s).
(b) Further effect of ideal low-pass digital filter of cut- $f_r << f_s$.
(c) As (b) but for first-order output filter, as for counter.

Time-domain view

With the above the frequency-domain view of the effect of the quantisation noise, for the corresponding time-domain view recall that the values of the filter output V_{oq} are the two-point differences of V_q. Thus the final sum obtained by the counter consists only of the difference between the last and first values of V_q, With all of the other $\approx T_c/T_s$ other samples are wasted in the average, this gives an increase in noise error by $\times(T_c/T_s)^{1/2}$.

Ideal filter noise amplitude

Taking G_q as uniform, as above, then

$$\overline{V_q^2} = \int_0^{f_s/2} G_q df \qquad (20.18)$$

giving

$$G_q = 2\overline{V_q^2} / f_s \qquad (20.19)$$

With

$$\overline{V_{oq}^2} = \int_0^{f_r} G_q A_d^2 .df \qquad (20.20)$$

and A_d as in [20.17] then

$$\overline{V_{oq}^2} = \frac{2}{3}\overline{V_q^2}(2\pi)^2\left(\frac{f_r}{f_s}\right)^3 \qquad (20.21)$$

With the typical peak-to-peak value of V_q equal to $\pm E$ then

$$V_{oq}(\text{peak-to-peak}) = 2E \times 2^{3/2}(\pi/3^{1/2})(f_r/f_s)^{3/2} \qquad (20.22)$$

Counter noise amplitude

With $A_d \approx 2\pi f/f_s$ for $f << f_s$ we take this as for the whole range. Thus the combined response in the counter stop-band $A = 4\pi f_r/f_s$. With

$$\overline{V_{oq}^2} = \int_0^{f_s/2} G_q A^2 .df$$

then

$$\overline{V_{oq}^2} = \overline{V_q^2}(4\pi)^2\left(\frac{f_r}{f_s}\right)^2 \qquad (20.23)$$

With the f_r-independent terms differing by ~×4 due to the triangular decrease for the ideal filter, then the added factor f_r/f_s for the ideal filter represents the increase in area from (b) to (c).

Digital filter

Fig. 20.17(b) shows a digital implementation of the simple first-order RC low-pass of (a). While a higher order digital filter is needed, the general topic of digital filtering is beyond the scope of this text (Proakis and Manolakis, 1962). The following is given as an introductory example of how an analogue filter can be translated to digital form. For the capacitor in (a)

$$I = CdV_o/dt \quad \text{and} \quad I = (V_{in} - V_o)/R \text{ giving}$$

$$dV_o/dt = (V_{in} - V_o)/T_r \qquad (20.24)$$

where $T_r = CR$ as usual. For the sampled equivalent the relation becomes

$$\Delta V_o/\Delta t = (V_{in} - V_o)/T_r$$

where ΔV_o is the change in V_o per sample and Δt is the sampling interval, here T_s. Thus

$$\Delta V_o = (V_o - V_{in})/N_r \qquad (20.25)$$

where $N_r = T_r/T_s$, with $T_s << T_r$ for effective approximation to the response of the RC filter. The corresponding algorithm is

$$D_o \Leftarrow D_o + (D_{in} - D_o)/N_r \qquad (20.26)$$

where D_o corresponds to V_o, etc. (b) is the implementation of this algorithm, with the input now D_{in}.

For $N_r = 2^N$, with N integer, the multiplication is most simply obtained by a shift down by N bits. Otherwise, and more often, a full parallel digital multiplier is needed.

The resolution of the subtractor and accumulator needs to be commensurate with the extent to which V_{oq} is reduced from V_q.

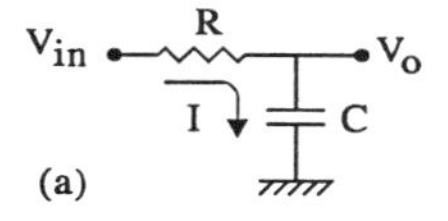

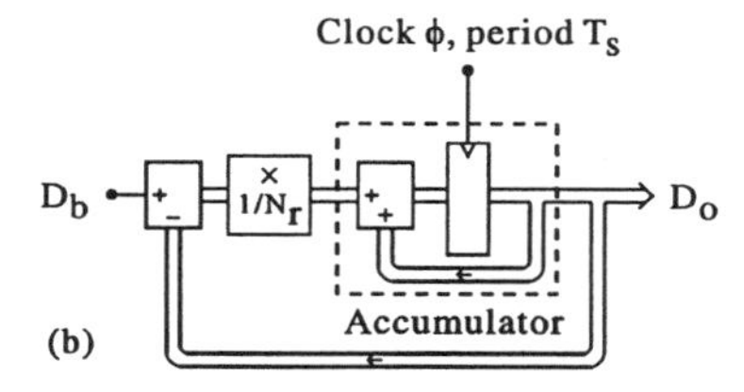

Fig. 20.17 Digital implementation of single-section low-pass RC filter.
(a) RC filter. Response time $T_r = RC$.
(b) Digital equivalent. $T_r = N_r T_s$. $N_r >> 1$.

Second-order circuit

The reason that the main feedback loop in the circuit model reduces the quantisation noise for $f < f_s$ is that the loop effectively contains an integrator, giving the loop gain $A_L \approx f_s/f$ and thus

$$G_{oq} \approx G_q(f_s/f)^2.$$

If a second integrator is added then $A_L \approx (f_s/f)^2$ giving $G_{oq} \approx (f_s/f)^4$, a yet greater decrease.

With $A_d \approx 2\pi f/f_s$ as in [20.17] for $f << f_s$ for the single integrator, then with the second added $A \approx 2\pi(f/f_s)^2$. With

$$\overline{V_{oq}^2} = \int_0^{f_r} G_q A^2 df$$

as in [20.20] then

$$\overline{V_{oq}^2} = \frac{2}{5}\overline{V_q^2}(2\pi)^2\left(\frac{f_r}{f_s}\right)^5 \qquad (20.27)$$

Stability

To retain stability of the loop, the response of the second integrator must level at f_2 a little before f_s to become unity for $f > f_2$. The term $(f_r/f_s)^5$ in [20.27] then becomes

$$(f_r/f_s)^3 \times (f_r/f_2)^2.$$

For experimental work supporting the oversampling ADC see Exercise 20.7.

Sigma-Delta

This term arose as the circuit evolved, as follows.

- An early form of digital communications link used the feedback loop of Fig. 20.14 as the modulator, but with the integrator in the feedback branch, following the one-bit DAC.
- With the integrator output V_i thus maintained equal to V_{in}, and with the single-bit D_b transmitted down the line, then a further integrator at the receiving station reconstructs V_i at its output. Here the modulator was termed a 'delta modulator', because of the way that D_b represented the increments required of V_i to track V_{in}.
- A limitation of this arrangement was the inevitable difference in input offset for the two integrators, causing the output of the second to drift to the point of saturation. This required a low frequency check to the demodulator integrator, giving an essentially AC-coupled (high-pass) overall response.
- The next step was to add a further integrator at the modulator input. Then, instead of D_b representing the differential of V_{in}, it became the representation of V_{in} directly. Thus no demodulator integrator with its offset drift was needed, only the one-bit DAC followed by a low-pass filter to remove the quantisation noise.
- The modulator was then called a 'sigma-delta modulator', the 'sigma' referring to the summing action of the added integrator, followed by the delta modulator as before.
- Unfortunately this simply transfers the saturation problem to the added integrator. However, it is possible to replace the two integrators with the one in the position in Fig. 20.14, for which the single integrator does not saturate. Thus the loop in Fig. 20.14 is the normal implementation of the sigma-delta modulator, which is why it is so labelled.
- Finally with the low-pass digital filter added we have the 'sigma-delta ADC'.

20.11 Oversampling DAC

As for the oversampling ADC, the value of the oversampling DAC is that it avoids the need for a highly accurate resistor network. As follows, apart from the interpolation filter at the input, the first-order oversampling DAC in Fig. 20.18 is the complement of the corresponding ADC in Fig. 20.14.

- The sigma-delta modulator is now digital rather than analogue.
- The low-pass output filter is now analogue rather than digital.

Interpolation filter

With f_{in} the frequency of the input signal, and the update (sampling) rate $f_{s.in}$ for the values of D_{in} possibly only a little above the Nyquist limit of $2f_{in}$, then this falls far short of the $f_s >> f_n$ required by the oversampling converter. With it fundamentally possible to fully reconstruct V_{in} from the D_{in} values, then the digital interpolation filter does so, providing the full set of samples at the increased rate f_s required by the oversampling DAC.

Sigma-delta modulator

This is a parallel digital implementation of the accumulator in the circuit model of Fig. 20.15, incorporating also the input subtractor.

With the one-bit DAC exactly as for the ADC, the analogue output filter is possibly the third-order circuit of Fig. 20.20(a), though probably of higher order.

For experimental work on the oversampling DAC, and a digital implementation of the first-order sigma-delta modulator see Exercise 20.8.

For a full development of both types of oversampling converter see Norsworthy, Schreier and Temes, 1967.

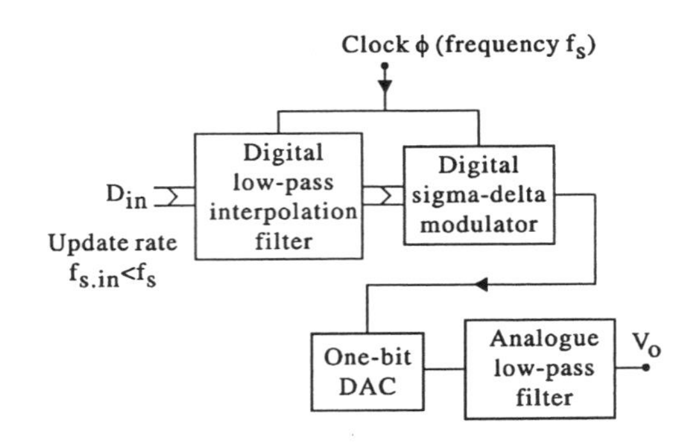

Fig. 20.18 Oversampling DAC with input interpolation filter.

20.12 Sampling gate and anti-aliasing

Where any kind of digital filtering is to be used it is essential that the sampling of the analogue input signal be at a constant rate. For example, for simple digital differentiation, where the previous sample is subtracted from the present, then a 10% variation in sampling period T_s will give a 10% error in the result.

With the sampling point somewhat indeterminate for say the successive-approximation type ADC, then it is usual to precede the ADC by a sampling gate, as in Fig. 20.19. With the switch opened at the required sampling point, this is kept open for the duration of the conversion, then to be closed ready for the next sampling point.

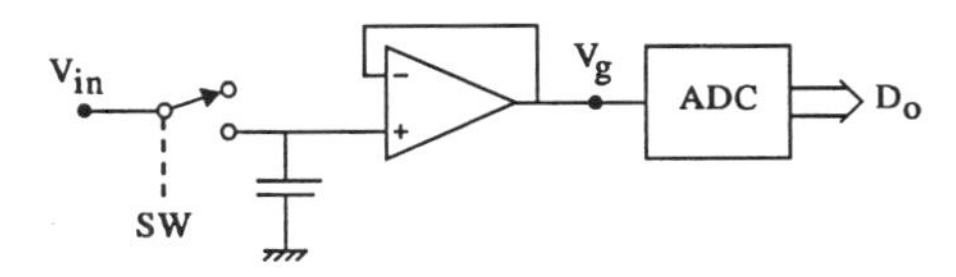

Fig. 20.19 Sampling gate preceding ADC.

Noise time averaging

A single point-sample is usually unfavourable for a signal accompanied by noise. As in Section 17.2, for white noise averaged over T_{av}, and with σ_n the standard deviation of the average, then

$$\sigma_n \propto T_{av}^{-1/2}.$$

Thus the samples represented by the ADC output should not be point samples of the noisy V_{in}, but averages of V_{in} over the sampling interval T_s.

Where there is no deliberate extension, the averaging period for the sample is the amplifier response-time T_r. For $T_r < T_s$ then the noise error is higher than need be, by

$$\times(T_s/T_r)^{1/2}$$

The dual-ramp converter does tolerably well here, because the output is the average of the input, but even here for only for half of T_s, giving σ_n is $2^{1/2}$ times higher than necessary.

Anti-aliasing

The need to time average over T_s in order to reduce noise error is the time-domain view of what is known in the frequency-domain view as 'anti-aliasing'.

As shown in Section 19.8, and now in Fig. 20.20(a), the sampling process is as if V_{in} were multiplied by a periodic train of unit impulses H_i, at the sampling rate T_s.

With the spectrum for H_i as in (b), and with the component at f_{in} that required, then any of the other input components shown will give a component at the multiplier output at f_{in}, and indeed of the same amplitude. These unwanted components are termed 'aliases' of the required component.

Here a low-pass filter having the cut-off $f_s/2$ as also shown will remove the unwanted component, also accepting any required input component of frequency up to the Nyquist limit of $f_s/2$. This is the anti-aliasing filter.

From the noise viewpoint, in the absence of the anti-aliasing filter, and with the remaining cut-off f_c, then $2f_cf_s$ times as many noise components as necessary are accepted, giving an increase in noise amplitude of $\times(2T_s/T_r)^{1/2}$ where T_r is the response-time of the remaining filter. This is approximately consistent with the above result for the time-domain view.

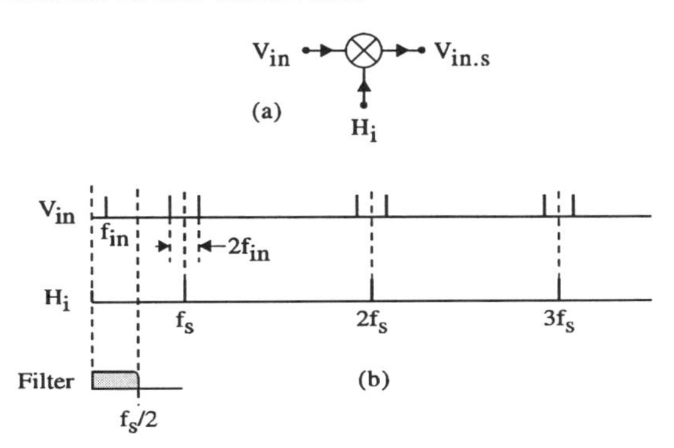

Fig. 20.20 Formation of aliases.
(a) Periodic point sampling viewed as multiplication by impulse train H_i to produce sampled input $V_{in.s}$.
(b) Spectra: V_{in} at f_{in} with aliases; H_i spectrum. Frequency response of anti-aliasing filter.

References

Norsworthy, S. R. Schreier, R. Temes, G. C. 1997: *Delta-Sigma Data Converters*. IEEE Press.

Proakis, J. G. and Manolakis, D. G. 1992, *Digital Signal Processing, Principles, Algorthms and Applications*. Macmillan.

20.13. Exercises and experiments

Exercise 20.1 Bipolar operation of DAC (Fig. 20.6)

Calculation For the eight-bit version, give the decimal equivalent D of the eight-bit B=1010 1110 when B is unsigned and signed.

Unsigned 1010=12, 1110=14, and $2^4=16$.
Thus $D_{unsigned}=12\times16+14$ giving **$D_{unsigned}=206$.**

Signed For MSB=0 $D_{signed}=D_{unsigned}$
while for MSB=1 $D_{signed}=D_{unsigned}-256$.

Here MSB=1 giving $D_{signed}=206-256$ and thus
$D_{signed}=-50$.

Exercise 20.2 MDAC bow-ties

Experiment Using two eight-bit single-quadrant DACs, a suitably biased sine wave signal-generator, an eight-bit counter and clock oscillator, and an analogue subtractor, devise some experiments to display the bow-ties equivalent to those in Fig. 15.3 showing

(i) the variation of V_o with V_{in}, for V_{in} varying slowly and D_{in} varying rapidly,
(ii) the variation of V_o with D_{in} for D_{in} varying slowly and V_{in} varying rapidly.

Sketch the 'range diagrams' for the various configurations. This is the D_{in} vs V_{in} plane with the areas shaded for which the device functions correctly.

Exercise 20.3 DAC sampling gate. (Fig. 20.10)

Explanation For the DAC output impedance $R_s=0$ and the DAC output voltage V_d constant, explain why the switch capacitance C_{sw} of the sampling gate gives a small DC output component.

• As the switch opens the charge $Q_{sw} \approx -V_{cc}C$ is transferred to C, causing V_c to be decreased from V_d by

$$\Delta V_c \approx -V_{cc}C_{sw}/C.$$

This remains for as long as the switch is open.
• As the switch closes, Q_{sw} passes through $R_s=0$ and not into C, returning V_c to V_d.

Thus V_c is a square-wave with the limits V_d and $V_d-\Delta V_c$. The added DC component is thus $\Delta V_c/2$.

Exercise 20.4 DAC glitches

Experiment Drive the ZN428 DAC or equivalent with an older type ripple counter such as the 74LS93 (or the even slower 7493 if you still have one!), driven by a suitable square-wave signal generator with a TTL type output. Observe the ripple-induced glitches, particularly the one at the transition from 0111 111 to 1000 0000.

Substitute the synchronous 74LS161 or the HCMOS 74HC93 to see the improvement.

Use an HC4316 to implement the sampling gate of Fig. 20.10 to avoid the irregular glitches. Finally add the low-pass filter shown to eliminate the periodic glitches incurred by the gate switch.

Adaptation of the scope probe as in Fig. 18.7(b) may well be necessary to see these effects reliably, cutting out coupling due to ground impedances etc.

Exercise 20.5 Flash ADC

Calculation For the eight-bit type, the total resistance of the potential divider is 2.5kΩ, and C in Fig. 20.12(b) is 100pF. Calculate the approximate value of the time T_d needed to elapse following application of the DC supply, for the converter to operate correctly.

The arrangement constitutes a lossy transmission line, with the distributed R and C. We shall not attempt this view but simply lump together the central 128 capacitors and regard then as charging through the impedance presented at the centre of the resistor chain, which is R/4=625Ω.

The corresponding capacitance is 128 × 100pF=12.8nF. Thus the charging time T_c=625Ω × 12.8nF=8µs. With the clock switching, the C are only connected to the divider for half of the time. Thus T_c becomes 16µs.

But the settling must be to within the resolution ΔV of the ADC, adding the factor

$$\ln(2^N)=\ln(256)=5.55.$$

Thus T_d=16µs×5.55 giving **T_d=88.7µs.**

Exercise 20.6 Oversampling ADC with counter as digital filter (Fig. 20.14)

Demonstration With the N-bit counter resolution $1/2^N$, show that this accuracy is realised.

For V_{in} constant and within range then $|V_e| < 2E$. With

$$dV_i/dt=V_e/T_i$$

and ΔV_i the change in V_i per clock period T_s then

$$|\Delta V_i|<2E\times(T_s/T_i).$$

With this the maximum possible amplitude of the V_i waveform as it cycles in semi-regular fashion about zero, and with ΔV_{ic} the difference between the initial and final values of V_i over the conversion period T_c then

$$|\Delta V_{ic}| <= |\Delta V_i| \qquad (20.28)$$

With

$$\Delta V_{ic}=\frac{1}{T_i}\int_0^{T_c}(V_{in}-V_d).dt=(T_c/T_i)\times(V_{in}-\overline{V_d})$$

then

$$\Delta V_{ic}=(T_c/T_i)\times(V_{in}-\overline{V_d}) \qquad (20.29)$$

With $T_c=T_s\times 2^N$ and the above relations

$$|V_{in}-\overline{V_d}|<2E/2^N \qquad (20.30)$$

With 2E the range of $\overline{V_d}$, and with $\overline{V_d} \propto D_o$, this accuracy is equal to that for the counter resolution.

Exercise 20.7. Oversampling ADC (experiments)

Fig. 20.21 shows an experimental circuit which demonstrates the operation of the oversampling ADC of Fig. 20.14. Here the digital output filter is replaced by the third-order Butterworth analogue equivalent, of cut-off $f_r=150Hz$.

Low levels of quantisation noise are encountered so the scope-probe should be connected as shown, with the final filter C directly across the BNC socket into which the probe adapter is plugged.

- Examine the waveforms for $V_{in}=0$ and for other DC values.
- With the overall high-frequency limit the cut-off f_r of the output filter, apply a sine wave and confirm the above $f_r=150Hz$.
- Apply a square wave to confirm the corresponding response-time $T_r=1ms$.
- Observe the quantisation noise for various values of DC input and then for a sine wave input. Compare the V_{oq} with that calculated from [20.22].
- To show that, apart from the amplitude of the integrator output V_i, the operation is independent of the integrator time-constant T_i, Reduce C_i to 680pF to give

$$T_i=T_s=20\mu s$$

as for the circuit model of Fig. 20.15. How far are we from being slew-limited?

- For the second-order loop, add the further integrator of (d) and repeat the above, noting the reduced quantisation noise. The comparator inputs need to be reversed.
- Sketch the expected frequency response confirming this is as required.
- Try reducing C_2 and note the resulting oscillation of the feedback loop. With the level of quantisation noise obtained, we have close to 16-bit accuracy. However the $f_r=150Hz$ is woefully inadequate for digital audio applications.
- Try $f_r=15kHz$ for a 5MHz clock

$$(C=1nF,\ C_i=100pF,\ C_2=47pF).$$

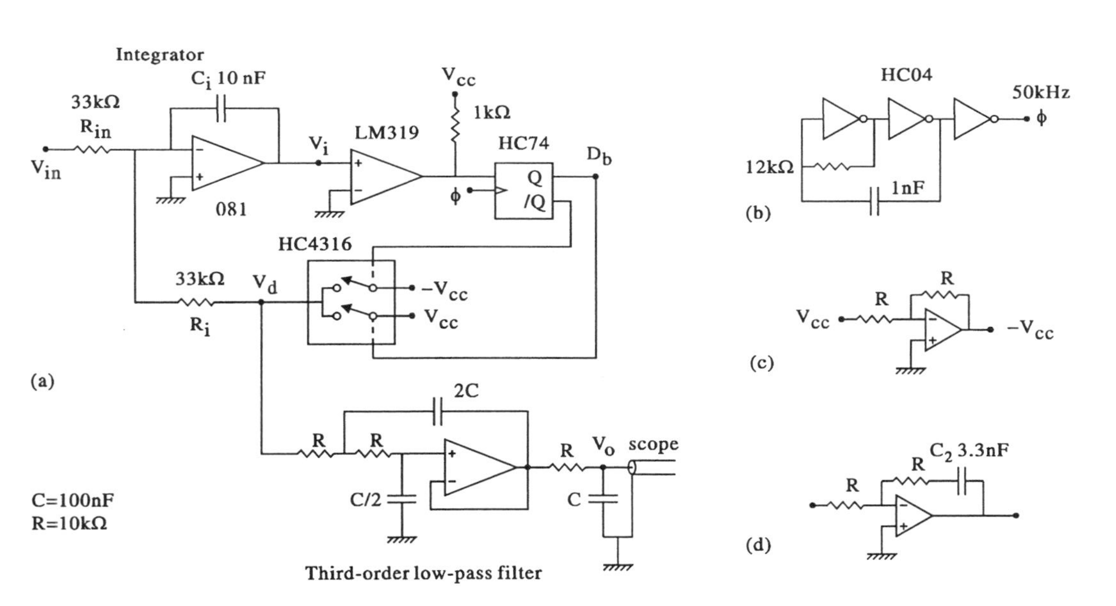

Fig. 20.21 Experimental circuit for oversampling ADC.
(a) Overall circuit.
(b) Clock oscillator ($f_s=50kHz$).
(c) Arrangement for developing $-V_{cc}=-5V$ from $V_{cc}=+5V$.
(d) Second integrator to follow first for second-order feedback loop.

Exercise 20.8 Oversampling DAC (experiments)

Fig. 20.22 shows the experimental arrangement used for demonstrating the operation of the oversampling DAC. Here

Clock	f_ϕ=800kHz,	T_ϕ=1.25μs.
DAC	f_s=50kHz,	T_s=20μs.
ADC	$f_{s.in}$=6.25kHz,	$T_{s.in}$=160μs.

Interpolation

While $T_{s.in}=T_s/8$, no interpolation filter is included, for the following reasons.

- $T_{s.in}$=160μs is still comfortably below the third-order final filter T_r=1ms.
- The ADC maintains the D_{in} value throughout the whole of the $T_{s.in}$ interval, instead of D_{in} being presented as a single impulse.

The effect of this is as if the V_{in} impulses of Fig. 20.19(b) were subject to a running-average filter of averaging period $T_{s.in}$. This is a first-order approximation to the interpolation filter.

Accumulator

The adder and clocked bistable in (b) constitute an accumulator, which is the digital counterpart of the integrator in the oversampling ADC of Fig. 20.14. The circuit is simpler than might be expected because the one-bit subtraction is an inherent feature of the overflow, with the carry-out bit used instead as the output.

Using the diagnostic DAC in (d) to monitor the Σ 'waveform' repeat the tests for the oversampling ADC. Experiment, using suitable outputs from the clock divider for the three clock lines.

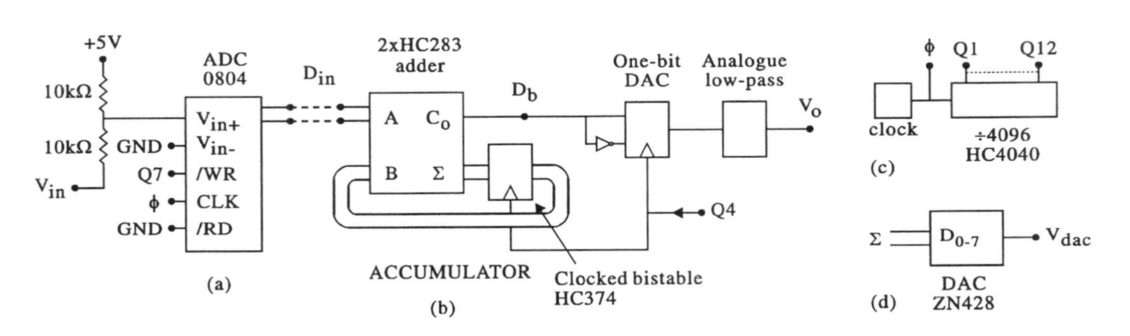

Fig. 20.22 Experimental circuit for oversampling DAC.
(a) Eight-bit successive approximation ADC providing input D_{in} to DAC in (b).
(b) Oversampling DAC. One-bit DAC and analogue low-pass filter as for oversampling ADC of Fig. 20.21(a).
(c) 800kHz clock and divider. Clock as in Fig. 20.20(b) with 1.2kΩ and 470pF.
(d) Diagnostic DAC for monitoring Σ.

21

SPICE and the audio power amplifier

Summary

As in the title, this chapter has two main topics. The first is the detailed consideration of one of the many types of audio power amplifier currently in use. The second is the use of the first as a typical example of where the SPICE type of simulation is a better way of obtaining a prediction of the circuit operation.

Most of the circuits covered so far have been simple enough for the nodal analysis to be of either one- or two-nodes. Even the two-node analysis is tedious, and strenuous efforts are made to restrict the number to one. For a circuit with as many nodes as the present, nodal analysis of the whole system in one would be impossibly complex.

Further, nodal analysis is applicable only to the incremental circuit model. For large-signal operation, as in the assessment of linearity, one must resort to the drawing of transistor characteristics, making graphic constructions to show the interaction with load-lines, etc. For circuits with the number of transistors in the present amplifier, this kind of approach is also too complex to be of any real help. Again SPICE comes to the rescue.

Two versions of SPICE are used, MicroSim PSpice and Those SpiceAge. Any of the experimental exercises given in the book can be backed by the appropriate SPICE simulation, most of them simple enough to be executable using the free evaluation version of PSpice. This is strongly recommended. Tuinenga (1988) is a suitable guide for the evaluation version.

The chapter includes the main features of the PSpice model for the bipolar transistor. This is more advanced than that in Chapter 6, and is based on Massobrio and Antognetti (1993), a comprehensive coverage of the many types of device modelled by PSpice.

Calculations in the main body of the chapter are cumulative as usual, with the numeric results set bold to the right-hand margin. Detailed calculations and derivations are placed in the Exercises section. With the emphasis on SPICE, there is no experimental work.

For the power amplifier the features discussed are

- supply requirements,
- dissipated heat and thermal stability,
- open-loop gain and frequency response,
- closed-loop gain, frequency response and stability,
- linearity and distortion.

21.1 Power amplifier

The basic requirements of an audio power amplifier, for domestic use, are considered to be as follows.

- Ability to deliver 50W of audio output power to an 8Ω speaker.
- Efficiency $\eta = P_o/P_{dc}$ as close to unity as possible. P_o=audio power output. P_{dc}=power provided by the supply.
- Frequency response flat over the audio range (20Hz to 25kHz) with comparable slew rate.
- Total distortion as low as possible.
- Input noise voltage as low as possible.
- Good thermal stability.
- Input voltage level for full output about 500mV.
- Input impedance high.

Maplin kit

The example chosen is based on a 50W audio power amplifier designed and marketed by Maplin Electronics, in kit form for assembly and use by non-expert customers. Some of the device types and circuit parameters used differ a little from the original. This is merely a matter of convenience.

Simplified circuit diagram

The circuit in Fig. 21.1. is basically that of a simple op-amp based non-inverting feedback amplifier. With R_a and R_b the feedback components, and R_L the 8Ω loudspeaker load, the remainder of the circuit constitutes the 'op-amp'.

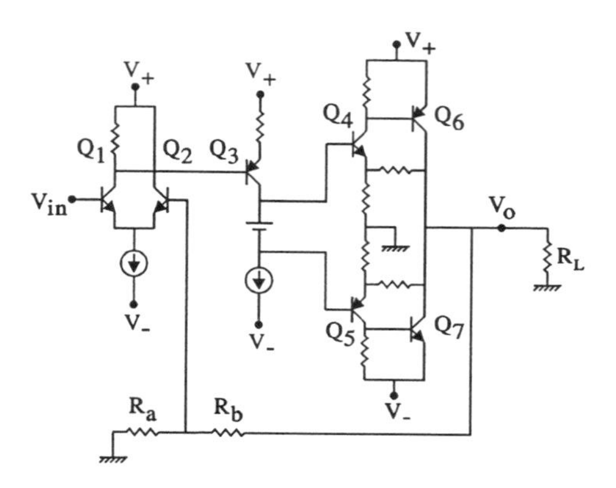

Fig. 21.1 Simplified circuit diagram for 50W audio power amplifier.

Indeed, apart from the last stage being of a higher power rating than for the usual op-amp, the make-up is much the same: Q_1 and Q_2 constitute the usual LTP differential input stage; Q_3 provides the largest part of the voltage gain; Q_4 to Q7 constitute the push-pull final power-amplifier stage.

Supply requirements

For the quoted audio output power of 50W, and the load resistance $R_L=8\Omega$, Exercise 21.1 gives the peak output voltage and current as $\quad$ $\mathbf{\hat{V}_o = 28.3V\ \hat{I}_o = 3.53A}$.

Supply voltages sufficient to avoid saturation of $Q_{6,7}$ are thus $\quad$ $\mathbf{V_+ = 30V,\ V_- = -30V}$.

Classes A and B

There are two main classes of operation of an audio power amplifier, Class A and Class B. Most commonly, and for the present circuit, it is Class B which is used. Here Q_6 supplies the positive half-cycles of the output and Q_7 the negative. For Class A, in contrast, both Q_6 and Q_7 would be conducting fully for the whole time.

Class B is clearly the more efficient, with Class A free from the potential 'crossover' distortion of Class B, as one transistor takes over from the other.

Efficiency

With $\overline{P_o}$ and $\overline{P_{dc}}$ the average values of the instantaneous output and DC supply powers P_o and P_{dc}, the efficiency

$$\eta = \overline{P_o} / \overline{P_{dc}}.$$

Class B For ideal transistors which do not saturate, η is highest when the peak output voltage is equal to that of the supply. For this ideal, and for the present push-pull output stage, Exercise 21.2 gives $\quad$ $\mathbf{\eta_{max}(Class\ B) = 78.5\%}$.

Class A Here neither transistor is driven below the point of cut-off. With the maximum output amplitude again that for which the peak output voltage is equal to that of the supply, and with $\hat{I}_o$ the corresponding peak output current, each I_c swings between $\hat{I}_o$ and zero. As shown in Exercise 21.3 $\quad$ $\mathbf{\eta_{max}(Class\ A) = 50.0\%}$.

For speech, music, etc., where the signal level is well below maximum for most of the time, the difference in efficiency for the two classes is much greater than the above figures suggest. While for Class B the two supply currents I_{dc} fall to zero for zero signal input, for Class A the I_{dc} remain at the value for maximum output. With the measures described below able to reduce the Class B crossover distortion to a very low level, these are the considerations leading to the normal use of Class B.

Transistor dissipation

While for Class B the supply power is highest for maximum output, the heat dissipated by the output transistors is highest at a lower level.

Consider the case where the signal amplitude is decreased by 5% from that where the peak output voltage is equal to the supply. Prior to the decrease, with the voltage at the peak zero, then the instantaneous power dissipation at the peak is also zero.

After the decrease the peak transistor voltage becomes 5% of the supply, while the peak current is only reduced by 5%.

With the peak transitor power thus increasing with reduced signal amplitude, the effect obtains to a lower degree for the rest of the cycle, giving the overall effect as stated.

More precisely, with $\overline{P_{th}}$ the average value of the transistor thermal power dissipation for a sine wave input, Exercise 21.4 gives the maximum value of $\overline{P_{th}}$ as occurring when

$$\hat{V}_o = 2V_+/\pi,$$

for which

$$\overline{P_{th}} = 2\,V_+^2/\pi^2 R_L \tag{21.1}$$

For the maximum power dissipation for the present two transistors together, then $\quad$ $\mathbf{\overline{P_{th}} = 22.8W}$.

With this for ideal non-saturating transistors, the small increase in supply voltage needed to allow for saturation increases the above figure a little.

Linearity

This is determined largely by the output stages, shown initially without the resistors in Fig. 21.2, and with R_s the output impedance of the previous stage.

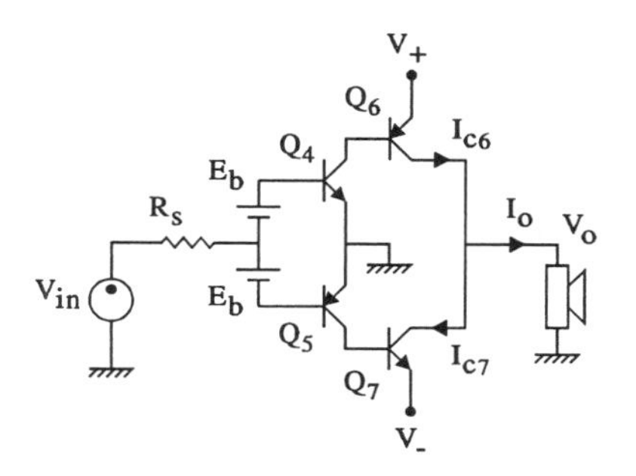

Fig. 21.2 Final stages of power amplifier in Fig. 21.1, with resistors omitted (R_s=75kΩ).

With the Pspice-derived voltage transfer function as in Fig. 21.3(a), for E_b=0, gross crossover distortion is evident. For say Q_4 and Q_6 conducting, with I_o=I_{c6} and with $I_{b6} \propto I_{c4}$, then significant current at any point requires V_{be4} ~ 600mV. Thus for V_{in} between ±600mV there is the flat region where I_o is negligible.

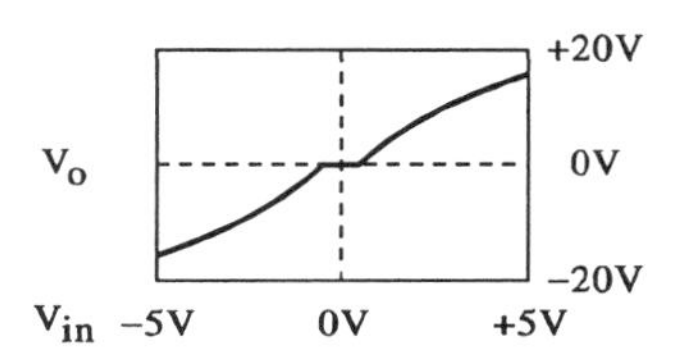

Fig. 21.3(a) PSpice-derived transfer functions for circuit of Fig. 21.2 with E_b=0, showing gross crossover distortion.

With the dual E_b applied the crossover distortion is largely removed, as in Fig. 21.3(b). However, the transfer function is still far from linear, with the dominant effect now the 'high-level injection' discussed in the next section. Here the current gain β decreases with increasing transistor collector current.

The reason for E_b=500mV being adequate, rather than the ususal ~600mV is the combined current gain, e.g. $\beta_4 \times \beta_6$.

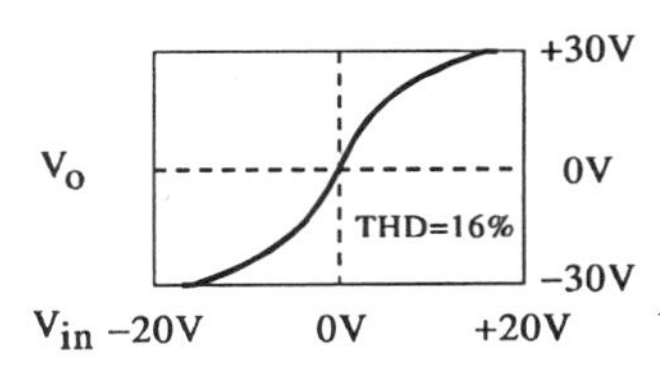

Fig. 21.3(b) As for (a) but with E_b=500mV.

Quiescent current

E_b=500mV in (b) gives a small 'quiescent' current I_q flowing through Q_6 and Q_7 for zero V_{in}. For $I_{b4} \propto V_{in}$ as for full linearity, this requires $r_{\pi 4} << R_s$. With

$$r_{\pi 4} = (kT/q)/I_{b4} \quad \text{and} \quad I_{b4} \propto I_{c6},$$

this requires the small I_{c6}=I_q at crossover.

Example

For $r_{\pi 4}$=R_s then $I_{c6} = (kT/q)\beta_4\beta_6/R_s$.

With R_s=75kΩ, and with β_4=140 and β_6=480 at crossover, then I_{c6}=22.4mA. The final value of I_q adopted is 50mA. With the non-zero I_q the class of operation is strictly Class AB, a fine point usually ignored.

Fourier decomposition

This PSpice facility resolves a distorted sine wave into its Fourier components. Table 21.1 shows the result for a 1Hz sine wave applied to the circuit of Fig. 21.2. This is with the above peak output voltage, which is a little short of the saturation limits.

Notice how only the odd harmonics are significant. With the circuit anti-symmetric, the amplitudes of the even harmonics should be zero. The reason they are not is that the parameters for the NPN and PNP transistors differ slightly. The 'total harmonic distortion' is given by

$$\text{THD} = \frac{\sqrt{V_2^2 + V_3^2 + V_4^2 \ldots}}{V_1} \tag{21.2}$$

where V_1 is for the fundamental, V_2 is for the second harmonic, and so on.

The value of 15.7% given is commensurate with the degree of non-linearity seen in the transfer function. We shall be looking for much better.

HARMONIC NO	FREQUENCY (HZ)	FOURIER COMPONENT
1	1.000E+00	3.346E+01
2	2.000E+00	3.129E-03
3	3.000E+00	4.839E+00
4	4.000E+00	6.752E-03
5	5.000E+00	1.773E+00
6	6.000E+00	8.393E-03
7	7.000E+00	9.148E-01
8	8.000E+00	1.033E-02
9	9.000E+00	4.914E-01

TOTAL HARMONIC DISTORTION
1.571094E+01 PERCENT

Table 21.1 Fourier decomposition for circuit of Fig. 21.2. E_b=500mV as for transfer function of Fig. 21.3(b).

21.2 SPICE bipolar transistor model

Variation of I_c and I_b with V_{be}

This is as in Fig. 21.4 (Massobrio and Antognetti, 1993) and gives a good account of the reasons why the current gain $\beta = I_c/I_b$ decreases, both for abnormally low values of collector current I_c and for the higher current densities occurring with the power transistor.

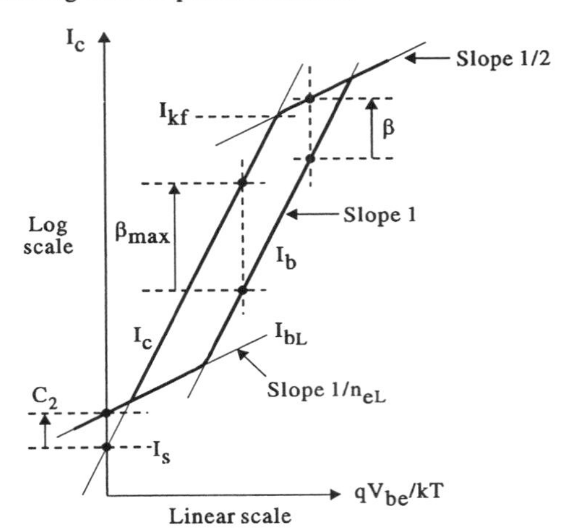

Fig. 21.4 Diagram showing how I_c and I_b vary with V_{be} for forward operation of transistor.

In the diagram β_{max} represents the normal value of β, with β decreasing at the extremities.

High-level injection

This is the effect causing the decrease of β at the upper extremity, as a result of the following sequence.

- As I_c increases then so does the stored base charge Q_{s-} for the NPN transistor.
- To maintain charge neutrality the compensating Q_{s+} increases by the same amount, in the form of extra holes in the base.
- With the concentration n_b of the passing electrons normally much less than the concentration p_b of holes in the base, the recombination rate R in the base is proportional to I_c, making $I_b \propto I_c$ as usual.
- However, as I_c increases there comes a point where no longer $n_b << p_b$. Here the concentration of the charge compensating holes has become comparable with the initial p_b.
- With $R \propto p_b \times n_b$ then R increases more rapidly than the previous $R \propto I_c$, causing I_b to increase more rapidly than I_c, hence the decrease in β.
- Ultimately p_b increases to a value large compared with the initial value, giving $p_b \approx n_b$.

With $\quad R \propto p_b \times n_b$

then in the limt $\quad R \propto n_b^2$.

With $\quad I_b \propto R \quad$ and $\quad n_b \propto I_c$

then $\quad I_c \propto I_b^{1/2}$

instead of the previous $\quad I_c \propto I_b$.

With I_b varying with V_{be} as before then the slope of the lin-log I_c plot is halved as shown.

Leakage current

This is the cause of the decrease in β for lower values of I_c. I_{bL} shown is the 'leakage' component of I_b, which varies less rapidly with V_{be} than the normal component.

With the normal $\quad I_b \propto \exp(qV_{be}/nkT)$

where n=1. For I_{bL} typically $n \approx 1/2$.

With I_{bL} arising from recombination at the surface and in the depletion layers, then only the surface recombination is truly 'leakage'.

Reverse transistor operation

Fig. 21.4 has a complement for reversed operation, where the collector and emitter connections are reversed but the transistor is otherwise biased as usual. Here the form of the I_c and I_b curves is the same, while the levels and transition points are different. Thus $\beta_f = I_c/I_b$ for normal 'forward' operation, and β_r for reverse, similarly I_{kf} and I_{kr}, etc.

PSpice parameters

With PSpice symbols only upper-case with no subscripting, Table 21.2 shows the corresponding symbols used in the present text.

PSpice	Text	
BF	$\beta_{f.max}$	Forward maximum β
BR	$\beta_{r.max}$	Reverse maximum β
IKF	I_{kf}	As diagram
IKR	I_{kr}	As IKF but reverse operation
VAF	V_{ea}	Early voltage, forward operation
VAR		Reverse VAF
IBL	I_{bl}	As diagram
MC	m_c	Junction capacitance parameter [21.7]
FC	ϕ_c	"
IS	I_s	As diagram
ISE	C_2I_s	As diagram (IS-emitter)
ISC		Reverse ISE (IS-collector)
NE	N_{el}	As diagram (N-emitter)
NC		Reverse NE (N-collector)

Table 21.2 PSpice parameter equivalents.

Early effect

As in Section 6.3, this is the effect accounting for the small increase of I_c with V_{cb}.

- An increase in V_{cb} causes the width W_d of the collector-base depletion layer to increase by ΔW_d.
- The width W_b of the undepleted region of the base is decreased by ΔW_d.
- With the base recombination thus decreased, then I_c is increased.

With the Early voltage V_{ea} the value of V_{bc} at which W_b is reduced to zero, then approximately

$$W_b \propto (V_{ea} - V_{cb}) \qquad (21.3)$$

For the NPN transistor, with n_b the concentration of electrons in transit across the base, with $x=0$ at the emitter end of the undepleted region, and with $n_b=0$ at the collector end,

$$dn_b/dx = n_b(x=0)/W_b.$$

With $I_c \propto dn_b/dx$ then

$$I_c \propto n_b(x=0)/W_b.$$

With $n_b(x=0)$ independent of V_{cb}, being determined only by V_{be}, then $I_c \propto 1/W_b$ for varying V_{cb}. With W_b as in [21.3] then

$$I_c(V_{cb}) = I_c(0)/(1 - V_{cb}/V_{ea}) \qquad (21.4)$$

Being reciprocal, this relation does not fully accord with the set of linear relations plotted in Fig. 6.6, and as normally observed in practice. In particular, $I_c=\infty$ for $V_{cb}=V_{ea}$ and the attendant $W_b=0$, does not accord. The assumed linearity of [21.3] is possibly the reason for the discrepancy. For $V_{cb} << V_{ea}$ in [21.4]

$$I_c(V_{cb}) \approx I_c(0)(1 + V_{cb}/V_{ea}) \qquad (21.5)$$

which, being linear, better accords with practice. PSpice therefore uses [21.5] rather than [21.4]. With the incremental output impedance of the common-emitter stage

$$r_o = dV_{cb}/dI_c,$$

then for [21.5] r_o is independent of V_{cb}.
Typically **$V_{ea}=70V$.**

Variation of β with base-collector voltage

As in Section 6.4, $I_b = I_{b1} + I_{b2}$

where I_{b1} is the component of I_b due to recombination in the base, while I_{b2} is that due to hole flow across the base-emitter junction. With

$$I_{b1} \propto W_b, \quad I_c \propto 1/W_b, \quad \text{and} \quad I_c = \beta_1 I_{b1}$$

then $\beta_1 \propto 1/W_b^2$.

With I_{b2} independent of W_b, and with $I_c \propto 1/W_b$, then

$$\beta_2 \propto 1/W_b.$$

SPICE assumes the linear relation, giving

$$\beta(V_{cb}) = \beta(0)(1 + V_{cb}/V_{ea}) \qquad (21.6)$$

Example

With $\beta_6(0)=336$ for the present Q_6, and with $V_{cb6} \approx 30V$ at crossover, then

$$\beta_6 = 336(1+30V/70V)$$

giving the considerable increase **β_6(crossover)=480,**

not as great, however, as the reduction due the high-level injection at peak, for which **β_6(peak)=80.**

Current crowding

The voltage I_bR_x developed by I_b passing through the spreading resistance R_x causes the base-emitter voltage at the centre of the transistor to be less than that at the edges. For a given I_c this increases the current density at the edges, giving an earlier onset of high-level injection.

This is allowed for in PSpice by suitably lowering I_{kf}. The reduction in R_x due to the crowding is also allowed for.

Crowding is reduced by making the area close to the edge as large a proportion of the total as possible. This is essentially a matter of making a power transistor in the form of a large number of much smaller transistors connected in parallel.

Depletion layer capacitance

As for Fig. 19.16 for the diode, the depletion layer capacitance is voltage-dependent, due to the widening of the layer when reverse bias is applied. For Pspice

$$c_\mu(V_{cb}) = \frac{c_\mu(0)}{(1 + V_{cb}/\phi_c)^{m_c}} \qquad (21.7)$$

where $\phi_c \sim 750mV$ is the collector-base built in junction voltage, and $m_C \sim 0.33$ (default values), similarly for c_π.

21.3 Thermal stability

Heat sinks

A suitable heat-sink is required to dissipate the above 22.8W of heat developed by the output transistors. Also the circuit must be designed to accommodate the resulting temperature rise.

A wide variety of sinks is available, ranging from a small clip placed as a cap over a small transistor, to large finned, and sometimes fan blown, systems occupying the whole of one side of the enclosure. Specifications range from H=55°C/W to 0.5°C/W for those in the Maplin catalogue.

The power transistor is mounted flat on to the sink, using thermally conducting grease. Where the transistor may not be grounded then a thin mica washer is interposed, or a suitable spongy insulating gasket.

These too have an H-value, ~0.4°C/W for the TO3 size mica washer. With H_s and H_w the values for the sink and washer then the total $H=H_s//H_w$, i.e.

$$H=1/(1/H_s+1/H_w) \qquad (21.8)$$

Thermal runaway

With I_c increasing by an amount corresponding to a 2mV/° increase in temperature, as in Section 6.7, then we have a form of positive feedback.

Increased $I_c \Rightarrow$ more heat $\Rightarrow$ further increase in I_c

This can soon destroy the transistors.

Thermal loop gain

Consider a single transistor of collector current I_c powered by the supply V_+. With θ_p the increase in temperature θ due to the thermal power P_{th}, and with H the heat-sink constant, then

$$\theta_p=P_{th}H.$$

With $\quad P_{th}=I_cV_+ \quad$ then $\quad d\theta_p/dI_c=V_+H.$

For the thermal loop gain A_{th}, working from input to output,

$$A_{th}=\frac{dV_{be}}{d\theta}\times g_m\times\frac{d\theta_p}{dI_c} \qquad (21.9)$$

With $\quad g_m=I_c/(kT/q), \quad dV_{be}/\theta=2mV/°C$

and $\quad kT/q=25mV$

then $\quad A_{th}=P_{th}H/12.5° \quad ...(a)$

or $\quad A_{th}=\theta_p/12.5° \quad ...(b) \qquad (21.10)$

Thus a mere $\theta_p=12.5°$ will incur runaway.

Example

With the required $H=12.5°/P_{th}$, and for the present $P_{th}=22.8W$, then $H=12.5°/22.8W$ giving $\mathbf{H=0.55°/W}$, a large heat sink.

Emitter degeneration

With the more complete circuit diagram for the last stages as in Fig. 21.5, and with initially only the R_e included (R_1 and R_f open-circuit) then the R_e have the effect of reducing the thermal sensitivity. This is as in Fig. 6.20 for the single-stage common-emitter amplifier, where it is shown that R_e develops local negative feedback for which the loop gain

$$A_L=-g_mR_e \qquad (21.11)$$

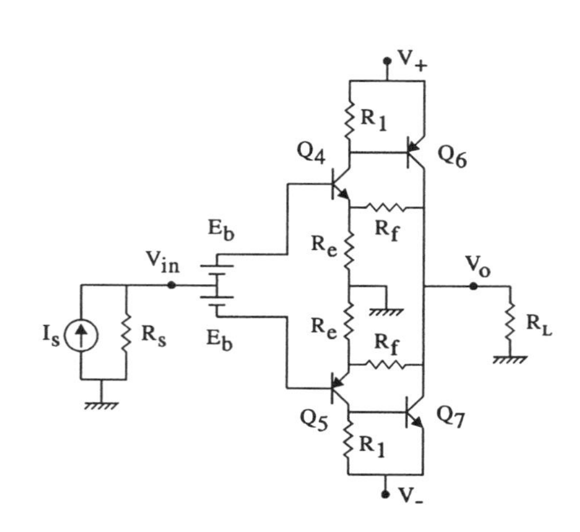

Fig. 21.5 Output circuit of Fig. 21.2 with added resistors R_e, R_1 and R_f.

With the thermal sensitivity thus reduced $\times 1/|A_L|$, and for the present $R_e=60\Omega$, Exercise 21.5 gives

$$\mathbf{A_L|(crossover)=0.25, \quad |A_L|(peak)=106.}$$

That the resulting reduction at crossover is negligible is of little importance. It is at peak that the thermal power is developed, and here the reduction is considerable.

Example

We design for $\theta_p=40°C$. Without R_e and with A_{th} as in [21.10] then $A_{th}=40°C/12.5°C$ giving $A_{th}=3.2$.

With R_e included then A_{th} is reduced to 3.2/106, which is well below runaway.

Quiescent

While there is little danger of runaway here, for a quiescent period following one of full power, during which θ has been increased by the 40°C, Exercise 21.6 shows that I_q is increased to ~650mA. This has no adverse effect on performance but adds to the general temperature rise and wastes power. For this and other reasons $R_1=100\Omega$ is included.

With $V_{be6}\approx 600mV$ then I_{c4} at crossover is increased to

$$I_{c4}\approx 600mV/R_i=6mA.$$

With $\quad g_{m4}=I_{c4}/(kT/q)$

then $\quad \mathbf{g_{m4}=240mS.}$

With A_L as in [21.11] and $R_e=60\Omega$ then $A_L=14.4$ which much reduces the thermal increase in I_q.

With $V_{e4}=I_{c4}R_e$

and with I_{c4} at crossover now ≈6mA then $V_{e4}\approx 360$mV, requiring E_b to be increased by this amount, giving

$$E_b \approx 860\text{mV}.$$

Thermal compensation

A further way of avoiding thermal effects is to arrange for the E_b to be temperature sensitive, but in the opposite direction to that for Q_4 and Q_5.

A simple method is to use two diodes in series in place of the single bias supply shown in the collector circuit for Q_3 in Fig. 21.1, with these fixed to the same heat sink as Q_4 and Q_5. With I_{c3} essentially constant, then for each diode

$$dV_d/d\theta=-2\text{mV}/^o\text{C},$$

exactly cancelling the effect of θ on $Q_{4,5}$.

Adjustable E_b

With the addition of the R_e requiring E_b to be increased to 860mV then the ≈600mV provided by the diode pair is not enough. While a third diode as in Fig. 21.6(a) may well suffice, adjustment of I_q to the optimum value requires the circuit of (b).

With
$$V_v \approx V_{be}(R_1+R_2)/R_2 \tag{21.12}$$

for $I_b(R_1//R_2) << V_{be}$ then $V_v=2E_b$ can be adjusted by the resistor values.

Overcompensation

With
$$dV_v/d\theta=-V_v/V_{be}\times(2\text{mV}/^o\text{C})$$

with the required $V_v=2\times 860$mV, and with $V_{be}\approx 600$mV, then $dV_v/d\theta \approx 5.73\text{mV}/^o$ which is in excess of the $2\times 2\text{mV}/^o$ required.

Here, instead of I_q being increased for a quiet passage following a loud one, it is reduced, giving the even more undesirable effect of temporary crossover distortion.

Fortunately the effect is moderated by the less than total thermal coupling of the compensating device to the output transistors. With the coupling suitably adjusted then the required degree of correction can be obtained. Exercise 21.7 gives an example calculation for the adjustable bias supply.

Current hogging

Consider two identical power transistors Q_1 and Q_2 connected in parallel to a common DC current source I_S. If Q_1 gets hotter than Q_2 then a higher proportion of I_s is steered into I_1, which then gets hotter still. This is 'current hogging'.

With a single power transistor actually a large number of smaller transistors connected in parallel (avoidance of current crowding) then a 'hot spot' becomes more so, hogging I_c to the point of destruction.

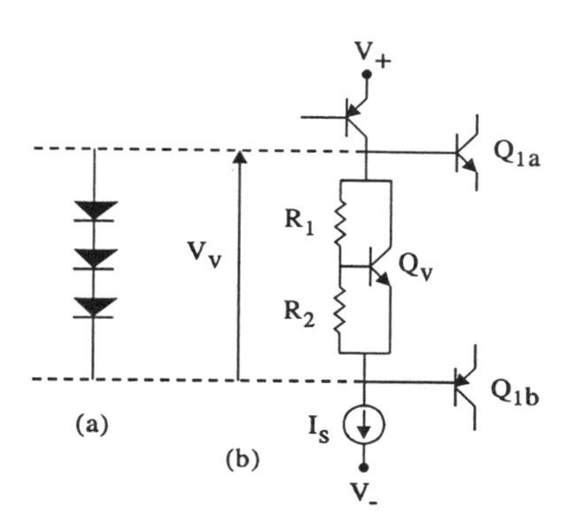

Fig. 21.6 Temperature compensating implementations of bias E_b in Fig. 21.2 and 21.5.
(a) Three-diode fixed bias supply.
(b) Adjustable bias supply.

Emitter degeneration as above is used here, with a small R_e added to each component transistor.

Similarly when two power transistors are connected in parallel on the same heat sink (as forming Q_6 in the present Maplin design) then small individual emitter resistors are added, ~0.5Ω.

21.4 Final stage linearity

The next step is to add the R_f in Fig. 21.5. This converts each half of the circuit into what is known as a 'voltage feedback pair', for which the incremental voltage gain $a_{4\text{-}7}$ (Q_4 to Q_7) becomes ideally $1/\beta_{fe}$ (fe: R_f, R_e) where the voltage feedback fraction

$$\beta_{fe}=R_e/(R_f+R_e) \tag{21.14}$$

as for the voltage feedback amplifier of Section 2.2. The values chosen are $R_e=R_f=120\Omega$, leaving the effective R_e as before, and giving ideally

$$a_{4\text{-}7}=2.$$

With the incremental loop gain a_L finite then

$$a_{4\text{-}7}=(1/\beta_f)\times a_L/(a_L+1) \tag{21.15}$$

Fig. 21.12 shown later gives the incremental circuit, from which the expression for a_L, and thus its value, can be derived.

But with the emphasis here on the SPICE approach, the details of this and other such calculations are left as exercises. Instead the numeric results will be given, as a check for the calculation, and also for comparison with the SPICE result.

Even here the parameters, g_m, r_π, etc., used in the incremental calculation require the SPICE-derived transistor β values in Table 21.3.

	Crossover	Peak
$Q_{4,5}$	140	113
$Q_{6,7}$	480	80

Table 21.3 PSpice-derived current gain β for transistors used in final stage of power amplifier.

From the calculation

$$\mathbf{a_L(crossover) \approx 16, \quad a_L(peak) \approx 5.}$$

With $a_{4\text{-}7}$ as in [21.15] then

$$\mathbf{a_{4\text{-}7}(crossover) \approx 1.88, \quad a_{4\text{-}7}(peak) \approx 1.67.}$$

With R_f included the PSpice-derived voltage transfer function becomes as in Fig. 21.7(a), with the much improved THD=1.7% shown. Here the non-linearity within the saturation points is barely discernible by eye, even with visual foreshortening.

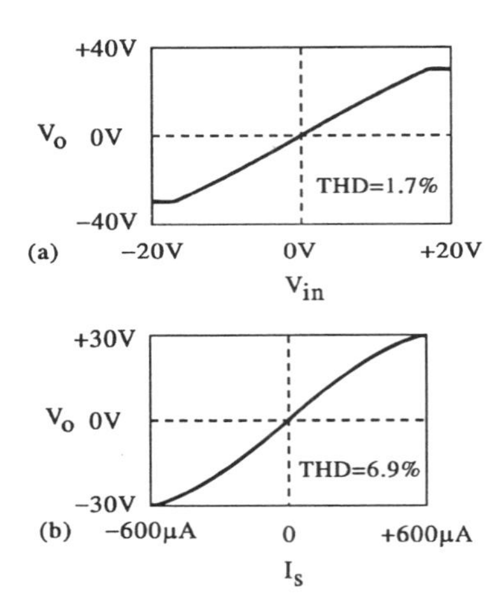

Fig. 21.7 Transfer functions for $R_e=R_f=120\Omega$ and $R_l=100\Omega$ in Fig. 21.5. (a) V_o vs V_{in}. (b) V_o vs I_s for $R_s=75k\Omega$.

Input impedance

Unfortunately the signal source of Fig. 21.5 is far from the ideal voltage source for which the voltage transfer function is appropriate. With the signal source impedance R_s included, the improvement in linearity, now as in (b), is a good deal compromised.

With the variation of V_{in} with I_{in} for $Q_{4,5}$ as in Fig. 21.8, this accords with the conventionally calculated

$$\mathbf{r_{in}(peak)=41k\Omega, \quad r_{in}(crossover)=69k\Omega.}$$

Here $(R_s=75k\Omega) \sim r_{in}$ rather than the required $R_s << r_{in}$.

Also closer consideration shows that, while the transfer from Q_6 to Q_7 is relatively close to the crossover point, both Q_4 and Q_5 remain conducting for about half of the output range. With this maintaining r_{in} at half the value that would otherwise obtain, in the absence of the effect the slope of the transfer function in (b) would be increased in the central region, further increasing the non-linearity. It is mainly this rather than the feedback via R_f that accounts for the modest improvement relative to the 16% in Fig. 21.3(b).

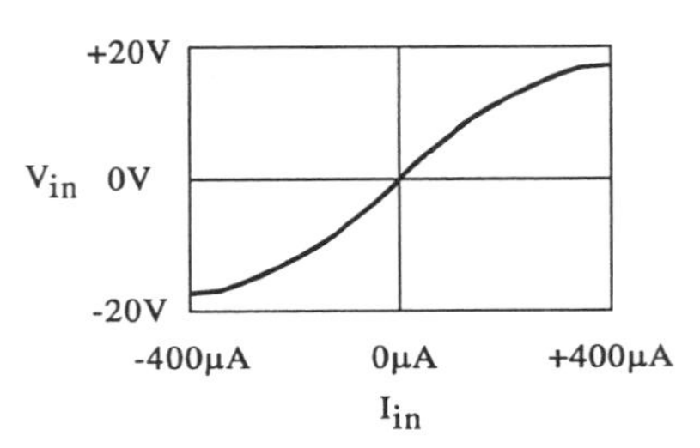

Fig. 21.8 PSpice-derived variation of V_{in} with associated I_{in}

Choice of quiescent current

With the crossover distortion discernible by ear much too small to be seen on a direct display of the transfer function, the method of Fig. 21.9(a) is required for the SPICE simulation. Here the ideal output AV_{in} is subtracted from that subject to the distortion, giving the distortion component only, which is then suitably expanded.

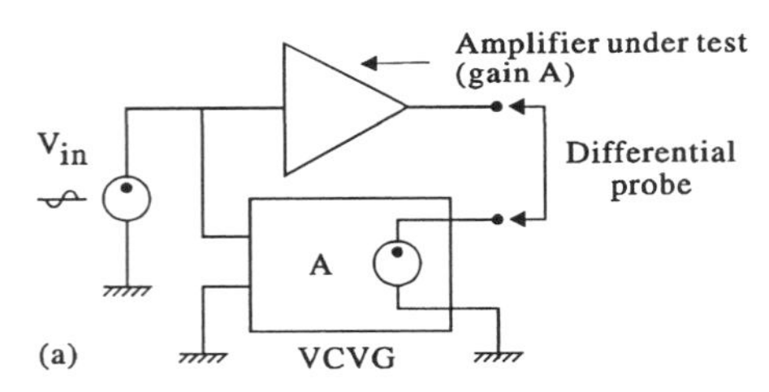

Fig. 21.9(a) SPICE configuration for enhanced display of low-level distortion

The experimental equivalent of (b) has long been used. Here one must be sure that it is not distortion in the scope differential amplifier that is being seen. Why is this less of a problem if the amplifier under test inverts?

In either case the A has to be adjusted to remove the tilt that otherwise occurs. With the results as in Fig. 21.10

$$\mathbf{I_q=50mA.}$$

was found to be adequate. The feedback through the R_f has no common-mode component and so leaves the thermal stability unchanged.

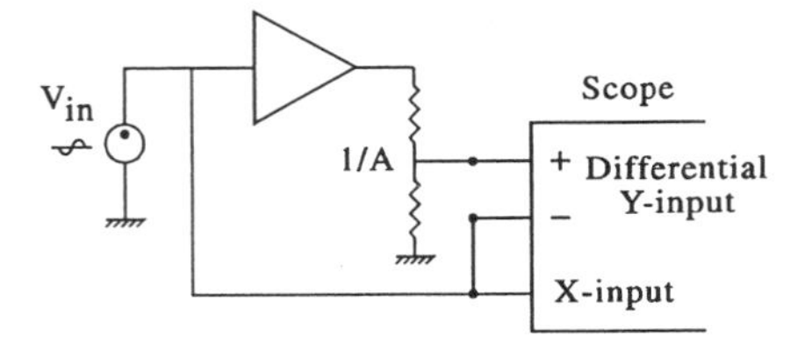

Fig. 21.9(b) Experimental equivalent of (a).

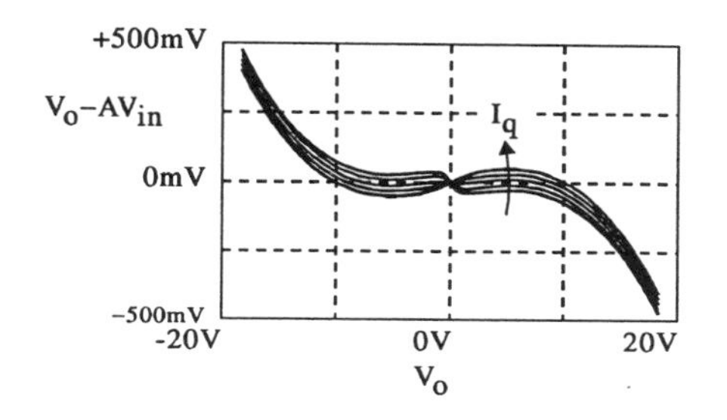

Fig. 21.10 Results obtained with PSpice method of Fig. 21.9(a). I_q (mA)=17, 28, 44, 67, 96.

21.5 Final stage frequency response

Fig. 21.11 shows the PSpice-derived frequency response for the present circuit. As before, the results of conventional analysis based on the incremental circuit of Fig. 21.12 are compared. Here it must be remembered that at crossover both halves of the circuit are operating, while at peak only one. The features to be accounted for are

- the two zero frequency values (crossover and peak),
- the two cut-off frequencies,
- the aberration between 10MHz and 100MHz.

Zero-frequency gain

Here the plotted values accord with the previously quoted calculation results

$$a_{4\text{-}7}(\text{crossover}) \approx 1.88 \quad \text{and} \quad a_{4\text{-}7}(\text{peak}) \approx 1.67$$

Cut-off frequencies

Here we first calculate the two open-loop cut-off frequencies. With the corresponding values of zero-frequency loop gain quoted above then the closed-loop cut-frequencies are derived. By 'open loop' we mean with no feedback through R_f, i.e. with the right-hand end of R_f grounded, making the effective R_e again 60Ω.

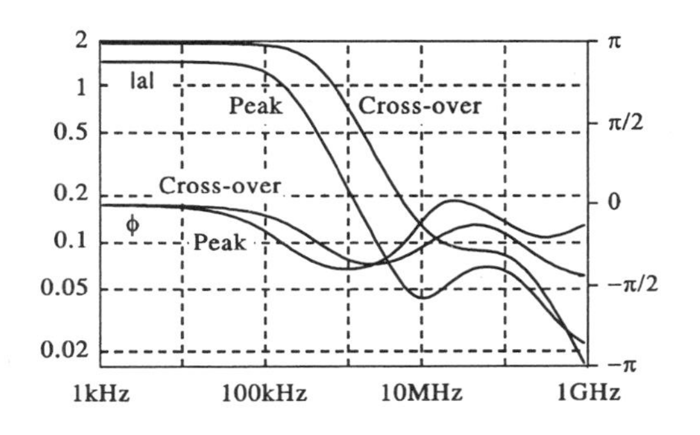

Fig. 21.11 PSpice frequency response of incremental voltage gain $a_{4\text{-}7}$ for circuit in Fig. 21.5.

Open loop It is shown in Section 7.2 for the emitter-follower, that for $R_e >> 1/g_m$ the effect of z_π is negligible. For Q_4 at crossover

$$I_{c4} \approx 600\text{mV}/R_1 = 6\text{mA}$$

With $$g_{m4} = I_{c4}/(kT/q)$$

then $$1/g_{m4} = 25\text{mV}/6\text{mA} \approx 4\Omega.$$

With R_e effectively 60Ω then indeed $R_e >> 1/g_{m4}$ so it becomes $z_{\pi 6}$ which determines the open-loop cut-off. With I_{c4} yet higher at peak then $1/g_{m4}$ is yet lower, so the same obtains.

With $c_{\pi s}$ determined by the base 'stored charge' then, whether I_c is above or below I_{kf},

$$I_c = Q_+/T_\tau$$

where T_τ is the base transit time.

With $c_{\pi s} = dQ_+/dV_{be}$ and $dI_c/dV_{be} = g_m$

then the normal relation $c_{\pi s} = g_m T_\tau$ obtains.

With PSpice giving $T_{\tau 6} = 53\text{ns}$,

with $g_m = I_c/(kT/q)$ and with $I_{c6} = 50\text{mA}$ and 3.5A

then $\mathbf{c_{\pi 6}(crossover) = 106nF}$, $\mathbf{c_{\pi 6}(peak) = 8\mu F}$.

Here $c_{\pi d6}$ is ignored, being only 600pF.

With $r_\pi = \beta/g_m$ and with $\beta_6 = 480$ and 80 at crossover and peak then

$$\mathbf{r_{\pi 6}(crossover) = 240\Omega, \quad r_{\pi 6}(peak) = 0.53\Omega.}$$

With the open-loop cut-off

$$f_L = 1/2\pi c_{\pi 6}(r_{\pi 6}//R_1) \tag{21.16}$$

then

$$\mathbf{f_L(crossover) = 22kHz, \quad f_L(peak) = 37.5kHz.}$$

Closed-loop With the closed-loop cut-off $f_{4\text{-}7} \approx f_L \times a_L$ and with the above $a_L \approx 16$ and ≈5 then

$$\mathbf{f_{4\text{-}7}(crossover) = 350kHz, \quad f_{4\text{-}7}(peak) = 188kHz.}$$

These values also accord with the PSpice plots.

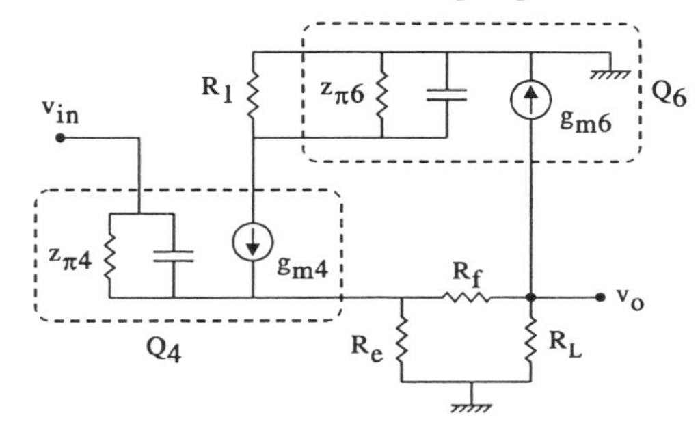

Fig. 21.12 Incremental equivalent circuit, upper half of Fig. 21.5.

Feed-forward

The temporary check at about 10MHz in the $\propto 1/f$ section of the plotted frequency response is due to 'feed-forward' transfer from say the emitter of Q_4 to the V_o point through R_f. Here Q_4 acts as an emitter-follower giving $v_{e4} \approx v_{in}$.

With $$v_o/v_{e4}=2R_L/(2R_L+R_f)$$

at crossover (both halves working) and

$$v_o/v_{e4}=R_L/(R_L+R_f)$$

at peak (one half only working), and with

$$R_L=8\Omega \quad \text{and} \quad R_f=120\Omega$$

then the overall feed-forward coefficients become

$$\boldsymbol{\alpha_{ff}(\text{crossover})=0.12, \quad \alpha_{ff}(\text{peak})=0.063.}$$

These values accord roughly with those for the semi-flat regions in the plot. The brief reduction in phase lag over the flat region is also as expected.

The final decay at about 100MHz is well beyond our range of interest and is probably as much due the transistor r_x as it is to $c_{\pi 4}$.

Input impedance z_{in}

Having noted above that the calculated zero-frequency input impedances r_{in} for the present circuit are 69kΩ and 41kΩ at crossover and peak, these values accord with the SpiceAge-derived plots of z_{in}.

With these the closed-loop values then there is enhancement by the loop gain a_L. With a_L subject to the open-loop cut-offs

$$f_L(\text{crossover})=22\text{kHz} \quad \text{and} \quad f_L(\text{peak})=37.5\text{kHz},$$

then these will be the transition frequencies for z_{in}, again in accordance with the SpiceAge plots.

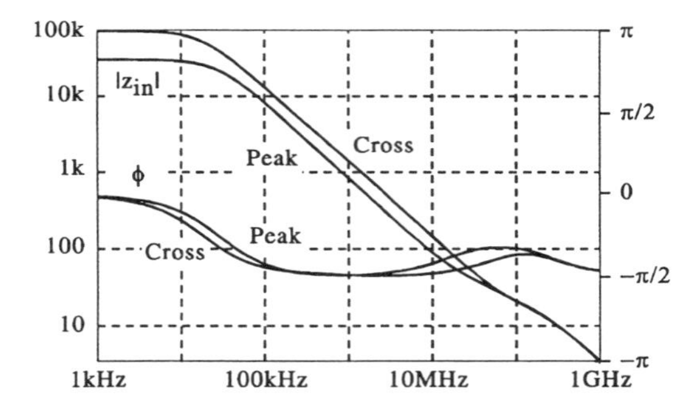

Fig. 21.13 SpiceAge-derived frequency response of input impedance z_{in} for circuit of Fig. 21.5 at peak and crossover.

21.6 Overall frequency response

Fig. 21.14 shows the full power amplifier comprising the active 'op-amp' section of voltage gain a_o and the feedback network R_a and R_b. For the present we take C_b to be omitted.

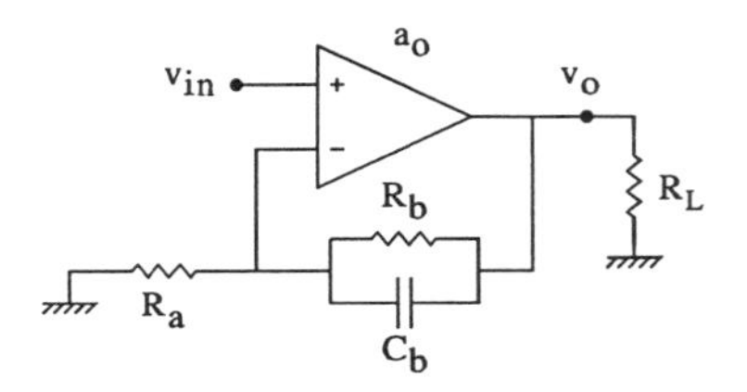

Fig. 21.14 Audio power amplifier in Fig. 21.1 showing feedback network.

'Op-amp' frequency response

Fig. 21.15 shows the straight-line approximation to the magnitude-frequency response of the 'op-amp' section of Fig. 21.14. Here the zero-frequency value of 30kΩ is as derived by PSpice, as also is $f_{1\text{-}2} \approx 3$MHz the cut-off for the LTP input stage in Fig. 21.1.

However, while the labelling of f_3 suggests it to be the response of f_3 this is a classic example of the inability to derive the frequency response of a multi-stage circuit by simply cascading the responses for the component stages considered separately.

In fact f_3 is determined by the loading of Q_3 by the input impedance z_{in} of the final push-pull feedback pair.

With $f_L \sim 30$kHz the transition frequency for z_{in}, and with f_L in fact the cut-off frequency for the loop-gain of the feedback pair, this is largely the transition for say $z_{\pi 6}$, a long way from Q_3.

With this allowed for, the plot shown gives helpful insight into the overall SPICE-derived open- and closed-loop frequency responses in Fig. 21.16 and following.

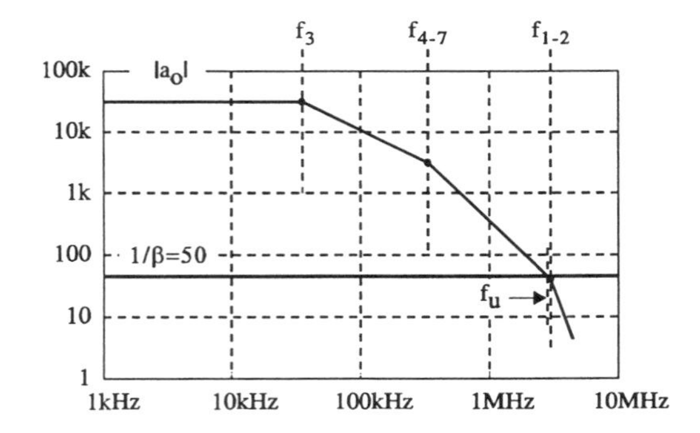

Fig. 21.15 Bode plot for open-loop frequency response of power amplifier in Fig. 21.1, simplified in Fig. 21.14.

Feedback resistors

The required input sensitivity is ≈500mV peak. Setting the feedback fraction β to 1/50 gives 25V peak output for 500mV input, which is adequate. $R_b=10k\Omega$ and $R_a=200\Omega$ are suitable.

Loop gain

With the overall loop gain $a_L=a_o\beta$ then $a_L=30k\Omega/50$ giving the zero frequency **$a_L=600$.**

Potential instability

With the straight-line plots of Fig. 21.15 also showing the feedback fraction $1/\beta$, then the loop-gain A_L is given as usual by the distance between this and the $|a_o|$ plot. Thus with the phase-lag advancing by $\pi/2$ at each transition, at the intersection where $|A_L|=1$, the lag is well in excess of the $2\times\pi/2$ giving oscillation.

SpiceAge plot

The open- and closed-loop SpiceAge frequency-response plots of Fig. 21.16 show the above tendency to instability to be much moderated. There is no oscillation, only the pronounced resonance at the unity-gain frequency f_u in (b), with this removed by adding the C_b in the feedback network as in Fig. 21.14.

For the open-loop response of Fig. 21.16(a), it is convenient to show the variation of the loop-gain $A_o\beta$ for β without C_b rather than just a_o as in the straight-line predictions of Fig. 21.15.

Apart from this, the two accord for the first two transitions but differ thereafter. This is because the straight-line prediction does not include the effect of the feed-forward component in the closed-loop response for the final stage, as in Fig. 21.11. Here the effect of the resulting zero at ≈10MHz is to move the phase lag at f_u back from near $3\pi/2$ to just a little less than $2\pi/2$, giving the resonance instead of the oscillation.

Feedback capacitor C_b

The addition of C_b to the feedback network removes the resonance as shown in the linear plots of Fig. 21.17. With f_a and f_b in (a) the transition frequencies for the feedback network then

$$f_a=1/(2\pi C_b R_b) \qquad (21.18)$$

and

$$f_b=1/(2\pi C_b R_a//R_b) \qquad (21.19)$$

First for (a) C_b is larger than desirable. With this value the circuit will certainly oscillate, because the $|a_o|$ curve crosses the $1/\beta$ (here 1) line at a point in the $|a_o|$ where it is falling $\approx 1/f^3$, with a phase lag $\approx 3\times 90°$.

Reducing C_b to give (b) probably avoids the oscillation but, if so, certainly gives a sharp resonance. Here ϕ_o is still close to $-3\times 90°$ but now $\phi_{1/\beta}\approx -90°$. Thus the phase lag of $A_o\beta \approx 2\times 90°$.

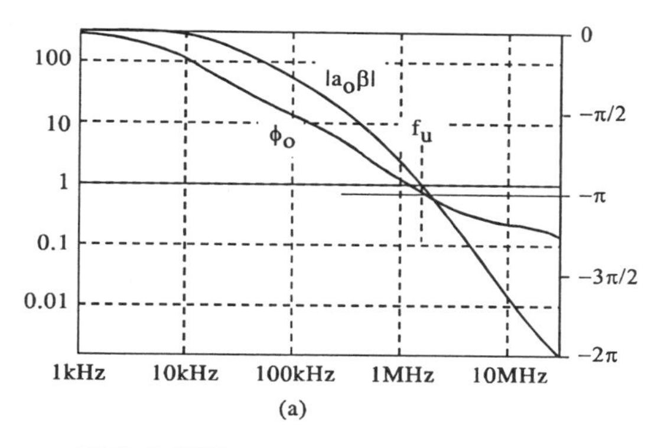

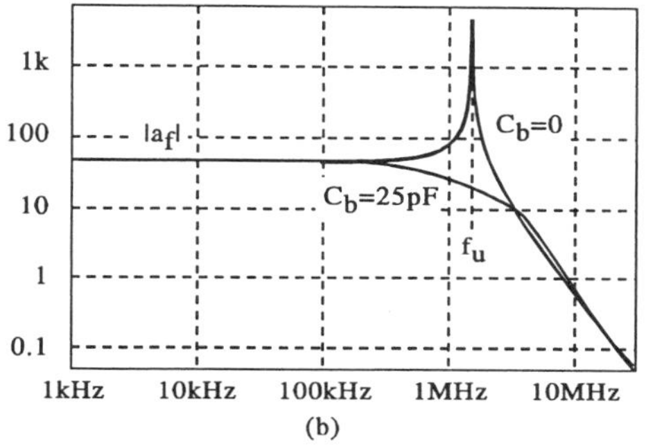

Fig. 21.16 Frequency response for SpiceAge simulation of compete audio power amplifier of Fig. 21.14.
(a) Open loop $|a_o|\beta$.
(b) Closed loop $|a_f|$.

(c) is the most stable case, but even then marginal. The transition of ϕ_o from −180° to −270° is about half-way down, but this is cancelled by the almost coincident transition in β, leaving the overall shift for the loop at ≈ −180°.

Apart from the zero, this would be highly marginal; the zero saves the situation.

Finally (d) shows the situation much as if C_b were absent, definitely giving oscillation apart from the zero.

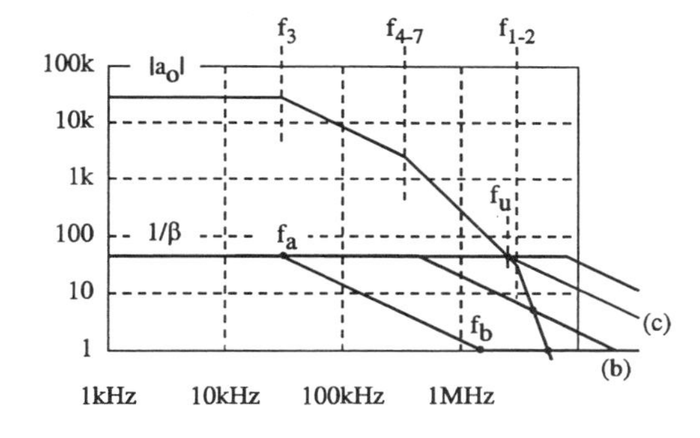

Fig. 21.17 Effect of C_b in feedback network of Fig. 21.15 on feedback fraction β. (a) to (d) decreasing C_b.

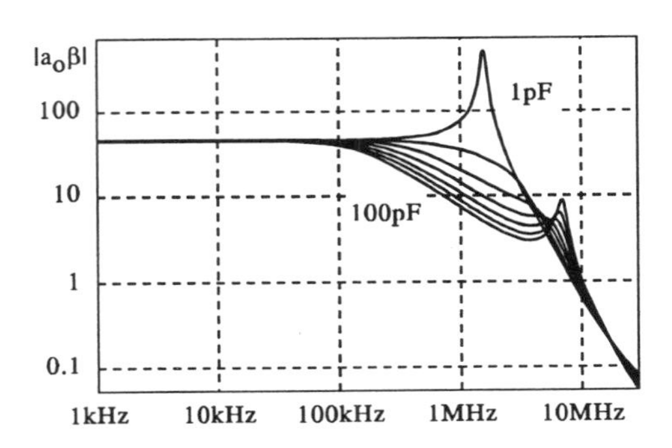

Fig. 21.18 Closed-loop response for Fig. 21.16(b). C_b=1, 2.2, 4.7, 10, 22, 47 and 100pF.

Fig. 21.18 shows the SpiceAge plots for the range of C_b roughly corresponding to cases (b) to (d), case (a) causing oscillation. Here the frequencies of the two peaks accord well with the above account, with the lower frequency peak for the lower C_b, and the higher frequency peak increasing slightly in frequency as C_b is increased.

Toning it down

While the response for C_b=2.2pF is stable, the resulting 1.5MHz cut-off is higher than required, leading to problems of instability resulting from unsuitable wiring layouts, etc. Remember that the design is for assembly and use by the inexperienced.

Suitable measures to reduce the cut-off to the required 150kHz are the connection of a 200pF capacitor across the load of Q_1 in Fig. 21.1 and 1nF capacitors across the R_f in Fig. 21.5. C_b can then be suitably increased to 100pF with the system still stable.

21.7 Overall linearity

Table 21.4 gives the PSpice analysis of the final open and closed-loop distortion, at the full 50W output. This is with a 1Hz sine wave and using the Transient Analysis.

With the open-loop THD=6.97% essentially that in Fig. 21.7(b) for the last stage, and with the zero-frequency loop gain a_L=600, then the closed-loop distortion should be less by $\times 1/A_L$, to give THD ~ 0.01%.

With the tabulated value approximately four times larger, this is possibly due to common-mode distortion in the LTP, which is not corrected by the negative feedback. As expected for the essentially anti-symmetric output stage, the dominant distortion components are those of the odd-order harmonics.

OPEN LOOP

FREQUENCY (HZ)	NORMALISED COMPONENT
1.000E+00	1.000E+00
2.000E+00	3.699E-03
3.000E+00	6.933E-02
4.000E+00	9.984E-04
5.000E+00	5.717E-03
6.000E+00	3.725E-04
7.000E+00	9.552E-04
8.000E+00	1.364E-04
9.000E+00	5.690E-05

THD = 6.968225E+00%

CLOSED LOOP

FREQUENCY (HZ)	NORMALISED COMPONENT
1.000E+00	1.000E+00
2.000E+00	2.188E-05
3.000E+00	3.981E-04
4.000E+00	6.067E-06
5.000E+00	4.571E-05
6.000E+00	5.014E-06
7.000E+00	2.022E-05
8.000E+00	6.068E-06
9.000E+00	3.989E-05

THD = 4.039658E-02 %

Table 21.4 Harmonic distortion components resulting from PSpice analysis for open- and closed-loop configurations.

Zobel impedance

The purpose of the Zobel impedance Z_z in Fig. 21.19 is to maintain stability when R_L is disconnected.

With the final-stage cut-off frequency $f_{4\text{-}7} \approx$ 300kHz as in Fig. 21.11, this is the frequency at which the local loop gain falls to unity. With r_o the output impedance of the final transistor, then at $f_{4\text{-}7}$ this is the output impedance of the local loop. With $r_o \sim 10\times R_L$ then when R_L is removed the main loop-gain is increased ~×10, usually causing oscillation.

With Z_z replacing R_L over this range then the oscillation is avoided. With C_z included then Z_z has negligible effect over the audio range. For the values shown

$$f_z = 1/(2\pi f C_z R_z)$$

gives **f_z=150kHz.**

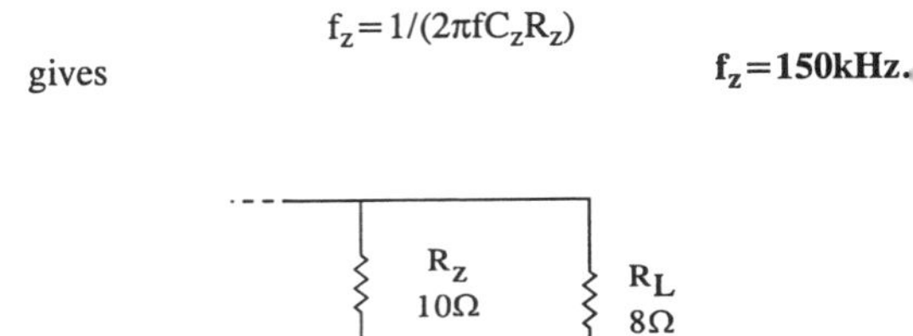

Fig. 21.19 R_ZC_Z Zobel impedance, R_L speaker load.

Inductive load

Z_z is also needed when the amplifier drives an inductive hearing-aid loop. With these usually of two-turns, there is a resonance with the wiring capacitance, normally at $f_o \sim$ 100kHz with $Q >> 1$. The author is hard of hearing and has had to deal with this on more than one occasion (see Exercise 21.8).

Alternative configurations

There is a wide variety of configurations used for the final stage of the audio power amplifier (Sinclair, 1998) and the more usual is where the final pair are in the form of a complementary emitter-follower (common base).

References

Massobrio, G. and Antognetti, P. 1993: *Semiconductor Device Modelling with SPICE.* McGraw Hill.
Sinclair, I. R. (ed.) 1998: *Audio & Hi-Fi Handbook.* Newnes.
Tuinenga, P. W. 1988: *SPICE a Guide to Circuit Simulation and Analysis using Using PSpice.* Prentice Hall.

21. 8 Exercises

Exercise 21.1 Peak load voltage and current

> **Calculate** the value of these for the output power 50W into the load $R_L = 8\Omega$.

With P_o the instantaneous output power then the quoted 50W is the mean $\overline{P_o}$. With the output voltage

$$V_o = \hat{V}_o \sin(\omega t) \quad \text{and} \quad P_o = V_o^2/R_L$$

then $P_o = \hat{P}_o \sin^2(\omega t) = (\hat{P}_o/2) \times (1 - \cos(2\omega ft)$

giving $\overline{P_o} = \hat{P}_o/2$.

With $\hat{P}_o = \hat{V}_o^2 / R_L$ then $\hat{V}_o = \sqrt{2\overline{P_o}R_L}$

With $\hat{P}_o = \hat{I}_o^2 R_L$ then $\hat{I}_o = \sqrt{2\overline{P_o} / R_L}$

giving **$\hat{V}_o = 28.3$V, $\hat{I}_o = 3.53$A.**

Exercise 21.2. Class B efficiency

> **Calculation** Show that the maximum efficiency η for the push-pull Class B power amplifier is 78.5%.

Output power As for the last exercise

$$\overline{P_o} = V_+^2/2R_L$$

Supply power With

$$\overline{P_{dc}} = \frac{2}{T}\int_0^{T/2} P_{dc}\,.dt\,,$$

$$P_{dc} = V_+ I_o \quad \text{and} \quad I_o = \hat{V}_o \sin(\omega t)/R_L$$

then $\overline{P_{dc}} = 2V_+\hat{V}_o / \pi R_L$

With $\eta = \overline{P_o} / \overline{P_{dc}}$ and $\hat{V}_o = V_+$

then for η maximum $\eta_{max} = \pi/4$

giving **$\eta_{max} = 78.5\%$.**

Exercise 21.3 Class A efficiency

> **Calculation** Show that the maximum efficiency η for the push-pull Class A power amplifier is 50%.

Output power From the last exercise

$$\overline{P_o} = \hat{P}_o/2 \quad \text{and} \quad \hat{P}_o = \hat{V}_o^2 / R_L\,.$$

With V_+ the supply voltage and $\hat{V}_o = V_+$ for maximum η

then $\overline{P_o} = V_+^2/2R_L$

Supply power With the upper supply

$$I_u = (V_+/R_L) \times (1 + \sin(2\pi ft))/2$$

then $\overline{I_u} = V_+/2R_L$.

With $\overline{P_u} = V_+ \times \overline{I_u}$ then $\overline{P_u} = V_+^2/2R_L$.

For both supplies together then

$$\overline{P_{dc}} = V_+^2/R_L$$

With $\eta = \overline{P_o} / \overline{P_{dc}}$ then **$\eta_{max} = 50\%$.**

Exercise 21.4 Thermal power

> **Calculate** the maximum transistor thermal power P_{th} for the amplifier of Fig. 21.1.

Over the conducting half-cycle $\overline{P_{th}} = \overline{P_{dc}} - \overline{P_o}$.

With $\overline{P_{dc}}$ and $\overline{P_o}$ as above then

$$\overline{P_{th}} = \frac{2V_+\hat{V}_o}{\pi R_L} - \frac{\hat{V}_o^2}{2R_L}\,.$$

For maximum P_{th} $\quad d\overline{P_{th}} / d\hat{V}_o = 0$

giving $\hat{V}_o = 2V_+/\pi$

for which $\overline{P_{th}} = 2V_+^2 / \pi^2 R_L$

With $V_+ = 30V$ and $R_L = 8\Omega$ then **$\overline{P_{th}}$ = 22.8W.**

Exercise 21.5 Thermal stability

> **Calculate** the magnitude of the loop gain A_L for the local feedback loop introduced by $R_e = 60\Omega$ in Fig. 21.5, at crossover and peak. R_1 and R_f omitted.

With A_L the local negative-feedback loop gain, with

$$|A_L| = g_{m4}R_e, \quad g_{m4} = I_{c4}/(kT/q),$$

and with $I_{c4} = I_{c6}/\beta_6$,

then $|A_L| = [(I_{c6}/\beta_6)/(kT/q)] \times R_e$.

Crossover With $I_{c6} = 50mA$ and $\beta_6 = 480$
then **$|A_L|$ (crossover) = 0.25.**

Peak With $I_{c6} = 3.53A$ and $\beta_6 = 80$
then **$|A_L|$ (peak) = 106.**

Exercise 21.6 Excessive quiescent current I_q

> **Calculation** For the circuit as above, calculate the effect of $\theta_p = 40°C$ on the quiescent $I_q = 50mA$.

Without R_e the effective $dV_{be}/d\theta = 2mV/°C$.

With $\Delta V_{be4} = 40°C \times 2mV/°$ then $\Delta V_{be4} = 80mV$.

With R_e then ΔV_{be} is reduced $\times 1/(1+A_L)$

With $A_L = 0.25$ then $\Delta V_{be4} = 64mV$.

With $I_{b4} \propto \exp(V_{be4}/(kT/q))$, $I_q \propto I_{c4}$,

and initially $I_q = 50mA$ then **$I_q = 645mA$.**

Exercise 21.7 Variable bias supply (Fig. 21.6(b))

> **Calculation. Loop gain** Derive the expression for the local loop gain a_L given
>
> $I_{c3} = 10mA$, $\beta_v = 200$, $V_v = 2\times 860mV$.

For $(R_1 + R_2) << r_o$

$$a_L \approx \beta_v R_2/(r_\pi + R_2) \tag{21.80}$$

giving $a_L \approx \beta_v$ for $R_2 >> r_\pi$

> **Calculation. Feedback resistor values** Derive suitable values for R_1 and R_2.

With the requirement $R_1//R_2 << V_{be}/I_b$

for negligible error due to I_b, and with $I_b = I_{c3}/\beta_v$,

then $R_1//R_2 << V_{be}\beta_v/I_{c3}$

giving $R_1//R_2 << 12k\Omega$.

With $R_2 >> (kT/q)/I_b$

for full loop gain then

$$R_2 >> (kT/q)/(I_{c3}/\beta_v)$$

giving $R_2 >> 500\Omega$.

We choose the geometric mean $(12k\Omega \times 500\Omega)^{1/2}$

giving **$R_2 = 2.5k\Omega$.**

With $V_v \approx V_{be}(R_1 + R_2)/R_2$ as in [21.12]

then $R_1 = R_2[(V_v/V_{be}) - 1]$

giving **$R_1 = 4.67k\Omega$.**

Exercise 21.8 Zobel impedance

> **Calculation** A hearing-aid loop in a public building has $R = 5\Omega$ and $L = 500\mu H$ and a shunt wiring capacitance of $C = 5nF$. Calculate the resonant frequency f_o and Q factor and thus show the need for the Zobel impedance even with the loop connected.

With the resonant frequency

$$f_o = 1/[2\pi(LC)^{1/2}]$$

then **$f_o = 101kHz$.**

With $Q = 2\pi f_o L/R$ then **$Q = 63.5$.**

With the impedance at resonance

$$Z = Q^2 R$$

then $Z = 20.1k\Omega$.

This is well in excess of $R = 8\Omega$ in a region where the open-loop gain of the main amplifier is well above unity. Thus the Zobel damping with f_z rather less than 100kHz is needed.

22

Transmission lines and transformers

Summary

A transmission line is essentially any pair of conductors conveying a signal from one location to another, with the following as examples.

- Twisted-pair telephone line.
- Coaxial cable.
- Ribbon cable, as frequently used to carry digital signals from one printed circuit board (PCB) to another.
- Tracks from one part of a PCB to another, with the ground-plane often the return.
- Tracks from one part of an on-chip design to another.

For a uniform structure such as coaxial cable the distributed series inductance L and shunt capacitance C combine to form a 'travelling wave structure'. Here a voltage step applied as in Fig. 22.1 travels down the line at a speed approaching that of light. Apart from the delay T_x, there is ideally no change in the wave reaching the load.

While such waves may travel in either direction, for the case shown there is only the 'forward wave', comprising the voltage and current components V_f and I_f. The line has a 'characteristic impedance' R_o such that for the forward wave $R_o=V_f/I_f$, with R_o normally in the approximate range from 30Ω to 600Ω.

Unless the load impedance R_L is equal to R_o, there is a reflection when the forward wave reaches the load, giving the 'reflected wave' V_r with I_r, as in Fig. 22.2 for $R_L=\infty$. Similarly, unless the source impedance $R_s=R_o$, there is a further reflection when the reflected wave reaches the source. For a high degree of 'mismatch' at each end of the line the resulting sequence of reflections can be as in Fig. 22.3(b). This is a point of particular importance for the digital circuit designer, with the time taken for the received signal to settle to the intended value a good deal increased from the inevitable T_x.

The above is the 'time-domain' view of the line function. Equally important is the response to a sine wave input, for example in the coupling the output of a radio transmitter to the antenna. Here for a load mismatch the forward and reflect waves interfere to give a 'standing-wave pattern', increasing the effect of losses due to the ohmic component of the line resistance.

A 'transmission line transformer' is one where the primary and secondary are in the form of the two conductors of a transmission line. With the load matched, this much increases the upper frequency limit set by the inter-winding capacitance of a conventional transformer. These are the topics of the chapter.

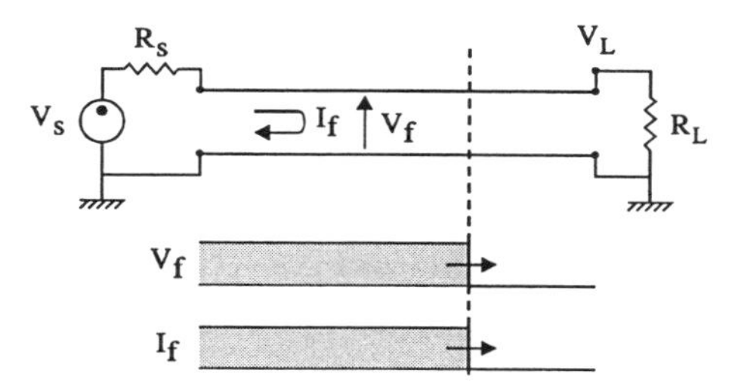

Fig. 22.1 Step travelling down line from source to load.

22.1 Line response to step input

With the forward and reflected waves as in Fig. 22.2, and with the components at the load end as in (a), then

$$V_f/I_f=R_o \ldots(a) \qquad V_r/I_r=R_o \ldots(b)$$
$$V_L=V_f+V_r \ldots(c) \qquad I_L=I_f-I_r \ldots(d) \qquad (22.1)$$

These the conventional symbols can be confusing. Be clear that 'L' signifies 'load' not 'line', and that 'o' signifies 'line characteristic impedance', not 'output'.

Reflection coefficient

With the load 'reflection coefficient'

$$\rho_L = V_r/V_f,$$

and with the above relations then

$$V_r = \rho_L V_f \ldots(a) \quad I_r = \rho_L i_f \ldots(b) \qquad (22.2)$$

giving

$$\rho_L = (R_L - R_o)/(R_L + R_o) \qquad (22.3)$$

and after the reflection,

$$V_L = V_f(1+\rho_L) \ldots(a) \quad I_L = I_f(1-\rho_L) \ldots(b) \quad (22.4)$$

With Fig. 22.2 as for $R_L = \infty$, Exercise 22.1 examines the effect of other values of R_L.

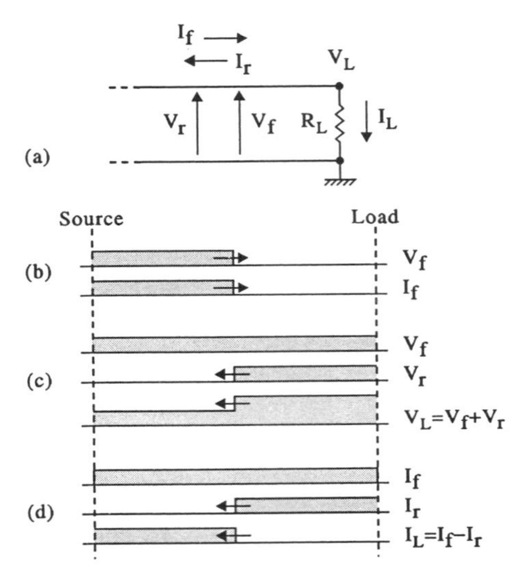

Fig. 22.2 Reflection at load for $R_L = \infty$.
(a) Voltage and current components.
(b) Forward wave before reflection.
(c) Voltages after reflection.
(d) Currents after reflection.

Multiple reflections

Fig. 22.3(b) to (d) show the V_L waveforms for R_L still infinite, and for R_s a few times less than R_o, equal to R_o, and a few times greater than R_o.

These compare with the waveforms in (e) to (g), for the circuit in (a), where the distributed line inductance and capacitance are each 'lumped' together into the single components L and C.

(e) to (g) are the waveforms of Fig. 12.16, which are for the same circuit, underdamped, critically damped, and overdamped.

The development of (b) to (d) is best understood by writing, debugging, and running the very short computer program describing the successive reflections at each end of the line.

This is detailed in Exercise 22.2 but you will learn much more by writing your own. Remember that V_f starts as

$$E_s \times R_o/(R_s + R_o),$$

not just as E_s. Here

- With [22.3] for the load reflection, this is also valid for that at the source, with R_s in place of the present $R_L = \infty$.
- For $R_s = 0$ the oscillatory transient of (b) becomes fully undamped, with the step bouncing between the ends of the line for ever.
- For $R_s = \infty$, i.e. an ideal current source, the stepped exponential of (c) becomes a never-ending staircase, as for a stepped integrator.

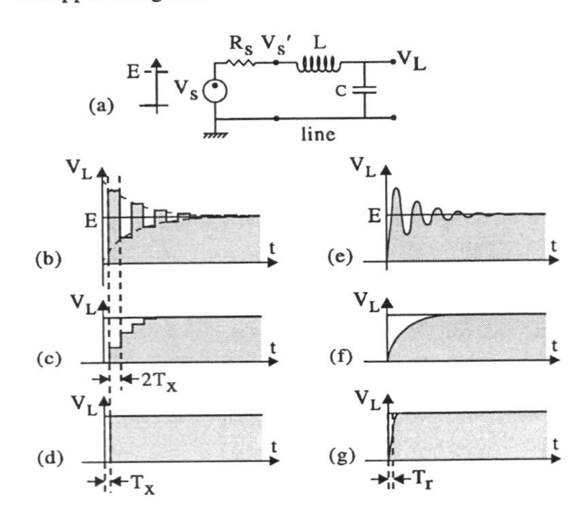

Fig. 22.3 Multiple reflections between source and load. $R_L = \infty$ as before, various R_s.
(a) Lumped equivalent LC network.
(b-d) Transmission line with $R_s \ll R_o$, $R_s \gg R_o$, $R_s = R_o$.
(e-g) As for (a), under-, over-, and critically-damped.
R_o = line impedance. T_x = line transit time.

22.2 Digital signal transmission

Long-distance transmission

Here, for just one source and one destination, the above principle is easily implemented. With the line correctly terminated at the load then there is no initial reflection and it becomes unnecessary for R_s to equal R_o. Indeed unless $R_s \ll R_o$ there will be a reduction in the level of the transmitted signal by $\times R_o/(R_s + R_o)$ which must be allowed for.

Matters become less straightforward when there are branching points, for the service of more than one receiving unit. Here ideally there needs to be a buffer stage presenting the matching impedance to the input line, and with a suitably low output impedance for the several output lines.

n-board transmission

ere the same principles apply, with the transmission line rmed by the PCB track and the common ground-plane.

With the concept of a relay stage at each branching oint impractical, and with the line impedance comparable ith the normal gate output impedance, matched termina- on at each gate input would give unacceptable attenuation ' the received signal. Thus the inevitable reflection from e gate input has to be accepted.

The obvious measure then is to make the lines as short possible, being careful with a multi-layer board to keep ose lines for which the response time needs to be short- t, closest to the ground plane.

With the terminations now just the gate input capaci- nce then once this has charged it is of less consequence the source impedance is no longer small compared with at of the line. Thus it can be advantageous to add to R_s including a series resistor of impedance comparable ith that of the line. While for a complex branching ructure or linear 'bus' this will not entirely eliminate condary reflections, the overall time for the signal to ttle should be reduced.

apacitive load

'e now consider the case in Fig. 22.4, where the source 'back-terminated' as above with $R_s=R_o$, and with the ad capacitance C_L the gate input capacitance $\mathbf{C_{in} \approx 3pF}$.

'ith $\rho_L=(R_L-R_o)/(R_L+R_o)$

in [22.3] for a resistive load, and now

$$Z_L(s)=1/sC_{in},$$

en
$$\rho_L(s)=\frac{s_o-s}{s_o+s} \qquad (22.5)$$

here
$$s_o=-1/C_{in}R_o \qquad (22.6)$$

'ith
$$V_L=V_f(1+\rho_L) \qquad \text{as in [22.4]}$$

en
$$V_L(s)=V_f(s)\times 2s_o/(s_o+s).$$

'ith $V_f=V_s/2$ for $R_s=R_L$ then

$$V_L(t)=V_s[1-\exp(-t/C_{in}R_o)] \qquad (22.7)$$

llowing also the time T_x taken for the step to reach the iven gate, then the V_L waveform becomes as in (d). his is exactly as if C_L had been connected directly to the urce, but with the delay T_x added.

eflected wave

'ith $V_r(s)=V_f(s)\rho_L(s)$

the load and with $\rho_L(s)$ essentially as in [22.5] for ρ_L then

$$V_r(t)=V_f(t)\times[1-2\exp(-t/CR_o)] \qquad (22.8)$$

initially giving $V_r=-V_f$ as for $R_L=0$,

and finally $V_r=V_f$ as for $R_L=\infty$.

With the essentially 100% reflection then, unless R_s is at least comparable with R_o, the above multiple reflections will result. With typically $\mathbf{R_s=30\Omega}$

for the HCMOS gate, if R_o can be made equal to R_s then the response time apart from T_x becomes $\mathbf{R_oC_L=90ps}$.

This is insignificant compared with the gate propagation time $\mathbf{T_p=10ns}$

so the main consideration becomes the delay T_x, and its possible multiplication by repeated reflections.

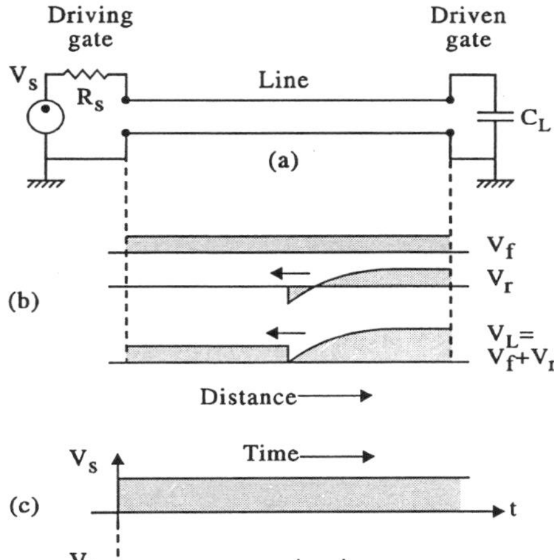

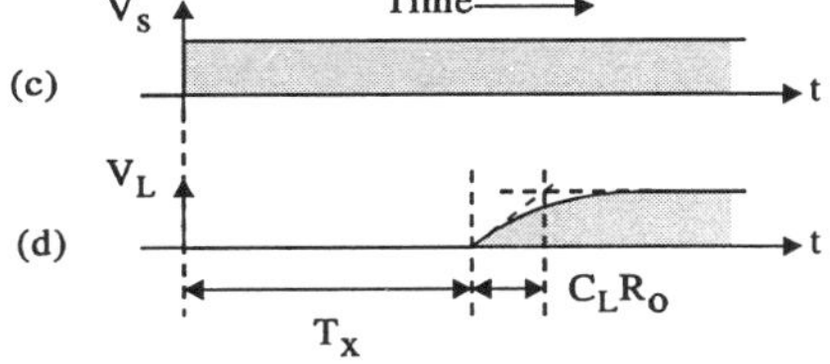

Fig. 22.4 Reflection from digital MOS logic gate.
(a) Source, line and load.
(b) Reflected wave.
(c) Source step waveform.
(d) Waveform at input of driven gate, response time C_LR_o.

Board-to-board transmission

Here the transmission line is either a ribbon cable or a printed circuit 'mother board'. With R_o~100Ω for both then (R_o ~ 30Ω) and R_s are comparable, causing the multiple reflections to decay relatively rapidly. Exercise 22.3 determines the time taken for the load voltage to settle to within the range for which the response of the load gate is guaranteed to be correct. This is shown to be $5T_x$, becoming $7T_x$ for a small noise margin.

Elimination of this extension of T_x by back termination is only of value for line lengths for which $7T_x$ is no longer small compared with the gate propagation time T_p. With

- x_L the length for $7T_x = T_p$,
- the velocity of propagation for ribbon cable measured as $\dot{x} = 180 \times 10^6$m/s,
- $T_p = 10$ns as above,

then **$x_L = 257$mm.**

With some board sizes comparable with this dimension then there can be a case for back termination for on-board transmission too.

PCB track impedance

With W the track width for standard 1/16" fibre-glass printed circuit board of $\varepsilon_r = 4.5$, the relation

$$R_o = -75.6 \times \log(0.074 \times W) \qquad (22.9)$$

fits experimental data reasonably well for W from 0.35mm to 4mm. However, for $R_o = R_s = 30\Omega$ then W needs to be 8mm. With this taking up far too much room on the board, then the more likely

W=0.5mm gives **$R_o = 108\Omega$.**

On-chip transmission

With the above $C_{in} = 3$pF mainly the integrated circuit pad capacitance then this component is eliminated for on-chip transfer.

With $c_{gs} = 0.2$pF as in Chapter 8, for the typical analog circuit MOS, with the digital device somewhat smaller, and with two in parallel for the CMOS gate, then $c_{in} = 0.1$pF is representative. With still $R_s = 30\Omega$ as above then **$c_{in}R_s = 3$ps.**

With the chip size sometimes over 10mm, x_L the line length, and allowing for transfer between the multi-layers on the chip, then $x_L = 15$mm is possible.

With the relative dielectric constant for $SiO_2 = 3.9$ then the wave velocity $\approx 150 \times 10^6$m/s. With the transit time $T_x = x_L / \dot{x}$ then **$T_x = 100$ps,**

giving T_x as the dominant value.

With c_{in} insignificant then the reflected wave is again much as in Fig. 22.2 for the load an open-circuit, still in principle requiring back-termination for multiple reflections to be avoided. However, with the connection again normally to several gate inputs, the only degree of back-termination practical is again the inherent output impedance of the driving gate.

Track impedance

With the rapidly advancing development in this area, this impedance can vary widely. For the now pedestrian W ~ 0.8μm, and with the SiO_2 thickness $t_{ox} \sim 25$nm the W >> t_{ox} giving $R_o << R_s$. Here the response is as in Fig 22.3(c), little different from the line capacitance plus c charging through R_s.

But with W comparable with t_{ox} now on target then R ~ R_s becomes feasible.

Track ohmic resistance

Also as W is reduced then the ohmic resistance R_Ω of th track becomes significant. This is primarily disadvanta geous in that it effectively increases the response time t $(R_s + R_\Omega) \times C$ where C is largely the track capacitance But for lower values the effect is advantageous in that tends to damp the repeated reflections that might other wise occur. The resistance of the layer-to-layer 'via' ir terconnection can also be significant.

It is probably true to say that for on-chip transmissio the need for the transmission line view is not yet s pressing as for the other cases considered.

22.3. Transmission line equations

To verify the above statements, and to give quantitativ precision, the requirements are to show

- that the applied signal travels down the line as in Fig 22.1,
- that the characteristic line impedance $R_o = \sqrt{\mathcal{L}/\mathcal{C}}$ where $\mathcal{L} = dL/dx$ and $\mathcal{C} = dC/dx$, with x the distanc along the line,
- that the speed at which the wave travels down the line i given by $dx/dt = 1/\sqrt{\mathcal{L}\mathcal{C}}$.

Characteristic impedance

Fig. 22.5 shows the last incremental section of the line terminated in the load R_L. With Z(s) the impedance pre sented then

$$Z(s) = s\delta L + R_o/(1 + s\delta C R_L) \qquad (22.10)$$

With $\mathcal{L}$ and $\mathcal{C}$ as above

$$Z(s) = s\mathcal{L}\,\delta x + R_L/(1 + s\mathcal{C}\,\delta x R_o) \qquad (22.11)$$

giving

$$Z(s) \approx R_L + s(\mathcal{L} - \mathcal{C} R_L^2)\delta x \qquad (22.12)$$

As $\delta x \to 0$ then obviously $Z(s) \to R_L$. However, for given length of line the number of sections $\propto 1/\delta x$. Thu for Z(s) to equal R_L for the segment of length equal to th initial δx as $\delta x \to 0$ requires that $R_L = \sqrt{\mathcal{L}/\mathcal{C}}$. With thi the characteristic impedance of the line then

$$R_o = \sqrt{\mathcal{L}/\mathcal{C}} \qquad (22.13)$$

Velocity of propagation

Here
$$V(x+\delta x,t+\delta t)=V_x(x,t)+(\partial V/\partial x).\delta x+(\partial V/\partial t).\delta t \quad (22.14)$$

With
$$\delta V=-\mathcal{L}.(\partial I/\partial t).\delta x$$

then
$$\partial V/\partial x=-\mathcal{L}.(\partial I/\partial t).$$

With also
$$V=I/R_o$$

then
$$V(x+\delta x,t+\delta t)=V(x,t)+(\partial I/\partial t)\times(-\mathcal{L}\delta x+R_o\delta t) \quad (22.15)$$

giving
$$V(x+\delta x,t+\delta t)=V(x,t)$$

or
$$\mathcal{L}\delta x=R_o\delta t.$$

Thus the wave travels at the speed $dx/dt=R_o/\mathcal{L}$,.

With R_o as in [22.13]

then
$$dx/dt=1/\sqrt{\mathcal{L}\mathcal{C}} \quad (22.16)$$

For a line with no dielectric, dx/dt is equal to the speed of light $c=300\times10^6$ m/s. For a dielectric this is reduced by the factor $1/\sqrt{\varepsilon}$ where ε is the relative dielectric constant.

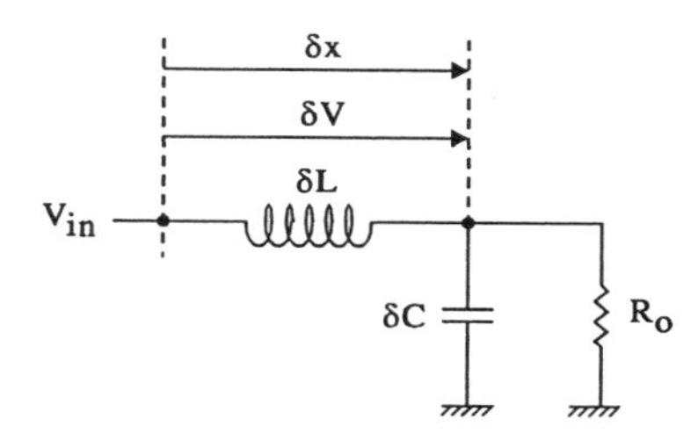

Fig. 22.5 Last incremental section of line with terminating load.

Line structures

Fig. 22.1 suggests a line in the form of a parallel twin cable. With this one of the forms used

$$R_o=120\frac{1}{\sqrt{\varepsilon}}\ln\left(w/d+\sqrt{(w/d)^2-1}\right)$$

$$\approx120\frac{1}{\sqrt{\varepsilon}}\ln\frac{2w}{d} \text{ for } w>>d \quad (22.17)$$

where d is the diameter of the wire, w the distance between line centres and ε is the relative dielectric constant of the surrounding media.

For the standard $R_o=300\Omega$ twin feeder the lines are about 1 cm apart and supported by a plastic web. The much more widely spaced $R_o=600\Omega$ 'open-wire' feeder has 15 cm insulating rod 'spreaders'. This tends to be used for a dipole antenna that is to be used at a wide variety of frequencies. With the resulting mismatch for most frequencies then 'standing waves' as in Fig. 22.7 occur.

Here 600Ω is the value R_o for which the resulting enhanced line ohmic losses are least. For coaxial cable $R_o=50\Omega$ is most common.

22.4 Sine wave transmission

Where the applied signal is a sine wave then the forward and reverse waves at time t and position x are given by

$$V_f(t,x)=V_f(t=0,\ x=0).\sin(\omega t-\beta x) \quad \ldots(a)$$
$$V_r(t,x)=V_r(t=0,\ x=0).\sin(\omega t+\beta x) \quad \ldots(b) \quad (22.18)$$

where ω is the signal frequency, $\beta=2\pi/\lambda$, λ is the wavelength, x=0 at the load, and x advances from source to load. Draw a sine wave and then move the paper to the left or to the right. This is what the waves are doing, only about 10^9 times faster, at about the speed of light.

Crank diagram

With ρ_L positive as for the resistive load $R_L>R_o$, then at t=0 the phasors at the load (x=0) are as in Fig. 22.6(a). As the point of observation is backed away from the load then V_f is advanced in phase as in (b), while V_r is retarded, still at t=0. Next (c) shows (b) but with V_f as the reference. For obvious reasons, this is termed a 'crank diagram'. With V the total line voltage V_r+V_f, the crank diagram shows how $|V|$ varies with distance from the load.

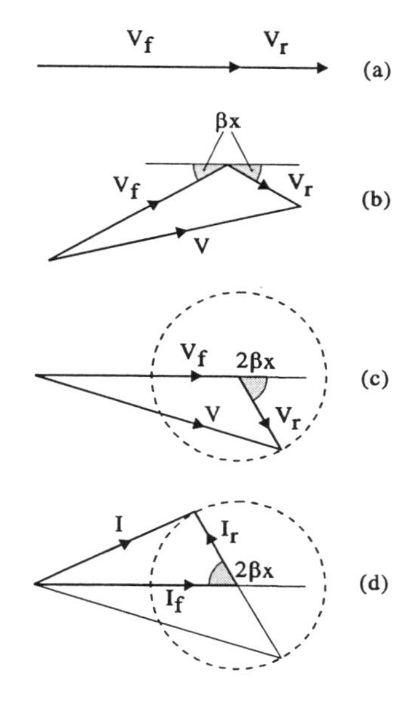

Fig. 22.6 Phasors for forward and reflected waves on line with $R_L>R_o$.
(a) At load.
(b) At a distance x before load.
(c) As (b) but with phase relative to V_f.
(d) As (c) but for current rather than voltage.

Complex reflection coefficient

With $\rho_L=(R_L-R_o)/(R_L+R_o)$

as in [22.3] for a resistive load R_L, and for the partly reactive Z_L, the now complex

$$\rho_L=(Z_L-R_o)/(Z_L+R_o) \qquad (22.19)$$

With V_f the reference phasor at the load, then the complex ρ_L means that V_r in Fig. 22.6(a) is no longer aligned with V_f. However, as one backs away from the load the crank-diagram develops as before. Whether the load is purely resistive or not then

$$|V|=|V_f|(1+|\rho_L|) \qquad \text{at the maxima}$$

and

$$|V|=|V_f|(|1-|\rho_L|) \qquad \text{at the minima.}$$

This leads to the concept of the 'voltage standing-wave ratio'

$$\text{VSWR} = (1+|\rho_L|)/(1-|\rho_L|) \qquad (22.20)$$

From Exercise 22.4, for $R_L=\infty$

$$|V|= 2|V_f|\cos(\beta x) \qquad (22.21)$$

as in Fig. 22.7, while, as also shown, for R_L a little higher than R_o

$$|V|\approx|V_f|(1+\rho_L\cos(2\beta x)) \qquad (22.22)$$

Standing and travelling waves

With no reflection for $R_L=R_o$, the wave is termed a 'travelling wave', with $|V|$ the same all along the line.

$R_L=\infty$, in contrast, gives a fully 'standing wave', with the peaks and troughs of Fig. 22.7, and these not moving along the line but standing still.

So far the line has been considered lossless. For a real line the conductors have a small ohmic resistance and the supporting dielectric a small effective conductance. With the resulting line power loss increasing sharply with VSWR then care is normally taken to match the load to the line, usually involving some kind of matching unit, as discussed later.

Current standing wave

With $I_L=I_f-I_r$, rather than $V_L=V_f+V_r$, then the current phasors corresponding to those of the voltage crank diagram in Fig. 22.6(c) become as in (d). Thus the current standing-wave patterns will be displaced by $\lambda/4$ relative to those in Fig. 22.7 for V, placing a current maximum at the load for $R_L=0$, and the expected current zero for $R_L=\infty$.

Presented impedance

With ρ_L as in [22.19] then

$$Z_L=R_o(1+\rho_L)/(1-\rho_L) \qquad (22.23)$$

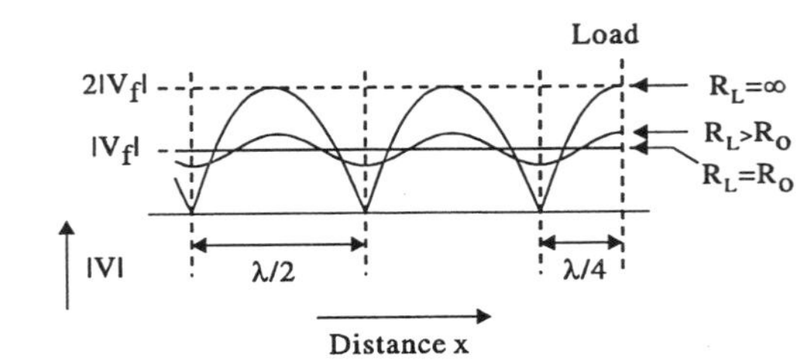

Fig. 22.7 Standing-wave pattern showing variation of $|V|$ with distance from load, where $V=V_f+V_r$.

Moving away from the load, as in Fig. 22.6, then the effective reflection coefficient ρ_x for the intervening li of length x and load is given by

$$\rho_x=\rho_L\exp(-2\beta x) \qquad (22.2$$

With Z_x the impedance presented by the line at the d tance x from the load then

$$Z_x=R_o(1+\rho_x)/(1-\rho_x) \qquad (22.2$$

With ρ_x as in [22.24] and ρ_L as in [22.19] then

$$Z_x = R_o\frac{Z_L + jR_o\tan(\beta x)}{jZ_L\tan(\beta x) + R_o} \qquad (22.2$$

Here $Z_x=R_o$ for the matched $Z_L=R_o$ as expected.

Line resonances

For $Z_L=0$ in [22.26]

$$Z_x=jR_o\tan(\beta x_L) \qquad (22.2$$

For the shorted line ($Z_L=0$) then, as shown in Exerc 22.5, Z_x varies with increasing distance x from the load follows.

(i) At $x=0$ then $Z_x=0$.

(ii) Z_x is inductive and increasing.

(iii) At $x=\lambda/4$, $Z_x=\infty$.

(iv) Z_x is capacitive and decreasing.

(v) At $x=\lambda/4$, $Z_x=0$.

Thereafter the sequence repeats. (iii) is akin to a paral resonance and (v) to a series. At either, as the frequen changes then so does λ, causing the frequency dependen to be much as usual for a resonator.

Tuning stubs

From the above, a length of line with a movable sh across it can be made to present any value of reac impedance. This can be used to match a line to an oth wise mismatched load as in Fig. 22.8. As one backs av from the load, the impedance Z_x may be considered a varying resistance R_p in parallel with a varying reacta X_p.

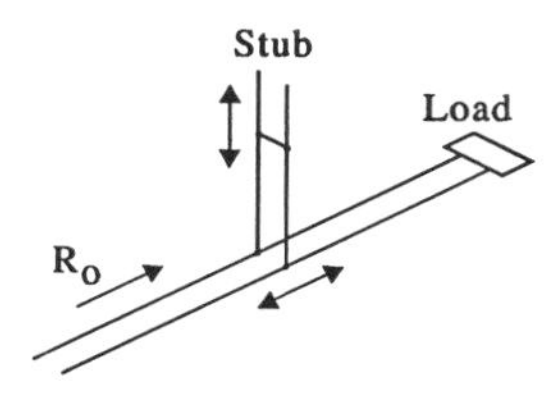

Fig. 22.8 Stub matching of load to line.

rom the voltage and current crank diagrams of Fig. 2.6(c,d), Z_x varies from

$$R_{max}=R_o(1+|\rho_L|)/(1-|\rho_L|$$

the maxima of $|V|$, to

$$R_{min}=R_o(1-|\rho_L|)/(1+|\rho_L|$$

the minima.

With $Y_x=G_x+B_x$ then G_x varies from G_{max} to G_{min},

here $\qquad G_{max}>G_o$ and $G_{min}<G_o$.

hus at some point $G_x=G_o$. With the stub placed there en B_x can be parallel resonated, leaving just the matchg G_o.

pen-end resonance

hould the section of the stub beyond the movable short be /4 then the resulting parasitic resonance can be coupled a the small inductance of the shorting link to draw ower and upset the match. The solution here is to conect a resistor of value R_o across the open end. Why isn't orting the ends as effective?

lternative arrangements

n open-ended stub can be used in much the same way, ith an extra $\lambda/4$ added to (or subtracted from) the stub ngth. However, this is less convenient to adjust.

Alternatively the adjustable stub can be placed in series ith the line. With $Z_x=R_x+B_x$, the stub is placed where $_x=R_o$. B_x is then series resonated with the stub impedce. Here, however, adjustment of position along the line far less convenient than for the parallel stub.

As shown in Exercise 22.6, [22.27] for the tuning stub actance can also be derived from the crank diagram.

2.5 Reflectometer

he reflectometer of Fig. 22.9 allows measurement of the nplitudes of the forward and reflected waves V_f and V_r, for the above matching process.

eflection coefficient and VSWR

/ith $\qquad VSWR=(1+|\rho_L|)/(1-|\rho_L|)$

as in [22.20] then VSWR can be calculated from the reflectometer reading. This is more convenient than the direct method of running a voltage or current probe along the line. Frequently the reflectometer display meter is calibrated in terms of VSWR rather than ρ_L, and is then termed a VSWR meter.

The direct VSWR measurement gives both the phase and magnitude of ρ_L, the magnitude from the measured voltages at the maxima and minima, and the phase from the placing of the maxima and minima relative to the load. In contrast, the reflectometer as shown only gives $|\rho_L|$, requiring the extension of Fig. 23.10 in the next chapter for $\phi(\rho_L)$ also.

Circuit

The transformer in (a) is implemented as in (b), with the line the a single-turn primary. With $I_2=I_1/n$ for the ideal transformer then the detector input voltages

$$V_{df}=\alpha_p(V_r+V_f)+(I_f-I_r)R/2n \quad ...(a)$$
$$V_{dr}=\alpha_p(V_r+V_f)-(I_f-I_r)R/2n \quad ...(b) \qquad (22.28)$$

where the potential divider $\quad \alpha_p=C_1/(C_1+C_2)$

For $\qquad \alpha_p=R/(2n)$

then $\quad V_{df}=2\alpha_pV_f$ ...(a) $\quad V_{dr}=2\alpha_pV_r$ ...(b) $\qquad$ (22.29)

as required. The two detector outputs are either displayed directly of processed to display the ratio $\propto|\rho_L|$.

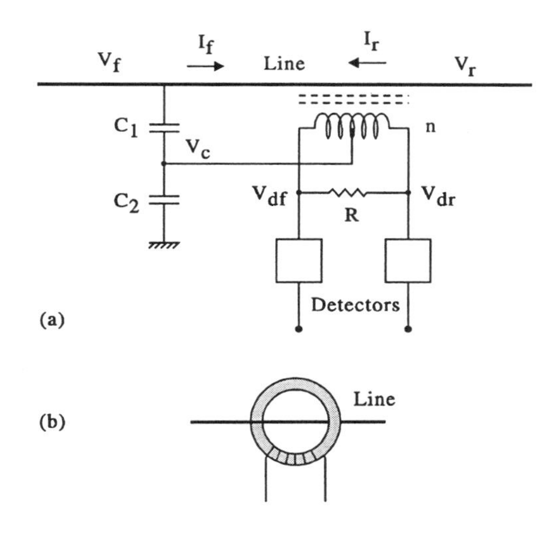

Fig. 22.9 Reflectometer based on toroidal current transformer.
(a) Circuit diagram.
(b) Transformer.

Insertion loss

With R' the impedance coupled into the line, and with α_{loss} the resulting signal loss factor then

$$\alpha_{loss}=2R_o/(R'+2R_o) \qquad (22.30)$$

requiring $R' << 2R_o$.

Low-frequency limit

With the impedance X_{L1} of the primary inductance L_1 in parallel with R' then the low-frequency limit f_{low} is where $|X_{l1}|=R'$, giving

$$f_{low}=R'/2\pi L_1 \qquad (22.31)$$

Potential divider

For a 100W transmitter $V_f \approx 70V$. With this excessive for the following circuitry, the potential divider provides suitable reduction, to a level suitably large relative to the 600mV needed for the detector diodes to conduct. Exercise 22.7 shows that for a 1% insertion loss ($\alpha_{loss}=0.99$) the required turns ratio n=5 and the $\alpha_p=0.05$. Also $f_{low}=141kHz$.

High-frequency limit

With f_{high} set by the core losses and the secondary winding capacitance then typically $f_{high}=50MHz$ giving

$$f_{high}/f_{low} \sim 300.$$

As below for the transmission line transformer, f_{high} can be increased to ~200MHz by using a grade of ferrite with lower losses. However, the lower relative permeability causes f_{low} to advance more rapidly, giving $f_{low} \sim 2MHz$ and thus the two-decade

$$f_{high}/f_{low}=100.$$

For yet higher frequencies the coaxial directional coupler of Fig. 23.11 is required.

Potential divider

The capacitive divider is preferred to the resistive equivalent because, for a given magnitude $|Z_p|$ of divider impedance Z_p, the signal loss is less. Exercise 22.8 shows that for $|Z_p|=10R_o$ and with α_{loss} the resulting signal loss factor then for the capacitive divider

$$\alpha_{loss} \approx 1-(1/8)(R_o/X_p)^2$$

giving $\alpha_p=1-1/800$, well below the target 1%. For the resistive divider $\alpha_{loss} \approx 1-R_o/2R_p$ giving $\alpha_{loss}=1-1/20$, a 5% loss. The exercise also considers the reason for the difference.

22.6 Balun

The 'balun' (balance-to-unbalance) is a device for allov ing a balanced load to be driven by an unbalanced sourc A common example is where a transmitter has to drive dipole antenna. The transmitter is in a grounded metal b while the antenna should be balanced relative to groun Fig. 22.10 shows the very simple device used, usually t or so turns (what will fit) wound round a ferrite ring. Th is also the basis of the transmission line transformer, b the balun provides a simpler introduction.

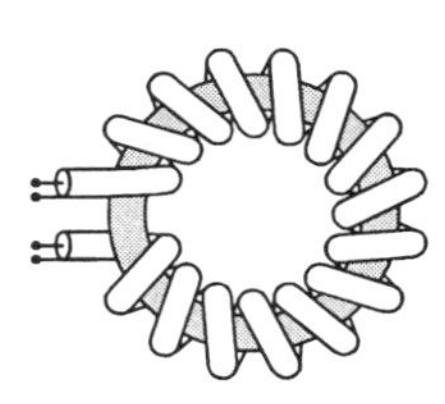

Fig. 22.10 Overwound toroid used in transmission line transformer and balun.

Fig. 22.11(a) shows a balanced source driving a balanc load, while (b) shows the present problem, where balun is needed. In the absence of line inductance lower $V_L=0$.

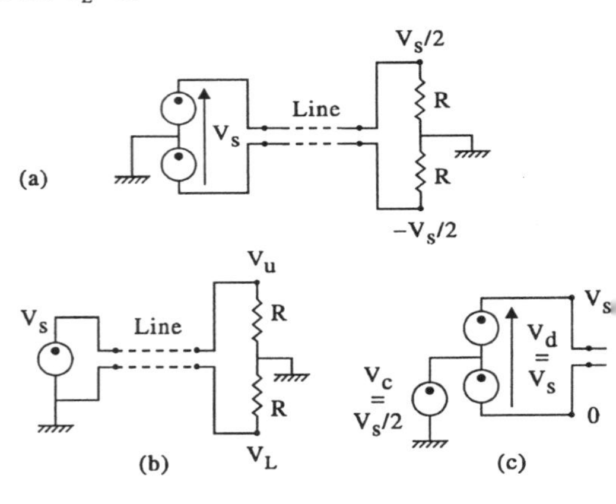

Fig. 22.11 Driving of balanced load.
(a) Balanced source correctly driving balanced load.
(b) Unbalanced source incorrectly driving balanced load.
(c) Resolution of source voltage V_s in (b) into common-mode a difference-mode components.

Here a conventional transformer with split seconda could be used. The reason why the balun is preferred that for the conventional transformer the range of oper ing frequencies is more limited. With the lower limit by the need for the inductive impedance to be $>>R_o$, upper limit is set by the winding capacitances. For balun and the transmission line transformer these are p of the line structure. Thus the typically two-decade ran of the conventional transformer is increased to typically three decades.

To understand the operation of the balun it is best to consider the source voltage in terms of its common-mode and difference-mode components as in (c). The effects of these are then considered separately.

V_d applied, $V_c=0$

Here the source becomes fully balanced as in (a), giving, apart from the line delay,

$$V_{ud}=V_s/2 \ldots(a) \qquad V_{Ld}=-V_s/2\ldots(b) \qquad (22.32)$$

where u: upper, L: lower, d: differential input. The fact that the line is wound on the toroid makes no difference. The cable is coaxial and the currents on inner and outer are equal and opposite. Thus the associated magnetic field is confined to within the cable.

V_c applied, $V_d=0$

With the line currents here equal and in the same direction then the line behaves much as a single wire wound round the core and terminated in R/2. With X_L the resulting line inductance and c: common-mode input then

$$V_{uc}=V_{Lc}=(V_s/2)R/(R+X_L) \qquad (22.33)$$

With the imbalance

$$V_{uc}/V_{ud}=R/(R+X_L) \qquad (22.34)$$

and with $R_L=R_o$ for the matched load, then we require $X_L >> R$. It is this which sets the lower operating frequency, as discussed below for use of the device as a transmission line transformer.

The inductance of a long transmission line will have something of the same effect, but with typically a considerably lower X_L than is afforded by the toroid.

22.7 Transmission line transformer

Fig. 22.12(a) shows the above balun connected as a transmission line transformer, with the line stretched out for clarity. Here both source and load are unbalanced and the function of the transformer is to give the 2:1 voltage step-up with the associated 4:1 impedance step-up.

Stated simply, the line transmits the voltage V_s to appear across its far end at B. This is added to the original V_s by the connection to point A, giving $V_L=2V_s$.

Apart from the line inductance L, the connection of A to B shorts the signal source, violating the above argument. But with L allowed for, and with the ends of the line brought together as in Fig. 22.10, then it is only the impedance of the connection from A to B which is of zero impedance. With L the inductance for each line, then for the inner conductor

$$V_L=V_s-j\omega L(I_1-I_2) \qquad (22.35)$$

while for the conducting sheath

$$V_s=-j\omega L((I_1-I_2) \qquad (22.36)$$

confirming

$$V_l=2V_s \qquad (22.37)$$

Thus

$$I_1=2V_s/R_L \quad \ldots(a)$$

$$I_2=V_s\times(1/j\omega L+2/R_L) \quad \ldots(b) \qquad (22.38)$$

giving $I_1=I_2$ for $X_L=\infty$, and thus $I_s=2I_L$ (22.39)

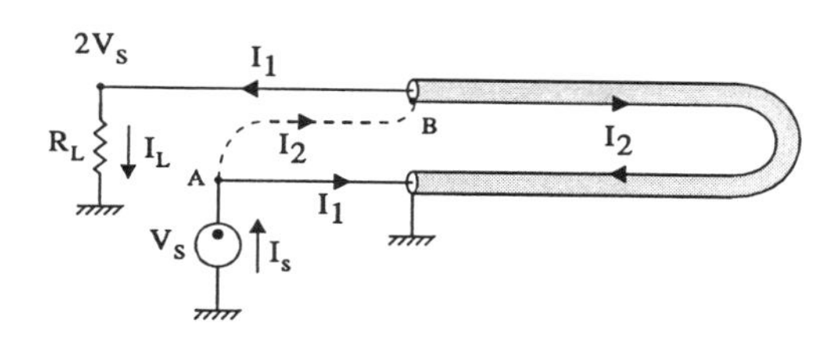

Fig. 22.12 Diagrammatic representation of use of overwound toroid in Fig. 22.10 as a 2:1 step-up transformer.

With $I_s=2I_L$ and $V_L=2V_s$ then the load R_a presented to the source is $R_L/4$. Thus the 2:1 voltage step-up and the 4:1 impedance step-up are confirmed. But avoidance of reflection at the line ends requires $V_s/I_1=R_o$ giving

$$R_L=2R_o \qquad (22.40)$$

and

$$R_a=R_o/2 \qquad (22.41)$$

Example A 50Ω line will be correctly matched for $R_s=25\Omega$ and $R_L=100\Omega$.

The device can clearly be reversed to give a 4:1 impedance step-down. Also Exercise 22.9 shows how it can be reconfigured for a balanced load.

Operating frequency limits

With ω as in [22.38](b) and with f_{low} the low-frequency limit for the transformer then

$$f_{low}=R_o/(2\pi L) \qquad (22.42)$$

The above analysis ignores the time taken for the signal to travel down the line. Taking f_{high} as the frequency for which the delay gives a 45° phase lag, and with the line length ≈30cm for ten turns round a ring of 4cm diameter, then Exercise 22.10 gives **$f_{high}\approx$ 75MHz**.

This value is lowered if the line is not correctly terminated, due to the resulting multiple reflections. With f_{low} as in [22.42], Exercise 22.11) gives **$f_{low}\approx$ 70kHz**

thus confirming the three-decade range.

The core losses also increase with frequency, so to move the range upwards the following measures are needed.

- Scale the toroid down $\times\alpha$.
- Reduce the number of turns N.
- Use a grade of ferrite having lower high-frequency losses.

Scaling

With $L \propto \alpha$ then $f_{low} \propto 1/\alpha$. With the line length $\propto\alpha$ then $f_{hich} \propto 1/\alpha$, leaving the three-decade f_{high}/f_{low} unaltered.

Number of turns

With the line length $\propto$N then

$$f_{high} \propto 1/N,$$

but with $\quad f_{low} \propto 1/L \quad \text{and} \quad L \propto N_2$

then $\quad f_{low} \propto 1/N^2.$

Thus as N is reduced then f_{low} advances more rapidly than f_{high}, closing the range.

Ferrite

For the lower loss ferrite μ_r is lower also, thus increasing f_{low} further. Typically for

$$f_{high} = 200\text{MHz}, \quad \text{then} \quad f_{low} = 2\text{MHz},$$

a two-decade range.

There are two types of ferrite, the present Nickel-Zinc, and also manganese-zinc. For the

$$\text{Ni-Zn } \mu_r \sim 500$$

while for $\quad \text{Mn-Zn } \mu_r \sim 1500$

However, the high-frequency losses of the Mn-Zn are much higher, making this only suitable for low-frequency applications like the core in the switched-mode power supply of Section 5.6.

Alternative line types

At least two alternatives exist to the coaxial cable used above for winding the toroid. A twin pair, twisted or otherwise, works well where the required line impedance R_o is larger than the typical 50Ω for coax. Conversely, parallel strip lines can be used where a lower R_o is needed.

Higher ratio configurations

Fig. 22.13(a) represents the 2:1 line transformer of Fig. 22.12 in a manner more like that of the conventional transformer.

With R_L disconnected, V_s drives the current V_s/X_L through the lower winding. The resulting core flux Φ develops the voltage V_s across both windings. Thus the output is $2V_s$ as shown.

When R_L is connected, equal and opposite additional currents I_w of magnitude $2V_s/R_L$ flow in the two windings. This leaves Φ unaltered and so the output is still $2V_s$. Also with the two I_w equal then

$$I_s = 2I_w.$$

Thus $\quad V_L = 2V_s \quad \text{and} \quad I_L = I_s/2$

to give the expected 4:1 impedance transformation ratio.

It is also possible to wind a bunch of three wires around the toroid. This can then be connected as in (b) to give the 3:1 step-up. The principle can be extended further, but 4:1 appears to be the normal limit.

For (b), the load current $3V_s/R_L$ flows through each of the windings in turn, to ground. To cancel the resulting flux, a current of three times this value needs to flow through the lowest winding, i.e. $I_s = 3 \times 3V_s/R_L$, giving $R_{in} = R_L/9$ as expected.

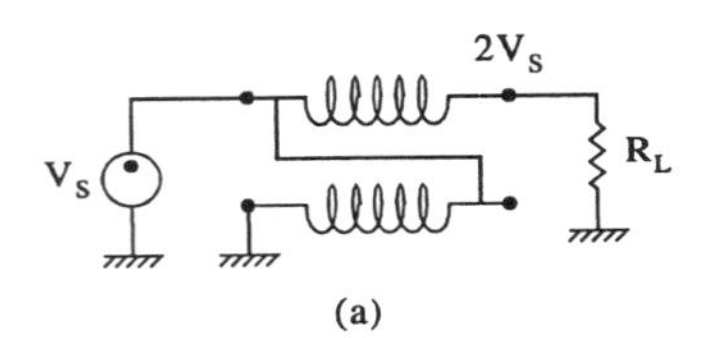

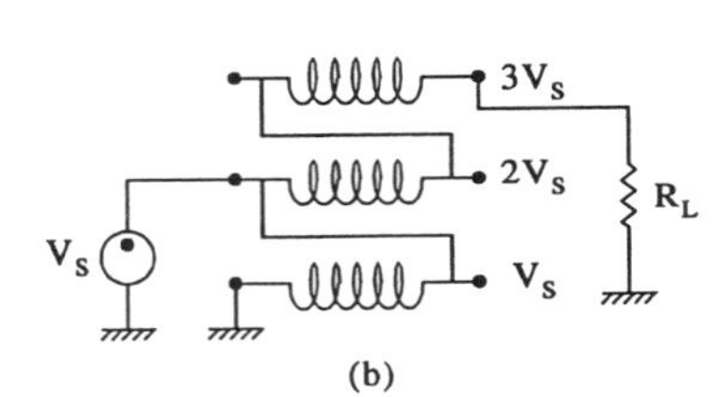

Fig. 22.13 Transformer style representation of transmission line transformers.
(a) 2:1 step-up of Fig. 22.12.
(b) Trifilar wound transformer giving 3:1 voltage step-up and 9:1 impedance step-up.

References

Jessop, J. P. 1985: *Radio Data Reference Book*. Radi[o] Society of Great Britain.
Large amounts of useful factual information, particular[ly] for RF and transmission line circuits and ferrites.
Sevick, J. 1987: *Transmission Line Transformers*. Amer[i]can Radio Relay League.
Short presentation of theory followed by many exper[i]mental systems and results obtained by the author.

22.8 Exercises

Exercise 22.1 Line reflections (manual)

> **Redraw** Fig. 22.2(b-d) for
> $R_L=0$, ΔR_o, $R_o(1-\Delta)$, R_o, $R_o(1+\Delta)$, ∞,
> where $\Delta << 1$.

$\mathbf{R_L=0}$ With $\rho_L=(R_L-R_o)/(R_L+R_o)$

as in [22.3] then $\mathbf{\rho_L=-1}$.

With $V_r=\rho_L V_f$ and $I_r=\rho_L I_f$

as in [22.2] then $\mathbf{V_r=-V_f,\ I_r=-I_f}$.

With $V_L=V_f(1+\rho_L)$ and $I_L=I_f(1-\rho_L)$

as in [22.4] then $\mathbf{V_L=0,\ I_L=2I_f}$.

$\mathbf{R_L=\Delta R_o}$ With $\rho_L=(R_L-R_o)/(R_L+R_o)$

as in [22.3] then $\mathbf{\rho_L=(\Delta-1)/(\Delta+1)\approx-1+2\Delta}$.

Thus $\mathbf{V_r=V_f\times(-1+2\Delta),\ I_r=I_f\times(-1+2\Delta)}$.

With $V_L=V_f(1+\rho_L)$ and $I_L=I_f(1-\rho_L)$

as in [22.4] then $\mathbf{V_L=V_f\times 2\Delta,\ I_r=I_f\times(2-2\Delta)}$.

Following this pattern

R_L	ρ_L	V_L/V_f	I_L/I_f
0	−1	0	2
ΔR_o	$-1+2\Delta$	2Δ	$2-2\Delta$
$R_o-\Delta R_o$	-2Δ	$1-2\Delta$	$1+2\Delta$
R_o	0	1	1
$R_o+\Delta R_o$	2Δ	$1+2\Delta$	$1-2\Delta$
∞	1	2	0

from which the waveforms may be redrawn.

Exercise 22.2 Line reflections (computer program)

> **Confirm** the transients of Fig. 22.3(b) and (e) for $R_s=0.25\times R_o$, and $R_s=4\times R_o$ with the initial V_s step of height $E=1V$.
>
> Then extend to $R_s=0$ and $R_s=\infty$.

$\mathbf{R_s=0.25\times R_o}$
Reflection coefficients: with

$\rho_L=(R_L-R_o)/(R_L+R_o)$ as in [22.3]

but here for reflection at the source then

$\rho_s=(0.25-1)/(0.25+1)$ giving $\mathbf{\rho_s=-0.6}$

With $R_L=\infty$ then $\rho_L=(\infty-1)/(\infty+1)$ giving $\mathbf{\rho_L=1}$

General algorithm for reflection at load

With $V_L\Leftarrow V_L+V_f+\rho_L V_f$ then $V_L\Leftarrow V_L+2V_f$
With $V_r=\rho_L V_f$ then $V_r=V_f$.

General algorithm for reflection at source

With $V_f\Leftarrow\rho_s V_r$ then $V_f\Leftarrow\rho_L V_f$.

Program sequence

Initial launch: with $V_f=ER_o/(R_o+R_s)$ then
$V_f=V\times R_o/(R_o+0.25\times R_o)$ giving $\mathbf{V_f=0.8V}$

Reflection at load: $V_L\Leftarrow V_L+2V_f$ giving
$V_L=0+2\times0.8V$ and thus $\mathbf{V_L=1.6V}$

Reflection at source: $V_f\Leftarrow\rho_s V_f$ giving
$V_f=-0.6\times0.8V$ and thus $V_f=-0.48V$.

Reflection at load: $V_L\Leftarrow V_L+2V_f$ giving
$V_L=1.6-2\times0.48V$ and thus $\mathbf{V_L=0.64V}$

Reflection at source: $V_f\Leftarrow\rho_s V_f$ giving
$V_f=0.6\times0.48V$ and thus $V_f=0.288V$.

Reflection at load: $V_L\Leftarrow V_L+2V_f$ giving
$V_L=0.64+2\times0.288V$ and thus $\mathbf{V_L=1.22V}$

and so on as below, giving the oscillatory sequence of Fig. 22.3(b).

V_L	0	1.6	0.64	1.22	0.870
→	1.08	0.953	1.028	0.983	1.01

$\mathbf{R_s=4\times R_o}$ As before but with the following changes

Source reflection coefficient: $\rho_s=(R_s-R_o)/(R_s+R_o)$
giving $\rho_s=(4-1)/(4+1)$ and thus $\mathbf{\rho_s=+0.6}$.

Initial launch: $V_f=ER_o/(R_o+R_s)$ giving
$V_f=1V\times R_o/(R_o+4\times R_o)$ and thus $\mathbf{V_f=0.2V}$.

With the results as below, this gives the stepped exponential sequence of Fig. 22.3(c).

V_L	0	0.4	0.64	0.784	0.870
→	0.922	0.953	0.972	0.983	0.99

$\mathbf{R_s=0}$ $V_L=0$, 2V, 0, 2V..., a raised square wave.

$\mathbf{R_s=\infty}$ (current source I_s). $V_L=0$, $2I_sR_o$, $4I_sR_o$, $6I_sR_o$... a staircase.

Exercise 22.3 Repeated reflections for ribbon-cable connection (Fig. 22.4)

Calculation For the HCMOS gate the output impedance $R_s=30\Omega$, and for the ribbon cable the line impedance $R_o=100\Omega$. With T_x the transit time for one pass along the line, calculate the number of passes that must elapse for a rising transition to be sure of registering the correct logic state.

For $R_s < R_o$ then the V_L waveform becomes the damped oscillatory transient of Fig. 22.3(b).

From Table 20.1 for the logic gate, E_s is guaranteed to be above 4.2V, while for the driven gate to correctly register logic 1 requires the load voltage $V_L > 3.6$V.

From the program of the last exercise, the lower peaks of the transient for are 2.98V, 3.85V, 4.10V, 4.17V, 4.19V..., etc., to the final 4.2V.

Thus only the first lower peak is below the required 3.6V, with the time needed therefore $5\times T_x$.

Problem A bus is driven at one end with a source that is correctly back terminated. The bus serves ten equally spaced gate inputs, each of negligible input capacitance. With T_x the transit time from one end of the bus to the other, what multiple of T_x will be needed for the gate closest to the source to respond correctly? (Nearly two).

What will be the voltages applied to the gate inputs for no back termination and with $R_s=0$?

Exercise 22.4. Crank diagram

Demonstration From the crank diagram in Fig. 22.6, and for $|V|$ as in Fig. 22.7 show

$$|V|\approx|V_f|(1+\rho_L\cos(2\beta x))$$

for R_L slightly greater than R_o, while for $R_L=\infty$.

$$|V| = 2|V_f|\cos(\beta x)$$

The results follow simply from Fig. 22.14.

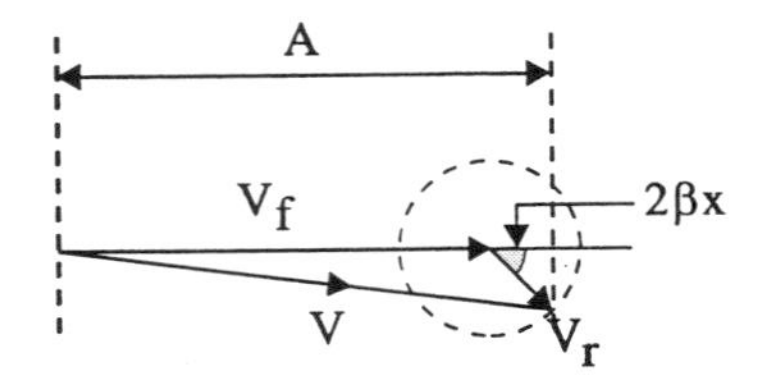

Fig. 22.14(a) Line crank diagram for R_L slightly > R_o.

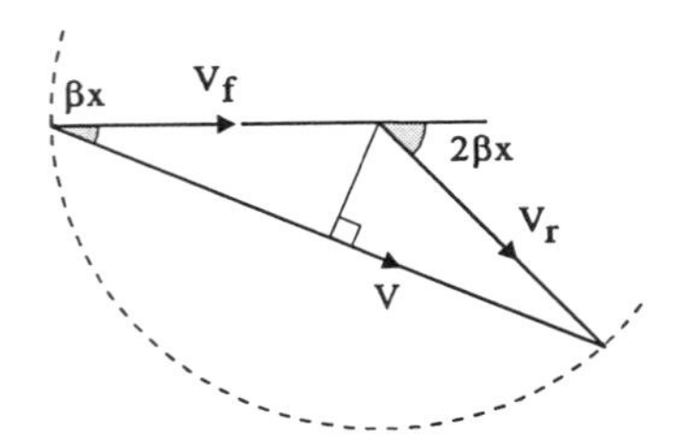

Fig. 22.14(b) Line crank diagram for $R_L=\infty$.

Exercise 22.5 Standing waves

Drawing From the crank diagrams of Fig. 22.6, draw the variation of $|I|$ and $|V|$ with x for $R_L=0$ (shorted line). Thus mark the points where the impedance $Z=V/I$ is zero and infinite.

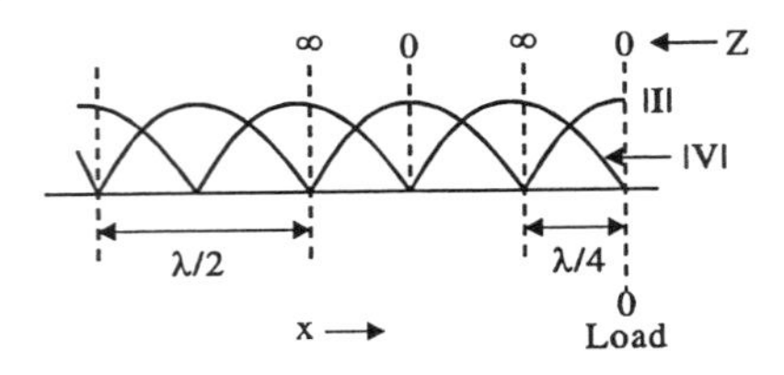

Fig. 22.15 Line voltage and current waves for $R_L=0$.

Exercise 22.6 Tuning stub impedances

Demonstration From the crank diagram of Fig. 22.6 show that, for $R_L=0$ then I and V are in exact quadrature at any point on the line, with the resulting impedance purely inductive as x is first increased from zero, to become purely capacitive after the first maximum, and so on. Thus show that for $R_L=0$ then $Z=jR_o\tan(\beta x)$ where $Z=V/I$.

With $\phi=90^o-\beta$ in Fig. 22.12 then I and V are in quadrature. From Fig. 22.16

$$|V|=2|V_f|\sin(\beta x) \quad \text{and} \quad |I|=2|I_f|\cos(\beta x)$$

giving

$$|Z|=R_o\tan(\beta x).$$

With the phasing shown then, $Z=jR_o\tan(\beta x)$.

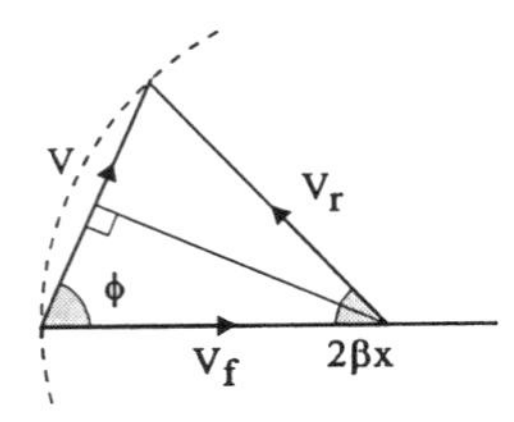

Fig. 22.16(a) Line crank diagrams for $R_L=0$, voltage phasors.

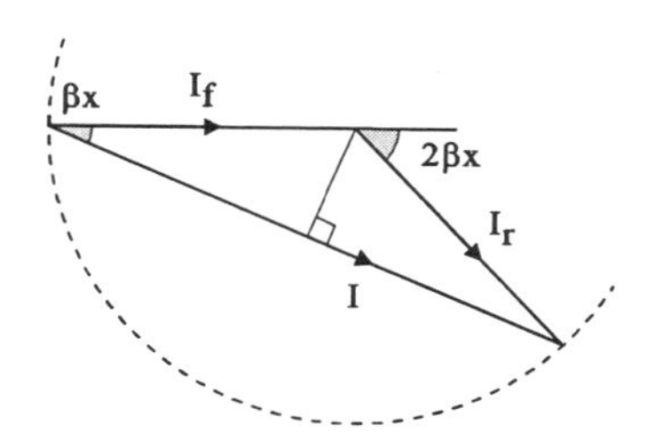

Fig. 22.16(b) Line crank diagrams for $R_L=0$, current phasors.

Exercise 22.7 Reflectometer parameters (Fig. 22.9)

Calculation For a 100W transmitter and a 1% insertion loss, calculate values of

- the transformer turns ratio n,
- the value of R,
- the potential division factor α_p,
- the low-frequency cut-off f_{low}.

The required detector voltage V_{df} is 7V.

For the ferrite ring used: $\mu_r=400$; diameter d=2.5cm; near square cross-section 7.5×7.5mm; permeability of free space $\mu_o=4\pi/10^{-7}$. The line impedance $R_o=50\Omega$.

Turns ration With $\alpha=0.01$ the factor by which V_L is reduced then

$$\alpha \approx R'/2R_o$$

giving $R'=1\Omega$

With the transmitter power $P=I_f^2R_o$ then $I_f=1.41A$.

With V_r' the voltage across R'

then $V_r'=I_fR'$
giving $V_r'=1.41V$.

With the total $V_{df}=2\times(n/2)\times V_r'$

then n=4.96 requring **n=5.**

Value of R With $R=n^2R'$
then **R=25Ω.**

Potential division factor With

$$V_{df}=2\times(n/2)V_r'$$

then $V_{df}=7.05V$.

With $P=V_f^2/R_o$ and $V_{df}=2\alpha_pV_f$ then

$$\alpha_p=V_{df}/[2(PR_o)^{1/2}]$$

giving $\mathbf{\alpha_p=0.05.}$

Low-frequency cut-off With the ring inductance

$$L=\mu_o\mu_rN^2w^2/d \quad \text{as in [1.20]}$$

then the primary $\mathbf{L_1=1.13\mu H.}$

With $f_{low}=R'/2\pi L$ as in [22.31] then $\mathbf{f_{low}=141kHz.}$

Exercise 22.8. Reflectometer potential divider (Fig. 22.9)

Calculation With

- $|Z_p|=10R_o$ where Z_p is the impedance of the potential divider,
- α_{loss} the signal loss factor due to the connection of the divider,
- the final load impedance $R_L=R_o$,
- Z_p first two capacitors and then two resistors, calculate

(i) the magnitude of the reflection coefficient ρ_p as a result of the connection of the potential divider without the transformer,
(ii) the resulting signal loss factor α_{loss}.

Reflection coefficient With

$$\rho_L=(Z_L-R_o)/(Z_L+R_o)$$

as in [22.19] for a load impedance Z_L, and with

$$Y_L'=Z_p//R_o$$

then $\rho_p=-Y_p/(2G_o+Y_p)$.

With $Y_p \ll G_o$ then

$$\rho_p \approx -Y_p/2G_o=-R_o/2Z_p$$

giving $|\rho_L|=R_o/(2|Z_p|)$
and thus $\mathbf{|\rho_p|=0.05.}$

Signal loss With source and load both matched to the line then the impedance presented to the divider is $R_o/2$.
With α_{loss} the signal loss factor then

$$\alpha_{loss}=Z_p/[(Z_p+(R_o/2)].$$

Capacitive divider With $Z_p=X_p$ then

$$|\alpha_{loss}|=|X_p|/[(|X_p|^2+(R_o/2)^2]^{1/2}.$$

With $|X_p| \gg R_o$

then $|\alpha_{loss}| \approx 1-(1/8)(R_o/|X_p|)^2$
giving $\mathbf{\alpha_{loss}=1-1/800.}$

Resistive divider With $Z_p=R_p$ then

$$\alpha_{loss}=R_p/[(R_p+(R_o/2)]$$

giving $\mathbf{\alpha_{loss}=1-1/20.}$

With the forward and back scattering approximately equal, this is consistent with $|\rho_p|=0.05$.

With the scattered waves in quadrature with the incident for the capacitive divider then the main effect at the final load is a phase shift.

Exercise 22.9 Impedance matching balun

> **Demonstration** Show how the 2:1 line transformer of Fig. 22.25 can be connected to give balun action but with a 4:1 impedance step-up ratio.

With the connection is as in Fig. 22.17, and with R_L' the impedance presented to the source, then

$$R_L'=V_s/2I_L$$

With the load voltages and currents as shown then

$$R_L=2V_s/I_L, \quad \text{giving} \quad R_L'=R_L/4.$$

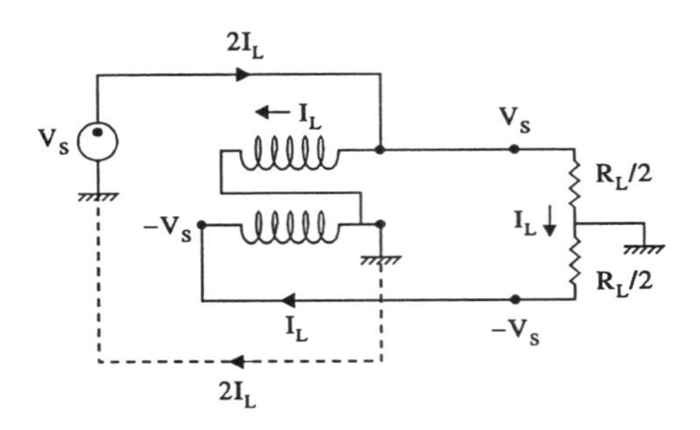

Fig. 22.17 Use of 2:1 line transformer as balun giving 4:1 impedance transformation.

Exercise 22.10 Transmission line transformer high-frequency cut-off due to line delay

> **Calculate** the frequency f_d at which the delay is sufficient to significantly violate the results of [22.35]ff. Assume coaxial line with polythene dielectric for which the relative permittivity $\varepsilon_r=2.7$. The velocity of light in free-space $=300\times10^6$m/s. For ten turns wound round a ferrite ring of 2.5cm diameter and near square cross-section $w=7.5\text{mm}\times7.5\text{mm}$ then the typical line-length $x=30$cm.

For a signal of frequency f and line wavelength λ

$$c\varepsilon^{-1/2}=\lambda f.$$

With the limiting phase shift ϕ_d and the corresponding $\lambda=\lambda_d$ then $\phi_d=2\pi x/\lambda_d$ giving

$$f_d=\phi_d c\varepsilon^{-1/2}/(2\pi x)$$

With $\phi_d=45^o$ then **$f_d=76.1$MHz.**

Exercise 22.11 Transmission line transformer low-frequency cut-off flow

> **Calculate** the low-frequency cut-off for the transformer in the above example. The line impedance $R_o=50\Omega$.
>
> For the ferrite used $\mu_r=400$ with the permeability of free space $\mu_o=4\pi/10^{-7}$.

With $L=\mu_o\mu_r N^2w^2/d$ as in [1.20]

then $L=113\mu$H

With $f_{low}=R_o/(2\pi L)$ as in [22.42]

then **$f_{low}=70.4$kHz**

23

S-parameters and the Smith chart

Summary

Vith the hybrid-π model of the bipolar transistor in hapter 6 ceasing to be accurate beyond ~30MHz, the evice, whether bipolar or other, will then be specified by set of four experimentally determined parameters, with ich given as a function of frequency. There are a number f possible parameter sets, z-, y-, h-, etc., all sound in rinciple. However, that most suited for measurement and igh-frequency application is the s-parameter set (s: scatring). Here the input and output components are considred as travelling waves on the transmission lines couing the signal to the input and output of the transistor.

With the coupling of the active device to the line usually ome form of matching LC section, resonant or otherwise, ien the required calculations are much simplified by the se of the Smith chart. These, with the usual exercises, e the topics of the chapter.

For the Smith chart, lower-case symbols signify impedices 'normalised' to that of the line R_o, e.g. $r=R/R_o$. or Section 23.4 onwards then the practice of using wer-case symbols to distinguish incremental from largegnal parameters is dropped. The distinction is not often levant for the present subject.

3.1 y- and h-parameters

he hybrid-π model of Section 6.6 is a small-signal model f the bipolar transistor which includes six incremental mponent values. With these values, and those for the mainder of the circuit, it is possible to calculate the sponse of the circuit at any frequency. However, with e accuracy of the hybrid-π model limited to ~30MHz ien a less compact and more experimentally determined presentation is needed.

At any one frequency a set of four parameter values is fficient to be able to calculate the response of the dece, but for any other frequency the values of the four parameters will be different. Thus instead of the mere six values for the hybrid-π, the set of four values must be given as a set of four graphs or tables, giving the frequency dependence of the complex value of each parameter.

We first review some of the earlier alternatives to the s-parameter set, starting with the y-parameter. With Y the standard symbol for complex admittance, and with the incremental value that of immediate concern, then for the present the practice of using lower-case for incremental values is retained.

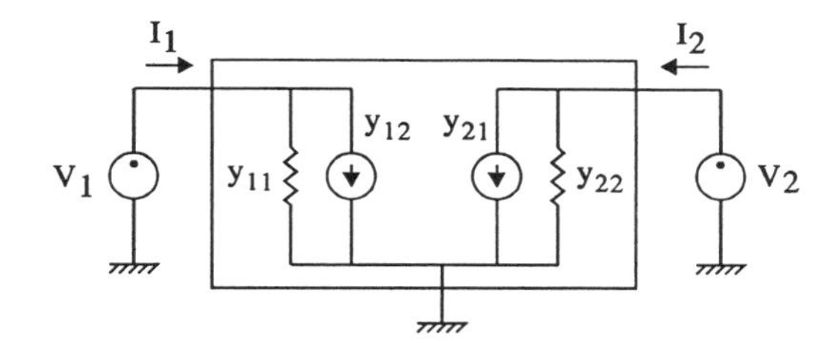

Fig. 23.1 Transistor y-parameters.

With the y-parameter representation as in Fig. 23.1, and with the ideal small-signal voltage sources v_1 and v_2 applied to the transistor input and output then, by superposition

$$i_1=y_{11}v_1+y_{12}v_2 \ldots(a) \quad i_2=y_{21}v_1+y_{22}v_2 \ldots(b) \qquad (23.1)$$

Amplifier

For the transistor used as an amplifier as in Fig. 23.2, then for the source and load

$$v_1=v_S-i_1R_S \ldots(a) \qquad v_2=-i_2R_L \ldots(b) \qquad (23.2)$$

With the voltage gain $a_v=v_2/v_s$ then

$$a_v=\frac{-y_{21}R_L}{(1+y_{11S}R_S)(1+y_{22}R_L)-y_{12}y_{21}R_LR_S} \qquad (23.3)$$

confirming that the four y-parameters are sufficient for calculation of the overall circuit performance.

Problem

For zero internal feedback $y_{12}=0$. Confirm [23.3] for this case by considering the input and output circuits separately. Exercise 23.1 gives the solution.

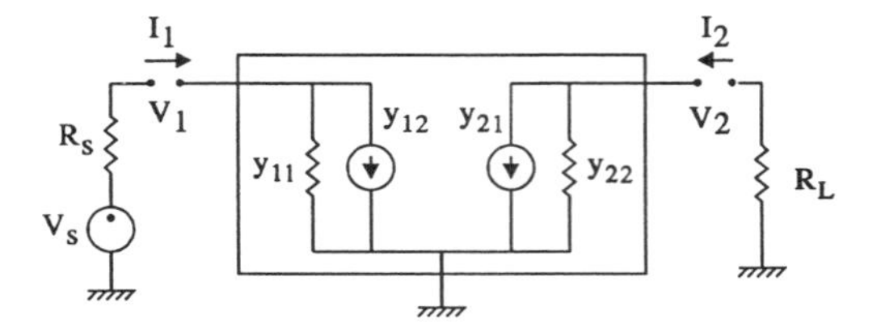

Fig. 23.2 y-parameters used to calculate amplifier gain $a_v=v_2/v_s$.

Measurement

For $v_2=0$ in [23.1] then $y_{11}=i_1/v_1$ and $y_{21}=i_2/v_1$. Thus the y-values are given by shorting the output and measuring the two currents. Similarly y_{12} and y_{22} are given by shorting the input and applying v_2. The published values are the result of the measurement of a large number of manufactured devices, with the production spread also stated.

h-parameters

An obvious alternative to the y-parameter is the z-parameter (z: impedance). Here say the parallel y_{11} and y_{12} is replaced by the series z_{11} and z_{12}.

The h-parameter set of Fig. 23.3 (h: hybrid) is favoured in that the parameters are closely related to those of the hybrid-π model. Here

$$v_1=h_{11}i_1+h_{12}v_2 \ldots(a) \quad i_2=h_{21}i_1+h_{22}v_2 \ldots(b) \qquad (23.4)$$

with the comparison as in Table 23.1.

$h_{11}=r_\pi$	$h_{12}=0$
$h_{21}=\beta$	$h_{22}=1/r_o$

Table 23.1 Comparison of h- and hybrid-π parameters.

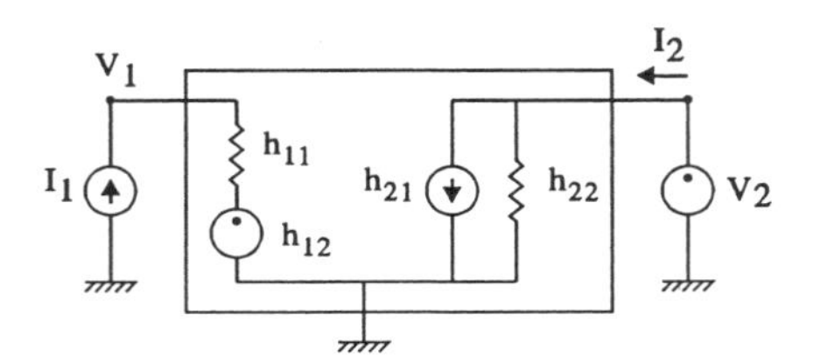

Fig. 23.3 Device h-parameters.

23.2 s-parameters

Each of the above parameter sets tends to fail f operating frequencies ~30MHz and above, simply becau it is not possible to make the measurements sufficient accurately. For example, measurement of h_{21} (β) in Fi 23.3 requires the source i_1 to be of infinite outp impedance, and i_2 to be measured using an instrument zero input impedance. The degree to which these idea can be approximated is lowered as the frequency increased.

Fig. 23.4 shows the corresponding method for mea uring the scattering s-parameters. In place of the abo idealised connections, connection is made to the device the standard 50Ω coaxial cable to which the paramete apply. With the a and b the travelling waves along the li then the directional couplers are able to resolve each the four components, somewhat as for the reflectometer Fig. 22.9. With the s the scattering parameters

$$b_1=s_{11}a_1+s_{12}a_2 \ldots(a) \quad b_2=s_{21}a_1+s_{22}a_2 \ldots(b) \qquad (23.$$

With the right-hand end terminated in a matched load th $a_2=0$ giving $s_{11}=b_1/a_1$ and $s_{21}=b_2/a_1$. Similarly with t left-hand end terminated in the matched load, and t right-hand end driven, then $a_1=0$ giving $s_{12}=b_1/a_2$ a $s_{22}=b_2/a_2$.

As for the other sets of four parameters, given the fo s-parameters for the operating frequency then it is possib to calculate the voltages and currents for any circuit in which the device is connected, in particular the gain of t resulting amplifier.

Wave labelling

The waves are labelled 'a' and 'b' rather than say v_a a

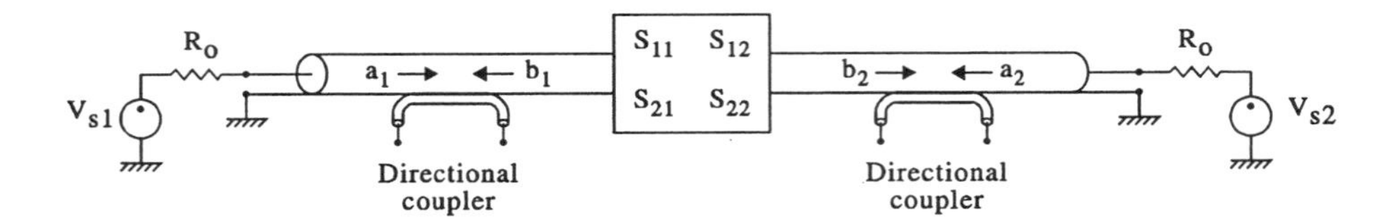

Fig. 23.4 Measurement of s-parameters (s: scattering).

v_b, partly for brevity and partly to give a simple relation to the power carried by the wave. For example, a_1 is scaled so that the associated power

$$P_{a1} = |a_1|^2.$$

With $$P_{a1} = \overline{v_{a1}^2} / R_o$$

with $\tilde{v}_{a1}$ the rms value of v_{a1}

and with $$\tilde{v}_{a1} = \hat{v}_{a1} / \sqrt{2}.$$

then $$a_1 = \hat{v}_{a1} / \sqrt{2R_o} \quad (23.6)$$

Amplifier

With the device connected as an amplifier in Fig. 23.5, the expression for the 'transducer power gain' G_T is to be derived. First G_T is defined as follows.

- The 'available source power' P_{av} is the power delivered to a load connected directly to the source, and with the load and source impedances matched.
- The load power P_L is that for the load in the diagram.

With these definitions $G_T = P_L/P_{av}$ (23.7)

To obtain the expression for G_T, the source and load are represented in terms similar to those of the s-parameters for the device, as follows.

b_s shown is the wave that would be transmitted down the line from the source with the source connected directly to the line (no matching unit) and with the line terminated in R_o (no reflection).

For the transistor in place of the termination, and with ρ_s the coefficient for the reflection of b_1 at the source,

$$a_1 = b_s + \rho_s b_1 \quad (23.8)$$

More simply for the load $$a_2 = \rho_L b_2 \quad (23.9)$$

These equations, together with [23.5] for the device, make up the four needed for the derivation of all the a and b, and hence the load power P_L. With

$$P_L = |b_2^2| - |a_2|^2$$

and a_2 as in (23.9) then

$$P_L = |b_2^2|(1 - |\rho_L|^2) \quad (23.10)$$

The factor $(1 - |\rho_L|^2)$ represents the loss of power due to the load mismatch. Similarly for the source

$$|b_s|^2 = P_{av}(1 - |\rho_s|^2) \quad (23.11)$$

From the above five relations, and with [23.5] for the device, then

$$G_T = \frac{|s_{21}|^2 (1 - |\rho_s|^2)(1 - |\rho_L|^2)}{|(1 - s_{11}\rho_s)(1 - s_{22}\rho_L) - s_{21}s_{12}\rho_L\rho_s|^2} \quad (23.12)$$

Stability

A device is said to be 'unconditionally stable' if there is no combination of passive impedance at its input and output terminals for which the device will oscillate. With G_T as in [23.12], instability requires den$[G_T] = 0$. For source and load passive in the sense that $|\rho_s| < 1$ and $|\rho_L| < 1$ then den$[G_T] \neq 0$ if all of the following conditions are satisfied for the device.

$$|s_{11}| < 1 \qquad |s_{22}| < 1 \qquad s_{12} = 0.$$

Here s_{12} is the reverse feedback factor, while s_{11} and s_{22} are the reflection coefficients for the device input and output.

Available power gain

The 'available power gain' G_{av} is the largest value of G_T that can be obtained by adjusting the source and load matching. This only has meaning for a device that is unconditionally stable. Otherwise as the threshold of oscillation is approached then G_T approaches ∞.

For the unconditionally stable device then $G_T = G_{av}$ if the source is matched to the transistor input and the load to the transistor output. The matching of any complex source and load Z_s and Z_L requires

$$Z_L = Z_s^*,$$

where Z_s^* is the complex conjugate of Z_s. With ρ_L as in [22.19] then

$$\rho_L = (Z_L - R_o)/(Z_L + R_o) \quad \ldots(a)$$
$$\rho_s = (Z_s R_o)/(Z_s + R_o) \quad \ldots(b) \quad (23.13)$$

For $$Z_L = Z_s^* \quad \text{then} \quad \rho_L = \rho_s^*.$$

Thus matching of the source and load to the transistor requires

$$\rho_s = s_{11}^* \quad \text{and} \quad \rho_L = s_{22}^*.$$

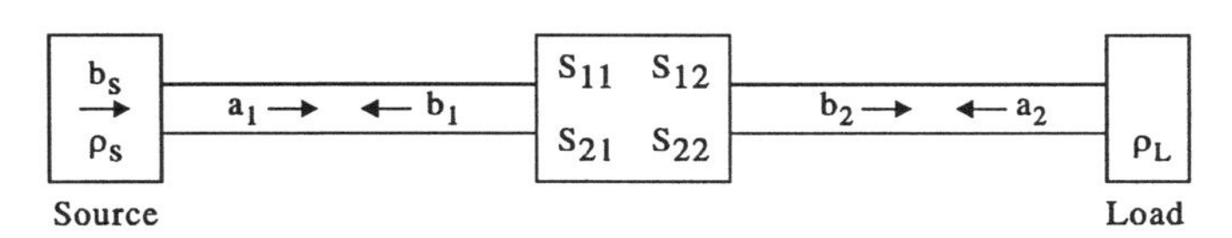

Fig. 23.5 Device of Fig. 23.4 connected as an amplifier.

With $s_{12}=0$ for the unconditionally stable amplifier, and with G_T as in (23.12), then

$$G_{av} = \frac{|s_{21}|^2}{(1-|s_{11}|^2)(1-|s_{22}|^2)} \qquad (23.14)$$

Confirmation

This result can be derived more simply. So far the matching has been of source to amplifier input, and of amplifier output to load, with the connecting lines regarded as part of the matching network. For short lines this is acceptable but for longer lines the resulting standing-waves increase the line losses. Then the matching needs to be in four stages.

(i) Source to line.
(ii) Line to amplifier input.
(iii) Amplifier output to line.
(iv) Line to load.

With all conditions satisfied then the G_T obtained is equal to G_{av}. For the present argument we start with (i) and (iv) satisfied, but not yet (ii) and (iii), giving $G_T < G_{av}$. With (i) then $|b_s|^2=P_{av}$. With no reflection at the source then $a_1=b_s$, giving

$$|a_1|^2=P_{av} \qquad (23.15)$$

With (iv)

$$P_L=|b_2|^2 \qquad (23.16)$$

With $\quad G_T=P_L/P_{av} \quad$ and $\quad b_2=s_{21}a_1$
then

$$G_T=|s_{21}|^2 \qquad (23.17)$$

The amplifier input mismatch incurs a power loss of

$$\times(1-|s_{11}|^2)$$

and the output mismatch

$$\times(1-|s_{22}|^2).$$

With the amplifier matching added then G_T is increased by

$$\times 1/(1-|s_{11}|^2) \quad \text{and} \quad \times 1/(1-|s_{22}|^2)$$

to confirm G_{av} as in [23.14].

23.3 Impedance matching

Fig. 23.6 shows a simple bipolar-transistor common-emitter stage that is to be matched to the normal $R_o=50\Omega$ of the coaxial connecting cable. Here L and C not only provide the required DC connection but also constitute the matching network. With the s-parameter representation of the transistor, and with the Smith chart of the next section, these are the methods to be used for designing the matching networks.

As a preliminary, however, the previous method of Section 12.6 will be reviewed. Here the transistor output is initially represented by the incremental output resistance r_o, subsequently allowing also for the parallel c_o.

With $r_o \sim 50k\Omega$ and $R_o=50\Omega$ giving $r_o >> R_o$, then this is a simple case of resonator matching. At any one frequency the series C and R_o transform to the parallel C_p and R_p.

With $Q=X_c/R_o$ and for $Q>>1$, then $R_p \approx Q^2R_o$ and $C_p \approx C$.

With L set to parallel resonate with C_p then the impedance presented to the transistor is the resistive R_p.

Where the required R_p and R_o do not differ so widely then, as also in Section 12.6, the calculations are a little more complex, but the basic effect the same. R_o is transformed to $R_p > R_o$, with the parallel C_p resonant with the parallel L.

With c_o allowed for, this simply comes in parallel with C_p, requiring only the appropriate reduction in L for parallel resonance.

Remember also that for a transistor RF power amplifier the required R_p bears no relation to the incremental r_o. Instead R_p is the value drawing the design power from the circuit at maximum output. Here normally $R_p < R_o$ requiring lateral reversal of the LC network.

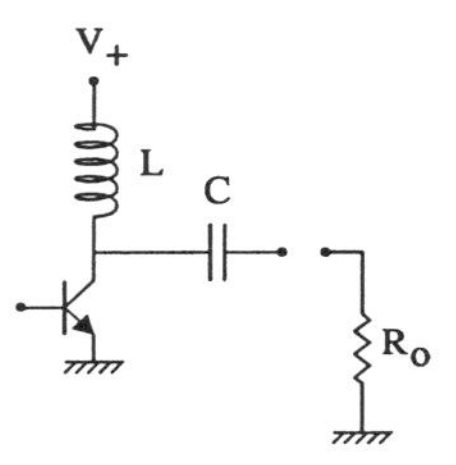

Fig. 23.6 LC matching of transistor amplifier to load.

23.4 Smith chart

While valuable as an aid to calculating matching networks, the Smith chart was originally devised as a means of calculating the impedance presented when a length of transmission-line is connected to the general load Z_L. With Z_x the impedance presented for a line of length x, then Z_x is as given by the somewhat complex [22.26]. With the load reflection coefficient

$$\rho_L=(Z_L-R_o)/(Z_L+R_o)$$

as in [22.19], and with $Z_L=R_L+X_L$, then the Smith chart of Fig. 23.7(a) is the polar plot of the general ρ, with the contours of constant R for varying X, and constant X for varying R, added. Thus the values of R and X for a given ρ (or vice versa) can be read from the chart directly.

Also for the added length of transmission-line the value of Z_L is simply rotated by $360^o \times x/(\lambda/2)$ about the chart centre to give the resulting R_x and X_x. Here λ is the wavelength on the line.

Normalised impedances

With the lower case r and x are used in the diagram, these are not incremental values but the normalised

$$r = R/R_0 \;\ldots(a) \qquad x = X/R_0 \;\ldots(b) \qquad (23.18)$$

allowing the chart to be used for any value of R_0, not just for the standard $R_0 = 50\Omega$. This therefore is the point at which we switch conventions.

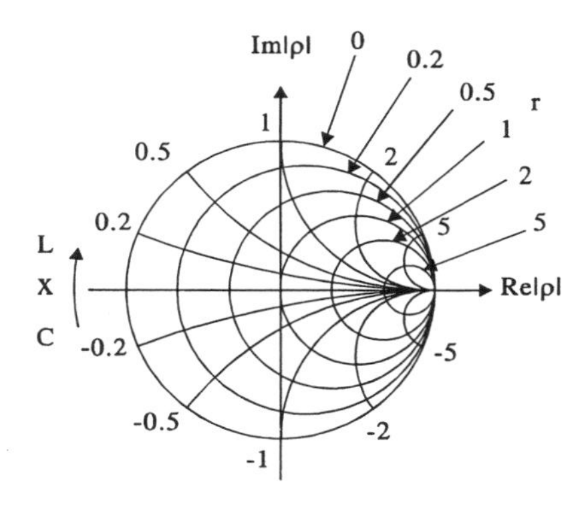

Fig. 23.7(a) Smith chart for plotting reflection coefficient ρ with impedance type grid showing normalised load impedance z=r+x.

Grid types

The chart uses two types of grid, that of Fig. 23.7(a) giving the values of the series R and X, and that of Fig. 23.8(a) for the equivalent parallel G and B. As for the normalised impedance grid, the normalised

$$y = g + b = Y/G_0,$$

giving

$$g = G/G_0 \;\ldots(a) \qquad b = B/G_0 \;\ldots(b) \qquad (23.19)$$

Thus

$$\rho = (z-1)/(z+1) \;\ldots(a) \quad \rho = (1-y)/(1+y) \;\ldots(b) \quad (23.20)$$

Impedance grid

Fig. 23.7(b) and (c) show some key contours, as derived from [23.20](a). The outer circle ($|\rho| = 1$) is that for r=0 as x varies from $-\infty$ to $+\infty$. At the two extremes, ρ=1, as previously established for an open circuit. Then for x=0 (short circuit) ρ=−1.

For all other values of x, $|\rho| = 1$, with ρ running round the circle in a clockwise direction for x increasing. Exercise 23.2 confirms.

For r=1 and x again varying from $-\infty$ to $+\infty$ then the circle is contracted to pass through the match point ρ=0 when x=0, as in (b).

(c) shows three significant cases where x is constant and r varies. First for x=0 the horizontal line is traced, from left to right as r increases from 0 through 1 (match) to ∞.

Thus the grid of (a) is completed, with one set for suitable increments of r, with x varying from $-\infty$ to $+\infty$ for each increment. The other set is for the corresponding increments of x, with r varying from 0 to ∞ for each increment.

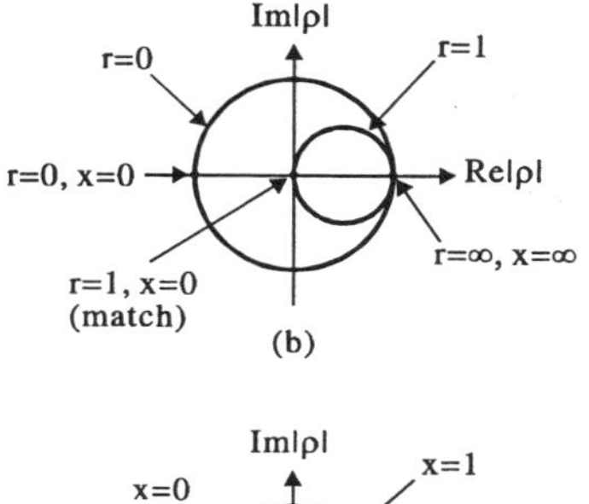

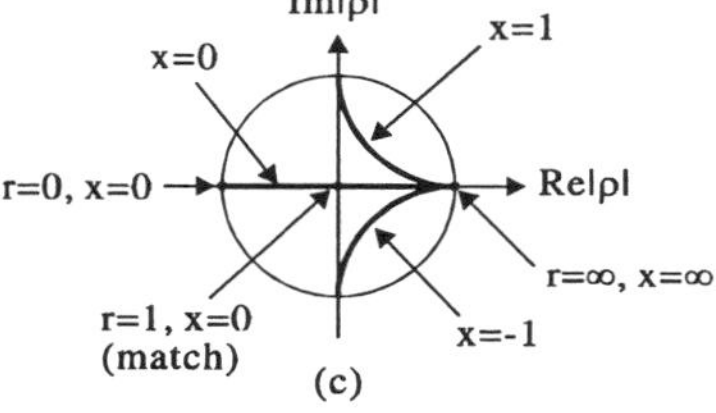

Fig. 23.7(b,c) Key constant resistance (b) and reactance (c) circles for Smith chart of (a).

Negative values of r give the off-grid $|\rho| > 1$. As shown in Exercise 23.3 this is where oscillation is possible.

Admittance grid

With ρ as in (23.20) $\qquad z = (1+\rho)/(1-\rho) \qquad (23.21)$

and $\qquad y = (1-\rho)/(1+\rho) \qquad (23.22)$

giving $\qquad y(\rho) = z(-\rho) \qquad (23.23)$

Thus the grid for y is obtained by rotating that for z by 180°. Here there is no change in ρ, only in the grid used. It remains for both that the impedance is partly or wholly inductive over the upper half of the diagram, and partly or wholly capacitive over the lower half. See Exercise 23.4 for exercises in reading the chart.

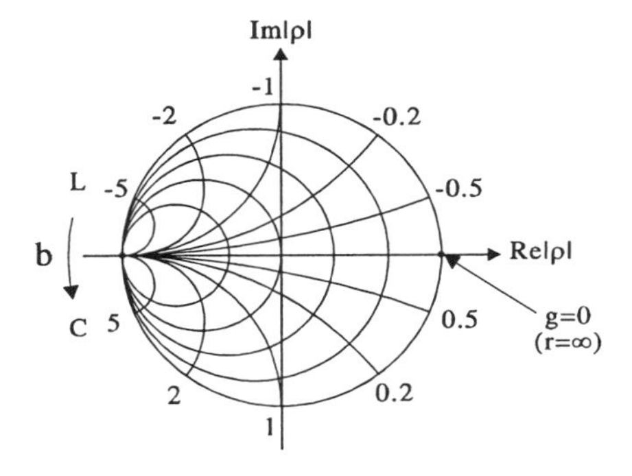

Fig. 23.8 Smith chart admittance grid showing y=g+b contours, and path for matching load to amplifier of Fig. 22.6.

23.5 Amplifier matching using the Smith chart

With the transistor amplifier of Fig. 23.6 redrawn in Fig. 23.9(a), it is now shown how the Smith chart in (b) can be used to derive the values of the matching L and C.

- For the reverse-feedback transistor scattering parameter s_{12} considered zero as above, the reflection coefficient at the transistor output is simply s_{22}. With the transformed R_o thus needing to be the conjugate s_{22}* then s_{22}* is marked on the admittance grid of (b).
- With the reflection coefficient for the load R_o zero, this is marked at the centre of the diagram.
- As well as the set of normalised admittance contours, (b) also shows the r=1 circle from the impedance grid. As the series C is added to R_o then the travel is around the r=1 circle as shown, up to the second point.
- Here the conductance is as required and the effect of the parallel L gives the further travel to the required final s_{22}* point.
- With the full impedance grid applied, the required value of x is read and converted to the appropriate series C for the frequency.
- With the admittance grid restored, the difference between the two b values on the constant g circle gives the required shunt L.

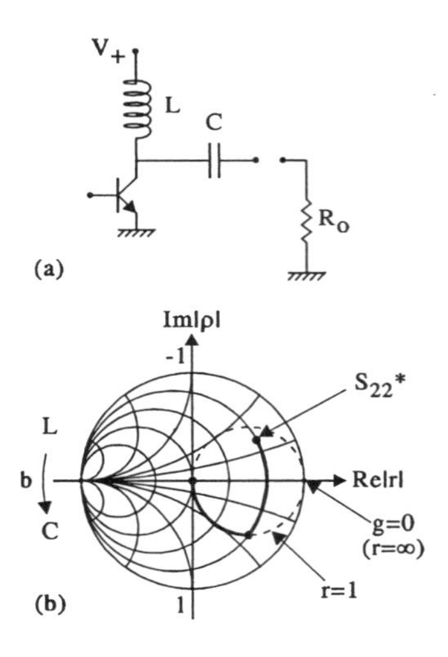

Fig. 23.9 Use of Smith chart for matching network design.
(a) Transistor amplifier with LC matching network.
(b) Admittance type Smith chart.

Series L

Starting again at the centre of the diagram, the initial movement around the r=1 circle is in the other direction. The final movement to s_{22}* now requires the shunt component to be a capacitor, as expected. With the required B smaller than for the previous shunt L then the bandwidth is higher.

Where the s_{22}* point lies outside of the r=1 circle then for the series L the required shunt B is an inductor also, making the difference in bandwidth for the series C and series L circuits yet higher.

Thus, overall, the series C is appropriate where an essentially band-pass response is required, and the series L for a wideband low-pass response.

Other load impedances

As for say the connection the output of one transistor to the input of the next, then the starting point is s_{22} for the first transistor and the end point s_{11}* for the second, or the other way round, as appropriate.

Matching using line section

Fig. 23.10 shows the series-L arrangement with the L provided by a section of the printed circuit board track.

For this example the transistor is the output stage of a transmitter. Here the impedance R_L' presented to the transistor is no longer that matching the transistor output impedance $r_o >> R_o$ but now the value normally $<<R_o$ needed to develop the design power in the load.

Thus the network is transposed laterally, here with the line and C_2 the matching components. At this stage the other components may be ignored.

Thus $R_o//C_2$ transforms to the series R_s and C_s, with now $R_s < R_o$. The effective inductance L of the line then series resonates with C_s, to leave only the R_s presented to the transistor as the required R_L'.

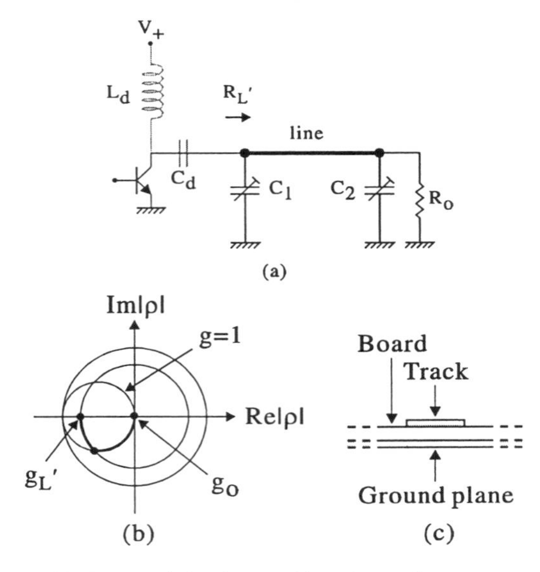

Fig. 23.10 Transmission line matching of transmitter output.
(a) Circuit diagram.
(b) Admittance type Smith chart.
(c) Cross-section of PCB track forming line in (a).

With the admittance type Smith chart as in (b), the sequence is as follows.

- R_o corresponds to the centre point marked as g_o.
- The shunt C_2 gives the movement shown, around the g=1 circle.
- Rotation due to the line gives the required g_L'.

C_1 is included to avoid the need for final adjustment of the line length. This is made a little less than that required and the required condition obtained by adjustment of C_1 and C_2. The transistor output capacitance becomes part of C_1.

Decoupling components

The DC blocking C_d needs to be large enough to present a negligible impedance at the operating frequency. Similarly the value of L_d needs to be large enough to present a negligible admittance. With the need for these components, the arrangement is less elegant than the series C network of Fig. 23.6. Worse, at a frequency well below that of the transmitter output, L_d and C_d form a parallel resonator. With the transistor often operating a good deal below its limit for maximum gain, the full gain obtains at the resonant frequency, with the resulting tendency for oscillation.

Lower frequency limit

As shown in Exercise 23.5, the frequency below which the required line length becomes inconveniently large is ~ 300MHz.

23.6 s-parameter and reflection coefficient measurement

Fig. 23.11(a) shows the complete arrangement for measuring the s-parameters, shown in part in Fig. 23.4. The connections are for the measurement of the forward transmission $s_{21}=b_2/a_1$ (the couplers have not been drawn the wrong way round - this is not water).

For the measurement of say s_{11} the second vector voltmeter is coupled to the V_{b1} point.

Vector voltmeters

With one of these generalised as in (b), the function of the circuit is to give V_{d0} and V_{d90} as the magnitudes of the components of V_a in phase and in quadrature with the primary reference V_{ro}. With

$$V_{ro}=\hat{V}_{ro}\cos(\omega t) \qquad \ldots(a)$$

$$V_{r90}=\hat{V}_{ro}\cos(\omega t+\pi/2) \qquad \ldots(b) \qquad (23.24)$$

and with

$$V_a=\hat{V}_a\cos(\omega t+\phi) \qquad (23.25)$$

where ϕ is the phase of V_a relative to V_{ro}, and with k_v the voltmeter constant, then

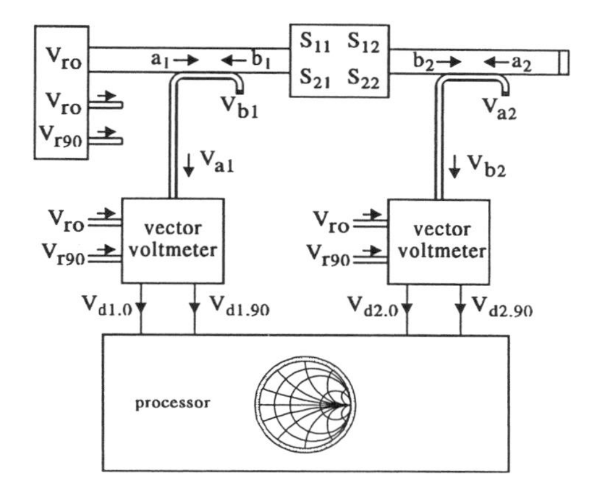

Fig. 23.11(a) Detailed arrangement for measuring S-parameters, configured for measurement of s_{21}. Block diagram.

$$V_{d0}=k_v\hat{V}_a\cos(\phi) \qquad \ldots(a)$$

$$V_{d90}=k_v\hat{V}_a\cos(\pi/2-\phi) \qquad \ldots(b) \qquad (23.26)$$

as required for displaying the in-phase and quadrature components of V_a on the complex plane.

With the two vector voltmeters giving the corresponding values for V_{a1} and V_{b2} then the processor can be programmed to calculate and plot $s_{21}=b_2/a_1$ on the Smith chart display shown.

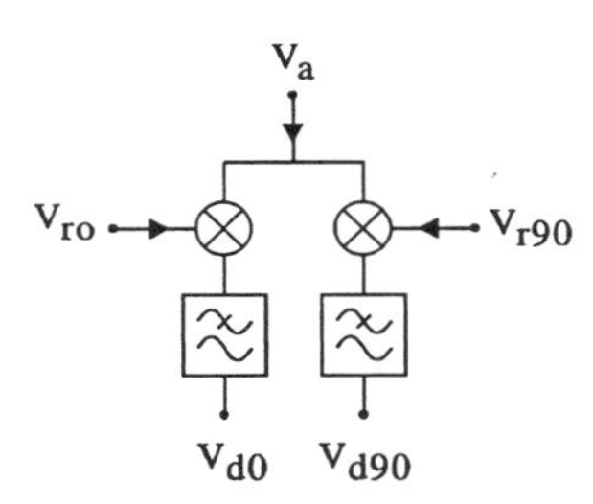

Fig. 23.11(b) Details of vector voltmeter in (a).

Line correction

The resulting s_{21} includes the combined phase shift for the two lengths of line between the device and the sampling points. Thus a preliminary measurement is required with the device removed, the lines coupled together, and the correction made. At the same time any departure from the expected $|s_{21}|=1$ for direct transfer can be noted and allowed for. For example, the coupler constants might not be quite the same.

For s_{11} and s_{22}, the line phase correction is obtained by replacing the device by a shorting connector.

Swept frequency

With the parameter values required over a range of frequencies then a swept frequency oscillator is needed. There are a good many factors to allow for here.

(i) The phase delay due to the lines will be proportional to the oscillator frequency ω.
(ii) The reference outputs V_{r0} and V_{r90} from the oscillator must maintain the required quadrature relation.
(iii) For the type of directional coupler used, the coupler constant k_c is proportional to ω.

Since each s-parameter is derived from the ratio of the two coupler outputs, (iii) is not a fundamental problem. Nor, for the same reason, is the possible variation in oscillator output amplitude with frequency. However, the processor has much to do to correct for these effects.

The arrangement shown is somewhat simplified, to make the principle clear. In reality at least one stage of frequency conversion would be used. Partly to overcome the problems of maintaining the two swept oscillator reference outputs in quadrature.

Coaxial directional coupler

This is shown in Fig. 23.12, with the sampled components of the forward and reverse waves now labelled V_f' and V_r'.

As for any measurement, the probe must not significantly perturb the signal that is to be measured. Thus the coupling capacitance C is made small enough to ensure $X_C >> R_0$ over the range of the swept oscillator. With the coupler output lines suitably terminated, this component of V_f' and V_r' is approximately $j\omega CR_0V/2$. With $j\omega MI/2$ the corresponding component due to the mutual inductance M then the total

$$V_f' = j\omega(CR_0V/2 + MI/2) \quad \ldots(a)$$

$$V_r' = j\omega(CR_0V/2 - MI/2) \quad \ldots(b) \qquad (23.27)$$

With

$$V = V_f + V_r, \quad I = I_f - I_r,$$

$$V_f = I_fR_0 \quad \text{and} \quad V_r = I_rR_0$$

then

$$V_f' = (j\omega/2)[V_f(CR_0 + M/R_0) + V_r(CR_0 - M/R_0)] \qquad (23.28)$$

$$V_r' = (j\omega/2)[V_f(CR_0 - M/R_0) + V_r(CR_0 + M/R_0)] \qquad (23.29)$$

For

$$R_0^2 = M/C$$

then

$$V_f' = j\omega V_fM/R_0 \ldots(a) \quad V_r' = j\omega V_rM/R_0 \ldots(b) \qquad (23.30)$$

Reflectometer

For higher operating frequencies (>300MHz) the coaxial directional coupler takes the place of the toroidal transformer in Fig. 22.9, the ferrite losses, etc., then being too high for its use. Unfortunately the present coupler lacks the primary independence of frequency of the transformer.

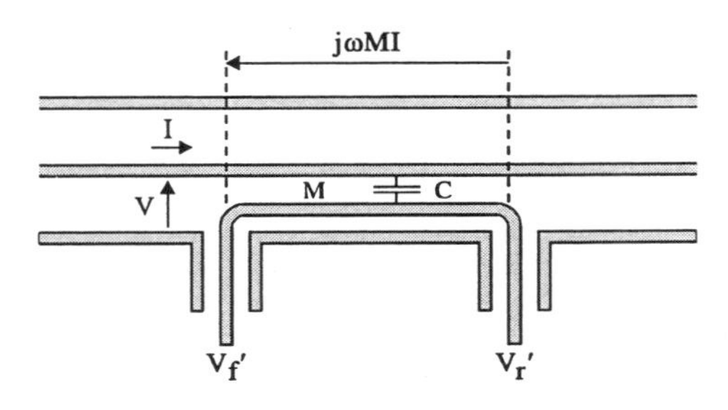

Fig. 23.12 Coaxial line directional coupler.

Reference

Hewlett Packard Application Note 95-1, *S-Parameter Techniques*, 1997.

Much of the material in this chapter is based on this note.

23.7 Exercises

Exercise 23.1. y-parameters

Confirmation [23.3] gives the output voltage V_2 for the amplifier of Fig. 23.2 in terms of the y-parameters. Confirm that the result for $y_{12}=0$ (no reverse feedback) accords with that given by the then much simpler direct analysis.

For $y_{12}=0$ [23.3] becomes

$$V_2 = V_s \frac{-y_{21}R_L}{(1 + y_{11}R_s)(1 + y_{22}R_L)} \qquad (23.31)$$

For $y_{12}=0$ in the diagram and for the potentiometer at the input

$$V_1 = V_s \frac{1/y_{11}}{1/y_{11} + R_s} = V_s \frac{1}{1 + y_{11}R_s} \qquad (23.32)$$

For $R_L//(1/y_{22})$ at the output

$$V_2 = -V_1y_{21} \frac{R_L(1/y_{22})}{R_L + (1/y_{22})} \qquad (23.33)$$

With V_1 as in [23.32] then [23.31] is confirmed.

Exercise 23.2 Smith chart, r = 0

Confirm the variation of ρ with x in Fig. 23.7 for z purely reactive.

Magnitude With $z=(1+\rho)/(1-\rho)$ as in [23.21]

then here $\rho=(x-1)/(x+1)$

giving $|\rho|=[x^2+(-1)^2]/[x^2+1^2]$

and thus $|\rho|=1.$

Phase For $x=-\infty$ to $+\infty$ then ϕ[num] varies clockwise from $-\pi/2$ to $\pi/2$.

For $x=-\infty$ to $+\infty$ then ϕ[den] varies anti-clockwise from $-\pi/2$ to $\pi/2$.

With $\phi=\phi[num]-\phi[den]$

then ϕ varies from clockwise $(-\pi/2+\pi/2)$ to $(\pi/2-\pi/2)$, i.e. clockwise from 0 to 0.

Exercise 23.3 Smith chart, r<0

Confirm that here $|\rho|>1$.

With $\rho=(z-1)/(z+1)$

as in (23.21) and $z=r+x$

then $|\rho|=[(r-1)^2+x^2]^{1/2}/[(r+1)^2+x^2]^{1/2}$.

With r<0 then |num|>|den| for any x giving $|\rho|>1$.

Exercise 23.4 Smith chart readings

Calculation Read the approximate values of r and x from the impedance grid of Fig. 23.7(a) for $\rho=0.5+j0.5$. Thus calculate R and X for the line impedance $R_o=50\Omega$.

From the chart $r=1$ and $x=2j$.

With $R=rR_o$ and $X=xR_o$

then **R=50Ω and X=j100Ω.**

Calculation Read the corresponding values of g and b from the admittance chart to give the parallel R_p and X_p.

From the chart $g=0.2$ and $b=-j0.3$,

giving $r_p=1/0.2=5$.

Thus $R_p=50\Omega\times5$ giving **$R_p=250\Omega$.**

Also $x_p=1/(-j0.3)=j3.3$

and thus $X_p=j50\Omega\times3.3$

giving **$X_p=j167\Omega$.**

Calculation Confirm the relation between z and y by calculation.

With $g+b=1/(r+x)$

then $g+b=(r-x)/(r^2+x^2)=(1-j2)/(1^2+2^2)$

giving **g=0.2** and **b=–j0.4,**

acceptable for the accuracy of chart reading

Exercise 23.5 Line-section matching (Fig. 23.9)

Calculate the convenient frequency limit.

With the required rotation typically ~1/10th of a full turn, this corresponds to 1/20th of the wavelength λ.

With the frequency $f=c\varepsilon^{-1/2}/\lambda$

where the velocity of light in free space $c=3\times10^8$ and the relative dielectric constant for the fibre-glass printedcircuit board $\varepsilon=4.5$ then

$$f=c\varepsilon^{-1/2}/20x$$

where x is the line length. With x=100mm the convenient limit then the corresponding **f=300Mz.**

24

Optoelectronics and other subjects

Summary

This the final chapter starts with an introduction to the developing topic of optoelectronics, particularly in connection with the fiber-optic digital data link. There follows one or two additional topics, with each summarised at the start of the section.

24.1 Fiber-optic data link

Here, as in Fig. 24.1, the usual connecting wire is replaced by a glass fiber. The semiconductor laser is a source of light which is switched on and off in response to the digital data stream D_{in}, while the photo-detector restores D_{in} as D_o. The advantage over the normal arrangement is a greatly increased bandwidth and thus data rate.

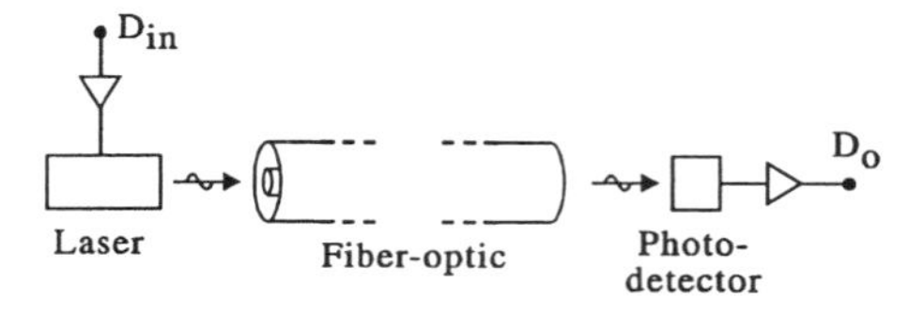

Fig. 24.1 Essentials of glass fiber optic-data link.

Light guide

The fiber shown is formed of silica glass (SiO_2) which must be extremely pure in order to avoid absorption of the light. Propagation is along the central core, with the diameter $d \approx 8\mu m$ comparable with the wavelength λ of the transmitted light. Confinement to the core is by total internal reflection. The core is a region of the silica doped to give the required higher refractive index. The standard outer diameter is 125μm, extended to 250μm by acrylic (not shown) for strength and flexibility.

Propagation

While the core is circular, the nature of the propagation is more simply understood for the hollow rectangular conducting guide of Fig. 24.2. Here the travelling wave in (a) can be viewed as the interference of a plane wave zigzagging across the guide at the speed of light. With β the propagation constant for the plane wave, for the progression along the guide $\beta_z=\beta \cos(\theta)$, while vertically across the guide $\beta_y=\beta \sin(\theta)$, where $\beta=2\pi/\lambda$, etc. For constructive interference across the guide $\lambda_y=2d$ giving

$$1/\lambda_z^2=1/\lambda^2 - 1/(2d)^2 \qquad (24.1)$$

as can also be derived from the geometry of (b). Thus λ_z increases to infinity as λ is increased to 2d. At this, the cut-off point, the plane wave bounces repeatedly between the faces with no lateral movement at all.

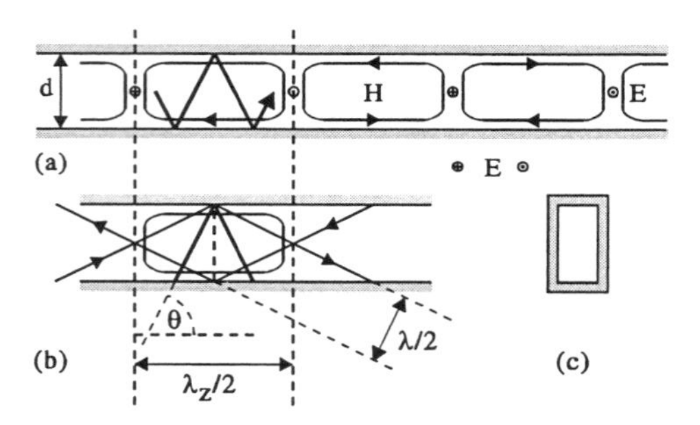

Fig. 24.2 (a) Waveguide propagation.
(b) Magnetic H field components, giving [24.1].
(c) Guide cross-section.

Multi-mode propagation

Conversely for $\lambda < d$ the pattern in (a) can be stacked two-high or more between the conducting faces. With the normal mode still available then the transmitted light becomes coupled into more than one mode. With $\beta_y=\beta \sin(\theta)$ and $\beta_y=2\pi/\lambda_y$ for the normal mode then $\sin(\theta)=\lambda/2d$. For the nth order mode this becomes $\sin(\theta)=\lambda/2(d/n)$. For a given value of λ then θ, and thus the rate at which the wave travels down the guide, is different for each of the available modes. Initially this smears the single received pulse, and for more extreme cases gives full resolution of the component pulses. The effect can be avoided while maintaining reasonably efficient propagation by setting d approximately mid way between $\lambda/2$ and λ.

Material dispersion

The above spreading of the received pulse is referred to as 'dispersion', the term previously used for the spreading of a ray of light by a prism into a band of rainbow colours. This is due to the light being composed of a range of spectral components, with each travelling through the prism glass at a different rate. For a pulse of white light transmitted through a block of the glass then again the component pulses will arrive at different times. For the prism the dispersion is spatial while for the pulse it is temporal. This type of temporal (pulse spreading) dispersion is termed 'material' while that above due to multimode propagation is termed 'inter-modal'.

Waveguide dispersion

This, the third type of temporal dispersion, exists for the optical fiber. With $\sin(\theta)=\lambda/(2d)$ and $f=c/\lambda$ for the hollow rectangular guide of Fig. 24.2, the rate of progression $c.\cos(\theta)$ along the guide, varies with frequency f, once more causing the components of the transmitted pulse to arrive at different times. Because material and waveguide dispersion most obviously obtain for light of more than one spectral component (colour), these are jointly termed 'chromatic dispersion'.

Of the various types of light source that might be used, white light has the full spread of visible colours, for LED light this is reduced to about 10% of the central frequency, and laser light is essentially of one frequency only. However, as soon as the single frequency source is pulsed it acquires a spectral width $\Delta f \sim 1/T_w$ where T_w is the pulse width. The same is so for the digital data stream with the data rate effectively $1/T_w$. With current rates approaching 10GHz, Δf is still only about $1/3 \times 10^4$ of the laser frequency. Thus use of the laser in place of the LED gives a very large reduction in the dispersive pulse spreading.

While the laser source and the single mode fiber are needed for the long-distance high-data-rate link, for shorter distances and/or lower data rates the lower cost LED source can be used. Also, where the resulting inter-modal dispersion is acceptable, not only is a fiber of larger diameter less costly, but coupling to the source and detector is simpler.

Phase and group velocity

For a full understanding of how chromatic dispersion spreads the pulse, the distinction between these two values needs to be clear. With λ_z as in [24.1] for the rectangular guide of Fig. 24.2 then $\lambda_z > \lambda$, inceasing to infinity as $\lambda/2$ approaches d. Thus while the speed of the zig-zag progression falls to zero at this limit, the speed $f\lambda_z$ at which the mode pattern moves rises to infinity. The latter, being the speed at which a point of constant phase moves, is termed the 'phase velocity' v_p and the former, for reasons we shall see directly, the 'group velocity' v_g.

Consider the extreme example of an impulse transmitted down a dispersive transmission line at time $t=0$. The impulse has an infinite spread of spectral components, all of the same amplitude, and all in phase only at the time of ocurrence of the impulse. Elsewhere the interference reduces the ampltude to zero.

Consider next two of the components, for simplicity $\cos(\omega_1 t - \beta_1 x)$ and $\cos(\omega_2 t - \beta_2 x)$ where $\beta=2\pi/\lambda$. With $x=0$ at the start of the line and the impulse applied at $t=0$, after the time t the value of x at which the impulse is placed is that for which the two phase angles are the same, giving $x=(\Delta\omega/\Delta\beta)t$ where $\Delta\omega=\omega_1-\omega_2$, etc. Thus the speed at which the impulse moves is $\Delta\omega/\Delta\beta$.

Returning to the light pulse, this is a group of such spectral components, with the speed $\Delta\omega/\Delta\beta_z$ at which the pulse moves aptly termed the 'group velocity' v_g. For the pulse not to disperse $v_g \equiv d\omega/d\beta_z$ must be constant over the spectral width of the pulse, a rather less demanding constraint than that $v_p \equiv \omega/\beta_z$ be constant. Much of the design of the optical fiber is thus directed to making $d^2\omega/d\beta^2=0$ at the laser frequency.

Where v_g and v_p differ, as for the rectangular guide, we see the envelope of the transmitted pulse moving at one speed (v_g) and the field pattern within it at another (v_p). With the propagation of waves across the surface of a pond also dispersive, the effect can be observed if a brick is oscillated briefly at the pond edge instead of just throwing it in.

As a final exercise, confirm that for $v_g \equiv d\omega/d\beta_z$ in [24.1], v_g is equal to the component of the zig-zag progression along the guide.

24.2 Light-emitting diode (LED)

While the LED is most widely used as a visual display device, the semiconductor laser normally used with the fiber-optic link is a suitably modified LED.

When the conventional semiconductor diode is forward biased the carriers flow across the junction to recombine on the far side of the depletion layer as in Fig. 5.10(a). Since the band-gap voltage E_g is the energy in electron-volts needed to free an electron from the silicon lattice, when recombination takes place that amount of energy must be released. With energy transfer quantised, this will either be in the form of a 'phonon' adding to the thermal vibration (temperature) of the lattice, or as a 'photon' of radiated light. It is the photons that constitute the light emitted by the LED.

From quantum theory, with f_p the photon frequency, E_p its energy and h Planck's constant then

$$E_p=hf_p \qquad (24.2)$$

With $c=\lambda f$, where c is the velocity of light and λ the wavelength then

$$\lambda=ch/E_g$$

Here $$h \approx 6.625 \times 10^{-34} \text{ joule-sec},$$

$$\text{one electron-volt}=1.602 \times 10^{-19} \text{ joule},$$

and $$c \approx 300 \times 10^6 \text{m/s}$$

giving $$\lambda_p \approx 1.24 \times 10^{-6}/E_g \qquad (24.3)$$

With $E_g=1.2V$ for Si then $\lambda_p \approx 1000nm$, well above $\lambda_{red}=625nm$. With the radiation emitted by the conventional Si diode thus well into the infra-red, a material with a larger bandgap is required for visible radiation.

Carbon and silicon carbide

Carbon is the first element in the periodic table to have four outer electrons and so is the usual example of covalent bonding. Here $E_g=5.2V$ giving $\lambda_p=250nm$ which is well into the ultra-violet and thus again invisible. Also the crystalline structure is diamond which can neither be grown nor doped. For SiC, $E_g=2.1eV$ giving $\lambda_p=590nm$ which matches $\lambda_{green}=569nm$ well. Crystals of SiC can be grown, but not easily.

Gallium arsenide

More easily grown is the crystal of the III-V compound Gallium arsenide (GaAs). Here Ga is an acceptor atom having three outer electrons, and As a donor having five. Thus the combined material is again able to form the covalent structure of eight shared outer electrons. But with $E_g=1.37eV$ giving $\lambda_p=905nm$, this is once more in the infra-red. While thus still not suitable for visual display, the GaAs devices are widely used for the digital data link. With the carrier mobility higher than that for Si, the improved speed of response gives data rates of several GHz.

Gallium arsenide phosphide

For this material Ga is again the acceptor, but the donor is a mix of the two donors As and phosphorous (P) giving GaAsP. By varying the mix, E_g can be made to give light ranging from red (625nm), through yellow (585nm) to green (569nm). Blue LEDs are not widely available. With the higher E_g and the need to allow for bulk resistance, the typical forward voltage for an LED is ~2V.

Spectral width

With the recombining electrons having some kinetic energy prior to recombination, that of the emitted photon will be somewhat higher than the band gap E_p. With the photon frequency $f_p=E_p/h$ then E_p will have a corresponding spread, to the above ~10% of the central frequency. See Senior 1992 for a fuller account.

24.3. Semiconductor laser

For a colliding hole and electron to recombine there has to be some means of dissipating the associated energy. Otherwise the two will just bounce apart like billiard balls. For the unmodified LED this is by 'spontaneous emission' of the photon. With the time at which the photon is emitted that at which the collision occurs, the phasing of the photons is randomly distributed. Also the radiation is essentially omni-directional. Finally, as previously noted, the photon frequencies have a spread of ~10%.

For reasons only explicable in terms of quantum mechanics, another means for the release of the photon exists if the colliding pair is subjected to radiation at the photon frequency. For the resulting 'stimulated emission', the emitted photon assumes the form of the stimulating field, phased so as to add to the field intensity.

If the stimulating field is a plane wave travelling the length of the LED recombination region, this will emerge amplified to give 'light amplification by stimulated emission of radiation' (LASER). Preceding the laser was the very low-noise microwave MASER used in the earliest satellite links.

Resonator

If the LED is cleaved as in Fig. 24.3 and the flat ends coated with reflective material then the above plane wave will bounce repeatedly between the two reflectors. If the distance D between the two reflectors is made equal to one half wavelength $\lambda_p/2$ for the photon frequency f_p then coherent oscillation will build up at this frequency, to give a 'light oscillator by stimulated emission of radiation' (the acronym was never adopted!).

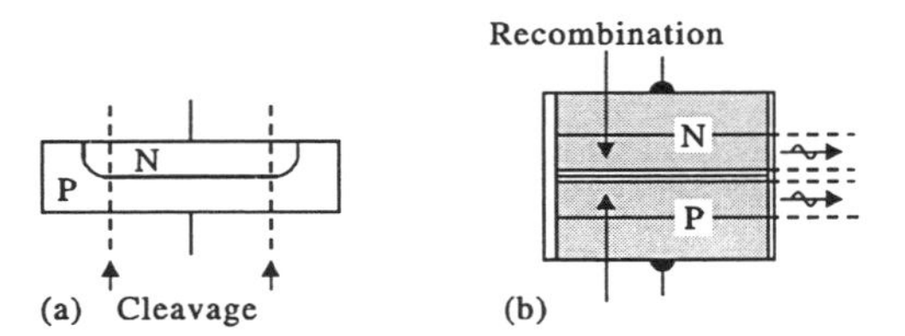

Fig. 24.3 Semiconductor laser.
(a) Formation of resonator by faces of cleaved crystal.
(b) Operation.

Multi and single mode operation

For reasons of mechanical stability, the length D of the laser cavity needs to be considerably more than the ideal $\lambda_p/2$. With oscillation possible at any of the frequencies over the LED output range for which D is an integral multiple of $\lambda_p/2$ then the laser will normally oscillate at a number of different frequencies at the same time.

With the fractional width of the LED output spectrum ~10%, it is possible to increase the cavity length to ten half wavelengths (~5μ) before more than one cavity resonances falls within the LED range. In practice single-mode operation can be maintained for D ~ 25μ. Alternatively the reflection of one of the laser end faces can be made frequency selective with a wider mode spacing by coupling to short resonator external to the semiconductor material.

Another reason why multi-mode operation of the laser is undesirable is that, as for broadcast radio waves, optical frequency domain multiplexing allows many channels to be available for one costly long-distance fiber. Multi-mode oscillation of the laser occupies more than one potential channel.

Coupling to optical fiber

For a coherent radiator in the form of a flat disc of diameter $d >> \lambda_p$ the radiation emerges as a nearly cylindrical beam with the divergence angle $\sim\lambda_p/d$. For the output of the semiconductor laser the larger dimension of the radiating surface is a number of times λ_p. For a circular radiating surface of this diameter a normal lens would reduce the beam to a spot of diameter comparable with λ_p at a distance from the lens equal to its focal length. The other dimension, however, is comparable with λ_p giving a large radiation angle. Here the lens needs to be such as to image the source at the core face. With the two required focal lengths different then a suitably asymmetric lens is required. With this provision, the light is concentrated on an area within that of the 8μm core diameter.

24.4 Photo-diode

The stability of the photo-detector amplifier was covered in Section 3.13, and the general noise aspects in Section 17.9. In this section we consider the photo-diode more closely. For incident light of frequency f_p such that the photon energy $E_p=hf_p$ is greater than the energy gap E_g, a hole-electron pair is generated for each photon absorbed. Those generated in the depletion layer are swept to the layer edges to constitute the required component of the diode current I_d. Other components of I_d are the 'dark current' and that due to the creation of hole-electron pairs by light absorbed in the remainder of the device.

Dark current

With $$I_d=I_{ds}[\exp(qV_d/kT)-1] \qquad \text{as in [5.2]}$$

for the normal diode, and with the photo-diode reverse biased in order to reduce the depletion layer capacitance then the dark current is essentially I_{ds}.

So far only one component of I_{ds} has been mentioned, that due to the flow of minority carriers across the depletion layer as in Fig. 5.10(b). In addition there are two components of 'leakage current', as in Fig. 24.4(a).

- The bulk component arises from the thermal generation of hole-electron pairs within the depletion-layer. Apart from the energy source being thermal, this is much as for the carriers created by the incident light.
- The surface component is the result of the relatively high concentration of defects at the surface.

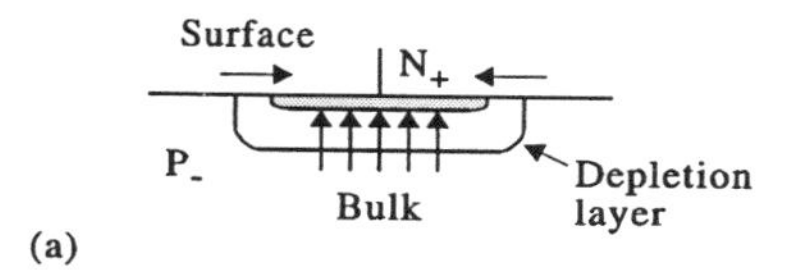

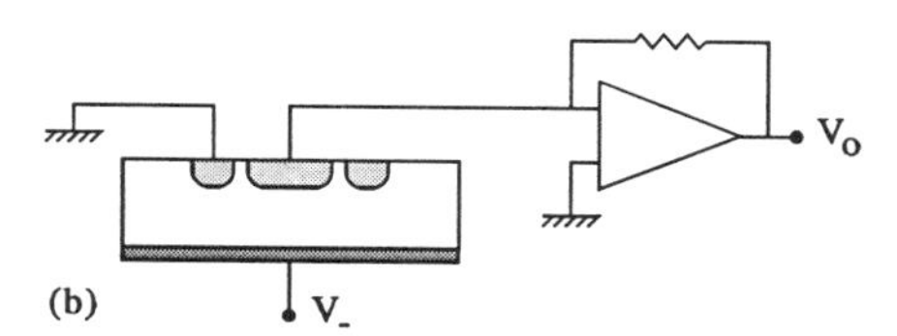

Fig. 24.4 Photo-diode leakage currents
(a) Bulk and surface components.
(b) Use of guard ring to divert surface component to ground.

Guard ring

All three components of I_{ds} are reduced by reducing the area of the photo-diode. With D the diameter of the active area, the minority-carrier and bulk leakage components are both $\propto D^2$. But with the surface component arising from a ring of constant width, this varies directly with D. For D reduced to suit the 8mm core diameter of the optical fiber then the surface component normally becomes dominant. The purpose of the guard ring in (b) is to divert this component to ground.

Integration

A further potential advantage of the reduced photo-diode area is the reduction in depletion layer capacitance, improving the speed of response. With it possible to make integrated MOS devices of comparable area, the limiting factor becomes the typically 3pF pad capacitance of the normal integrated circuit. This would be avoided by full integration of diode and preamplifier.

Extinction length

As light from above penetrates the photo-diode in Fig. 24.4(a) the intensity decays exponentially, with the decay constant the 'extinction length' λ_{ex}. To the extent that it is only the light absorbed in the depletion layer that gives the photo-diode current I_d, the thickness of the N_+ layer needs to be $<<\lambda_{ex}$ and that of the depletion layer $>>\lambda_{ex}$.

Stored charge tail

However, hole-electron pairs created by absorption of the light in the bulk regions above and below the depletion layers do add to I_d, but not very helpfully. As shown in Fig. 24.5(b), following abrupt cut-off of the light, the depletion layer component of I_d falls to zero relatively quickly, but the bulk component takes much longer, giving the 'tail' shown.

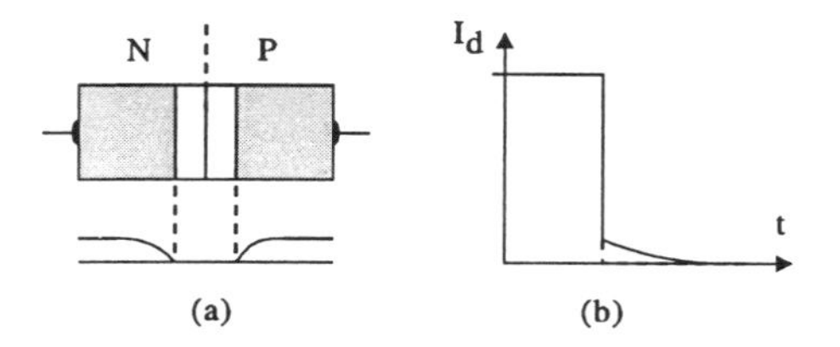

Fig. 24.5 Stored photo-detector charge tail.
(a) Minority carrier concentration.
(b) Tail to photo-detector output I_d following cut-off of light.

First (a) shows the minority carrier concentration for any reverse-biased diode. With the minority carriers at the edge of the depletion layer swept immediately across, the concentration there is effectively zero. The resulting concentration gradient is that giving the diffusion of minority carriers to the layer edge which constitutes I_{ds}.

When the diode is illuminated the hole-electron pairs generated in the bulk regions greatly increase the scale of the profile in (a), making the proportional current the few percent of the depletion layer component in (b).

When the light is extinguished the carriers in the depletion layer are swept away relatively quickly, but the time taken for the excess concentration in the bulk regions to decay is the much longer recombination time. The same obtains for any step transition in the light intensity, whether an increase or decrease.

For the tail as shown this does little more that slightly reduce the noise-margin for the resolution of D_o. The time taken to cross the threshold is essentially unaltered. Limitation of the tail height is thus the main reason for ensuring that most of the hole-electron generation is in the depletion layer.

PIN diode

With the same material (GaAs) used for laser and detector in the fiber-optic data link, it is ensured that the laser photon energy is above that of the energy gap E_g of the detector. But with E_g close to the lowest obtainable, for light of yet larger wavelength λ_p it becomes progressively more difficult to provide an effective photo-diode detector. With the initial effect here an increase in the extinction length λ_{ex}, the required increase in the effective depletion layer is obtained by adding the I-layer as in Fig. 24.6. This is a layer of undoped (intrinsic) semiconductor of the type doped to give the P and N regions.

With the larger distance across the E region, and E essentially as before, the required applied voltage is considerably increased. This proportionally increases the reverse breakdown voltage, causing the PIN structure to also be the basis of the high voltage class of normal diode.

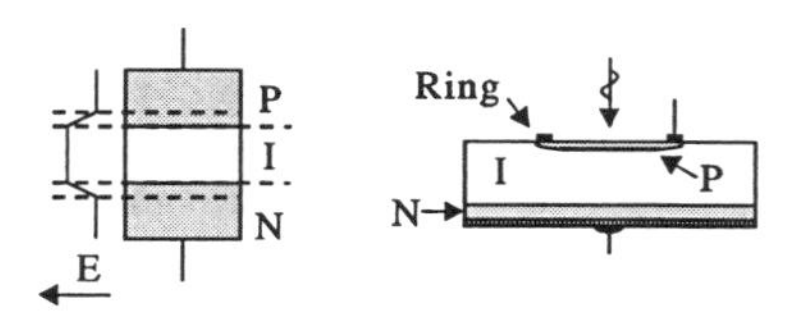

Fig. 24.6 PIN photo-diode
(a) Essentials with electric field E.
(b) Detailed structure with thin transparent P-region and conducting gold annular ring.

Avalanche photo-diode (APD)

This is essentially a PN diode with the reverse bias deliberately advanced to the verge of avalanche breakdown. As explained in Section 5.5, an electron in the depletion layer moves to the edge at a rate determined by the acceleration imposed by the E field in the layer, and limited by the collisions with the lattice. A proportion of such collisions cause a further electron to be released from the lattice. Avalanche breakdown is where the average number released per electron is greater than unity, perpetuating the process.

For the E field such that the average number is just less than one, say 0.99, then the charge passed from one edge of the layer to the other is increased from the q for the single hole-electron pair, to ≈100q.

With the mean value of I_d thus increased by the factor M, the above M≈100 can be routinely obtained with a suitably constructed diode and a well stabilised bias supply, with much higher values reported for research conditions.

With M obtaining for the dark current as well as the photo-current, the main advantage of the APD is for the low values of load resistor (~50Ω) needed for high-speed operation. Here even the multiplied dark current noise is dominated by that from the amplifier.

Guard ring

For the typical PN type APD of Fig. 24.7, the function of the guard ring is to prevent premature avalanche breakdown at the curved ends of the N_+ diffusion where the E field would otherwise be the highest.

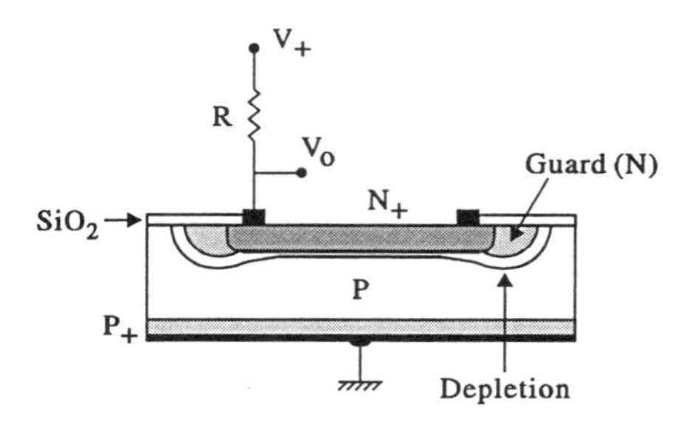

Fig. 24.7 Avalanche photo-diode with guard ring.

With the lighter doping N of the ring, the depletion layer is there wider. With the same potential difference for all parts of the layer, that for the ends is lower than at the centre ensuring that the charge multiplication is where required. It is also important that the doping and dimensions of the high field region be uniform. Otherwise the multiplication is only obtained at the points of maximum E.

A similar guard ring is used for the normal high voltage diode, again to prevent premature breakdown.

Avalanche PIN photo-diode

With the voltages needed to cause the PIN diode to avalanche impracticably high, the arrangement of Fig. 24.8 may be used. Here the photo-electrons are produced in the I region as for the normal PIN diode, to be swept towards the P region as usual. With the width of the P region made small compared with the recombination length then the remaining operation is much as for the bipolar transistor, with most of the photo-electrons passing through the thin P-layer base to reach the 'base-collector' PN depletion layer. With the applied voltage there able to be at the avalanche threshold, this is where the multiplication occurs.

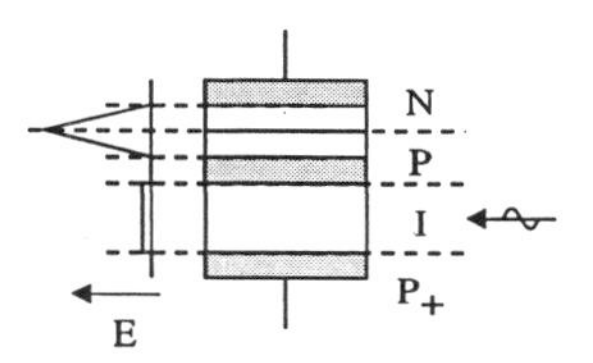

Fig. 24.8 Avalanche PIN photo-diode with separate photon absorption and charge multiplication regions.

24.5 Opto-isolator

The opto-isolator exists in a various forms, with three as in Fig. 24.9. With each an LED driving a photo-detector, the purpose is to couple an electrical signal from one point to another with no electrical contact. Applications are as follows.

- Breaking a ground loop, particularly where digital signals are concerned.
- Safety, e.g. coupling to a battery-operated system for monitoring body functions such as the EEG or skin resistance.

The LED to photo-diode in (a) is the simplest, while the other two give increasing sensitivity, by $\times\beta$ for the photo-transistor in (b) and by $\times\beta^2$ for the Darlington in (c). Each is usually encapsulated as an integrated circuit, for convenience and to shut out stray light. For another type both input and output are digital signals, giving the required 5V or 0V output levels.

The Darlington pair is used more widely, often as the pass transistor in the voltage-regulated DC supply of Fig. 5.21.

Linearity

As an analog link the isolator is far from linear. While the detector $I_d \propto$ light intensity W, W is by no means proportional to I_{led}, and even less to V_{led}. A solution is to use two HC4046 PLL chips. For one chip the included VCO provides a linear voltage-to-frequency converter, with the output driving the LED. For the other the full PLL forms the frequency-to-voltage converter of Section 19.5. An ADC and DAC is an evident alternative.

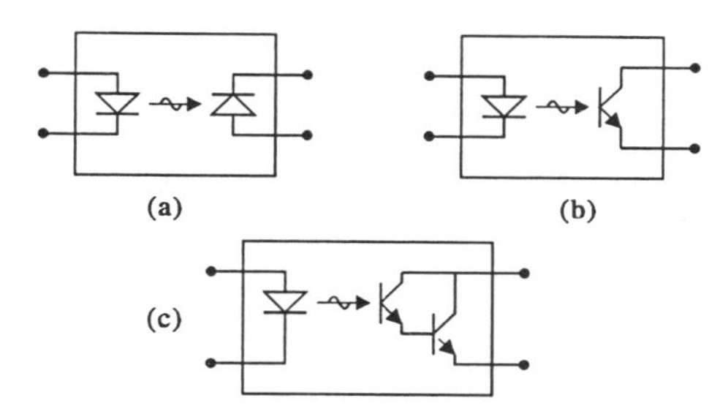

Fig. 24.9 Opto-isolators.
(a) LED to photo-diode. (b) LED to photo-transistor.
(c) LED to Darlington.

24.6 Piezo-electric filters

A piezo-electric material such as quartz is one for which applied mechanical stress deforms the crystal in such a way as to give the molecules an electric dipole moment. This places a positive charge on one face of the crystal and an equal negative charge on that opposite. Conversely, if conducting plates are placed on the crystal where the charge was formed and a voltage applied, then the dimension changes.

Resonator

For the material in the form of a rectangular slab with conducting plates on the two larger surfaces as in Fig. 24.10(a) the device becomes a resonator. The main resonance occurs where the largest dimension d is equal $\lambda/2$ where λ is the wavelength, with the associated series of 'overtones' for which $d=3\lambda/2$, $5\lambda/2$, etc.

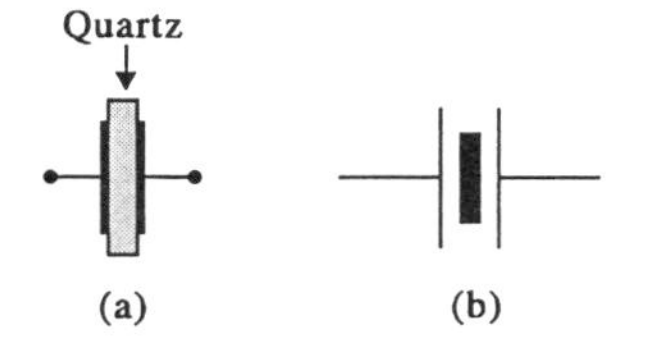

Fig. 24.10 Quartz crystal resonator.
(a) Construction.
(b) Circuit symbol.

With the Q-factor of the resonance very high ($\sim 20\times 10^3$), and the dimensions highly temperature independent, the quartz resonator becomes the basis of a frequency stable oscillator, or of a high-grade narrow-band band-pass or band-stop filter. Typical oscillator stability is one part in 10^5 improving ×10 for a temperature controlled enclosure.

Band-pass filter

Fig. 24.12 shows the kind of multi-section bandpass filter that would be used in the intermediate freqency section of a single-sideband transceiver. With 8MHz the normal f_{if} and 3kHz the required bandwidth for speech, we consider first the single section filter composed of the last three components. This is as in Fig. 24.11(a) with the crystal shown as its equivalent circuit in the region of the main resonance of the crystal. Here the series resonator C_m, L_m and R_m represents the piezo-electric (mechanical) impedance Z_m of the crystal, and C_o the inter-plate capacitance. Typically

$$R_m=400\Omega, \quad Q_m=20\times10^3, \quad C_o=200\times C_m.$$

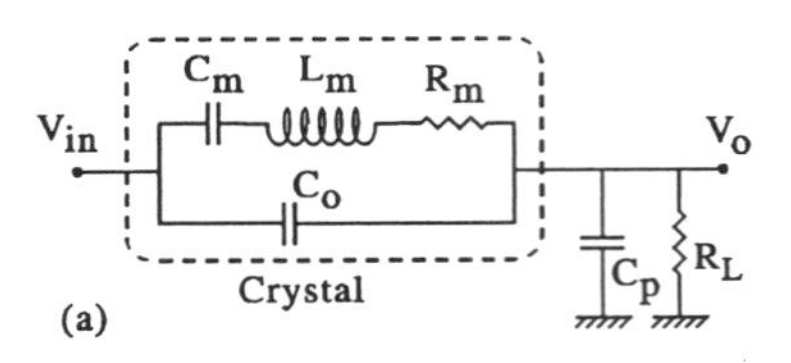

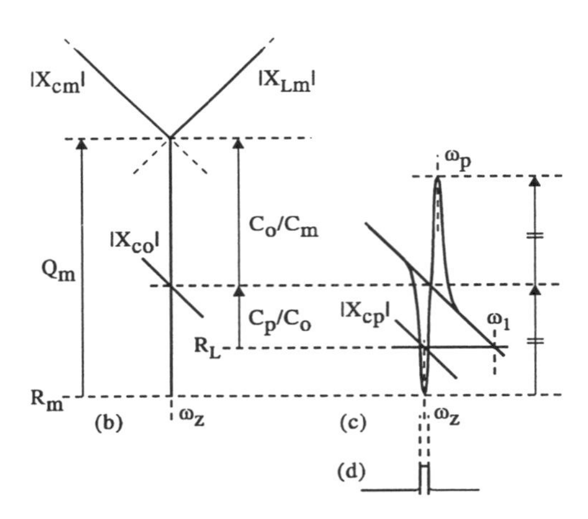

Fig. 24.11 Narrow-band band-pass filter based on quartz crystal resonator of Fig. 24.10.
(a) Equivalent circuit
(b) Summary representation of variation of crystal impedances with frequency.
(c) Interaction of crystal impedance in (b) with shunt network C_pR_L in (a).
(d) Resulting filter frequency response.

Apart from C_o, the frequency response of Z_m is the claasic series resonator response of Fig. 12.2(a). But, with the present Q_m so large, the essentials of the response are best shown in the summary 'Y' form of (b) here. With C_o included, this adds a parallel resonance at the frequency ω_p (pole) to the previous series resonance at $\omega_z=\omega_{if}$ (z: zero). The resonance occurs where $|X_{co}|=|Z_m|$, beyond ω_z where Z_m is inductive. Note the anti-symmetry of the resulting response in (c), relative to the symmetry of (b). Analysis of the circuit gives

$$\omega_p=\omega_z(1+C_m/C_o)^{1/2} \qquad (24.4)$$

where, as usual, $\qquad \omega_z=1/(L_mC_m)^{1/2} \qquad (24.5)$

With the above figures, and with Δf_m the natural bandwidth of the mechanical resonance

$$\Delta f_m=f_z/Q_m=400\text{Hz}.$$

Omitting C_p for the moment, and with Δf the filter bandwidth in (d)

$$\Delta f=\Delta f_m(R_L+R_m)/R_m$$

requring R_L somewhat larger than R_m as in (c) for $\Delta f=3$kHz.

Without C_p the filter transfer rises to unity beyond ω_1 shown. With C_p in place of R_L the transfer becomes flat at $\approx C_o/C_p$ away from the double-resonance on both sides, but gives a resonant peak at the upper limit of the filter bandpass in (d), where the magnitude of the total crystal impedance is equal to $|X_{cp}|$. Retention of R_L damps this resonance, and also causes the transfer to fall further $\propto\omega$ below the double resonance.

Multi-section quartz filter

With the stop-band transfer α_{stop} still $\approx 1/10$ (−20dB) above the double resonance, this is inadequate. The four-resonator filter of Fig. 24.12 increases α_{stop} to ≈−80dB, a much better figure. Also the values of ω_z and C_p can be adjusted to better approach the ideal flat response over the pass-band.

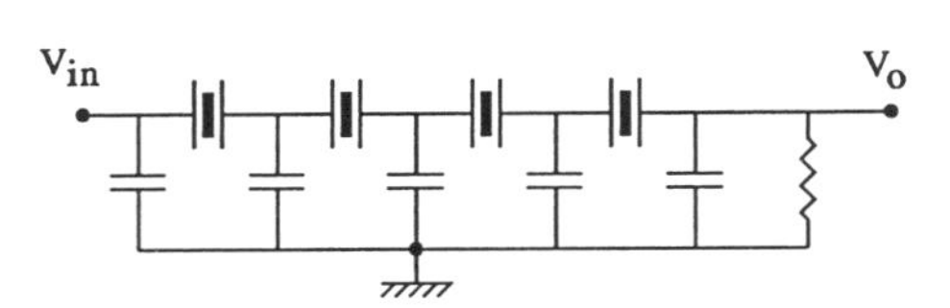

Fig. 24.12 Multi-section band-pass filter based on quartz resonators.

Frequency limits

The approximate frequency range for normal operation of the quartz resonator is from 100kHz to 20MHz, with the upper limit set by fragility. This, however, can be extended to ~200MHz by using up to the ninth overtone. For the oscillator, an additional LC resonator is then needed, to suppress the lower overtones and the main resonance.

Ceramic resonators

With quartz by no means the only piezo-electric material, the above principles can be implemented using the somewhat cheaper piezo-electric ceramics such as lead zirconate titanate (PZT). Compared with the quartz resonator, both the mechanical Q_m and the ratio C_m/C_o are lower, with

$$Q_m \sim 1000, \quad C_o/C_m \sim 10.$$

Here the lower Q_m can be offset by the ability to make better use of the parallel resonance with C_o. As in Fig. 24.13(a), the C_p of Fig. 24.11(a) is replaced by a second resonator.

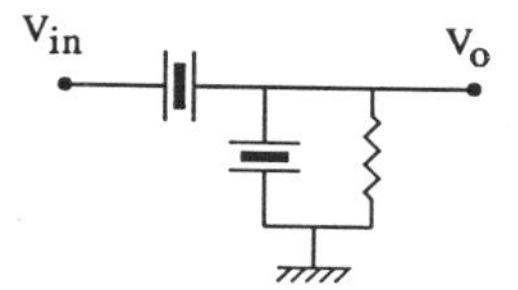

Fig. 24.13(a) Two ceramic resonators used in place of single quartz resonator and C_p in Fig. 24.11(a).

With the wider choice of materials, etc., it is possible to make the general impedance of the shunt resonator well below that of the series. By aligning the pole of the shunt resonator with the zero of the series then the band-pass response shown is obtained. The resulting two notches give well defined limits to the pass-band.

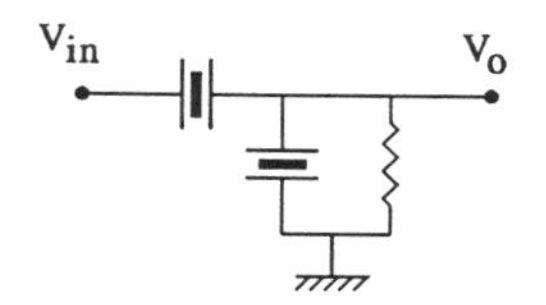

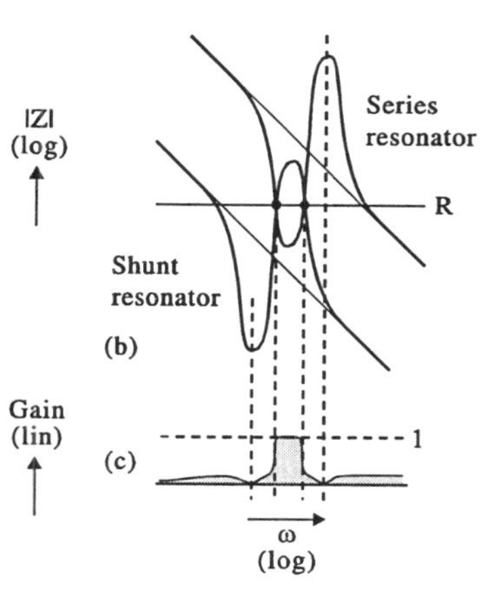

Fig. 24.13(b,c) Frequency response for filter in (a).
(b)Impedance of each component.
(c) Resulting $|V_o|/|V_{in}|$.

Again R_L is required to damp the two further resonances that would otherwise occur at the intersection points of the two curves.

Much as for the quartz equivalent, for frequencies well above the pass-band, the transmission factor $\alpha_{stop} \approx C_{os}/C_{op}$ where these are the C_o for the series and shunt (parallel) C_o. This is one reason why the general shunt impedance needs to be lower than the series.

The extension to the multi-section filter becomes as in Fig. 24.14. MuRata manufacture a wide range of such devices, with the CFS455 a popular choice for the 455kHz standard intermediate frequency.

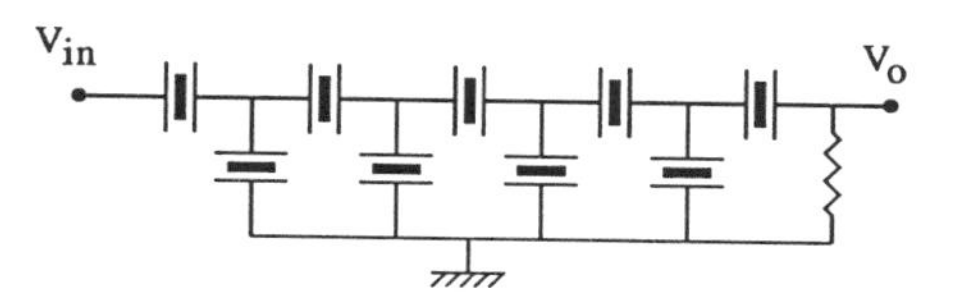

Fig. 24.14 Multi-section band-pass filter based on ceramic resonators.

SAW resonator

Beyond the limit of 20MHz to 200MHz for the normal quartz crystal resonator, the surface acoustic wave (SAW) resonator/filter takes over, either quartz or ceramic. Operation is up to ~2GHz, sometimes also using harmonic-mode operation.

Water waves

Everyone has thrown a pebble into a pond and watched the circular wave spread out. The more curious might have tried oscillating a brick up and down at the edge of the water to see the spreading of the continuous circular sinusoidal pattern. In much the same way so-called 'acoustic waves' can be made to travel across the surface of a piezo-electric material.

Two strip launch

If two adjacent conducting strips are laid across the top of a rectangular slab of the material, and an alternating voltage applied, then the acoustic waves will propagate away from the edge of the strip in each direction.

With the speed at which the waves travel determined by the material, with d the distance between the strips, and with λ the wavelength, the process works best for the driving frequency f such that $\lambda/2 = d$. As f decreases to zero then so does the coupling. Conversely as f is increased then the coupling becomes zero for every frequency for which d is an integral multiple of λ. With essentially

$$\text{coupling} \propto |\sin(\pi d/\lambda)| \qquad (24.6)$$

there is also a peak for each frequency for which d is an odd number of half wavelengths.

Resonator

These are travelling waves, but if reflecting objects are placed in the path of each, at the same distance from the launching site on either side, then the wave will be reflected back to the source. For any driving frequency for which the phase of the reflected wave arriving at the source is in phase with that being transmitted, a strong standing wave pattern will be built up. At the source this is experienced as a sharp resonant drop in the impedance presented by the two strips.

This, however, is a multi-mode resonator. For the distance of the reflector from the source large compared with the distance between the strips there will be many resonances even over the first half cycle of the above $|\sin(\pi d/\lambda)|$ coupling variation.

Inter-digital structure

If the number of launching strips is increased from the original two to give the inter-digital structure in Fig. 24.15 then the range of frequencies for effective launching is much reduced. While the coupling remains optimum for the frequency f_0 for which $\lambda/2=d$, this is only retained for the range centred on f_0 and of width

$$\Delta f_0 \sim f_0/N \qquad (24.7)$$

where N is the number of pairs of fingers. With the reflectors placed close to the structure so as to space the multiple resonances as widely as possible then the resonances adjacent to that at f_0 are suppressed.

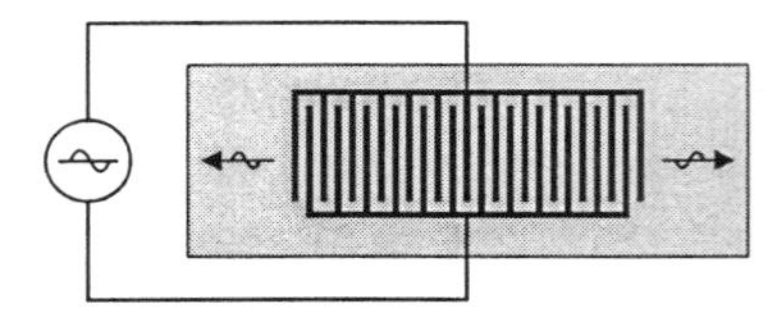

Fig. 24.15 Inter-digital structure used for launching acoustic surface waves.

For the SAW resonator it is possible to obtain variations in general impedance level such as to allow the two-resonator band-pass filter shown in Fig. 24.13(a) as for the normal ceramic filter, and giving the notch-bounded frequency response of (c).

Higher-order operation

Unfortunately or not, there are equivalent bands of width Δf_0 centred on each odd multiple of f_0, corresponding to the maxima of the $|\sin(\pi d/\lambda)|$ for the two-strip launcher. This is unfortunate if it is only the resonance at f_0 that is required, but fortunate in that the higher resonances are available for frequencies too high for it to be possible to arrange for the main resonance to be obtainable. In either case supporting lower-Q LC resonators will be required to suppress the unwanted resonances.

Transmission mode

The method can also be used in the transmission mode, as in Fig. 24.16, with the effective bandwidth the above Δf_0 in [24.7].

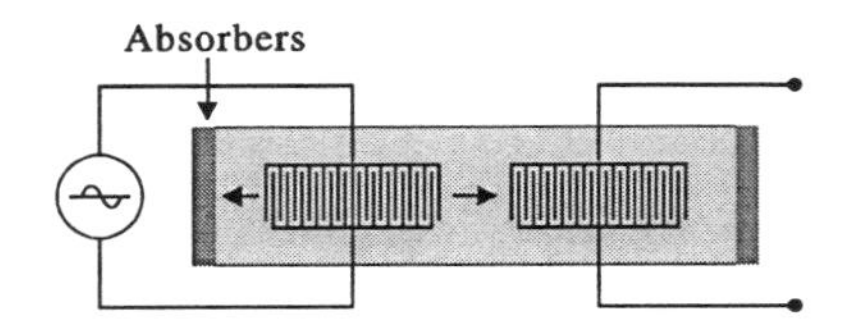

Fig. 24.16 Transmission type SAW band-pass filter.

24.7 Electrostatic microphones

Taking the operation of the moving coil microphone and loudspeaker as being well known, we now consider some microphones based on the electrostatic principle.

Capacitor microphone

This microphone consists of two conducting plates, one rigid and the other the diaphragm upon which the sound waves are incident. Connected as in Fig. 24.17, the resulting capacitor C is charged to give $Q=CV_+$ and $V_o=V_+$. For a step change Δd in d, Q will initially be unchanged. With $C \propto 1/d$ where d is the distance between the plates, and with $V_o=Q/C$, then the resulting $\Delta V_o \propto \Delta d$. Only after the time CR will Q adjust to restore V_o to V_+. Thus, to the extent that Δd is proportional to the sound pressure, the device functions as a microphone with a lower cut-off $f_c = 1/(2\pi CR)$.

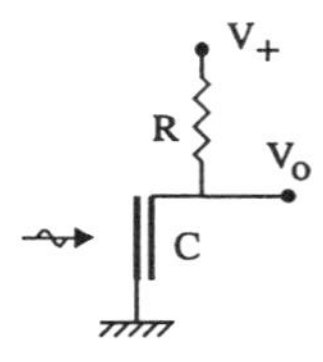

Fig. 24.17 Capacitor microphone

Electret microphone

This device, now extensively used in headsets, mobile phones, portable audio cassette tape recorders, etc., is an extension of the capacitor microphone principle.

As in Fig. 24.18, the capacitor is a sheet of plastic with a metal backing plate and a thin film of aluminium on the top. The molecules in the plastic are highly polar, having a fixed positive charge at one and the balancing negative charge at the other. Normally these are either orientated randomly or in closed groups, giving no charge imbalance for the bulk material. If, however, the plastic is softened by heating and a strong electric field applied then the molecules will change direction to become aligned with he field. As the plastic is cooled, with the applied field till present, the alignment remains after the field has been witched off. Thus opposite charges are left locked into he opposite faces of the sheet, somewhat as for a permaent magnet.

When the resistor R is connected sufficient charge will low from one conducting 'plate' to the other to balance hat on the faces of the dielectric, making the potential ifference between the plates zero. With the plates harged in this way then the device functions as a microhone much as before, but now with no need for the bias V_+. Moreover the charge is much higher than can coneniently be obtained for the capacitor microphone, giving much higher sensitivity.

With the electret capacitance C_e small, R needs to be igh to give the required low-frequency cut-off f_c. The ircuit is thus highly susceptible to capacitively coupled nterference, requiring the full screening shown. Note too ow the first amplifying stage is split in such a way as not requires a V_+ supply line to the JFET.

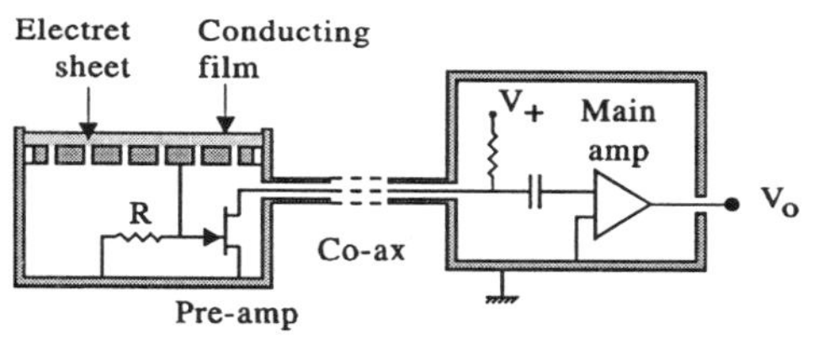

Fig. 24.18 Electret microphone with pre- and main amplifier.

24.8 Bandgap voltage reference

his is an alternative to the Zener voltage reference, with he advantages of a generally lower temperature coefficient, and lower noise. The name derives from the fact hat the output $V_{ref}=V_g$, the bandgap voltage of 1.2V for ilicon.

iode-connected transistor

onsider first the diode-connected transistor of Fig. 24.19 sed as an elementary voltage reference.

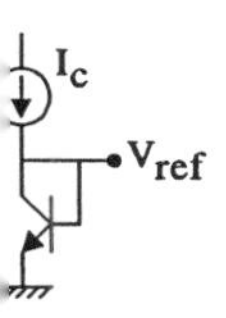

ig. 24.19 Diode-connected transistor as voltage reference.

Vith $V_{ce}=V_{be}\approx 600$mV well above the saturation voltage $V_{sat}\approx 150$mV then the device operates as normal, giving

$$I_c \approx k_s A \exp((V_{be}-V_g)/(kT/q)) \qquad (24.8)$$

where A is the device area, and k_s is relatively temperature-independent. For constant I_c then

$$dV_{be}/dT=(V_{be}-V_g)/T \qquad (24.9)$$

With $V_{be}\approx 600$mV and $T=300°C$ then dV_{be}/dT is the usual -2mV/°.

Bandgap

To see how the bandgap reference improves upon this, consider two such circuits with the I_c equal but with the device areas different, giving the difference

$$\Delta V_{be}=V_{be1}-V_{be2}$$

With dV_{be}/dt as in [24.9] then

$$d(\Delta V_{be})/dT=\Delta V_{be}/T \qquad (24.10)$$

For $\Delta V_{be}>1$ this is positive, unlike dV_{be}/dt in [24.9] which is negative. Thus by adding V_{be} and ΔV_{be} in suitable proportions it becomes possible to form a voltage V_{ref} for which $dV_{ref}/dT=0$. With K_g the scaling constant

$$V_{ref}=V_{be1}+K_g\Delta V_{be} \qquad (24.11)$$

then

$$\frac{dV_{ref}}{dT}=\frac{V_{be1}-V_g}{T}+K_g\frac{\Delta V_{be}}{T} \qquad (24.12)$$

which is zero for

$$K_g=\frac{V_g-V_{be1}}{\Delta V_{be1}} \qquad (24.13)$$

With this value of K_g in [24.11] then $V_{ref}=V_g=1.2$V as stated.

For differing device area A in [24.8] and for the two I_c equal

$$\Delta V_{be}=(kT/q)\ln(A_2/A_1) \qquad (24.14)$$

which for

$$A_2/A_1=10 \quad \text{and with} \quad kT/q=25\text{mV}$$

gives

$$\mathbf{\Delta V_{be}=57mV.}$$

With K_g as in [24.13], $V_g=1.2$V, and $V_{be1}\approx 600$mV then

$$\mathbf{K_g\approx 10.}$$

These values appear suitable. Too low a ΔV_{be} might incur drift error from the amplifier providing the gain factor K_g.

Balance upset

With K_g as in [24.13], V_{be1} changes with T while V_g does not, thus upsetting the balance in [24.12]. This second-order effect leaves the published variation in V_{ref} as

5mV for T=0→70°C.

At 0.4% of $V_{ref}=1.2$V, this is about ten times better than for the Zener, and 30 times better than for the single diode-connected transistor of Fig. 24.19.

Noise

The circuit is also preferable to the Zener in that the enhancement of shot-noise for avalanche breakdown discussed in Section 10.6 is avoided.

Temperature dependence of k_s

With I_d and k_s as in [24.8], it has been assumed that k_s is independent of temperature T. With, in fact, $k_s \propto T^n$ where $n \approx 3$, dV_{be}/dT in [24.9] becomes

$$dV_{be}/dT = (V_{be} - V_g - nkT/q)/T \qquad (24.15)$$

With dV_{ref}/dT as in [24.12] the added nkT/q cancels in the term ΔV_{be} but does alter the term V_{be1}. If not allowed for the error component of dV_{ref}/dT will equal

$$nk/q = 3 \times 25\text{mV}/300^{o}\text{K} \qquad \text{for } n=3.$$

This is $0.25\text{mV}/^{o}\text{K}$ which gives

$$17.5\text{mV for } T = 0 \rightarrow 70\ ^{o}\text{C}.$$

This is larger than the above 5mV and needs to be allowed for, requiring K_g to be increased by approximately 10%.

Feedback implementation

Fig. 24.20 shows a suitable implementation of the above principle, avoiding ideal current sources of the type in Fig. 24.19. None of the above calls for a particular or even a constant value of I_c, only that the two be the same. In Fig. 24.20 the negative feedback maintains the two I_c equal, with the area of Q_2 the larger, as the symbol suggests. For Q_1 and R_2

$$V_{ref} = V_{be1} + 2I_cR_2 \qquad (24.16)$$

while for R_1

$$\Delta V_{be} = I_cR_1 \qquad (24.17)$$

giving

$$V_{ref} = V_{be1} + 2R_2/R_1 \Delta V_{be} \qquad (24.18)$$

With V_{ref} as in [24.11] then

$$K_g = 2R_2/R_1 \qquad (24.19)$$

Example

With R_1 as in [24.17], with $\Delta V_{be1} = 57.5\text{mV}$ as above, and for $I_c = 100\mu\text{A}$ then **$R_1 = 57.5\Omega$.**

With $K_g = 10$ also as above then $R_2 = 5R_1$ giving **$R_2 = 287\Omega$.**

Supply independence

With the op-amp keeping the I_c equal, and thus ΔV_{be} as in [24.14], ΔV_{be} is independent of the supply V_+.

With I_c as in [24.17] and ΔV_{be} independent of V_+ then so also is I_c.

With V_{be} as in [24.8] and with I_c independent of V_+ then so also is V_{be1}.

With V_{ref} as in [24.11] and [24.18], and with ΔV_{be} a V_{be1} both independent of V_+ then so also is V_{ref}.

Feedback sense

With the $I_{c1} = I_{c2}$ then $g_{m1} = g_{m2}$. With R_1 adding to t emitter degeneration of Q_2 then Q_1 dominates the fee back, requiring the sense shown.

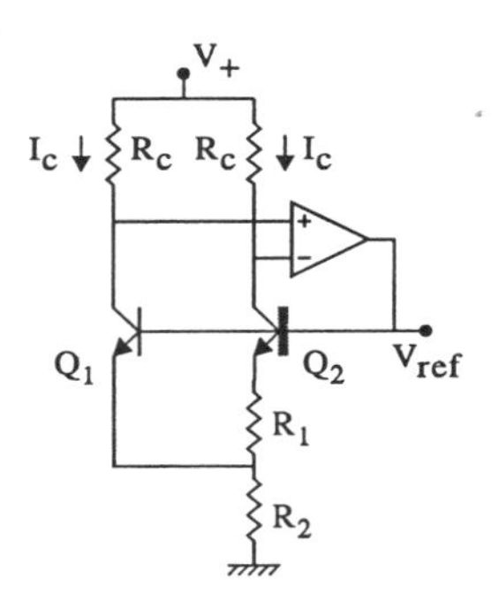

Fig. 24.20 Bandgap voltage reference.

Variants

There are a number of variants of Fig. 24.20, some which allow V_{ref} to be higher than $V_g = 1.2\text{V}$, with othe adapted to the limitations of the type of technology used the microcircuit fabrication, e.g. CMOS vs bipolar. In a cases the two key transistors must be bipolar.

24.9 Current feedback op-amp

With the circuit for this device as in Fig. 24.21, the nam alludes to inverting input impedance being small compare with that of the feedback components R_1 and R_2, rath than as large as possible as for the normal op-amp. I warned, however, that the term is highly misleading.

- It suggests that the device is more suitable than the no mal op-amp for current feedback, which it is not.
- It suggests that the fed back signal which is compare with V_{in} is the current I_{i-}, which it is not.

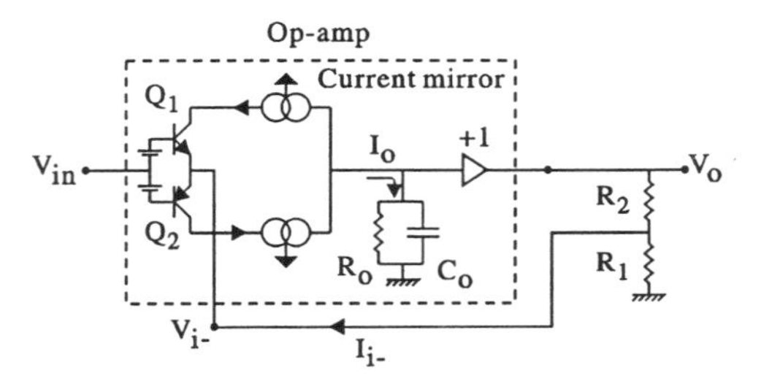

Fig. 24.21 Current-feedback op-amp used in non-inverting voltage feedback amplifier.

ı fact the low inverting input impedance R_{i-} is merely the rice paid for a simpification in the op-amp circuit which ives a much higher upper cut-off frequency than for the onventional op-amp, typically 50MHz is contrast to the MHz for the conventional bi-fet TL081. The diagram ıows the circuit connected in the normal, non-inverting oltage feedback amplifier configuration of Fig. 2.4, to hich it is fully suited. Here a useful feature not obtaining r the normal op-amp, is that as R_1 is varied in order to ter the feedback amplifier gain $A_f=(R_1+R_2)/R_1$ then the edback amplifier cut-off remains at the same high value for $A_f=1$.

With the inverting input impedance $R_{i-} >> R_1//R_2$, the ason that this does not upset the feedback fraction $=R_1/(R_1+R_2)$ for which $A_f=1/\beta$ is that, in the extreme f the op-amp voltage gain A_o being infinite, then $_{in} - V_{i-}=0$ making $I_{i-}=0$ also. Thus there is no loading f the feedback network.

oltage and current feedback

With the first of the above two cautionary remarks, we eed to be clear as to the difference between current and oltage feedback. The distinction is only meaningful for ases where the load resistance R_L (not shown in Fig. 4.21) varies. For example, as the electro-magnet in Fig. 4.22 heats, its resistance R_m increases. With the 'sam-ing resistor' R_s as shown then the voltage I_mR_s is main-ined equal to V_{in}, and thus I_m constant, in the face of arying R_m. This is true current feedback, also used by e electro-chemist where the current through an electro-emical cell needs to remain constant as the reaction roceeds.

Concerning the second caution, to omit R_s and simply-rive I_m directly into the op-amp as I_{i-}, apart from proba-ly destroying the op-amp, will fundametally not have the equired effect.

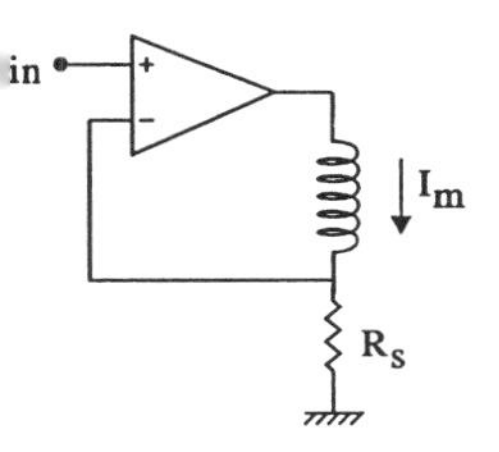

ig. 24.22 Current feedback stabilisation of electro-magnet I_m r thermally varying magnet resistance.

o add to the confusion, suppliers tend to add the OTA to ose listed under the heading of 'current feedback op-mp'. It is nothing of the kind, merely an amplifier for hich the output impedance is as high, rather than as low, s possible.

Upper frequency cut-off

Here the upper and lower halves of the circuit operate essentially in Class B, with only one half working at a time except in the cross-over region. Within the feedback loop then say Q_1 and the upper current-mirror function essentially as two cascaded common-base stages. With the associated cut-off f_p comparable with the base transit time cut-off f_τ, and with $f_\tau=150$MHz for the BC182, then $f_p=50$MHz is typical.

f_p, however, represents the two equal secondary poles; the main pole is set by R_o and C_o. With R_o the parallel combination of the output impedance of the current mirror and the input impedance of the voltage-follower then the associated cut-off f_o can be set well below f_p as required.

Thus, overall, we have an op-amp capable of giving a closed-loop cut-off $f_f=50$MHz, at least for unity feedback amplifier gain.

Input impedance

At cross-over the input impedance R_{i-} of the inverting input is given by

$$R_{i-}=1/2g_m \qquad (24.20)$$

which for $I_c=1$mA gives **$R_{i-}\approx 10\Omega$.**

Nodal analysis

Approximating $Q_{1,2}$ with the current-mirror by the single transconductance g_m, and from the resulting single-node (V_{i-}) analysis, the feedback voltage amplifier gain

$$A_f = \frac{1}{\beta} \times \frac{-A_L}{1-A_L} \qquad (24.21)$$

where the usual feedback fraction

$$\beta=(R_1+R_2)/R_1 \qquad (24.22)$$

and

$$A_L = -\frac{R_o}{R_2} \times \frac{g_m}{g_m + G_1 + G_2} \qquad (24.23)$$

With normally $g_m >> (G_1+G_2)$ then

$$A_L \approx -R_o/R_2 \qquad (24.24)$$

For confirmation of [24.24], with the current gain of $Q_{1,2}$ and the mirror unity, and with I_{i-} the current flowing into the inverting input, then $V_o \approx I_{i-}R_o$. With $R_{i-} << R_2$ then $I_{i-} \approx V_o/R_2$, giving A_L as above.

With A_L thus independent of R_1, while the lower R_{i-} initally lowers A_L a good deal, it is confirmed that the upper frequency cut-off of A_f remains independent of R_1 and thus A_f.

EL202C

It is for this device that the above typical $f_p=50$MHz. Also $R_o\approx 500$kΩ although considerably temperature dependent. With $A_L \approx 1/R_2$ as in [24.24], $R_2=1k\Omega$ is the value for which the frequency for which $|A_L|=1$ is equal to f_p, and therefore the minimum for stability. It is also lowest value giving the full range for V_o.

Frequency responses

These are shown in Fig. 24.23, for the loop gain A_L in (a) and the feedback voltage gain A_f in (b), with both for R_2 the limiting 1kΩ for stability. Thus it is confirmed that A_L and f_f remain independent of A_f until R_1 is reduced to be comparable with ($R_{i-} \approx 10\Omega$)

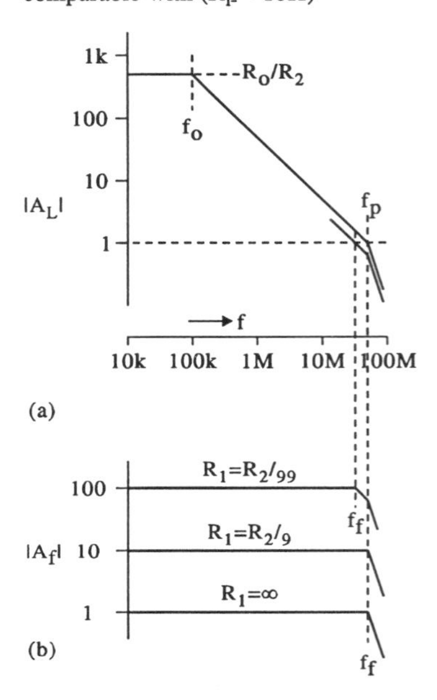

Fig. 24.23 Frequency responses for op-amp of Fig. 24.21.
(a) Open loop gain A_L with R_2 set to stability limit, showing small reduction for $R_1=R_2/99$ in (b).
(b) Closed-loop gain for various values of R_1.

24.10 Thyristor (SCR)

For the thyristor the alternative term 'silicon-controlled rectifier' (SCR) is more descriptive, since the device is a rectifier but one that will not conduct in the normal forward direction until triggered by a voltage applied briefly to a third electrode, the 'gate'.

Power control

In the extended form of the 'triac', the SCR is the control element in the familiar electric light dimmer. The alternative of a variable resistor placed in series with the light is unacceptable on account of the wasted power in the resistor, and the heat developed in the control box.

The triac works instead in the equally familiar manner by which the heating element in an electric cooker is controlled. This is by periodically switching the supply o[n] and off, and by varying the duty cycle. With essentially n[o] heat dissipated in the control element the arrangement i[s] 100% efficient.

For the cooker the switch is mechanical so, to avoi[d] wear, the switching cycle is a few tens of seconds. Wit[h] this unsuitable for the light dimmer, for the SCR contr[ol] the switching period is the 20ms period of the 50Hz main[s] supply, with the waveforms as in Fig. 24.25(b,c).

Four-layer structure, bistable operation

With the essential structure of the SCR as in Fig 24.24(a), (b) and (c) show how this is effectively a pair o[f] cascaded complementary bipolar transistors with 100% positive feedback (a 'bistable'). Taking the cathode K a[s] reference, and for a constant positive DC voltage applie[d] to anode A, then either

- neither transistor is conducting, or
- both are conducting, to the point of destruction if th[e] supply impedance is zero.

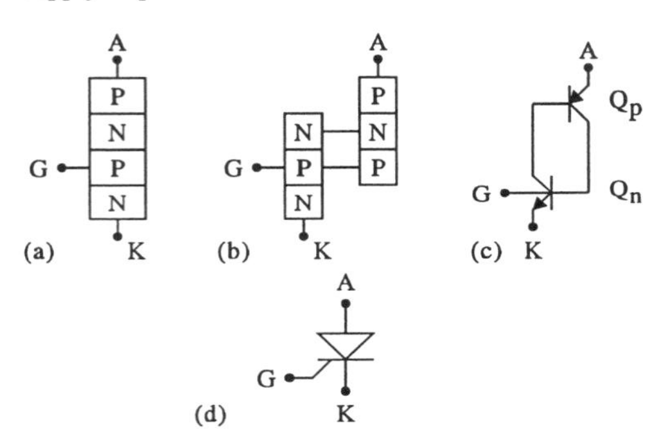

Fig. 24.24 Silicon controlled rectifier (SCR)
(a,b) Four-layer structure
(c) Equivalence to NPN, PNP positive feedback pair.
(d) Circuit symbol. A: Anode, K: cathode, G: gate

Sine wave switching

Fig. 24.25(a) shows the circuit diagram and (b) the wave forms for the following simplified switching sequence.

- V_{in} increases from zero with the SCR so far non conducting.
- With the RC network introducing the phase lag shown as V_G rises above zero then the SCR is triggered int[o] conduction.
- As V_{in} falls to zero then conduction stops, remaining s[o] for as long as V_{in} is negative.
- The cycle repeats.

The operation has two clear limitations

- The phase lag ϕ_{cr} is limited to 90°.
- The switch is 'off' for all of the second half-cycle.

With the second limitation resolved by use of the triac a[s] below, the first is overcome as follows.

Gate threshold voltage

With $V_{in} = \hat{V}_{in}\sin(\theta)$

and with the 'firing angle' ϕ the value of θ at which conduction starts, we seek to extend ϕ from the limiting $\phi_{cr} < 90^o$ to $\phi = 180^o$.

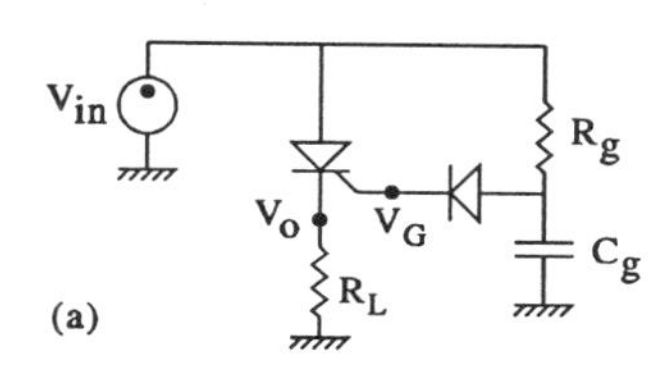

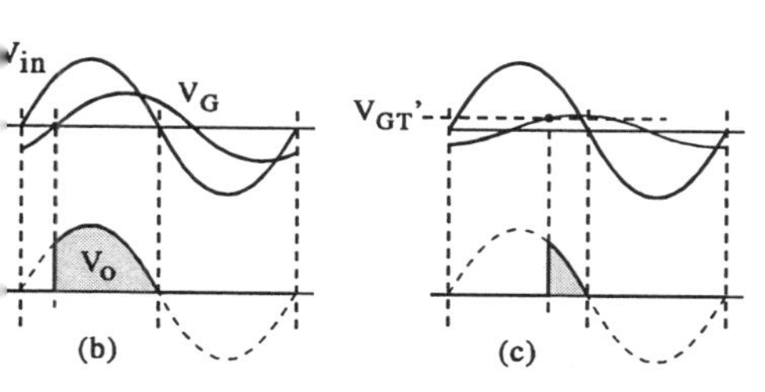

Fig. 24.25 Thyristor control of AC supply to load R_L.
(a) Circuit diagram.
(b) Waveforms for D absent and gate trigger threshold $V_{GT}=0$.
(c) As (b) but with $V_{GT} \sim 600mV$ as normal and D added to give total effective V_{GT}'. R_g increased to give RC phase-lag closer to 90° limit.

With the gate threshold V_{GT} that at which the SCR is triggered into conduction, the above sequence was for $V_{GT}=0$. Instead V_{GT} will need to be the ≈600mV needed for significant current to flow across the input junction, and in fact is somewhat higher. Also by the addition of the diode D the effective V_{GT}' is made yet higher.

With the waveforms now as in (c), ϕ is increased beyond the 90° limit. As R_g is increased, not only does ϕ_{rc} approach the limiting 90° but also the amplitude of the V_G sine wave decreases, eventually such that the peak is equal to V_{GT}. Here the addition to ϕ is equal to 90°, causing the limiting ϕ to approach 180° as required.

A slight consequent disadvantage is that the minimum value of ϕ is no longer zero. For $R_g=0$ for which $V_G=V_{in}$, triggering now occurs when $V_{in}=V_{GT}'$ rather than when $V_{GT}=0$.

For the usual $V_{in}=240V$ rms and $V_{GT} \sim 1V$ then the lower and upper limits to ϕ come very close to the ideal 0° and 180°. However, with the limits not needing to be quite this close, and with the precise value of V_{GT} somewhat variable with production spread, it is usual to make V_{GT}' considerably higher than V_{GT}. The addition of the diode D is the first step, but a Zener gives a further increase. $V_{GT}'=15V$ or so is typical.

Gate reverse breakdown

There is another reason for the inclusion of D. Without it, for $R_g=0$, and with the reverse breakdown voltage for the input junction well below the peak value of 340V for the 240V RMS mains, then V_{in} is coupled directly to the load for most of the negative half-cycle of V_{in}, in an uncontrollable manner. For the same reason D must be included in series with the Zener.

Load connection

One's first inclination might be to reverse to positions of load and thyristor. Apart from the frequent requirement that the lower terminal of R_L be grounded, the other advantage of the arrangement shown is that as soon as the device is triggered the gate-to-cathode voltage is immediately reversed, stopping the gate conduction and thus allowing the RC network to operate much as if unloaded.

Firing angle ϕ

With the above ϕ termed the 'firing angle', and the remainder of the half-cycle the 'conduction angle', we now derive the expressions for the limits to ϕ.

Minimum With $dV_{in}/d\theta = 2\pi\hat{V}_{in}$

and $\phi \approx V_{GT}'/(dV_{in}/d\theta)$

then $$\phi \approx V_{GT}'/(2\pi\hat{V}_{in}) \qquad (24.25)$$

For say $V_{GT}' = \hat{V}_{in}/20$

then $\boldsymbol{\phi_{min} \approx 3^o}$.

Maximum Here R_g has been increased from the previous zero to the point where

$$R_g/X_g \approx V_{GT}'/\hat{V}_{in}$$

for which the added delay is the full 90°.

With here $R_g >> X_g$

and with ϕ_{rc} the network phase lag then

$$\phi_{rc} = \tan^{-1}(R_g/X_g) \approx 90^o \qquad (24.26)$$

Thus $$\phi_{max} \approx 90^o + \tan^{-1}(\hat{V}_{in}/V_{GT}') \qquad (24.27)$$

giving $\boldsymbol{\phi_{max} \approx 180^o - 3^o}$.

a good approximation to the ideal of from 0° to 180°.

Zener voltage V_z With $\hat{V}_{in} \approx 340V$

as for the 240V RMS mains, and with the above

$$V_{GT}' = \hat{V}_{in}/20$$

then $V_{GT}'=17V$. For $V_{GT}=1V$ and with D included then

$$V_z=(\hat{V}_{in}/20)V_{GT}-600mV-V_{GT} \quad (24.28)$$

giving $$V_z=15.4V.$$

Triac

The switching control in Fig. 24.25 is extended to the second half-cycle by connecting two SCRs in parallel, with the second the complement of the first (from the bottom in Fig. 24.24(a), PNPN rather than NPNP). With the terms 'anode' and 'cathode' now ambiguous, the rather unimaginative 'main-terminal' 1 and 2 are used instead. Fig. 24.26(a) gives the circuit symbol.

Diac

The diac in Fig. 24.26(b) is essentially a triac with the twin gate terminal unused. Its function is to replace the Zener used with the SCR to effectively increase the gate threshold voltage V_{GT}. With the effective Zener now needing to break down equally for both polarities, the diac material is adjusted to break down at the required approximate ±15V

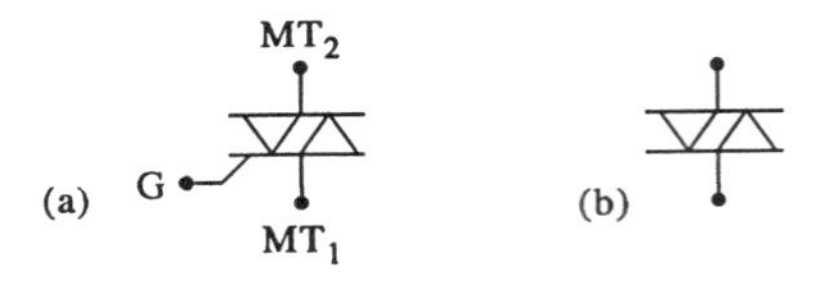

Fig. 24.26 Circuit symbols.
(a) Triac (MT: Main terminal).
(b) Diac.

Latch-up

This, a potential hazard in microcircuit design, is akin to the positive feedback in the SCR. Fig. 24.27 shows the relevant features of the CMOS digital inverter in Fig 8.16. Here it is seen how a parasitic lateral NPN and vertical PNP are formed with a path as for the closed loop in Fig. 24.24(c). As for the SCR, the loop has to be triggered to enter the conducting state, but once so triggered it will remain so, grossly inhibiting the required operation. Triggering may be by untoward events such as power line surges, however brief, unsuitable sequencing of dual supply application, and so on. Preventive measures are as follows.

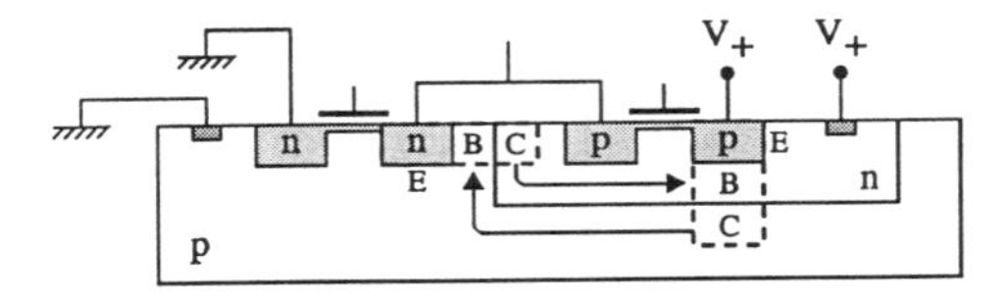

Fig. 24.27 Latch-up path for CMOS inverter.

- Re-positioning the V_+ and ground connections to diver the current paths away from the effective base regions possibly extending the connection structure.
- Addition of suitable guard rings.

References

Rhea, R.W. 1997: *Oscillator design and computer simu lation,* 2nd Ed, McGraw Hill.

Senior, J.M. 1992: *Optical fiber communications, princi ples and practice,* 2nd Ed, Prentice Hall.

Tagami, T. Ehara, H. Noguchi, K. and Komasaki, T 1997: *Resonator type SAW filter*, Oki Technical Review **63**, No 158.

Wilson, J. and Hawkes, J.F.B. 1989: *Optoelectronics, an introduction,* 2nd Ed, Prentice Hall.

ndex

300 type DAC, 242
'f-noise, 206
55 precision astable, 44
55 precision monostable, 44
)0mV model, diode, 59

C coupled feedback amp, 31
C resistance, inductor, 158
C theory, 11
cceptor, semiconductor, 57
ctive load, common-emitter amplifier, 74
ctive RC filters, 164
ctive tail, bipolar LTP, 94
DC, 245-250
dder, feedback, 22
dmittance grid, Smith chart, 287
GC control, 195, 198
ll-pass filter, ideal response, 137
mplifier matching, using Smith chart, 288
mplifier noise temperature, 208
mplifier s-parameters, 285
mplitude measurement, noise-optimised, 230
mplitude modulation, 180
nalog multiplier, 190
nalog shunt switch, bipolar transistor, 98
nalog-to-digital converter, *see ADC*
ntenna efficiency, 181
nti-aliasing, 251
stable, 42, 44
udio power amp, *see Power amplifier, audio*
uto zero, 49
utomatic gain control, *see AGC*
vailable power, 6
vailable power gain, 285
valanche breakdown, 61
valanche photo-diode, 296, 297

ack diode, 62
alls on a tray model, 55
alun, 276
andgap voltage reference, 301
and-pass filter,
  ideal response, 137
  OTA-C, 173
  quartz crystal, 298
  state-variable, 171
  incorporating notch, 155
and-stop filter, ideal response, 137
andwidth, series resonator, 150
ase, bipolar transistor, 72-78
asic filter responses, 143
ias current, op-amp, 28
ipolar DAC, 243

Bipolar transistor, 72-81
  analogue shunt switch, 98
  damping of input resonator, 194
  forward and reverse β, 98
  switching type, 98
  noise, 209
Bistable, 41, 42, 303, 305
Board-to-board transmission of digital signal, 271
Body effect, MOS, 103, 109
Bow-tie diagram, modulator, 188
Boxcar sampling gate, 49
Bridge rectifier, 64
Broad band band-pass and band-stop filters, 157
  coupled LCR resonators, 158
  quartz crystal filters, 158, 297
  ceramic, 299
  SAW, 300
Bulk effect, *see Body effect*
Bulk resistance, 59
Butterworth filter, 164-166

Capacitance, 3
Capacitive load, transmission line, 271
Capacitively coupled interference, 216
Capacitor, 3-4
  discharge, 138
  microphone, 300
Capture-range, PLL, 226
Cascaded low-pass sections, pole-zero analysis, 144
Cascode,
  bipolar transistor, 91
  MOS, 111
  OTA, 124
Cascode current mirror, basic, 123
  wide-swing, 124
Cascode tail, MOS LTP, 122
Ceramic resonators, 299
Characteristic impedance, transmission line, 269
  equations, 272
Charge control model, bipolar transistor, 78
Class A and B audio power amplifier, 256
CMOS switch, 113-115
CMOS, switching modulator, 200
CMOS, technology, 104
CMR, 31
  bipolar transistor LTP, 94
  MOS LTP, 120
Coaxial directional coupler, 290
Colpitts oscillator, 234
Comb filter, ideal response, 137
Common-base amplifier stage, bipolar transistor, 91
Common-emitter amplifier, 71-81
Common-gate amplifier, MOS, 111

Common-mode feedback, 121
Common-mode input, 31
Common-mode offset, OTA, 175
Common-mode rejection ratio, *see CMR*
Common-mode transconductance,
  MOS LTP, 119
Common-mode voltage gain, bipolar LTP, 93
Common-source amplifier stage, JFET, 132
  MOS, 105-108
Comparator, 42
Conductance, 1
Conductor, 55
Constant current source, 136
Convolution, 15
Corkscrew waves, 142
Correlation, noise, 207
Covalent bonding, 55, 305
Crank diagram, transmission line, 273
Critical damping, 156
Current bias, CE amp, 81
Current crowding, 259
Current drive, CE amp, 76
Current feedback op-amp, 302
Current follower, 21
Current mirror,
  bipolar transistor LTP, 94
  MOS LTP, 118
Current sampling transformer, 7
Current source, ideal, 10

DAC, 240-244
Dark current, photo-diode 295
Darlington, 297
DC power supplies, 63-65
DC resistance, inductor, 158
DC restorer, 50
DDS, 236-237
Decibel scale, 139
Decoupling, 221
Delyiannis-Friend filters, 168-169
Demodulator applications, 180-185
Depletion layer,
  capacitance, diode, 66
  capacitance, PSpice bipolar transistor, 259
  diode, 58
Diac, 306
Difference-mode input, 31
Difference-mode transconductance, MOS LTP, 118
Difference-mode voltage gain, bipolar LTP, 93
Differential amplifier, 18
  ground loop prevention, 220
Differentiator, feedback, 23
  resonance, 35
Digital signal line transmission, 270-272
Digital-to-analog converter,
  see DAC
Diode,
  limiter, 62
  ring modulator, 200
  semiconductor, 57-62, 66-68
Direct digital synthesis, *see DDS*
Donor, semiconductor, 56
Double-conversion frequency changing, 184
Drift, time averaging, 206
Dual ramp ADC, 246
Dual source noise model, 207
Dual-gate MOS, 112, 195, 198
Dynamic range, diode rectifier, 46

Early effect, 74
  Pspice, bipolar transistor, 259
Eddy current screening, 219
Efficiency, audio power amp, 256
Electret microphone, 300
Electrolytic capacitor, 3
Electrostatic screening, 216
Emitter degeneration, 260
Emitter-follower, 86-90
End connections, diode, 59
Envelope detector, 47
  high-efficiency, 50
  precision, 47
Envelope distortion, 195-198
Equivalent noise resistance, 207
Exclusive-OR modulator, 185
Exponential waves, 142
Extinction length, 295
Extrinsic semiconductor, 56

Farad, capacitance unit, 3
Feedback amplifier,
  limitations 26-34
  OTA, 128
  simple circuits, 18-21
Feedback bias, CE amplifier, 81
Feed-forward,
  audio power amplifier, 264
  feedback amplifier, 21
Feed-through capacitor, 222
Fibre-optic, 292
Filter responses, ideal, 137
Flash ADC, 245
Flyback switched-mode supply, 64
Folded cascode, OTA, 125
Forward and reflected waves, transmission line, 269
Forward β, bipolar transistor, 98
Fourier analysis, 13-17
Fourier decomposition, PSpice, 257
Four-quadrant modulator, 189
Frequency changer, 184, 199
Frequency counter, 40
Frequency doubler, 15
Frequency jitter, 233
Frequency measurement, noise-optimised, 231

Frequency-domain multiplexing, 181, 294
Full-wave rectifier,
  Gilbert cell, 191
  power supply, 63
  precision, 46
Fully differential,
  feedback-amplifier, MOS, 176
  feedback integrator, MOS, 176
  OTA, 175

Gain control, OTA, 118
Gallium arsenide, 294
Gated integrator phase-det, 231
Gilbert cell, 189
  full-wave rectifier, 191
Glitches DAC, 244
Ground loop prevention, 219-221
Ground strapping, 220
Guard ring, 295, 296

h-parameters, 283
HC4046 PLL chip, 227
Heat sink, 259
Henry, inductance unit, 5
High-k capacitor, 4, 306
High-level injection, 258
High-mu screen, 218
High-pass filter,
  Butterworth, 164-166
  first order, RC, 141
  ideal response, 137
  OTA-C, 173
  state-variable, 171
High-voltage diode, PIN, 296
  guard ring, 297
Holbrook third-order Butterworth
  filter, 166
Hybrid-π model, bipolar, 76
Hyper-abrupt junction varactor, 68
Hysteresis,
  Schmidt, 43

Ignition system, inductor based, 5
Image response, 194
Impedance grid, Smith chart, 287
Impedance matching, LC, 157, 286
  line section, Smith chart, 288
Incremental circuit model,
  bipolar transistor, 73
  semiconductor diode, 60
Inductance, 5
Inductive load, audio power amplifier, 267
Inductor, 4-5
  resistance, 154
  self-resonance, 155
  specifications, 158
Inter-digital structure, SAW, 300

Insulator, 55
Integrator, feedback, 22
Interference, 216-222
Interpolation filter, oversampling
  DAC, 250
Intrinsic semiconductor, 56
Inverter switch, bipolar transistor, 96-98
Inverting feedback amplifier, 20
  noise, 212

JFET (junction field-effect transistor), 132-136
j-operator, 12
Junction potential, diode, 58

Laplace transform, 145, 148
LASER, 294
Latch-up, microcircuit, 306
LC,
  impedance matching, 157, 286, 288
  oscillator, 159, 233
  resonator based filters, 149-158
Leakage current, base,
  bipolar transistor, 258
  photo-diode, 295
Limiter, diode, 62
Light emitting diode (LED), 293
Light guide, 292
Line resonances, transmission line, 274
Linearity, 11
  audio power amplifier, 257, 266
Linear-phase filters, 166
Link coupling, 157
Lock-range, PLL, 226
Log plot, 20
  using squared paper, 146
Logic levels, digtal gate, 241
Long-tailed pair, *see LTP*
Loop gain, feedback amplifier, 20
Low-pass filter,
  Butterworth, 164-166
  digital, 249
  first-order RC, 138-141
  ideal response, 137
  OTA-C, 173
  state-variable, 171
LTP,
  bipolar transistor, 92-96
  large signal analysis, 188
  modulator, bipolar, 187-189
  MOS, 118-122
  noise, 210
  series resonator, 152

Magnetic screening, 218
Majority carrier flow, 58
Mean-square detector, analog multiplier, 191
Memory-box model, noise, 205

Microphone, 300
Microwave modulator, 201
Miller effect, 80, 107
Minority carrier flow, 58
Missing pulse detector, 45
Modulator, 180, 200
  LTP, 187-191
Monostable, 42
  precision 555, 44
  retriggerable, 45
MOS, transistor 102-115
MOSFET-CR filter, 176
MOSFET-C filter, 175-177
Motor phase-locking, 227-230
MSF frequency reference, 235
Multi-mode guide propagation, 292
Multiplying DAC, 244
Mutual conductance, 73, 105
  MOS, 105
Mutual inductance, 7
Mutually coupled LCR resonators, 158

Narrow-band band-pass filters
  active RC, 167-169
  LC, 149-153
Narrow-band band-stop filter, 153
  See also Notch filters
Near-differentiation, RC high-pass filter, 141
Near-integration, RC low-pass filter, 138
Network theorems, 7
Nodal analysis, 10
Noise, 204-214
  Barkhausen, 306
  corner, 206
  factor/figure, 207
  margin, logic levels, 241
  matching, 210
  photo diode amplifier, 213
  resistance, equivalent, 207
  temperature, amplifier, 208
  time-averaging, 206
Noise-optimised measurements, 230
  amplitude, 230
  frequency, 231
  general timing, 232
  pulse timing, 231
Non-inverting feedback amp, 19
  noise, 213
Normalisation, 13
Normalised impedances, Smith chart, 287
Norton's theorem, 10
Notch filter,
  band-pass incorporating, 155
  Delyiannis-friend, 169
  LC, 154
  Q-enhanced, 154
Notch-enhanced RC low-pass active filter, 170

N-type semiconductor, 56
Nyquist limit, 237

Ohm's law, 1
On-chip digital signal, line and ohmic effects, 272
Operational amplifier,
  current feedback type, 302
  limitations, 26-34
  MOS, 117
  positive feedback, 41
  simple circuits, 18-23
Operational transconductance amplifier, *see OTA*
Opto-isolator, 297
Optoelectronics, 292-297
Oscillation hysteresis, 168
Oscilloscope, triggering, 42
Oscilloscope-probe, 146
OTA, 117-118
  cascode, 124
  feedback amplifier, 128
  folded cascode, 125
  linearity, 173
OTA-C filters, 172-175
Oversampling ADC, 247-250
Oversampling DAC, 250
Overwound toroid, 277

Parallel resistors, 8
Parallel resonator, 152
Parallel strip transmission line, 278
PCB track impedance, 272
Peak and valley detectors, 47
Permeability, 5
Permittivity, 3
Phase-detector,
  DAC type, 228
  for laser speckle signal, 231
  gated integrator, 231
  XOR-gate, 225
Phase-locked loop, *see PLL*
Phasor diagram, 11
Photo detector, resonance, 36
Photo diode, 295
Photo transistor, 297
Photo diode amplifier noise, 213
Piezo-electric effect, 297
PIN diode, 296
Pinch-off, JFET , 132
  MOS , 103
PLL, 224-230
  frequency synthesisers, 234-235
π-network, bipolar transistor, 78
Pole-zero diagram, 142-145
  Butterworth filters, 165
  narrow-band band-stop filter, 154
  second-order low and high-pass filters, 156
  series resonator, band-pass filter, 150

Polyester capacitor, 4
Positive feedback, op-amp, 41
Pot-cores, 218
Potentiometer, 7
Power, 2
Power amplifier,
  audio, 255-266
  RF, 195
  VHF, 195
Power matching, transformer, 6
Power spectral density, noise, 206
Power JFET, 135
Preferred values, 2
PSpice, 255
  bipolar transistor model, 258-259
P-type semiconductor, 57
Pulse timing measurement, noise-optimised, 231
Push-pull amplifier, 15

Q-enhanced notch filter, 154
Q-factor, series resonator band-pass filter, 150
  inductor specification, 158
Quantisation noise, 248
Quality factor, *see Q-factor*
Quartz crystal resonator, 297
  filters, 158, 297-298
Quiescent current, audio power amplifier, 257, 262

Ratemeter, conventional, 228
  detector, DAC type, 228
RC filter, single-section passive, 137-145
RC network, PLL, 226
Reactance, capacitor, 3
  inductor, 5
Rectifier, dynamic range and precision, 46
Reflection coefficient measurement, 289
Reflection coefficient, transmission line, 270
Reflectometer, 275-276
Reluctance, 5
Resistor, 1-2
Resistor-bridge strain gauge, 182
Resonance, differentiator, 35
  capacitive loading,
    emitter-follower, 90
    source-follower, 109
    voltage-follower, 36
  photo detector, 36
  LC, 149-163
  water model, 5
Reverse β, bipolar transistor, 98
Reverse breakdown, diode, 61
Reversing switch,
  demodulator, 182
  LTP modulator, 187
  modulator, 181
RF amplifier, receiver first stage, 194
Ribbon cables, 219

RMS value, 2
Roll-off, filter response, 139
Roofing filter, 184
Running-average, 140

Sallen-Key implementation of Butterworth filter, 165
Sample-and-hold gate, 48
Sampling gates, 48-49
  glitch avoidance, 245
Saturation voltage, bipolar transistor, 74
SAW resonator, 299
Sawtooth, Fourier analysis, 15, 17
Schmidt trigger, 40-43
Schottky diode, 68
SCR, *see Thyristor*
Screening, 216-219
Self-resonance, inductor, 155, 158
Semiconductor, 55-57
Sensitivity, filter, 167
Series resonator band-pass filter, 149-152
Series to parallel impedance transformation, 153
Settling time, DAC, 244
Short-base diode, 68
Shot noise, 205
Sigma-delta ADC, *see Oversampling ADC*
Sigma-delta modulator, oversampling DAC, 250
Sigma-delta nomenclature, 250
Silicon, 55
Silicon carbide, 294
Silicon controlled rectifier, *see Thyristor*
Sine wave, 2
Sine wave adapter, 43
Sine wave line transmission, 273-275
Single-sideband suppressed-carrier,
  carrier synchronisation, 183
  radio link, 183
Skin effect, 219
Slew-rate, 34
Smith chart, 286-288
Source-follower, 108-110
S-parameters, 284-291
SPICE, see PSpice
SpiceAge, 265
Spontaneous emission, 294
Square-wave, Fourier analysis, 14,16
Stability, feedback amplifier, 32
Stability criterion, s-parameters, 285
Standard deviation, noise, 206
Standing waves, transmission line, 274
State-variable filter, 170-171
Stimulated emission, 294
Stored charge,
  bipolar transistor LTP, 97
  diode, 66
  tail, photo-diode, 295
Strain gauge, resistor-bridge, 182
Strapped transistor diode, 68

Substrate, MOS , 104
Subtractor, 22
Successive approximation ADC, 245
Superposition, 9
Supply-independent bias source, MOS, 129
Surface acoustic wave, *see SAW*
Switch, JFET, 136
Switched input ladder DAC, 241
Switched output ladder DAC, 241
Switched-capacitor comparator, flash ADC, 246
Switched-capacitor,
  filter, 170
  integrator, 171
  VCO, 232
Switched-mode supply, flyback type, 64
Switching time, semiconductor diode, 66
Switching bipolar transistor, 98

$Tanh^{-1}$ converter, 190
Temperature coefficient,
  bipolar transistor, 81
  JFET, 135
  op-amp, 29
Thermal,
  compensation, 261
  frequency drift, oscillator, 233
  loop-gain, 260
  noise, 204
  runaway, 260
  stability, 259-261
Thevenin's theorem, 8
Thyristor, 304
Third-order intercept, 197
Time averaging, RC low-pass, 140
Timing measurement, noise-optimised, 232
Toroidal core, 5
Toroid, ferrite, 218
Track-and-hold, 48
Transceiver, 184
Transductor, *see OTA*
Transfer function,
  operational amplifier, 19
  symmetric and anti-, 14
Transformer, 5-7
  noise-matching, 210
Transistor dissipation, audio power amp, 256
Transmission lines, 269-275, 278
Transmission-line transformer, 276-278
Transposed input LTP modulator, 189
Travelling-waves, line, 274
Triac, 306
Triangular and square wave signal generator, 43
Trimmer capacitor, 4
Tuning stubs, transmission line, 274
Tuning wheel, PLL frequency synthesiser, 235
Tuning, OTA-C filter, 174-175
Tunnel diode, 62
Twin pair, twisted line, 278
Twin-tub technology, MOS, 104
Two-stage Miller op-amp, 126-128
Two-tone test, 197

Unit ramp and step, 143

Varactor diode, 68
  tuning, Colpitts oscillator, 234
VCO, 232
Velocity of propagation, transmission line, 273
VHF amplifier,
  receiver, 194
  transmitter, power, 195
Virtual ground, 20
Voltage,
  bias, CE amplifier, 81
  controlled oscillator, *see VCO*
  offset, op-amp, 27
  reference,
    bandgap, 301
    JFET, 136
    Zener, 62
  regulator, 64
  source, ideal, 10
Voltage-current law, diode , 58
Voltage-follower, 21
  resonance, 36
VSWR (voltage standing wave ratio), 275

Wein network, 167
  narrow-band band-pass filter, 167
  oscillator, 167
White noise, 205
Wide-swing cascode current mirror, 124

XOR phase-detector, 225
Y-parameters, 283
Zener,
  breakdown, 61
  diode, 61
  voltage reference, 62
Zero correction, two-stage Miller op-amp, 128
Zobel impedance, 266-267